SCHAUM'S OUTLINE OF

THEORY AND PROBLEMS

OF

BIOCHEMISTRY

•

PHILIP W. KUCHEL, Ph.D.
Coordinating Author

GREGORY B. RALSTON, Ph.D.
Coordinating Author

AUDREY M. BERSTEN, M.Sc.

SIMON B. EASTERBROOK-SMITH, Ph.D. **ALAN R. JONES, Ph.D.**

M. DAN MONTAGUE, Ph.D. **MICHAEL B. SLAYTOR, Ph.D.**

MICHAEL A. W. THOMAS, D.Phil. **R. GERARD WAKE, Ph.D.**

Department of Biochemistry
The University of Sydney
Sydney, Australia

•

SCHAUM'S OUTLINE SERIES

McGRAW-HILL, INC.

New York St. Louis San Francisco Auckland Bogotá Caracas
Hamburg Lisbon London Madrid Mexico Milan Montreal
New Delhi Paris San Juan São Paulo Singapore
Sydney Tokyo Toronto

Schaum's Outline of Theory and Problems of
BIOCHEMISTRY

3 4 5 6 7 8 9 10 11 12 13 14 15 16 17 18 19 20 SHP SHP 9 2 1 0

ISBN 0-07-035579-7

Sponsoring Editor, Elizabeth Zayatz
Production Manager, Nick Monti
Editing Supervisor, Marthe Grice
Cover design by Amy E. Becker.

Library of Congress Cataloging-in-Publication Data

Schaum's outline of theory and problems of biochemistry.

(Schaum's outline series)
Includes index.
1. Biological chemistry--Outlines, syllabi, etc.
2. Biological chemistry--Examinations, questions, etc.
I. Kuchel, Philip W.
QP518.3.S3 1985 574.19'2'0202 86-18154
ISBN 0-07-035579-7

Preface

This book is the result of a cooperative writing effort of approximately half of the academic staff of the largest university department of biochemistry in Australia. We teach over 1,000 students in the Faculties of Medicine, Dentistry, Science, Pharmacy, Veterinary Science and Engineering. So, for whom is this book intended and what is its purpose?

This book, as the title suggests, is an *Outline* of Biochemistry—principally mammalian biochemistry and not the full panoply of the subject. In other words, it is not an encyclopedia but, we hope, a guide to understanding for undergraduates up to the end of their B.Sc. or its equivalent.

Biochemistry has become the language of much of biology and medicine; its principles and experimental methods underpin all the basic biological sciences in fields as diverse as those mentioned in the faculty list above. Indeed, the boundaries between biochemistry and much of medicine have become decidedly blurred. Therefore, in this book, either implicitly through the solved problems and examples, or explicitly, we have attempted to expound *principles* of biochemistry. In one sense, this book is our definition of biochemistry; in a few words, we consider it to be the description, using *chemical* concepts, of the processes that take place in and by living organisms.

Of course, the chemical processes in cells occur not only in free solution but are associated with macromolecular structures. So inevitably, biochemistry must deal with the structure of tissues, cells, organelles, and of the individual molecules themselves. Consequently, this book begins with an overview of the main procedures for studying cells and their organelle constituents, with what the constituents are and, in general terms, what their biochemical functions are. The subsequent six chapters are far more chemical in perspective, dealing with the major classes of biochemical compounds. Then there are three chapters that consider enzymes and general principles of metabolic regulation; these are followed by the metabolic pathways that are the real soul of biochemistry.

It is worth making a few comments on the *style* of presenting the material in this book. First, we use so-called *didactic questions* that are indicated by the word *Question*; these introduce a new topic, the answers for which are not available from the preceding text. We feel that this approach embodies and emphasizes the inquiry in any research, including biochemistry: the answer to one question often immediately provokes another question. Secondly, as in other Schaum's Outlines, the basic material in the form of *general* facts is emphasized by what is, essentially, optional material in the form of *examples*. Some of these examples are written as questions; others are simple expositions on a particular subject that is a specific example of the general point just presented. Thirdly, the solved problems relate, according to their section headings, to the material in the main text. In virtually all cases, students should be able to solve these problems, at least to a reasonable depth, by using the material in this outline. Finally, the supplementary problems are usually questions that have a minor twist on those already considered in either of the previous three categories; answers to these questions are provided at the end of the book.

While this book was written by academic staff, its production has also depended on the efforts of many other people, whom we thank sincerely. For typing and word processing, we thank Anna Dracopoulos, Bev Longhurst-Brown, Debbie Manning, Hilary McDermott, Elisabeth Sutherland, Gail Turner, and Mary Walsh and for editorial assistance, Merilyn Kuchel. For critical evaluation of the manuscript, we thank Dr. Ivan Darvey and many students, but especially Tiina Iismaa, Glenn King, Kiaran Kirk, Michael Morris, Julia Raftos, and David Thorburn. Dr. Arnold Hunt helped in the early stages of preparing the text. We mourn the sad loss of Dr. Reg O'Brien, who died when this project was in its infancy. We hope, given his high standards in preparing the written and spoken word, that he would have approved of the final form of the book. Finally, we thank Elizabeth Zayatz and Marthe Grice of McGraw-Hill; Elizabeth for raising the idea of the book in the first place, and both of them for their enormous efforts to satisfy our publication requirements.

Philip W. Kuchel
Gregory B. Ralston
Coordinating Authors

Contents

CONTENTS

CONTENTS

Chapter 1

Cell Ultrastructure

1.1 INTRODUCTION

Question: What are the basic units of life?

All animals, plants, and microorganisms are composed of small units known as *cells*. Cells range in volume from a few attoliters among bacteria to milliliters for the giant nerve cells of squid; typical cells in mammals have diameters of 10 to 100 μm and are thus about five times smaller than the smallest visible particle. They are generally flexible structures with a delimiting membrane that is in a dynamic, undulating state. Different animal and plant tissues contain different types of cells, which are distinguished not only by their different structure but by their different metabolic activities.

EXAMPLE 1.1

Antonie van Leeuwenhoek (1632–1723), draper of Delft in Holland, ground his own lenses and made simple microscopes that gave magnifications of $\sim\times200$. In 40 years of work he wrote over 200 manuscripts and letters to the Royal Society of London and to other notable organizations and people. His first letter to the Royal Society in 1673 described his microscopical observations on mold and bee stings, but on October 9, 1676, he sent a $17\frac{1}{2}$-page letter to the Society, in which he described *animalcules* in various water samples. These small organisms included what are today known as protozoans and *bacteria*; thus Leeuwenhoek is credited with the first observation of bacteria. Later work of his included the identification of spermatozoa and red blood cells from many species.

The development of a *stem cell* into cells with specialized function is called the process of *differentiation*. This takes place most dramatically in the development of a fetus, from the single cell formed by the fusion of one *spermatozoon* and one *ovum* to a vast array of different tissues.

Cells appear to be able to recognize cells of like kind, and thus to unite into coherent organs, principally because of specialized glycoproteins (Chap. 2) on the cell membranes.

1.2 METHODS OF STUDYING THE STRUCTURE AND FUNCTION OF CELLS

Light Microscopy

Many cells and, indeed, parts of cells (*organelles*) react strongly with colored dyes such that they can be easily distinguished in thinly cut sections of tissue by using light microscopy. Hundreds of different dyes with varying degrees of selectivity for tissue components are used for this type of work, which constitutes the basis of the scientific discipline *histology*.

EXAMPLE 1.2

In the clinical biochemical assessment of patients, it is common practice to inspect a blood sample under the light microscope, with a view to determining the number and type of inflammatory white cells present. A thin film of blood is smeared on a glass slide, which is then placed in methanol to *fix* the cells; this process rigidifies the cells and preserves their shape. The cells are then dyed by the addition of a few drops each of two dye mixtures; the most commonly used ones are the *Romanowsky* dyes, named after their nineteenth-century discoverer. The commonly used hematological dyeing procedure is that developed by J.W. Field: A mixture of *azure I* and *methylene blue* is first applied to the cells, followed by *eosin*; all dyes are dissolved in a simple phosphate buffer. The treatment stains nuclei blue, cell cytoplasm pink, and some subcellular organelles either pink or blue. On the basis of different staining patterns, at least five different types of white cells can be identified. Furthermore, intracellular organisms such as the malarial parasite *Plasmodium* stain blue.

The exact chemical mechanisms of differential staining of tissues are poorly understood. This aspect of histology is therefore still empirical. However, certain features of the chemical structure of dyes allow some interpretation of how they achieve their selectivity. They tend to be multi-ring, heterocyclic, aromatic compounds, with the high degree of bond conjugation giving the bright colors. In many cases they were originally isolated from plants, and they have a net positive or negative charge.

EXAMPLE 1.3

Methylene blue stains cellular nuclei blue.

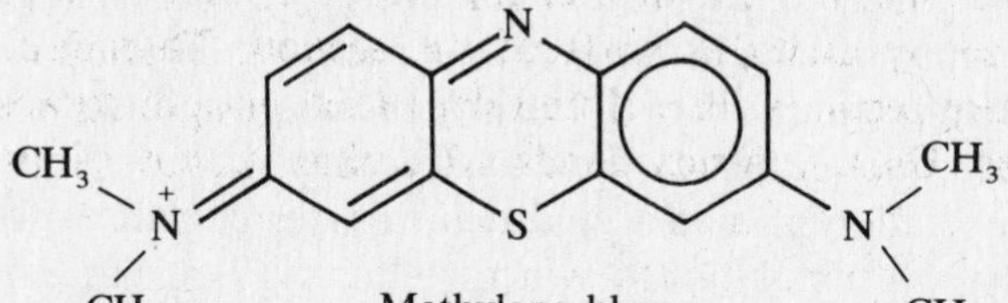

Methylene blue

Mechanism of staining: The positive charge on the N of methylene blue interacts with the anionic oxygens in the phosphate esters of DNA and RNA (Chap. 7).

Eosin stains protein-rich regions of cells red.

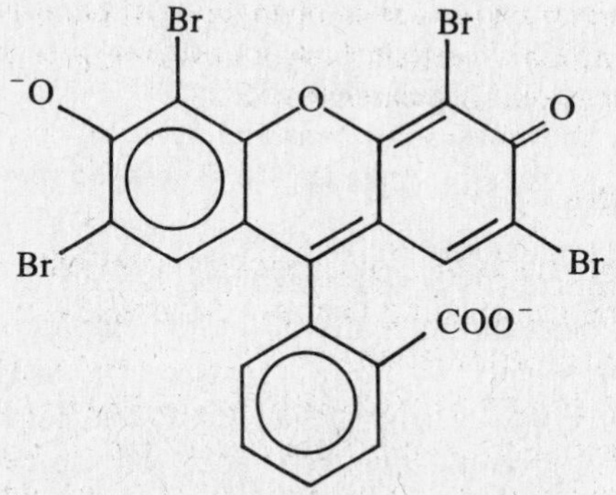

Eosin

Mechanism of staining: Eosin is a *dianion* at pH 7, and so it binds electrostatically to protein groups, such as arginyls, histidyls, and lysyls, that have a positive charge at this pH. Thus, this dye highlights protein-rich areas of cells.

PAS (*periodic acid Schiff*) stain is used for the histological staining of carbohydrates; it is also used to stain glycoproteins (proteins that contain carbohydrates; Chap. 2) in electrophoretic gels (Chap. 4). The stain mixture contains *periodic acid* (HIO_4), a powerful oxidant, and the dye *basic fuchsin*:

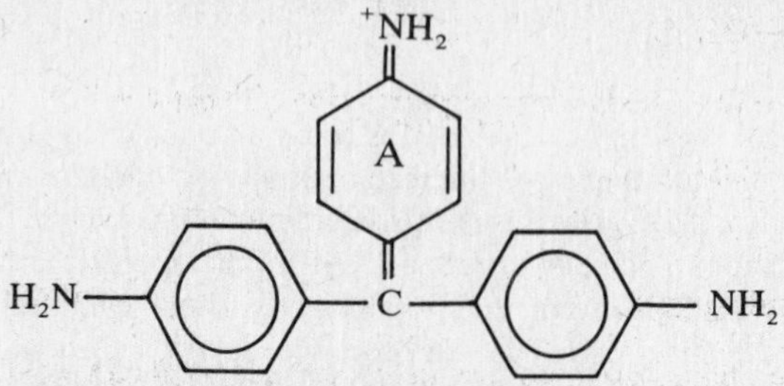

Basic fuchsin

Mechanism of staining: Periodic acid opens the sugar rings at cis-diol bonds (i.e., the C-2—C-3 bond of glucose) to form two aldehyde groups and *iodate* (IO_3^-). Then the $=^+NH_2$ group of the dye reacts to form a so-called *Schiff base* bond with the aldehyde, thus linking the dye to the carbohydrate. The basic reaction is:

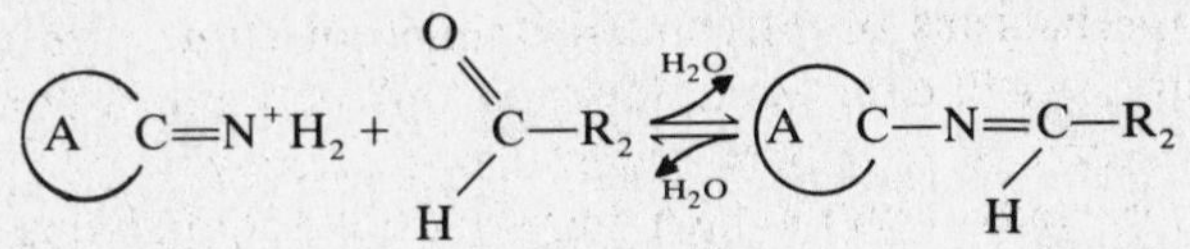

$$\text{(A}\ C{=}N^{+}H_2 + \overset{O}{\underset{H}{C}}{-}R_2 \underset{H_2O}{\overset{H_2O}{\rightleftharpoons}} \text{(A}\ C{-}N{=}\underset{H}{C}{-}R_2$$

The conversion of ring A of basic fuchsin to an aromatic one, with a carbocation (Chap. 8) at the central carbon, renders the compound pink.

Electron Microscopy

Image magnifications of thin tissue sections up to ×200,000 can be achieved using *electron microscopy*. The sample is placed in a high vacuum and exposed to a narrow beam of electrons that are differentially scattered by different parts of the section. Therefore, in staining the sample, differential electron density replaces the colored dyes used in light microscopy. A commonly used dye is *osmium tetroxide* (OsO_4), which binds to amino groups of proteins, leaving a black, electron-dense region.

EXAMPLE 1.4

The wavelength of electromagnetic radiation (light) limits the resolution attainable in microscopy. The *resolution* of a device is defined as the smallest gap that can be perceived between two objects when viewed with the device; resolution is approximately half the wavelength of the electromagnetic radiation used. Electrons accelerated to high velocities by an electrical potential of ~100,000 V have electromagnetic wave properties as well, with a wavelength of 0.004 nm; thus a resolution of about 0.002 nm is theoretically attainable with electron microscopy. This enables the distinction of certain features even on protein molecules, since the diameter of many globular proteins, e.g., hemoglobin, is greater than 3 nm.

Histochemistry and Cytochemistry

Histochemistry deals with whole tissues, and *cytochemistry* with individual cells. The techniques of these disciplines give a means for locating specific compounds or enzymes in tissues and cells. A tissue slice is incubated with the substrate of an enzyme of interest, and the product of this reaction is caused to react with a second, pigmented compound that is also present in the incubation mixture. If the samples are adequately fixed before incubation and the fixing process does not damage the enzyme, the procedure will highlight, in a thin section of tissue under the microscope, those cells which contain the enzyme or, at higher resolution, the subcellular organelles which contain it.

EXAMPLE 1.5

The enzyme *acid phosphatase* is located in the lysosomes (Sec. 1.3) of many cells, including those of the liver. The enzyme catalyzes the hydrolytic release of phosphate groups from various phosphate esters including the following:

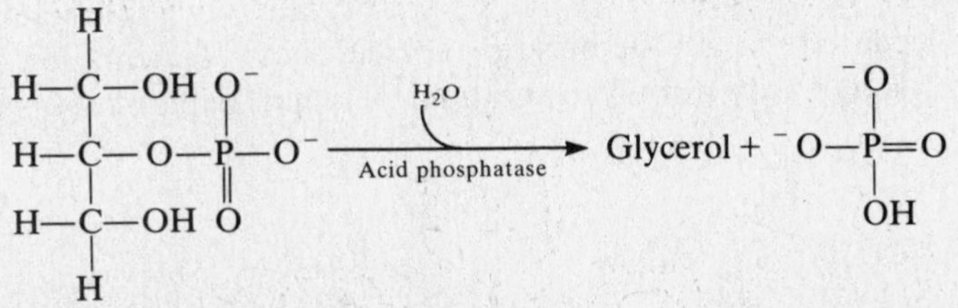

$$\begin{array}{c} \text{H} \\ | \\ \text{H—C—OH}\ \ \text{O}^- \\ |\qquad\quad | \\ \text{H—C—O—P—O}^- \\ |\qquad\quad \| \\ \text{H—C—OH}\ \ \text{O} \\ | \\ \text{H} \end{array} \xrightarrow[\text{Acid phosphatase}]{H_2O} \text{Glycerol} + \begin{array}{c} ^-\text{O} \\ | \\ ^-\text{O—P=O} \\ | \\ \text{OH} \end{array}$$

2-Phosphoglycerol　　　　　　Phosphate

In the *Gomori* procedure, tissue samples are incubated for ~30 min at 37°C in a suitable buffer that contains 2-phosphoglycerol. The sample is then washed free of the phosphate ester and placed in a buffer that contains lead nitrate. The 2-phosphoglycerol freely permeates lysosomal membranes, but the more highly charged phosphate does not, so that any of the phosphate released inside the lysosomes by phosphatase remains there. As the Pb^{2+} ions penetrate the lysosomes, they precipitate as lead phosphate. These regions of precipitation appear as dark spots in either an electron or a light micrograph.

Autoradiography

Autoradiography is a technique for locating radioactive compounds within cells; it can be conducted with light or electron microscopy. Living cells are first exposed to the *radioactive precursor* of some intracellular component. The labeled precursor is a compound with one or more hydrogen (^{1}H) atoms replaced by the *radioisotope tritium* (^{3}H); e.g., [^{3}H]thymidine is a labeled precursor of DNA, and [^{3}H]uridine is a labeled precursor of RNA (Chap. 7). Various tritiated amino acids are also available. The labeled precursors enter the cells and are incorporated into the appropriate macromolecules. The cells are then fixed, and the samples are embedded in a resin or wax and then sectioned into thin slices.

The radioactivity is detected by applying (in a darkroom) a photographic silver halide emulsion to the surface of the section. After the emulsion dries, the preparations are stored in a light-free box to permit the radioactive decay to expose the overlying emulsion. The length of exposure depends on the amount of radioactivity in the sample but is typically several days to a few weeks for light microscopy and up to several months for electron microscopy. The long exposure time in electron microscopy is necessary because of the *very* thin sections ($<1\ \mu$m) and thus the minute amounts of radioactivity present in the tiny samples. The preparations are developed and fixed as in conventional photography. Thus, the silver grains overlie regions of the cell that contain radioactive molecules; the grains appear as tiny black dots in light micrographs and as twisted black threads in electron micrographs. Note that this whole procedure works only if the precursor molecule can traverse the cell membrane and if the cells are in a phase of their life cycle that involves incorporation of the compound into macromolecules.

EXAMPLE 1.6

The sequence of events involved in the synthesis and transport of *secretory proteins* from glands can be followed using autoradiography. For example, rats were injected with [^{3}H]leucine, and at intervals thereafter they were sacrificed and autoradiographs of their *prostate glands* prepared. In electron micrographs of the sample obtained 4 min after the injection, silver grains appeared overlying the *rough endoplasmic reticulum* (RER) of the cells, indicating that [^{3}H]leucine had been incorporated from the blood into protein by the *ribosomes* attached to the RER. By 30 min the grains were overlying the Golgi apparatus and secretory vacuoles, reflecting intracellular transport of labeled secretory proteins from the RER to those organelles. At later times after the injection radioactive proteins were released from the cells, as evidenced by the presence of silver grains over the glandular lumens.

Ultracentrifugation

The *biochemical* roles of subcellular organelles could not be studied properly until the organelles had been separated by fractionation of the cells. George Palade and his colleagues, in the late 1940s, showed that *homogenates* of rat liver could be separated into several fractions using *differential centrifugation*. This procedure relies on the different velocities of sedimentation of various organelles of different shape, size, and density through a solution. A typical experiment is outlined in Example 1.7.

EXAMPLE 1.7

Liver is suspended in 0.25 *M* sucrose and then disrupted using a rotating, close-fitting Teflon plunger in a glass barrel (known as a *Potter-Elvehjem homogenizer*). Care is taken not to destroy the organelles by excessive homogenization. The sample is then spun in a centrifuge (see Fig. 1-1). The nuclei tend to be the first to sediment to the bottom of the sample tube at forces as low as 1,000 *g* for ~15 min in a tube 7 cm long.

High-speed centrifugation, such as 10,000 *g* for 20 min, yields a pellet composed mostly of mitochondria, but contaminated with lysosomes. Further centrifugation at 100,000 *g* for 1 h yields a pellet of ribosomes and so-called microsomes that contain endoplasmic reticulum. The soluble protein and other solutes remain in the *supernatant* (overlying solution) from this step.

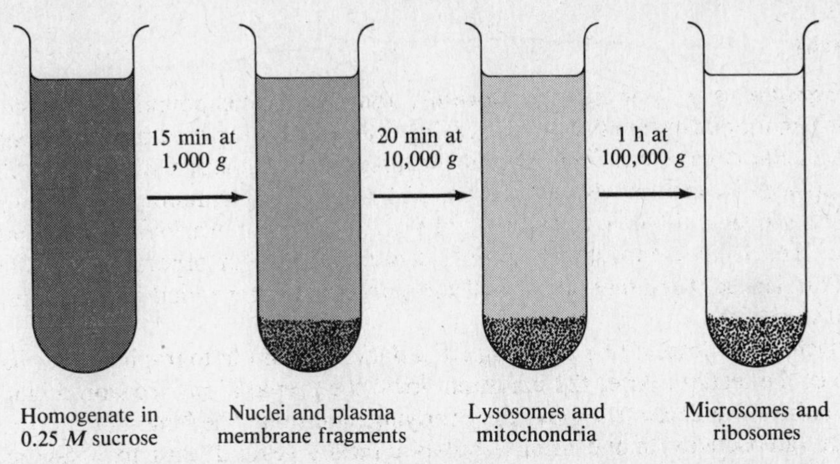

Fig. 1-1 Separation of subcellular organelles by differential centrifugation of cell homogenates.

Density-gradient centrifugation (also called *isopycnic centrifugation*) can also be used to separate the different organelles (Fig. 1-2). The homogenate is layered onto a discontinuous or continuous concentration gradient of sucrose solution, and centrifugation continues until the subcellular particles achieve density equilibrium with their surrounding solution.

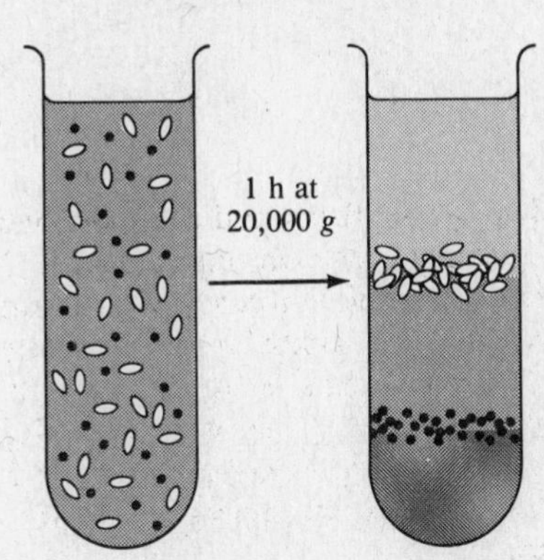

Fig. 1-2 Isopycnic centrifugation of organelles. The shading indicates increasing solution density.

Question: Can a procedure similar to isopycnic separation in a centrifugal field be used to separate different *macromolecules*?

Yes; in fact one way of preparing and purifying DNA fragments for genetic engineering uses density gradients of CsCl. Various proteins also have different densities and thus can be separated on sucrose density gradients; however, the time required to attain equilibrium is much longer, and higher centrifuge velocities are needed than is the case for organelles.

1.3 SUBCELLULAR ORGANELLES

Question: What does a typical animal cell look like?

There is no such thing as a typical animal cell, since cells vary in overall size, shape, and distribution of the various subcellular organelles. Fig. 1-3 is, however, a composite diagram that indicates the relative sizes of the various *microbodies*.

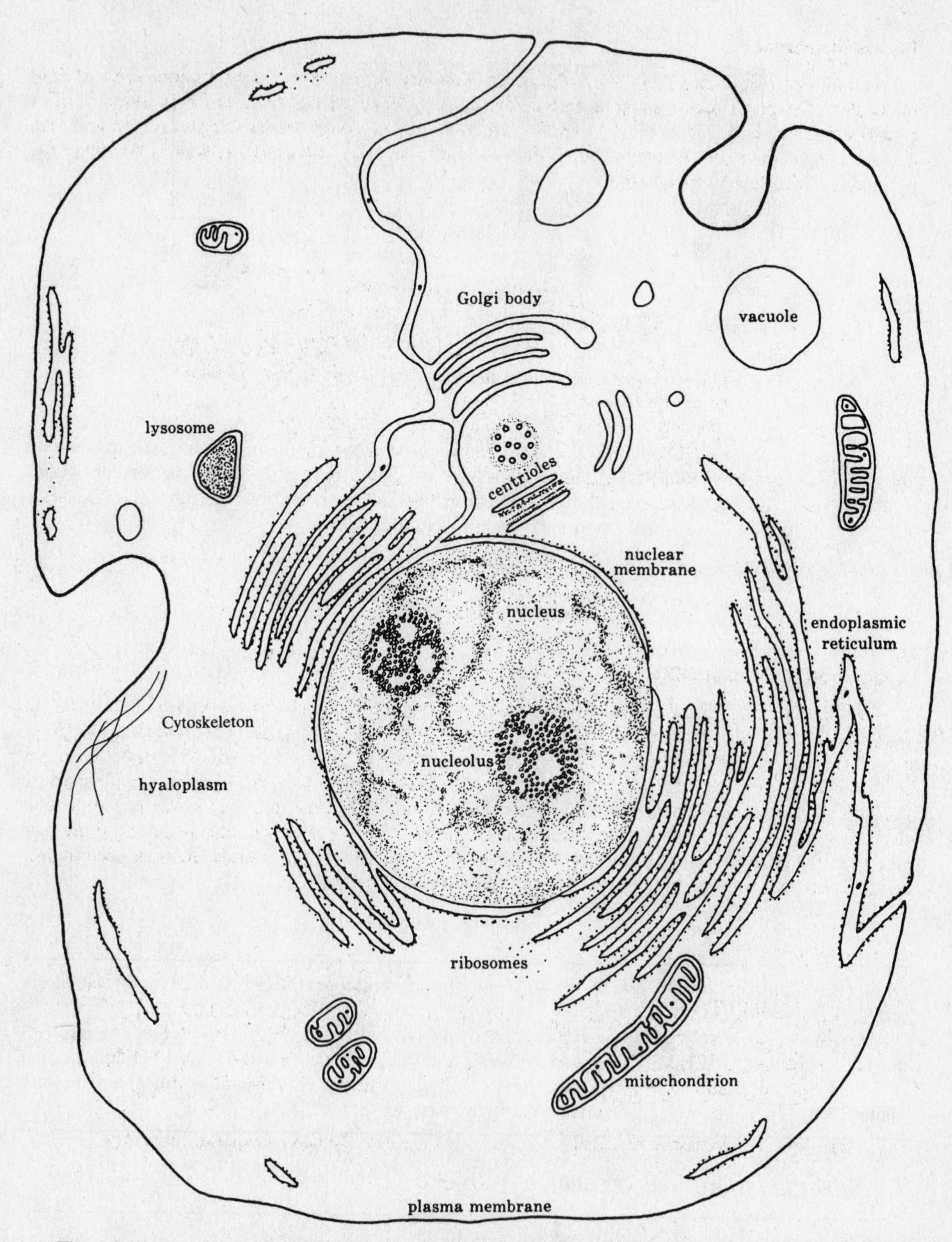

Fig. 1-3 Diagram of a mammalian cell. The organelles are approximately the correct relative sizes.

Plasma Membrane

The *plasma membrane* (Fig. 1-4) is the outer boundary of the cell; it is a continuous sheet of lipid molecules (Chap. 6) arranged as a molecular bilayer 4–5 nm thick. In it are embedded various proteins that function as enzymes (Chap. 8), structural elements, and molecular pumps and selective channels that allow entry of certain small molecules into and out of the cell, as well as receptors for hormones and cell growth factors.

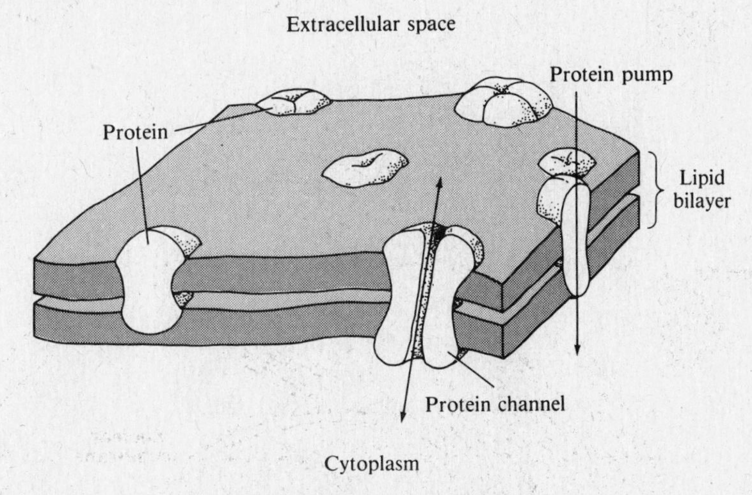

Fig. 1-4 Plasma membrane.

Endoplasmic Reticulum (ER)

The *endoplasmic reticulum* (ER) is composed of flattened sacs and tubes of membranous bilayers that extend throughout the cytoplasm enclosing a large intracellular space. The *luminal* space (Fig. 1-5) is continuous with the outer membrane of the *nuclear envelope* (Fig. 1-11). It is involved in the synthesis of proteins and their transport to the cytoplasmic membrane (via *vesicles*, small spherical particles with an outer bilayer membrane). The *rough* ER (RER) has flattened stacks of membrane that are studded on the outer (cytoplasmic) face with *ribosomes* (discussed later in this section) that actively synthesize proteins (Chap. 17). The *smooth* ER (SER) is more tubular in cross section and lacks ribosomes; it has a major role in lipid metabolism (Chap. 13).

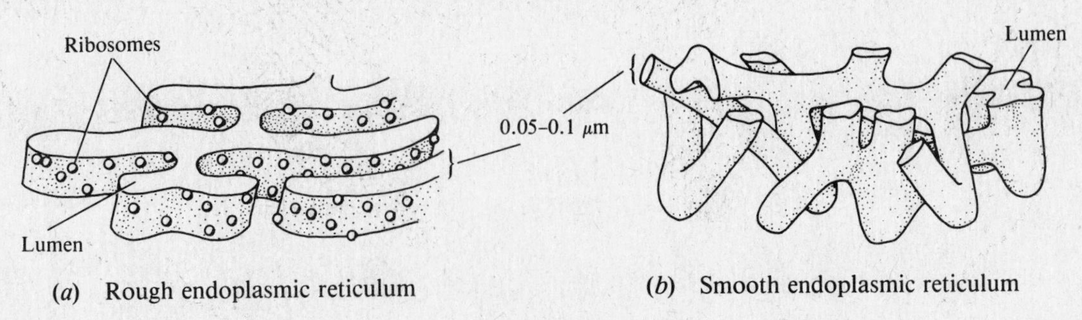

(*a*) Rough endoplasmic reticulum (*b*) Smooth endoplasmic reticulum

Fig. 1-5

EXAMPLE 1.8

What mass fraction of the lipid membranes of a liver cell is plasma membrane?

Only about 10 percent; the remainder is principally ER and mitochondrial membrane.

Golgi Apparatus

The *Golgi apparatus* is a system of *stacked*, membrane-bound, flattened sacs organized in order of decreasing breadth (see Fig. 1-6). Around this system are small *vesicles* (50-nm diameter and larger); these are the secretory vacuoles that contain protein that is *released* from the cell (see Example 1.6).

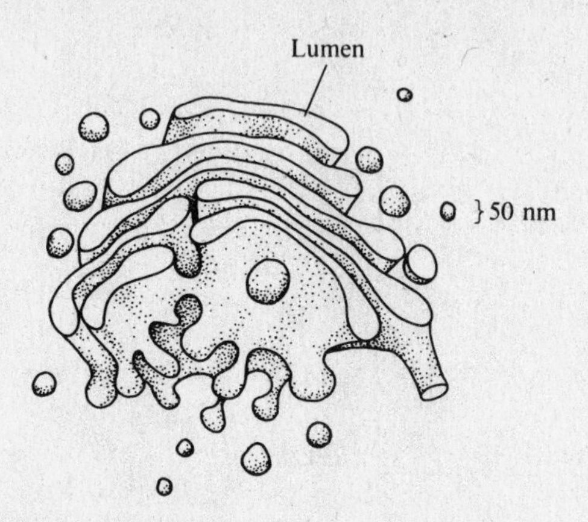

Fig. 1-6 Golgi apparatus and secretory vesicles.

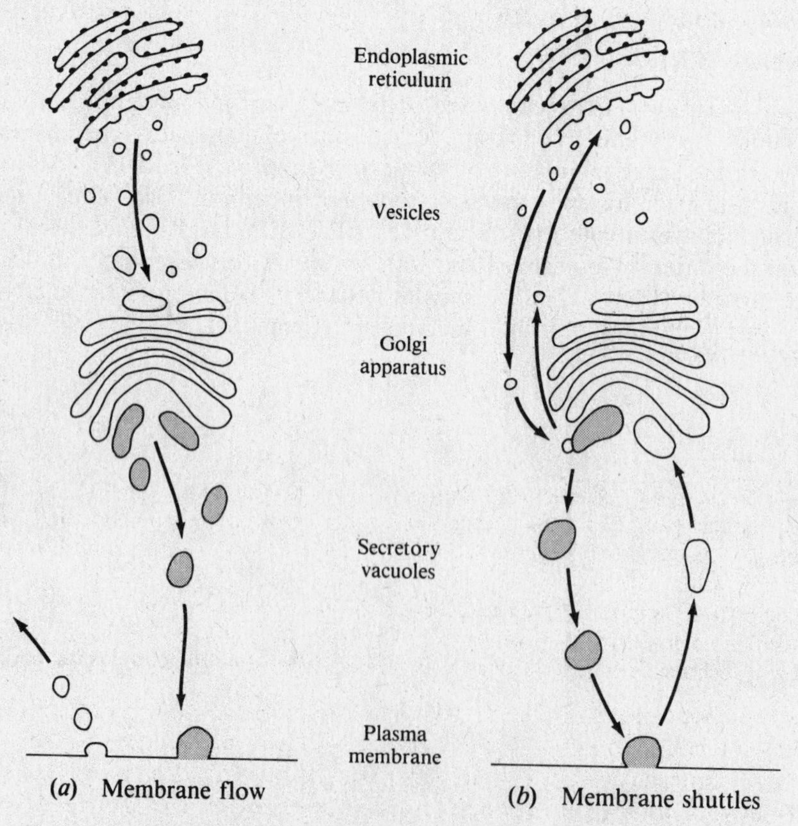

Fig. 1-7 Possible membrane-exchange pathways during secretion of protein from a cell.

The pathway of secretory proteins and glycoproteins (Chap. 4) through *exocrine* (secretory) gland cells in which *secretory vacuoles* are present is well established. However, the exact pathway of exchange of the membranes between the various organelles is less clear and could be either one or a combination of both of the schemes shown in Fig. 1-7.

In the *membrane flow* model of Fig. 1-7, membranes move through the cell from ER→Golgi apparatus→secretory vacuoles→plasma membrane. In the *membrane shuttle* proposal, the vesicles shuttle between ER and Golgi apparatus, while secretory vacuoles shuttle back and forth between the Golgi apparatus and the plasma membrane.

Question: What controls the *directed* flow of membranous organelles?

No one really knows; it is one of the great wonders of cell physiology yet to be fully understood.

Lysosomes

Lysosomes are membrane-bound vesicles that contain *acid hydrolases*; these are enzymes that catalyze hydrolytic reactions and function optimally at a pH (~5) found in these organelles. Lysosomes range in size from 0.2 to 0.5 μm. They are instrumental in intracellular digestion (*autophagy*) and the digestion of material from outside the cell (*heterophagy*). Heterophagy, which is involved with the body's removal of bacteria, begins with the invagination of the plasma membrane, a process called *endocytosis*; the whole digestion pathway is shown in Fig. 1-8.

Since lysosomes are involved in digesting a whole range of biological material, exemplified by the destruction of a whole bacterium with all its different types of macromolecules, it is not surprising to find that a large number of *different* hydrolases reside in lysosomes. These enzymes catalyze the breakdown of nucleic acids, proteins, cell wall carbohydrates, and phospholipid membranes (see Table 1.1).

Mitochondria

Mitochondria are membranous organelles (Fig. 1-9) of great importance in the energy metabolism of the cell; they are the source of most of the ATP (Chap. 14) and the site of many metabolic reactions. Specifically, they contain the enzymes of the citric acid cycle (Chap. 12) and the electron-transport chain (Chap. 14), which includes the main oxygen-utilizing reaction of the cell. A mammalian liver cell contains about 1,000 of these organelles; about 20 percent of the cytoplasmic volume is mitochondrial.

EXAMPLE 1.9

Mitochondria were first observed by R. Altmann in 1890. He named them *bioblasts*, because he speculated that they and chloroplasts (the green chlorophyll-containing organelles of plants) might be intracellular *symbionts* that arose from bacteria and algae, respectively. This idea lay in disrepute until the recent discovery of mitochondrial nucleic acids.

In histology, mitochondria can be stained *supravitally*; i.e., the metabolic activity of the functional (*vital* = living) organelle or cell allows selective staining. The *reduced* form of the dye Janus green B is colorless, but it is *oxidized* by mitochondria to give a light-green pigment that is easily seen in light microscopy.

Mitochondria are about the size of bacteria. They have a diameter of 0.2 to 0.5 μm and are 0.5 to 7 μm long. They are bounded by *two* lipid bilayers, the inner one being highly folded. These folds are called *cristae*. The innermost space of the mitochondrion is called the *matrix*. They have their own DNA in the form of at least one copy of a circular double helix (Chap. 7), about 5 μm in overall diameter; it differs from nuclear DNA in its density and denaturation temperature by virtue of being richer in guanosine and cytosine (Chap. 7). The different density from nuclear DNA allows

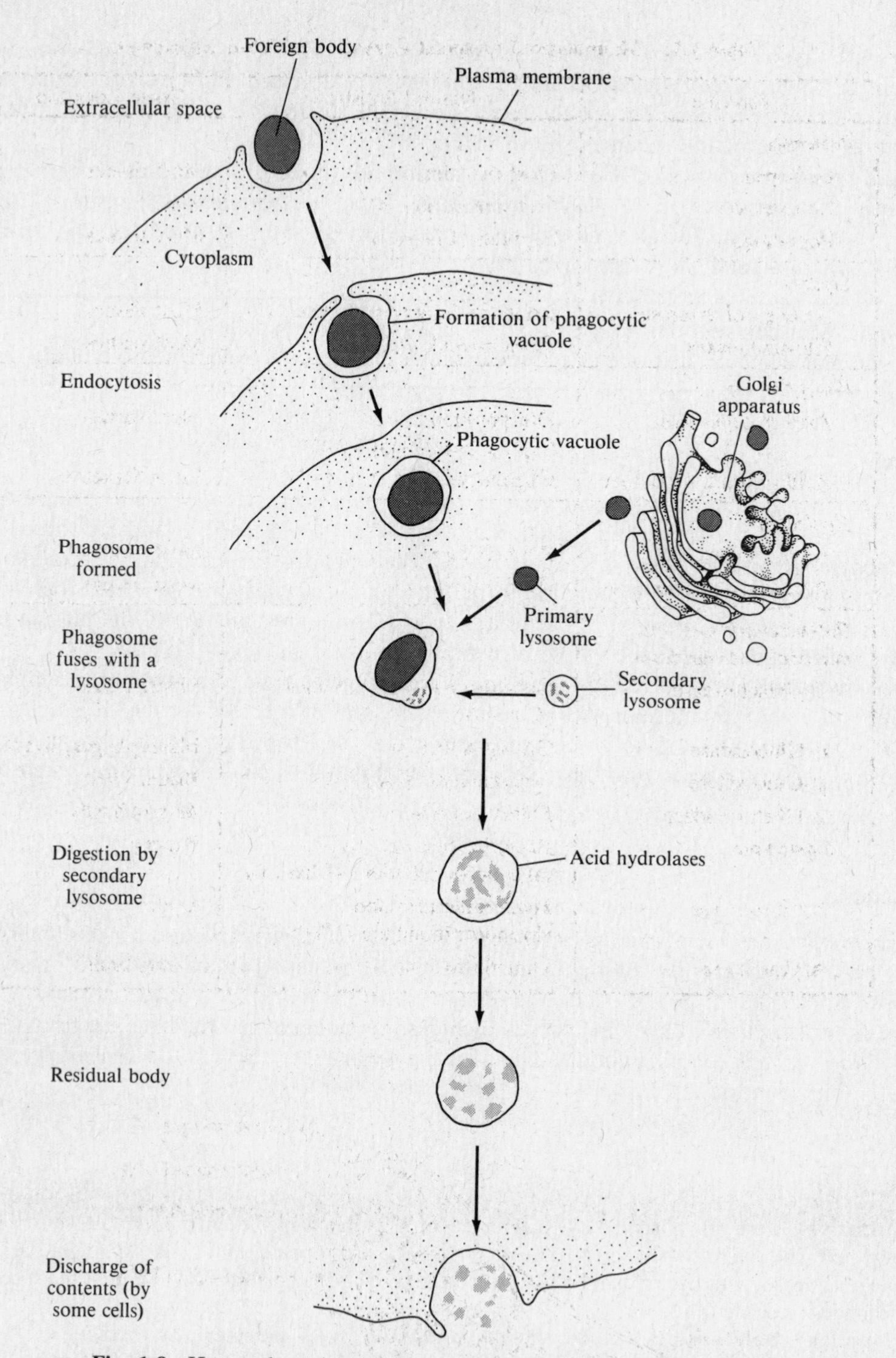

Fig. 1-8 Heterophagy in a mammalian cell, typically a *macrophage*.

its separation by isopycnic centrifugation. Mitochondria also have their own type of *ribosomes* that differ from those in the cytoplasm but are similar to those of bacteria.

Most of the enzymes in mitochondria are *imported* from the cytoplasm; the enzyme proteins are largely coded for by *nuclear* DNA (Chap. 17). The enzymes are disposed in various specific regions of the mitochondria (Table 1.2); this has an important bearing on the direction of certain metabolic processes.

Table 1.1. Mammalian Lysosomal Enzymes and Their Substrates

Enzyme	Natural Substrate	Tissue Location
Proteases		
Cathepsin	Most proteins	Most tissues
Collagenase	Collagen (Chap. 4)	Bone
Peptidases	Peptides (Chap. 3)	Most tissues
Lipases		
A range of esterases	Esters of fatty acids (Chap. 13)	Most tissues
Phospholipases	Phospholipids (Chap. 6)	Most tissues
Phosphatases		
Acid phosphatase	Phosphomonoesters (e.g., 2-phosphoglycerol)	Most tissues
Acid phosphodiesterase	Oligonucleotides (Chap. 7)	Most tissues
Nucleases		
Acid ribonuclease	RNA (Chap. 7)	Most tissues
Acid deoxyribonuclease	DNA (Chap. 7)	Most tissues
Polysaccharidases and mucopolysaccharidases		
β-Galactosidase	Galactosides of membranes (Chap. 6)	Liver, brain
α-Glucosidase	Glycogen (Chap. 11)	Macrophages, liver
β-Glucosidase	Gangliosides (Chap. 6)	Brain, liver
β-Glucuronidase	Polysaccharides	Macrophages
Lysozyme	Bacterial cell wall and mucopolysaccharides (Chap. 8)	Kidney
Hyaluronidase	Hyaluronic acid and chrondroitin sulfate (Chap. 2)	Liver
Arylsulfatase	Organic sulfates	Liver, brain

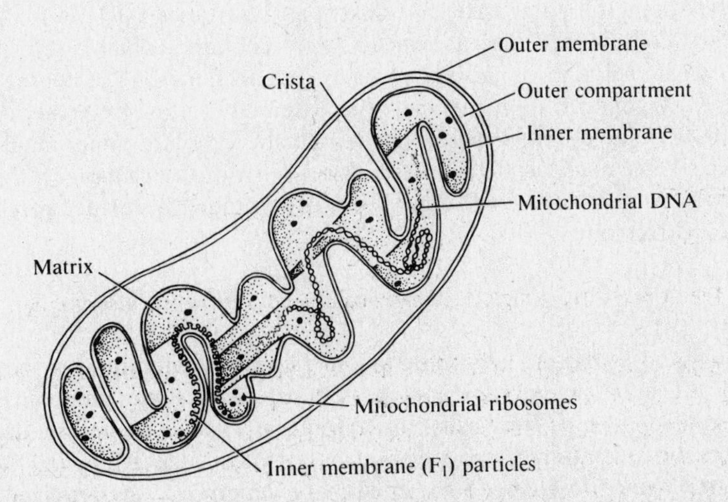

Fig. 1-9 Mitochondrion.

Table 1.2. Enzyme Distribution in Mitochondria

Location	Characteristics or Cross-Reference to Discussion
Outer membrane	
Monoamine oxidase	Neurotransmitter; catabolism
Rotenone-insensitive NADH–cytochrome c reductase	Chap. 14
Kynurenine hydroxylase	Tryptophan catabolism; Chap. 15
Fatty acid–CoA ligase	Chap. 13
Space between inner and outer membrane	
Adenylate kinase	$AMP + ATP \rightleftharpoons 2ADP$
Nucleoside diphosphokinase	$XDP + YTP \rightleftharpoons XTP + YDP$ where X and Y are any of several ribonucleosides
Inner membrane	
Respiratory chain enzymes	Chap. 14
ATP synthetase	Chap. 14
Succinate dehydrogenase	Chap. 14
β-Hydroxybutyrate dehydrogenase	Chap. 11
Carnitine–fatty acid acyltransferase	Chap. 13
Matrix	
Malate and isocitrate dehydrogenase	Chap. 12
Fumarase and aconitase	Chap. 12
Citrate synthase	Chap. 12
2-Oxoacid dehydrogenase	Chap. 13
β-Oxidative enzymes for fatty acids	Chap. 13
Carbamoyl phosphate synthetase I	Chap. 15
Ornithine carbamoyltransferase	Chap. 15

Peroxisomes

Peroxisomes are about the same size and shape as lysosomes (0.3 to 1.5 μm in diameter). However they do *not* contain hydrolases; instead, they contain *oxidative* enzymes that generate *hydrogen peroxide* by catalyzing the combination of oxygen with a range of compounds. The various enzymes present in *high* concentration (even to the extent of forming crystals of protein) are (1) urate oxidase, Chap. 15; (2) D-amino acid oxidase, Chap. 15; (3) L-amino acid oxidase; and (4) α-hydroxy acid oxidase (includes lactate oxidase). Also, most of the *catalase* in the cell is contained in peroxisomes; this enzyme catalyzes the conversion of hydrogen peroxide, produced in the other reactions, to water and oxygen.

Cytoskeleton

In the cytoplasm, and especially subjacent to the plasma membrane, are networks of protein filaments that stabilize the lipid membrane and thus contribute to the maintenance of cell shape. In cells that grow and divide, such as liver cells, the cytoplasm appears to be organized from a region near the nucleus that contains the cell's pair of *centrioles* (Fig. 1-10). There are three main types of cytoskeletal filaments: (1) *microtubules*, 25 nm in diameter, composed of organized aggregates of the protein *tubulin* (Chap. 5); (2) *actin* filaments, 7 nm in diameter (Chap. 5); and (3) so-called *intermediate filaments*, 10 nm in diameter (Chap. 5).

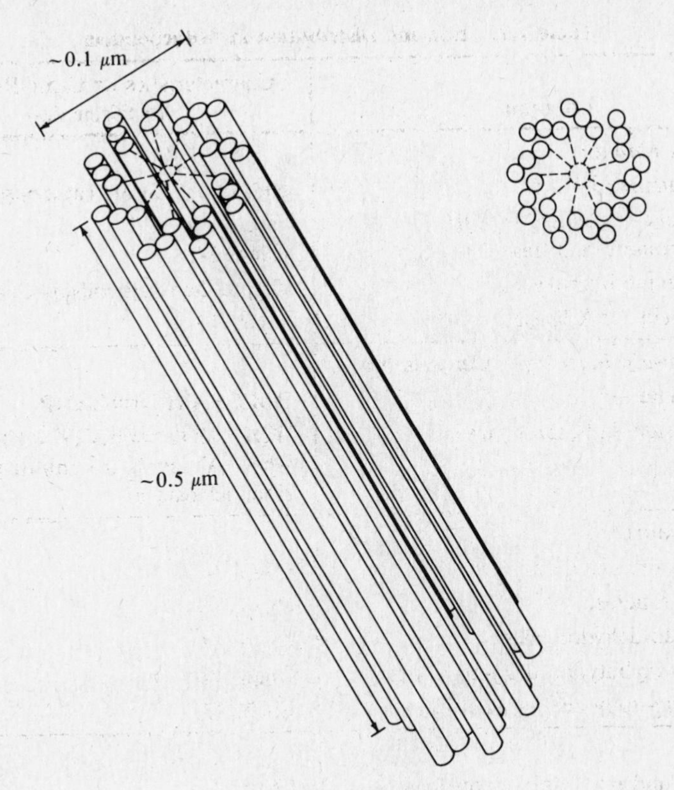

Fig. 1-10 Centriole.

Centrioles

Centrioles are a pair of hollow cylinders that are composed of nine triplet tubules of protein (Fig. 1-10). The members of a pair of centrioles are usually positioned at right angles to each other. Microtubules form the fine weblike protein structure that appears to be attached to chromosomes during cell division (mitosis); the web is called the *mitotic spindle* and is attached to the ends of the centrioles. While centrioles are thought to function in chromosome segregation during mitosis, it is worth noting that cells of higher plants, which clearly undergo this process, lack centrioles.

Ribosomes

Ribosomes are the site of protein synthesis and exist (1) in the cytoplasm as rosette-shaped groups called *polysomes* (in immature red blood cells there are usually five per group), (2) on the outer face of the RER, or (3) in the mitochondrial matrix, although this last type is different in size and shape from ribosomes in the cytoplasm. Ribosomes are composed of RNA and protein and range in size from 15 to 20 nm. Their central role in protein synthesis is described in Chap. 16.

EXAMPLE 1.10

Ribosomes were first isolated by differential centrifugation and then examined by electron microscopy. This and related work by George Palade in the early 1950s earned him the Nobel prize in 1975. For a time ribosomes were known to electron microscopists as *Palade's granules*.

Nucleus

The *nucleus* is the most conspicuous organelle of the cell (see Fig. 1-11). It is delimited from the cytoplasm by a membranous envelope called the *nuclear membrane*, which actually consists of two membranes forming a flattened sac. The nuclear membrane is perforated by *nuclear pores* (60 nm in

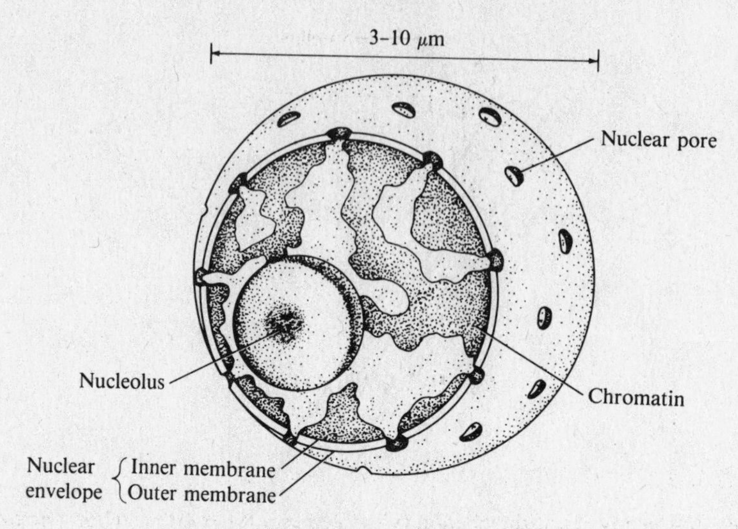

Fig. 1-11 Mammalian cell nucleus.

diameter), which allow transfer of material between the *nucleoplasm* and the cytoplasm. The nucleus contains the *chromosomes*, which consist of *DNA* packaged into *chromatin* fibers by association of the DNA with an equal mass of *histone* proteins (Chap. 16).

Nucleolus

The *nucleolus* is composed of 5 to 10 percent RNA, and the remainder of the mass is protein and DNA. In light microscopy it appears to be spherical and *basophilic* (Prob. 1.1). Its function is the synthesis of ribosomal RNA (Chap. 7). There may be more than one per nucleus.

Chromosomes

Chromosomes are the bearers of the hereditary instructions in a cell; thus they are the overall regulators of cellular processes. Important features to note about chromosomes are:

(*a*) *Chromosome number*. In animals, each *somatic* cell (body cells, excluding sex cells) contains one set of chromosomes inherited from the female parent and a comparable (*homologous*) set from the male parent. The number of chromosomes in the dual set is called the *diploid number*; the suffix *-ploid* means "a set" and the *di* refers to the multiplicity of the set (in this case, "two"). Sex cells (called *gametes*) contain *half* as many chromosomes as found in somatic cells and are therefore referred to as *haploid* cells. A *genome* is the set of chromosomes that corresponds to the haploid set of a species.

EXAMPLE 1.11

Human somatic cells contain 46 chromosomes, cattle 60, and fruit fly 8. Thus, the diploid number bears no relationship to the species' positions in the phylogenetic scheme of classification.

(*b*) *Chromosome morphology*. Chromosomes become visible under the light microscope only at certain phases of the nuclear division cycle. Each chromosome in the genome can usually be distinguished from the others by such features as (1) relative length of the whole chromosome; (2) the position of the *centromere*, a structure that divides the chromosome into a crosslike structure with two pairs of arms of different length; (3) the presence of knobs of chromatin called *chromomeres*; and (4) the presence of small terminal extensions called *satellites* (Fig. 1-12).

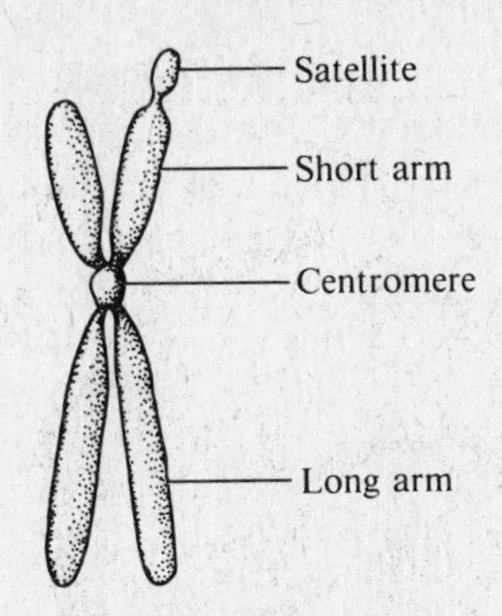

Fig. 1-12 Mammalian chromosome.

EXAMPLE 1.12

In the clinical investigation of infants or fetuses with possible inborn errors of metabolism or morphology, it is common practice to prepare a *karyotype*. Usually, white cells are cultured and then stimulated to divide. The predivision cells are squashed between glass slides, causing the cellular nuclei to disgorge their chromosomes, which are then stained with a blue dye. The chromosomes are photographed and then ordered according to their length, the longest pair being numbered 1. The sex chromosomes do not have a number.

The inherited disorder *Down's syndrome* (also called mongolism) involves mental retardation and distinctive facial features. It results from the inclusion of an extra chromosome 21 in each somatic cell of the body. Hence, the condition is called *trisomy 21.*

(*c*) *Autosomes and sex chromosomes*. In humans, gender is associated with a morphologically dissimilar pair of chromosomes called the *sex chromosomes*. The two members of the pair are labeled *X* and *Y*, X being the larger. Genetic factors on the Y chromosome determine maleness. All chromosomes, exclusive of the sex chromosomes, are called *autosomes*.

1.4 CELL TYPES

There are over 200 different cell types in the human body. These are arranged in a variety of different ways, often with mixtures of cell types, to form *tissues*. Among this vast array of types are some highly specialized ones.

Red Blood Cell (Erythrocyte)

Erythrocytes are small compared with most other cells and are peculiar because of their biconcave disk shape (see Fig. 1-13). They have no nucleus, because it is extruded just before the

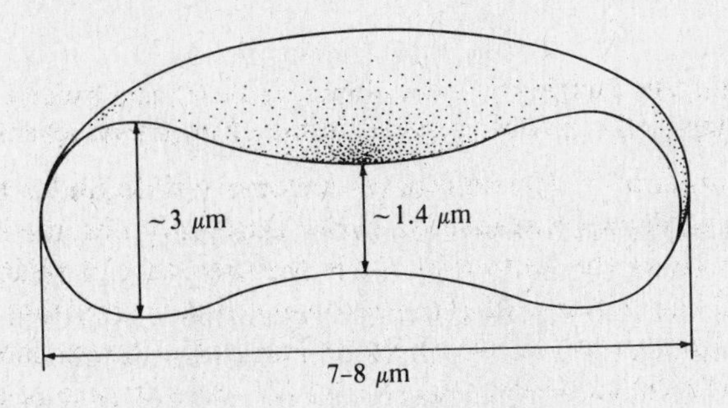

Fig. 1-13 Human erythrocyte.

release of the cell into the blood stream from the bone marrow, where the cells develop. Their cytoplasm has no organelles and is full of the protein *hemoglobin* that binds O_2 and CO_2. In the cytoplasm are other proteins also, namely, (1) the submembrane cytoskeleton, (2) enzymes of the glocolytic and pentose phosphate pathways (Chap. 11), and (3) a range of other hydrolytic and special-function enzymes that will not be discussed here. In the membrane are specialized proteins associated with (1) *anion transport* and (2) the carrying of the carbohydrate cell-surface antigens (blood group substances).

Adipocyte

Adipocytes are the specialized cells of fat tissue (Fig. 1-14). The cells range in size from 60 to 120 μm in diameter and have the characteristic feature of a huge vacuole that is full of triglycerides (Chap. 6). The nucleus and mitochondria are flattened on one inner surface of the plasma membrane, and there is only a small amount of endoplasmic reticulum.

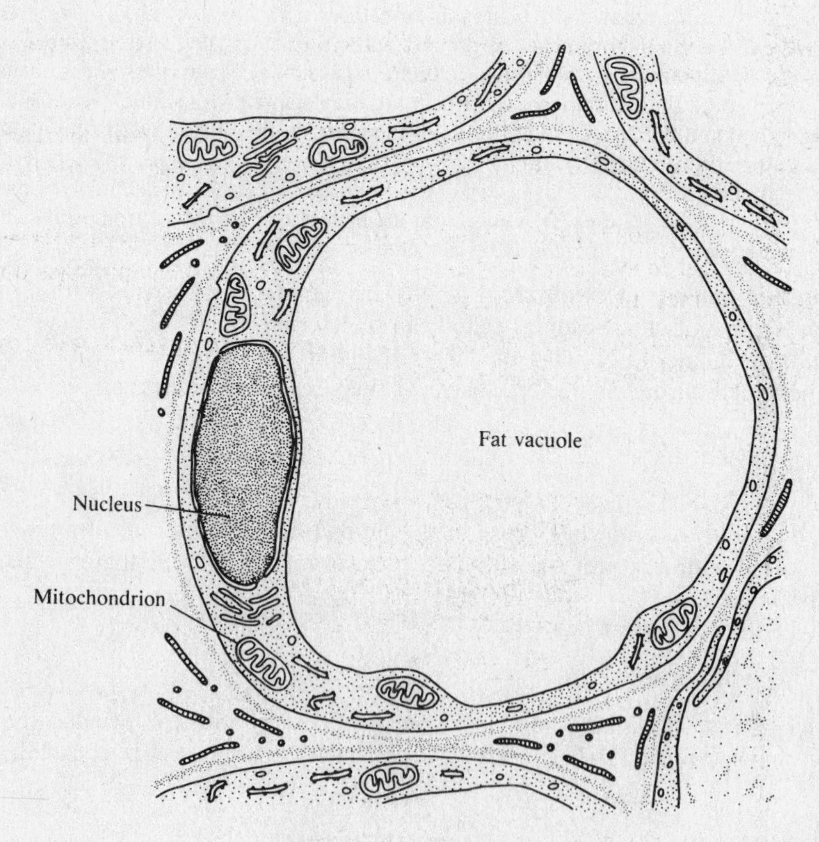

Fig. 1-14 Adipocyte.

Liver Cell (Hepatocyte)

Liver tissue contains an array of cell types, but the preponderant one is the *hepatocyte*. It has an overall structure much like that of the cell in Fig. 1-3. The cells are arranged in long, branching columns of about 20 cells in a cross section around a central *bile cannaliculus* (channel). Into the cannaliculus the cells secrete bile. The liver is the main producer of urea (Chap. 15), stores glycogen (Chap. 11), synthesizes many of the amino acids used by other tissues (Chap. 3), and produces serum proteins, among many other metabolic roles.

Muscle Cell (Myocyte)

Muscle cells produce mechanical force by contraction. In vertebrates there are three basic types:

1. *Skeletal muscle* moves the bones attached to joints. These muscles are composed of bundles of long, multinucleated cells. The cytoplasm contains a high concentration of a special macromolecular contractile-protein complex, *actomyosin* (Chap. 5). There is also an elaborate membranous network called the *sarcoplasmic reticulum* that has a high Ca^{2+} content. The contractile-protein complex has a banded appearance under microscopy.
2. *Smooth muscle* is the type in the walls of blood vessels and the intestine. The cells are long and spindle-shaped, and they lack the banding of skeletal muscle cells.
3. *Cardiac muscle* is the main tissue of the heart. The cells are similar in appearance to those of skeletal muscle but in fact have a different biochemical makeup.

Epithelia

Epithelial cells (Fig. 1-15) form the coherent sheets that line the inner and outer surfaces of the body. There are many specialized types, but the main groups are as follows:

1. *Absorptive cells* have numerous hairlike projections called *microvilli* on their *outer* surface; they increase the surface area for absorption of nutrients from the gut lumen and other areas.
2. *Ciliated cells* have small membranous projections (*cilia*) that contain interior contractile proteins; cilia beat in synchrony and serve to sweep away foreign particles on the surface of the respiratory tract, i.e., in the lungs and the nasal lining.
3. *Secretory cells* occur in most epithelial surfaces; e.g., sweat gland cells in the skin and mucus-secreting cells in the intestine and respiratory tract.

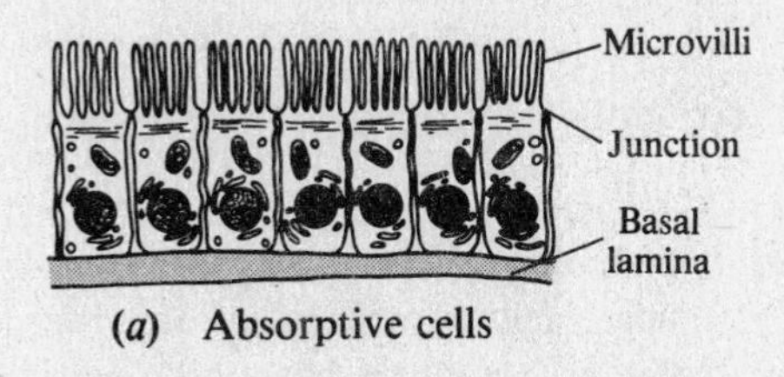

(*a*) Absorptive cells

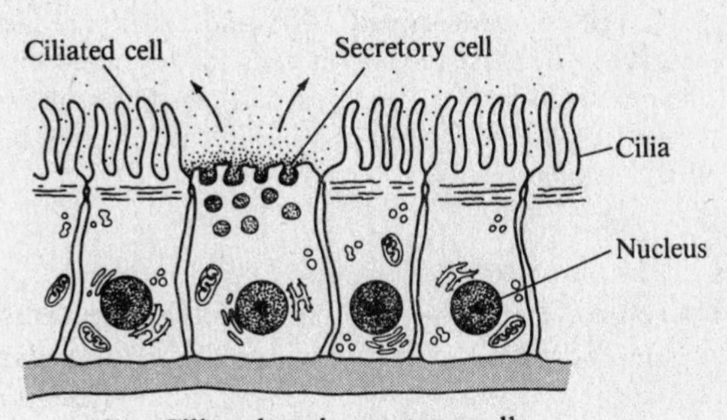

(*b*) Ciliated and secretory cells

Fig. 1-15 Epithelial cells.

1.5 THE STRUCTURAL HIERARCHY IN CELLS

The organic molecules that are building blocks of biological macromolecules are very small; e.g., the amino acid alanine is only 0.7 nm long, whereas a typical globular protein, hemoglobin (Chap. 5), which consists of 574 amino acids, has a diameter of ~6 nm. In turn, protein molecules are small compared with the ribosomes that synthesize them (Chap. 17); these macromolecular aggregates are composed of over 70 different proteins and four nucleic acid strands (Chap. 16). They have a molecular weight (M_r) of around 2.8×10^6 and a diameter of ~20 nm. In contrast, mitochondria contain their own ribosomes and DNA and range in length up to 7 μm. Intracellular vesicles are often seen to be about the same size as mitochondria, and yet the Golgi apparatus or the lipid vacuole of an adipocyte is much larger. The nucleus may be even larger and also contains some ribosomes and other macromolecular aggregates, including, most importantly, the chromosomes.

Even though the building blocks of macromolecules are small in relation to the size of the cell (e.g., the ratio of the volume of one molecule of alanine to that of the red blood cell is $1:10^{11}$), a defect in the order of sequence of one amino acid in a protein can profoundly affect not only the protein but also the cell structure. Furthermore, an altered enzymatic activity or binding affinity can greatly influence the survival of not only the cell but the whole being.

EXAMPLE 1.13

In humans having the inherited disease called *sickle-cell anemia*, the hemoglobin molecules of the erythrocytes are defective; 2 of the 574 amino acids in the protein are substituted for another. Specifically, glutamate in position 6 of each of the two β chains of the hemoglobin tetramer (see Chap. 5) is replaced by valine. This single change increases the likelihood of the molecules aggregating when they are deoxygenated. The aggregated protein forms large *paracrystalline structures* (called *tactoids*) inside the cells and distorts the cells into a relatively inflexible sickle shape. These cells tend to clog small blood vessels and capillaries and lead to poor oxygen supply in many organs. Also, they are more fragile and thus rupture, reducing the number of cells and causing anemia.

Solved Problems

METHODS OF STUDYING THE STRUCTURE AND FUNCTION OF CELLS

1.1. Basic dyes such as methylene blue or toluidine blue are positively charged at the pH of most staining solutions used in histology. Thus the dyes bind to acidic (negatively charged) substances in the cell. These acidic molecules are therefore referred to as basophilic substances in cells. Give some examples of basophilic substances.

SOLUTION

Examples of basophilic cell components are DNA and RNA; the latter includes messenger RNA (Chap. 7) and ribosomes. The youngest red blood cells in the blood circulation contain a basophilic *reticulum* (network) in their cytoplasm; this is composed of messenger and ribosomal RNA. The network slowly dissolves over the first 24 hours of the cell's life in the circulation. This readily identifiable red blood cell type is called the *reticulocyte*.

1.2. Acidic dyes such as eosin and acid fuchsin have a net negative charge at the pH of usual staining solutions. Therefore, they bind to many cellular proteins that have a net positive charge. Give some regions of a liver cell that might be *acidophilic*.

SOLUTION

The cytoplasm, mitochondrial matrix, and the inside of the smooth endoplasmic reticulum are acidophilic; all these regions are almost exclusively protein in content.

1.3. Describe a possible means for the cytochemical detection and localization of the enzyme *glucose 6-phosphatase*; it exists in liver and catalyzes the following reaction:

$$\text{Glucose 6-phosphate} \xrightarrow{H_2O} \text{Glucose} + \text{Phosphate}$$

SOLUTION

Incubate a tissue slice at 37°C with glucose 6-phosphate in a suitable buffer solution. Wash the tissue free of the substrate, and then precipitate the phosphate ions by the addition of lead nitrate to the tissue slice. The remainder of the preparation is as described in Example 1.5. In liver cells the reaction product is found *within* the endoplasmic reticulum, thus indicating the location of the enzyme.

1.4. How may cells be disrupted in order to obtain subcellular organelles by centrifugal fractionation?

SOLUTION

There are several ways of disrupting cells:

1. *Osmotic lysis*: The plasma membranes of cells are water-permeable but are impermeable to large molecules and some ions. Thus, if cells are placed into water or dilute buffer, they swell owing to the *osmotically* driven influx of water. Since the plasma membrane is not able to stretch very much (the red blood cell membrane can stretch only up to 15 percent of its normal area before disruption), the cells burst. The method is effective for isolated cells but is not so effective for tissues.
2. *Homogenizers*: One of these is described in Example 1.7.
3. *Sonication*: This involves the generation of shear forces in a cell sample in the vicinity of a titanium probe (0.5 mm in diameter and 10 cm long) that vibrates at ~20,000 Hz. The device contains a crystal of lead zirconate titanate that is *piezoelectric*, i.e., it expands and contracts when an oscillatory electric field is applied to it from an electronic oscillator. The ultrasonic pressure waves cause *microcavitation* in the sample, and this disrupts the cell membranes, usually in a few seconds.

SUBCELLULAR ORGANELLES

1.5. On the basis of the pathway of heterophagy (Fig. 1-8), make a proposal for the pathway of *autophagic* degradation of a mitochondrion.

SOLUTION

Figure 1-16 shows the scheme for autophagic degradation of a mitochondrion. Note that once the so-called *phagosome* has been formed, the process of digestion, etc., is the same as for heterophagy (Fig. 1-8).

1.6. There is an inherited disease in which a person's lysosomes lack the enzyme *β-glucosidase* (Table 1.1). What are the clinical and biochemical consequences of this deficiency?

SOLUTION

The disease is called *Gaucher's disease* and is the most common of the so-called *sphingolipidoses*; its incidence in the general population is ~1:2,500. This class of disease results from defective hydrolysis of membrane components, called *sphingolipids* (Chap. 6) that are normally turned over in the cell by hydrolytic breakdown in the lysosomes. The sphingolipids are lipid molecules with attached carbohydrate groups. An inability to remove glucose from these molecules results in their accumulation in the lysosomes. In fact over a few years, the cells that have rapid membrane turnover, such as in the liver and spleen, become engorged with this lipid breakdown product. Clinically, patients have an enlarged liver and spleen and may show signs of mental deterioration if much of the lipid accumulates in the brain as well.

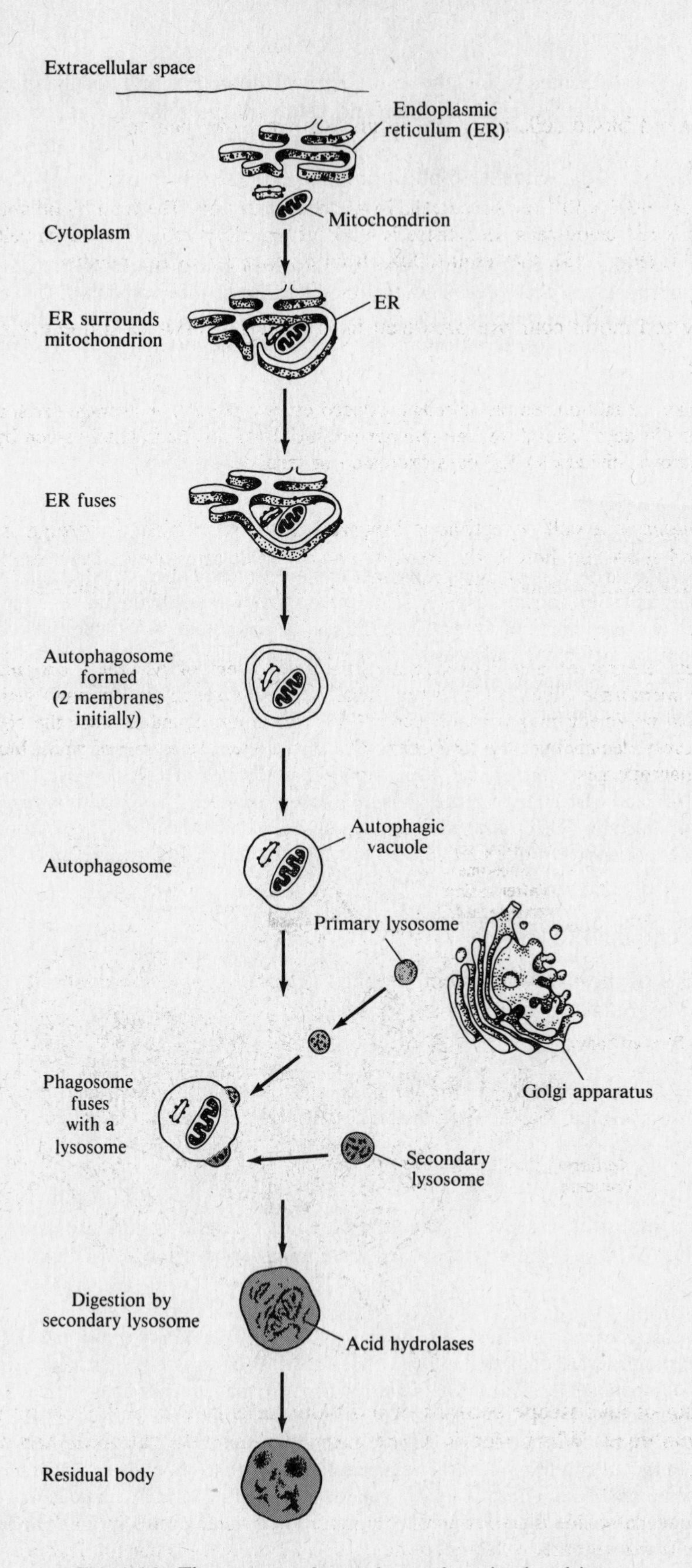

Fig. 1-16 The process of autophagy of a mitochondrion.

CELL TYPES

1.7. How many red blood cells are there in an average 70-kg person?

SOLUTION

There are $\sim 3.3 \times 10^{13}$ red blood cells in an average person. The total blood volume is ~6 L, and half of that is red blood cells; i.e., there is ~3 L of red blood cells. Since each cell has a volume of $\sim 90 \times 10^{-15}$ L (Fig. 1-13), the result follows from dividing 3 L by this number.

1.8. How many red blood cells are produced in an average 70-kg person every second?

SOLUTION

There are 3.2 million red blood cells produced every second! The average life span of a human red blood cell is 120 days. Therefore, the number produced per second is simply given by the answer from Prob. 1.7, above, divided by 120 days expressed in seconds.

1.9. A *macrophage* is a cell type that is involved in the engulfing of foreign material, such as bacteria and damaged host cells. In view of this specialized phagocytic function, draw what you think an electron microscopist would see in a cross section of the cell.

SOLUTION

The key features of a macrophage are its large system of *lysosomes* and *invaginations of the cytoplasmic membrane* (Fig. 1-17). Also, there is a rich rough endoplasmic reticulum where the lysosomal hydrolytic enzymes are produced. Mitochondria are abundant since the highly active protein synthesis is very demanding of ATP (Chap. 17). Monocytes are a type of white blood cell which are related to macrophages.

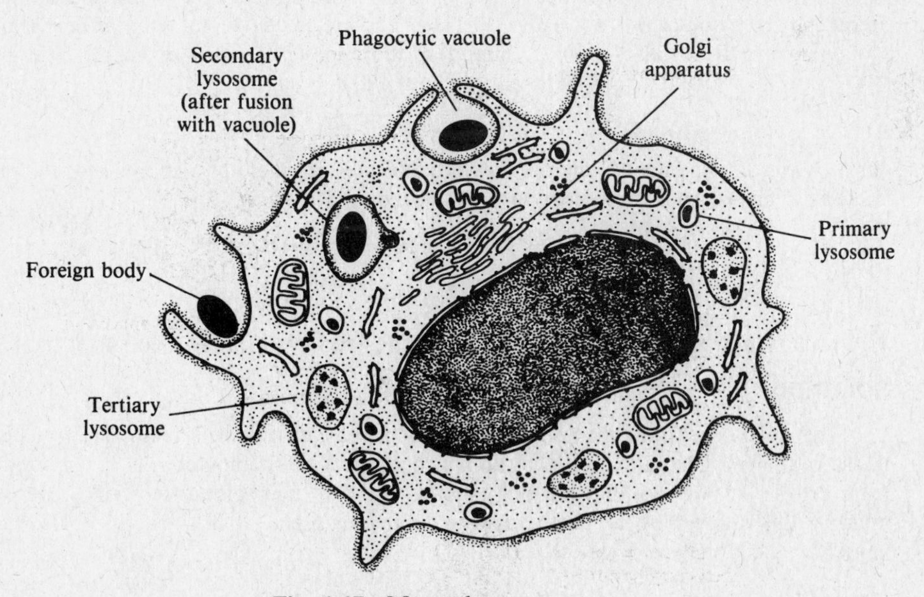

Fig. 1-17 Macrophage.

1.10. PAS staining of microscope sections of red blood cells gives a pink stain on one side only of the cell membrane. Which side is it, the extracellular or the intracellular side?

SOLUTION

The extracellular side is stained pink. All glycoproteins and glycolipids of the plasma membrane of red blood cells and *all* other cells are on the outside of the cell. No oligosaccharides are present on the inner face of the plasma membrane.

1.11. Why do the vesicles of *mast* cells, which contain large quantities of *histamine*, stain red with eosin?

SOLUTION

Eosin is negatively charged, and histamine has the structure

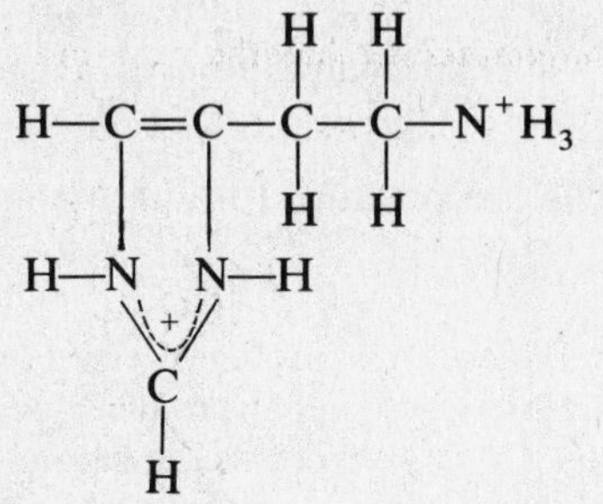

```
               H   H
               |   |
   H—C=C—C—C—N⁺H₃
      |   |   |   |
      |   |   H   H
      |   |
   H—N   N—H
      \ + /
       \ /
        C
        |
        H
```

Histamine

i.e., it has a net charge of +2. The two types of molecules interact electrostatically inside the vesicles, and thus the red eosin stains the vesicles red.

THE STRUCTURAL HIERARCHY IN CELLS

1.12. How many molecules of hemoglobin are there in one human red blood cell?

SOLUTION

There are $\sim 3 \times 10^8$ molecules of hemoglobin in one erythrocyte. The concentration of hemoglobin in human red blood cells is normally 330 g L^{-1}. The molecular weight (M_r) of hemoglobin is 64,500, and the volume of a red cell is ~90 fL. Therefore the number of moles of hemoglobin in one cell is

$$\frac{330 \times 90 \times 10^{-15}}{64{,}500} = 4.6 \times 10^{-16}$$

Since Avogadro's number is the number of molecules per mole of a compound, the previous number is multiplied by Avogadro's number to give the required estimate:

$$4.6 \times 10^{-16} \times 6.02 \times 10^{23} \simeq 3 \times 10^8$$

1.13. How many hemoglobin molecules are synthesized per human red blood cell per second?

SOLUTION

The mean generation time of a red cell, from the stem cell to a mature reticulocyte, is ~90 h. The phase in the cell generation pathway in which most of the hemoglobin is synthesized is ~40 h. Since, from Prob. 1.12, we saw that the cell contains $\sim 3 \times 10^8$ hemoglobin molecules, we proceed by simply dividing this number by the time taken to generate them, 40 h. This gives the rate of production, namely, ~2,000 molecules per second.

1.14. It has been estimated that it takes ~1 min to synthesize one hemoglobin subunit from its constituent amino acids. Using this fact, calculate how many hemoglobin molecules are produced on average at any one time in the differentiation of the cell.

SOLUTION

From Prob. 1.13, ~2,000 hemoglobin molecules are produced per second; this is equal to $\sim 1.2 \times 10^5$ per minute. However, hemoglobin is a *tetrameric* protein (four subunits; Chap. 5), so four times 1.2×10^5 chains are produced per minute: 4.8×10^5.

Supplementary Problems

1.15. A commonly used test of the viability of cells in tissue culture is whether or not they exclude a so-called supravital dye such as toluidine blue. If the cells exclude the dye, they are considered to be viable. What is the biochemical basis of this test?

1.16. The chemical compound *glutaraldehyde* has the structure

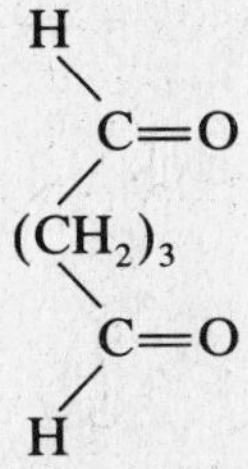

$$\mathrm{H{-}C(=O){-}(CH_2)_3{-}C(=O){-}H}$$

It is used as a fixative of tissues for light and electron microscopy. What chemical reaction is involved in this fixation process?

1.17. Outline the design of a histochemical procedure for the localization of the enzyme arylsulfatase in tissues; the enzyme catalyzes the following reaction type:

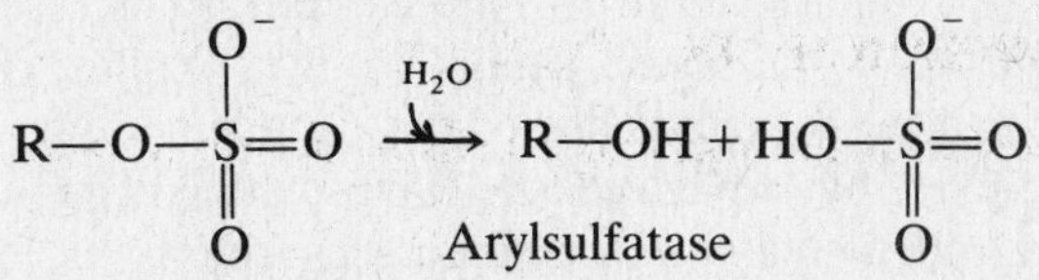

$$\mathrm{R{-}O{-}S(O^-)(=O){=}O} \xrightarrow[\text{Arylsulfatase}]{H_2O} \mathrm{R{-}OH + HO{-}S(O^-)(=O){=}O}$$

1.18. In an attempt to determine the localization of glycogen in the liver, could there be any problems of interpretation of the electron microscopic autoradiographic images if $[^3H]$glucose were used as the radioactive precursor molecule of glycogen?

1.19. *Microsomes* are small, spherical, membranous vesicles with attached ribosomes. During differential sedimentation, they sediment only in the late stages of a preparation, when very high centrifugal forces are used. They don't appear in electron micrographs of a cell. From where do they arise?

1.20. There are two forms of the enzyme carbamoyl phosphate synthetase, one in the mitochondrial matrix and the other in the cytoplasm. What might be the consequence and role of this so-called *compartmentation* of enzymes?

1.21. Human reticulocytes (Prob. 1.1) continue to synthesize hemoglobin for approximately 24 hours after release into the circulation. Design an electron microscopic experiment using autoradiography to identify *which* of the cells are actively synthesizing the protein.

1.22. (*a*) Who is the *primary* source of the DNA in *your* mitochondria—your mother or your father?

(*b*) Speculate on possible inheritance patterns if there were a defect in one or the other parent's mitochondria.

1.23. Given that mitochondria do not have the same aggressive autolytic capacity as lysosomes, what might be the significance of having such a complex membranous structure? After all, the endoplasmic reticulum and the plasma membrane could potentially support those enzymes found in mitochondrial membranes.

1.24. The disease *epidermolysis bullosa* involves severe skin ulceration and even loss of the ends of the ears, nose, and fingers. It is possibly the result of a primary defect in the stability of lysosomal membranes.

(*a*) How does this lead to the signs, just mentioned, of the disease?

(*b*) What biochemical procedure might you suggest to treat the disorder?

1.25. In some sufferers of Down's syndrome, the somatic cell nuclei do not contain three chromosomes 21. However, there is a chromosomal defect relating to chromosome 21; what might it be?

Chapter 2

Carbohydrates

2.1 INTRODUCTION AND DEFINITIONS

It is not possible to give a simple definition of the term *carbohydrate*. The name was applied originally to a group of compounds containing C, H, and O that gave an analysis of $(CH_2O)_n$, i.e., compounds in which n carbon atoms appeared to be hydrated with n water molecules. These compounds possessed reducing properties because they contained a carbonyl group as either an aldehyde or a ketone, as well as an abundance of hydroxyl groups. The use of the term *carbohydrate* was extended to describe the derivatives of these simple compounds, although the derivatives failed to give the simple analysis shown above. Moreover, many naturally occurring compounds proved to be derivatives in which the reducing group (aldehyde or ketone) had undergone reaction.

The simplest definition of carbohydrates that can be given is that they are *polyhydroxy aldehydes* or *ketones*, or compounds derived from these. They range in M_r from less than 100 to well over 10^6. The smaller compounds, containing three to nine carbon atoms, are called *monosaccharides*. The larger compounds are formed by condensation of the smaller ones via *glycosidic bonds*. A *disaccharide* consists of two monosaccharides linked by a single glycosidic bond; a *trisaccharide* is three monosaccharides linked by two glycosidic bonds, etc. *Oligo-* and *polysaccharides* are terms describing carbohydrates with few and many monosaccharide units, respectively.

Because many mono- and oligosaccharides have a sweet taste, carbohydrates of low M_r are often called *sugars*.

EXAMPLE 2.1

Of the compounds shown, (1) and (2) are not carbohydrates because they have only one hydroxyl group each; (3), (4), and (5) are carbohydrates because they have the general formula $(CH_2O)_n$ and are polyhydroxylic.

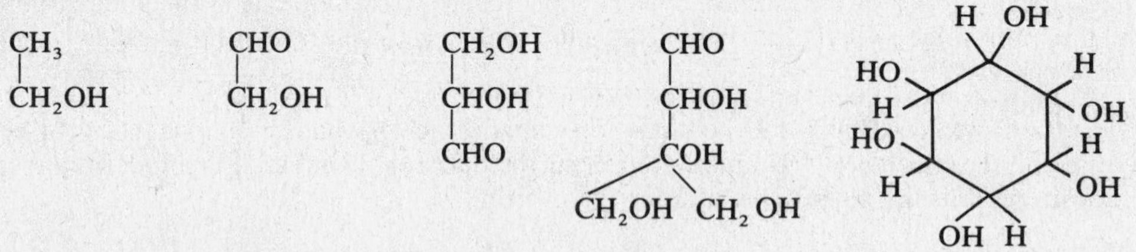

(1) CH_3–CH_2OH

(2) CHO–CH_2OH

(3) CH_2OH–CHOH–CHO

(4) CHO–CHOH–COH(CH_2OH)(CH_2OH)

(5) Six-membered carbon ring, each ring carbon bearing H and OH (ring substituents as printed: H OH; HO, H; H, OH; HO, H; H, OH; OH H)

Let us for now restrict discussion to simple monosaccharides, that is, polyhydroxy compounds containing a carbonyl function. There are two series—*aldoses*, containing an aldehyde group, and *ketoses*, containing a ketone group. Simple monosaccharides can also be classified according to the number of carbon atoms they contain—*trioses*, *tetroses*, *pentoses*, *hexoses*, etc., containing three, four, five, and six carbon atoms, respectively. The two systems can be combined. Thus, glucose, the most common sugar, is an *aldohexose*; i.e., a six-carbon monosaccharide with an aldehyde group.

EXAMPLE 2.2

The classification of each of the following monosaccharides is given below the structure.

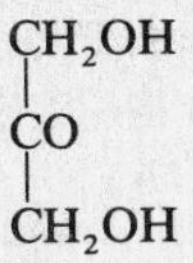

Ketotriose	Aldopentose	Ketoheptose	Ketopentose	Aldopentose
$CH_2OH-CO-CH_2OH$	$CH_2OH-CHOH-CHOH-CHOH-CHO$	$CH_2OH-CO-CHOH-CHOH-CHOH-CHOH-CH_2OH$	$CH_2OH-CHOH-CO-CHOH-CH_2OH$	$CHO-CHOH-COH(CH_2OH)_2$

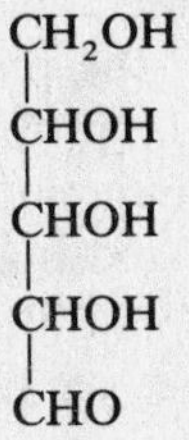

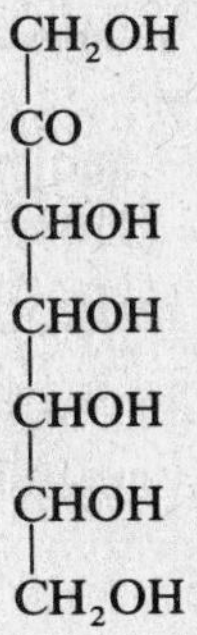

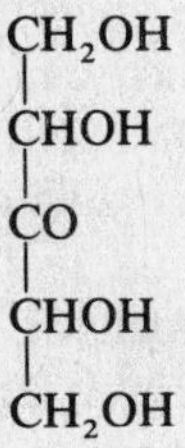

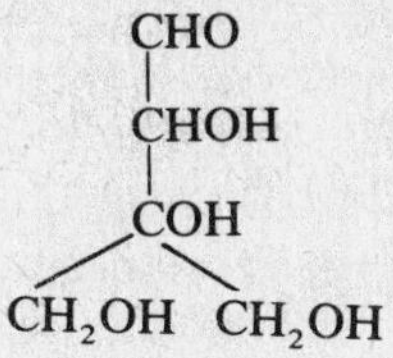

2.2 GLYCERALDEHYDE

The simplest aldose is glyceraldehyde:

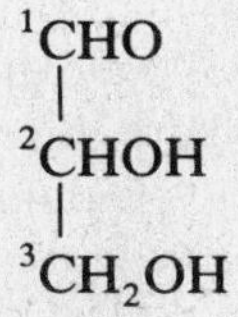

$$ {}^{1}CHO - {}^{2}CHOH - {}^{3}CH_2OH $$

It has reducing properties because it is an aldehyde. The C-2 of glyceraldehyde is a *chiral center* (also known as an asymmetric center), i.e., there are two possible isomers, known as *enantiomers*, of glyceraldehyde.

EXAMPLE 2.3

The structures of the two enantiomers of glyceraldehyde are

$$ \begin{array}{c} CHO \\ | \\ H-C-OH \\ | \\ CH_2OH \end{array} \qquad \begin{array}{c} CHO \\ | \\ HO-C-H \\ | \\ CH_2OH \end{array} $$

These structures, when written as shown below (left), are called *Fischer projection formulas*, which are attempts to represent three-dimensional molecules in two dimensions. For example, the two molecules would appear in three-dimensional space as shown below (right)

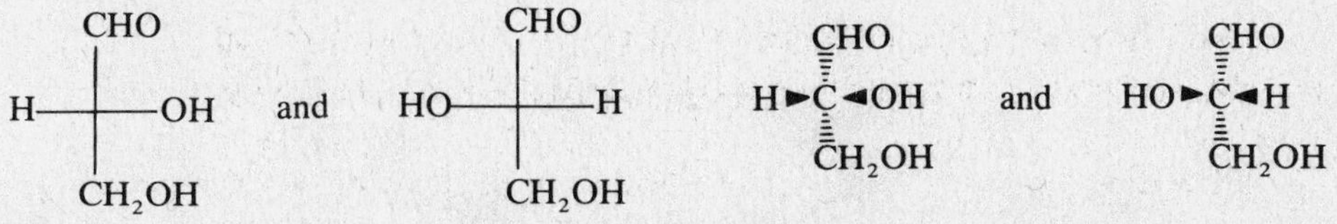

$$ \begin{array}{c} CHO \\ H-\!\!\!+\!\!\!-OH \\ CH_2OH \end{array} \text{ and } \begin{array}{c} CHO \\ HO-\!\!\!+\!\!\!-H \\ CH_2OH \end{array} \qquad \begin{array}{c} CHO \\ H \blacktriangleright C \blacktriangleleft OH \\ CH_2OH \end{array} \text{ and } \begin{array}{c} CHO \\ HO \blacktriangleright C \blacktriangleleft H \\ CH_2OH \end{array} $$

where the C-2 is in the plane of the paper, the aldehyde and hydroxymethyl groups are behind the paper (as denoted by the dashed bond lines), and the hydrogen and hydroxyl groups are in front of the paper (as denoted by the solid bond lines). Enantiomers are *mirror images* of each other: when placed in front of a plane mirror, one structure will give an image that is identical with the structure of the other.

In the pairs of figures above, the structure on the left is called D-glyceraldehyde, and that on the right L-glyceraldehyde. The prefixes D- and L- refer to the overall shape of the molecules; more specifically the letters refer to the *configuration*, or arrangement, of groups around the chiral center.

Generally, with compounds containing a single chiral carbon atom, there are only minor differences in the physical and chemical properties of the pure enantiomers. There is one physical property, however, in which enantiomers are markedly different—the property of *optical activity*. This refers to the ability of a solution of an enantiomer to rotate the plane of plane-polarized light. One of a pair of enantiomers will rotate the plane in a *clockwise* direction and is given the symbol (+). The other will rotate the plane in a *counterclockwise* direction and is given the symbol (−). The D enantiomer of glyceraldehyde is (+) and is described as D-(+)-glyceraldehyde; the other is L-(−)-glyceraldehyde. Mixtures of D and L enantiomers will have a net rotation depending on the proportions of the enantiomers; equal proportions give a net rotation of zero, in which case the solution is said to be *racemic*.

Optical activity is measured in a *polarimeter*. The magnitude of the optical activity is measured as an angle of rotation, given the symbol α. The units are degrees or radians (SI).

Question: On what factors will the α of a solution of an optically active compound depend?

The factors are the concentration of the compound, the length of the cell in which the solution is placed, the wavelength of the polarized light, the temperature, and the solvent.

Because of the dependencies noted above, the experimentally measured α is always converted to and expressed as the *specific rotation* $[\alpha]_D^T$, where the super- and subscripts refer, respectively, to the temperature and the wavelength of the light (D referring to the D lines of sodium vapor, about 589 nm).

$$[\alpha]_D^T = \frac{\alpha}{\text{length of cell (dm)} \times \text{concentration (g cm}^{-3})} \tag{2.1}$$

The name of the solvent is given in parentheses after the value is given; e.g., $[\alpha]_D^{25} = +17.5°$ (in water).

2.3 SIMPLE ALDOSES

The simple aldoses are related to D- and L-glyceraldehyde in that structurally they may be considered to be derived from glyceraldehyde by the introduction of hydroxylated chiral carbon atoms between C-1 and C-2 of the glyceraldehyde molecule. Thus, two tetroses result when CHOH is introduced into D-glyceraldehyde:

```
    CHO
     |
 H—C—OH
     |
    CH2OH
```

D-Glyceraldehyde

```
    CHO               CHO
     |                 |
 H—C—OH          HO—C—H
     |                 |
 H—C—OH           H—C—OH
     |                 |
    CH2OH             CH2OH
```

D-Erythrose D-Threose

Question: Two tetroses can be formed from L-glyceraldehyde. These tetroses are called L-*erythrose* and L-*threose*. Why is it unnecessary to invent new names for the tetroses derived from L-glyceraldehyde?

If the structures of the two tetroses are written alongside those of D-erythrose and D-threose using Fischer projection formulas, it is seen that two pairs of mirror images are given. That is, the four aldotetroses constitute two pairs of enantiomers:

D-Erythrose	L-Erythrose	D-Threose	L-Threose
CHO	CHO	CHO	CHO
H—C—OH	HO—C—H	HO—C—H	H—C—OH
H—C—OH	HO—C—H	H—C—OH	HO—C—H
CH_2OH	CH_2OH	CH_2OH	CH_2OH

Two simple aldopentoses can be derived structurally from each of the four aldotetroses described, making a total of eight aldopentoses. Therefore, there will be 16 aldohexoses.

EXAMPLE 2.4

Simplified structures for the 8 aldopentoses and 16 aldohexoses are shown (O represents aldehyde; — represents an OH group; H atoms on carbons are omitted).

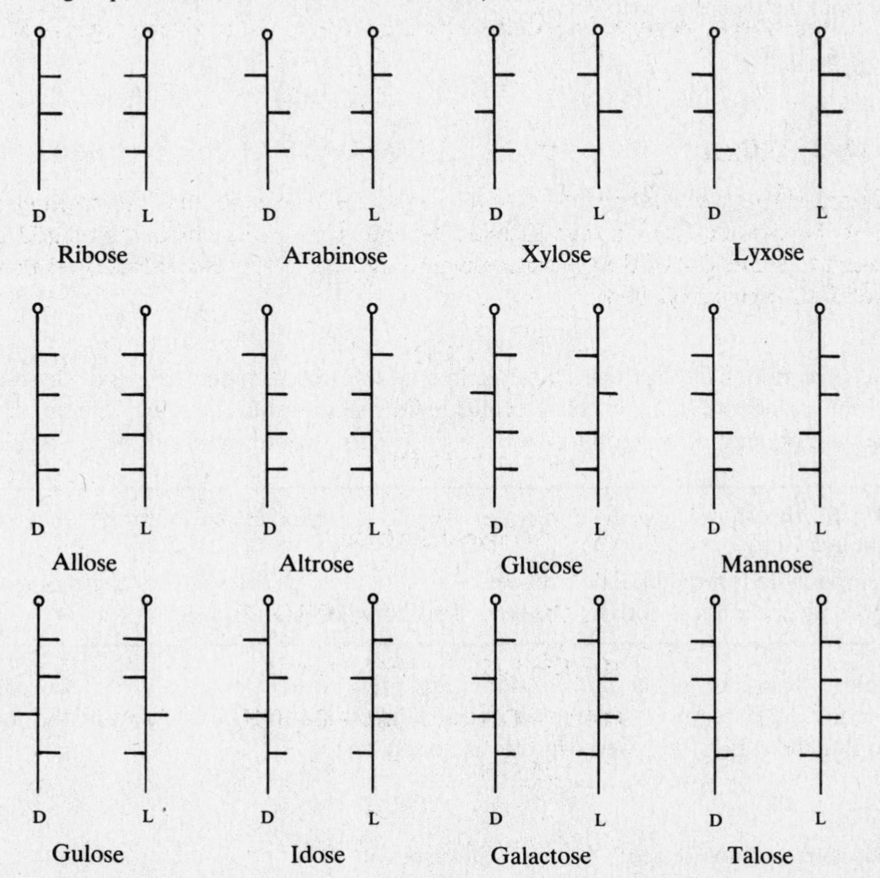

Question: Glyceraldehyde is sometimes known as *glycerose*. Why?

The names of all the aldotetroses, -pentoses, and -hexoses end in *-ose*. Glyceraldehyde is the parent aldose; thus, its name, glycerose, is valid.

There are two series of simple aldoses: a D series and an L series. To determine to which series an aldose belongs, locate the chiral carbon atom most remote from the reducing group and determine its relationship to glyceraldehyde; e.g., the sugar shown below, glucose, is called D-glucose:

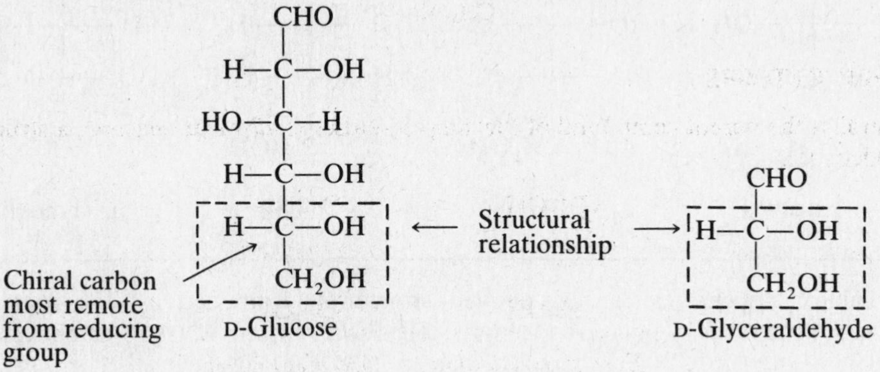

```
             CHO
              |
          H—C—OH
              |
         HO—C—H
              |
          H—C—OH                                       CHO
       - - - -|- - - -                              - - -|- - - -
      |   H—C—OH  |  ←— Structural       —→  |H—C—OH  |
      |   ↗   |      |      relationship           |     |         |
Chiral carbon  CH2OH |                               |  CH2OH   |
most remote - - - - - -                              - - - - - - -
from reducing   D-Glucose                       D-Glyceraldehyde
group
```

EXAMPLE 2.5

In the sugars shown below, (1) is L and (2), (3), and (4) are D. Notice that in (3) the *chiral* carbon atom most remote from the reducing group is C-2.

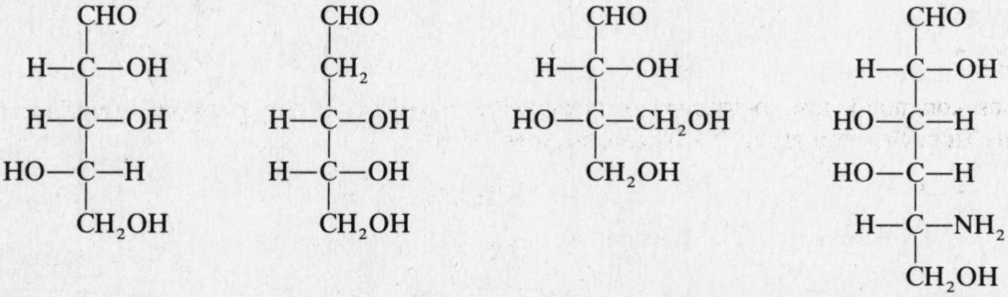

```
     CHO              CHO                CHO                  CHO
      |                |                  |                    |
  H—C—OH             CH2             H—C—OH              H—C—OH
      |                |                  |                    |
  H—C—OH          H—C—OH       HO—C—CH2OH          HO—C—H
      |                |                  |                    |
 HO—C—H           H—C—OH             CH2OH              HO—C—H
      |                |                                       |
    CH2OH            CH2OH                                H—C—NH2
                                                               |
                                                             CH2OH

     (1)              (2)                (3)                  (4)
```

Whereas glyceraldehyde has one chiral center, aldotetroses, -pentoses, and -hexoses have two, three, and four, respectively. Each chiral center gives rise to optical activity. The net optical activity of an aldose will depend on contributions from each chiral center and will be (+) or (−).

Question: D-Erythrose was dissolved in water. Would you predict the solution to have a (+) or (−) optical rotation?

This is impossible to predict. The prefixes D- and L- refer to the *shape* of the molecule and imply nothing regarding the optical activity. In fact, a solution of D-erythrose is (−).

When more than four chiral carbon atoms are present, an aldose is given two configurational prefixes, one for the four lowest-numbered chiral centers and one for the rest of the molecule. The configuration of the *highest*-numbered group is stated first.

EXAMPLE 2.6

The aldooctose shown is named D-*erythro*-L-*galacto*octose.

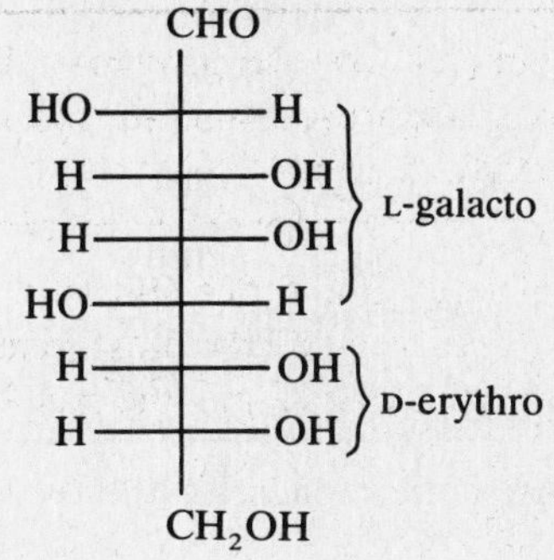

```
          CHO
           |
   HO------+------H  \
   H-------+------OH  |
   H-------+------OH  } L-galacto
   HO------+------H  /
   H-------+------OH \
   H-------+------OH / D-erythro
           |
         CH2OH
```

2.4 SIMPLE KETOSES

Structurally, the parent compound of the simple ketoses is *dihydroxyacetone*, a structural isomer of glyceraldehyde.

```
CH2OH            CHO
 |                |
CO               CHOH
 |                |
CH2OH            CH2OH
```

Dihydroxyacetone   Glyceraldehyde

Although dihydroxyacetone does not possess a chiral carbon atom, the simple ketoses are related to it structurally by the introduction of hydroxylated chiral carbon atoms between the keto group and one of the hydroxymethyl groups. Thus there are two ketotetroses, four ketopentoses, and eight ketohexoses.

EXAMPLE 2.7

The most common ketose, D-*fructose*, is shown below. Compare the configuration of the chiral carbon atom most remote from the keto group (C-5) with D-glyceraldehyde.

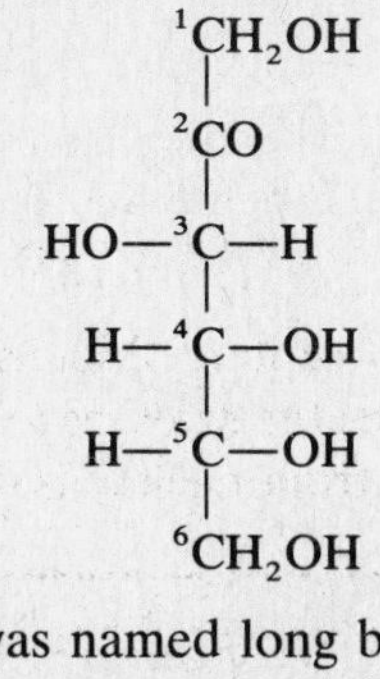

```
        1CH2OH
         |
        2CO
         |
   HO—3C—H
         |
    H—4C—OH
         |
    H—5C—OH
         |
        6CH2OH
```

Fructose (shown in Example 2.7) was named long before its structure was known. The same is true for most aldoses. Names like glucose, mannose, ribose, and fructose are called *trivial names*; i.e., they are *nonsystematic*. However, ketoses, which are isomers of aldoses, were isolated or synthesized after the corresponding aldoses and were given names based on the names of the isomeric aldoses. Such names are misleading because they do not reflect the structural elements present in the ketoses.

EXAMPLE 2.8

The ketose shown below is known universally as D-*ribulose* because it is an isomer of D-ribose. The name is incorrect; the compound has only two chiral centers, *not* three as the prefix *rib-* would imply. The compound is related to D-erythrose, and its correct name is D-*erythro*-pentulose.

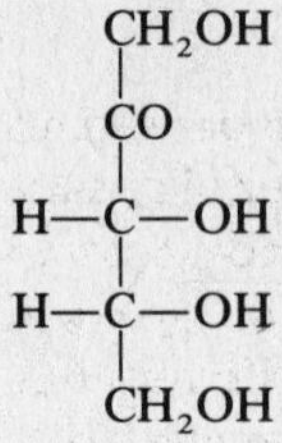

```
   CH2OH
    |
    CO
    |
H—C—OH
    |
H—C—OH
    |
   CH2OH
```

Question: What are the correct systematic names for the two sugars shown?

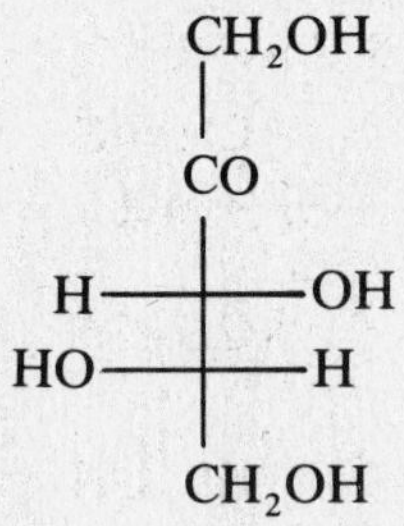

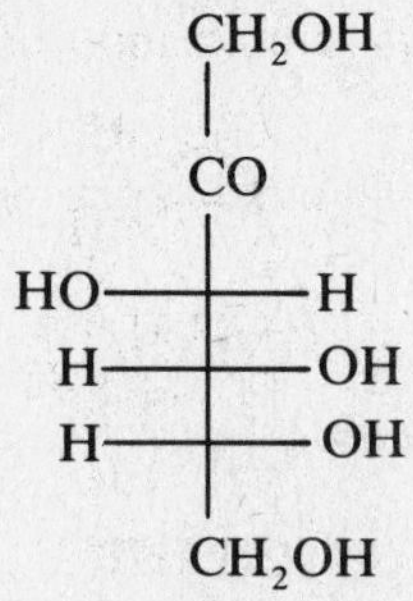

```
     CH2OH            CH2OH
       |                |
       CO               CO
       |                |
  H————+————OH    HO————+————H
 HO————+————H      H————+————OH
       |           H————+————OH
     CH2OH              |
                      CH2OH
      (1)              (2)
```

(1) L-*Threo*-pentulose; commonly called L-xylulose
(2) D-*Arabino*-hexulose; commonly called D-fructose

Question: How can you tell if a monosaccharide is a ketose from its name?
The systematic name always ends in *-ulose*; apart from fructose, so do the trivial names.

Some ketoses are not related structurally to dihydroxyacetone. They are named by considering the configurations of all the chiral carbon atoms as a unit, ignoring the carbonyl group.

EXAMPLE 2.9

Consider the ketose shown. It has three chiral carbon atoms in the D-arabino configuration (even though interrupted by a keto group). The keto group is at position 3. The ketose has six carbon atoms. It is therefore called D-*arabino*-3-hexulose.

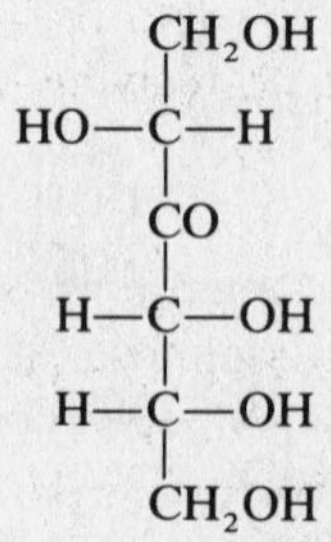

```
     CH2OH
      |
HO—C—H
      |
      CO
      |
 H—C—OH
      |
 H—C—OH
      |
     CH2OH
```

If the name of a ketose contains no number, it is assumed the ketose is related to dihydroxyacetone and the keto group is at position 2.

Question: Why is there no need to use numbers to indicate the position of the carbonyl group in aldoses?
The aldehyde group must always be terminal to a chain of carbon atoms.

2.5 THE STRUCTURE OF D-GLUCOSE

D-Glucose is the most common of the monosaccharides, occurring in the free state in the blood of animals and in the polymerized state, inter alia, as starch and cellulose. Tens of millions of tons of these polysaccharides are made by plants and photosynthetic microbes annually. A detailed study of the structure of glucose is justified on these grounds, and many of the structural features of all monosaccharides can be illustrated using glucose as an example.

The Fischer projection formula for D-glucose (Sec. 2.3) is also known as the *open-* or *straight*-chain structure. This structure occurs only in solution. There are two crystalline forms of D-glucose, known as α and β, which also have different optical activities when dissolved. X-ray diffraction studies have confirmed chemical evidence that α- and β-D-glucose are structures containing a ring of five carbon atoms and one oxygen atom:

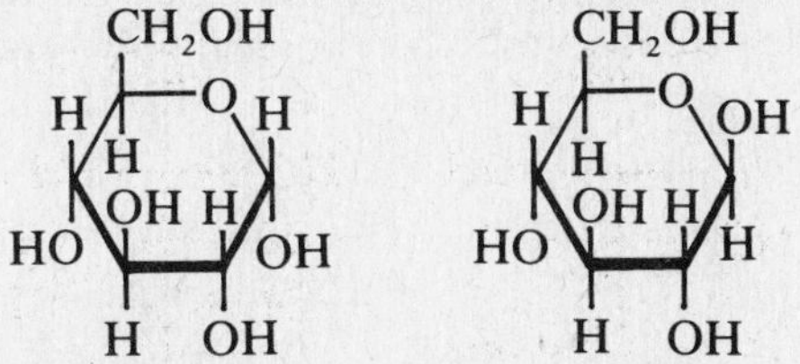

α-D-Glucose β-D-Glucose

These structures are known as the *Haworth projection formulas*. They do not represent the true shapes of the molecules (Sec. 2.6), but they do show the configuration at each of the chiral atoms. The formulas are meant to represent the ring as a generalized plane standing at right angles to the plane of the paper and an element of perspective is given to the structure by thickening the three bonds of the ring that are meant to appear to be in front of the paper; the remaining three bonds of the ring and the oxygen atom in the ring lie behind the paper.

When α-D-glucose or β-D-glucose is dissolved in water, the ring opens and the open-chain structure is formed. The reaction is *reversible*, and an equilibrium is established between the open form and the two ring forms. The chemistry of the process is understood in terms of the chemistry of the aldehyde group. In general, aldehydes react reversibly with alcohols to give *hemiacetals* and then, in the presence of an acid catalyst (Chap. 8), *acetals*:

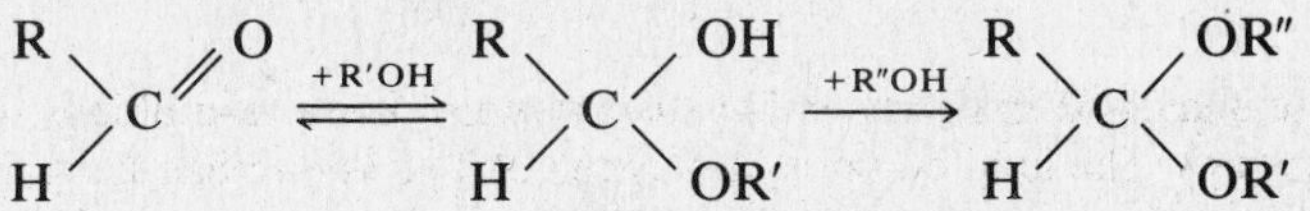

The formation of a ring by the open-chain form of D-glucose can be considered to be the result of a reaction between the hydroxyl group on C-5 and the aldehyde group to give a hemiacetal. The aldehyde carbon becomes chiral as a result, thus giving rise to two hemiacetals, α- and β-D-glucose.

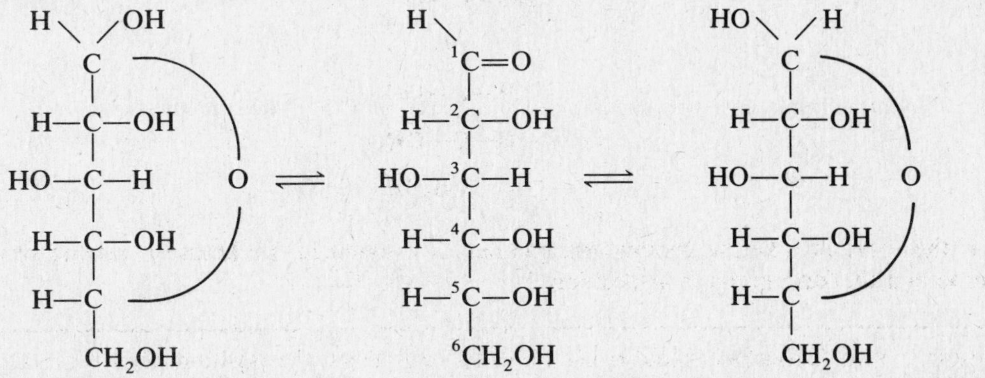

A hemiacetal (α-D-glucose) A hemiacetal (β-D-glucose)

These hemiacetals are known as *anomers*, and C-1 of the ring form is called the *anomeric* carbon.

Question: Why is the anomeric carbon sometimes called the *potential reducing carbon*?

When the ring opens in solution, C-1 becomes the carbon atom of the aldehyde group, which has reducing properties. Only the open-chain form has reducing properties; the ring forms are nonreducing because they lack an aldehyde group.

The Haworth projection formulas are neater ways of writing the ring forms shown in the equilibria above and yet preserving the configuration shown at each chiral carbon. It is not difficult to translate the open-chain structure for a monosaccharide into the Haworth ring structure.

EXAMPLE 2.10

The translation of the open-chain structure into the Haworth ring structure is best achieved in a step-by-step procedure, since a direct translation is tricky. D-Glucose is used in this example:

1. Write the open-chain structure.

CHO
H—C—OH
HO—C—H
H—C—OH
H—C—OH
CH_2OH

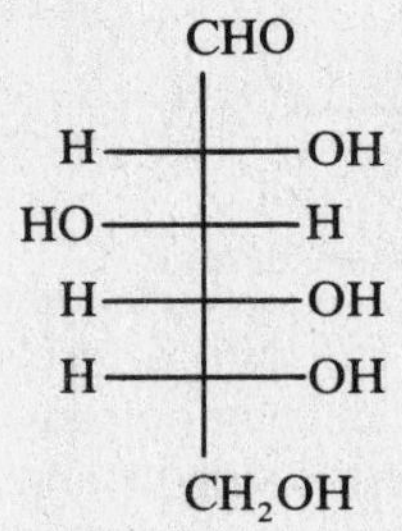

2. Turn the structure clockwise on its side and bend it round to almost form a ring.

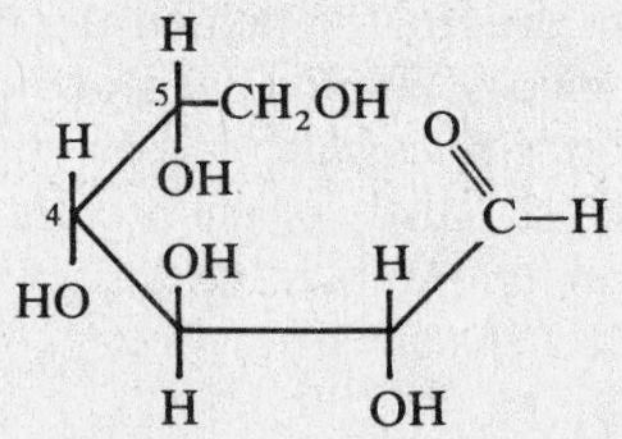

H, 5, CH_2OH, O, H, OH, 4, C—H, OH, H, HO, H, OH

3. Rotate the C-4—C-5 bond to bring the hydroxyl on C-5 close to the carbonyl group.

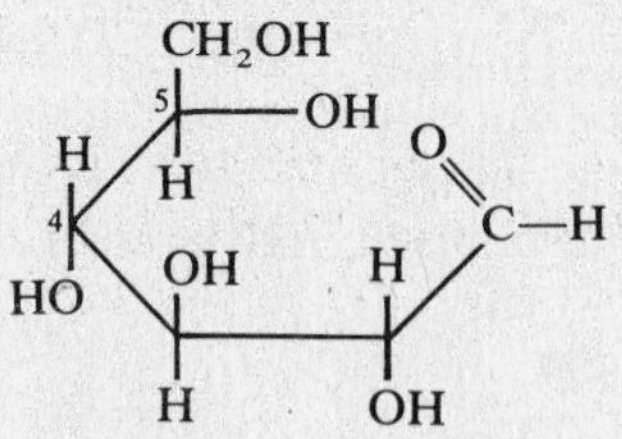

CH_2OH, 5, OH, O, H, H, 4, C—H, OH, H, HO, H, OH

4. Form the hemiacetal(s) by bonding of the hydroxyl on C-5 to the carbonyl group.

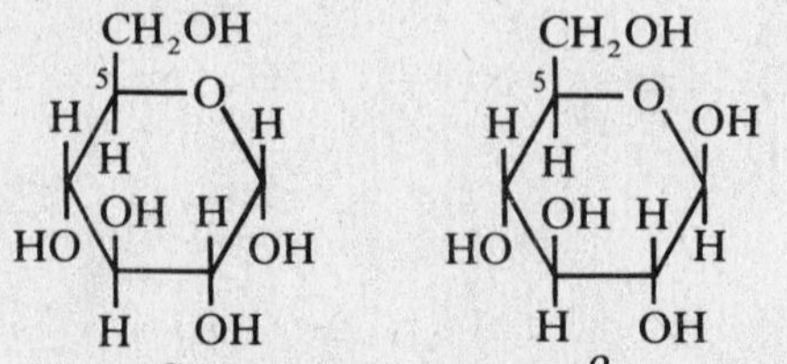

CH_2OH, 5, O, H, H, H, OH, H, HO, OH, H, OH — α

CH_2OH, 5, O, H, OH, H, OH, H, HO, H, H, OH — β

Note: It is normal in writing the ring structures of sugars to omit the H atoms attached to carbon, in which case β-D-glucose is written as:

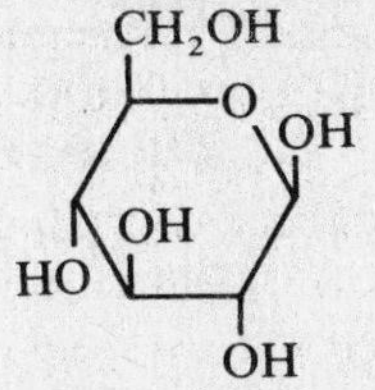

CH_2OH O OH OH HO OH

Question: What determines which of the anomers shown in Example 2.10 is called α and which β?

This depends on the configuration of the anomeric carbon relative to the configuration of the chiral atom that establishes whether the monosaccharide belongs to the L or the D series (in the case of glucose, C-5). If, in the Fischer projection formula, the hydroxyl groups on these carbons are cis, the anomer is called α; if the hydroxyl groups are trans, the anomer is called β. A comparison of the structure in step (2) of Example 2.10 with those in step (4) shows that the hydroxyl group on C-5 was below the plane of the ring before the rings were closed. Thus, the anomer with the anomeric hydroxyl below the plane of the ring is called α.

Although the open-chain form of glucose has four chiral centers, formation of the ring creates a fifth chiral center, which is, of course, why there are two anomers that differ in optical rotation. When solid α-D-glucose is dissolved in water, $[\alpha]_D^{25}$ is $+112°$. When solid β-D-glucose is dissolved in water, $[\alpha]_D^{25}$ is $+19°$.

EXAMPLE 2.11

The optical rotations of freshly made solutions of α- and β-D-glucose change with time. This is called *mutarotation*. A change in optical rotation must mean a change in structure. When α-D-glucose is dissolved in water, the α-D anomer is the only structure present at the instant of dissolution. The opening of the ring is a slow reaction, but since the reaction is reversible, some β-D anomer as well as the open-chain form will appear. Ultimately, an equilibrium between the open-chain form and the two anomers is established with an $[\alpha]_D^{25}$ of $+52°$. For this reason, the structure shown below is used to describe the state of glucose in solution, and the same device is used to depict the structure of a sugar in which the configuration of the anomeric carbon is unknown.

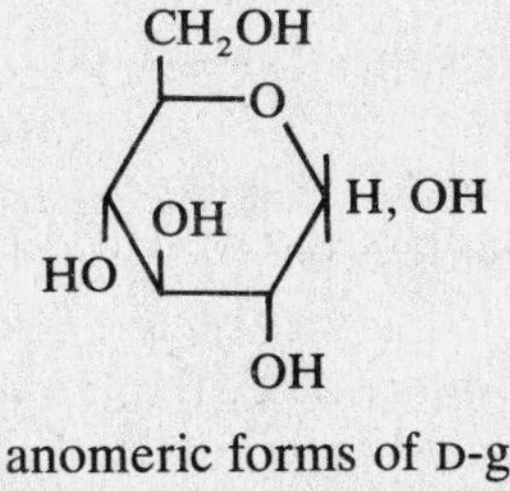

CH_2OH O H, OH OH HO OH

Although we have considered the anomeric forms of D-glucose to consist of six-membered rings, it is possible for two other anomeric forms to arise from the open-chain form by the addition of the hydroxyl group on C-4 to the carbonyl group of the aldehyde. These are five-membered ring forms.

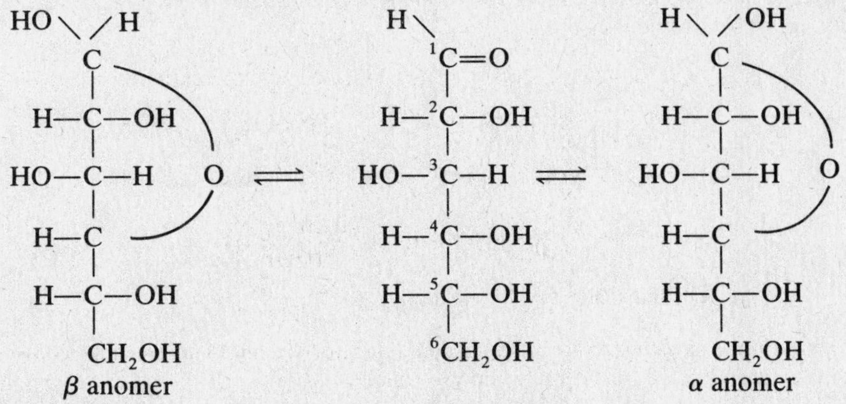

β anomer ⇌ open-chain form ⇌ α anomer

β anomer

α anomer

Question: How do we translate the open-chain form of glucose into five-membered Haworth projection formulas and assign the symbols α and β to the appropriate formula?

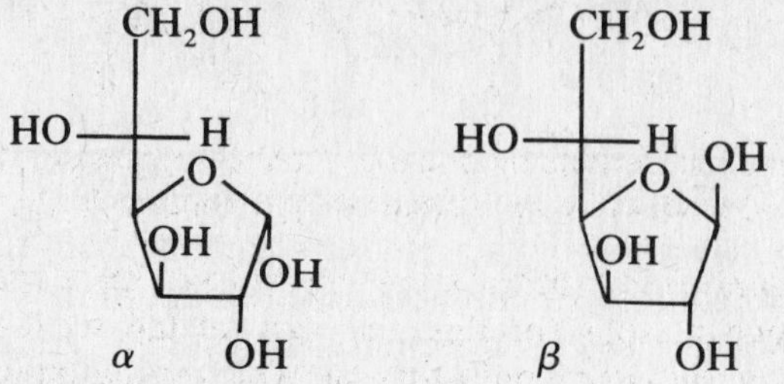

α $\qquad$ β

Clearly, it is necessary to distinguish between six- and five-membered rings. This is done by expanding the name *glucose* to *glucopyranose* for the six-membered anomers and to *glucofuranose* for the five-membered anomers.

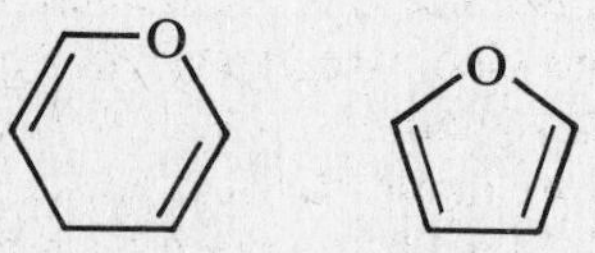

Pyran $\qquad$ Furan

Question: Would you expect furanose or pyranose ring forms to be the more stable for a given sugar?

Furanoses are *generally* less stable because the area of the ring is smaller and the opportunity for destabilizing the structure through steric interference between the H, hydroxyl, and hydroxymethyl groups on different carbons is greater. With glucose, the α- and β-glucofuranoses contribute very little to the equilibrium mixture.

2.6 THE CONFORMATION OF GLUCOSE

It was stressed in the previous section that the Haworth structures for the anomers of D-glucopyranose do not represent the true shape of the rings. The carbon atoms of glucose are all saturated, and the most stable form of a ring will be one that is *strain free*, i.e., where the angles formed by the bonds at each carbon atom are 109°, the tetrahedral angle.

EXAMPLE 2.12

For simplicity, consider cyclohexane. Only two strain-free conformations are possible: a *chair* and a *boat*, which, for clarity, are shown without H atoms:

Cyclohexane $\qquad$ Chair $\qquad$ Boat

There are 12 H atoms in the structure; 6 are in the general plane of the rings and are called *equatorial* (e); 6 are perpendicular to the general plane of the rings and are called *axial* (a).

The chair and boat forms are interconvertible by rotation of the C—C bonds. Therefore, they are not isomers, and the term *conformers* is used to describe the various shapes a molecule can possess.

Although both conformers of cyclohexane are strain-free, the chair conformer is more stable than the boat. There are two reasons for this. (1) Two of the six C—C bonds in the boat are *eclipsed*, whereas none are in the chair conformation. This can be illustrated by imagining the eye placed to look along the C-2—C-3 bond, as shown in Fig. 2-1. The eye, positioned as in (*a*), will see the configuration shown in (*b*): C-2 with its two H atoms and the bond leading to C-1; however, the eye cannot see C-3 or the two H atoms attached to it, because these three atoms are immediately behind C-2 and its two H atoms. Likewise, the C-5—C-6 bond is eclipsed, but all other C—C bonds in the boat and *all* C—C bonds in the chair are *staggered*, e.g., (*c*) shows the eye looking along C-3–C-4. Thus in the boat, the four axial H atoms on C-2, C-3, C-5, and C-6 are as close as possible, and interactions between them tending to distort the ring are maximal. (2) The axial H atoms on C-1 and C-4 of the boat come closer together (0.18 nm) than the sum of their van der Waals radii (0.24 nm), and with these two atoms there is an unfavorable interaction tending to distort the ring.

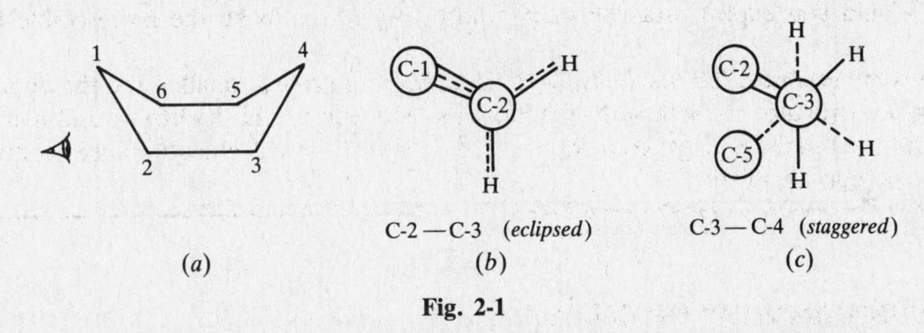

Fig. 2-1

The difference in energy between the boat and chair conformers of cyclohexane is about 25 kJ mol^{-1}, which means that at 25°C only 1 in 1,000 molecules exists in the boat conformation.

Question: What are two ways in which a monosaccharide in a ring form, like glucose, differs from cyclohexane in conformation?

(1) Monosaccharides have hydroxyl and hydroxymethyl groups replacing some of the H atoms on the carbon atoms of the ring. These are much more bulky than H atoms and would tend to be in equatorial positions around the edge of the ring rather than in axial positions on the faces of the ring, where they would be closer together and thus distort the ring. (2) Monosaccharides have an oxygen atom in the ring in the place of a carbon. This has little effect on the shape of the ring since, although the C—O bond length is slightly shorter than the C—C bond length, the valencies of oxygen are at 109°. However, the presence of an oxygen atom in the ring does mean there are *two* possible chair conformers, known as *C1* and *1C*.

C1 1C

These cannot be superimposed, but C1 can be converted reversibly to 1C by rotation of the bonds, passing through a boat conformer in the process.

A Haworth structure for a monosaccharide is translated readily into a structure showing the true shape of the molecule.

EXAMPLE 2.13

The Haworth structure tells us that a substituent that is above the general plane of the monosaccharide ring must also appear above the general plane of the chair; whether this is axial or equatorial will depend on the carbon atom being considered. For example, looking at each carbon of β-D-glucopyranose in turn, we find

Substituent	Haworth Structure—Position Relative to Ring	Chair Conformers* C1	Chair Conformers* 1C
OH of C-1	above	e	a
H of C-1	below	a	e
OH of C-2	below	e	a
H of C-2	above	a	e
OH of C-3	above	e	a
H of C-3	below	a	e
OH of C-4	below	e	a
H of C-4	above	a	e
CH_2OH of C-5	above	e	a
H of C-5	below	a	e

*e = equatorial plane; a = axial plane.

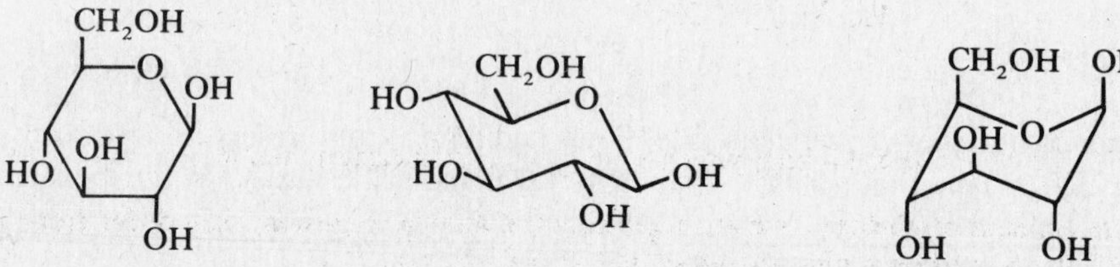

CH_2OH O OH OH HO OH

Haworth

HO CH_2OH O HO OH OH

C1

CH_2OH OH OH O OH OH

1C

The preferred conformer is C1, where all the bulky substituents are equatorial, on the edge of the ring. There will be considerable distortion in the 1C conformer, where three bulky groups on the upper face of the ring produce considerable steric hindrance, as do two bulky groups on the lower face.

Question: Which is the preferred conformer for α-D-glucopyranose?

Proceeding as in Example 2.13, we see that C1 is again the preferred conformer:

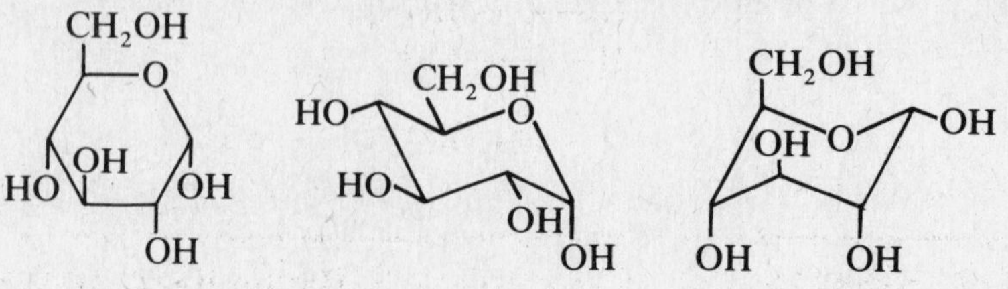

CH_2OH O OH HO OH OH

HO CH_2OH O HO OH OH

C1

CH_2OH OH O OH OH OH

1C

If the preferred forms of α- and β-D-glucopyranose (both C1) are compared, it is seen that the β form is the more stable since it has no bulky axial substituent, whereas the α form has one. This helps explain why β-D-glucopyranose is the dominant anomer in an aqueous solution of glucose.

2.7 MONOSACCHARIDES OTHER THAN GLUCOSE

Aldohexoses

Two structural isomers of glucose are mannose and galactose.

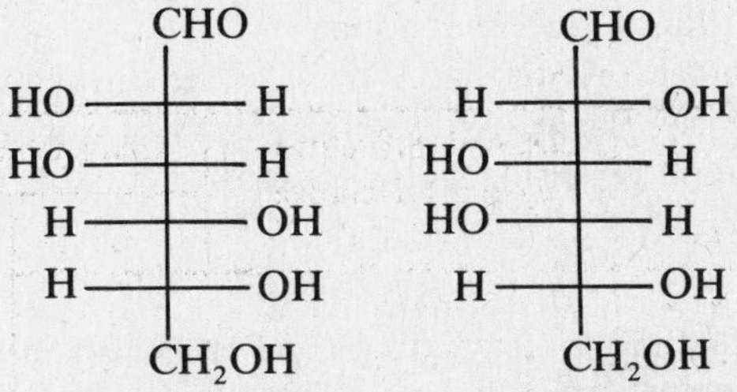

D-Mannose		
	CHO	
HO	—	H
HO	—	H
H	—	OH
H	—	OH
	CH_2OH	

D-Galactose		
	CHO	
H	—	OH
HO	—	H
HO	—	H
H	—	OH
	CH_2OH	

D-Mannose D-Galactose

EXAMPLE 2.14

D-Mannose and D-galactose are readily translated into six-membered Haworth structures. The α anomers only are shown.

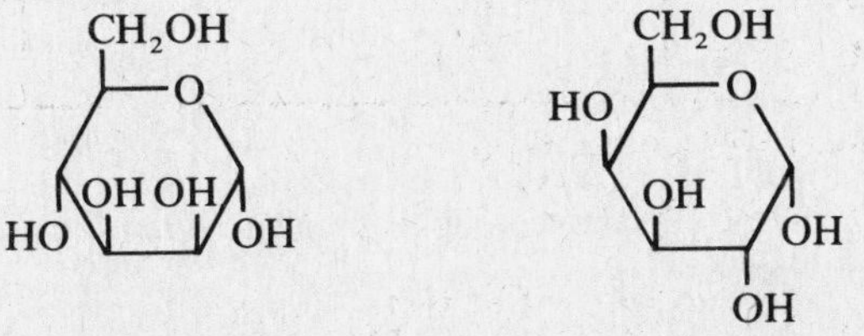

CH_2OH O OH OH HO OH

CH_2OH HO O OH OH OH

α-D-Mannopyranose α-D-Galactopyranose

Both of these are *epimers* of glucose; that is, neglecting the anomeric carbon, they differ from glucose in the configuration at just one carbon atom—mannose at C-2, galactose at C-4.

Aldopentoses

The most common aldopentose is D-ribose, the sugar present in RNA.

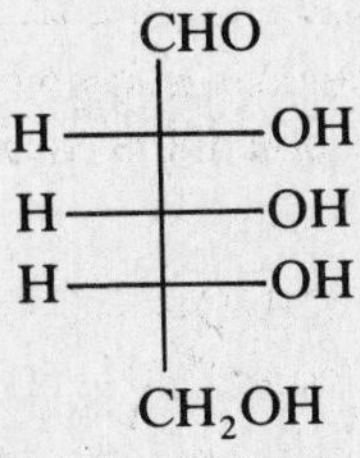

	CHO	
H	—	OH
H	—	OH
H	—	OH
	CH_2OH	

Question: Translate D-ribose into five- and six-membered Haworth structures. Give the β anomers only.

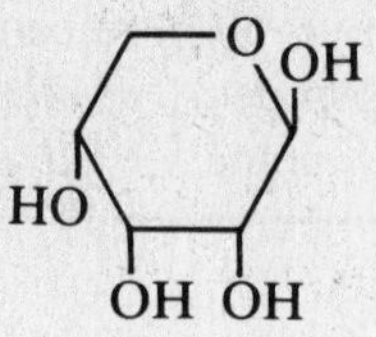

β-D-Ribopyranose β-D-Ribofuranose

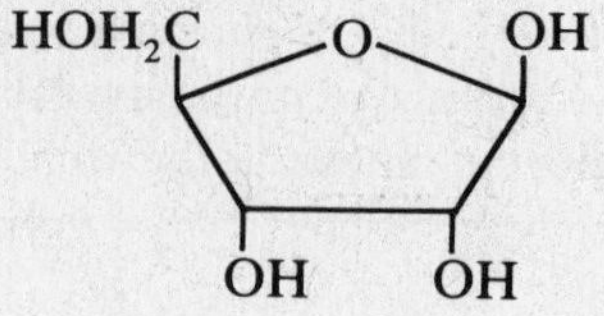

The only known *crystalline* form of D-ribose is β-D-ribopyranose, but in solution the α and β anomers of both pyranose and furanose forms occur.

The angle of a regular pentagon is 108°, very close to the tetrahedral angle, suggesting that the shape of a furanose ring is nearly planar. With some sugars having a furanose ring, the structure cannot be planar because of steric repulsion between substituents on the ring.

EXAMPLE 2.15

In β-D-ribofuranose, notice that the hydroxyl groups and H atoms attached to the carbon atoms of the C-2—C-3 bond are eclipsed. Repulsions between these eclipsed groups would be relieved by twisting the ring out of the plane, i.e., by raising C-3 above the plane and dropping C-2 below the plane. This conformer is known as T_2^3 (T for *twist*, with C-3 raised and C-2 lowered). An alternative way of relieving steric repulsion is to raise or lower C-3 out of the plane of the ring. This gives E (for *envelope*) conformers.

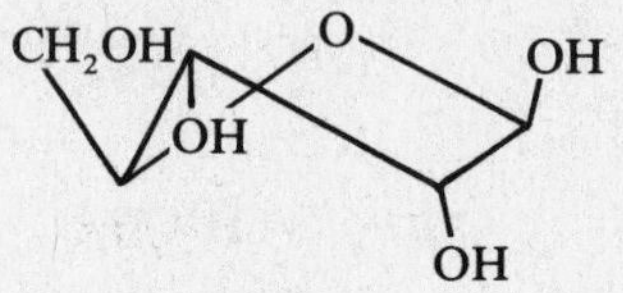

T_2^3 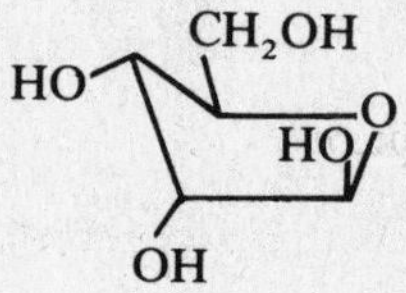 E^3

The loss of stability of the ring when C-3 is moved out of the plane is more than compensated by the relief from steric repulsion.

Deoxy Sugars

Deoxy sugars are reduced forms of sugars in which a hydroxyl group is replaced by a hydrogen atom. The most widely distributed deoxy sugar is known as 2-deoxy-D-ribose and is present in DNA.

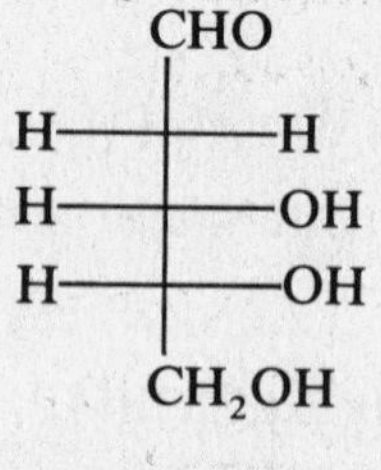

EXAMPLE 2.16

As with ketoses, many deoxy sugars are incorrectly named. In so-called 2-deoxy-D-ribose, there are only two chiral carbon atoms and the sugar is related to D-erythrose. The systematic name is 2-deoxy-D-*erythro*-pentose.

The other deoxy sugars commonly found are L-fucose, particularly in animals, and L-rhamnose, occurring in plants and bacteria.

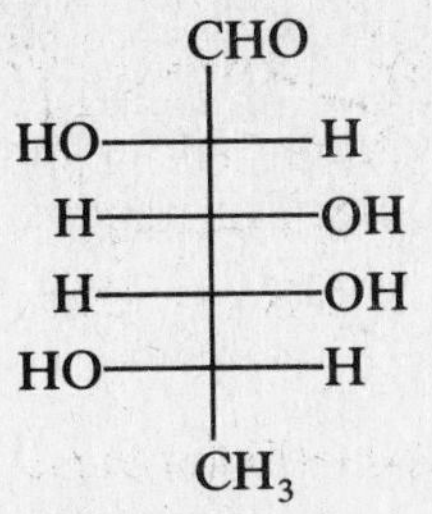

$$\begin{array}{rcl} & \mathrm{CHO} & \\ \mathrm{HO}- & | & -\mathrm{H} \\ \mathrm{H}- & | & -\mathrm{OH} \\ \mathrm{H}- & | & -\mathrm{OH} \\ \mathrm{HO}- & | & -\mathrm{H} \\ & \mathrm{CH_3} & \end{array} \qquad \begin{array}{rcl} & \mathrm{CHO} & \\ \mathrm{H}- & | & -\mathrm{OH} \\ \mathrm{H}- & | & -\mathrm{OH} \\ \mathrm{HO}- & | & -\mathrm{H} \\ \mathrm{HO}- & | & -\mathrm{H} \\ & \mathrm{CH_3} & \end{array}$$

L-Fucose 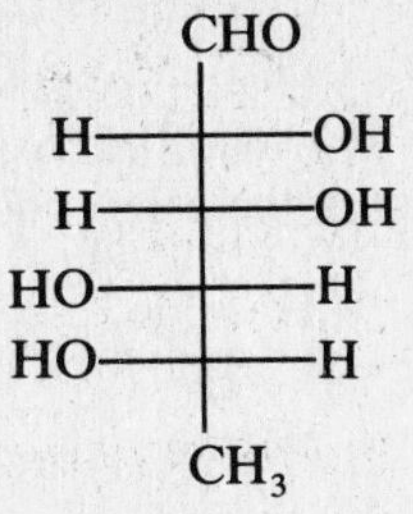 L-Rhamnose

---

Question: What are the systematic names for these sugars?

6-Deoxy-L-galactose (L-fucose) and 6-deoxy-L-mannose (L-rhamnose).

Alditols

A different type of reduced sugar is an *alditol*, in which the aldehyde group of an aldose has been reduced. For example, the alditol produced from D-glucose is D-glucitol (the trivial name is sorbitol). The name of an alditol is obtained by adding *-itol* to the root of the name of the aldose (except for glycerol, a reduction product of glyceraldehyde).

EXAMPLE 2.17

Two alditols are shown below.

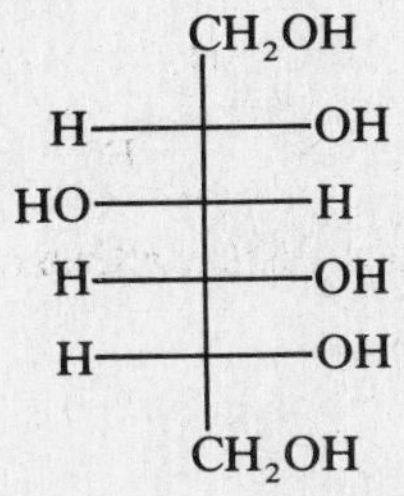

$$\begin{array}{rcl} & \mathrm{CH_2OH} & \\ \mathrm{H}- & | & -\mathrm{OH} \\ \mathrm{HO}- & | & -\mathrm{H} \\ \mathrm{H}- & | & -\mathrm{OH} \\ \mathrm{H}- & | & -\mathrm{OH} \\ & \mathrm{CH_2OH} & \end{array} \qquad \begin{array}{rcl} & \mathrm{CH_2OH} & \\ \mathrm{H}- & | & -\mathrm{OH} \\ \mathrm{H}- & | & -\mathrm{OH} \\ \mathrm{H}- & | & -\mathrm{OH} \\ & \mathrm{CH_2OH} & \end{array}$$

D-Glucitol 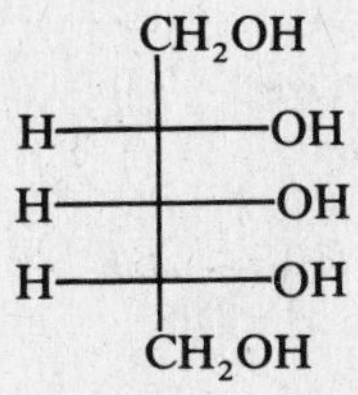 D-Ribitol

---

Question: What are the names of the compounds obtained when the carbonyl group of D-fructose is reduced to an alcohol?

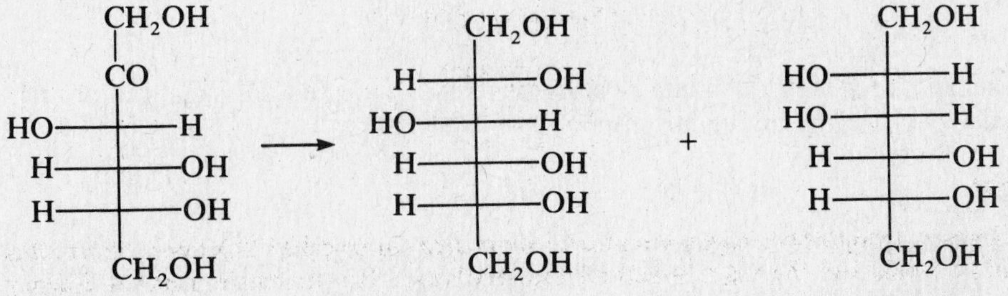

$$\begin{array}{rcl} & \mathrm{CH_2OH} & \\ & \mathrm{CO} & \\ \mathrm{HO}- & | & -\mathrm{H} \\ \mathrm{H}- & | & -\mathrm{OH} \\ \mathrm{H}- & | & -\mathrm{OH} \\ & \mathrm{CH_2OH} & \end{array} \longrightarrow \begin{array}{rcl} & \mathrm{CH_2OH} & \\ \mathrm{H}- & | & -\mathrm{OH} \\ \mathrm{HO}- & | & -\mathrm{H} \\ \mathrm{H}- & | & -\mathrm{OH} \\ \mathrm{H}- & | & -\mathrm{OH} \\ & \mathrm{CH_2OH} & \end{array} + \begin{array}{rcl} & \mathrm{CH_2OH} & \\ \mathrm{HO}- & | & -\mathrm{H} \\ \mathrm{HO}- & | & -\mathrm{H} \\ \mathrm{H}- & | & -\mathrm{OH} \\ \mathrm{H}- & | & -\mathrm{OH} \\ & \mathrm{CH_2OH} & \end{array}$$

D-Fructose D-Glucitol D-Mannitol

Uronic and Aldonic Acids

Uronic acids are sugars in which the hydroxymethyl group of an aldose has been oxidized to a carboxylic acid. These occur as the salts, known as *uronates*, at physiological pH.

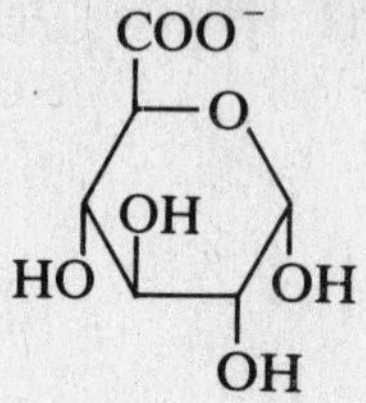

α-D-Glucuronate

Oxidation of the aldehyde group of an aldose to a carboxylic acid group gives a derivative known as an *aldonic acid*. This occurs as a salt, *aldonate*, at physiological pH. If an aldonic acid contains five or more carbon atoms, a *δ-lactone* is formed spontaneously by the condensation of the carboxylic acid group and the hydroxyl group on C-5.

EXAMPLE 2.18

With glucose, chemical oxidation of C-1 occurs with the open-chain form, whereas enzyme-catalyzed oxidation of C-1 occurs with the ring form of the sugar. The aldonic acid (gluconic acid) that is formed is an equilibrium mixture of the free acid (open chain) and the δ-lactone (ring).

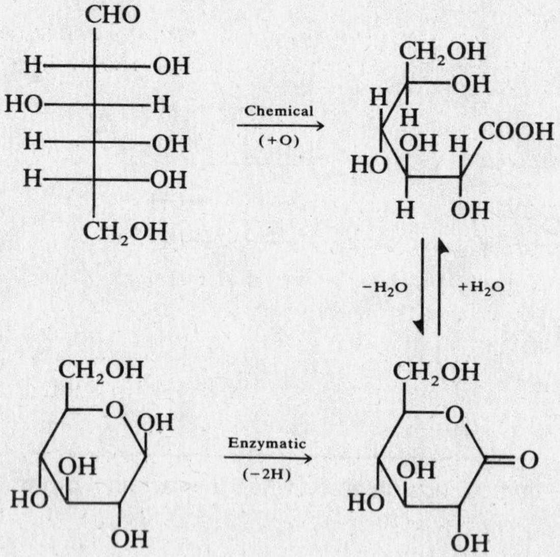

Chemical (+O)

D-Gluconic acid ($C_6H_{12}O_7$)

$-H_2O$ $+H_2O$

Enzymatic (−2H)

D-Gluconolactone ($C_6H_{10}O_6$)

Amino Sugars

Amino sugars are widely distributed naturally. Generally, they are sugars in which a hydroxyl group has been replaced by an amino group.

EXAMPLE 2.19

The most common amino sugars are:

D-Glucosamine (2-amino-2-deoxy-D-glucose)

D-Galactosamine (2-amino-2-deoxy-D-galactose)

Neuraminic acid, a derivative of D-mannosamine (2-amino-2-deoxy-D-mannose)

The amino groups in these compounds are usually acetylated.

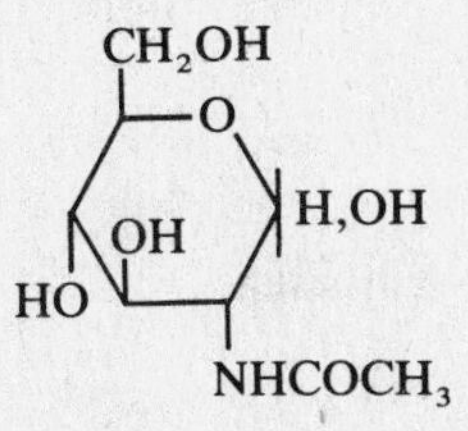

N-Acetyl-D-glucosamine (ring structure: CH_2OH, O, H,OH, OH, HO, $NHCOCH_3$)

N-Acetyl-D-galactosamine (ring structure: CH_2OH, HO, O, H,OH, OH, $NHCOCH_3$)

N-Acetylneuraminic acid (sialic acid) (Fischer projection: COOH; CO; H—H; H—OH; H_3COCHN—H; HO—H; H—OH; H—OH; CH_2OH)

N-Acetyl-D-glucosamine 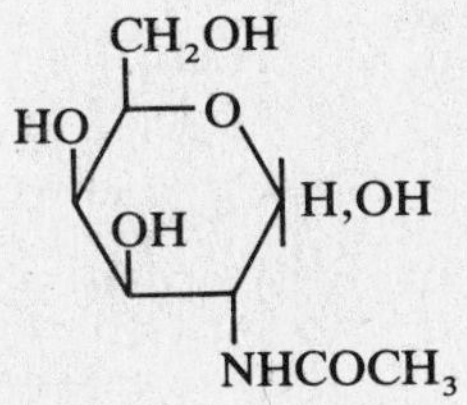   *N*-Acetyl-D-galactosamine 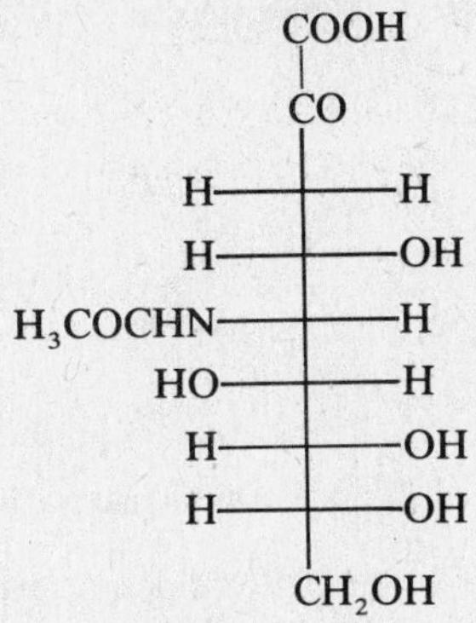   *N*-Acetylneuraminic acid (sialic acid)

Phosphate and Sulfate Esters

Many monosaccharides and their derivatives occur naturally in a form in which one or more of the hydroxyl groups has been substituted by a phosphate or a sulfate group. These are known as *esters*. In general, the phosphate esters are found as components of metabolic pathways within cells, whereas the sulfate esters are found in oligosaccharides and polysaccharides occurring outside cells.

EXAMPLE 2.20

Two phosphate esters and two sulfate esters of monosaccharides that occur naturally are fructose 1,6-diphosphate, 6-phosphogluconate (for both, see Chap. 11); and D-galactose 4-sulfate, *N*-acetylgalactosamine 4-sulfate.

2.8 THE GLYCOSIDIC BOND

All monosaccharides and their derivatives that possess aldehyde or ketone groups (that is, excepting derivatives such as alditols and aldonic acids) will have reducing properties. Moreover, those with the appropriate number of carbon atoms can form rings occurring in two forms (anomers) and in which the potential reducing carbon is called the anomeric carbon.

Question: How reactive is the anomeric hydroxyl group compared with the other hydroxyl groups in a monosaccharide?

It is much more reactive than a typical primary or secondary alcohol. This reactivity is due to the electron-withdrawing influence of the ring oxygen. The situation may be compared to the structure written for a carboxylic acid in which the four atoms, C-O and O-H, constitute the chemical group. With a monosaccharide the anomeric C atom, the H and O-H atoms on the anomeric carbon, and the ring oxygen constitute the corresponding chemical grouping.

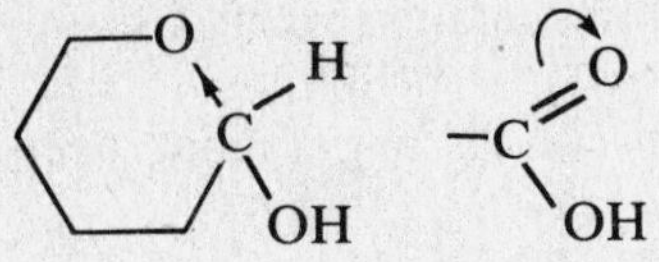

(Structures: ring O–C(H)(OH); –C(=O)OH)

EXAMPLE 2.21

The reactivity of the anomeric hydroxyl group is illustrated by the ease with which monosaccharides react with alcohols and with amines. The normal hydroxyl groups in the molecule do not react, but the anomeric hydroxyl does. The process is known as *glycosylation* (of the alcohol or amine), and the products as O-*glycosides* and N-*glycosides*.

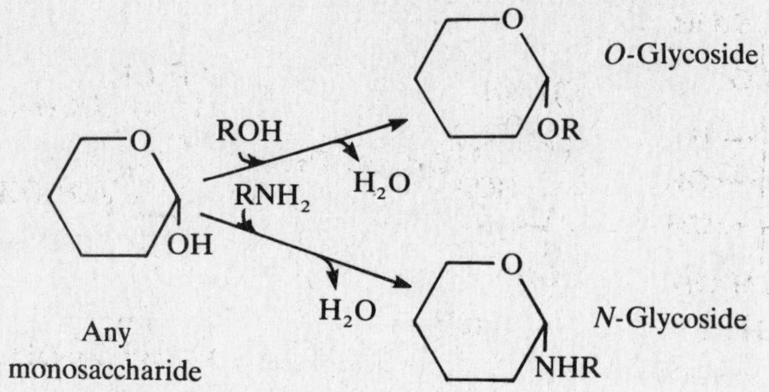

The R group in these glycosides is referred to as the *aglycone*.

As with the parent monosaccharide, the anomeric carbon can have either the α or the β configuration since it remains a chiral center.

Question: In what respect does a glycoside ring differ from the ring structure of a monosaccharide?

The ring of a glycoside cannot open to give a straight-chain structure. Consequently, glycosides have no reducing properties.

A glycoside is, in chemical terms, an acetal (see also Sec. 2.5):

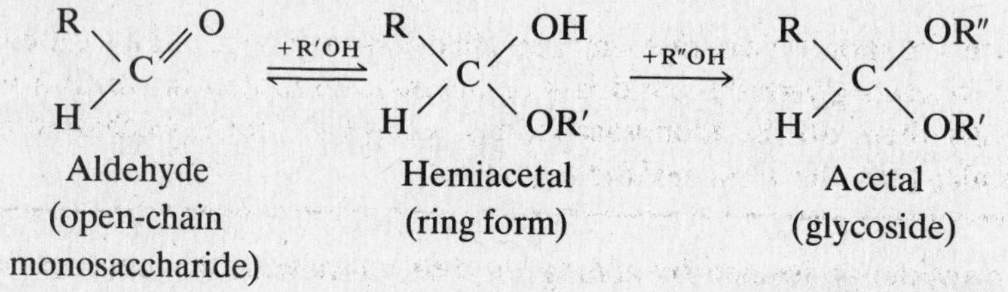

Aldehyde (open-chain monosaccharide) $\xrightleftharpoons{+R'OH}$ Hemiacetal (ring form) $\xrightarrow{+R''OH}$ Acetal (glycoside)

The term *glycoside* is a generic one, and glycosides derived from glucose, fructose, and ribose are known as *glucosides*, *fructosides*, *ribosides*, etc.; i.e., *-oside* is the suffix for glycosides. Glycosides, like the parent monosaccharides, can have either five- or six-membered rings known as *furanosides* or *pyranosides*, respectively.

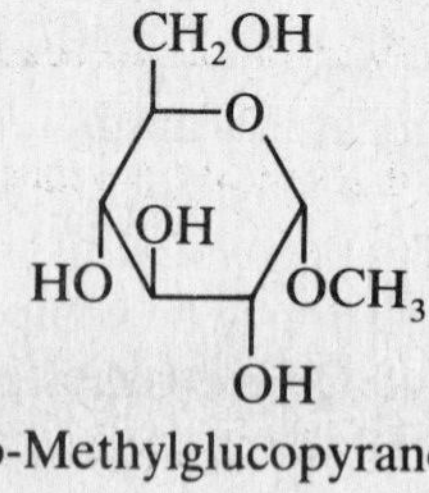

α-D-Methylglucopyranoside 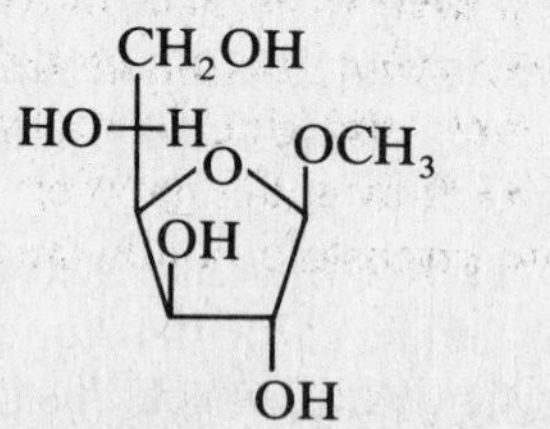 β-D-Methylglucofuranoside

---

Question: What happens to the optical rotation when a freshly prepared solution of α-D-methylglucopyranoside is allowed to stand?

It remains constant. Since a glycosidic ring cannot open, these structures do not display mutarotation.

EXAMPLE 2.22

Give the structure for β-D-methylfructofuranoside.

1. Write the structure for D-fructose.
2. Convert this to the five-membered Haworth ring form (use Example 2.10 as a guide).
3. Take the β anomer and substitute the anomeric hydroxyl with a methyl group.

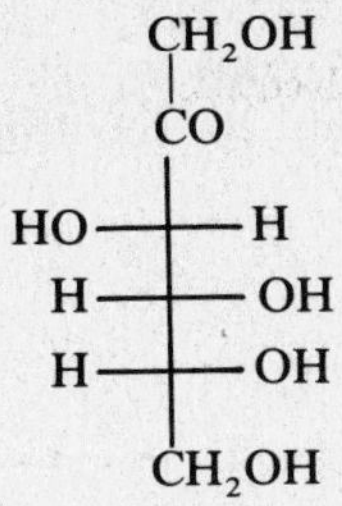

CH_2OH / CO / HO—H / H—OH / H—OH / CH_2OH
(1)

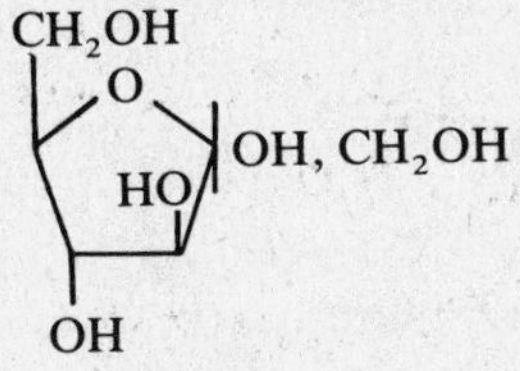

(2): CH_2OH, O, HO, OH, CH_2OH, OH

(3): CH_2OH, O, OCH_3, HO, CH_2OH, OH
(3)
β-D-Methylfructofuranoside

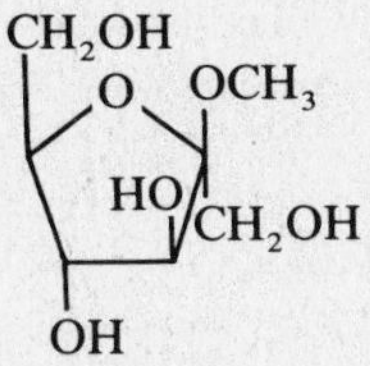

The glycosidic bond is found in a very wide range of biological compounds. In addition, there is a particularly important group of *O*-glycosides in which the glycosidic bond links two monosaccharides, i.e., the aglycone is a sugar. Such compounds are called *disaccharides*. The anomeric hydroxyl group of the second monosaccharide can itself glycosylate a hydroxyl group in a third monosaccharide to give a trisaccharide, and so on. *Polysaccharides* are polymers in which a large number of monosaccharides are linked by glycosidic bonds.

Question: What is an oligosaccharide?

An *oligosaccharide* is a compound consisting of an undefined but small number of monosaccharides linked by glycosidic bonds.

Question: Why are the monosaccharide units in oligo- and polysaccharides called *residues*?

The formation of each glycosidic bond is a *condensation reaction* in which a water molecule is produced. Effectively then, all the monosaccharides, except the monosaccharide at one end, have lost a water molecule; thus the term *residue* is justified.

The name of a residue is formed by adding *-osyl* to the root of the name of the sugar. Thus, a trisaccharide made from three glucose molecules is glucosylglucosylglucose. In order to abbreviate the written descriptions for oligo- and polysaccharides, a shorthand method for naming residues has been introduced; e.g., Glc (glucosyl), Gal (galactosyl), Fru (fructosyl), GlcN (glucosaminyl), GlcNAc (*N*-acetylglucosaminyl), GlcA (glucuronyl), NeuNAc (*N*-acetylneuraminyl).

The term *glycan* is used as an alternative to the word *polysaccharide*. This is a generic term, and names such as *glucan*, *xylan*, *glucomannan* describe polymers composed, respectively, of glucose residues, xylose residues, and glucose and mannose residues.

The greatest problem in understanding the structures of oligo- and polysaccharides is related to the ways in which the glycosidic bonds are written.

EXAMPLE 2.23

The structures of two different disaccharides, both composed of glucose, are shown.

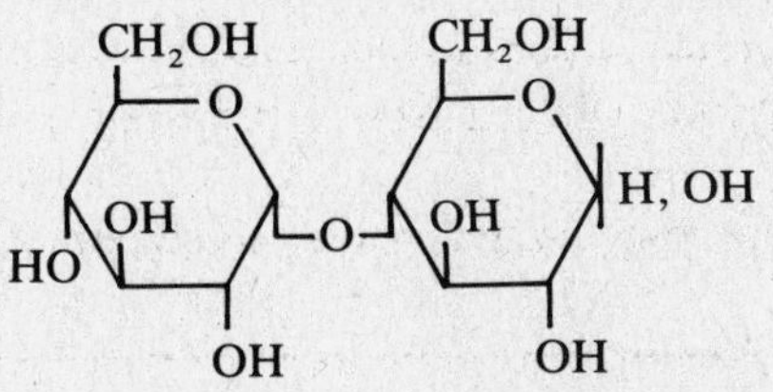

CH_2OH O CH_2OH O; OH, HO, O, OH, H, OH; OH OH

Maltose
α-Glc-(1→4)-Glc

CH_2OH O CH_2OH O; OH, HO, O, OH, H, OH; OH OH

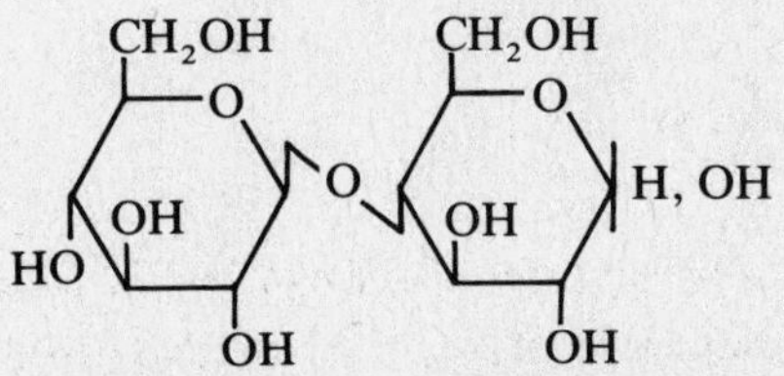

Cellobiose
β-Glc-(1→4)-Glc

The glycosidic bonds joining the glucose molecules are printed the way they are for the following reason. Consider a disaccharide formed between two monosaccharides, A and B, in which the anomeric hydroxyl of A is used to glycosylate the hydroxyl on C-4 of B:

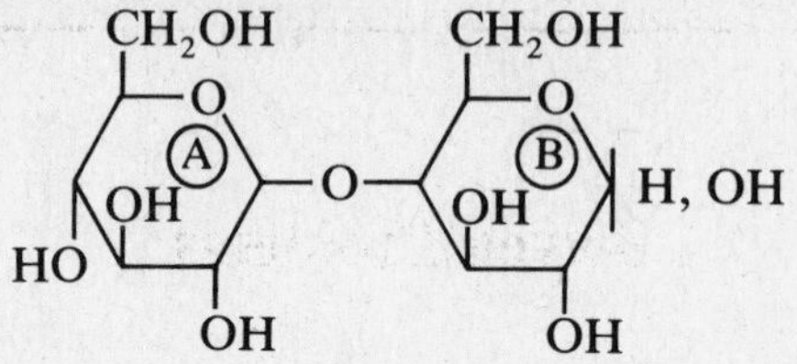

CH_2OH CH_2OH
O O
(A) (B)
OH O OH H, OH
HO
OH OH

This structure is ambiguous in two respects. It does not show the configuration at C-1 of A—i.e., whether the glycosidic bond is α or β. Nor does it show the configuration at C-4 of B, and therefore the identity of B is not revealed. B could be either glucose (with the glycosidic O below the ring B) or galactose (with the glycosidic O above ring B). Thus maltose and cellobiose are written as shown to indicate clearly (1) whether the glycosidic bond is α or β with respect to the glycosyl component and (2) what the identity of the other sugar is. The printing of the glycosidic bonds as C C and C O C is a pictorial device: it gives clearly the correct configuration of the chiral carbon atoms. The bonds in reality are not bent.

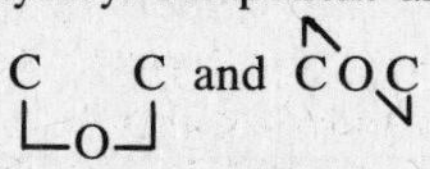

Question: What are the systematic names for maltose and cellobiose?

Maltose is α-D-glucosyl-(1→4)-D-glucose or, more specifically, α-D-glucopyranosyl-(1→4)-D-glucopyranose. Cellobiose is β-D-glucosyl-(1→4)-D-glucose.

In the structure for maltose shown in Example 2.23, no configuration is given for the anomeric carbon of the glucose unit on the right; the structure represents the state of maltose in solution—a mixture of α and β anomers. In crystalline maltose, the anomeric hydroxyl is α, and maltose can be described as α-D-glucosyl-(1→4)-α-D-glucose.

Question: Is maltose a reducing sugar?

Yes. Although it is a glycoside, the second glucose unit possesses an anomeric carbon atom and its ring can open to give an aldehyde. For the same reason, solutions of maltose display mutarotation.

Question: Why doesn't sucrose, another disaccharide, have reducing properties?

Sucrose, or cane sugar, is a disaccharide in which the anomeric hydroxyl of α-D-glucose is condensed with the anomeric hydroxyl of β-D-fructose (Example 2.22). It is therefore both an α-glucoside and a β-fructoside. Neither unit possesses an anomeric hydroxyl and neither ring can open to give an aldehyde.

CH_2OH
O
OH
HO
HO
O
CH_2OH
O
HO CH_2OH
OH

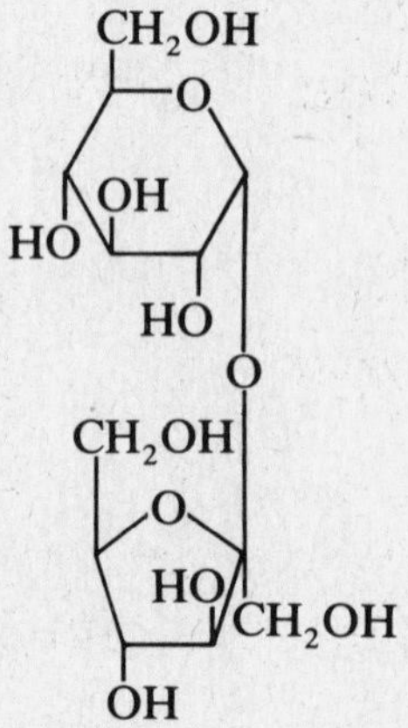

Question: What is the systematic name for sucrose?

Either α-D-glucosyl-(1→2)-β-D-fructoside (α-D-glucopyranosyl-(1→2)-β-D-fructofuranoside) or β-D-fructosyl-(2→1)-α-D-glucoside (β-D-fructofuranosyl-(2→1)-α-D-glucopyranoside). In abbreviated form, these are α-Glc-(1→2)-β-Fru and β-Fru-(2→1)-α-Glc, respectively.

Solved Problems

THE STRUCTURE OF D-GLUCOSE

2.1. A solution of D-glucose contains predominantly the α and β anomers of D-glucopyranose, both of which are nonreducing. Why is a solution of D-glucose a strong reducing agent?

SOLUTION

Because there is some open-chain glucose present with reducing properties. As this reacts, the equilibria between it and the nonreducing ring forms are disturbed, causing more of the open-chain form to appear. Ultimately, all the glucose will have reacted via the chain form.

2.2. What percentage of D-glucose in solution at equilibrium exists as the β anomer?

SOLUTION

By definition, 1 g of β-D-glucose in 1 cm^3 will give a rotation of +19° in a 1-dm polarimeter tube. Likewise 1 g of α-D-glucose will give a rotation of +112° (see Sec. 2.5). At equilibrium, let there be b g cm^{-3} of β-D-glucose. There will be $(1-b)$ g cm^{-3} of α-D-glucose at equilibrium. The rotation at equilibrium (+52°; Example 2.11) will be due to a contribution from b g of β-D-glucose and $(1-b)$ g of α-D-glucose; i.e., $b(+19°)+(1-b)(+112°)=+52°$, from which $b=0.64$ g. Therefore the percentage of glucose present as the β anomer at equilibrium is 64 percent. (This answer assumes that the open-chain form of glucose makes a negligible contribution to the equilibrium mixture, an assumption supported by spectral analysis of the solution at equilibrium.)

THE CONFORMATION OF GLUCOSE

2.3. Write the Haworth structure for β-L-glucopyranose.

SOLUTION

By definition, this is the mirror image of β-D-glucopyranose. The structure can be drawn in six ways by imagining a mirror placed (1) to the left, (2) to the right, (3) above, (4) below, (5) in front of, or (6) behind the structure of β-D-glucopyranose. Three of these images are shown.

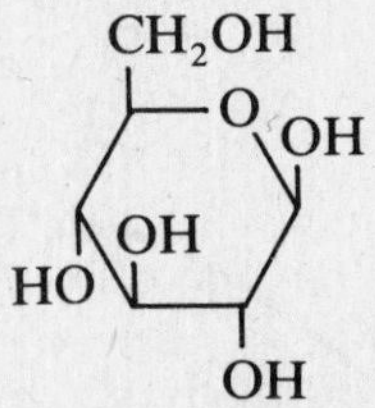

β-D-Glucopyranose

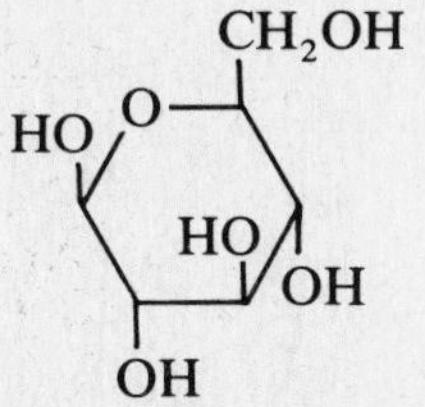

(2)

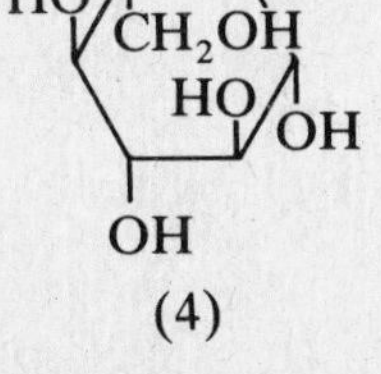

(4)

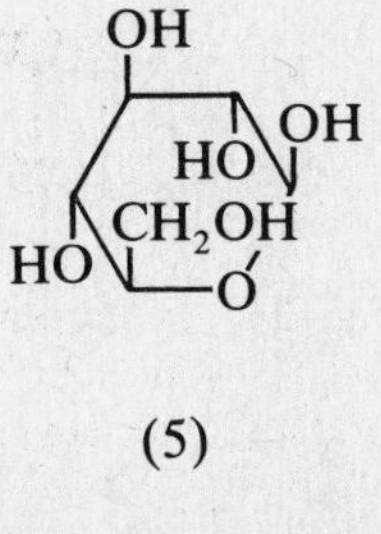

(5)

All six images represent the same structure and can be seen to be so by rotating any image about an appropriate axis; e.g., rotating (4) toward you through 180° about a horizontal axis in the plane of the paper will give (5).

2.4. What is the preferred conformation for 3,6-anhydro-D-glucose?

SOLUTION

3,6-Anhydro-D-glucose (1) is a glucose molecule in which a water molecule is lost between C-3 and C-6. Although the preferred conformation of D-glucose is C1, this conformation is not possible with (1) because the oxygen on C-3 is too far from C-6 for a bond to be formed between them (2). Consequently, 1C is a more likely conformation (3), but the large number of axial hydroxyl groups and the strain in the ring bounded by C-3, C-4, C-5, C-6 and the O between C-3 and C-6 make this conformation unstable. The preferred conformation for 3,6-anhydro-D-glucose is (4), with two fused, planar five-membered rings.

(1) (2) (3) (4)

MONOSACCHARIDES OTHER THAN GLUCOSE

2.5. Are D-mannose and D-galactose epimers?

SOLUTION

No. They differ in configuration at *two* carbon atoms.

2.6. Figure 2-2 shows the mutarotation that occurs when β-D-ribose is dissolved in water. How can this curve be interpreted?

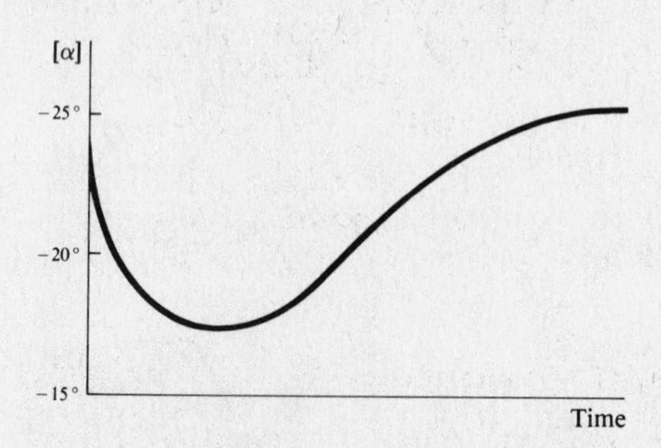

Fig. 2-2 Mutarotation of β-D-ribose dissolved in H_2O.

SOLUTION

Crystalline β-D-ribose is β-D-ribopyranose (Sec. 2.7), and when this ribose is dissolved in water, the pyranose ring opens. The resulting open-chain form will close to give furanose and pyranose ring forms. Formation of the furanose ring is much faster since the random motion of the open-chain form allows the hydroxyl group on C-4 to approach C-1 far more frequently than the hydroxyl group on C-5 can. However, the pyranose ring is more stable because it is larger and the steric repulsion is less. Thus, the initial changes in rotation shown in Fig. 2-2 are due to the appearance of a relatively high concentration of furanose forms, and the later changes in rotation reflect the reappearance of increasing concentrations of pyranose forms.

2.7. Show that L-ribitol 1-phosphate is identical with D-ribitol 5-phosphate.

SOLUTION

All carbon atoms of ribitol are in the same state of oxidation, and the carbon chain can be numbered from either end. If numbered as shown in (*a*) below, and assuming C-1 was derived from an aldehyde, then the compound would be related to D-ribose and can be called D-ribitol. D-Ribitol 5-phosphate would be as shown in (*b*). However, if C-5 was derived from an aldehyde, then the compound would be related to L-ribose [turn the structure in (*a*) upside down] and called L-ribitol. L-Ribitol 1-phosphate would be as shown in (*c*), clearly the same as the phosphate ester shown in (*b*).

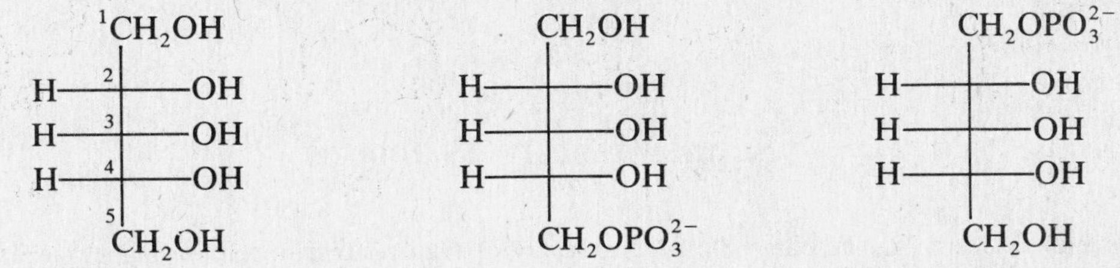

(*a*)	(*b*)	(*c*)
$^{1}CH_2OH$	CH_2OH	$CH_2OPO_3^{2-}$
H—^{2}C—OH	H—C—OH	H—C—OH
H—^{3}C—OH	H—C—OH	H—C—OH
H—^{4}C—OH	H—C—OH	H—C—OH
$^{5}CH_2OH$	$CH_2OPO_3^{2-}$	CH_2OH

THE GLYCOSIDIC BOND

2.8. The disaccharide shown is lactose, the carbohydrate of mammalian milk. Give (*a*) its full name and (*b*) its abbreviated name.

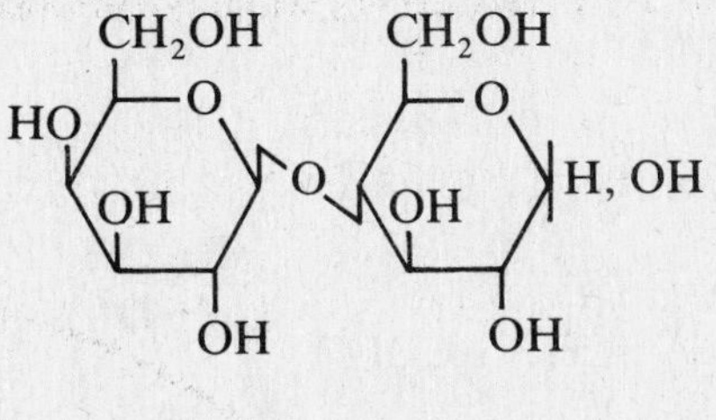

CH_2OH CH_2OH
HO O O
O
OH OH H, OH
OH OH

Lactose

SOLUTION

(*a*) β-Galactopyranosyl-(1→4)-glucopyranose

(*b*) β-Gal-(1→4)-Glc

2.9. Give the structure for a disaccharide, containing only glucose, that is nonreducing.

SOLUTION

For the disaccharide to be nonreducing, the anomeric carbons of both glucose molecules must be linked via a glycosidic bond. The product is called *trehalose*, and it exists in three isomeric forms in which the glucose molecules are linked α,α; β,β; or α,β (shown). α,β-Trehalose is α-D-glucopyranosyl-$(1 \rightarrow 1)$-β-D-glucopyranose.

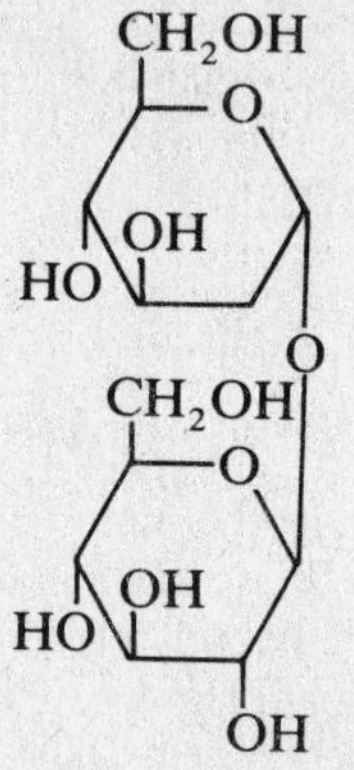

α,β-Trehalose

Supplementary Problems

2.10. Identify by class and by name the monosaccharides that occur in the following glycosides. Give the configuration of each glycosidic bond.

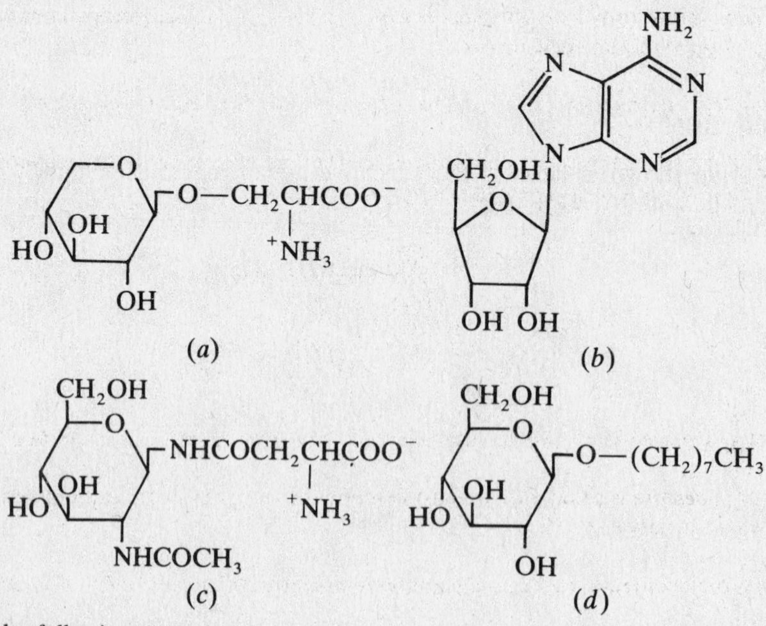

2.11. Which of the following are reducing sugars?

(*a*) Galactose
(*b*) β-Methylgalactoside
(*c*) Maltose
(*d*) Mannose
(*e*) Xylose
(*f*) Fructose
(*g*) Rhamnose
(*h*) Ribose
(*i*) Glucosamine
(*j*) Gluconic acid
(*k*) Glucitol

2.12. (*a*) Draw the structure of any β-D-aldoheptose in the pyranose form. (*b*) Draw the structures of the anomer, the enantiomer, an epimer.

2.13. A sugar ($C_5H_{10}O_5$) was treated by a method that reduces aldehyde groups and gave a product that was optically inactive. Assuming the sugar was D, what are two possible structures of the product?

2.14. An aqueous solution of D-galactose has an $[\alpha]_D^{25}$ of +80.2° after standing for some hours. The specific rotations of pure α-D-galactose and β-D-galactose are +150.7° and +52.8°, respectively. Calculate the proportions of α- and β-D-galactose in the equilibrium mixture.

2.15. The specific rotation of maltose in water is +138°. What would be the concentration of a maltose solution that had an optical rotation of +23° if a polarimeter tube 10 cm long was used?

2.16. An enzyme known as *invertase* catalyzes the hydrolysis of sucrose to an equimolar mixture of D-glucose and D-fructose. During the hydrolysis, the optical rotation of the solution changes from (+) to (−). What can you conclude from this observation?

2.17. Nojirimycin (5-amino-5-deoxy-D-glucose) is an antibiotic used in studies of the biosynthesis of glycoproteins. Write its open-chain and pyranose ring structures.

2.18. There are two possible chair conformers of β-D-glucopyranose. How many boat conformers are possible?

2.19. Show that in β-L-glucopyranose all the substituents on the ring carbon atoms are equatorial.

2.20. Write the preferred conformation for α,α-trehalose (see Prob. 2.9), the carbohydrate found in the hemolymph of insects.

2.21. Erythritol is the reduction product of D-erythrose. Why is the prefix D- omitted from its name?

2.22. Galactitol, the reduced form of D-galactose, is the toxic by-product that accumulates in persons suffering from galactosemia. Write its structure.

2.23. Write the open-chain structures for (*a*) D-gluconic acid and (*b*) D-glucuronic acid.

2.24. L-Iduronic acid is part of the structure of some polysaccharides found in connective tissue. How could this sugar be formed in a one-step reaction from D-glucuronic acid?

O
COOH H, OH
OH
HO
OH

2.25. Write the Fischer projection formulas for the sugar derivatives named in Example 2.20.

2.26. The name *methylpentose* is sometimes used to describe the sugars L-fucose and L-rhamnose. Is this name valid? Explain your answer.

2.27. Are α-methyl-D-glucopyranoside and β-methyl-D-glucopyranoside anomers, isomers, or conformers?

Chapter 3

Amino Acids and Peptides

3.1 AMINO ACIDS

Amino Acids Found in Proteins

All proteins are composed of *amino acids* linked into a linear sequence by *peptide bonds* between the amino group of one amino acid and the *carboxyl* group of the preceding amino acid. The amino acids found in proteins are all *α-amino acids*; i.e., the amino and carboxyl groups are both attached to the same carbon atom (the α-carbon atom; Fig. 3-1). The α-carbon atom is a potential chiral center, and except when the —R group (or *side chain*) is H, amino acids display optical activity. All amino acids found in proteins are of the L configuration, as indicated in Fig. 3-1.

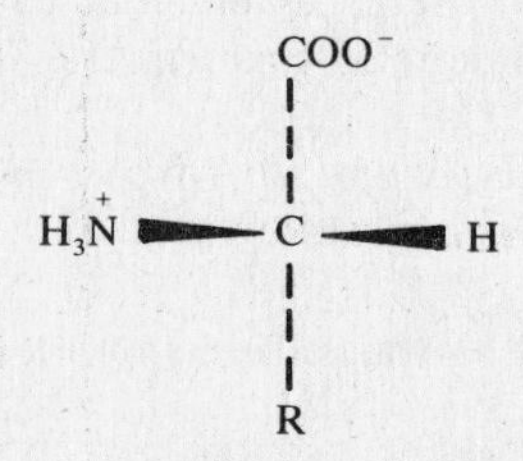

Fig. 3-1 General structure of α-amino acids in the L configuration; R is one of over 20 different chemical groups.

There are 20 different amino acids used in the synthesis of proteins; these amino acids are listed in Table 3.1, which also contains the two commonly used symbols for each amino acid. The three-letter symbols are easier to remember, but the single-letter symbols are often used in writing long sequences. In many proteins some of the amino acids are modified after incorporation into proteins; e.g., in collagen, a hydroxyl group is added to each of several proline residues to yield *hydroxyproline* residues. With the exception of proline, the α-amino acids that are incorporated into proteins can be represented by the formula shown in Fig. 3-1.

The side chains of the amino acids do not form a natural series, and thus, there is no easy way to learn their structures. It is useful to classify them according to whether they are polar or nonpolar, aromatic or aliphatic, or *acidic* or *basic*, although these classifications are not mutually exclusive. Tyrosine, for example, can be considered to be both aromatic and polar, although the polarity introduced by a single hydroxyl group in this aromatic compound is somewhat feeble.

The aromatic amino acids absorb light strongly in the ultraviolet region. Use may be made of this in determining the concentration of these amino acids in solution. The *Beer-Lambert* relationship states that the absorbance of light at a given wavelength by a substance in solution is proportional to its concentration C (in mol L^{-1}) and the length l (in cm) of the light path in the solution:

$$A = \varepsilon C l \tag{3.1}$$

where A is the absorbance of the solution and ε is the molar absorbance coefficient. The absorbance

Table 3.1. Amino Acids Used in Protein Synthesis, Grouped according to Chemical Type

Structure	Name	Abbreviation	
1. Neutral Amino Acids			
(*a*) *Nonpolar, aliphatic*			
$H_3\overset{+}{N}-\underset{COO^-}{\overset{H}{C}}-H$	Glycine	Gly	G
$H_3\overset{+}{N}-\underset{COO^-}{\overset{H}{C}}-CH_3$	Alanine	Ala	A
$H_3\overset{+}{N}-\underset{COO^-}{\overset{H}{C}}-CH{<}^{CH_3}_{CH_3}$	Valine	Val	V
$H_3\overset{+}{N}-\underset{COO^-}{\overset{H}{C}}-CH_2-CH{<}^{CH_3}_{CH_3}$	Leucine	Leu	L
$H_3\overset{+}{N}-\underset{COO^-}{\overset{H}{C}}-\overset{CH_3}{CH}-CH_2-CH_3$	Isoleucine	Ile	I
(*b*) *Polar, aliphatic*			
$H_3\overset{+}{N}-\underset{COO^-}{\overset{H}{C}}-CH_2-OH$	Serine	Ser	S
$H_3\overset{+}{N}-\underset{COO^-}{\overset{H}{C}}-\overset{OH}{CH}-CH_3$	Threonine	Thr	T
$H_3\overset{+}{N}-\underset{COO^-}{\overset{H}{C}}-CH_2-\underset{O}{\overset{}{\underset{\Vert}{C}}}-NH_2$	Asparagine	Asn	N
$H_3\overset{+}{N}-\underset{COO^-}{\overset{H}{C}}-CH_2-CH_2-\underset{O}{\underset{\Vert}{C}}-NH_2$	Glutamine	Gln	Q

Table 3.1. (*Cont.*)

Structure	Name	Abbreviation	
(*c*) *Aromatic*			
$H_3\overset{+}{N}-C(H)(COO^-)-CH_2-C_6H_5$	Phenylalanine	Phe	F
$H_3\overset{+}{N}-C(H)(COO^-)-CH_2-C_6H_4-OH$	Tyrosine	Tyr	Y
$H_3\overset{+}{N}-C(H)(COO^-)-CH_2-C$ (indole ring, N–H)	Tryptophan	Trp	W
(*d*) *Sulfur-containing*			
$H_3\overset{+}{N}-C(H)(COO^-)-CH_2-SH$	Cysteine	Cys	C
$H_3\overset{+}{N}-C(H)(COO^-)-CH_2-CH_2-S-CH_3$	Methionine	Met	M
(*e*) *Containing secondary amino group*			
$H_2\overset{+}{N}$, CH_2, CH_2, CH_2, $H-C$ ring; COO^-	Proline	Pro	P
2. Acidic Amino Acids			
$H_3\overset{+}{N}-C^{\alpha}(H)(COO^-)-\overset{\beta}{C}H_2-COO^-$	Aspartate	Asp	D
$H_3\overset{+}{N}-C^{\alpha}(H)(COO^-)-\overset{\beta}{C}H_2-\overset{\gamma}{C}H_2-COO^-$	Glutamate	Glu	E

Table 3.1. (*Cont.*)

Structure	Name	Abbreviation	
3. Basic Amino Acids			
$H_3\overset{+}{N}-\underset{COO^-}{\overset{H}{C^{\alpha}}}-\overset{\beta}{C}H_2-\overset{\gamma}{C}H_2-\overset{\delta}{C}H_2-\overset{\varepsilon}{C}H_2-\overset{+}{N}H_3$	Lysine	Lys	K
$H_3\overset{+}{N}-\underset{COO^-}{\overset{H}{C}}-CH_2-CH_2-CH_2-NH-\overset{NH_2}{C}\overset{+}{=}NH_2$	Arginine	Arg	R
$H_3\overset{+}{N}-\underset{COO^-}{\overset{H}{C}}-CH_2-C$ (imidazole ring: C=CH–NH–CH=N)	Histidine	His	H

The Greek symbols indicate the nomenclature of the carbon chains in certain amino acids. The carbon atom carrying (i.e., next to) the carboxyl group is labeled α.

is defined as the logarithm of the ratio of the intensity of incident light (I_o) to that of transmitted light (I):

$$A = \log_{10} \frac{I_o}{I} \qquad (3.2)$$

EXAMPLE 3.1

Given that the *molar absorbance coefficient* of tyrosine in water is 1,420 L mol^{-1} cm^{-1} at 275 nm, what is the concentration of tyrosine in a solution of path length 1 cm for which the absorbance is 0.71?

By use of the Beer-Lambert relationship:

$$C = \frac{A}{\varepsilon \cdot l} = \frac{0.71}{1{,}420 \times 1.0} = 5 \times 10^{-4} \text{ mol L}^{-1}$$

Nonprotein Amino Acids

A large number of amino acids involved in metabolism are not found in proteins; e.g., β-alanine, $^-OOC-CH_2-CH_2-NH_3^+$, is an intermediate in the synthesis of the B vitamin *pantothenic acid*, but it is not found in proteins. Although most naturally occurring amino acids are of the L configuration, some D-amino acids are found in certain antibiotics and in the cell walls of some bacteria.

3.2 ACID-BASE BEHAVIOR OF AMINO ACIDS

Amino acids are *amphoteric* compounds; i.e., they contain both acidic and basic groups. Because of this, they are capable of bearing a net electrical charge, which depends on the nature of the solution.

The charge carried by a molecule influences the way in which it interacts with other molecules; use can be made of this property in the isolation and purification of amino acids and proteins. Therefore, it is important to have a clear understanding of the factors that influence the charge carried on amino acids.

The Ionization of Water

The major biological solvent is water, and the acid-base behavior of dissolved molecules is intimately linked with the dissociation of water. Water is a weak electrolyte capable of dissociating to a proton and a hydroxyl ion. In this process, the proton binds to an adjacent water molecule to which it is *hydrogen-bonded* (Chap. 4) to form a *hydronium ion* (H_3O^+):

$$\mathrm{H_2O \cdots H{-}OH} \overset{K_e}{\rightleftharpoons} \mathrm{H_2O^+{-}H} + \mathrm{OH^-}$$

In pure water at 25°C, at any instant there are 1.0×10^{-7} mol L^{-1} of H_3O^+ and an equivalent concentration of OH^- ions. It must be stressed that the proton is hardly ever "bare" in water because it has such a high affinity for water molecules. Hydrated complexes other than H_3O^+ have been suggested, but since water is so extensively hydrogen-bonded, it is difficult to identify these species experimentally, and as a simplification the hydrated proton is often written as H^+.

The dissociation of water is a rapid equilibrium process for which we can write an equilibrium constant:

$$K_e = \frac{a_{H_3O^+} \times a_{OH^-}}{(a_{H_2O})^2} \tag{3.3}$$

Since, for dilute solutions, the *activity a* (Chap. 10) of water is considered to be constant and very close to 1.0 and the activities of the solutes may be represented by their concentrations, we can define a practical constant, K_w, called the *ionic product of water*:

$$K_w = [H_3O^+] \cdot [OH^-] \tag{3.4}$$

or, as is often seen, $K_w = [H^+][OH^-]$, with the hydration of the proton ignored for simplicity. Note that the square brackets denote concentrations of species in mol L^{-1}.

At 25°C in pure water, $K_w = 10^{-14}$. Since, in pure water, $[H^+] = [OH^-]$

$$[H^+] = \sqrt{10^{-14}} = 10^{-7} \text{ mol L}^{-1}$$

In acid solution, $[H^+]$ is higher, and $[OH^-]$ is correspondingly lower, because the ionic product is constant. The value of K_w is temperature-dependent; at 37°C, for example, $K_w = 2.4 \times 10^{-14}$.

EXAMPLE 3.2

Calculate $[OH^-]$ in aqueous solution at 25°C when $[H^+] = 0.1$ mol L^{-1}.
Since $[H^+] \cdot [OH^-] = 10^{-14}$

$$[OH^-] = \frac{10^{-14}}{10^{-1}} = 10^{-13} \text{ mol L}^{-1}$$

Acidity and pH

The Danish chemist S. P. L. Sørensen defined pH (*potentia Hydrogenii*) as:

$$\mathrm{pH} = -\log_{10}[H^+] \tag{3.5}$$

Neutral solutions are defined as those in which $[H^+] = [OH^-]$, and for pure water at 25°C

$$\mathrm{pH} = -\log_{10}(10^{-7}) = 7.0 \tag{3.6}$$

Note that *distilled* water normally used in the laboratory is not absolutely pure. Traces of CO_2 dissolved in it produce carbonic acid that increases the hydrogen-ion concentration to about 10^{-5} mol L^{-1}, thus rendering the pH around 5.

EXAMPLE 3.3

Calculate the pH of a 4×10^{-4} mol L^{-1} solution of HCl.

At this low concentration we may consider HCl to be completely dissociated to H^+ and Cl^-. Therefore

$$[H^+] = 4 \times 10^{-4} \text{ mol L}^{-1}$$

and

$$\text{pH} = -\log(4 \times 10^{-4}) = 3.40$$

Note that acid solutions (high H^+ concentrations) have *low* pH and alkaline solutions (high OH^- concentration and *low* H^+ concentration) have *high* pH. A 10-fold increase in $[H^+]$ corresponds to a decrease of 1.0 in pH.

Weak Acids and Bases

Acids

An *acid* is a compound capable of donating a proton to another compound (this is the so-called *Brønsted definition*). The substance CH_3COOH is an acid, *acetic acid*. However, because the dissociation of all the carboxyl groups is not complete when acetic acid is dissolved in water, acetic acid is referred to as a *weak* acid. The dissociation reaction for any weak acid of type HA in water is

$$HA + H_2O \rightleftharpoons A^- + H_3O^+$$

The dissociation reaction for acetic acid is therefore

$$\underset{\substack{H^+ \text{ donor} \\ \text{(acid)}}}{CH_3COOH} + \underset{\substack{H^+ \text{ acceptor} \\ \text{(base)}}}{H_2O} \rightleftharpoons \underset{\substack{\text{Conjugate} \\ \text{base}}}{CH_3COO^-} + \underset{\substack{\text{Conjugate} \\ \text{acid}}}{H_3O^+}$$

The donating and accepting of the proton is a two-way process. The H_3O^+ ion that is formed is capable of donating a proton back to the acetate ion to form acetic acid. This means that the H_3O^+ ion is considered to be an acid and the acetate ion is considered a base. Acetate is called the *conjugate base* of acetic acid.

The two processes of association and dissociation come to equilibrium, and the resulting solution will have a higher concentration of H_3O^+ than is found in pure water; i.e., it will have a pH below 7.0.

A measure of the *strength* of an acid is the *acid dissociation constant*, K_a:

$$K_a = \frac{a_{H_3O^+} \cdot a_{A^-}}{a_{HA} \cdot a_{H_2O}} \tag{3.7}$$

For acetic acid, K_a is therefore

$$K_a = \frac{a_{H_3O^+} \cdot a_{CH_3COO^-}}{a_{CH_3COOH} \cdot a_{H_2O}}$$

where a denotes the thermodynamic activity (Chap. 10) of the chemical species.

In dilute solutions, the concentration of water is very close to that of pure water, and the activity of pure water, by convention, is taken to be 1.0. Furthermore, in dilute solutions, the activity of solutes may be approximated by their concentrations; so we may write an expression for a *practical* acid dissociation constant:

$$K_a = \frac{[H^+][A^-]}{[HA]} \tag{3.8}$$

For acetic acid, this becomes

$$K_a = \frac{[H^+][CH_3COO^-]}{[CH_3COOH]}$$

The *larger* the value of K_a, the greater the tendency of the acid to dissociate a proton, and so the *stronger* the acid.

In a manner similar to the definition of pH, we can define

$$pK_a = -\log K_a \tag{3.9}$$

Thus, the *lower* the value of the pK_a of a chemical compound, the higher the value of K_a, and the *stronger* it is as an acid.

EXAMPLE 3.4

Which of the following acids is the stronger: boric acid, which has a $pK_a = 9.0$, or acetic acid, with a $pK_a = 4.6$?

For boric acid, $K_a = 10^{-9}$, while for acetic acid, $K_a = 10^{-4.6} = 2.5 \times 10^{-5}$. Thus, acetic acid has the greater K_a and is therefore the stronger acid.

Bases

A *base* is a compound capable of accepting a proton from an acid. When methylamine (CH_3—NH_2) dissolves in water, it accepts a proton from the water, thus leading to an increase in the OH^- concentration and a high pH.

$$\underset{\text{Base}}{CH_3\text{—}NH_2} + \underset{\text{Acid}}{H_2O} \rightleftharpoons \underset{\substack{\text{Conjugate}\\ \text{acid}}}{CH_3\text{—}NH_3^+} + \underset{\substack{\text{Conjugate}\\ \text{base}}}{OH^-}$$

As in the case of acetic acid, as the concentration of OH^- increases, the reverse reaction becomes more significant and the process eventually reaches an equilibrium.

We can write an expression for a *basicity constant*, K_b:

$$K_b = \frac{[CH_3\text{—}NH_3^+][OH^-]}{[CH_3\text{—}NH_2]} \tag{3.10}$$

However, the use of this constant can be confusing, as we would need to keep track of two different types of constant, K_a and K_b. Since chemical equilibrium is a two-way process, it is perfectly correct, and more convenient, to consider the behavior of bases from the point of view of their *conjugate acid*. The latter can be considered to donate a proton to water:

$$CH_3\text{—}NH_3^+ + H_2O \rightleftharpoons CH_3\text{—}NH_2 + H_3O^+$$

and

$$K_a = \frac{[CH_3\text{—}NH_2][H_3O^+]}{[CH_3\text{—}NH_3^+]} \tag{3.11}$$

Of course, K_a and K_b are related as follows:

$$K_a \cdot K_b = \frac{[CH_3\text{—}NH_2][H_3O^+]}{[CH_3\text{—}NH_3^+]} \cdot \frac{[CH_3\text{—}NH_3^+][OH^-]}{[CH_3\text{—}NH_2]} = [H_3O^+][OH^-]$$

i.e.,

$$K_a \cdot K_b = K_w \tag{3.12}$$

In other words, if we know K_a for the conjugate acid, we can calculate K_b for the base. A base is thus characterized by a *low* value of K_a for its conjugate acid.

Buffers

A mixture of an acid and its conjugate base is capable of resisting changes in pH when small amounts of additional acid or base are added. Such a mixture is known as a *buffer*.

Consider again the dissociation of acetic acid:

$$CH_3COOH + H_2O \rightleftharpoons CH_3COO^- + H_3O^+$$

Additional acid causes recombination of H_3O^+ and CH_3COO^- to form acetic acid, so that a buildup of H_3O^+ is resisted. Conversely, addition of NaOH causes dissociation of acetic acid to acetate, reducing the fall in H_3O^+ concentration.

This behavior can be quantified by taking logarithms of both sides of Eq. (*3.8*):

$$K_a = \frac{[\mathrm{H}^+][\mathrm{A}^-]}{[\mathrm{HA}]}$$

$$\log K_a = \log[\mathrm{H}^+] + \log[\mathrm{A}^-] - \log[\mathrm{HA}]$$

Multiplying both sides by -1, we get

$$-\log K_a = -\log[\mathrm{H}^+] - \log[\mathrm{A}^-] + \log[\mathrm{HA}]$$

From Eq. (*3.9*), $-\log K_a = \mathrm{p}K_a$, and from Eq. (*3.5*), $-\log[\mathrm{H}^+] = \mathrm{pH}$. By substitution, we therefore get

$$\mathrm{p}K_a = \mathrm{pH} + \log\frac{[\mathrm{HA}]}{[\mathrm{A}^-]} = \mathrm{pH} + \log\frac{[\text{acid}]}{[\text{conjugate base}]} \qquad (3.13a)$$

or

$$\mathrm{pH} = \mathrm{p}K_a + \log\frac{[\text{base}]}{[\text{conjugate acid}]} \qquad (3.13b)$$

These are two forms of the *Henderson-Hasselbalch equation*. This useful relationship enables us to calculate the composition of buffers that have a specified pH. Note that if $[\mathrm{HA}] = [\mathrm{A}^-]$, then $\mathrm{pH} = \mathrm{p}K_a$.

EXAMPLE 3.5

(*a*) Calculate the pH of a solution containing 0.1 mol L^{-1} acetic acid and 0.1 mol L^{-1} sodium acetate. The $\mathrm{p}K_a$ of acetic acid is 4.7. (*b*) What would be the pH value after adding 0.05 mol L^{-1} NaOH? (*c*) Compare the latter value with the pH of a simple solution of 0.05 mol L^{-1} NaOH.

(*a*) Using the Henderson-Hasselbalch equation (*3.13b*),

$$\mathrm{pH} = \mathrm{p}K_a + \log\frac{[\text{base}]}{[\text{acid}]}$$

Therefore

$$\mathrm{pH} = 4.7 + \log\frac{0.1}{0.1} = 4.7 + 0 = 4.7$$

(*b*) On adding 0.05 *M* NaOH, the concentration of undissociated acetic acid falls to 0.05 *M*, while the acetate concentration rises to 0.15 *M*. Therefore,

$$\mathrm{pH} = 4.7 + \log\frac{0.15}{0.05} = 4.7 + \log 3 = 4.7 + 0.48 = 5.18$$

(*c*) The pH of 0.05 mol L^{-1} NaOH is $-\log[\mathrm{H}^+]$ in the solution. If we assume the NaOH is fully dissociated in water, the value of $[\mathrm{OH}^-]$ is 0.05 mol L^{-1}. The known value of the ionic product of water is 10^{-14} $(\mathrm{mol\ L}^{-1})^2$; therefore,

$$[\mathrm{H}^+] = 10^{-14}/0.05$$

and

$$\mathrm{pH} = 12.7$$

In comparison with this, even after adding the same amount of alkali, the buffered solution changes pH only slightly.

Acid-Base Behavior of Simple Amino Acids

Many biological molecules have more than one dissociable group. The dissociation of one group can have profound effects on the tendency for dissociation of the other groups. Amino acids, containing both carboxyl and amino groups, illustrate this phenomenon. In water, the carboxyl group tends to dissociate a proton, while the amino group binds a proton. Both reactions can therefore proceed largely to completion, with no buildup either of $\mathrm{H_3O^+}$ or $\mathrm{OH^-}$. An important result is that amino acids carry both a negative and a positive charge in solution near neutral pH; in this state the compound is said to be a *zwitterion*.

Titration of Amino Acids

A way of examining the zwitterionic behavior of amino acids is to study their *titration*. Suppose, for example, that we begin with a solution of glycine hydrochloride, in which both groups are in their acidic forms. Addition of sodium hydroxide brings about changes in the pH of the solution as shown in Fig. 3-2.

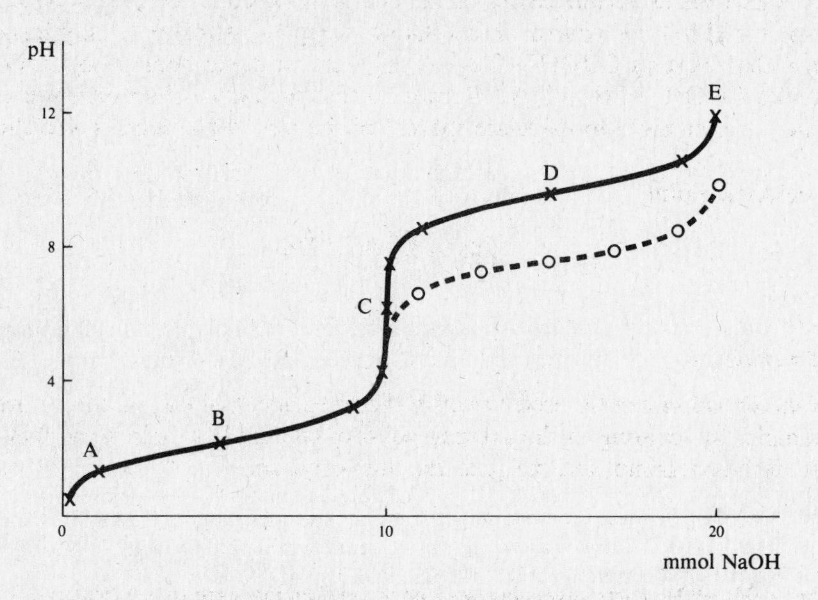

Fig. 3-2 Titration of 10 mmol glycine · HCl with NaOH solution. The broken line shows the titration in the presence of formaldehyde.

The curve in Fig. 3-2 shows two distinct branches, one for each acidic group on glycine hydrochloride. At point A, both groups are in the acid (protonated) form: $HOOC—CH_2—NH_3^+$, and the charge carried by the molecule is +1. After addition of 10 mmol of NaOH (point C, near pH 6.0), one group is almost completely deprotonated, and the molecule carries no net charge. After an additional 10 mmol of NaOH, the second group is also deprotonated (point E), and the form of the molecule will be $^-OOC—CH_2—NH_2$, with a charge of −1.

In the vicinity of points B and D, the pH changes least for a given amount of added NaOH; i.e., the solution is acting as a buffer. At point B, pH 2.3, half of the first group has been titrated to its conjugate base and half remains in its acid form; i.e., [acid] = [conjugate base]. At this point, then, from Eq. (*3.13*), $pK_a = pH$, or for this group, $pK_a = 2.3$. The first group to titrate must be a relatively strong acid, and this group would be expected to be the carboxyl group. Similar considerations lead to a value of 9.6 for the pK_a of the second group, the amino group.

EXAMPLE 3.6

What fraction of the first group (the carboxyl) has been titrated by pH 6.3?

Using the Henderson-Hasselbalch equation (*3.13b*):

$$pH = pK_a + \log \frac{[base]}{[acid]}$$

At pH 6.3, for a group with a pK_a of 2.3:

$$\log \frac{[\text{base}]}{[\text{acid}]} = 6.3 - 2.3 = 4.0$$

Therefore, [base] = 10^4 × [acid]. This result indicates that only 1 in 10,000 of the carboxyl groups (0.01 percent) is protonated at pH 6.3.

The pH at which the molecule carries no net charge is called the *isoelectric point*. For glycine the isoelectric point is pH 6.0. Of course, in a solution of glycine at pH 6, at any instant there will be some molecules that exist as COOH—CH_2—NH_3^+, and an equal number as COO^-—CH_2—NH_2, and even fewer of COOH—CH_2—NH_2. Because of the symmetry of the titration curve around the isoelectric point, it is possible to calculate the pH of the isoelectric point, given the individual pK_a values.

At the isoelectric point:

$$\text{pH}_I = \frac{\text{p}K_{a1} + \text{p}K_{a2}}{2} \qquad (3.14)$$

The Formol Titration

Chemical modification can be used to verify the assignment of pK_a values in titrations of amino acids. For example, by titrating in the presence of formaldehyde, we can show that pK_{a1} in Fig. 3-2 belongs to the carboxyl group, and pK_{a2} to the amino group.

EXAMPLE 3.7

Formaldehyde (HCHO) reacts reversibly with uncharged amino groups as follows:

$$\text{R—NH}_2 + 2\text{HCHO} \rightleftharpoons \text{R—N}\begin{matrix}\diagup \text{CH}_2\text{—OH} \\ \diagdown \text{CH}_2\text{—OH}\end{matrix}$$

Thus, the pK_a of amino groups is altered in formaldehyde solutions, while the carboxyl groups remain unaffected. It is found that the pK_{a1} of amino acids is unchanged in formaldehyde and therefore represents the carboxyl group, while pK_{a2} is lowered (see the broken line in Fig. 3-2). The extent of lowering depends on the concentration of formaldehyde. Thus, pK_{a2} must reflect the dissociation of the amino group.

In general, among simple compounds, the carboxyl group is a stronger acid than the —NH_3^+ group. Carboxyl groups tend to have pK_a values below 5, and —NH_3^+ groups above 7.

Question: Why is the pK_a of the glycine carboxyl (2.3) less than the pK_a of acetic acid (4.7)?

In glycine solutions at pH values below 6, the amino group is present in the positively charged form. This positive charge stabilizes the negatively charged carboxylate ion by electrostatic interaction. This means that the carboxyl group of glycine will *lose* its proton more readily and is therefore a stronger acid (with a lower pK_a value).

Acidic and Basic Amino Acids

Some of the amino acids carry a *prototropic* side chain: e.g., aspartic acid and glutamic acid have an extra carboxyl, histidine has an imidazole, lysine has an amino, and arginine has a guanidino group. The structures of these side chains are shown in Table 3.1, and their pK_a values are listed in Table 3.2.

Table 3.2. pK_a Values of Some Amino Acids

Amino Acid	pK_{a1} (α-COOH)	pK_a (α-NH_3^+)	pK_{aR} (side chain)
Glycine	2.3	9.6	—
Serine	2.2	9.2	—
Alanine	2.3	9.7	—
Valine	2.3	9.6	—
Leucine	2.4	9.6	—
Aspartic acid	2.1	9.8	3.9
Glutamic acid	2.2	9.7	4.3
Histidine	1.8	9.2	6.0
Cysteine	1.7	10.8	8.3
Tyrosine	2.2	9.1	10.1
Lysine	2.2	9.0	10.5
Arginine	2.2	9.0	12.5

The titration curves of these amino acids have an extra inflection, as shown for glutamic acid in Fig. 3-3.

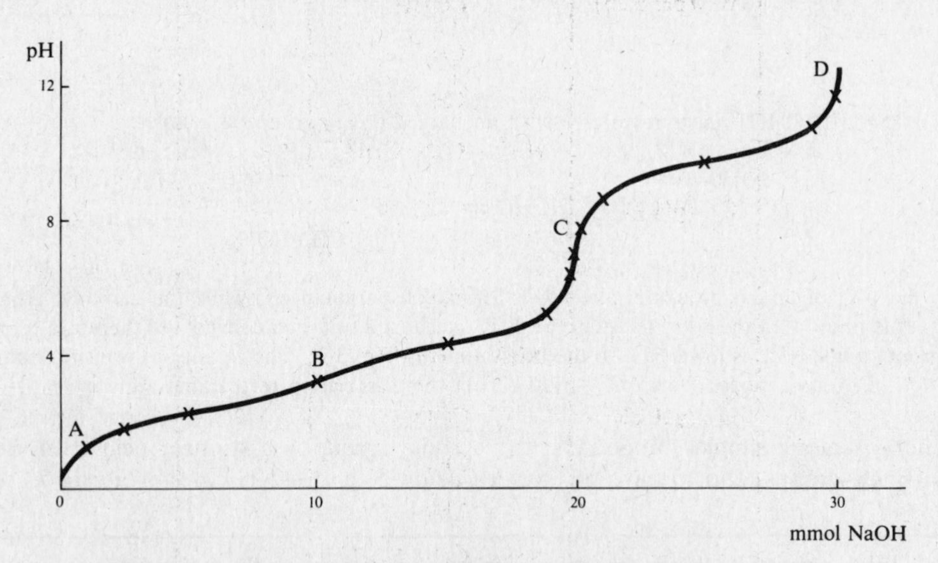

Fig. 3-3 Titration of 10 mmol glutamic acid hydrochloride by NaOH.

Charge on Polybasic Amino Acids

Recall that for amino acids, the various groups have a charge that depends on the pH of the solution:

1. The protonated forms of the carboxyl groups and the tyrosine side chain are *uncharged*, while the deprotonated forms are negatively charged, or *anionic*.
2. The protonated forms of the amino group, the *imidazolium* side chain of histidine and the *guanidinium* group of arginine, are positively charged (or *cationic*), while the deprotonated forms are uncharged.

At pH values *below* the pK_a of a group, the solution is more acidic, and the protonated form predominates. As the pH is raised above the pK_a of a group, that group loses its proton; i.e., the $—COOH$ group becomes the *negative* $—COO^-$, while the positively charged acid form of the amino group, $—NH_3^+$, becomes *uncharged* NH_2.

Isoelectric Point

By using the above arguments, we can consider the example of the isoelectric point of glutamic acid, with the aid of the pK_a values given in Table 3.2.

At pH values less than 2.2, *all* the groups are protonated. The predominant forms of glutamic acid, present at different points during the titration shown in Fig. 3-3, are given below:

Point	Charged Form	Charge
A (pH 1)	$COOH—CH_2—CH_2—CH(NH_3^+)—COOH$	+1
B (pH 3.25)	$COOH—CH_2—CH_2—CH(NH_3^+)—COO^-$	0
C (pH 7)	$COO^-—CH_2—CH_2—CH(NH_3^+)—COO^-$	−1
D (pH 12)	$COO^-—CH_2—CH_2—CH(NH_2)—COO^-$	−2

As the pH is increased through 2.2, the α-carboxyl dissociates to yield the zwitterion near pH 3. As the pH is further increased through 4.3, the side-chain carboxyl dissociates to yield a molecule with two negative and one positive charges (i.e., a net charge of −1). Note that near pH 7 (i.e., in neutral solution), both glutamic acid and aspartic acid exist as glutamate and aspartate, respectively, with ionized side chains. Finally, as the pH rises through the pK_{a2}, the amino group dissociates to yield the species with a charge of −2. Thus, the isoelectric point is flanked by the two pK_a values of 2.2 and 4.3.

So, it can be shown that for glutamic acid,

$$pH_I = (pK_{a1} + pK_{aR})/2 = (2.2 + 4.3)/2 = 6.5/2 = 3.25.$$

Thus, as a general rule, the isoelectric point of an amino acid is the average of the pK_a values of the protonation transitions on *either side* of the isoelectric species; this implies a knowledge of the charges of the various forms.

3.3 AMINO ACID ANALYSIS

After hydrolysis of proteins to amino acids (usually in concentrated HCl), the amino acids can be separated from each other, by means of *ion-exchange chromatography*. Three buffers of successively higher pH are used to *elute* the amino acids from the chromatography column. The order of elution depends on the charge carried by the amino acid. The basic amino acids (lysine, histidine, and arginine) bind most tightly to the negatively charged ion-exchange resin. By using this technique, it is possible to determine which amino acids occur in a given protein. Their *relative abundance* can also be determined by measuring the concentration of each amino acid. The compound *ninhydrin* reacts with amino acids to form a purple derivative. By measuring the absorbance at 570 nm of the purple solution so formed, the relative concentrations of each amino acid can be determined (Fig. 3-4).

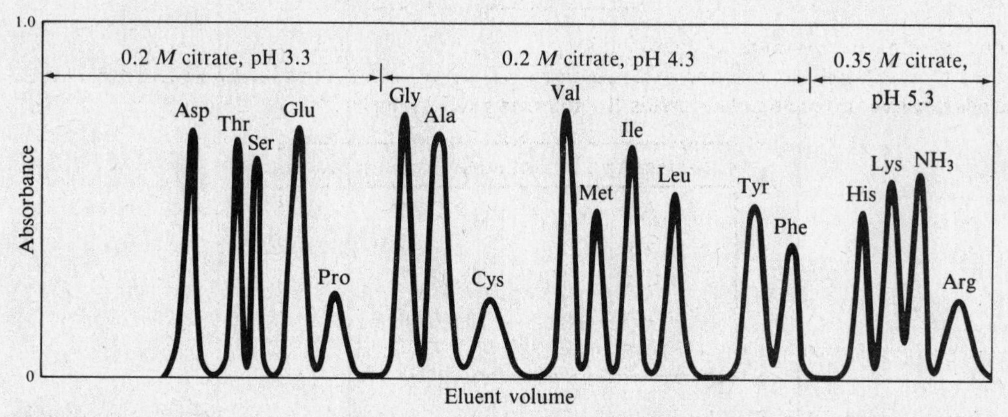

Fig. 3-4 Separation of amino acids by ion-exchange chromatography. The areas under the peaks are proportional to the amounts of amino acids in the solution.

Reaction of amino acids with ninhydrin:

$$2\ \underset{\text{Ninhydrin}}{\text{(ninhydrin)}} + \underset{\text{Amino acid}}{R{-}\overset{H}{\underset{\overset{+}{N}H_3}{C}}{-}COO^-} \longrightarrow \underset{\text{Purple pigment}}{\text{(purple pigment)}} + R{-}\underset{O}{\overset{}{C}}{-}H + CO_2 + 3H_2O$$

Ninhydrin

Amino acid

Purple pigment

Although proline does not give a purple color when treated with ninhydrin, it does yield a weaker yellow pigment, which can be quantified. Proline can be located readily in a chromatogram during amino acid analysis because the relative heights of the 570- and 440-nm absorbance profile are reversed compared with the other amino acids.

EXAMPLE 3.8

A peptide was hydrolyzed, and the resulting solution was examined by amino acid analysis. The following data were obtained:

Amino Acid	μmol
Asp	1.21
Ser	0.60
Gly	1.78
Leu	0.58
Lys	0.61

Determine the empirical formula of the peptide.

We may take leucine as a convenient reference compound and determine the proportion of each of the amino acids relative to leucine. This yields the following stoichiometry:

Amino Acid	Relative Concentration
Asp	2.086
Ser	1.03
Gly	3.07
Leu	1.00
Lys	1.05

Since there must be a whole number for each amino acid in the peptide, these values can be rounded off to integers to give the most likely empirical formula: Asp_2, Ser, Gly_3, Leu, Lys.

3.4 THE PEPTIDE BOND

In protein molecules, α-amino acids are linked in a linear sequence. The α-carboxyl group of one amino acid is linked to the α-amino group of the next through a special amide bond known as a *peptide bond*. The peptide bond is formed by a *condensation* reaction, requiring an input of energy:

$$\underset{\text{Glycine}}{NH_3^+{-}CH_2{-}COO^-} + \underset{\text{Alanine}}{NH_3^+{-}\overset{\displaystyle CH_3}{\overset{|}{C}H}{-}COO^-} \underset{H_2O}{\overset{H_2O}{\rightleftharpoons}} \underset{\text{Glycylalanine}}{NH_3^+{-}CH_2{-}\underset{\underset{\displaystyle O}{\|}}{C}{-}\overset{\displaystyle H}{\overset{|}{N}}{-}\overset{\displaystyle CH_3}{\overset{|}{C}H}{-}COO^-}$$

Note that the acidic and basic character of the carboxyl and amino groups taking part in forming the bond are lost after condensation. The hydrolysis of the peptide bond to free amino acids is a spontaneous process, but is normally very slow in neutral solution.

Nomenclature

In naming peptides, we start with the amino acid that has the free α-NH_3^+ (the *N terminal*) and replace the *-ine* endings (except the last one) with the ending *-yl*. The amino acids in the peptide are called *residues*, since they are the residues left after the removal of water during peptide bond formation.

EXAMPLE 3.9

Distinguish between the dipeptides glycylalanine and alanylglycine.

We start naming a peptide from the amino terminus; thus, glycylalanine has a free α-amino group on the glycine residue, while the free carboxyl group is that of the alanine residue.

Glycylalanine and alanylglycine are examples of *sequence isomers*; they are composed of the same amino acids, but they are combined in different sequences.

Peptides

Compounds of two amino acids linked by a peptide bond are known as *dipeptides*; those with three amino acids are called *tripeptides*, and so on. *Oligopeptides* contain an unspecified but small number of amino acid residues, while *polypeptides* comprise larger numbers. Natural polypeptides of 50 or more residues are generally referred to as *proteins*.

EXAMPLE 3.10

The following peptides are not derived from proteins. How could this be deduced from an inspection of their structures?

(*a*) Glutathione (γ-L-glutamyl-L-cysteinylglycine)

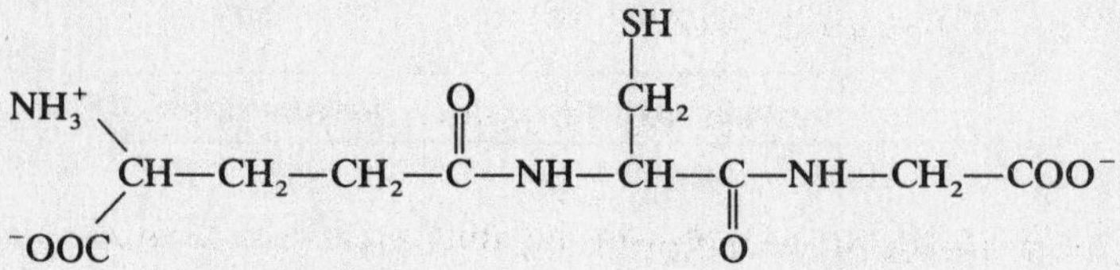

$$\begin{array}{l} NH_3^+ \\ \quad\diagdown \\ \quad\; CH{-}CH_2{-}CH_2{-}\overset{O}{\overset{\|}{C}}{-}NH{-}\overset{SH}{\overset{|}{\underset{}{\overset{CH_2}{\overset{|}{CH}}}}}{-}\underset{O}{\underset{\|}{C}}{-}NH{-}CH_2{-}COO^- \\ \quad\diagup \\ {}^-OOC \end{array}$$

The peptide bond between glutamic acid and cysteine is through the γ-carboxyl of glutamic acid, not the α-carboxyl group. All natural proteins are composed of α-peptide links (unless there have been postsynthetic modifications).

(*b*) Carnosine

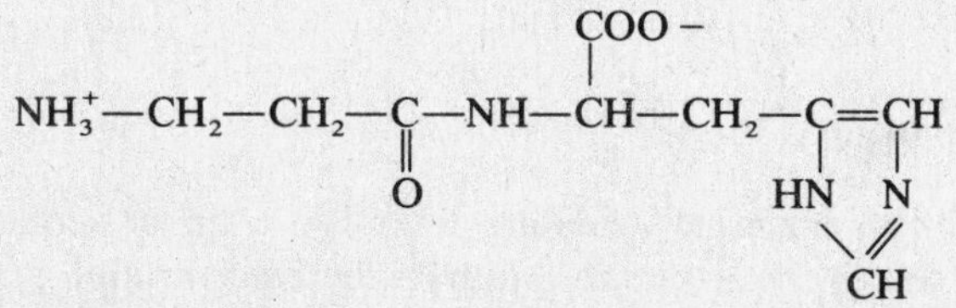

$$NH_3^+{-}CH_2{-}CH_2{-}\underset{O}{\underset{\|}{C}}{-}NH{-}\overset{COO^-}{\overset{|}{CH}}{-}CH_2{-}\underset{\substack{| \quad\; | \\ HN \quad N \\ \diagdown\;\diagup\!\!\!\diagup \\ CH}}{C{=}CH}$$

This compound contains β-alanine. Naturally occurring proteins are composed of α-amino acids only.

3.5 AMINO ACID SEQUENCE

The structure and properties of peptides and proteins depend critically upon the sequence of amino acids in the peptide chain. In 1953, F. Sanger determined the sequence of the 51 amino acid residues in insulin. The first step in determining the sequence was to determine the composition, i.e., the number of residues of each type in the protein molecule, by means of amino acid analysis.

N-Terminal Analysis

The N-terminal residue, i.e., the first amino acid in the sequence of a peptide, can be determined by reaction with *phenylisothiocyanate*. At neutral pH, this compound reacts with the α-amino group. After mild acid hydrolysis, the reaction product cyclizes, releasing the terminal residue as a *phenylthiohydantoin* (PTH) derivative (the Edman degradation, Fig. 3-5). The derivative can be analyzed to determine its parent amino acid and its quantity.

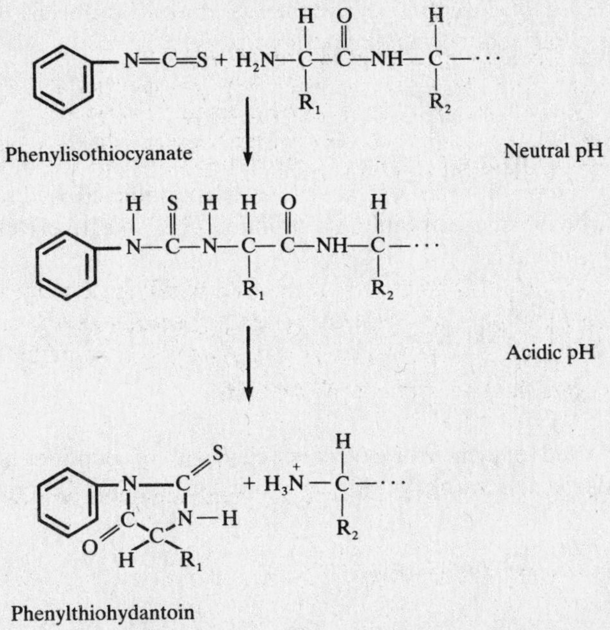

Fig. 3-5 The Edman degradation.

Other reagents capable of reacting with the amino-terminal residues, such as *dansyl chloride* and *fluoro-2,4-dinitrobenzene*, may also be used to identify the N terminus.

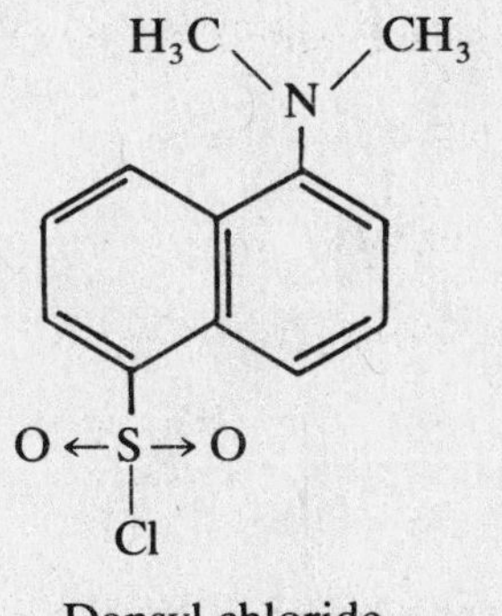

Dansyl chloride

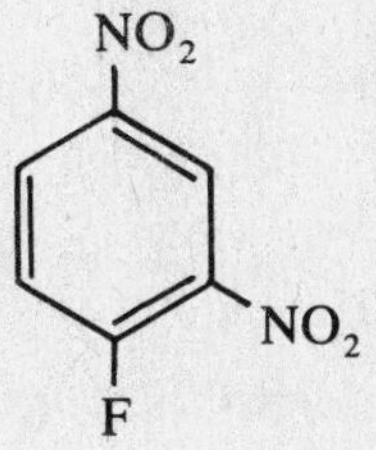

Fluoro-2,4-dinitrobenzene

EXAMPLE 3.11

Given a means of identifying and quantifying the N-terminal residue, describe a way of determining the entire sequence of a protein.

If the phenylisothiocyanate method is used, the cyclization and release of the N-terminal derivative occurs under mild conditions that leave the rest of the chain intact. It is therefore possible to take the protein chain, now without its original N-terminal residue, and repeat the procedure to determine the second residue in the sequence, and so on. Unfortunately, at each step, there is a finite chance of additional peptide hydrolysis or incomplete reaction, and uncertainty tends to accumulate after 10 to 20 cycles.

Specific Cleavage of Peptides

Many proteins contain hundreds of amino acid residues, and it may not be feasible to determine the sequence of the whole molecule in one go, because of the buildup of uncertainty at each step. It is therefore convenient first to cleave the protein into smaller, more manageable pieces. Also the protein may contain *disulfide bridges* linking cysteine residues in different parts of the chain; these links must be broken to allow sequencing to continue.

Trypsin

One of the most frequently used means of breaking polypeptides into smaller fragments is hydrolysis by the enzyme *trypsin*. This digestive enzyme, produced by the pancreas, hydrolyzes peptide bonds on the carboxyl side of lysine and arginine (i.e., positively charged) residues:

$$\overset{\text{Trypsin}}{\downarrow}$$
$$R_1\text{-Lys-Ala-}R_2 \longrightarrow R_1\text{-Lys-COO}^- + NH_3^+\text{-Ala-}R_2$$

Chymotrypsin

Another commonly used enzyme for selective cleavage of peptides is *chymotrypsin*, which hydrolyzes peptide bonds on the carboxyl side of aromatic residues (phenylalanine, tyrosine, and tryptophan):

$$\overset{\text{Chymotrypsin}}{\downarrow}$$
$$R_1\text{-Phe-Ser-}R_2 \longrightarrow R_1\text{-Phe-COO}^- + NH_3^+\text{-Ser-}R_2$$

Cyanogen Bromide Cleavage

A nonenzymatic, chemical method for specific cleavage of polypeptides involves the reaction of *cyanogen bromide*, CNBr, with methionine residues:

```
          H         H                           H         H
          |         |                           |         |
   R1—C—N—CH—C—N—R2                   R1—C—N—CH—C=N—R2
      ||    |   ||                         ||    |   |  +
      O    CH2  O                          O    CH2  O
            |           ———→                      \ /
           CH2             \                      CH2
            |               ↓                      |
           S—CH3        CH3—S—CN                   |  H2O
            +              +                       ↓
          Br—C≡N          Br−
                                                H          O
                                                |         //
                                          R1—C—N—CH—C              +
                                             ||    |   |       + H3N—R2
                                             O    CH2  O
                                                    \ /
                                                    CH2
```

Overlapping Peptides

With a protein cleaved into manageable peptides and the sequence of each peptide determined, the next problem is that of ordering the peptides themselves in the correct sequence. For this, at least two sets of peptide sequences from different selective cleavage methods are required.

EXAMPLE 3.12

A peptide was broken into two smaller peptides by cyanogen bromide (CNBr), and into two different peptides by trypsin (Tryp). Their sequences were as follows: CNBr 1, Gly-Thr-Lys-Ala-Glu; CNBr 2, Ser-Met; Tryp 1, Ser-Met-Gly-Thr-Lys; Tryp 2, Ala-Glu. Determine the sequence of the parent peptide.

By arranging the sequences in an overlapping set, as follows, it is possible to determine the parent sequence:

CNBr 2:	Ser-Met
Tryp 1:	Ser-Met-Gly-Thr-Lys-
CNBr 1:	-Gly-Thr-Lys-Ala-Glu
Tryp 2:	Ala-Glu
	Ser-Met-Gly-Thr-Lys-Ala-Glu

Note: Some of this information is redundant; peptides Tryp 1 and CNBr 1 would be sufficient for unambiguous sequence determination. In addition, in such a simple case, with only two peptides from each cleavage reaction, we could determine the sequence from either one of these cleavages alone using the fact that the carboxyl sides of Met and Lys are the sites of cleavage with CNBr and trypsin, respectively.

3.6 REACTIONS OF CYSTEINE

The side chain of cysteine is important because of the possibility of its oxidation to form the disulfide-bridged amino acid *cystine*:

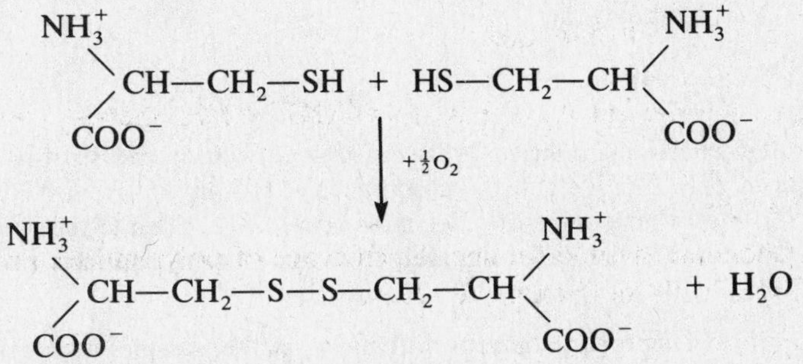

$$(NH_3^+)(COO^-)CH—CH_2—SH + HS—CH_2—CH(NH_3^+)(COO^-) \xrightarrow{+\frac{1}{2}O_2} (NH_3^+)(COO^-)CH—CH_2—S—S—CH_2—CH(NH_3^+)(COO^-) + H_2O$$

Cystine

Disulfide bridges are often found in proteins if the cysteine side chains are close enough in the *tertiary structure* (Chap. 4) to form a bridge. In addition, oxidation of free —SH groups on the surface of some proteins can cause two different molecules to be linked covalently by a disulfide bridge. This process may be biologically undesirable, and cells frequently contain reducing agents that prevent or reverse this reaction. The most common of these agents is *glutathione*; it can reduce the oxidized disulfide back to the sulfhydryl form, becoming itself oxidized in the process. Cells contain other reducing systems linked to metabolism that can then re-reduce the glutathione.

Disulfide links can be broken in the laboratory by reagents similar to glutathione that carry a free —SH group. The most common of these is *mercaptoethanol*, HO—CH_2—CH_2—SH.

There are more powerful disulfide reducing agents (lower *standard redox potential*, Chap. 10); e.g., *dithiothreitol* carries two sulfhydryl groups, and oxidation of these causes ring closure, forming a very stable disulfide. As a result, dithiothreitol is several orders of magnitude more powerful a reducing agent than mercaptoethanol.

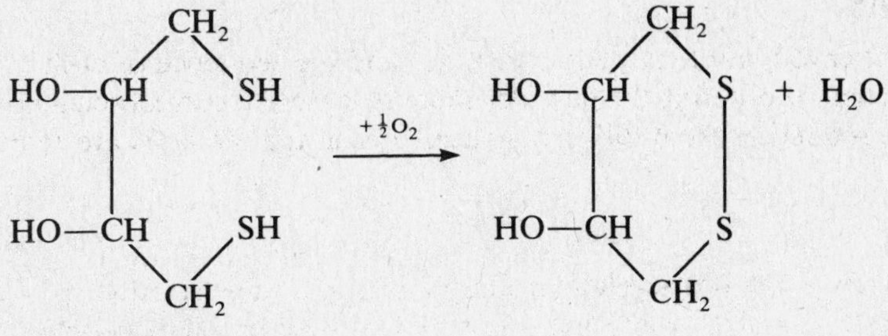

$$HS—CH_2—CH(OH)—CH(OH)—CH_2—SH \xrightarrow{+\frac{1}{2}O_2} \text{cyclic } CH_2—S—S—CH_2 \text{ (ring with two HO—CH)} + H_2O$$

Dithiothreitol

Often it is necessary to prevent the oxidation of sulfhydryl groups, as the presence of disulfide links could lead to insolubility of a protein or to inability to determine its sequence. It is possible to *block* the reactive —SH groups with a range of chemical reagents.

1. *Iodoacetate*. This reagent forms an *S*-carboxymethyl derivative of cysteine residues:

$$R—SH + I—CH_2—COO^- \longrightarrow RS—CH_2—COO^- + HI$$

2. N-*Ethylmaleimide*. The reaction with *N*-ethylmaleimide results in a loss in absorbance of the reagent at 305 nm; this characteristic can be used to measure the extent of reaction:

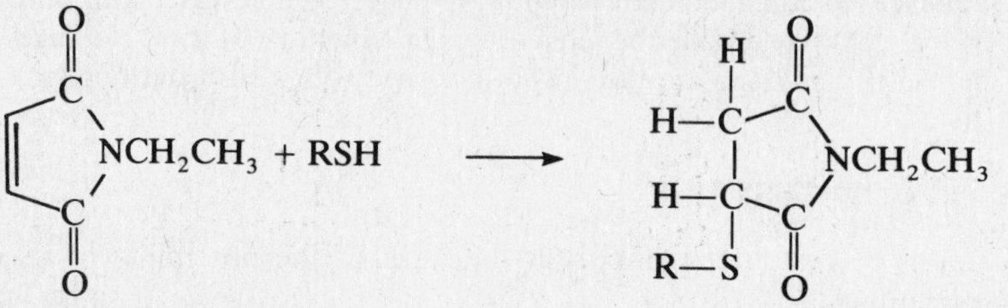

N-Ethylmaleimide

EXAMPLE 3.13

A solution of a protein containing 2 mg in 1 mL was treated with an excess of *N*-ethylmaleimide. During the reaction in a cuvette of path length 1 cm, the absorbance at 305 nm fell from 0.26 to 0.20. Given a molar absorbance coefficient for *N*-ethylmaleimide of 620 L mol^{-1} cm^{-1}, (*a*) calculate the concentration of sulfhydryl groups in the solution; (*b*) what is the molecular weight of the protein, assuming there is only one cysteine residue per molecule?

(*a*) The concentration of sulfhydryl groups originally in the sample is equal to the decline in *N*-ethylmaleimide concentration. The Beer-Lambert law (Eq. *3.1*) can be used to determine the concentration change, given the change in absorbance:

$$A = \varepsilon C l$$

$$\Delta C = \frac{\Delta A}{\varepsilon l}$$

$$= \frac{0.26 - 0.20}{620 \times 1.0} = \frac{0.06}{620} = 9.8 \times 10^{-5} \text{ mol L}^{-1}$$

(*b*) The protein concentration is 2 mg mL^{-1}, or 2 g L^{-1}. If there were a single —SH per protein molecule, then the molar protein concentration would also be 9.8×10^{-5} mol L^{-1}. Therefore, molar weight = $2/(9.8 \times 10^{-5})$ g mol^{-1}; i.e., $M_r = 20{,}400$.

Solved Problems

AMINO ACIDS

3.1. Which of the following compounds are α-amino acids?

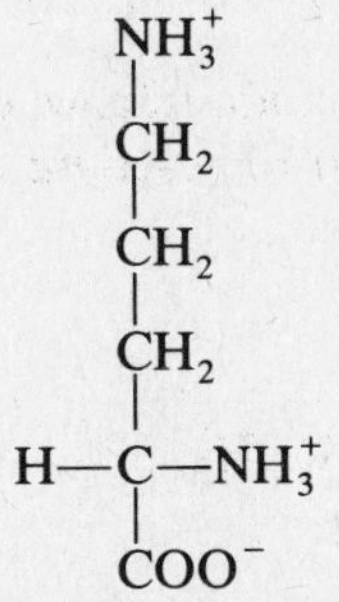

$$NH_3^+-CH_2-CH_2-CH_2-\underset{COO^-}{\underset{|}{C}}H-NH_3^+$$

(*a*) Ornithine

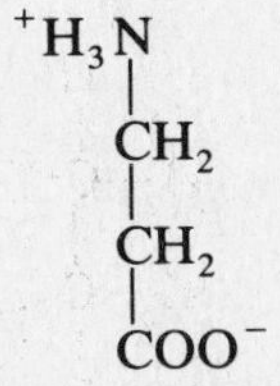

$$^+H_3N-CH_2-CH_2-COO^-$$

(*b*) β-Alanine

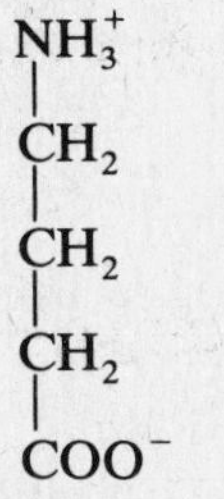

$$NH_3^+-CH_2-CH_2-CH_2-COO^-$$

(*c*) γ-Aminobutyrate

$$^+H_2N-CH(COO^-)-CH_2-C(H)(OH)-CH_2-\ (\text{ring})$$

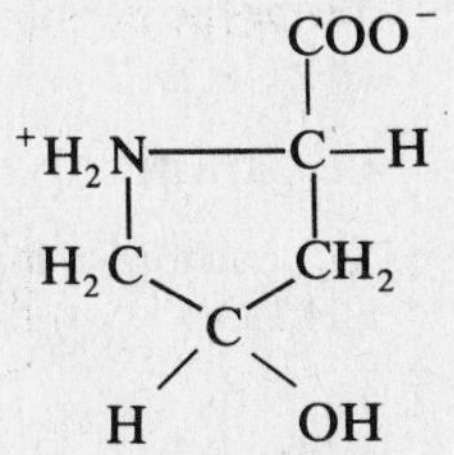

(*d*) Hydroxyproline

SOLUTION

Ornithine (*a*) and hydroxyproline (*d*) are both α-amino acids, because an amino group is attached to the same carbon atom that carries a carboxyl. Although ornithine is not used in protein synthesis, it is an intermediate in the urea cycle (Chap. 15). β-Alanine (*b*) and γ-aminobutyrate (*c*) have their amino group attached to a carbon atom different from that bearing the carboxyl, and are a β- and γ-amino acid, respectively.

3.2. Which amino acids can be converted into different amino acids by mild hydrolysis, with the liberation of ammonia?

SOLUTION

Both glutamine and asparagine have *amide* side chains. Amides can be hydrolyzed to yield the carboxyl and free ammonia.

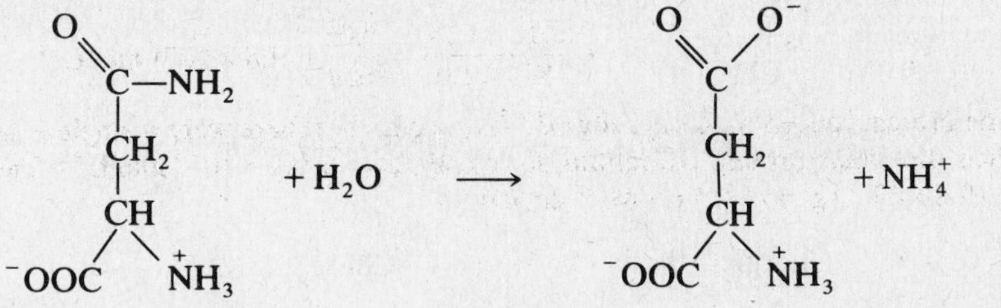

$$\underset{\text{Asparagine}}{H_2N-C(=O)-CH_2-CH(COO^-)(\overset{+}{N}H_3)} + H_2O \longrightarrow \underset{\text{Aspartate}}{^-O-C(=O)-CH_2-CH(COO^-)(\overset{+}{N}H_3)} + NH_4^+$$

3.3. Why is phenylalanine very poorly soluble in water, while serine is freely water-soluble?

SOLUTION

The aromatic side chain of phenylalanine is nonpolar, and its solvation by water is accompanied by a loss of entropy and is therefore unfavorable. On the other hand, the side chain of serine carries a polar hydroxyl group that allows hydrogen bonding with water.

3.4. Many proteins absorb ultraviolet light strongly at 280 nm, yet gelatin does not. Suggest an explanation.

SOLUTION

The ultraviolet absorbance of many proteins is due to the presence of amino acids with aromatic side chains. Gelatin, a derivative of collagen (Chap. 4), has an unusual composition, with a low proportion of aromatic amino acids.

ACID-BASE BEHAVIOR OF AMINO ACIDS

3.5. Write the conjugate bases of the following weak acids: (*a*) CH_3COOH; (*b*) NH_4^+; (*c*) $H_2PO_4^-$; (*d*) $CH_3NH_3^+$

SOLUTION

Each acid forms its conjugate base by loss of a proton. The conjugate bases are thus: (*a*) CH_3COO^-; (*b*) NH_3; (*c*) HPO_4^{2-}; (*d*) CH_3NH_2

3.6. Write the equations for the ionization equilibria of aspartic acid. Indicate the net charge carried on each species.

SOLUTION

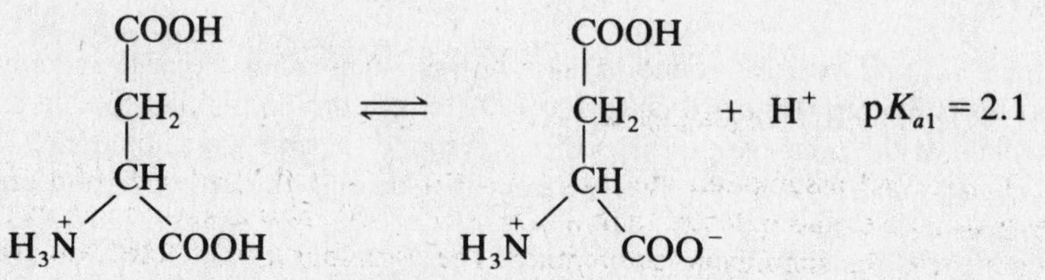

$$HOOC{-}CH_2{-}CH(\overset{+}{N}H_3){-}COOH \rightleftharpoons HOOC{-}CH_2{-}CH(\overset{+}{N}H_3){-}COO^- + H^+ \qquad pK_{a1} = 2.1$$

Charge: +1 Charge: 0

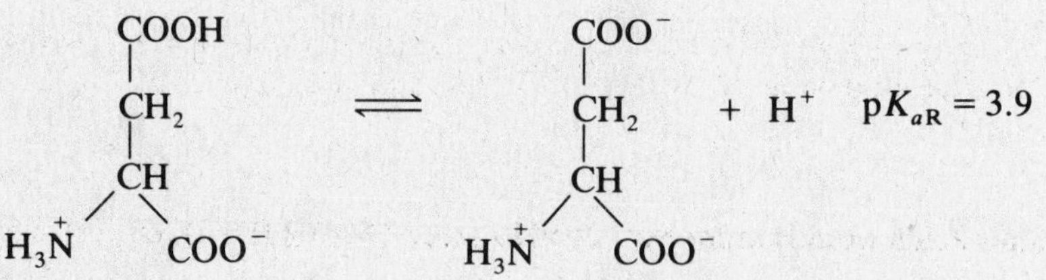

$$HOOC{-}CH_2{-}CH(\overset{+}{N}H_3){-}COO^- \rightleftharpoons {}^-OOC{-}CH_2{-}CH(\overset{+}{N}H_3){-}COO^- + H^+ \qquad pK_{aR} = 3.9$$

Charge: 0 Charge: −1

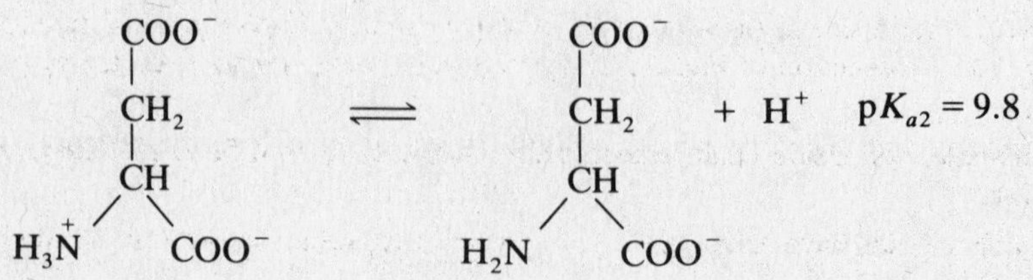

$$^-OOC{-}CH_2{-}CH(\overset{+}{N}H_3){-}COO^- \rightleftharpoons {}^-OOC{-}CH_2{-}CH(NH_2){-}COO^- + H^+ \qquad pK_{a2} = 9.8$$

Charge: −1 Charge: −2

3.7. Calculate the isoelectric point (pH_I) of aspartic acid, using the information from Prob. 3.6.

SOLUTION

Since the isoelectric species is flanked by transitions whose pK_a values are 2.1 and 3.9, respectively,

$$\mathrm{pH}_I = \frac{2.1 + 3.9}{2} = 3.0$$

3.8. Calculate the pH of 0.1 M acetic acid (pK_a = 4.7).

SOLUTION

Acetic acid, although a weak acid, is a stronger acid than water. Therefore, as a first approximation, we can ignore the protons from water and represent the dissociation as follows:

$$CH_3COOH \rightleftharpoons CH_3COO^- + H^+$$

Now we can approximate $[H^+] = [CH_3COO^-] = x$.

In addition, because acetic acid is a weak acid, the concentration of the undissociated acid will not be appreciably lowered by dissociation. We can assume then that $[CH_3COOH] = 0.1\ M$.

Thus
$$K_a = \frac{[CH_3COO^-][H^+]}{[CH_3COOH]} = \frac{x^2}{0.1}$$

$$x^2 = 0.1 \times K_a = 0.1 \times 10^{-4.7}$$

Therefore
$$x = 0.0014\ \mathrm{mol\ L^{-1}} \quad \text{and} \quad \mathrm{pH} = 2.85$$

Note: The second assumption made above could be tested by setting $[CH_3COOH] = 0.1 - x$ and solving the quadratic equation for x. In that case, pH = 2.80. This answer is not very different from that obtained employing the simplifying assumption. The assumption, therefore, was reasonable.

3.9. At what pH values would glutamate be a good buffer?

SOLUTION

By inspection of Fig. 3-3, it can be seen that the pH of a glutamate solution changes least rapidly with NaOH concentration in the vicinity of the pK_a values, i.e., near pH 2.2, 4.3, and 9.7.

3.10. What amino acids would buffer best at pH values of (*a*) 2.0; (*b*) 6.0; (*c*) 4.5; (*d*) 9? (Refer to Table 3.2.)

SOLUTION

(*a*) All amino acids have a carboxylic group whose pK_a is close to 2.0. So all amino acids would be good buffers in this pH range.

(*b*) Histidine has a side chain whose pK_a value is close to 6. Histidine would therefore be a good buffer at pH 6.0. Inspection of Fig. 3-2 shows that the simple amino acids are very poor buffers at this pH.

(*c*) Glutamate has a side chain whose pK_a value is close to 4.5, so glutamate would buffer in this region.

(*d*) All amino acids have an α-amino group whose pK_a value is near 9, so all would buffer well at this pH value.

3.11. Derive a general equation that describes the average net charge on a prototropic group in terms of the solution pH and the pK_a of the group.

SOLUTION

By rearranging the Henderson-Hasselbalch equation (*3.13*), we obtain:

$$\frac{[\text{acid}]}{[\text{conjugate base}]} = 10^{(\text{p}K_a - \text{pH})}$$

If we define α as the fraction of the group in the acid form, then

$$\alpha = \frac{[\text{acid}]}{[\text{acid}] + [\text{conjugate base}]} = \frac{10^{(\text{p}K_a - \text{pH})}}{1 + 10^{(\text{p}K_a - \text{pH})}}$$

$$= \frac{1}{10^{(\text{pH} - \text{p}K_a)} + 1}$$

Therefore, if the group (such as the amino group) has a cationic acid form, α represents the fractional positive charge:

$$Z = +\alpha = \frac{+1}{1 + 10^{(\text{pH} - \text{p}K_a)}}$$

Note: When $\text{pH} \ll \text{p}K_a$, Z approaches +1.

On the other hand, for groups such as the carboxyl, with a neutral acid form and an anionic conjugate base, α represents the fraction uncharged. The fractional charge is then:

$$Z = -(1 - \alpha) = \alpha - 1 = \frac{-1}{1 + 10^{(\text{p}K_a - \text{pH})}}$$

In this case, when $\text{pH} \ll \text{p}K_a$, Z approaches zero.

In general, when the pH is more than 2 units below the $\text{p}K_a$ value, the group can be considered completely protonated.

3.12. Using the $\text{p}K_a$ values for aspartic acid listed in Table 3.2, determine the average net charge on the molecule at pH values of 1.0, 3.9, 6.8.

SOLUTION

From the solution to Prob. 3.11, the average net charge on each of the three groups can be determined at each of the pH values. The results are as follows:

	Average Charge		
Group	pH 1.0	pH 3.9	pH 6.8
α-COOH	−0.07	−0.98	−1.0
β-COOH	0	−0.5	−1.0
α-NH_3^+	+1.0	+1.0	+1.0
Average net charge	+0.93	−0.48	−1.0

3.13. In most amino acids, the $\text{p}K_a$ of the α-carboxyl is near 2.0, and that of the α-amino is near 9.0. However, in peptides, the α-carboxyl has a $\text{p}K_a$ of 3.8, and the α-amino has a $\text{p}K_a$ of 7.8. Can you explain the difference?

SOLUTION

In the free amino acids, the neighboring charges affect the $\text{p}K_a$ of each group. The presence of a positive $\text{—}NH_3^+$ group stabilizes the charged $\text{—}COO^-$ group, making the carboxyl a stronger acid; conversely, the negative carboxylate stabilizes the $\text{—}NH_3^+$ group, making it a weaker acid and thus raising its $\text{p}K_a$. When peptides are formed, the free α-amino and carboxylate are farther apart, and exert less mutual influence.

A secondary effect is that the carbonyl of the first peptide group has an inductive electron-withdrawing effect on the terminal α-amino, decreasing the stability of the $\text{—}NH_3^+$ and lowering the $\text{p}K_a$ slightly.

AMINO ACID ANALYSIS

3.14. Serine phosphate

$$\begin{array}{c} \qquad\qquad\quad O \\ \qquad\qquad\quad \| \\ CH_2\text{—}O\text{—}P\text{—}O^- \\ |\qquad\qquad \| \\ CH \qquad\quad O \\ / \quad \backslash \\ ^-OOC \qquad NH_3^+ \end{array}$$

can be found in enzymatic hydrolysates of casein, a protein found in milk, yet this is not one of the 20 amino acids coded for in protein synthesis. Explain.

SOLUTION

In the synthesis of casein, serine is incorporated into the protein sequence. Subsequently, some of these serine residues are phosphorylated to form serine phosphate.

3.15. The elution profile shown in Fig. 3-4 indicates that glycine, alanine, valine, and leucine can be separated clearly by ion-exchange chromatography; yet these four amino acids have almost indistinguishable sets of pK_a values (Table 3.2). How can you account for this behavior?

SOLUTION

These amino acids all differ in their polarity. The polystyrene matrix that carries the charged groups of the ion-exchange resin is relatively nonpolar, and some separation will be achieved by means of partition effects, as well as by means of ion exchange. Amino acids such as leucine interact more strongly with the resin than do more polar amino acids such as serine, and this interaction retards passage through the column.

3.16. A small peptide was subjected to hydrolysis and amino acid analysis. In addition, because acid hydrolysis destroys tryptophan, the tryptophan content was estimated spectrophotometrically. From the following data, determine the empirical formula of the peptide.

Amino Acid	μmol
Ala	2.74
Glu	1.41
Leu	0.69
Lys	2.81
Arg	0.72
Trp	0.65

SOLUTION

Taking leucine as a convenient reference, we determine the molar ratios of the amino acids, which are 3.97, 2.04, 1.00, 4.07, 1.04, 0.94. Rounding these numbers off to integers yields the formula: Ala_4, Glu_2, Leu, Lys_4, Arg, Trp.

AMINO ACID SEQUENCE

3.17. (*a*) What peptides would be released from the following peptide by treatment with trypsin?

Ala-Ser-Thr-Lys-Gly-Arg-Ser-Gly

(*b*) If each of the products were treated with fluoro-2,4-dinitrobenzene (FDNB) and subjected to acid hydrolysis, what DNP-amino acids could be isolated?

SOLUTION

(*a*) Trypsin hydrolyzes peptides at the carboxyl side of lysine and arginine residues. The resulting peptides would be Ala-Ser-Thr-Lys, Gly-Arg, and Ser-Gly.

(*b*) Treatment with FDNB and hydrolysis will liberate DNP derivatives of the N-terminal amino acids: DNP-Ala, DNP-Gly, and DNP-Ser. Note that the ϵ-amino group of lysine can also react with FDNB; however, the ϵ-DNP derivative of lysine can be distinguished from the α-DNP derivative by its chromatographic behavior.

3.18. Predict the migration direction (anodal, cathodal, or stationary) for the following peptide at pH 6.0: Lys-Gly-Ala-Glu.

SOLUTION

In peptides, the α-NH_3^+ and α-COO^- groups involved in peptide bond formation have been lost; only the N-terminal α-NH_3^+ and C-terminal α-COO^- groups remain. In addition, in the peptide above, lysine carries a positively charged side chain at pH 6.0, and glutamate carries a carboxylate group. The two amino groups, both on lysine, have pK_a values above 8. They will therefore both be nearly completely *protonated* and positively charged at pH 6.0. The two carboxyl groups on glutamate have pK_a values near 4. They will therefore be nearly totally *deprotonated* at pH 6.0. Thus, the peptide can be written

$$\begin{array}{ccccc} & NH_3^+ & & & COO^- \\ & | & & & | \\ {}^+NH_3— & \text{Lys} & —\text{Gly}— & \text{Ala}— & \text{Glu}—COO^- \end{array}$$

and the net charge is zero. The peptide will therefore be stationary.

3.19. The following data were obtained from partial cleavage and analysis of an octapeptide:

Composition:	Ala, Gly_2, Lys, Met, Ser, Thr, Tyr
CNBr:	(1) Ala, Gly, Lys, Thr
	(2) Gly, Met, Ser, Tyr
Trypsin:	(1) Ala, Gly
	(2) Gly, Lys, Met, Ser, Thr, Tyr
Chymotrypsin:	(1) Gly, Tyr
	(2) Ala, Gly, Lys, Met, Ser, Thr
N terminus:	Gly
C terminus:	Gly

Determine the sequence of the peptide.

SOLUTION

A set of overlapping peptides can be prepared from the above data by making use of the fact that one CNBr peptide must end in methionine, one tryptic peptide must end in either lysine or arginine, and one chymotryptic peptide must end in an aromatic amino acid. The sequence of the peptide is, therefore, Gly-Tyr-Ser-Met-Thr-Lys-Ala-Gly.

3.20. The enzyme *carboxypeptidase A* hydrolyzes amino acids from the C-terminal end of peptides, provided that the C-terminal residue is not proline, lysine, or arginine. Fig. 3-6 shows the sequential release of amino acids from a protein by means of treatment with carboxypeptidase A. Deduce the sequence at the C terminus.

SOLUTION

The sequence at the C terminus is:

-Ser-His-Ile

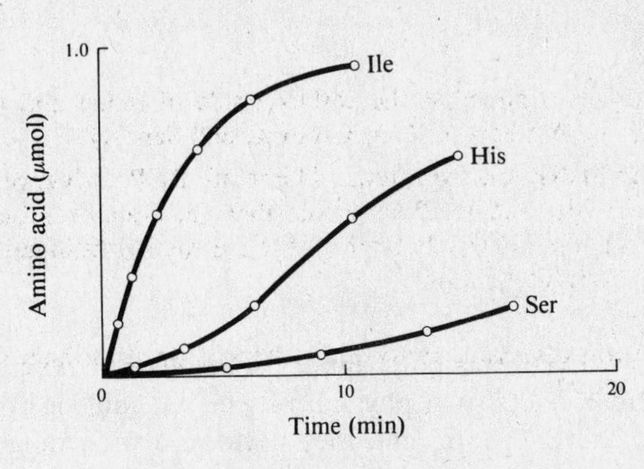

Fig. 3-6

Supplementary Problems

3.21. List the amino acids with side chains capable of acting as (*a*) H-bond donors; (*b*) H-bond acceptors.

3.22. List those amino acids that absorb ultraviolet light strongly.

3.23. Which amino acid found in proteins does not show optical activity?

3.24. Write the conjugate acids of the following weak bases:

(*a*) $HO-CH_2-CH_2-S^-$

(*b*) pyridine ring (N)

(*c*) imidazole ring: N=CH—N—H, CH=CH

3.25. Calculate the isoelectric point of arginine.

3.26. Calculate the isoelectric point of histidine.

3.27. Calculate the pH of 1 *M* acetic acid.

3.28. What is the average net charge on lysine at the following pH values: (*a*) 2.0; (*b*) 5.0; (*c*) 10.0?

3.29. Using the information given in Prob. 3.13, determine the average net charge on the peptide *glycylglycylglycine* at pH: (*a*) 2.0; (*b*) 5.0; (*c*) 9.0.

Chapter 4

Proteins

4.1 INTRODUCTION

Proteins are naturally occurring polypeptides of molecular weight greater than 5,000. These *macromolecules* show great diversity in physical properties, ranging from water-soluble enzymes to the insoluble keratin of hair and horn, and they perform a wide range of biological functions.

4.2 FUNCTIONS

Proteins fulfill the following biochemical roles:

1. *Enzymatic catalysis*. Enzymes are protein catalysts, capable of enhancing rates of reactions by factors of up to 10^{12}.
2. *Transport and storage*. Many small molecules and ions are transported in the blood and within cells by being bound to *carrier proteins*. The best example is the oxygen-carrying protein *hemoglobin*. Iron is stored in various tissues by the protein *ferritin*.
3. *Mechanical functions*. Proteins often fulfill structural roles. The protein *collagen* provides tensile strength in skin, teeth, and bone. The membranes surrounding cells and cell organelles are also partly composed of proteins, having both functional and structural roles.
4. *Movement*. Muscle contraction is accomplished by the interaction between two types of protein filaments—*actin* and *myosin*. Myosin also possesses an enzymatic activity for facilitating the conversion of the chemical energy of ATP into mechanical energy.
5. *Protection*. The antibodies are proteins, aided in mammals by *complement*, a complex set of proteins involved in the destruction of foreign cells.
6. *Information processing*. Stimuli external to a cell, such as hormone signals or light intensity, are detected by specific proteins that transfer a signal to the interior of the cell. A well-characterized example is the visual protein *rhodopsin*, located in membranes of retinal cells.

Question: What is a common feature among those functions listed above that may be explained in terms of protein structure?

In all the above examples, the phenomenon of *specific binding* is involved. For example, hemoglobin specifically binds molecular oxygen, antibodies bind to specific foreign molecules, enzymes bind to specific substrate molecules and in doing so bring about selective chemical bond rearrangement. The function of a protein, then, is understood in light of how the structure of the protein allows the specific binding of particular molecules.

4.3 COMPOSITION OF PROTEINS

Heating proteins in 6 *M* HCl hydrolyzes them to their component amino acids. The task of understanding the chemical structure of a protein involves determining the sequence of its amino acids. Some proteins also contain other compounds, apart from amino acids. These proteins are known as *conjugated proteins*, and the non-amino acid part is known as a *prosthetic group* (Chap. 8);

the amino acid part is termed the *apoprotein*. *Glycoproteins* and *proteoglycans* contain covalently bound carbohydrates, while *lipoproteins* (Chap. 6) contain lipids as the prosthetic groups.

EXAMPLE 4.1

The oxygen-binding proteins *hemoglobin* and *myoglobin* possess a prosthetic group known as *heme*. Heme consists of an organic molecule, known as *porphyrin*, in the center of which an iron atom is bound. Heme gives blood its red color and endows hemoglobin and myoglobin with their oxygen-binding ability.

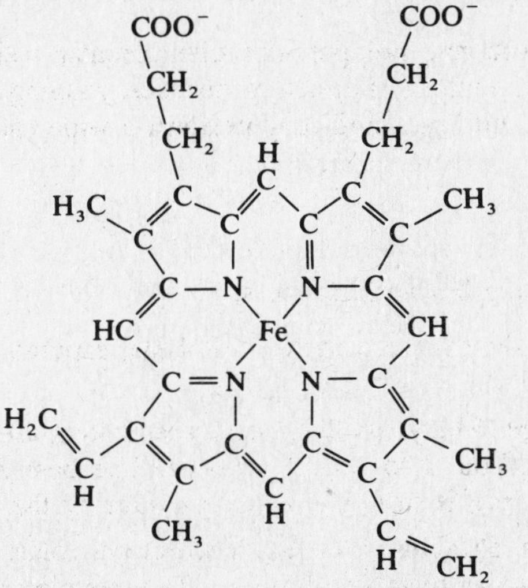

Heme [Fe(II)-protoporphyrin IX]

4.4 PROTEIN ISOLATION AND PURIFICATION

The first step in protein purification often involves separating the protein molecules from low-molecular-weight solutes. Some degree of separation of different proteins can then be achieved on the basis of physical properties such as electrical charge, molecular size, and differential solubility in various solvents. Finally, high purification can be attained on the basis of specific affinity for certain compounds that are linked to some form of solid support, the process known as *affinity chromatography*.

EXAMPLE 4.2

How can protein molecules be separated from low-molecular-weight solutes?

Protein molecules are of high molecular weight, generally (by definition) greater than 5,000. Thus, they will not pass through a cellulose membrane that allows free passage to smaller molecules. This selective permeability is the basis of the process termed *dialysis*, which is usually carried out by placing the protein solution inside a cellophane bag and immersing the bag in a large volume of buffer. The small molecules diffuse across the bag into the surrounding buffer, while the proteins remain within the bag.

Selective Solubility

Proteins may be selectively precipitated from a mixture of different proteins by adding (1) neutral salts such as ammonium sulfate (*salting out*), (2) organic solvents such as ethanol or acetone, or (3) potent precipitating agents such as trichloroacetic acid. Proteins are *least* soluble in any given solvent when the pH value is equal to their *isoelectric point* (PH_I). At the isoelectric point, the protein carries no net electric charge, and therefore electrostatic repulsion is minimal between protein molecules. Although a protein may carry both positively and negatively charged groups at its isoelectric point, the *sum* of these charges is zero. At pH values *above* the isoelectric point, the net charge will be *negative*, while at pH values *below* the pH_I, the net charge will be *positive*.

Electrophoresis

The movement of electrically charged protein molecules in an electrical field, termed *electrophoresis*, is an important means of separating different protein molecules. In electrophoresis, proteins become banded into narrow zones, providing that the effects of convection are minimized, for example, by the use of a stabilizing support such as paper or acrylamide gel.

The terms *anode* and *cathode* are often used in electrophoresis and are often confused. Remember that an *anion* is a negatively charged ion; *anions* move toward the *anode*.

EXAMPLE 4.3

In what direction will the following proteins move in an electric field [toward the anode, toward the cathode, or toward neither (i.e., remain stationary)]: (*a*) egg albumin ($pH_I = 4.6$) at pH 5.0; (*b*) β-lactoglobulin ($pH_I = 5.2$) at pH 5.0; at pH 7.0?

(*a*) For egg albumin, pH 5.0 is above its isoelectric point, and the protein will therefore carry a *small* excess negative charge. It will thus migrate toward the anode.

(*b*) For β-lactoglobulin, pH 5.0 is below its isoelectric point; at this pH, the protein will be positively charged and will move toward the cathode. At pH 7.0, on the other hand, the protein will be negatively charged and will move toward the anode.

Ion-Exchange Chromatography

Ion-exchange chromatography relies on the electrostatic interaction between a charged protein and a stationary ion-exchange-resin particle carrying a charge of opposite sign. The strength of attraction between the protein and the resin particle depends on the charge on the protein (and thus on the solution pH) and on the dielectric constant of the medium. The interaction can be modified in practice by altering the pH or the salt concentration.

EXAMPLE 4.4

A solution containing egg albumin ($pH_I = 4.6$), β-lactoglobulin ($pH_I = 5.2$), and chymotrypsinogen ($pH_I = 9.5$) was loaded onto a column of diethylaminoethyl cellulose (DEAE-cellulose) at pH 5.4. The column was then *eluted* with pH 5.4 buffer, with an increasing gradient of salt concentration. Predict the elution pattern.

$$
\begin{array}{c}
CH_3 \\
| \\
CH_2 \\
| \\
-CH_2-CH_2-N^+-H \\
| \\
CH_2 \\
| \\
CH_3
\end{array}
$$

Diethylaminoethyl (DEAE) group

DEAE-cellulose carries a positive charge at pH 5.4. At this pH value, chymotrypsinogen also carries a positive charge, and therefore does not bind. β-Lactoglobulin, carrying a very small negative charge, binds only weakly and will be displaced as the salt concentration is raised. However, egg albumin, carrying a larger negative charge, binds tightly and will be displaced only at high salt concentrations, or on lowering the pH to a value at which its negative charge is reduced.

Gel Filtration

Separation of proteins on the basis of size can be achieved by means of *gel filtration*. This technique relies on diffusion of protein molecules into the pores of a *gel matrix* in a column. A commonly used type of gel material is *dextran*, a polymer of glucose, in the form of very small beads. This material is available commercially as *Sephadex* in a range of different pore sizes.

When the protein is larger than the largest pore of the matrix, no diffusion into the matrix takes place and the protein is eluted rapidly. Molecules smaller than the smallest pore can diffuse freely into all gel particles and elute later from the column, after a larger volume of buffer has passed through.

EXAMPLE 4.5

In what order would the following proteins emerge upon gel filtration of a mixture on Sephadex G-200: myoglobin ($M_r = 16\,000$), catalase ($M_r = 500{,}000$), cytochrome *c* ($M_r = 12{,}000$), chymotrypsinogen ($M_r = 26{,}000$), and serum albumin ($M_r = 65{,}000$)?

Catalase is the largest protein in this set and would be excluded completely from the Sephadex beads (exclusion limit approx. 200,000). That means that to the catalase molecules, the beads appear to be solid, and so the catalase molecules can only dissolve in the fluid outside the beads (the *void volume*). They would be eluted from the column when a volume of effluent equal to this void volume had passed through the column.

Cytochrome *c* is the smallest protein of the set and would be able to diffuse freely into all the space within the beads. The column would therefore appear to have a larger volume available to cytochrome *c*, which would not emerge until the column had been flushed with a volume almost equal to its total volume.

The order of elution would thus be: catalase, serum albumin, chymotrypsinogen, myoglobin, and cytochrome *c*.

4.5 MOLECULAR WEIGHT

Each protein has a unique *molecular weight*. Furthermore, the size or molecular weight of a particular protein, under specified conditions, distinguishes it from many other proteins.

Proteins that possess a *quaternary* structure are composed of several separate polypeptide chains held together by noncovalent interactions. When such proteins are examined under dissociating conditions (e.g., 8 *M* urea to weaken hydrogen bonds and hydrophobic interaction, 1 m*M* mercaptoethanol to disrupt disulfide bonds), the molecular weight of the component polypeptide chains can be determined. By comparison with the *native* molecular weight, it is often possible to determine how many polypeptide chains are involved in the native structure.

The molecular weight of a protein may be determined by the use of thermodynamic methods, such as *osmotic pressure* measurement and *sedimentation equilibrium* in the ultracentrifuge. Osmotic pressure is sensitive to the *number* of molecules in solution, and if one knows the total mass of protein in solution, the molar weight can be calculated. The sedimentation of a protein molecule in a centrifuge depends directly on the *mass* of the molecules.

EXAMPLE 4.6

The molar mass (M) of a solute can be determined from measurements of the *sedimentation coefficient* s and the *diffusion coefficient* D (see Prob. 4.18), according to the Svedberg equation:

$$M = \frac{RTs}{D(1 - \bar{v}\rho)} \tag{4.1}$$

where R is the gas constant, T the absolute temperature, $\bar{v}$ the *partial specific volume* of the solute (the reciprocal of its density), and ρ the solution density. The sedimentation coefficient (s)—which depends on the molar weight of the solute, as well as its size and shape—is determined from the rate at which the solute sediments in the gravitational field of an ultracentrifuge. The diffusion coefficient (D), dependent on size and shape, is determined from the rate of spreading of the boundary between a solution and the pure solvent.

The sedimentation coefficient is often reported in Svedberg units (S), named in honor of Thé Svedberg, the inventor of the ultracentrifuge: $1S \equiv 10^{-13}$ seconds. Sedimentation coefficients of some selected proteins are listed below.

Protein	M_r	s
Lysozyme	14,000	1.9S
Hemoglobin	64,500	4.5S
Catalase	248,000	11.3S
Urease	480,000	18.6S

Sedimentation techniques are often difficult to apply, especially if the sample is not absolutely pure. Frequently, techniques are used to measure molecular weight that are not so theoretically sound, since they are based on a prior *calibration* with proteins of known molecular weight.

Gel Filtration

The volume required to elute a globular protein from a gel-filtration column is a monotonically decreasing function of molecular weight. Thus, by comparing the elution behavior of an unknown protein with that of a series of standards of known molecular weight, an estimation of molecular weight can be interpolated (Fig. 4-1).

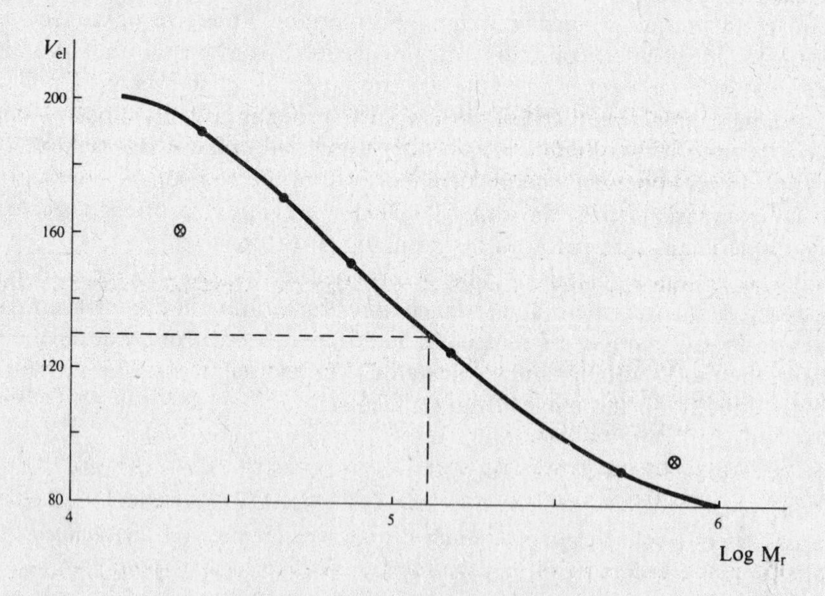

Fig. 4-1 Gel filtration of proteins. Elution volumes V_{el} are plotted for a series of proteins of different M_r applied to a column of Sephadex G-200. The two data points denoted by ⊗ were not used in constructing the calibration curve.

EXAMPLE 4.7

A series of standard proteins and an unknown enzyme were examined by means of gel filtration on Sephadex G-200. The elution volume, V_{el}, for each protein is given below.

Protein	M_r	V_{el} (mL)
Blue dextran*	10^6	85
Lysozyme	14,000	200
Chymotrypsinogen	25,000	190
Ovalbumin	45,000	170
Serum albumin	65,000	150
Aldolase	150,000	125
Urease	500,000	90
Ferritin†	700,000	92
Ovomucoid†	28,000	160
Unknown	——	130

*Blue dextran is a high-molecular-weight carbohydrate with a covalently bound dye.
†Do not use for calibration.

(*a*) Plot the data in the form of log M_r versus elution volume. (*b*) From the line of best fit through the points for the standards, determine the molecular weight of the unknown enzyme. (*c*) Explain why ferritin and ovomucoid behave anomalously.

(*a*) Figure 4-1 shows the line of best fit for the data and indicates coordinates for the unknown.

(*b*) From Fig. 4-1, M_r of the unknown enzyme $\simeq$ 138,000.

(*c*) Ferritin contains an iron hydroxide core, and therefore its density is higher than that of the standards. Ovomucoid is a glycoprotein of different density and probably different shape from the standards.

Gel Electrophoresis

In the presence of the detergent *sodium dodecyl sulfate* (SDS), many proteins are disaggregated and unfolded. As with elution volume, the electrophoretic mobility of the polypeptide chains is a monotonically decreasing function of molecular weight, and the molecular weight of an unknown protein may be inferred from the mobilities of standards.

Question: What assumptions are inherent in both of the above techniques for the determination of molecular weight?

1. In both cases, the hydrodynamic behavior depends on the *shape* of the molecule. We must, therefore, assume that the shape (i.e., spherical, ellipsoidal, etc.) of the unknown protein is the same as that of the standards.
2. The gel-filtration behavior really depends on the *effective size*, not mass. Therefore, if the protein differs from the standards in *density*, an incorrect estimate of molecular weight will result.
3. Proteins bind SDS, and as a first approximation it is assumed that the amount of SDS bound per gram is the same for all proteins. Differences in the amount bound per molecule will result in differences in the total charge, leading to differences in the electrophoretic mobility, and an incorrect value of molecular weight will be inferred.

4.6 PROTEIN STRUCTURE

In order to understand the functions of proteins, we need to know something about the *conformation*, or the three-dimensional folding pattern, that the polypeptide chain adopts. Although many artificial polyamino acids have no well-defined conformation and seem to exist in solution as nearly random coils, most biological proteins adopt a well-defined *folded structure*. Some, such as the keratins of hair and feathers, are *fibrous* and organized into linear or sheetlike structures with a regular, repeating folding pattern. Others, such as most enzymes, are folded into compact, nearly spherical, *globular* conformations.

The fibrous structure of keratin is sufficiently regular to scatter a beam of x-rays in a manner that reveals the regularity of folding. Measuring the intensities and positions of spots in the resulting *x-ray diffraction pattern* provides estimates of the distance between regularly repeating features of the folding pattern.

Linus Pauling and Robert Corey first realized that the peptide bond was planar and rigid. With this structural restriction, the number of folding patterns available for a protein is limited to two basic forms. One of these corresponded to the *α pattern*, an x-ray diffraction pattern observed with keratin from hair; the other corresponded to the *β pattern*, observed with silk fibroin, the fibrous protein of silk, and with keratin that had been stretched.

EXAMPLE 4.8

Why is the peptide group planar?

The C—N bond has a partial double-bond character owing to resonance between the two forms shown below:

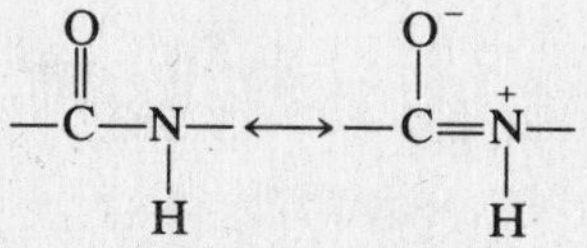

$$\text{—}\overset{\overset{\displaystyle O}{\|}}{C}\text{—}\underset{\underset{\displaystyle H}{|}}{N}\text{—} \longleftrightarrow \text{—}\overset{\overset{\displaystyle O^-}{|}}{C}\text{=}\underset{\underset{\displaystyle H}{|}}{\overset{+}{N}}\text{—}$$

The length of the C—N bond (0.132 nm) is intermediate between that of a C—N single bond (0.149 nm) and a C=N double bond (0.129 nm). The partial double-bond character restricts rotation around the C—N bond, such that the favored arrangement is for the O, C, N, and H atoms to lie in a plane, with the O and H atoms trans.

EXAMPLE 4.9

In the dipeptide *glycylalanine*, which bonds of the backbone allow free rotation?

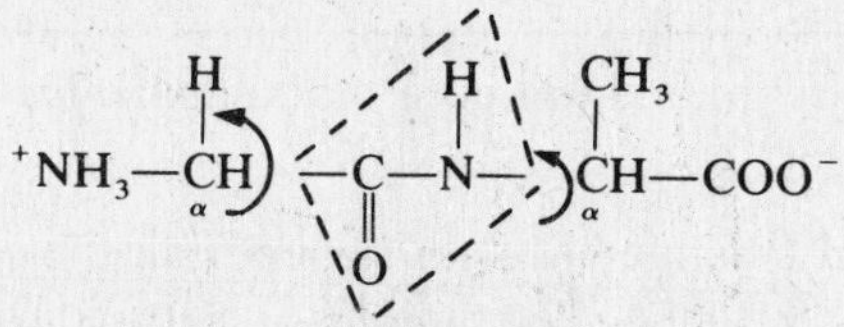

$$^{+}NH_3\text{—}\underset{\alpha}{\overset{\overset{\displaystyle H}{|}}{CH}}\text{—}\underset{\underset{\displaystyle O}{\|}}{C}\text{—}\overset{\overset{\displaystyle H}{|}}{N}\text{—}\underset{\alpha}{\overset{\overset{\displaystyle CH_3}{|}}{CH}}\text{—}COO^-$$

Glycylalanine

In this structure, the peptide group is indicated by the dashed lines. Because the peptide group itself is rigid and planar, there is no rotation around the bond between the carbonyl carbon atom and the nitrogen atom (the C′—N bond). However, free rotation is possible around the bond between the α carbon and the carbonyl carbon atom (the C_α—C′ bond) and about the bond between the nitrogen atom and the alanyl α-carbon atom (the N—C_α bond). Thus, for every peptide group in a protein, there are two rotatable bonds, the relative angles of which define a particular backbone conformation.

Regular, Repeating Structures

Given the normal van der Waals radii of the atoms, expected bond angles, and the planarity of the peptide bond, only two regular, repeating structures exist without distortion and with maximum hydrogen-bond formation:

1. The α helix, found in the α-keratins
2. The β pleated sheets (parallel and antiparallel), as exemplified by the β form of stretched keratin and silk protein

These regular, repeating structures are also commonly found as elements of folding patterns in the globular proteins. For many globular proteins, a significant proportion of the polypeptide chain displays no regularity in folding. These regions are often referred to as having a *random-coil conformation*. However, in most cases, these regions are well-defined, even if they are not regular, and they are better referred to as *disordered* regions.

The α Helix

In the α helix, the polypeptide *backbone* is folded in such a way that the —C=O group of each amino acid residue is *hydrogen-bonded* to the —N—H group of the *fourth* residue along the chain: i.e., the —C=O group of the first residue bonds to the —N—H group of the fifth residue, and so on.

Question: Do the —N—H groups on residues 1 to 4 also hydrogen bond to other groups in the α helix?

There are no available —C=O groups with which these amide groups can interact. Consequently, they remain unbonded. Similarly, at the other end of the chain, four —C=O groups remain unbonded. If the polypeptide chain is very long, the lack of bonding at the ends has a negligible effect on the overall stability. However, short α-helical chains are less stable because the end effects are *relatively* more important.

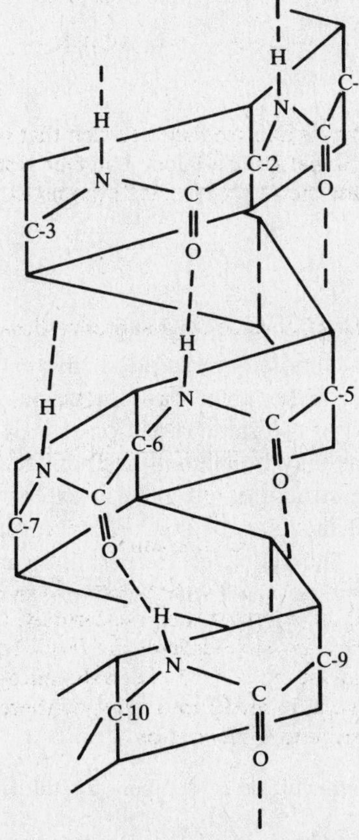

Fig. 4-2 The right-handed α helix. This diagram shows the peptide groups represented as planar segments, with the α-carbon atom at the junctions of successive planes.

The backbone of the α helix winds around the long axis, as shown in Fig. 4-2. The hydrogen bonds are all aligned approximately parallel to this axis, and the side chains protrude outward. Each residue is spaced 0.15 nm from the next along the axis, and 3.6 residues are required to make a complete turn of the helix. Although both left and right screw senses are possible, the right-hand screw sense is energetically favored with L-amino acids.

Because each —C=O and —N—H group is hydrogen-bonded (except for the four at each end), the α helix is strongly stabilized. However, for some amino acids, interactions involving the side chains may weaken the α helix, making this conformation less likely in polypeptide chains containing high proportions of such helix-destabilizing amino acids (Table 4.1).

Table 4.1. Tendency of Amino Acid Residues to Form α Helices

Helix Formers	Helix Breakers	Indifferent Residues
Glu	Pro	Asp
Ala	Gly	Thr
Leu	Tyr	Ser
His	Asn	Arg
Met		Cys
Gln		
Trp		
Val		
Phe		
Lys		
Ile		

The β-Sheet Structures

The second major regular, repeating structure, the β structure, differs from the α helix in that the polypeptide chains are almost completely extended, as in Fig. 4-3(a), and hydrogen bonding occurs *between* polypeptide strands, rather than *within* a single strand, as shown in Fig. 4-3(c).

Adjacent chains can be aligned in the same direction (i.e., N terminal to C terminal) as in the *parallel β sheets*, or alternate chains may be aligned in opposite orientations as in the *antiparallel β sheets*, shown in Fig. 4-3(c). These structures often form extensive sheets, as shown in Fig. 4-3(b). Sometimes it is possible for several sheets to be stacked upon one another. Because the side chains tend to protrude above and below the sheet in alternating sequence, as shown in Fig. 4-3(b), the β-sheet structures are favored by amino acids with relatively small side chains, such as alanine and glycine. Large, bulky side chains can lead to steric interference between the various parts of the protein chain.

EXAMPLE 4.10

Predict which regular, repeating structure is more likely for the two polypeptides (a) poly(Gly-Ala-Gly-Thr); (b) poly(Glu-Ala-Leu-His).

Polypeptide (a) is composed largely of amino acid residues with small side chains. Except for Ala, none of the residues favors helix formation, and Gly destabilizes the α helix. Thus, this polypeptide is more likely to form β structures.

Polypeptide (b) is composed of amino acid residues with bulky side chains that would destabilize β structures. However, all the amino acids are helix-stabilizing, and thus polypeptide (b) is more likely to form an α helix.

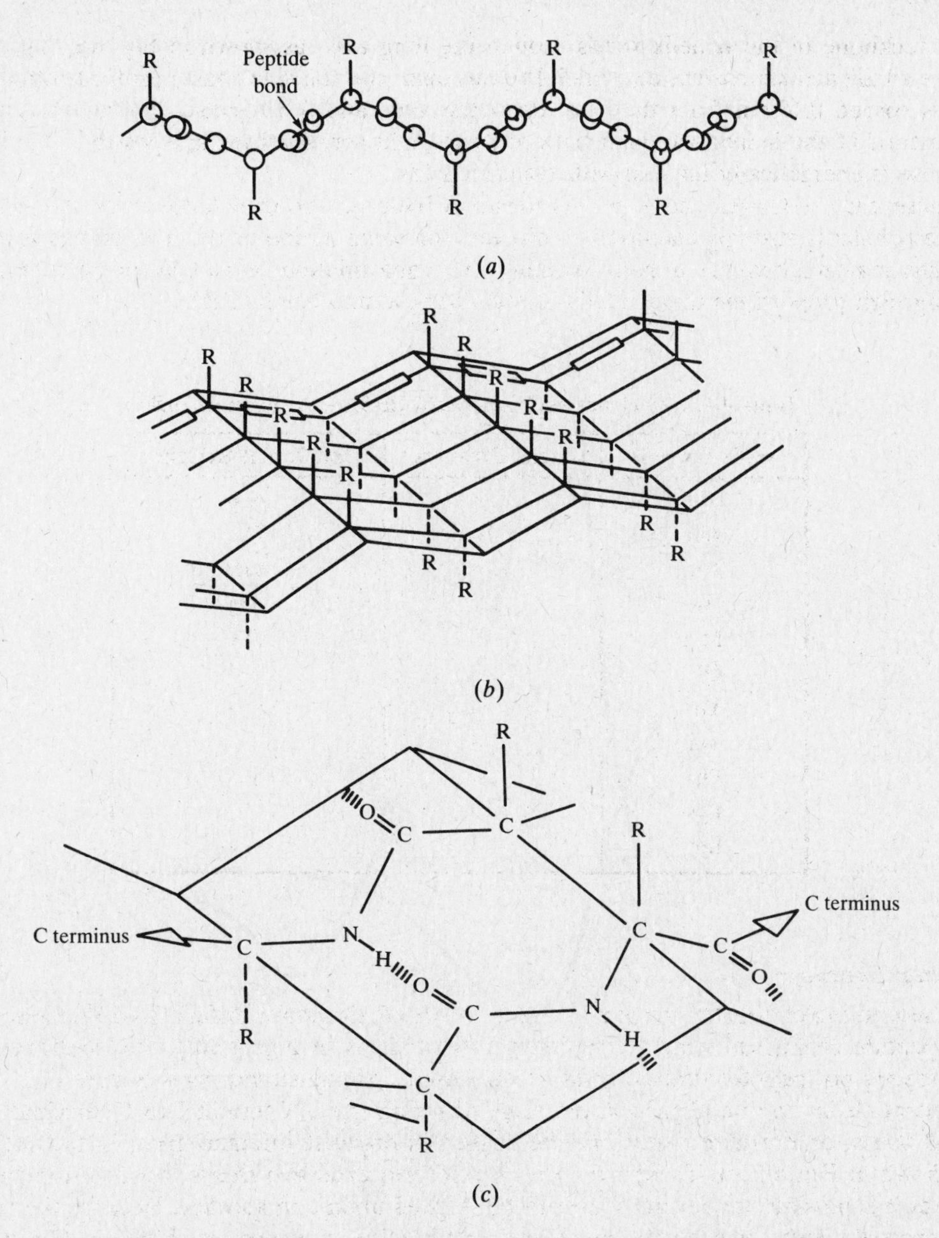

Fig. 4-3 β-Sheet structures: (*a*) polypeptide segment in an extended conformation; (*b*) sheet formed by the assembly of extended polypeptide chains side by side; (*c*) detail showing H bonding between adjacent polypeptide chains in an antiparallel β sheet.

Optical Activity

Asymmetric molecules such as carbohydrates, amino acids, and proteins rotate the plane of polarization of plane-polarized light. The amount of rotation depends on the concentration of the substance and the path length of light in the sample (Chap. 2), in much the same way as for optical absorbances (Chap. 3). The amount of rotation (and, in fact, even the *direction* of rotation) also depends on the wavelength of the light. The dependence of the *specific rotation* $[\alpha]$ (the measured rotation per unit concentration and path length) on the wavelength of light is known as the *optical rotatory dispersion* (ORD).

The conformation of a protein introduces an additional source of asymmetry that affects the ORD spectrum. Helical regions in soluble proteins give rise to a particular ORD spectrum (see Fig. 4-4) quite distinct from that of disordered regions. By comparing the ORD spectrum of an unknown protein with suitable standards of known conformation, it is possible to calculate the approximate proportions of the different types of structure in the protein.

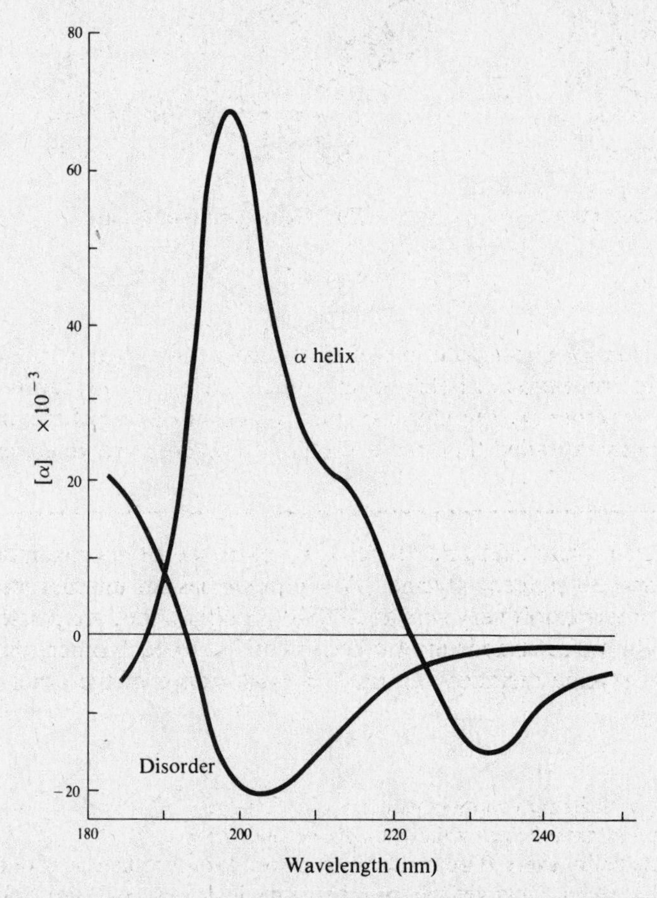

Fig. 4-4 ORD spectrum of poly-D-lysine showing the α-helical and disordered conformation regions. (Units of [α] are degrees mL dm^{-1} g^{-1}.)

EXAMPLE 4.11

The structures of many globular proteins are composed of elements of α helix, β structures, and disordered regions. Some examples are shown in stylized form in Fig. 4-5. The α helices are represented by coiled ribbons, and β structures are represented by arrows pointing in the N⟶C direction. Parallel β sheets have their arrows pointing in the same direction; antiparallel β sheets have their arrows alternating.

The Collagen Triple Helix

Question: Why is proline rarely found within α-helical segments?

The α-amino group of proline is a *secondary* amino group. When proline participates in peptide bonds through its amino group, there is no longer an amide hydrogen to participate in the hydrogen-bond stabilization of the α helix. In addition, because the side chain of proline is attached to the α-amino group, there is no free rotation about the N—C_α bond and proline cannot take up the correct conformation for an α-helical residue.

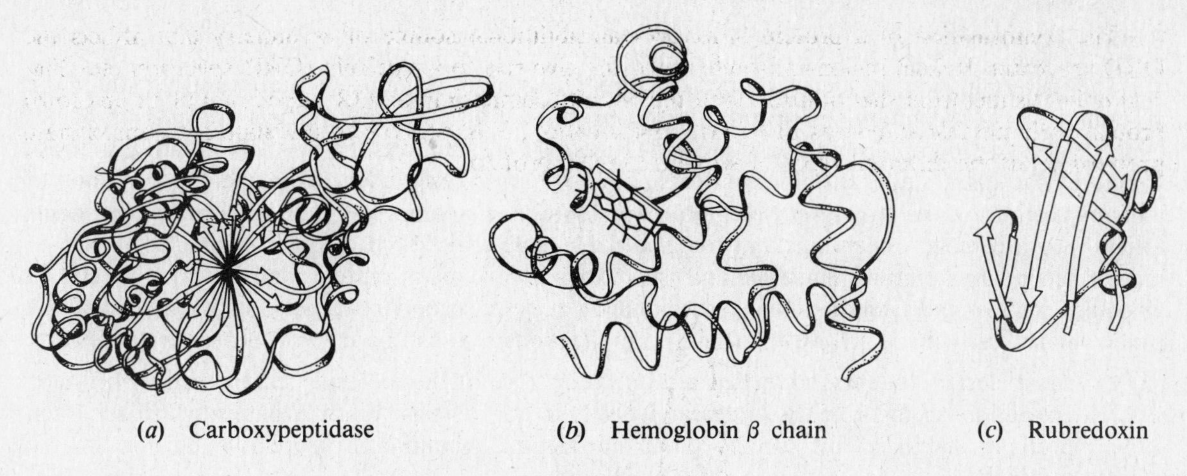

(*a*) Carboxypeptidase (*b*) Hemoglobin β chain (*c*) Rubredoxin

Fig. 4-5.

Although proline cannot participate in α-helical conformations, polypeptides composed only of proline can adopt a different type of helical conformation. This *polyproline helix* is *not* stabilized by hydrogen bonding, but rather by the steric mutual repulsion effects of the prolyl side chains. The polyproline helix is more extended than the α helix, with adjacent residues separated along the axis by 0.31 nm.

The protein *collagen*, from skin and tendons, is composed of approximately 30 percent proline and hydroxyproline and 30 percent glycine. This protein has an unusual structure in which *three* chains, each with a conformation very similar to that of polyproline, are twisted about each other to make a triple helix. The three strands are hydrogen-bonded to each other, through hydrogen bonds between the —NH of glycine residues and the —C═O groups of the other amino acids.

EXAMPLE 4.12

Why does collagen, with its polypeptide sequence that is largely (Gly-Pro-x-)$_n$, where x is another type of amino acid residue, form a triple helix, while polyproline does not?

In the collagen triple helix, every third residue is positioned toward the center of the helix and comes into close contact with another chain. Only glycine, with its simple hydrogen atom side chain, is small enough to fit into this crowded space.

Levels of Structure

The Danish protein chemist Kai Linderstrøm-Lang suggested that one could consider the structure of a protein on several levels:

(*a*) *Primary structure*: the sequence of amino acids.

(*b*) *Secondary structure*: the regular, repeating folding pattern (such as the α helix and β structures), stabilized mostly by hydrogen bonds between peptide groups close together in the sequence.

(*c*) *Tertiary structure*: for a globular protein, the way that segments of secondary structure fold together in three dimensions, stabilized by interactions often far apart in the sequence. For those proteins with little or no detectable α helix or β structure, the tertiary structure can be considered to be the way the protein folds in three dimensions, stabilized by interactions between distant parts of the sequence.

(*d*) *Quaternary structure*: the interaction between different polypeptide chains to produce an *oligomeric* structure, stabilized by noncovalent bonds only.

Tertiary Structure

The determination of the tertiary structure of soluble proteins is more difficult than the determination of secondary structure in fibrous proteins. Most of our structural information comes from x-ray crystallographic analysis of protein *crystals*. Myoglobin was the first soluble protein whose structure was determined at sufficient resolution to enable researchers to locate the positions of individual atoms in the structure. Myoglobin was found to be composed of short α-helical segments folded into a compact, near spherical (or *globular*) shape and linked by short segments having no regular secondary structure. Since the analysis of myoglobin, many proteins and enzymes have been examined by x-ray crystallography, and in many, a clear secondary-tertiary structure distinction cannot be so easily made. However, some generalizations that can be made from these studies are:

1. Most electrically charged groups are on the surface of the molecule, interacting with water. Exceptions to this rule are often catalytically important residues in enzymes, which are often partially stabilized by specific polar interactions within a hydrophobic portion of the molecule.
2. Most nonpolar (e.g., hydrocarbon) groups are in the interior of the molecule, avoiding the thermodynamically unfavorable contact with water. Exceptions to this may function as specific binding sites on the surface of the molecule for other proteins or ligands.
3. Maximal hydrogen bonding occurs within the molecule.
4. Proline often *terminates* α-helical segments.

Allowed Conformations

The conformation of an amino acid residue can be defined by specifying the angles of rotation ϕ (around the N—C_α bond) and ψ (around the C_α—C′ bond; see also Example 4.9). The zero position for ϕ is defined with the —N—H group trans to the C_α—C′ bond, and for ψ with the C_α—N bond trans to the C═O bond (Fig. 4-6).

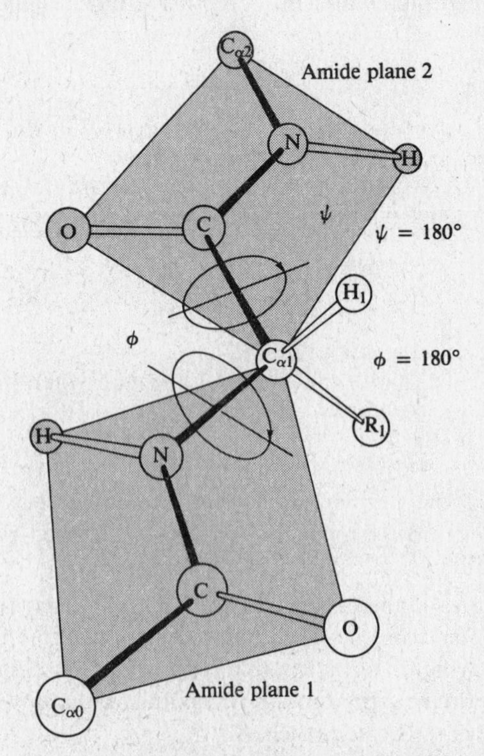

Fig. 4-6 Rotation of adjacent amide planes in a polypeptide.

Not all combinations of ϕ and ψ angles are possible, however, as many lead to clashes between atoms in adjacent residues. For all residues except glycine, the existence of such steric restriction involving side-chain atoms reduces drastically the number of possible conformations. The possible combinations of ϕ and ψ angles that do not lead to clashes can be plotted on a conformation map (also known as a *Ramachandran plot*, named after the chemist who did much of the pioneering work in this field). Figure 4-7 shows a Ramachandran plot for the allowed conformations of alanylalanine. The double-hatched areas represent conformations (combinations of ϕ and ψ) for which no hindrance exists. The single-hatched areas represent conformations for which some hindrance exists, but which may be possible if the distortion can be compensated for by interactions elsewhere in the protein.

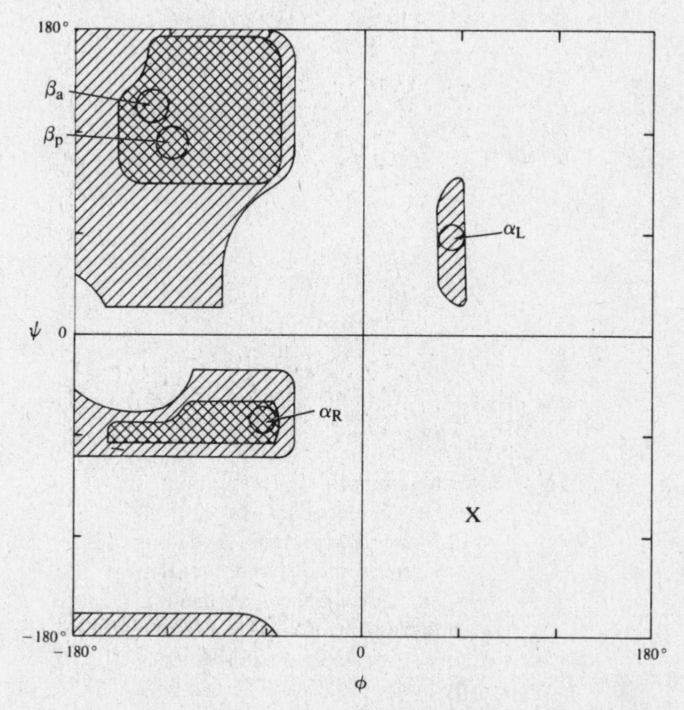

Fig. 4-7 Ramachandran plot for alanylalanine, showing the fully allowed regions (double-hatched) and partially allowed regions (single-hatched) of ϕ and ψ angles (see Fig. 4-6). The coordinates for the parallel and antiparallel β structures (β_p and β_a, respectively) and for the left-handed and right-handed α helices (α_L and α_R, respectively) are indicated.

EXAMPLE 4.13

The right-handed α helix has ϕ and ψ values of $-57°$ and $-47°$, respectively. Would you expect an α helix to be a stable structure?

Yes. These values put the right-handed α helix into a particularly favorable area of the Ramachandran plot, indicated in Fig. 4-7 with the symbol α_R.

Structure Hierarchy

With the large number of protein structures now known, it is possible to draw some generalizations; there are common, repeating patterns of secondary structure that occur in many proteins. For example, one common, recurring pattern is β-α-β, which has a segment of extended chain, an intervening helical loop, and a second segment of extended chain hydrogen-bonded to the first (Fig. 4-8). In the enzyme triosephosphate isomerase, this motif is repeated several times to make a barrel-like structure. Such recurring patterns are referred to as *supersecondary structures*.

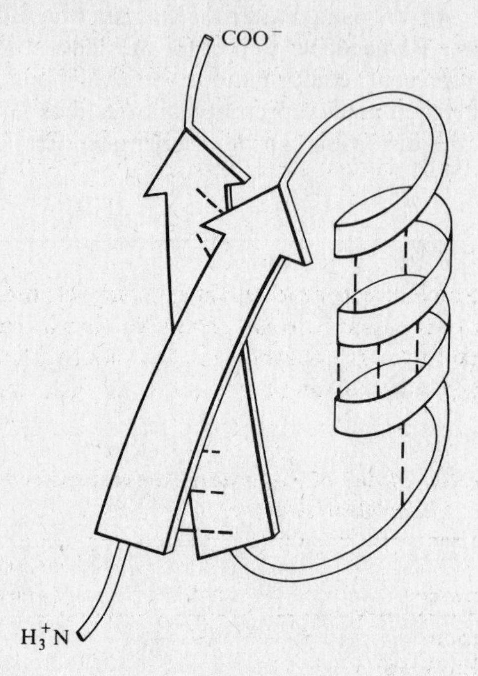

Fig. 4-8 Diagrammatic representation of the supersecondary β-α-β folding unit of a protein. β regions are represented by the arrows, while the α-helical segment is indicated by the coiled structure. Approximate positions of hydrogen bonds are shown with dashed lines.

Domain Organization

EXAMPLE 4.14

For a spherical protein, how do (*a*) volume and (*b*) surface area depend on molecular weight?

(*a*) Given a uniform density, the volume V would be directly proportional to the mass, and therefore to the molecular weight; i.e., $V \propto M_r$.

(*b*) The surface area of a sphere is proportional to the square of the radius, while the volume is proportional to the cube of the radius. Thus, $A \propto V^{2/3}$. We have just seen that $V \propto M_r$; therefore, $A \propto M_r^{2/3}$.

The above calculation shows that the volume of a spherical protein will increase more rapidly with molecular weight than will the surface area. Thus, in order to accommodate all charged groups on the surface and all nonpolar groups in the interior, three strategies are possible:

1. To overcome the increasing volume-to-surface-area ratio, larger proteins may show altered composition, with increasing proportions of nonpolar amino acid side chains occupying the increased interior volume.

2. Larger proteins may fold into separate *domains*, each domain being a *globular folding unit* with its own interior and surface, and with an interconnecting strand of backbone linking the domains.
3. Large proteins may fold into more elongated or rodlike shapes.

Domains are a common feature of many globular proteins, particularly where the molecular weight is above 20,000. Larger proteins are often folded so that each domain is approximately 17,000 daltons. For example, the enzyme glyceraldehyde 3-phosphate dehydrogenase is folded into two domains, with each domain having a separate function: one domain binds the cofactor NAD^+, while the other, catalytic, domain binds the substrate glyceraldehyde 3-phosphate (Chap. 11).

Forces Determining Structure

The folding of a chain of amino acid residues into a compact, ordered structure is accompanied by a large *decrease* in *entropy*; this decrease is thermodynamically unfavorable. The native folded state of a protein is maintained by a large number of weak, noncovalent interactions (Table 4.2) that act cooperatively; i.e., although any individual interaction is weak, the effect of a large number can be considerable.

Table 4.2. Types of Noncovalent Interactions Stabilizing Protein Structure

Interaction	Example	Bond energy* ($kJ\ mol^{-1}$)
Dispersion (van der Waals)	C—H···H—C	1–5
Electrostatic	$—COO^- \cdots H_3\overset{+}{N}—$	12–20
Hydrogen bond	—N—H···O=C—	10–20
Hydrophobic†		12–15

*The bond energy is the energy required to break the interaction.
†This value represents the free energy required to transfer a $—CH_2—$ group of a nonpolar side chain from a protein's interior to water.

Hydrogen Bonds

A hydrogen bond results from an electrostatic interaction between a hydrogen atom, covalently bound to an *electronegative* atom (such as O, N, or S), and a second electronegative atom with a *lone pair* of *nonbonded* electrons:

$$\underset{\text{Donor}}{—O—H} \quad \cdots \quad \underset{\text{Acceptor}}{O{=}C}$$

The hydrogen atom is partly shared between the *donor* group and the *acceptor*. These bonds are highly directional and are strongest when all three participating atoms lie in a straight line.

Hydrophobic Interactions

The placing of a nonpolar group in water leads to an energetically unfavorable organization of the water molecules around it, i.e., a lowering of the entropy of the solution. Transfer of nonpolar groups from water to a nonpolar environment is thus accompanied by an increase in entropy and is spontaneous (Chap. 10). The folding of a protein chain to a compact globular conformation removes nonpolar groups from contact with water; the increase in entropy arising from the liberation of water molecules compensates for the decrease in entropy of the folded polypeptide chain.

EXAMPLE 4.15

If the overall folding energy of a particular protein is only 80 kJ mol^{-1}, how many H bonds would have to be broken in order to disrupt the structure?

Since each H bond contributes 10 to 20 kJ mol^{-1} of stabilizing energy, the breaking of between four and eight such bonds would lead to a loss of 80 kJ mol^{-1}, sufficient to disrupt the native structure.

Because the stabilization energy of most proteins is so small, many proteins show rapid, small fluctuations in structure, even at normal temperatures. In addition, it is fairly easy to cause protein molecules to unfold, or *denature*. Common denaturation agents are:

(*a*) High temperature

(*b*) Extremes of pH

(*c*) High concentrations of compounds such as urea or guanidine hydrochloride:

$$NH_2{-}\underset{\displaystyle O}{\underset{\|}{C}}{-}NH_2 \qquad\qquad NH_2{-}\underset{\displaystyle {}^{+}NH_2Cl^-}{\underset{\|}{C}}{-}NH_2$$

Urea Guanidine hydrochloride

(*d*) Solutions of detergents such as sodium dodecyl sulfate, $CH_3(CH_2)_{10}CH_2OSO_3^-Na^+$

Protein Structure Dictated by Amino Acid Sequence

Several proteins and enzymes, completely unfolded by urea and with disulfide bridges reduced, are capable of refolding to the active, native state on removal of urea. This demonstrates that the *information* for the correct folding pattern exists in the sequence of amino acids.

EXAMPLE 4.16

Ribonuclease, the enzyme that hydrolyzes *ribonucleic acids* (Chap. 7), contains four disulfide bonds that help to stabilize its conformation. In the presence of 6 *M* guanidine hydrochloride, to weaken hydrogen bonds and hydrophobic interactions, and 1 m*M* mercaptoethanol, to reduce the disulfide bonds, all enzymatic activity is lost, and there is no sign of residual secondary structure. On removing the guanidine hydrochloride by dialysis or gel filtration, enzymatic activity is restored, the native conformation is regained, and correct disulfide bonds are reformed.

4.7 SEQUENCE HOMOLOGY AND PROTEIN EVOLUTION

Myoglobin

The protein myoglobin serves as an oxygen binder in muscle. This protein was among the first to be studied with the aid of x-ray crystallography, which revealed a compact globular structure comprising eight segments of α helix linked by short nonhelical segments. The helical segments were named in sequence from A to H. A heme group is bound in a nonpolar pocket between helices B, E, F, and G. The iron atom of the heme group is coordinated to the nitrogen atom of histidine F8 (the eighth residue in helix F), also known as the *proximal histidyl*. See Fig. 4-9. Another histidine residue, E7, or the *distal histidyl*, is close to, but not directly coordinated with, the other side of the iron atom. When oxygen binds, it occupies the sixth coordination position on the iron atom. The protein part of myoglobin (the *apoprotein*) stabilizes the heme group in the Fe(II) state, which is capable of reversible oxygenation. In the absence of the protein, free heme is rapidly oxidized by oxygen to the Fe(III) state, which no longer binds oxygen.

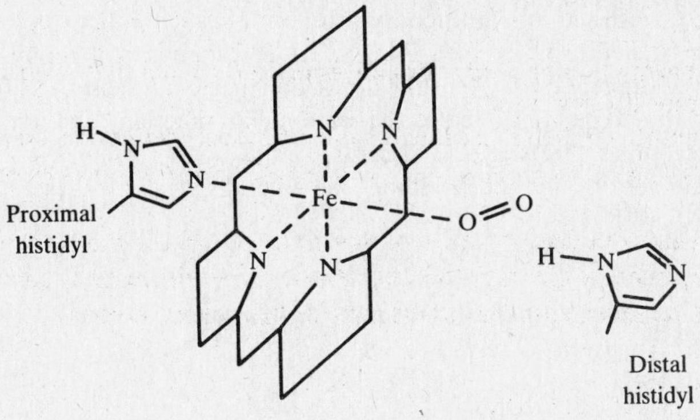

Fig. 4-9 The disposition of the heme between the functionally important histidine residues in myoglobin.

Hemoglobin

Hemoglobin, the oxygen-carrying protein of vertebrate blood, is similar in structure to myoglobin. However, hemoglobin is composed of four chains; i.e., it has a *quaternary structure*. The four chains are held together in a particular geometrical arrangement by noncovalent interactions. There are two major types of hemoglobin polypeptide chains in normal adult hemoglobin, the α chain and the β chain.

The polypeptide chain of myoglobin and the two chains of hemoglobin are remarkably similar, both in primary and in tertiary structure. The two proteins are said to be *homologous*. Myoglobin has 153 residues; the hemoglobin α chain has 141, and the hemoglobin β chain 146. The sequence is identical for 24 out of 141 positions for the human proteins, and many of the differences show *conservative replacement*. This means that an amino acid residue in one chain has been replaced by a chemically similar residue at the corresponding position in another chain, for example, the replacement of glutamate for aspartate.

Species Differences

Comparison of the amino acid sequences of hemoglobin and myoglobin chains from different species of animals shows that the chains from related species are similar. The number of differences increases with phylogenetically more separated species.

On the assumption that proteins evolve at a constant rate, the number of differences between two homologous proteins will be proportional to the time of divergence in evolution of the species.

EXAMPLE 4.17

Draw an evolutionary tree for the human, rabbit, silkworm, and fungus *Neurospora* by using the data in the following table for differences in the respective cytochrome *c* sequence.

	Number of Sequence Differences			
	Human	Rabbit	Silkworm	*Neurospora*
Human	0	11	36	71
Rabbit		0	35	70
Silkworm			0	69
Neurospora				0

These data allow us to construct an evolutionary tree, with branch lengths approximately proportional to the number of differences between species (see Fig. 4-10). The human and rabbit show most similarity and therefore are connected by short branches. The silkworm cytochrome *c* is closer to both mammalian forms than it is to *Neurospora*, and so should be connected to the mammalian junction. The length of the silkworm branch will be approximately three times the length of the rabbit and human branches. Finally, *Neurospora* shows approximately the same number of differences with all the animal species, and therefore can be joined to their common branch.

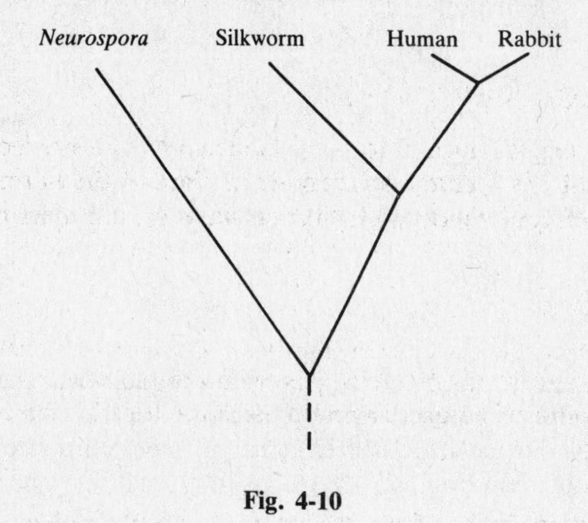

Fig. 4-10

Solved Problems

COMPOSITION OF PROTEINS

4.1. A pure heme protein was found to contain 0.426 percent of iron by weight. What is its minimum molecular weight?

SOLUTION

The minimum molecular weight is the molecular weight of a molecule containing only a single iron atom. Thus, if 0.426 g of iron is contained in every 100 g of protein, then 1 mol (56 g) iron is contained in

$$\frac{100 \times 56}{0.426} = 13{,}145 \text{ g of protein}$$

The mass of protein containing one gram atom of iron is 13,145 g, which therefore represents the molar mass of the protein. The minimum molecular weight is thus 13,145. This is close to the molecular weight of the heme protein cytochrome *c*, and is thus a reasonable value. However, had the molecule of protein contained more than one atom of iron, the molecular weight would have been a multiple of 13,145.

4.2. Threonine constitutes 1.8 percent by weight of the amino acid content of insulin. Given that the molecular weight of threonine is 119, what is the minimum molecular weight of insulin?

SOLUTION

Since water is lost during the condensation of amino acids to form peptide bonds, the *residue weight* of threonine will be $119 - 18 = 101$. If we assume a single threonine residue per molecule, then 101 represents 1.8 percent of the molecular weight, and

$$M_r = \frac{101}{0.018} = 5{,}600$$

Physical measurements in dissociating solvents confirm this value.

PROTEIN ISOLATION AND PURIFICATION

4.3. Hemoglobin A (the major, normal form in humans) has an isoelectric point of pH 6.9. The variant hemoglobin M has a glutamate residue in place of the normal valine at position 67 of the α chain. What effect will this substitution have on the electrophoretic behavior of the protein at pH 7.5?

SOLUTION

Since pH 7.5 is above the isoelectric point of hemoglobin A, the protein carries a negative charge and will migrate to the anode. At pH 7.5 the glutamate side chain has a negative charge, while valine is uncharged. Hemoglobin M, therefore, carries an additional negative charge at pH 7.5 and will migrate *faster* toward the anode.

4.4. The proteins ovalbumin ($pH_I = 4.6$), urease ($pH_I = 5.0$), and myoglobin ($pH_I = 7.0$) were applied to a column of DEAE-cellulose at pH 6.5. The column was eluted with a dilute pH 6.5 buffer, and then with the same buffer containing increasing concentrations of sodium chloride. In what order will the proteins be eluted from the column?

SOLUTION

At pH 6.5, both ovalbumin and urease are negatively charged, and will bind to the DEAE-cellulose. Myoglobin has a positive charge at pH 6.5 and will be eluted immediately. As the salt concentration is raised, electrostatic interactions are weakened; urease will be eluted next, and ovalbumin will be eluted last.

4.5. An enzyme of $M_r = 24{,}000$ and $pH_I = 5.5$ is contaminated with a protein of similar molecular weight, but with $pH_I = 7.0$, and another protein of $M_r = 100{,}000$ and $pH_I = 5.4$. Suggest a purification strategy.

SOLUTION

Gel filtration will allow the high-molecular-weight contaminant to be removed. The remaining mixture of lower-molecular-weight proteins can be separated by ion-exchange chromatography, as described in Prob. 4.4.

PROTEIN STRUCTURE

4.6. Can you explain why protein amino acids are all α-amino acids?

SOLUTION

In peptides made from α-amino acids, the only bonds in the peptide backbone that allow free rotation are the C_α—C′ and N—C_α bonds. The C′—N bond is rigid and planar. If a β-amino acid were to participate in peptide bond formation, two additional bonds (one on either side of the β-carbon atom) would allow free rotation. The additional flexibility introduced would prevent the folding into secondary structures.

4.7. The artificial polypeptide poly-L-glutamate forms α helices in solution at pH 2, but not at pH 7. Suggest an explanation.

SOLUTION

At pH 2, the side chains of poly-L-glutamate are largely uncharged and protonated. However, at pH 7 they are negatively charged. The negative charges lead to mutual repulsion and destabilization of the α helix.

4.8. Poly-L-glutamate in 100 percent α-helical form shows a trough in the ORD spectrum with $[\alpha]_{233} = -15{,}000°$; in the random-coil form, $[\alpha]_{233} = -1{,}000°$ (the subscript 233 refers to the wavelength of light used). Calculate the proportion of α helix in a protein for which $[\alpha]_{233} = -7{,}160°$, assuming the presence of α-helical and disordered regions only.

SOLUTION

100 percent α helix corresponds to $[\alpha]_{233} = -15{,}000°$; 0 percent α helix corresponds to $[\alpha]_{233} = -1{,}000°$. Therefore

$$\% \text{ helix} = \frac{-[\alpha]_{233} - 1{,}000}{15{,}000 - 1{,}000} \times 100\% = \frac{6{,}160}{14{,}000} \times 100\% = 44\%$$

4.9. Myoglobin in solution at pH values above 6.0 has a value of specific rotation $[\alpha]_{233} = -12{,}000°$. However, on adjusting the solution to pH 2.0, the specific rotation changes to $-2{,}000°$. Explain the change.

SOLUTION

Using the method of Prob. 4.8, the helical content of myoglobin can be calculated at both pH values, if we assume the presence of α-helical and disordered regions only. At pH 6.0, the helical content is 79 percent. However, at pH 2.0 the helical content is only 7 percent. We may conclude, then, that at low pH, the protein has lost its α-helical conformation; i.e., it has been denatured.

4.10. Partial sequence determination of a peptide gave the following:

-Gly-Pro-Ser-Gly-Pro-Arg-Gly-Leu-Hyp-Gly-

What conclusions can be made about the possible conformation of the protein from which this peptide was derived?

SOLUTION

This sequence closely resembles that of collagen. In particular, the repeating (Gly-Pro-x) pattern and the occurrence of hydroxyproline are characteristic of collagen. It is possible, then, that the peptide was derived from a protein resembling collagen in its triple-helical conformation.

4.11. Insulin possesses two polypeptide chains, A and B, linked by disulfide bonds. On the denaturation and reduction of insulin, followed by reoxidation, only 7% recovery of activity was obtained. This is the level of activity expected for random pairing of disulfide bridges. How can these data be reconciled with the hypothesis that the amino acid sequence directs protein folding?

SOLUTION

Insulin is synthesized as *proinsulin*. After synthesis and folding, a section of the molecule (the C peptide) is excised, leaving the A and B peptides connected via disulfide bridges. Thus, native insulin, lacking the C peptide, lacks some of the information necessary to direct the folding process.

4.12. Insulin and hemoglobin are both proteins that comprise more than one polypeptide chain. Contrast the interactions between the component polypeptide chains of the two proteins.

SOLUTION

The two chains of insulin, fragments of what was originally a single chain, are held together by covalent disulfide bonds. Hemoglobin has four polypeptide chains, held together by noncovalent interactions only.

4.13. There are four disulfide bonds in ribonuclease. If these are reduced to their component sulfhydryl groups and allowed to reoxidize, how many different combinations of disulfide bonds are possible?

SOLUTION

In forming the first disulfide bond, a cysteine residue may pair with any one of the remaining seven. In forming the second bond, a cysteine residue may pair with one of the remaining five, and for the third bond, a cysteine residue may pair with one of three. Once three bonds have formed, there is only one way to form the last bond. Consequently, the number of possibilities is

$$7 \times 5 \times 3 = 105$$

Thus, the likelihood of forming the correct disulfide bonds by chance alone is

$$\frac{1}{105} = 0.0095, \text{ or } 0.95\%$$

4.14. The glycine residue at position 8 in the sequence of insulin has torsion angles $\phi = 82°$, $\psi = -105°$, which lie in the unfavorable region (marked X) of the Ramachandran plot in Fig. 4-6. How is this possible?

SOLUTION

Glycine has a very small side chain—a single hydrogen atom. The plot of Fig. 4-6 was determined for alanine, which has a methyl group as the side chain. Thus, there are conformations allowed for glycine that are not possible for the methyl group of alanine.

MOLECULAR WEIGHT

4.15. Using the data of Example 4.7, determine the molecular weight of an enzyme for which the elution volume was 155 mL.

SOLUTION

From Fig. 4-1, the value of log M_r corresponding to 155 mL elution volume is 4.8. The molecular weight is therefore 63,000, providing that the density and shape of the enzyme are similar to those of the calibration standards.

4.16. An enzyme examined by means of gel filtration in aqueous buffer at pH 7.0 had an apparent molecular weight of 160,000. When examined by gel electrophoresis in SDS solution, a single band of apparent molecular weight 40,000 was formed. Explain these findings.

SOLUTION

The detergent SDS causes the dissociation of quaternary structures and allows the determination of molecular weight of the component subunits. The data suggest that the enzyme comprises four identical subunits of $M_r = 40{,}000$, yielding a tetramer of $M_r = 160{,}000$.

4.17. During an attempt to determine the molecular weight of the milk protein β-lactoglobulin, by means of gel filtration, the following data were obtained with different sample concentrations:

Protein Concentration	Apparent M_r
$10\ g\ L^{-1}$	36,000
$5\ g\ L^{-1}$	35,000
$1\ g\ L^{-1}$	32,000
$0.1\ g\ L^{-1}$	25,000

Electrophoresis in acrylamide gels containing SDS led to an apparent molecular weight of 18,000, consistent with the known amino acid sequence of this protein. Explain these data.

SOLUTION

These data show that the polypeptide chain of β-lactoglobulin has a molecular weight of 18,000, and that in high concentration, the protein exists as a dimer of $M_r = 36{,}000$. However, on dilution, the dimer, maintained by reversible, noncovalent interactions, undergoes a partial dissociation: $A_2 \rightleftharpoons 2A$.

4.18. Calculate M_r of a protein, given the following experimental data obtained at 20°C.

$$s = 4.2 \times 10^{-13}\ s$$
$$D = 1.2 \times 10^{-10}\ m^2\ s^{-1}$$
$$\bar{v} = 0.72\ mL\ g^{-1}$$
$$\rho = 0.998\ g\ mL^{-1}$$

SOLUTION

From Eq. (*4.1*)

$$M = \frac{RTs}{D(1 - \bar{v}\rho)}$$

where R (gas constant) = $8.314\ J\ K^{-1}\ mol^{-1}$
T (absolute temperature) = 293 K

$$M = \frac{8.314\ J\ K^{-1}\ mol^{-1} \times 293\ K \times 4.2 \times 10^{-13}\ s}{1.2 \times 10^{-10}\ m^2\ s^{-1} \times (1 - 0.72 \times 0.998)}$$
$$= \frac{10{,}231 \times 10^{-13}\ kg\ m^2\ s^{-2}\ mol^{-1}\ s}{0.338 \times 10^{-10}\ m^2\ s^{-1}}$$
$$= 30\ kg\ mol^{-1} \text{(to two significant figures)} \quad \text{or} \quad 30{,}000\ g\ mol^{-1}$$

The molecular weight is thus 30,000.

Supplementary Problems

4.19. List (*a*) the major proteins of muscle, (*b*) the major protein of skin, connective tissue, and bone, and (*c*) the major protein of hair and feathers.

4.20. A pure heme protein was found to contain 0.326 percent iron. If the molecule contains only one iron atom, what is its molecular weight?

4.21. In what direction (toward the anode, toward the cathode, or toward neither) will the following proteins move in an electric field?

(*a*) Serum albumin (I.P. = 4.9) at pH 8.0

(*b*) Urease (I.P. = 5.0) at pH 3.0; pH 9.0

(*c*) Ribonuclease (I.P. = 9.5) at pH 4.5; pH 9.5; pH 11

(*d*) Pepsin (I.P. = 1.0) at pH 3.5; pH 7.0; pH 9.5

4.22. In what order would the following globular proteins emerge on gel filtration of a mixture on Sephadex G-200: ribonuclease ($M_r = 12{,}000$); aldolase ($M_r = 159{,}000$); hemoglobin ($M_r = 64{,}000$); β-lactoglobulin ($M_r = 36{,}000$); and serum albumin ($M_r = 65{,}000$)?

4.23. Distinguish between the terms *primary*, *secondary*, and *tertiary* structures.

4.24. Of the following amino acid residues—methionine, histidine, arginine, phenylalanine, valine, glutamine, glutamic acid—which would you expect to find on the (*a*) surface of a protein and which would you expect to find (*b*) in the interior?

4.25. What functions would you expect to be served by residues such as (*a*) phenylalanine at the protein surface or (*b*) aspartic acid in the interior?

4.26. What is meant by a *domain* of protein structure? Give examples of domains in real proteins.

4.27. What is meant by the statement that a particular conformation of an amino acid residue lies in an *unfavorable region* of the Ramachandran plot?

4.28. (*a*) Why does urea cause denaturation? (*b*) Why do our kidneys not denature in the presence of urinary urea?

4.29. (*a*) What are the important noncovalent interactions within proteins? (*b*) How do weak interactions result in a stable structure?

4.30. The *pitch* (p) of a helix is defined as $p = dn$, in which n is the number of repeating units per turn and d is the distance along the helix axis per repeating unit. Therefore, the pitch is a measure of the distance from one point on the helix to the corresponding point on the next turn of the helix.

(*a*) What is the pitch of an α helix and the distance per residue?
(*b*) How long would myoglobin be if it were one continuous α helix?
(*c*) How long would myoglobin be if it were one strand of a β sheet?
(*d*) How long would myoglobin be if it were fully extended (distance/residue = 0.36 nm)?

4.31. Predict which of the following polyamino acids will form α helices and which will form no ordered structures in solution at room temperature.

(*a*) Polyleucine, pH = 7.0
(*b*) Polyisoleucine, pH = 7.0
(*c*) Polyarginine, pH = 7.0
(*d*) Polyarginine, pH = 13.0
(*e*) Polyglutamic acid, pH = 1.5
(*f*) Polythreonine, pH = 7.0

4.32. What forces hold protein subunits in a quaternary structure?

4.33. Poly-L-proline can form a single-strand helix that is similar to that of a single strand of the collagen triple helix, but it cannot form a triple helix. Why not?

4.34. Poly(Gly-Pro-Pro) is capable of forming triple helices. Why?

4.35. Compare and contrast the structures of (*a*) insulin, (*b*) hemoglobin, and (*c*) collagen, all of which are proteins consisting of several chains but held together by different types of bonds.

4.36. What are the reasons for the marked stability of an α helix?

4.37. (*a*) In what important ways do the α helix and β structure differ? (*b*) How are they similar?

Chapter 5

Proteins: Supramolecular Structure

5.1 ASSEMBLY OF SUPRAMOLECULAR STRUCTURES

Although proteins are large molecules, they are small compared with a cell or even with structures that may be found within a cell, such as membranes, ribosomes, filaments, enzyme complexes, and viruses. Such structures are formed by the stepwise, noncovalent association of preformed macromolecules, largely proteins, according to the principles discussed in the previous chapter. The processes of assembly of *supramolecular structures* are governed by the same chemical and physical principles that govern protein folding and the association into quaternary structure.

The assembly process is often brought about by the self-association of a number of *identical subunits*, forming a complex structure held together by a multitude of relatively weak noncovalent bonds. This stepwise assembly process has certain advantages: it reduces the amount of genetic information needed to code for a complex structure, and it allows mistakes to be circumvented by the exclusion of faulty subunits or by the dissociation of any spurious aggregations (see Prob. 5.1).

The driving force for the assembly process is primarily *hydrophobic*, but *specificity* is provided by hydrogen and ionic bonds between polar groups in the largely hydrophobic binding sites. Binding occurs between complementary sites that are in van der Waals contact with each other; a projection on one site must be opposite an indentation on the other. Similarly, a positively charged group must be opposite a negatively charged group and a hydrogen-bond donor opposite a hydrogen-bond acceptor. The specificity of complementary binding sites means that the subunits in the aggregate bear a fixed orientation to each other.

In some cases, all the information for assembly is contained within the component molecules; such a process can be termed *self-assembly*.

EXAMPLE 5.1

Ribosomes are large macromolecular complexes, the components of which contain all the information necessary for their self-assembly. The *E. coli* ribosome has a sedimentation coefficient of 70S and consists of two subunits (50S and 30S) with a total mass of 2.8×10^6 daltons and with 58 different components. Three of these components are RNA molecules that together constitute 65 percent of the mass, and they act as a framework (template) for the ordering of the different proteins. When the pure dissociated components are mixed together in the proper order under the correct conditions, they spontaneously reassemble to form a fully active ribosome (Fig. 5-1).

Question: Are all subcellular organelles capable of self-assembly from isolated components?

Many complex organelles are not able to assemble spontaneously from isolated components, but require the provision of additional information for assembly. Mitochondria fall into this category; new mitochondria arise from the growth and division of preexisting mitochondria.

Although certain viruses, notably tobacco mosaic virus (TMV), can be dissociated into their component proteins and nucleic acids and then reassembled into infective virus particles on mixing the components together again, the assembly of other viruses may be far more complex.

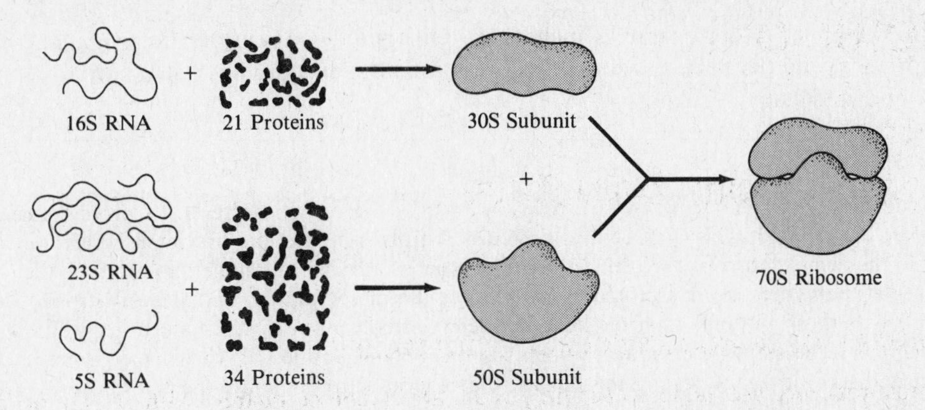

Fig. 5-1 Steps in the assembly of an *E. coli* ribosome.

EXAMPLE 5.2

Tobacco mosaic virus (TMV) consists of a cylindrical coat of 2,130 identical protein subunits enclosing a long RNA molecule of 6,400 nucleotides. Thirty years ago, it was shown that the dissociated coat protein subunits and the RNA would, under appropriate conditions, spontaneously self-assemble to form fully active virus particles. This process is multistage—the critical intermediate being a 34-unit, two-layered disk, which when added to the RNA forms a helical structure with 16.33 protein subunits per turn (Fig. 5-2). In the absence of the RNA, the protein may be polymerized into helical tubes of indefinite length. The presence of the RNA aids the polymerization and results in a virus particle of fixed length.

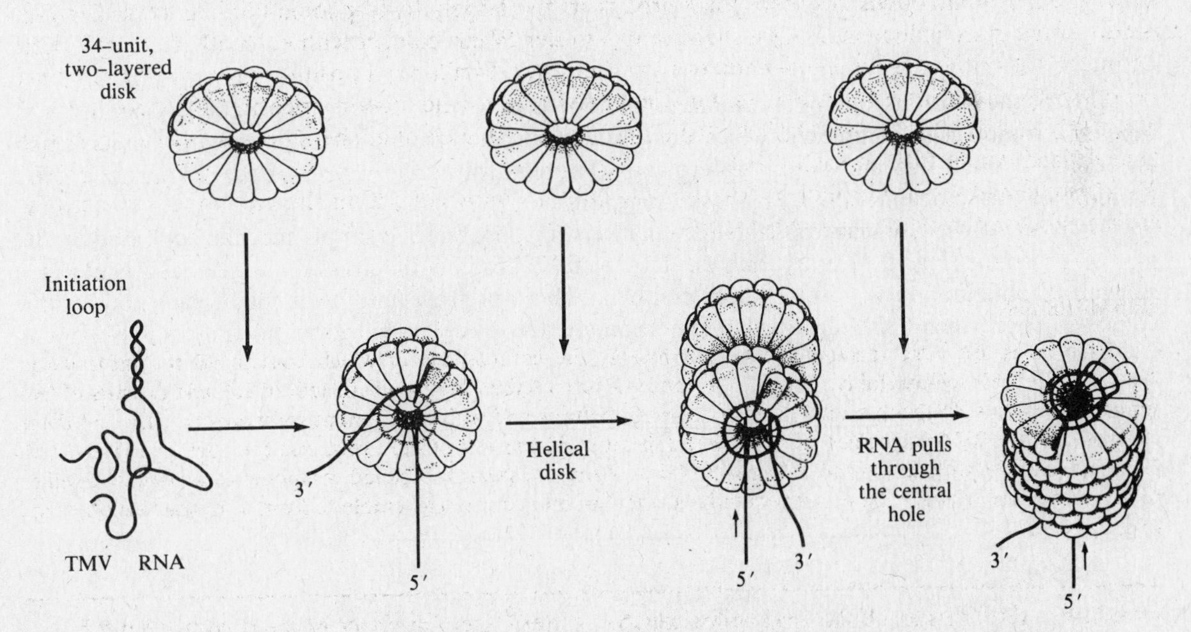

Fig. 5-2 The assembly of tobacco mosaic virus.

The construction of a viral coat from multiple copies of a comparatively small protein represents a huge saving in genetic information. Even when the virus is more complex, with several coat proteins, the saving is still large. For instance, in the poliovirus coat, there are four proteins formed by proteolytic cleavage of a 115,000-dalton precursor, which first aggregates to form a pentamer. Twelve of these cleaved pentamers then associate to form the complete coat. *Processing* by enzymes such as proteases is one example where information from outside the system is needed for the assembly to continue. Other examples include the provision of a *scaffold*, or *template*, to order the assembly process, and the necessity for the information already stored in a structure to be used in the formation of a new copy.

EXAMPLE 5.3

The bacteriophage T4 is a complex virus capable of infecting certain bacteria. Its protein coat (*head*, *tail*, and *tail fibers*; Fig. 5-3) contains 40 structural proteins. A further 13 proteins are required for assembly but do not appear in the completed virus particle: three of these act as a transient template to promote the formation of the tail baseplate; another one is a protease that cleaves the major protein of the head (from 55,000 daltons down to 45,000 daltons), but only after the head has been partially assembled. T4 illustrates well the spatial and temporal control of the stepwise assembly process, as it is only when the tail reaches its correct length that the cap protein is placed on top, allowing the head to become attached. Finally, the tail fibers are added at the opposite end.

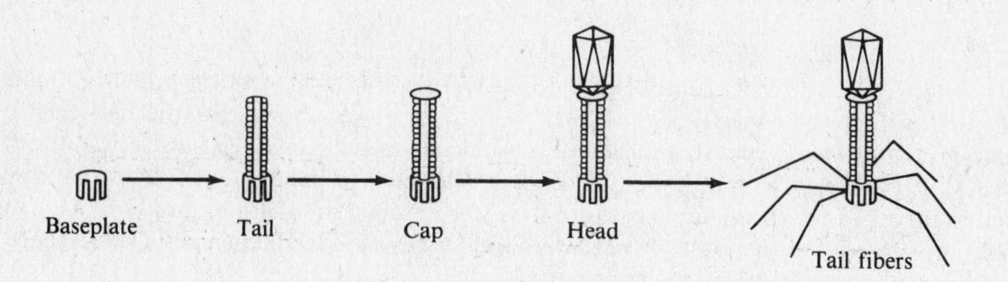

Fig. 5-3 The assembly sequence of bacteriophage T4.

One of the most fascinating questions in biology is how organs, cells, organelles, filaments, and other macromolecular complexes, which are often constructed from a series of repeating units, reach an exactly defined size and then stop growing. The control of the length of the T4 tail is a good example of this problem.

There are obviously many possible explanations for this, but one simple mechanism based on the *vernier principle* (Fig. 5-4) is as follows. Two rodlike proteins of different lengths aggregate in a staggered configuration to form a linear complex. This will grow until the ends of each rod exactly coincide. There then being no overlapping segment, the complex will grow no further.

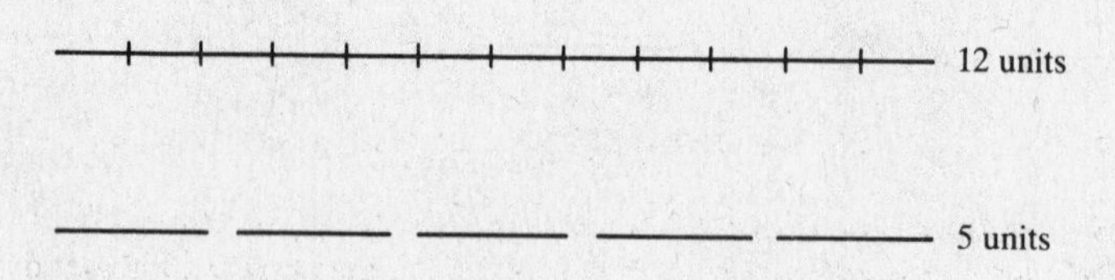

Fig. 5-4 The vernier principle.

5.2 PROTEIN SELF-ASSOCIATION

A large number of biologically active proteins exist in solution as large complexes. The simplest cases of self-assembly are those in which all the molecules are identical. Many proteins possess a quaternary structure in which multiple identical subunits are assembled into geometrically regular structures: e.g., the enzyme aldolase is a tetramer, while the iron-storage protein ferritin has 24 subunits.

Symmetrical Dimers

Question: In what ways can proteins self-associate to form geometrically regular oligomers?

If the binding site is complementary to itself, then a *symmetrical dimer* will be formed. There will be a diad (or twofold rotational) axis of symmetry between the two subunits, such that a rotation of one subunit by 180° about this axis will superimpose it onto the other subunit. The dimer so formed may itself act as a subunit of larger aggregates; e.g., two dimers may associate through a different binding interface to generate a tetramer, with two axes of symmetry. In such cases, where two different axes of symmetry exist, the symmetry is described as *dihedral*.

The term *protomer* is often used to describe the basic unit taking part in a self-association reaction. In the above question, the monomer could be described as the protomer of the dimer. Similarly, since two identical dimers associate to form the tetramer, the dimer could be described as the protomer of the second association reaction (Fig. 5-5).

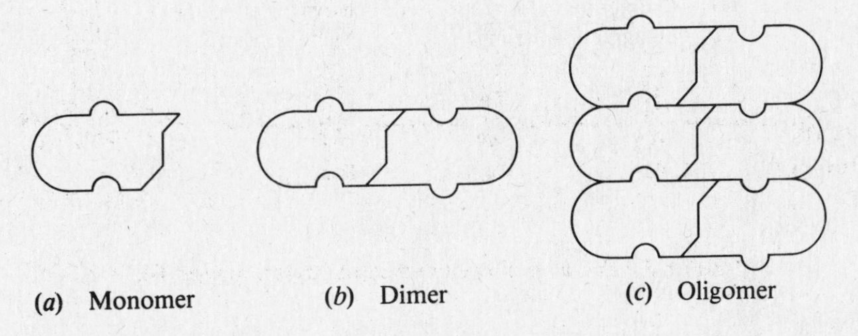

Fig. 5-5 Examples of simple self-association of a protomer.

Rotational Symmetry

If the binding site is complementary to a site elsewhere on the protein, then a *chain* will be formed. For certain angles between the protomers, the chain will close upon itself and form a *ring*. This ring may vary in size from a dimer to high oligomers. Such a regular ring possesses *rotational symmetry*, and this type of symmetry is commonly found in proteins having 3, 5, or other uneven numbers of protomers, although it may also be found among oligomers with an even number of protomers.

EXAMPLE 5.4

The hemagglutinin membrane protein of the influenza virus is a *trimer*. Each protomer is a single polypeptide chain, folded into two segments: an α-helical segment that forms a left-hand *superhelix* (coiled coil) with the two helices from the other two subunits and a globular head region constituting an eight-stranded β-sheet structure, which contains the binding site for receptors on the host cells. The major forces stabilizing the trimer arise from the nonpolar amino acid residues within the triple-stranded coiled coil. The axis of symmetry passes through both regions (Fig. 5-6), producing a threefold cyclic symmetry.

Indefinite Self-Association

In many cases, so long as the binding between the monomers is identical, an open-ended continuous chain in the form of a helix will be produced (Fig. 5-7). Since additional protomers can be added to an oligomer of any size, this process is termed an *indefinite self-association*.

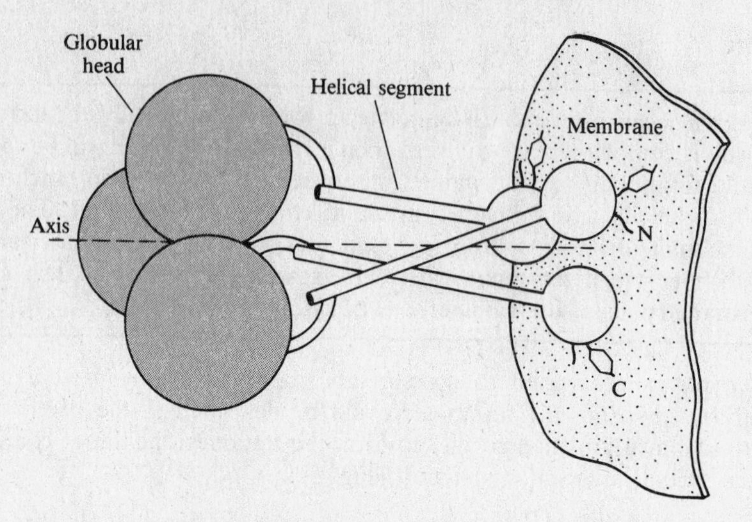

Fig. 5-6 A diagrammatic representation of the influenza virus hemagglutinin trimer.

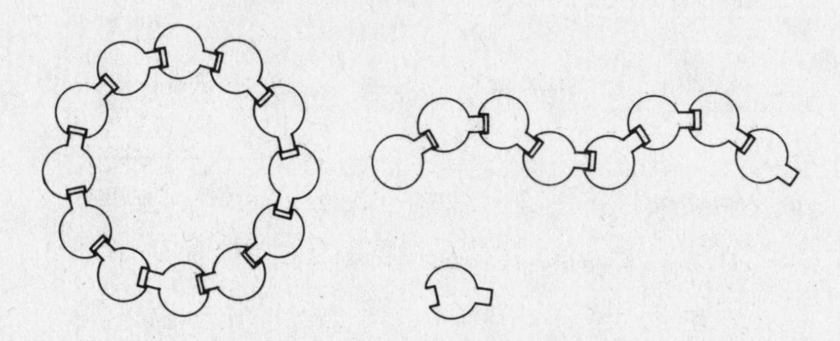

Fig. 5-7 Rings or helices can be formed when protein subunits interact with each other in identical manners.

Sheets and Closed Surfaces

If two helical strands are wound around each other to form a double helix or if succeeding turns of the helix are in contact, there is a considerable increase in stability due to each monomer's interacting with two monomers in the opposite strand as well as with its two neighbors in its own strand. Multiple interaction in the same plane can lead to the formation of *sheets*, in which each monomer interacts with six neighbors in *hexagonal packing*. Sheets can, with a slight readjustment, be converted into cylindrical *tubes* or even into *spheres*. These closed structures provide greater stability since they maximize the number of interactions that can be made; the protein coats of certain viruses are excellent examples of this. (See Fig. 5-8.)

EXAMPLE 5.5

In *tomato bushy stunt virus*, 180 identical protein subunits ($M_r = 41{,}000$) form a shell that surrounds a molecule of RNA containing 4,800 nucleotides. For geometrical reasons, no more than 60 identical subunits can be positioned in a spherical shell in a precisely symmetrical way. This limits the volume; so, in order to accommodate a greater amount of RNA, a larger shell is needed. This can be achieved by *relaxing* the symmetry: the subunits are divided into three sets of 60, each of which packs in strict symmetry. The relationship between each set is different, so that the overall packing is only *quasi-equivalent* (Fig. 5-9).

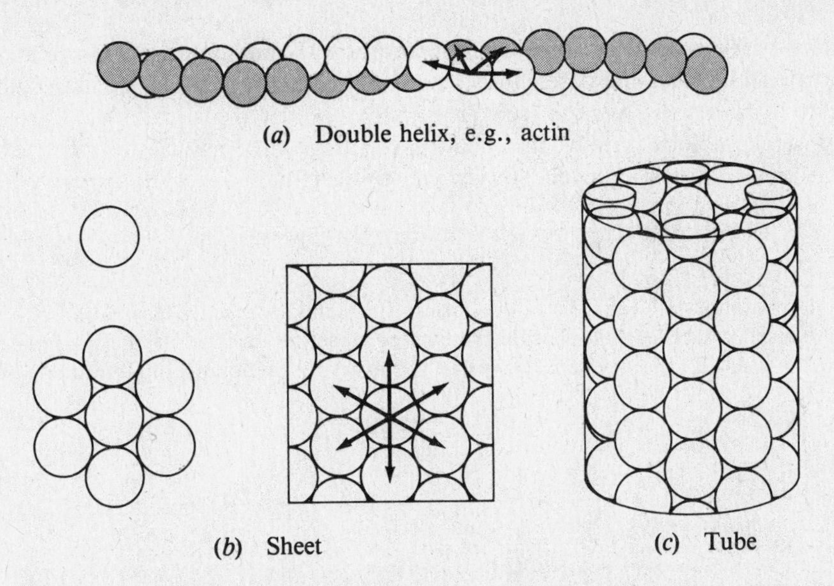

Fig. 5-8 The formation of double helices, sheets, and tubes showing multiple interactions between subunits.

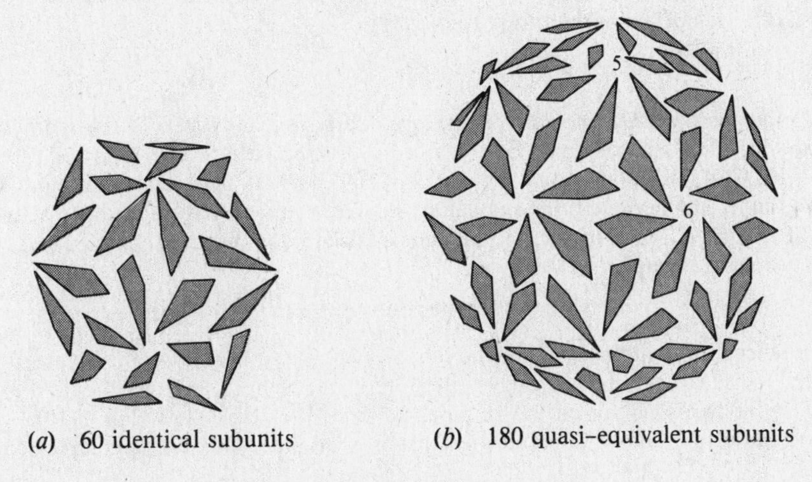

Fig. 5-9 Spherical protein shells.

Even in fairly simple oligomeric structures, all subunits do not necessarily function identically; in crystals of malate dehydrogenase, only one of the two otherwise identical chains in the dimer will bind a molecule of the cofactor NAD^+; the other is altered reversibly in some way that prevents binding.

Equilibria

The protomer association reaction is characterized by an equilibrium constant. For example, the dimerization reaction

$$2A \rightleftharpoons A_2$$

can be characterized by a *dimerization constant*:

$$K = \frac{[A_2]}{[A]^2} \tag{5.1}$$

where the square brackets represent molar concentration. Examination of this relationship reveals that the proportion of dimer increases with the total concentration of the molecule in question; conversely, dilution favors *dissociation* (see Prob. 5.4).

Binding of small molecules may also change the degree of association. For example, if the associated form binds a small molecule (or *ligand*) preferentially, then the presence of that ligand will favor the associated state.

EXAMPLE 5.6

Rabbit muscle phosphorylase can exist in two forms: an essentially *inactive dimer*, phosphorylase *b*, and an *active tetramer*, phosphorylase *a*. On phosphorylation of a serine residue of phosphorylase *b* or on the noncovalent binding of AMP, the enzyme is converted to the active, predominantly tetrameric, form (see Chap. 11):

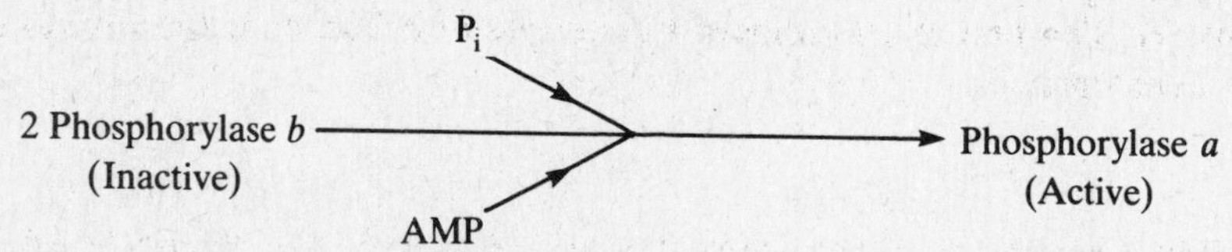

Heterogeneous Self-Association

There are many complexes in which the subunits are different. Frequently, they have different functional roles, often *catalytic*, sometimes *regulatory*.

EXAMPLE 5.7

The enzyme *aspartate transcarbamoylase* catalyzes an early regulated step in the synthesis of pyrimidine nucleotides (Chap. 15). The enzyme from *E. coli* can be dissociated into two kinds of subunits: a *catalytic* subunit, C_3 ($M_r = 100{,}000$), and a *regulatory* subunit, R_2 ($M_r = 34{,}000$). The catalytic subunit is a trimer and is catalytically active, but is not regulated; the regulatory subunit, a dimer, has no catalytic activity but will bind the regulators ATP and CTP. When mixed, the subunits associate, the resulting complex being fully active and regulated by the two nucleotides:

$$2C_3 + 3R_2 \rightleftharpoons R_6C_6$$

X-ray diffraction studies show that the two catalytic subunits lie back to back, with the regulatory subunits fitting into grooves on the outside. The catalytic (active) sites are in the center, near the threefold axis of symmetry, while the regulatory binding sites are on the outside, far from the active sites (Fig. 5-10). The binding of a regulator molecule causes a conformational change that is transmitted across the complex to the active site.

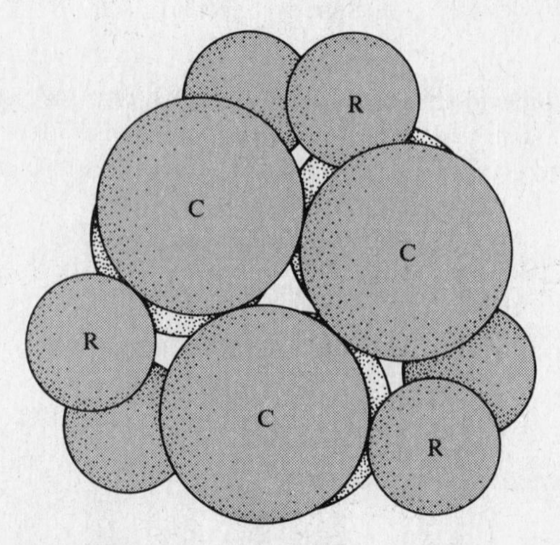

Fig. 5-10 The arrangement of the catalytic (C) and regulatory (R) subunits in aspartate transcarbamoylase.

Rarely, the specificity of an enzyme may be altered by its association with a *modifier* subunit. An example is *lactose synthetase*, the enzyme that catalyzes the linking of glucose and galactose to form lactose. It consists of a catalytic subunit and a modifier subunit. By itself, the catalytic subunit catalyzes the addition of a galactose residue to a carbohydrate chain of a glycoprotein; in the presence of the modifier, the specificity of the catalytic subunit is changed and lactose is synthesized. Another example of altered specificity occurs at the termination of protein synthesis on a ribosome where the peptidyltransferase component of the 50S ribosomal subunit is modified by the binding of a *release factor* (Chap. 17): the specificity is changed so that water becomes the peptide acceptor rather than an amino acid attached to tRNA.

A further catalytic advantage arises when a series of enzymes catalyzing a sequence of reactions in a metabolic pathway is assembled to form a *multienzyme complex*. This assembly increases the efficiency of the pathway in that the *product* of one enzymatic reaction is in place to be the *substrate* of the next (Chap. 9). The limitation imposed by the rate of diffusion of the reactants in solution is therefore largely overcome.

EXAMPLE 5.8

The *pyruvate dehydrogenase complex* catalyzes the conversion of pyruvate to acetyl-CoA. This conversion links the breakdown of carbohydrates to the processes of respiration and oxidative phosphorylation (Chap. 12). The overall reaction is

$$\text{Pyruvate} + \text{CoA} + \text{NAD}^+ \longrightarrow \text{Acetyl-CoA} + \text{CO}_2 + \text{NADH} + \text{H}^+$$

This is the sum of five reactions, catalyzed by three enzymes and requiring five cofactors, including CoA and NAD^+:

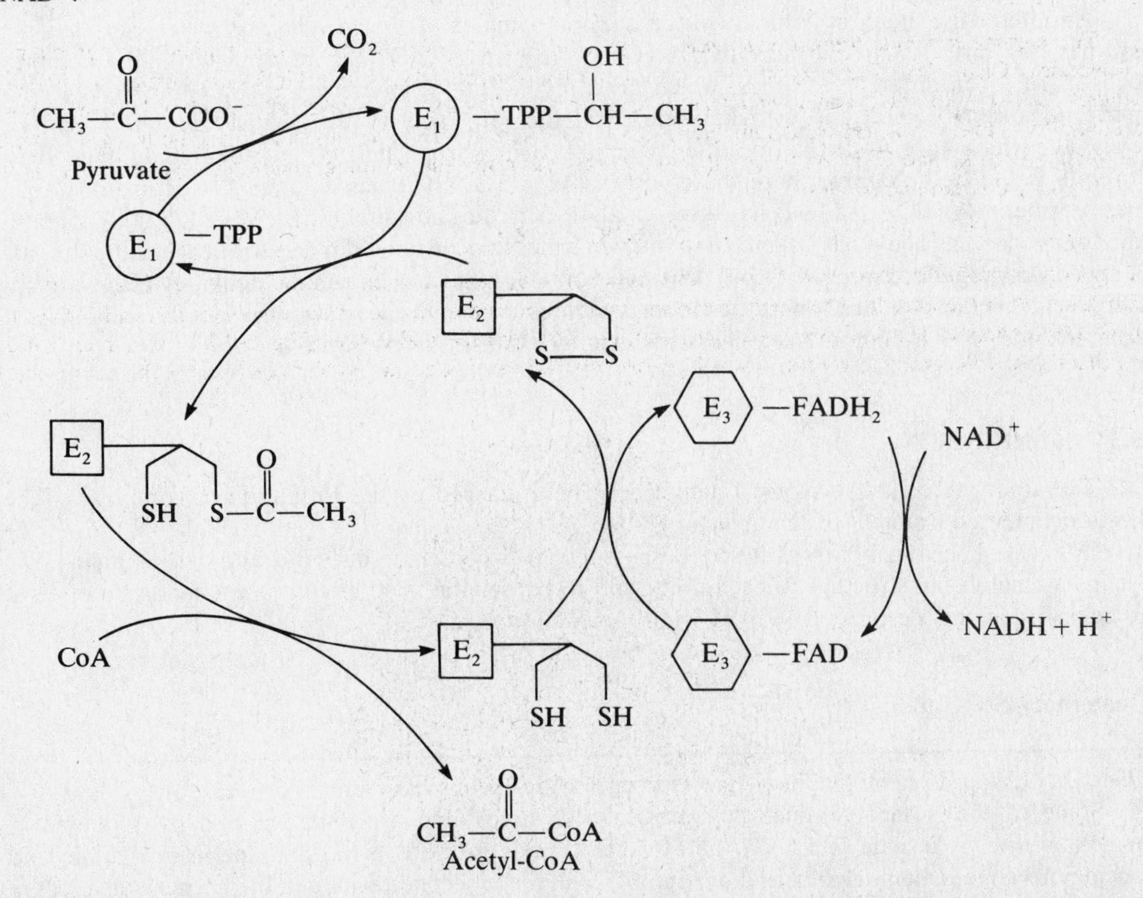

In *E. coli*, the complex has a mass of about 4×10^6 daltons and consists of 60 polypeptide chains of the three different enzymes. In the center of the complex, there is a core of eight trimers of *transacetylase* (E_2)

arranged in cubic symmetry. Dimers of *dihydrolipoyl dehydrogenase* (E_3) are bound on each of the six faces of the cube. Finally, pairs of *pyruvate decarboxylase* molecules (E_1) bind to each edge of the cube, encircling the dehydrogenase dimers (Fig. 5-11). The central position of the transacetylase (E_2) allows the flexible lipoyl groups to transfer reactants from E_1 to E_3 or to coenzyme A (CoA).

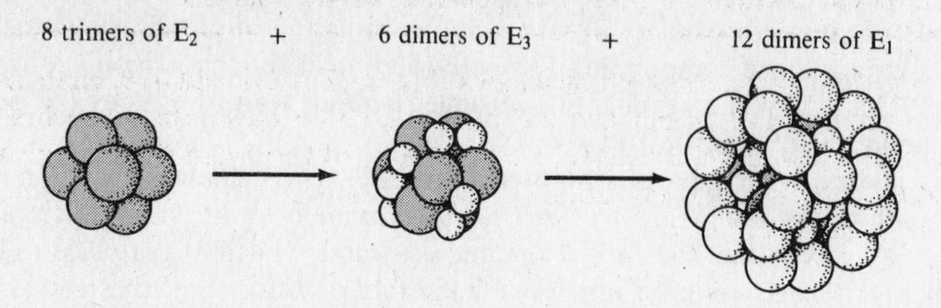

Fig. 5-11 Model of the *E. coli* pyruvate dehydrogenase complex.

The mammalian enzyme complex is even larger, containing almost 200 polypeptide chains, to give a molecular weight of over 7×10^6. In addition to performing the enzymatic activities described above, the mammalian complex contains two other enzymes that act as regulators by catalyzing a phosphorylation and dephosphorylation of the decarboxylase subunit in response to the metabolic demand; i.e., the complex is activated when there is a need for the product acetyl-CoA and inactivated by phosphorylation when there is no need. Thus, this multienzyme complex reduces waste from side reactions or from the unnecessary formation of product.

An interesting situation occurs with the enzyme complex *fatty acid synthase*, which catalyzes the synthesis of fatty acids from acetyl-CoA (Chap. 13). In *E. coli* and most bacteria, the complex consists of seven different enzymes, which probably exist as a multienzyme complex. However, in more advanced bacteria and in eukaryotic cells, the fatty acid synthase exists as *multifunctional enzymes*, which have several different enzymatic activities in a single polypeptide chain. The yeast enzyme is the best example; here the complex ($M_r = 2.3 \times 10^6$) has a subunit structure of A_6B_6, where subunit A ($M_r = 185{,}000$) has three catalytic activities and subunit B ($M_r = 175{,}000$) contains the four remaining activities. Mammalian liver enzymes appear to be dimers, with each subunit being a single polypeptide chain ($M_r = 240{,}000$) and containing all the enzymatic activities necessary for fatty acid synthesis. These multifunctional enzymes have all the advantages of a multienzyme complex and have a built-in equivalent stoichiometry and assembly process.

5.3 HEMOGLOBIN

The ability to control biological activity can be enhanced by the formation of complexes. The best-documented example of this is hemoglobin.

In Chap. 4, hemoglobin was described as a tetramer consisting of two α and two β chains. The chains are similar in structure to each other and to myoglobin. The *globin fold* is made up of eight α-helical segments designated A to H from the N terminus.

Quaternary Structure

Question: Why do hemoglobin chains associate while myoglobin chains do not?

Some of the surface residues that are polar in myoglobin are nonpolar and hydrophobic in hemoglobin; e.g., residue B15 (i.e., the fifteenth residue in the B helix) is lysine in myoglobin and leucine or valine in hemoglobin, and residue FG2 (i.e., the second residue in the corner between the F and G helices) is histidine in myoglobin and leucine in both chains of hemoglobin. In hemoglobin both these residue sites are part of the contact between the α and β chains. Approximately 20 percent of the surface area of the isolated α and β chains is buried in the process of forming the hemoglobin tetramer. These contacts are mainly hydrophobic, but about one-third of the contacts involve polar side chains in hydrogen bonds and electrostatic interactions (sometimes called *salt links*). It is these polar interactions that give specificity to the contacts.

Of the two types of contact between the chains (Fig. 5-12), $\alpha_1\beta_1$ and $\alpha_2\beta_2$ contacts involve the B, G, and H helices and the GH corner and are known as the *packing contacts*, while $\alpha_1\beta_2$ and $\alpha_2\beta_1$ contacts involve the C and G helices and the FG corner and are called the *sliding contacts*, because movement between dimers may occur here.

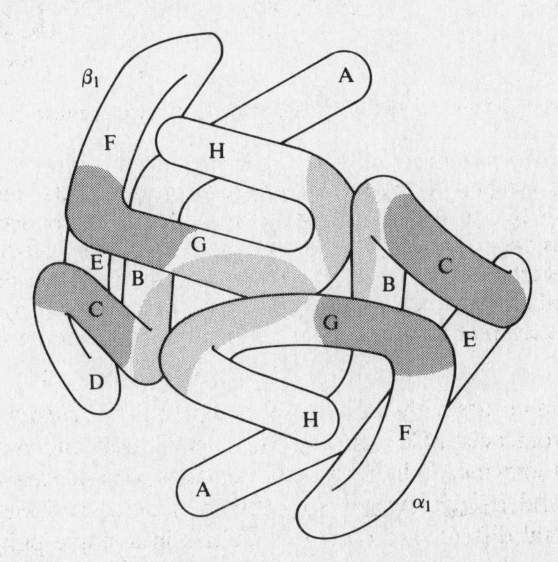

Fig. 5-12 A view of the $\alpha_1\beta_1$ dimer of hemoglobin. The packing contacts shown here in grey hold the dimer together. The sliding contacts (darker shading) form interactions with the $\alpha_2\beta_2$ dimer.

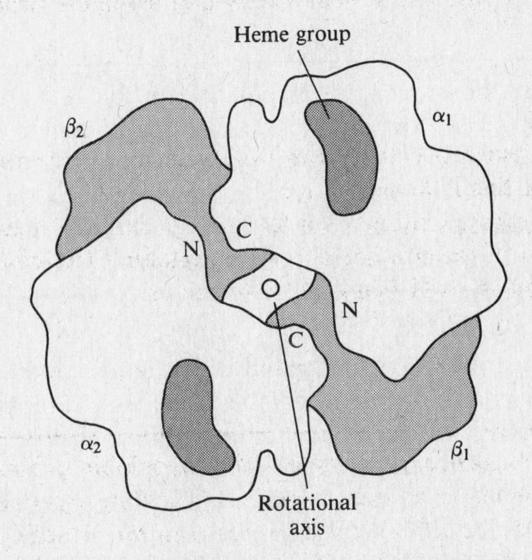

Fig. 5-13 Diagrammatic representation of hemoglobin showing its cyclic symmetry.

The tetramer of hemoglobin may be considered to be a symmetrical molecule made up of two asymmetrical, but identical, *protomers*, the $\alpha_1\beta_1$ and $\alpha_2\beta_2$ dimers. They are related by a twofold rotational or cyclic axis of symmetry; i.e., if one dimer is rotated by 180°, it will superimpose on the other (Fig. 5-13).

On the other hand, if the α and β chains are considered to be identical, then hemoglobin has dihedral symmetry with two rotational axes, and with the four subunits arranged at the apexes of the tetrahedron. When hemoglobin combines with oxygen to form oxyhemoglobin, there is a change in the quaternary structure due to the relative movement of the subunits. The $\alpha_1\beta_1$ dimer rotates by 15° on the $\alpha_2\beta_2$ dimer, sliding upon the $\alpha_1\beta_2$ and $\alpha_2\beta_1$ contacts, and the two β chains come closer together by 0.7 nm.

Allosteric Behavior

DPG

The compound *2,3-diphosphoglycerate* (*DPG*) is produced within the red blood cell of many animal species and acts to modify the oxygen-binding affinity of hemoglobin:

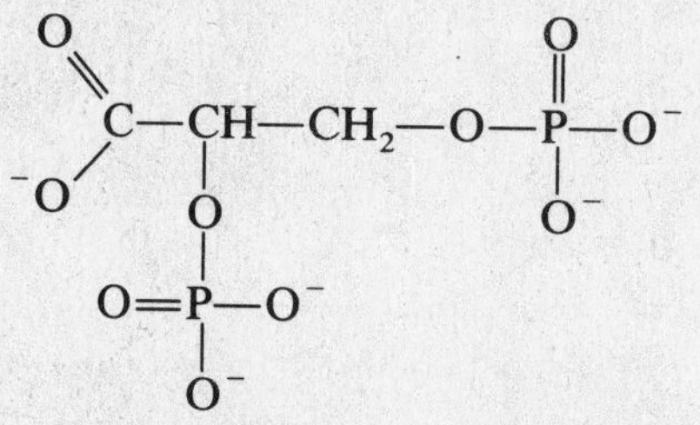

```
 O                    O
  \\                  ||
    C—CH—CH2—O—P—O⁻
  /    |             |
⁻O     O             O⁻
       |
   O=P—O⁻
       |
       O⁻
```

2,3-Diphosphoglycerate (DPG)

This compound binds in the central cavity (the so-called *β-β cleft*) of deoxyhemoglobin, making ionic interactions with cationic groups on the β chains of the hemoglobin molecule (Fig. 5-14). In oxyhemoglobin, the cavity is too small to accept DPG. The binding of oxygen and DPG are thus mutually exclusive, and the effect is that DPG reduces the oxygen affinity of hemoglobin:

$$\text{Hb}(\text{O}_2)_4 + \text{DPG} \rightleftharpoons \text{Hb-DPG} + 4\text{O}_2$$

Question: At high altitudes, the concentration of DPG in the red blood cell can increase by 50 percent (after 2 days at 4,000 m). How does this assist adaptation to a lower partial pressure of oxygen?

As the concentration of DPG increases, the products depicted on the right-hand side of the preceding reaction are favored and more oxygen is released from the blood to the tissues. This compensates for the decreased arterial oxygen concentration.

The change in quaternary structure is associated with changes in the tertiary structure triggered by the movement of the proximal histidine when the heme iron combines with oxygen (Chap. 4). These changes cause the breakage of the constraining salt links between the terminal groups of the four chains. Hemoglobin is thus a structure with a low affinity for oxygen (the so-called *tense structure*), while oxyhemoglobin is a structure with a high affinity for oxygen (the so-called *relaxed structure*). As oxygen successively binds to the four heme groups of the hemoglobin molecule, the oxygen affinity of the remaining heme groups increases. This *cooperative* effect (Chap. 9) produces a sigmoidal oxygen dissociation curve (Fig. 5-15).

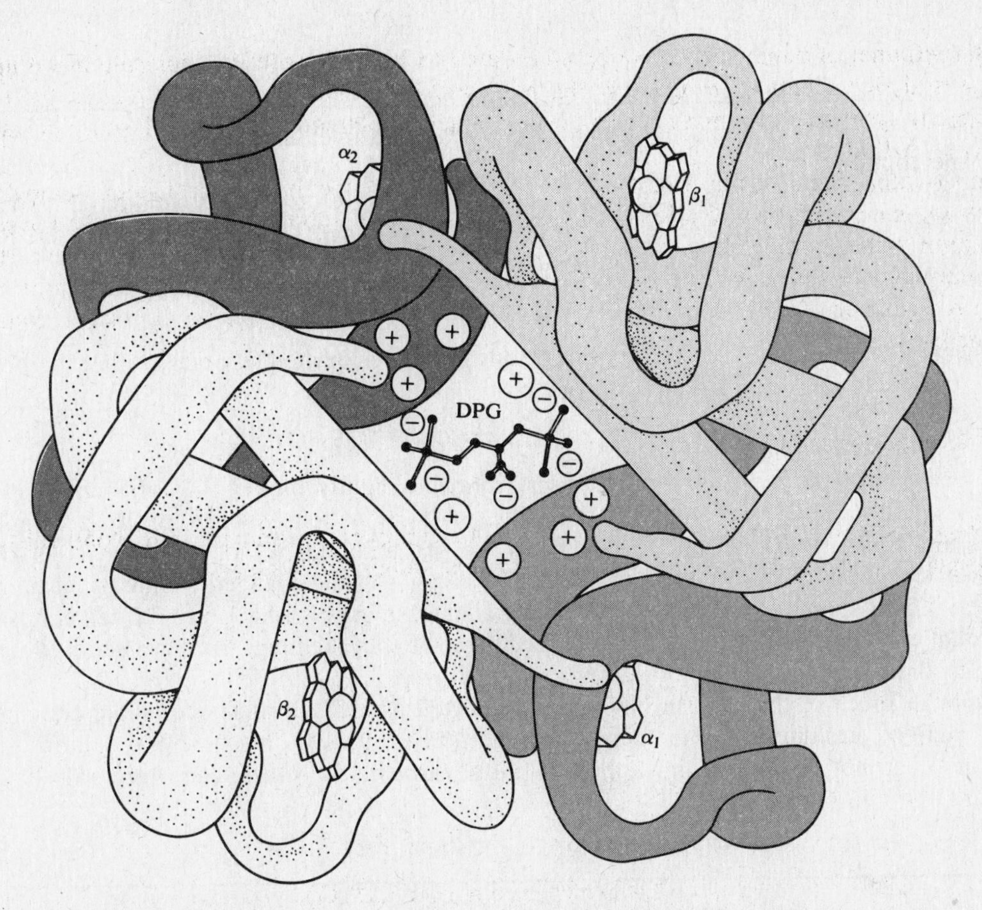

Fig. 5-14 Binding of 2,3-DPG in the β-β cleft of deoxyhemoglobin.

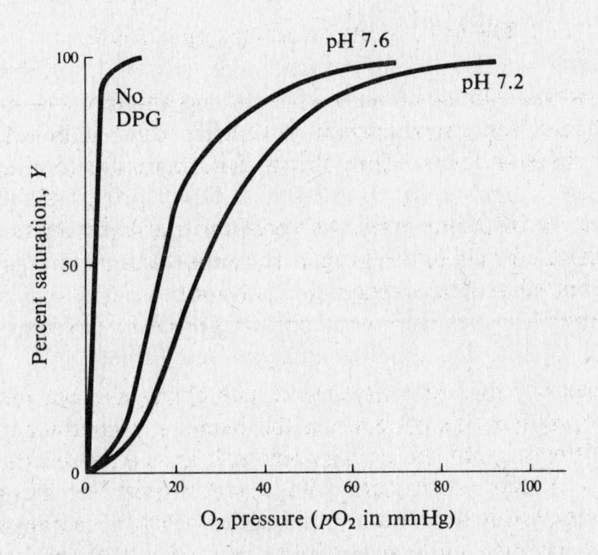

Fig. 5-15 Effect of DPG and pH on the oxygen affinity of hemoglobin.

pH and CO_2

Question: What is the mechanism by which oxyhemoglobin releases more oxygen in tissues that are rapidly metabolizing?

In rapidly metabolizing tissues, there is a fast buildup of CO_2 from the oxidation of fuels such as glucose. This causes an increase in proton concentration (decrease in pH) through the following reaction:

$$H_2O + CO_2 \rightleftharpoons HCO_3^- + H^+$$

Deoxyhemoglobin has a higher affinity for protons than does oxyhemoglobin, so that the binding of protons competes with the binding of oxygen:

$$Hb(O_2)_4 + 2H^+ \rightleftharpoons Hb(H^+)_2 + 4O_2$$

This effect, known as the *Bohr effect*, arises from the slightly higher pK_a of ionizing groups in deoxyhemoglobin. One such group is histidine β146 as a consequence of its proximity to a neighboring aspartate. The higher pK_a reflects the greater tendency of histidine β146 to bind a proton in deoxyhemoglobin. In oxyhemoglobin, as a result of the changed geometry, histidine β146 is free, and has a more normal pK_a value. Thus, a decrease in pH from 7.6 to 7.2 can almost double the amount of oxygen released in the tissues (see Fig. 5-15). In the lungs, the reverse reactions occur, leading to the release of CO_2.

Although much of the CO_2 in blood is transported as HCO_3^-, as above, some combines with hemoglobin and acts directly as an *allosteric effector* (Chap. 9). CO_2 reacts with the un-ionized form of α-amino groups in deoxyhemoglobin to form *carbamates*, which can form salt links, thus stabilizing the tense structure:

$$\text{Hb-NH}_2 + CO_2 \rightleftharpoons \text{Hb(NHCOO}^-) + H^+$$

The interactions between the subunits of hemoglobin allow the release of oxygen to be fine-tuned to the physiological needs. The allosteric effectors DPG, H^+, and CO_2 all lower the affinity of hemoglobin for oxygen by increasing the strength of the subunit interactions.

5.4 THE EXTRACELLULAR MATRIX

The *extracellular matrix* consists of an intricate network of interacting macromolecules that surround, support, and regulate the behavior of cells and tissues. The term *connective tissue* is commonly used to describe this extracellular matrix and the cells, such as *fibroblasts*, located in it.

The extracellular matrix is made up of fibrous proteins, such as *collagen* and *elastin*, embedded in a gel-like ground substance of *proteoglycan* and water. Other proteins, such as *fibronectin*, play a specialized role in binding the matrix components together. In calcified tissue, a *filler* of *hydroxyapatite* crystals, $Ca_{10}(PO_4)_6(OH)_2$ replaces the ground substance to enable high load bearing. Both the amount of connective tissue and the composition of the macromolecular components vary greatly from one organ to another. This gives rise to an enormous diversity of forms and functions (compare the marblelike bones and teeth to the ropelike tendons; see Table 5.1).

Collagen

Structure

The *collagens* are a very widely distributed family of proteins; in mammals they make up over 25 percent of the total body protein. The basic collagen molecule consists of a stiff, inextensible, triple-stranded helix (Chap. 4). Eighteen genetically different individual polypeptide chains have now been described, and these combine in triplets to form ten collagen types (Table 5.2). Of these, type I, the most abundant, has been most fully investigated.

Table 5.1. Approximate Composition of Connective Tissues (% of Dry Weight*)

	Hydroxyapatite	Collagen	Elastin	Proteoglycan
Bone	80	19	0	1
Cartilage	0	50	0	50
Tendon	0	90	9	1
Skin	0	50	5	45
Ligament	0	23	75	2

*The water content of bone is low (~10%), but high in the others (~70%).

Table 5.2. Types and Properties of Collagen

Type	Chain Designations and Approximate Triplet Composition	Structural Form	Distribution
I	2 $\alpha1$(I) and 1 $\alpha2$(I); 300 nm	Broad fibrils with 67-nm bands	Most abundant; found in skin, bone, tendon, and cornea
II	3 $\alpha1$(II); 300 nm	Small fibrils with 67-nm bands	Abundant, in cartilage, vitreous humor, and intervertebral disks
III	3 $\alpha1$(III); 300 nm	Small reticulin fibrils with 67-nm bands	Abundant; found in skin, blood vessels, and internal organs
IV	2 $\alpha1$(IV), 1 $\alpha2$(IV); 390 nm	Nonfibrillar network	Found in all basement membranes (a delicate layer under the epithelium of many organs)
V	$\alpha1$(V), $\alpha2$(V), $\alpha3$(V); 300 nm	Small fibers	Widespread in interstitial tissues
VI	$\alpha1$(VI), $\alpha2$(VI), $\alpha3$(VI); 105 nm	Fibrils with 100-nm bands	In most interstitial tissues
VII	3 $\alpha1$(VII); 450 nm	End-to-end dimers that assemble laterally	Anchoring fibrils between basement membrane and stroma
VIII	3 $\alpha1$(VIII)	?	Some endothelial cells
IX	$\alpha1$(IX), $\alpha2$(IX), $\alpha3$(IX)	?	Minor cartilage protein
X	3 $\alpha1$(X); 150 nm	?	In hypertrophic or mineralizing cartilage

Type I collagen has three α chains, with a mass of 95,000 daltons. Each chain contains about 1,000 amino acids, of which the glycine content is one-third and proline one-fifth. (Early investigators denatured collagen to form gelatin and found molecular species of approximately 100,000, 200,000, and 300,000 daltons, which they called α, β, and γ chains. In due course, it was discovered that the β chains were dimers and the γ chains trimers of the α chains, but the nomenclature has remained.)

Question: By what processes do three single α chains line up to form a regular triple helix?

The individual chains are synthesized as *procollagen* α chains, which, with a mass of 150,000 daltons, have additional *extension peptides* at both the amino and carboxyl termini. The amino- and carboxyl-terminal regions from the three α chains each fold to form globular structures, which then interact to guide the formation of the triple helix. Disulfide bonds stabilize the carboxyl-terminal domain (Fig. 5-16).

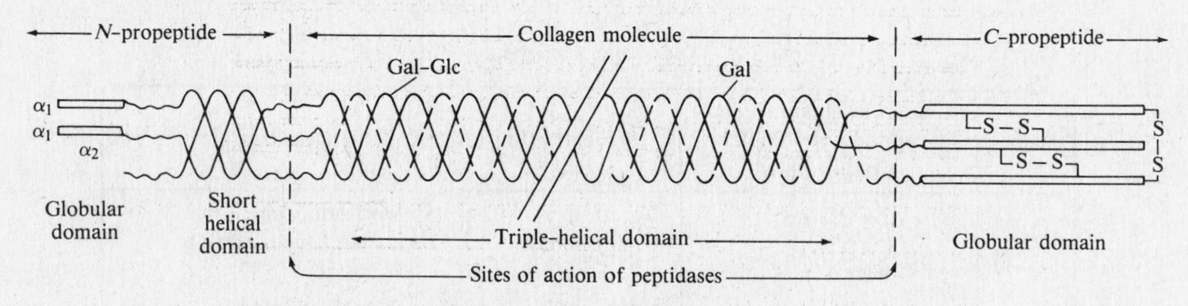

Fig. 5-16 Model of type I procollagen.

Question: Collagen is unusual in that it contains residues of both hydroxyproline and hydroxylysine. As these amino acids cannot be directly incorporated into the polypeptide chain, how are they formed in collagen?

The enzyme *prolylhydroxylase* catalyzes the hydroxylation of proline residues, which lie immediately before the repetitive glycine residues in the pro-α-chains:

$$\text{—Prolylglycyl— + 2-Oxoglutarate} + O_2 \xrightarrow[\text{Ascorbate}]{Fe^{2+}} \text{—Hydroxyprolylglycyl— + Succinate} + CO_2 + H^+$$

The enzyme *lysylhydroxylase* acts in a similar way on lysine residues. Note that the triple-helical portion of collagen consists of $(x\text{-}y\text{-Gly})_n$, where x is often proline and y is commonly hydroxyproline.

Question: In scurvy, caused by a lack of ascorbate (vitamin C), the skin and the blood vessels become extremely fragile. Why is that?

Without ascorbate, hydroxylation of proline and lysine residues does not take place. This has three effects: (1) It prevents the formation of interchain hydrogen bonds involving the hydroxyl groups of hydroxyproline, and thus the triple helix is less stable. (2) It prevents *glycosylation*, which is the addition of galactose units to hydroxylysine residues, followed in most cases by the addition of a glucose to the galactose. (3) It limits the extent of cross-linking of mature collagen molecules (see Fig. 5-19). Most of the nonhydroxylated pro-α-chains are degraded more rapidly within the cell.

Question: Once the procollagen molecule is hydroxylated, glycosylated, and coiled into a triple helix, how is it processed to form higher structures?

The procollagen molecule is secreted from the cell, and the extension propeptides are excised by two specific procollagen peptidases to form the *tropocollagen* molecule. The removal of the peptides ($M_r = 20{,}000$ and 35,000) allows the tropocollagen molecules to self-assemble to form *fibrils*. This assembly is regulated to some extent by the cells and by other extracellular components to produce the wide variety of structures found in collagen fibers.

For electron microscopy, collagen fibers are fixed and stained with heavy-metal reagents; they show cross-striations of dark and light bands every 67 nm (Fig. 5-17). This effect arises from accumulation of the heavy metal in the gap regions, thus producing the dark-stained bands. The individual tropocollagen molecules assemble in such a way that adjacent molecules are staggered, i.e., displaced longitudinally by approximately one-quarter of their length (67 nm) and leaving a gap between the ends of each molecule.

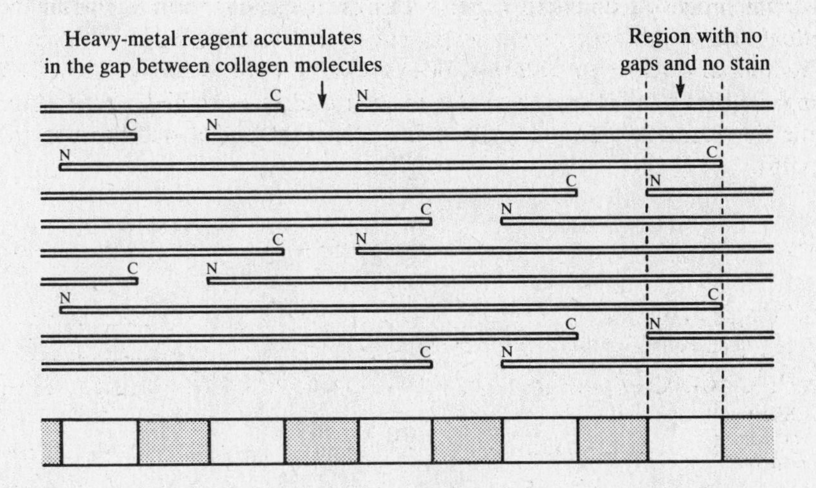

Fig. 5-17 Diagram showing how the staggered arrangement of tropocollagen molecules gives rise to cross-striations in a fibril negatively stained with phosphotungstic acid.

This packing of the tropocollagen molecules with a displacement of 67 nm is caused by repeating clusters of charged and uncharged residues along the polypeptide chains with a periodicity of 67 nm. Hence the maximum number of intermolecular interactions (electrostatic and hydrophobic) is formed when the tropocollagen molecules are displaced by multiples of 67 nm.

However, it is more difficult to explain how the tropocollagen molecules pack three-dimensionally in the quarter-staggered array to produce a cylindrical fibril. One favored model is a pentamicrofibril consisting of groups of five tropocollagen molecules packing together in the 67-nm staggered pattern (Fig. 5-18).

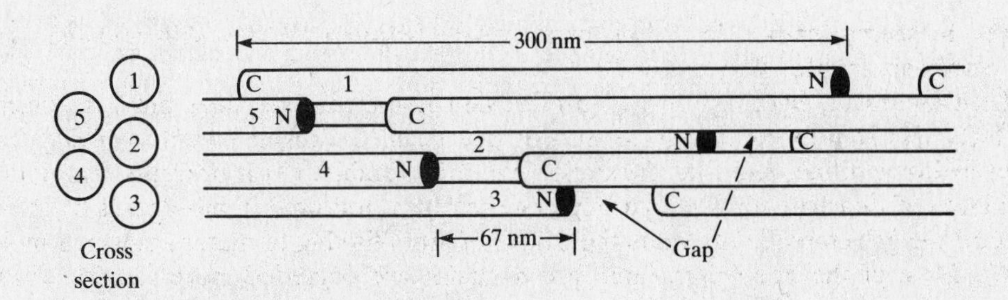

Fig. 5-18 A possible model for the three-dimensional packing arrangement of collagen molecules in a pentamicrofibril.

The size and arrangement of the collagen fibrils vary considerably from tissue to tissue, the diameter ranging from 10 to 300 nm and the packing from being apparently random, as in mammalian skin, to being in strict parallel bundles, as in tendon.

Cross-Linking of Collagen

The collagen fibrils are further strengthened to withstand high tensile forces through the introduction of cross-links. Covalent cross-links are formed both within a tropocollagen molecule and between different molecules. The first step is the enzymatic *oxidative deamination* of the ϵ-amino group of lysine residues to form the aldehyde groups of *allysine* residues. These highly reactive aldehydes then spontaneously react with each other or with lysine residues (Fig. 5-19) to form covalent bonds, which can often react further with other residues such as histidine.

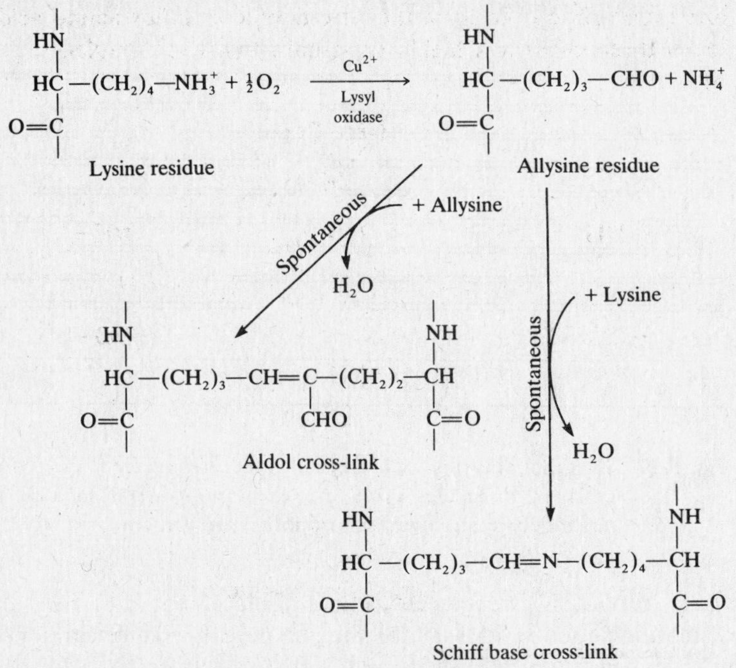

$$\underset{\text{Lysine residue}}{\text{HN}-\text{HC}(\text{O=C})-(\text{CH}_2)_4-\text{NH}_3^+} + \tfrac{1}{2}\text{O}_2 \xrightarrow[\text{Lysyl oxidase}]{\text{Cu}^{2+}} \underset{\text{Allysine residue}}{\text{HN}-\text{HC}(\text{O=C})-(\text{CH}_2)_3-\text{CHO}} + \text{NH}_4^+$$

Allysine residue + Allysine $\xrightarrow{\text{Spontaneous}}$ Aldol cross-link + H_2O

$$\underset{\text{Aldol cross-link}}{\text{HN}-\text{HC}(\text{O=C})-(\text{CH}_2)_3-\text{CH}=\underset{\text{CHO}}{\text{C}}-(\text{CH}_2)_2-\text{CH}(\text{C=O})-\text{NH}}$$

Allysine residue + Lysine $\xrightarrow{\text{Spontaneous}}$ Schiff base cross-link + H_2O

$$\underset{\text{Schiff base cross-link}}{\text{HN}-\text{HC}(\text{O=C})-(\text{CH}_2)_3-\text{CH}=\text{N}-(\text{CH}_2)_4-\text{CH}(\text{C=O})-\text{NH}}$$

Fig. 5-19 Reactions that produce cross-links in collagen.

Collagen types I, II, and III and possibly types V and VI form fibrillar structures. The others all aggregate to form supramolecular structures, but with a variety of forms (Table 5.2).

EXAMPLE 5.9

The secreted form of type IV collagen contains a globular region at the C terminus and a bend in the triple helix near the N terminus. It polymerizes without any further processing and cannot associate laterally, but it still polymerizes to form a two-dimensional sheetlike network. The asymmetric monomers aggregate only through mutual association of their identical ends; the globular regions of two monomers bind together, while four monomers are held together by their triple-helical N termini (Fig. 5-20).

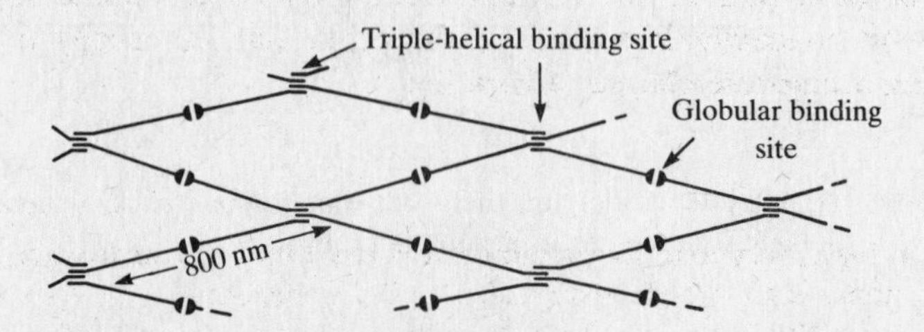

Fig. 5-20 A model of the association of type IV collagen in basement membrane.

Elastin

The other major protein in the extracellular matrix is *elastin*, which is the main component of elastic fibers found in ligaments, large arteries, and lungs. After synthesis and partial hydroxylation of proline residues, a 72,000-dalton molecule of *tropoelastin* is secreted into the matrix. This protein is rich in nonpolar amino acids and contains repeating sequences, such as (Val-Pro-Gly-Val-Gly). These sections form an amorphous, random-coiled structure with frequent *reverse turns*. Other recurrent sequences are rich in alanine with paired lysine residues, e.g., -Ala-Ala-Ala-Ala-*Lys*-Ala-Ala-*Lys*-Ala. The action of lysyl oxidase to produce allysine allows three of these modified residues to condense with one lysine residue to form the heterocyclic complex amino acid *desmosine*, which cross-links two or even three chains. A highly cross-linked network results.

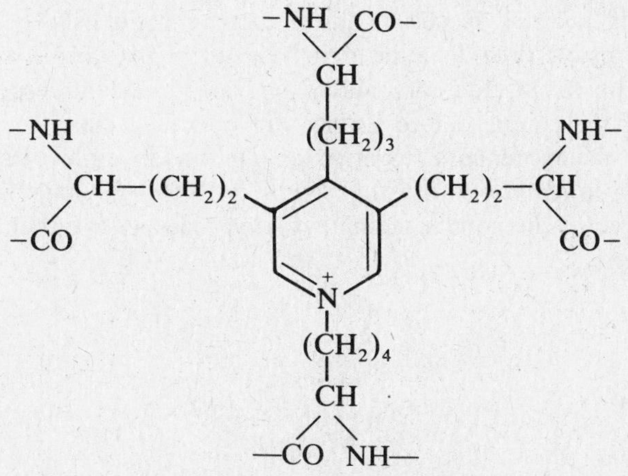

-NH CO-
CH
$(CH_2)_3$
-NH NH-
CH—$(CH_2)_2$ $(CH_2)_2$—CH
-CO CO-
$\overset{+}{N}$
$(CH_2)_4$
CH
—CO NH—

Desmosine residue

EXAMPLE 5.10

Elastin is not a true *rubber* as it is not self-lubricating. It has elastic properties only in the presence of water. At rest, elastin is tightly folded, stabilized by hydrophobic interactions between nonpolar residues; this has been termed an *oiled coil*. On stretching, these hydrophobic interactions are broken, and the nonpolar residues are exposed to water. This conformation is thermodynamically unstable, and once the stretching force is removed, the elastin recoils to its resting state.

Proteoglycans

The ground substance of the extracellular matrix is a highly hydrated gel containing large polyanionic *proteoglycan* molecules, which are about 95 percent polysaccharide and 5 percent protein. The polysaccharide chains are made up of repeating disaccharide units and are called *glycosaminoglycans* because half of the disaccharide is always an amino sugar derivative, either *N*-acetylglucosamine or *N*-acetylgalactosamine; the other half is usually a uronic acid, such as glucuronic acid, giving a negative charge (Chap. 2).

EXAMPLE 5.11

Hyaluronic acid is a single, very long glycosaminoglycan chain having from five hundred to several thousand repeating disaccharide units, e.g., $[\beta\text{-}(1\rightarrow4)\text{-GlcA-}\beta\text{-}(1\rightarrow3)\text{-GlcNAc-}\beta\text{-}(1\rightarrow4)]_n$. (See Fig. 5-21.) These molecules have molecular weights between 0.2×10^6 and 10×10^6 and may exist without being bound covalently to a protein. Because of the carboxyl groups, this substance carries a large negative charge at neutral pH values. It therefore exists as the anionic *hyaluronate* in vivo.

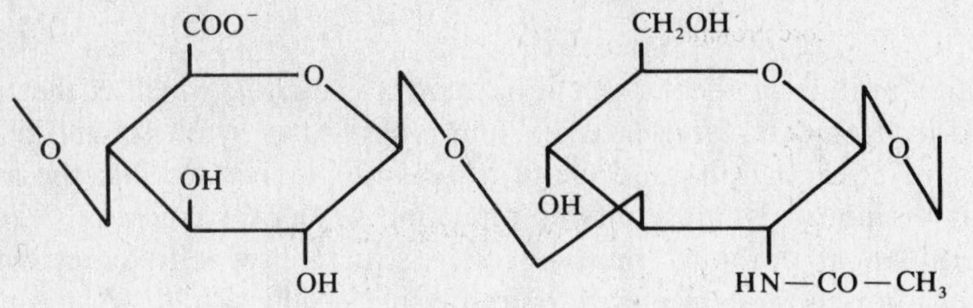

Fig. 5-21 Structural formula of the repeating disaccharide of hyaluronate.

The *core protein* of the proteoglycan is unusually large ($M_r \simeq 300{,}000$), with a globular head ($M_r \simeq 75{,}000$) and a long tail rich in serine and threonine residues. The glycosaminoglycans are covalently attached to these hydroxyl amino acids via an *oligosaccharide linkage* (see Fig. 5-22). During synthesis, the monosaccharides are added one at a time by a specific *glycosyltransferase*; during the elongation of the chain, sulfate groups are added—usually to a hydroxyl group of the amino sugar—to add to the more high negative charge of the molecules. In cartilage, the most abundant glycosaminoglycan is *chondroitin sulfate* ($M_r \simeq 10{,}000$–$30{,}000$), and there may be over a hundred such chains linked to the core protein. *Sulfation* may occur either on the 4′ or 6′ hydroxyl group of the galactosamine.

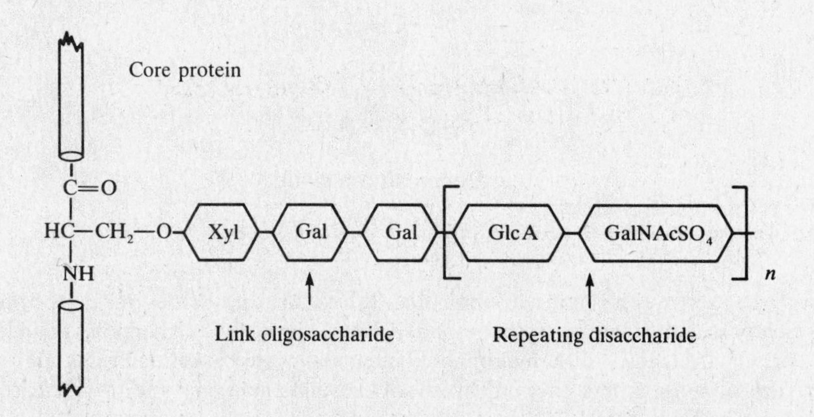

Fig. 5-22 Schematic representation of the linkage between the glycosaminoglycan chondroitin sulfate and a serine residue of the core protein.

EXAMPLE 5.12

As well as chondroitin sulfate, other glycosaminoglycans, such as *keratan sulfate*, with shorter chains ($M_r \simeq 2{,}500$–$5{,}000$) are found in cartilage (Fig. 5-23). N-linked oligosaccharides similar to those found in glycoproteins are also present. Note that keratan sulfate in other tissues, e.g., intervertebral disks or corneas, has a longer chain ($M_r \simeq 10{,}000$–$25{,}000$) and may have a different oligosaccharide link to the core protein.

There is still a further level of assembly, for, in the presence of hyaluronate, most cartilage proteoglycan molecules aggregate to form massive complexes, containing some 50 monomers, to give a total mass around 10^8 daltons. The globular heads of the core proteins interact with segments of five disaccharide units along the length of the hyaluronate. This noncovalent interaction is stabilized by the binding of a *link protein* ($M_r \simeq 50{,}000$) to both the hyaluronate and the core protein (Fig. 5-24).

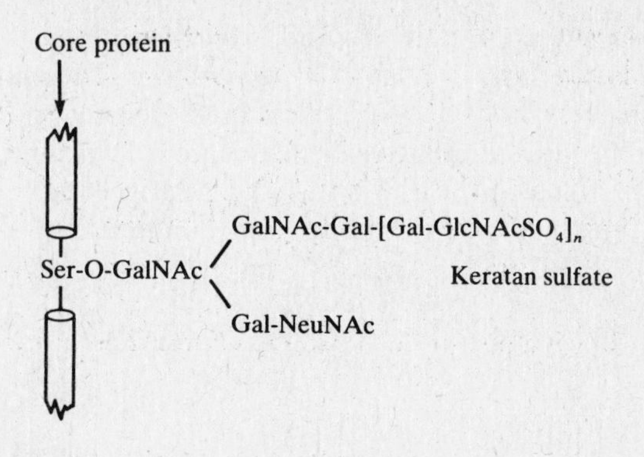

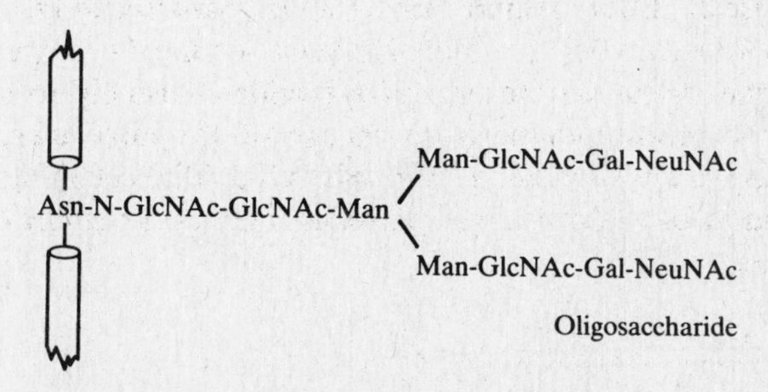

Fig. 5-23 Additional linked carbohydrates found in cartilage.

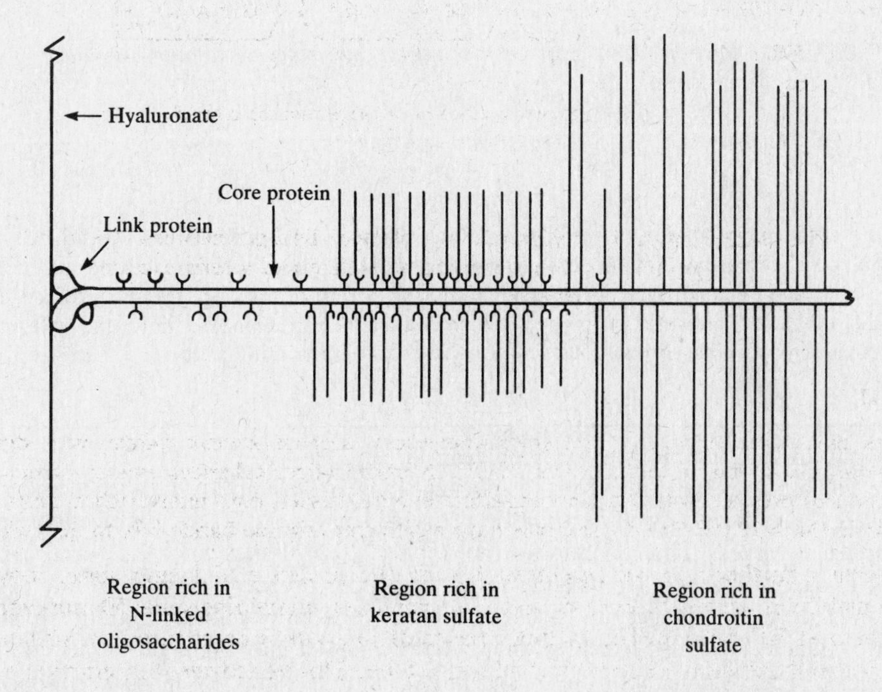

Fig. 5-24 A diagrammatic representation of part of a proteoglycan aggregate from cartilage.

The physical properties of proteoglycans are due almost entirely to the glycosaminoglycan components; the core proteins act largely as *spatial organizers*. The number and length of the glycosaminoglycan chains can vary considerably. The high density of negative charges causes electrostatic repulsion, so that the glycosaminoglycan chains are fully extended and separated to form a *bottlebrush*-like structure (see Fig. 5-25), which occupies a relatively large volume. The hydrophilic groups of the glycosaminoglycans bind and immobilize large numbers of water molecules.

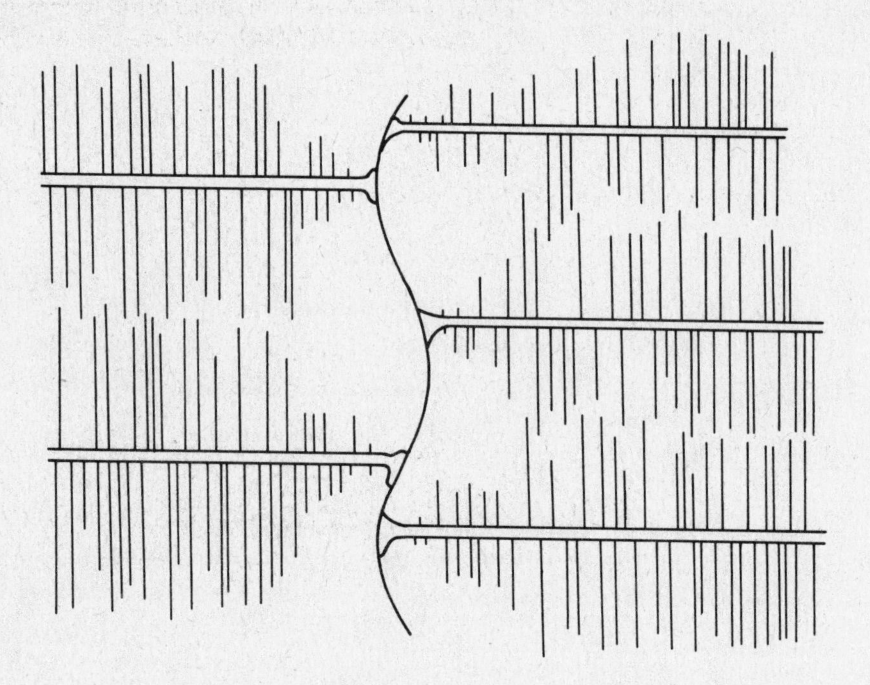

Fig. 5-25 Portion of a proteoglycan aggregate showing bottlebrush-like structure.

EXAMPLE 5.13

Hyaluronate is the most studied of the glycosaminoglycans; it is also the largest. In aqueous solution, it forms a random coil and occupies a huge domain, whose volume is filled with immobilized water and to which small molecules and ions, but not large ones, have access. In a 0.01% solution, the hyaluronate domains will occupy the total volume of the water; at higher concentrations, the hyaluronate molecules will interpenetrate neighboring domains, producing very viscous solutions that have lubricating properties, such as are needed in joints.

Question: What are the functions of proteoglycans?

The large volume-to-mass ratio of proteoglycans, due to their ability to attract and retain large amounts of water, produces a swelling osmotic pressure, or turgor, in the extracellular matrix that resists compressive forces. This is demonstrated clearly in tissues such as joint cartilage and intervertebral disks. In *degenerative diseases* (e.g., arthritis), proteoglycan is partially depleted and disaggregated, leading to changes in tissue resilience. The polyanionic bottlebrush structure of proteoglycans produces a sieving effect, so that the diffusion of macromolecules through connective tissue is restricted while small molecules, especially if anionic, may even show *enhanced rates* of diffusion.

The space-filling character of proteoglycans appears to be important in morphogenesis, particularly in the development of the skeleton. During these developmental processes, the presence of hyaluronate appears to facilitate the migration of cells. This effect is stopped by the removal of hyaluronate by hyaluronidase and by its replacement with aggregating proteoglycans.

Fibronectin

Of the several other protein components of the extracellular matrix, the best understood is *fibronectin*. This glycoprotein is a *heterodimer* composed of two very similar, but not identical, disulfide-bonded polypeptide chains ($M_r \simeq 220{,}000$). It is found in several forms: as a bound complex on the surface of cells such as fibroblasts, as large aggregates in the extracellular space, and in modified form as the so-called *cold-insoluble globulin* (i.e., it readily precipitates at 0°C) in plasma.

Fibronectin is a *multifunctional* protein consisting of a series of globular domains connected by flexible segments that are sensitive to proteolytic cleavage. These domains carry binding sites for components of the extracellular matrix and for cell surfaces (see Fig. 5-26). Fibronectin can thus form multiple interactions that are important in tissue homeostasis. For example, sulfated glycosaminoglycans increase the rate of binding of fibronectin to collagen, thus increasing the stability of the complex and allowing it to precipitate, forming large aggregates in the matrix.

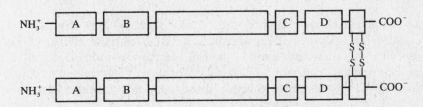

Fig. 5-26 Diagrammatic representation of the domains of fibronectin:

Domain A binds to heparin, fibrin, bacteria (staphylococci).
Domain B binds to collagen and fibronectin.
Domain C binds to cell surfaces.
Domain D binds to heparin.

EXAMPLE 5.14

An important function of fibronectin is that of cell *adhesion*. Fibronectin forms a network that connects the cell to other components of the extracellular matrix, especially collagen, anchoring the cell in position. The connection to the cell is through a membrane protein that connects with the cytoskeleton. *Transformed* cells produce less fibronectin and fail to adhere. The addition of fibronectin to cultures of such cells causes adherence, and the cells change to a more normal appearance. *Metastasizing* cancerous cells lack fibronectin or cause its proteolytic breakdown.

Structural studies have indicated that fibronectin contains at least three types of internal amino acid sequence homologies that have not yet been found in other known protein sequences. Two other *attachment glycoproteins* have been described recently, and their properties are compared with fibronectin in Table 5.3.

5.5 CYTOSKELETON

The complex network of protein filaments found in the cytoplasm of eukaryotic cells is called the *cytoskeleton*. It organizes the cell contents so that the organelles and subcellular particles are oriented and directed in their patterns of movement. It also enables the eukaryotic cell to change shape and in many cases to move from one position to another.

Table 5.3. Comparison of Attachment Proteins

Property	Fibronectin	Laminin	Chondronectin
Molecular weight	440,000	~900,000	~170,000
Subunits	2 × 220,000	2 or 3 × ~200,000 1 × ~400,000	3 × ~55,000
Carbohydrate content	5–9%	12–15%	~8%
Shape	Extended V shape	Cross shape; 3 short arms, 1 long arm	Compact
Tissue distribution	Fibrous connective tissue	Basement membranes	Cartilage, vitreous body
Collagen-binding	Types I–V; best III and I	Type IV	Type II
Glycosaminoglycan-binding	Heparin, heparan sulfate	Heparan sulfate, heparin	Chondroitin sulfate, heparin
Cell-binding	Fibroblasts	Epithelial and endothelial cells	Chondrocytes

Three types of filaments are found in the cytoskeleton; the two most important are *actin filaments* (also called *microfilaments*) and *microtubules*. Both of these consist of globular protein subunits that assemble and disassemble within the cell and are evolutionarily highly conserved. The third type, the *intermediate filaments*, are so called because they are intermediate in diameter between the other two. These filaments, generally more stable than the other two types, are made of fibrous subunits and show greater phylogenetic variation.

The subunits of these three types of filaments have several features in common. They all form *polar* helical filaments, with chemically distinct heads and tails. Helical structures possess multiple and equivalent binding sites, so the filaments can exist in several polymeric forms and can bind repeatedly and specifically with a variety of other proteins. Calcium ions play a highly significant role in the behavior of the cytoskeleton, related possibly to the high content of acidic amino acids and to the presence of many phosphorylated residues.

In addition to these three major types of protein filaments, the cytoskeleton contains a large number of *auxiliary* proteins; these are involved either in the formation of the filaments or with the linkage between the filaments themselves or between the filaments and other cell components such as the plasma membrane. Two types of complex cytoskeletal assemblies are unusually ordered and stable and thus have been studied extensively. One of these is that of the *myofibrils*, which produce muscle contraction and are based on actin filaments; the other is the collection of *cilia*, which beat rhythmically and are based on microtubules. Both these movements obtain their energy from the hydrolysis of ATP and depend on the sliding of one protein filament relative to another.

Actin Filaments

Actin is widely distributed in eukaryotic cells, often being the most abundant protein, commonly making up about 10 percent of the total cell protein. Actin is an ancient, highly conserved protein; e.g., there are only 17 differences out of 375 amino acids between slime mold actin and rabbit muscle actin. The actin monomer ($M_r = 41,800$) is a single polypeptide, often known as globular actin or *G actin*, which contains binding sites for divalent cations (Ca^{2+} or Mg^{2+}) and for nucleotides (ATP). Under physiological conditions, there is an equilibrium between G actin and its polymer, filamentous actin, or *F actin*. Under certain conditions, the rate of polymerization may be extremely rapid, leading to a sudden change of cellular shape and the production of thin projections or *filopodia*, which may grow at the rate of 1 μm per second.

EXAMPLE 5.15

Platelets are small cellular fragments in the blood that lack a nucleus and ribosomes. In response to an injury, platelets rapidly change from a disk shape to one with numerous thin projections that are involved in the formation of the platelet plug and the subsequent blood clot. The rapid polymerization of G actin from a cellular pool to form a large number of actin filaments produces these projections.

F actin consists of double strands of G actin (4 nm in diameter) that are twisted into a double helix of diameter 7 nm with 13.5 molecules per turn; this gives a helical repetition of 36 nm (cf. Fig. 5-8). During polymerization, hydrolysis of ATP occurs:

$$n\text{G actin} \cdot \text{ATP} \longrightarrow \text{F actin} \cdot (\text{ADP})_n + n\text{P}_\text{i}$$

This hydrolysis is not absolutely necessary, as polymerization proceeds equally well in the presence of nonhydrolyzable analogs of ATP. The asymmetry of the actin monomers leads to the filament having a direction, or *polarity*. The *plus* or barbed end of the filament is the fast-assembly end with a lower *critical concentration* of monomer. The *minus* or pointed end is slower growing and requires a higher critical concentration. (The critical concentration is that concentration of monomer at which addition of monomers just balances dissociation; above the critical concentration, net association occurs; below, there is net dissociation.)

Question: How can the polarity of an actin filament be determined?

The protein *myosin* or its fragments, *heavy meromyosin* and S_1 (Fig. 5-32), interact specifically with each actin molecule in a filament, which becomes "decorated" with a pattern of arrowheads, all pointing the same way.

In a cell, several distinct processes may be occurring simultaneously: association of monomers to form trimers, which may act as *nucleation sites*; addition of monomers at either end; dissociation at either end; breakage of the filament due to thermal agitation; and *annealing* (or joining together) of the filaments (Fig. 5-27).

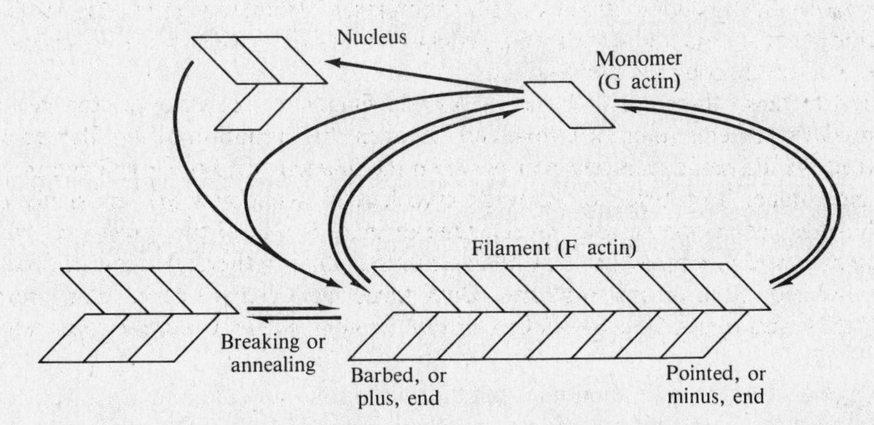

Fig. 5-27 Diagrammatic representation of the assembly and disassembly of actin filaments.

Owing to the *polarity*, the opposite ends of the actin filament grow and dissociate at very different rates. If the concentration of actin monomer is between the critical concentrations of the two ends, a *steady state* may be produced in which G actin molecules largely dissociate from the minus end and assemble at the plus end. This phenomenon is called *treadmilling* and has the effect of moving the actin filament longitudinally in space from the minus to the plus end. The energy required for this process is provided by the hydrolysis of ATP, which occurs on polymerization. It should be noted that, in vivo, normally one or both ends would be blocked by specific binding proteins.

Actin filaments in the cell often form cross-linked bundles that serve two main functions: they provide mechanical support for various cellular structures, and together with myosin, they form various contractile systems. In both these cases, it is the role of the actin-binding protein that becomes significant to the function.

EXAMPLE 5.16

On the surface of the hair cells of the inner ear are found large numbers of specialized microvilli called *stercocilia*. These are rigid structures that sensitively respond to sound frequencies by making tiny movements about the taper point at the surface of the cell. In stercocilia, hundreds of cross-linked actin filaments are found packed longitudinally in strict alignment. The cross-linking proteins bind to precise positions so that the crossover points of the actin helices all lie in the same plane (Fig. 5-28).

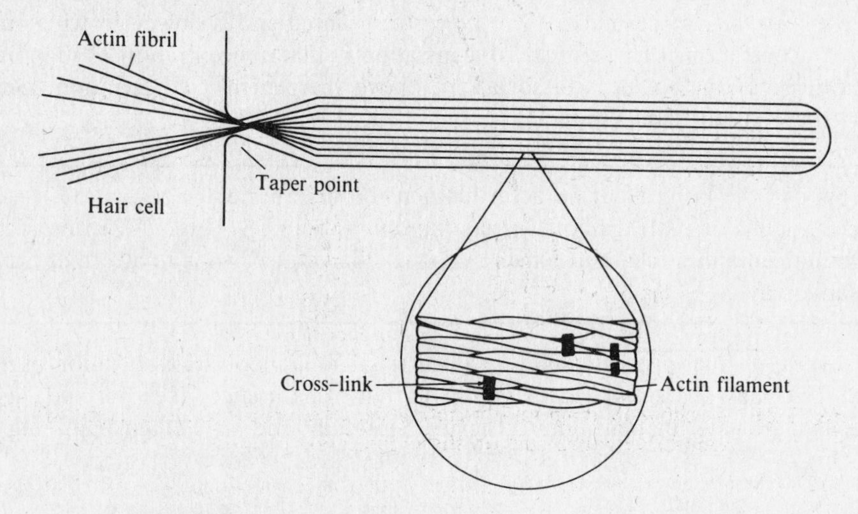

Fig. 5-28 Diagram of a stercocilium with the expanded section showing actin helices in register.

Actin-Binding Proteins

A multitude of actin-binding proteins have been described, but they can be conveniently divided into five groups according to their sites of binding. They also vary in their sensitivity to calcium. The effects of the various types of actin-binding proteins on the organization of actin are illustrated in Fig. 5-29.

Monomer Binding

Proteins that bind to the actin monomer inhibit nucleation by weakening the interaction between monomers. In most cases, this retards the polymerization process. Thus the calcium-sensitive protein *fragmin*, when bound to G actin, suppresses the elongation of the filament.

EXAMPLE 5.17

In the disk-shaped, inactivated form of platelets, the protein *profilin* ($M_r = 16{,}000$) forms a rapidly reversible one-to-one complex by binding to a large portion of the G actin molecules. This has the effect of keeping the concentration of the *free* actin constant and low. Any mechanism that could inhibit the actin-profilin interaction would produce a large increase in the concentration of monomer, triggering nucleation and the whole polymerization process (see Example 5.15). A sharp change in Ca^{2+} concentration is one possible trigger.

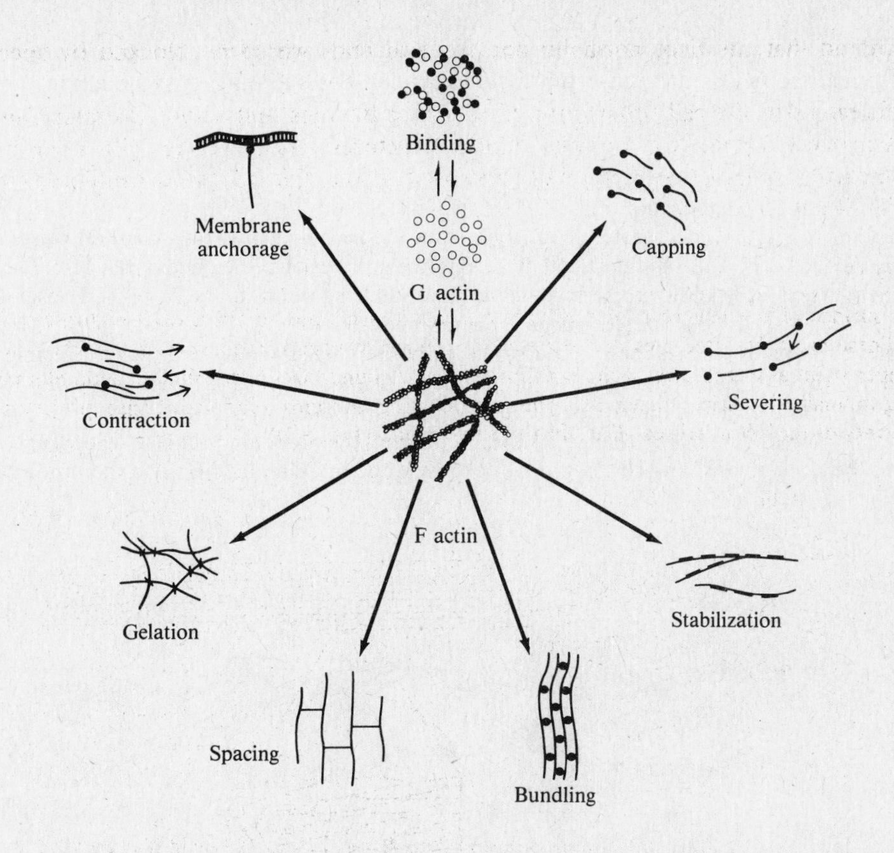

Fig. 5-29 A schematic diagram illustrating the effect of the different types of actin-binding proteins on the organization of actin.

Binding to the Plus End of the Filament

The plus, barbed, or fast-assembly end of actin filaments can be blocked by the *capping* proteins. In some cases, they act as fragmenting proteins before capping. The addition of one such protein for every 400 or more actin monomers causes a marked drop in the viscosity as the gelled actin filament solution undergoes a gel-sol (discontinuous phase) transition. The best known of these proteins is *gelsolin*, which inhibits elongation and annealing but promotes nucleation. The overall effect is to shorten the filament length when added to a gel of long filaments, but to promote polymerization when added to a solution of monomer. Gelsolin and similar proteins will bind to actin only when the concentration of Ca^{2+} is above 10^{-6} M; such a concentration occurs only briefly in response to some change in the environment of the cell.

Binding to the Minus End of the Filament

The effect of a protein binding to the minus, pointed, or slow-assembly end of an actin filament is indistinguishable from binding to the plus end, except under special circumstances. Only one such protein has been described so far, namely, *acumentin* from macrophages.

Binding to the Side of the Filament

There are some proteins that bind to the side of an actin filament without interacting with other filaments. These proteins may either stabilize or destabilize the filament. *Villin*, a major protein of the microvilli of the intestinal brush border, is a calcium-sensitive capping protein that will also fragment (sever) the filament by inserting itself between the subunits in the filament and forming a cap on the new positive end.

EXAMPLE 5.18

Tropomyosin is a rod-shaped protein consisting of two α-helical chains ($M_r = 35{,}000$) twisted around each other to form a coiled coil. This rigid molecule lies along the grooves on either side of the actin filament, thereby stabilizing the filament. It is known to be involved in the stabilizing of the *stress fibers*, first observed in cultured cells, that lie parallel to the plasma (cell-surface) membrane.

The role of tropomyosin is best understood in vertebrate skeletal muscle. In resting muscle, the tropomyosin rod is held out of the groove of the actin filament by the *troponin complex* ($M_r = 18{,}000 + 21{,}000 + 37{,}000$); this blocks the interaction of actin with the heads of myosin molecules (Fig. 5-30). When the level of Ca^{2+} is raised, there is a change in the binding of the troponin complex with Ca^{2+} that causes a movement of the tropomyosin toward the groove, exposing the myosin-binding site and allowing interaction to occur.

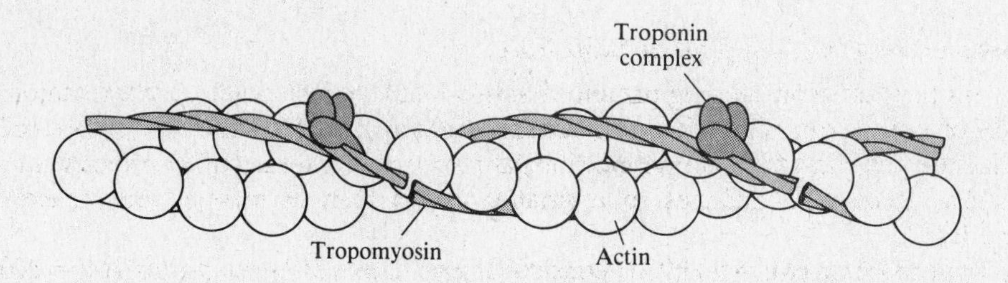

Fig. 5-30 Model of the interaction between actin, tropomyosin, and the troponin complex in resting muscle.

Filament Cross-Linking

The proteins that form cross-links between actin filaments can be divided into three groups. *Fimbrin*, a compact monomer, is an example of a protein that links parallel actin filaments into dense bundles. Another group consists of rodlike spacing proteins, which form bridges, separated by distances of ~200 nm, between parallel actin filaments. The third group of proteins consists of long, flexible molecules, such as *filamin* ($M_r = 250{,}000$), with an extensive binding site for actin filaments. In solution, filamin forms dimers, which form flexible links between filaments; the links are close together but at any orientation to each other. The addition of a relatively small number of filamin molecules to a solution of F actin can produce a solid gel.

EXAMPLE 5.19

The cytoskeleton of red blood cells provides the strength and flexibility to allow their passage through the extremely fine capillaries. The main protein of the cytoskeleton is *spectrin*, an actin-binding protein that cross-links very short lengths of actin, forming a two-dimensional mesh. The spectrin protomer is a heterodimer consisting of two large polypeptides ($M_r = 220{,}000\beta$ and $240{,}000\alpha$) that lie side by side and then associate head to head to form the functionally important linear tetramer. The tails of several tetramers interact with the sides of each small F actin oligomer, consisting of about a dozen G actin monomers. This meshwork is held in position by the spectrin dimer's being bound to the protein *ankyrin* ($M_r = 200{,}000$), which is itself attached to the *anion-channel protein* within the membrane of the red blood cell. In addition, a protein known as *band 4.1* is thought to bind to the spectrin-actin complex, increasing the strength of the spectrin-actin interaction and providing an additional link with the membrane. Some idea of the interactions within the cytoskeleton may be seen in Fig. 5-31.

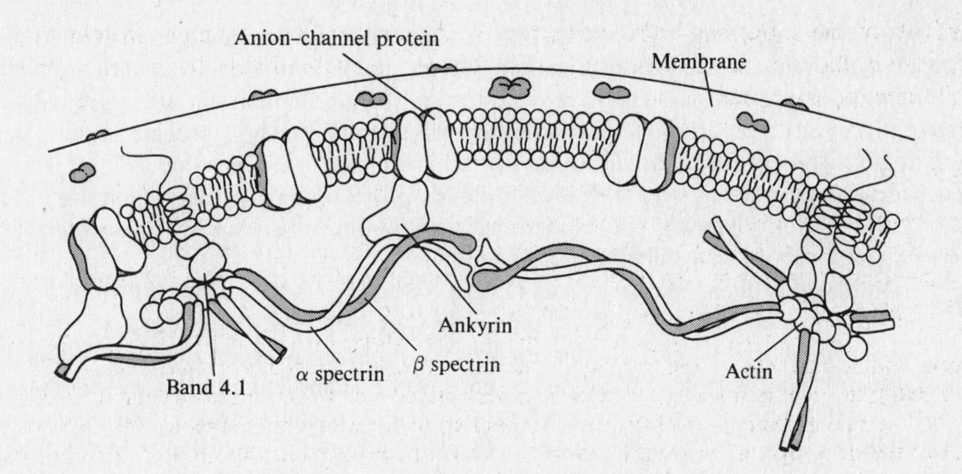

Filament Cross-Linking Producing Mechanical Force

Myosin was the first actin-binding protein described and characterized. It is the major protein of the myofibrils of muscle but is also found in smaller amounts in many other vertebrate cell types. In all cases it has the potential to generate mechanical force when interacting with F actin; however, in nonmuscle cells, myosin polymerizes to a smaller extent than in muscle, and hence the forces produced are less.

The myosin protomer ($M_r = 500{,}000$) consists of two identical *heavy chains* ($M_r = 200{,}000$) and two pairs of light chains ($M_r = 16{,}000$ to $27{,}000$, depending on the source). This protomer can be split by the action of the protease *papain* to produce two globular *subfragments* (S_1), known as heads, which contain the light chains, and one long α-helical coiled-coil rod, referred to as the *tail* (see Fig. 5-32). Alternatively, the protease *trypsin* cleaves a sensitive region in the tail to produce highly α-helical *light meromyosin* (LMM) and *heavy meromyosin* (HMM), which consists of a

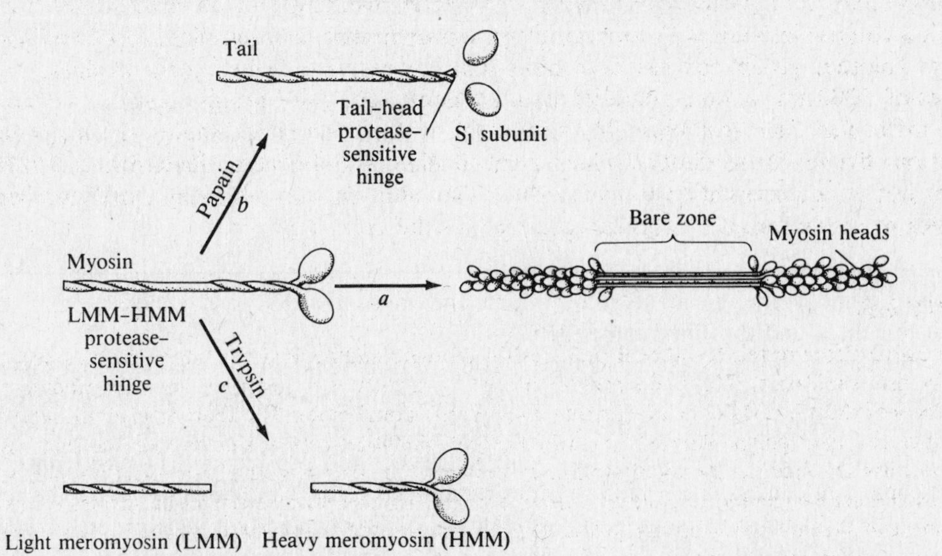

Fig. 5-32 Schematic diagram of a myosin molecule showing (*a*) how it assembles to form the thick filament and how it may be cleaved by the action of (*b*) papain and (*c*) trypsin.

segment of the tail linked to the two heads. The myosin heads, and hence S_1 and HMM, contain the F actin binding site and the enzymatic site that hydrolyzes ATP to ADP and inorganic phosphate. The rodlike tails and LMM associate in solution to form filamentous structures that tend to precipitate. In the same way, the myosin protomer will spontaneously aggregate under physiological conditions to form the myofibrillar *thick filaments* by the assembly of some 400 myosin tails in a staggered, side-by-side packing with the myosin heads projecting at regular repeating intervals in a helical array. The thick filaments are bipolar with a 150-nm bare zone in the center, where two oppositely oriented sets of myosin tails come together.

The myofibrillar *thin filaments* consist of F actin (although tropomyosin and troponin are found in striated muscle thin filaments; see Example 5.18). The thin filament is bipolar; when *decorated* with HMM or S_1, as described earlier, the arrowheads point away from a central region where the protein *α-actinin* ($M_r = 2 \times 95{,}000$) binds to the ends of the actin filaments, anchoring them firmly in the myofibril. Six thin filaments surround each thick filament, partially overlapping it. On contraction the filaments slide past each other, increasing the overlap without any change in the length of the filaments themselves (Fig. 5-33). This is the *sliding filament model* (Fig. 5-33), which is supported by a wide variety of experimental evidence.

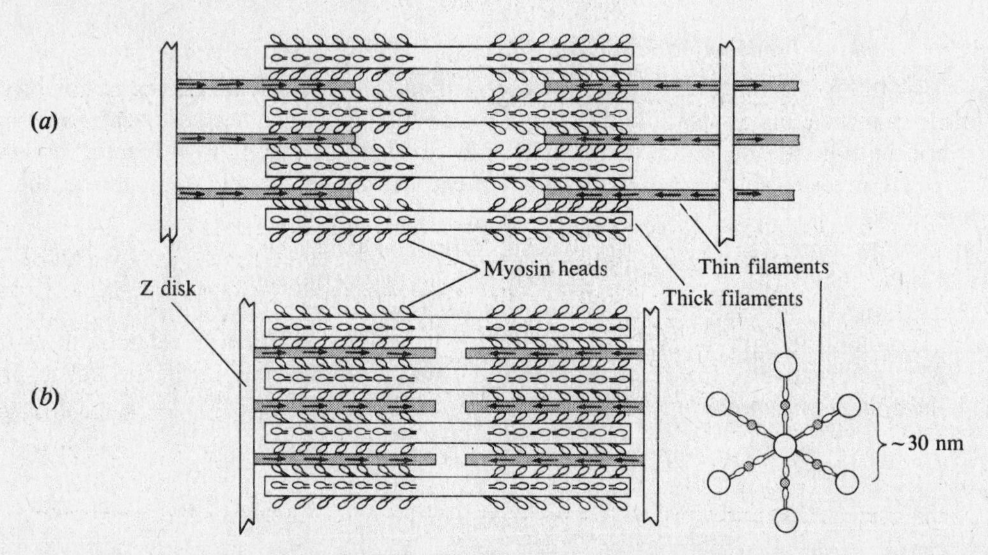

Fig. 5-33 Schematic diagram showing the arrangement of the filaments in a *sarcomere* (the basic contractile repeat unit of muscle fibers) when (*a*) relaxed and (*b*) contracted. The polarities of the filaments are indicated by arrowheads. The arrangement of the filaments in cross section is shown at the overlap regions.

Question: What is the mechanism by which the myosin cross-links generate mechanical force between the thick and the thin filaments?

The mechanical force is generated by a cyclic reaction between the heads of myosin on the thick filaments and the F actin, or thin filaments. The energy for this force comes from the hydrolysis of ATP; myosin by itself is an *ATPase*. In the presence of actin, the rate of the reaction increases 200-fold; the binding of actin accelerates the release of the products—ADP and phosphate—and causes a change of conformation around the myosin head. This is the *power stroke*. The myosin head is released from the actin filament by the binding of ATP, which is then hydrolyzed. A new cycle can now begin with the binding of the myosin-ADP to actin. Each cycle can take 0.2 s during rapid contraction. This process is illustrated in Fig. 5-34. Note the importance of the two *hinge* areas in the myosin, which allow flexibility of movement. The exact details of the conformational changes are uncertain. In the myofibril, there are hundreds of myosin heads in each thick filament, and during contraction, approximately half are attached at any one time, so pressure is always maintained.

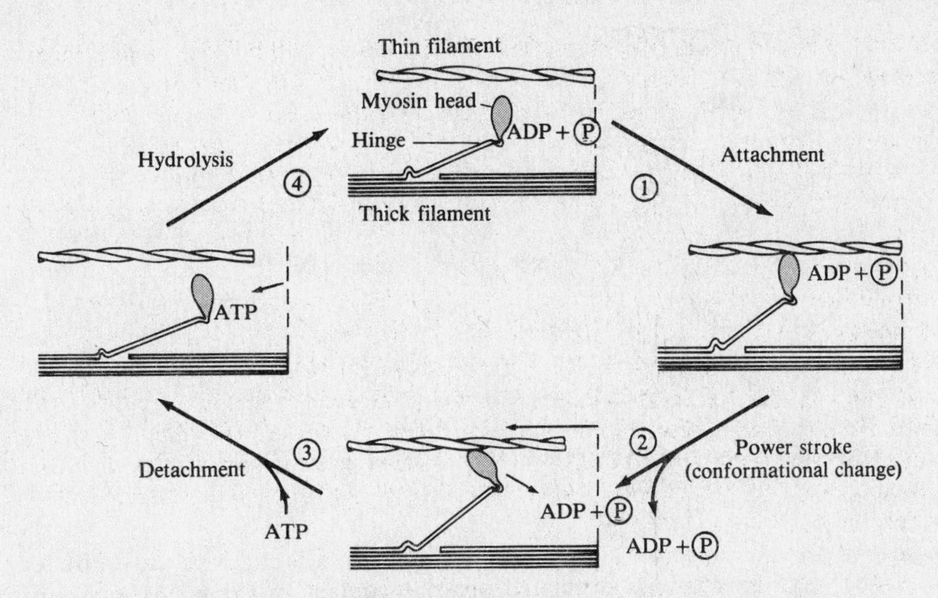

Fig. 5-34 A diagrammatic representation of the four steps in the contraction cycle showing the association and dissociation of myosin and actin driven by the hydrolysis of ATP.

Question: How is this generation of mechanical force initiated and regulated?

An increase in calcium ions (e.g., from 10^{-8} M to 10^{-5} M) acts as the trigger. In striated muscle, calcium interacts with the troponin complex, causing a movement of tropomyosin to expose the myosin binding sites on the thin filaments, as described in Example 5.18. In smooth muscle, calcium activates an enzyme that *phosphorylates* the light chains of myosin; hence the control effect is on the thick filament. In some nonmuscle cells, the action of calcium is at the level of the assembly of myosin molecules into filaments. In all these cases, a subsequent drop in Ca^{2+} concentration will reverse the process. It should be noted that a continuing supply of ATP is required for the contraction process. When ATP is completely depleted, *rigor* sets in.

Microtubules

The largest filaments in the cytoskeleton are the microtubules, which are long, hollow cylinders with an external diameter of 24 nm and an internal diameter of 14 nm. Each cylinder is composed of 13 circularly arranged, parallel *protofilaments*, each of which is a head-to-tail linear polymer of α,β protomers. The protomer is a dimer of α- and β-tubulin (each $M_r \simeq 50{,}000$), distinct but homologous globular proteins that have been highly conserved in evolution (Fig. 5-35).

The polymerization process has many similarities to that of actin. Each tubulin molecule has a binding site for a *guanine nucleotide*. The protomers associate to form a nucleus, which then grows in both directions, with hydrolysis of GTP. However, this hydrolysis is not essential, as nonhydrolyzable analogs of GTP and even pyrophosphate will support polymerization. Calcium ions and low temperature (4°C) inhibit polymerization, while magnesium ions and heat (37°C) stimulate it. The polymerization process is *biased*, as each end has a different critical concentration, allowing treadmilling to occur when the subunit concentration lies between the two critical concentrations. However, it is believed that the concentration of free tubulin subunits is usually so low in the cytoplasm that depolymerization will occur unless the minus end is capped.

The microtubule-associated proteins could be classified in the same way as the actin-binding proteins, but at the present time, they are less well defined and have been largely studied in terms of their effects upon the polymerization process. Some have been shown to induce polymerization by

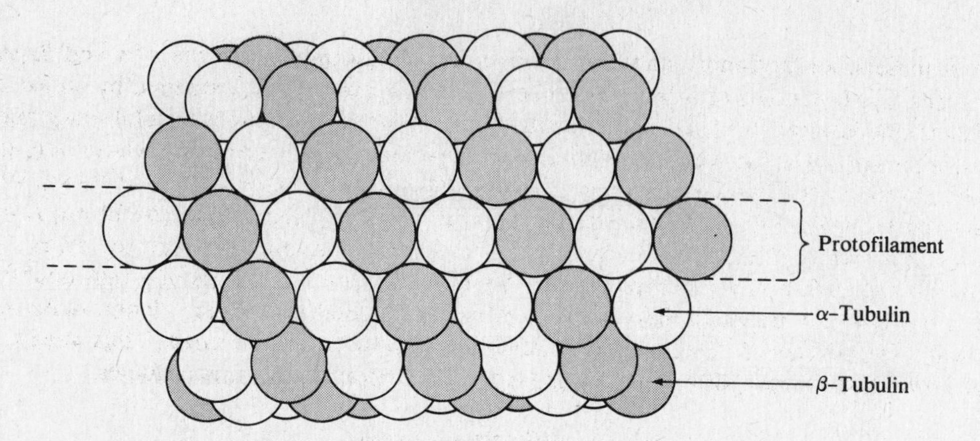

Fig. 5-35 A short segment of microtubule showing the tubulin dimers aligned into protofilaments.

binding to several tubulin molecules simultaneously, often associating with the entire length of the microtubule. The large microtubule-associated proteins contain two domains, one binding to the microtubule, the other extending outward, possibly to form cross-links with other cellular components. Thereby microtubules appear to be involved in establishing the geometry of the cell, often radiating out from organizing centers.

EXAMPLE 5.20

During the *interphase* of almost all animal cells, microtubules have been shown to fan out from the *centrosome*, or cell center, which consists of a pair of *centrioles* surrounded by densely staining *pericentriolar* material (Chap. 1). Each centriole is a small cylinder, 0.1 μm in diameter by 0.5 μm in length, made up of nine triplet microtubules. The *basal body* of cilia and flagella has a similar structure (Fig. 5-36). Here, the long microtubules making up the cilia and flagella arise directly from the basal body; however, the extensive microtubules radiating from the cell center do not appear to arise directly from the centrioles but, rather, from the surrounding pericentriolar material. In both cases, the plus ends of the microtubules point away from the organizing center and may grow at the rate of about 1 μm per minute.

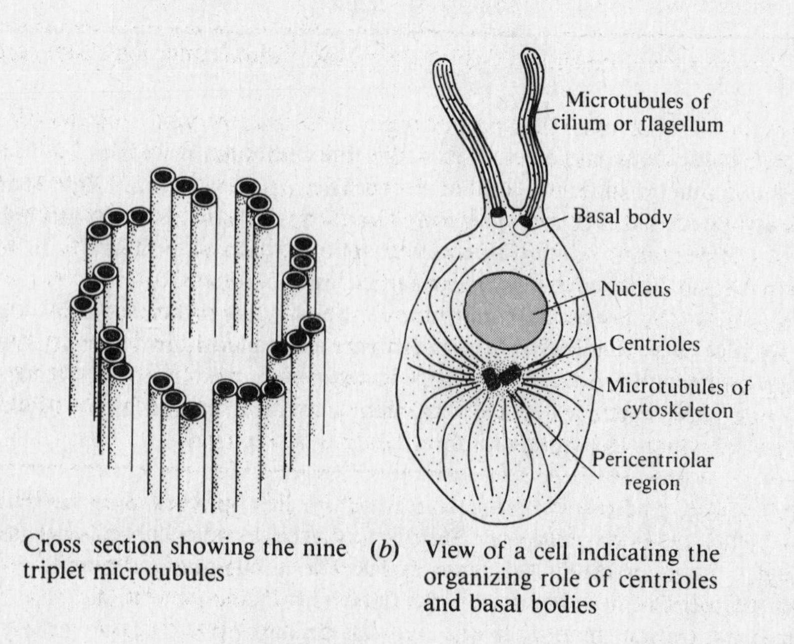

(*a*) Cross section showing the nine triplet microtubules (*b*) View of a cell indicating the organizing role of centrioles and basal bodies

Fig. 5-36 Diagrammatic representations of a centriole and basal body.

The best understood and most stable of the microtubule assemblies are the *cilia* and *flagella* of eukaryotic cells. Each consists of a bundle of fibers called an *axenome*, surrounded by an extension of the plasma membrane. The axenome is made up of a ring of nine doublet microtubules, surrounding a pair of single microtubules and associated with many other proteins whose interactions generate the power for the oarlike movement. Figure 5-37 is a representation of the cross section of a cilium, showing the 9 + 2 array and the positions of the various proteins that extend out from the microtubules at periodic intervals. The most important of the proteins are pairs of *dynein arms*, which project from each doublet at intervals of 24 nm and extend toward the adjacent doublet. At greater intervals (86 nm) are *nexin* links between adjacent microtubules. Every 29 nm, *radial spokes* project inward from each of the nine outer doublets toward the *inner sheath*; this sheath itself consists of protein arms extending every 14 nm from the central pair of microtubules.

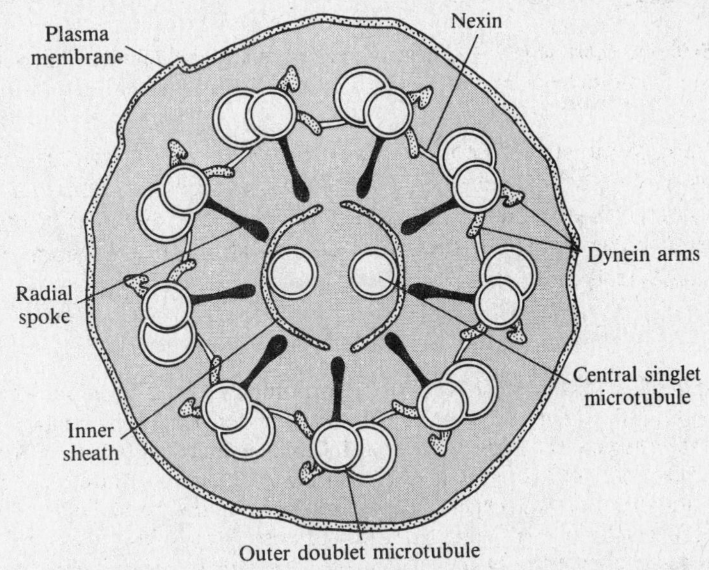

Fig. 5-37 Diagram of a cross section of a cilium.

Question: What is the mechanism by which the cilium and flagellum produces its bending movement?

There is considerable evidence supporting a *sliding microtubule mechanism*, in which adjacent doublet microtubules slide past each other within the axenome. Because of the presence of the nexin cross-links, the extent of sliding is limited. Bending is produced by a difference in the extent of the sliding on the opposite sides of the axenome. The force is generated by the dynein arms, which point downward in the presence of ATP; however, in the absence of ATP, they are oriented at right angles, making contact with the neighboring microtubule. Dynein itself is an ATPase whose activity is much increased by the presence of microtubules, analogous to myosin in the presence of actin. The force generated by each dynein arm is similar to that of a myosin molecule in muscle. The regulation of the sliding doublet microtubules appears to be carried out by the inner sheath and the central microtubule pair. It has been suggested that chemical signals relayed by the radial spokes activate the dynein arms to produce the regular cyclical beating of the cilia.

The determination of the spatial geometry of the cell appears to be under the influence of the microtubules. This is clearly seen in the long processes of nerve cells, whose asymmetry is maintained by microtubules. In addition, microtubules can support movements within the cell; vesicles can move along axons at speeds of 400 mm per day compared with the growth of a neurite of 1 mm per day. The presence of ATP and undefined microtubule-associated proteins is necessary for this movement. The separation of the daughter chromosomes at mitosis is another example of microtubule function.

EXAMPLE 5.21

During mitosis, a *mitotic spindle* is formed, which enables the two daughter chromosomes to separate. This spindle is a highly ordered, bipolar, fibrous structure consisting of polar microtubules radiating out from each of the two cell centers (*centrosomes*) and overlapping with each other at the equator of the spindle. During *prometaphase*, the chromosomes condense and produce two opposing sets of microtubular *kinetochore* fibers, which interact with the polar fibers of the spindle. These interactions produce oscillating movements, which, at metaphase, align the chromosomes at the equator of the spindle, balanced by the opposing poleward forces. When the daughter chromosomes split apart at *anaphase*, the two chromosomes are pulled toward the poles; at the same time, the two poles are pushed apart. In both cases, the mechanism involves microtubular sliding combined with elongation of the microtubules. A dyneinlike ATPase appears to generate the force that pushes the poles apart. The details of the chromosomal movement are different, however; movement is limited by the rate of disassembly of the kinetochore microtubules as they reach a pole. These processes are illustrated in simplified form in Fig. 5-38.

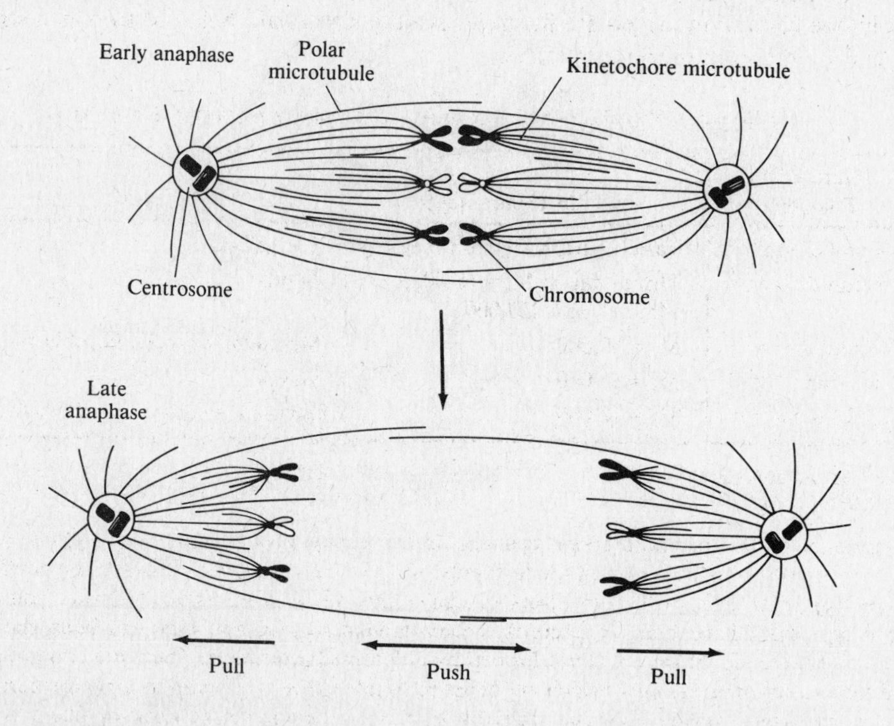

Fig. 5-38 Diagram showing the daughter chromosomes being separated by being pulled toward the centrosome poles while the poles are pushed apart.

Intermediate Filaments

Intermediate filaments form a family of tough, insoluble protein fibers with diameters of about 10 nm, found in most eukaryotic cells. The protein subunits (Fig. 5-39) vary greatly in size ($M_r = 40{,}000$ to $200{,}000$), but they all appear to have a common coiled-coil, α-helical region in the center of each subunit. The rodlike region is of constant length, with considerable sequence homology, 30–70 percent of the amino acids being identical between the various types; it also appears to be the basis for the filamentous structure, about 25 subunits interlocking side by side to form the cross section of one filament. The terminal globular regions are the variable segments, which differentiate the various types of intermediate filaments and which project from the filament into the cytoplasm to play an as yet ill-defined role in the cell.

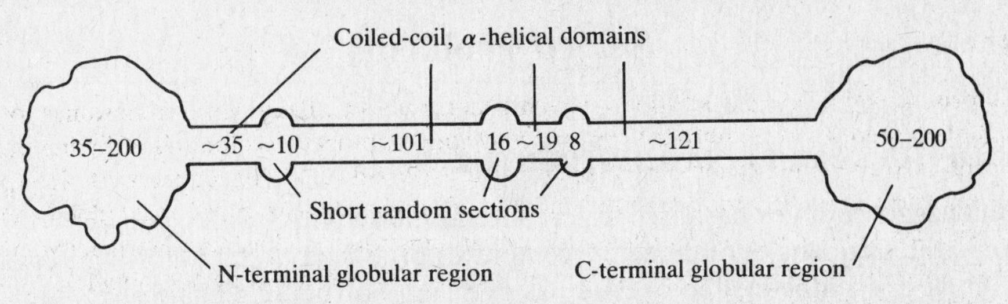

Fig. 5-39 A diagrammatic representation of the subunit of an intermediate filament. (The figures represent the number of amino acid residues in each structural domain.)

There are five classes (see Table 5.4) of intermediate filaments, which can be differentiated by biochemical, immunological, and molecular biological techniques. In general, one type of cell contains only one class of intermediate filament. All the classes are believed to play a structural role within the cell by resisting mechanical stress.

Table 5.4. The Classes of Intermediate Filaments in Vertebrate Cells

Intermediate Filaments	Protein Components	Cell Type
Cytokeratins	Various: $M_r = 40{,}000$ to 70,000	Epithelium
Neurofilaments	Three major: $M_r = 68{,}000$, 145,000, and 220,000	Neurons
Vimentin	$M_r = 55{,}000$	Mesenchyme, e.g., fibroblasts
Desmin	$M_r = 53{,}000$	Muscle
Glial filaments	$M_r = 50{,}000$	Some glial cells, e.g., astrocytes

EXAMPLE 5.22

The most-studied class of intermediate filaments and the most diverse is the *cytokeratins*, also called the *prekeratins*, which are cross-linked by disulfide bonds. On reduction, several different polypeptide chains are obtained from epithelial cells, with a spectrum of chains that varies from tissue to tissue. Some 19 different cytokeratin polypeptides have been characterized from human tissues, and each has been shown to be the product of a distinct individual gene and not formed by differential processing. There are two groups, the *acidic keratins* and the *neutral-basic keratins*, which are believed to interact in a 1:1 ratio to form the keratin filaments. These play an important role in epithelial cells by forming a continuous network of fibers across a whole epithelial sheet, giving it *tensile strength*. The cells are fastened by rivetlike connections, called *spot desmosomes*, which act as the anchoring point for the keratin filaments. Keratin fibers are abundant in the tough outer protective covering of higher animals, and they become increasingly cross-linked as the cell matures and finally dies, producing the keratinized outer skin layer, nails, and hair.

The intermediate filaments are durable and, once formed, do not assemble and disassemble under physiological conditions. There is no pool of unpolymerized subunits, no nucleotide is involved in assembly, and treadmilling does not occur.

Question: In view of the stability of the intermediate filaments, how can the cell control the number and length of these filaments?

One of the few intermediate filament–associated proteins to have been characterized is a calcium-ion-activated protease, which has been found associated with desmin, neurofilaments, and vimentin. Proteolysis rapidly disassembles the filaments, by cleaving their polypeptides into smaller units, and this degradation may be the only option available to the cell for control of the intermediate filament component of the cytoskeleton.

Solved Problems

ASSEMBLY OF SUPRAMOLECULAR STRUCTURES

5.1. A multienzyme complex has three different catalytic activities with eight sites for each activity. Compare the frequencies of defective complexes produced in the following two situations: (*a*) The complex is synthesized in one step as one long polypeptide chain containing 8,000 amino acid residues. (*b*) The complex is constructed in three steps. First, 24 polypeptides are synthesized: 8×200, 8×300, and 8×500 amino acid residues. Next, trimers consisting of one of each chain type are formed. Last, these eight trimers are assembled to form the complex. (Assume in both cases that the error frequency is 10^{-5} for each operation and that a single mistake will cause complete rejection.)

SOLUTION

(*a*) The probability of incorporating an incorrect amino acid is 1 in 10^5. Therefore, the frequency of defective complexes would be $8 \times 10^3 \times 10^{-5}$, i.e., eight of every 100 complexes are defective.

(*b*) Because defective polypeptides will be rejected as being unable to form trimers and any faulty trimers will not assemble further, the frequency of defective complexes will be related only to the seven steps required for the final assembly of the trimers to form the complex, and this would be equal to 7×10^{-5}. Thus, the three-step process produces about 1,000 times fewer defective complexes than the single-step process and, incidentally, requires one-eighth as much genetic information.

PROTEIN SELF-ASSOCIATION

5.2. For the dimerization reaction $2\text{A} \rightleftharpoons \text{A}_2$, the equilibrium constant in the mol L^{-1} scale is 10^5. What is the concentration in mol L^{-1} of dimer in equilibrium with 10^{-3} mol L^{-1} monomer?

SOLUTION

Rearranging Eq. (*5.1*), $[\text{A}_2] = K[\text{A}]^2$.
Thus, at equilibrium,

$$[\text{A}_2] = 10^5(10^{-3})^2 \text{ mol L}^{-1} = 0.1 \text{ mol L}^{-1}$$

5.3. For the dimerization reaction $2\text{A} \rightleftharpoons \text{A}_2$, in which A is a protein of molar weight $40{,}000 \text{ g mol}^{-1}$, the equilibrium constant in the mol L^{-1} scale is 10^6. Calculate the percentage by weight of dimer when the total concentration of the protein is 1 g L^{-1}.

SOLUTION

It is important to convert all units to a consistent set. Here, it is most convenient to use the mol L^{-1} scale. Thus, the molar concentrations are $[\text{A}] = c_\text{A}/M$ and $[\text{A}_2] = c_{\text{A}2}/2M$, where c_A and $c_{\text{A}2}$ are the concentrations in g L^{-1} of A and A_2 respectively, and M is the molar weight of A.

Now, the total concentration of A is

$$c_\text{T} = c_\text{A} + c_{\text{A}2}$$

Hence,

$$c_{\text{A}2} = c_\text{T} - c_\text{A}$$

By substituting in Equation (*5.1*), we get

$$K = \frac{[\text{A}_2]}{[\text{A}]^2} = \frac{(c_\text{T} - c_\text{A})/2M}{(c_\text{A}/M)^2}$$

Rearranging and then solving the quadratic in c_A gives two roots, one negative, and thus physically meaningless, and the other positive. The positive root is given by

$$c_A = \frac{-1 + (1 + 8Kc_T/M)^{1/2}}{4K/M}$$

Substituting for K, c_T, and M, we get

$$c_A = 0.13 \text{ g L}^{-1}$$

Therefore, the percentage by weight of monomer is 13, and that of the dimer is 87.

5.4. Repeat the above calculation for a total concentration of 10 g L^{-1}.

SOLUTION

Using the same procedure as in Prob. 5.3 and substituting into the expression for c_A, we get

$$c_A = \frac{-1 + [1 + (8 \times 10^6 \times 10)/(4 \times 10^4)]^{1/2}}{4 \times 10^6/4 \times 10^4} \text{ g L}^{-1}$$

$$= \frac{-1 + (2{,}001)^{1/2}}{100} \text{ g L}^{-1} = 0.437 \text{ g L}^{-1}$$

This represents only 4.4 percent of the total at this new concentration, and thus the percentage by weight of the dimer is 95.6 percent.

Note the general principle that can be drawn from this: As the total concentration of a self-associating molecule increases, the proportion of the associated form increases. A corollary is that dilution favors the monomeric species. This is an example of Ostwald's dilution principle.

HEMOGLOBIN

5.5. Fetal hemoglobin (HbF) is a tetramer of two α chains and two γ chains. The γ chains are similar to the β chains of HbA, but there are many sequence differences between them. One significant difference is that the residue H21 is the positively charged histidine in the β chain but the neutral serine in the γ chain. (*a*) Explain why HbF has a higher oxygen affinity than normal adult hemoglobin (HbA). (*b*) Why is this effect physiologically important?

SOLUTION

(*a*) The amino acid substitution at H21 affects the β-β cleft, where DPG is bound, there being two fewer positive charges in HbF. Thus, DPG is bound less strongly to HbF, and oxygen is more tightly bound and therefore is less readily released from HbF.

(*b*) This effect allows HbF to draw O_2 across the placenta from the maternal HbA for the use of the fetus.

5.6. In vertebrate hemoglobins, it is the tetramer that shows cooperativity on binding with oxygen; the dimer has no cooperative effect. Recently, the blood clam has been shown to contain two hemoglobins: a tetramer consisting of two pairs of unlike chains and a homodimer of a third, each of these chains having a tertiary structure similar to the globin fold. In the quaternary structure of the blood clam hemoglobins, the E and F helices, which lie on the proximal and distal sides of the heme group, are involved in subunit interactions forming an extensive contact between the chains in the dimer. This is "back to front" when compared with the vertebrate hemoglobins, where the E and F helices are on the outside of the molecule. Interestingly, both the dimer and the tetramer clam hemoglobins bind oxygen cooperatively. Explain how the mollusk dimer hemoglobin could demonstrate a cooperative effect.

SOLUTION

On combining with oxygen, the F helix, containing the proximal histidine residue, moves about 0.5 Å. In vertebrate hemoglobin, this movement, being on the surface of the molecule, has no direct effect, acting only on the tetrameric structure (the $\alpha_1\beta_1$ dimers are quite rigid). In the clam, such a movement will directly affect the structure of the dimer and may lead to an increase in the oxygen affinity, producing cooperativity.

5.7. There are many abnormal or mutant hemoglobins, some of which cause pathological conditions. One is the sickle-cell hemoglobin (HbS), in which the glutamate residue in the sixth position of the normal human hemoglobin (HbA) β chain has been replaced by valine. This position, referred to as $\beta 6$, is on the outside of the hemoglobin molecule. Individuals who are homozygous for HbS suffer from circulatory problems and anemia. This is because the red blood cells become sickle-shaped in the venous circulation, causing blockages in the capillaries, and are principally removed by the spleen. Why should such a small change—two amino acids in a total of 574—produce such a drastic effect?

SOLUTION

Amino acid residues on the outside of soluble proteins are almost without exception polar. The replacement of the strongly polar glutamate in HbA by the nonpolar valine in HbS leads to a hydrophobic area on the exterior of the molecule. Such a "sticky patch" would readily interact with a complementary binding site on another molecule. HbS, but not oxyHbS, has such a site, and this allows HbS to polymerize into long fiberlike chains that distort the red blood cell and produce the sickle shape. It is of interest that the HbS gene is common in tropical Africa, because the heterozygous individual, who rarely has sickling episodes, is protected against the most dangerous form of malaria. Should infection with malaria occur, the red cells would tend to sickle because of the increased oxidative stress imposed on them; these cells would then be removed from the circulation by the spleen.

5.8. Mutations in the α_1,β_2 sliding contact of hemoglobin generally cause impairment of the cooperative effect. In $Hb_{KEMPSEY}$, the aspartate residue G1 in the β chain is replaced by asparagine. This removes a hydrogen bond from the sliding contact (Fig. 5-12) that stabilizes the tense structure of deoxyhemoglobin but is not involved in the relaxed structure of oxyhemoglobin. What effect will this amino acid substitution have on the oxygen affinity of $Hb_{KEMPSEY}$?

SOLUTION

As the stability of the tense structure is reduced by the loss of this interaction in the two sliding contacts α_1,β_2 and α_2,β_1, less energy is required for oxygen to bind to $Hb_{KEMPSEY}$. Thus, $Hb_{KEMPSEY}$ has a higher oxygen affinity ($p50 = 15$ torr) and a lower Hill coefficient (see Chap. 9 and Prob. 5.21). Individuals with this hemoglobin compensate by producing more red blood cells.

THE EXTRACELLULAR MATRIX

5.9. There is a continual turnover of collagen molecules in the body, e.g., in the remolding of bone and in wound healing. For many years it was believed that the triple helix was resistant to attack by all proteolytic enzymes. In the last 20 years, many mammalian collagenases have been found that cleave each of the three chains of the triple helix between a glycine and a large hydrophobic residue (e.g., leucine, isoleucine, or phenylalanine) roughly three-quarters of the distance from the N terminus. This is an unstable region, low in proline and hydroxyproline. Explain how this cleavage of the triple helix allows the degradation of collagen to proceed.

SOLUTION

The specific cleavage of collagen by mammalian collagenases produces two fragments that are unstable at body temperature and begin to uncoil into individual chains. These are then susceptible to attack by other proteolytic enzymes.

5.10. The monomers of type VI collagen are the shortest collagen molecules (105 nm), with only one-third of the mass being the triple-helical domain, the other two-thirds being the two terminal globular regions. On dispersion and rotary shadowing for the electron microscope, the following structures are the most abundant of those observed:

Explain the construction of the structures, and suggest how further assembly could occur to form extended microfibrils.

SOLUTION

The structure on the left shows a dimer with two antiparallel collagen type VI molecules assembling in a staggered manner, with the triple-helical segments overlapping by 75 nm. The dimers align symmetrically with the free (nonoverlapping) triple-helical ends crossing over like scissors to form the tetramer shown on the right. This tetramer then acts as the protomeric form, which aggregates end to end, with further crossing over of the ends to form the linear polymers of microfibrils.

5.11. The genes for the chains of fibrillar collagens (types I, II, III, and probably V) contain about 50 short exons (expressed regions), separated by long introns (nonexpressed regions); the introns are removed before the genetic information is translated into the protein chain (see Chap. 17). The exons for the triple-helical domain are unusual in that they all contain multiples of 9 base pairs, 54 being the most common, followed by 108. If three nucleotide bases code for one amino acid, what is the significance of this gene organization?

SOLUTION

On translation each exon produces a length of oligopeptide that is a multiple of three amino acids; lengths of 18 and 36 amino acids are the most common, being equivalent to 6 or 12 turns of the triple helix. Each exon begins with a codon for glycine, and this is followed every nine nucleotides by another. It has been suggested that the collagen gene evolved from a small primordial gene and that the organization of the gene is now conserved to an extraordinary extent.

5.12. Defects in collagen synthesis can cause severe connective tissue disorders. In Ehlers-Danlos syndrome VII, those affected suffer from thin, velvety skin, which heals poorly, and from hypermobile joints and weak ligaments, a condition which leads to multiple joint dislocations. It has been reported that in this syndrome, exon 46 (numbering from the C-terminal end) is missing from the gene for the $\alpha 2(\mathrm{I})$ chain. This sequence includes the cleavage site for the procollagen N-peptidase and a lysine residue near the normal N terminus of the collagen $\alpha 2(\mathrm{I})$ chain. Describe the effect of the loss of this exon on the structure of type I collagen.

SOLUTION

The loss of the N-peptidase cleavage site would prevent the proper processing of the $\alpha 2(\mathrm{I})$ procollagen chain, and thus the mature chain would have a large N-terminal extension of about 50 amino acids, which would disrupt the proper assembly of the triple-helical collagen molecules into a fibril. The lack of the lysine residue would prevent the formation of one of the four major interchain cross-links that the $\alpha 2(\mathrm{I})$ chain makes with its quarter-staggered neighbors. The overall effect is to weaken tissues that depend on type I collagen for tensile strength.

5.13. How many specific glycosyltransferases are required for the sequential addition of monosaccharide units in the synthesis of chondroitin sulfate? Note that substrate specificity is strict.

SOLUTION

Six glycosyltransferases are required (Fig. 5-22) and they act in the following order: (1) xylyltransferase, (2) galactyltransferase I (specific for xylose), (3) galactyltransferase II (specific for galactose), (4) glucuronate transferase I (specific for Gal-Gal), (5) *N*-acetylglucosamine transferase, (6) glucuronate transferase II (specific for *N*-acetylglucosamine). Finally, a sulfotransferase will catalyze the addition of the sulfate groups.

CYTOSKELETON

5.14. If the spectrin heterodimer can associate only head to head, explain the observation of higher polymers as large as the dodecamer.

SOLUTION

The head of each spectrin chain interacts with the head of the complementary chain of another heterodimer. In the tetramer, there are paired interactions (Fig. 5-31), while in higher oligomers, a closed loop is formed, as the head region is quite flexible. This is shown for the hexamer in Fig. 5-40, but this self-association may continue indefinitely.

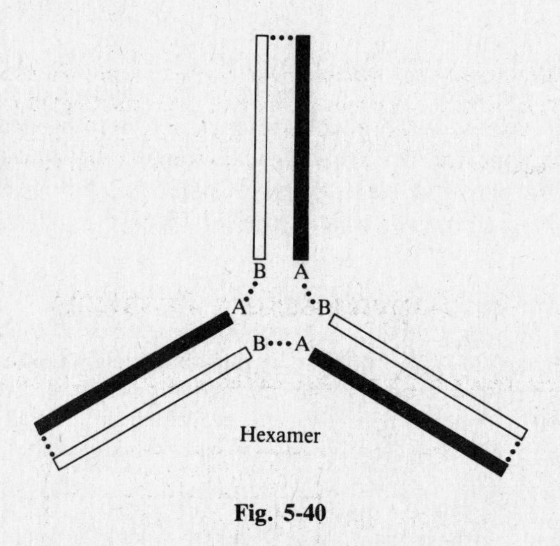

Fig. 5-40

5.15. Several cytoskeletal proteins contain long coiled-coil domains in which two or even three α helices coil around each other. What sequence criteria would be necessary for the α helices to form stable coiled coils in aqueous solutions?

SOLUTION

The α helix contains 3.6 amino acid residues per turn, and in a coiled coil, this produces seven residues for every two turns of the α helix. If these seven residues are denoted a, b, c, d, e, f, and g, then residues a and d (with e as an alternative) would be expected to be hydrophobic. These residues would be next to each other on one side of the α helix and in the center of the coiled coil, which is thus stabilized by hydrophobic interactions. This is known as the *heptad repeat*.

5.16. Many assemblies of actin and myosin in cells are temporary structures. One example is the beltlike *contractile ring* that appears during cell division just below the plasma membrane. As the ring contracts, the center of the cell is constricted until two daughter cells are produced. What interactions and processes must occur for this constriction to take place, noting that the ring remains a constant thickness?

SOLUTION

To allow the force generated between the actin and myosin to act upon the membrane, the contractile ring must be anchored to the membrane by specific actin-binding proteins. As the cell constricts, the contractile ring must disassemble little by little to remain a constant thickness.

5.17. Explain how the alkaloid drug *colchicine*, which binds tightly to tubulin dimers, blocks mitosis.

SOLUTION

The binding of the tubulin dimer by colchicine not only will prevent polymerization and the formation of microtubules but also, by reducing the concentration of free tubulin below the critical concentration, will cause depolymerization and breakdown of the mitotic spindle. Similar drugs have been used as antitumor agents.

5.18. The nuclear envelope is a supramolecular assembly that lines the inner surface of the nucleus. The major proteins of the envelope are the A, B, and C *lamins*, and it has been suggested that they may be members of the intermediate filament family. What characteristic features should the lamins possess for this to be so?

SOLUTION

The intermediate filament family consists of proteins with a large central region with considerable sequence homology, which folds up into an α helix, with the heptad repeat allowing the formation of coiled-coil domains. The positions of the short random sections between the coiled-coil domains are conserved. Strong sequence divergence in the two globular terminal regions would be expected.

Supplementary Problems

5.19. For the dimerization reaction $2A \rightleftharpoons A_2$, calculate the concentration in mol L^{-1} of dimer for the following conditions:

	K	[A] (mol L^{-1})
(*a*)	10^3	10^{-3}
(*b*)	10^4	10^{-2}
(*c*)	10^6	10^{-4}

5.20. For the dimerization reaction $2A \rightleftharpoons A_2$, in which A is a protein of molar weight 50,000 g mol^{-1}, calculate the percentage by weight of the dimer when the total concentration of the protein is 1 g L^{-1} and the equilibrium constant in the mol L^{-1} scale is (*a*) 10,000; (*b*) 10^8.

5.21. Early this century, A.V. Hill derived a useful equation that describes the oxygen dissociation curve of hemoglobin fairly accurately. It is

$$Y = \frac{(pO_2)^n}{(pO_2)^n + (p50)^n}$$

where Y is the fraction of hemoglobin oxygen-binding sites occupied by oxygen, pO_2 the partial pressure of O_2, $p50$ the partial pressure of oxygen when $Y = 0.5$ (i.e., 50 percent of the sites are filled), and n the *Hill coefficient*, which is a measure of cooperativity. The value of n for human HbA is 2.8, while that of myoglobin, which is not cooperative, is 1. Calculate the change in saturation ΔY for the situation in which hemoglobin goes from lung to working muscle and in which pO_2 for lung is 100 torr and pO_2 for muscle is 20 torr under the following circumstances:

	Protein	$p50$ (torr)	n
(*a*)	HbA	26	2.8
(*b*)	Noncooperative Hb	26	1
(*c*)	Mb-like Hb	1	1
(*d*)	HbA with high DPG	31	2.8
(*e*)	Clamlike dimer Hb	26	1.5

5.22. One of the salt links stabilizing deoxyhemoglobin involves a chloride ion. Explain how this ion could act as an allosteric effector.

5.23. In the abnormal hemoglobin Hb_{KANSAS}, residue G4 on the β chain is threonine instead of asparagine in HbA. This change weakens the interactions in the sliding contact in the relaxed state of oxyhemoglobin without affecting the tense state. How would the oxygen affinity of Hb_{KANSAS} compare with that of HbA?

5.24. Which of the following synthetic polypeptides would be expected to form a triple helix similar to collagen?

(*a*) $(\text{Pro-Gly})_n$
(*b*) $(\text{Pro-Pro-Gly})_n$
(*c*) $(\text{Pro-Gly-Gly})_n$
(*d*) $(\text{Pro-Gly-Pro-Pro})_n$

Of those which form a triple helix, which would be the most stable?

5.25. Why is the triple helix of collagen a better structure for providing tensile strength than the α helix?

5.26. The following is a sequence of part of a pro-α-chain of collagen: -Gly-Pro-Met-Gly-Pro-Ser-Gly-Pro-Arg - Gly - Leu - Pro - Gly - Pro - Pro - Gly - Ala - Pro - Gly - Pro - Gln - Gly - Pro - Arg - Gly - Pro-Pro-Gly-Glu-Pro-Gly-Glu-Pro-Gly-Ala-Ser-. (*a*) How many residues would be hydroxylated by the enzyme *prolylhydroxylase*? (*b*) How many peptides would be produced after digestion by the collagenase from one of the gas gangrene organisms *Clostridium histolyticum*? The specificity of the enzyme is to cleave the bond before glycine in the sequence -Pro-*x*-Gly-Pro-

5.27. In the calcification process of bone formation, the initial crystals of hydroxyapatite are found at intervals of 67 nm along the collagen fiber. What is the reason for this?

5.28. In osteoarthritis, fragments of proteoglycans are produced. It is believed that those fragments are formed by the action of proteases produced in response to inflammation. Where is the action of the proteases on the proteoglycan most likely to occur?

5.29. In a 0.01% solution, the domains of hyaluronate occupy the total volume. What would be the ratio of the volume occupied by one molecule of hyaluronate ($M_r = 1.5 \times 10^6$) to the volume occupied by five molecules of collagen ($M_r = 300{,}000$; diameter 1.5 nm; length 300 nm)?

5.30. How might the identification of the type and subtype of intermediate filament help in the diagnosis and classification of cancer?

Chapter 6

Lipids, Membranes, Transport

6.1 INTRODUCTION

Lipids are defined as water-insoluble compounds extracted from living organisms by weakly polar or nonpolar solvents. This definition is based on a *physical* property, in contrast to the definitions of proteins, carbohydrates, and nucleic acids, which are based on chemical structure. Consequently, the term *lipid* covers a structurally diverse group of compounds, and there is no universally accepted scheme for classifying lipids.

EXAMPLE 6.1

Consider the following compounds, most of which are lipids:

$CH_3(CH_2)_{14}COO^-$
(1) Palmitate

$CH_3CH_2COO^-$
(2) Propionate

$CH_3(CH_2)_{14}CH_2OH$
(3) Cetyl alcohol

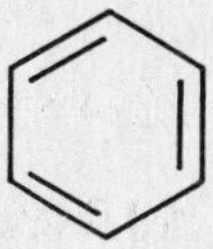

(4) Benzene

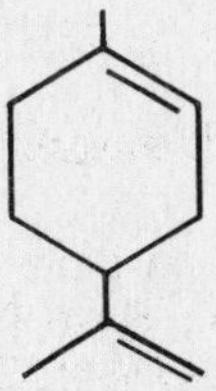

(5) Limonene

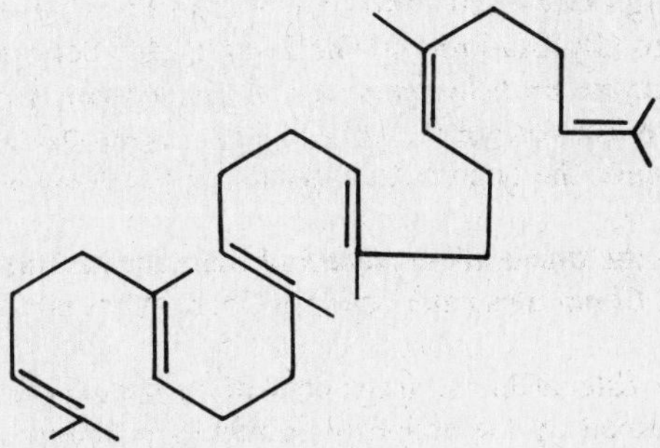

(6) Squalene

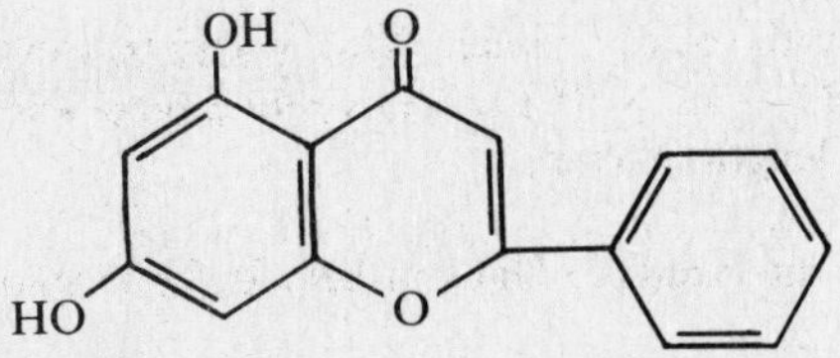

OH O

HO O

(7) Chrysin

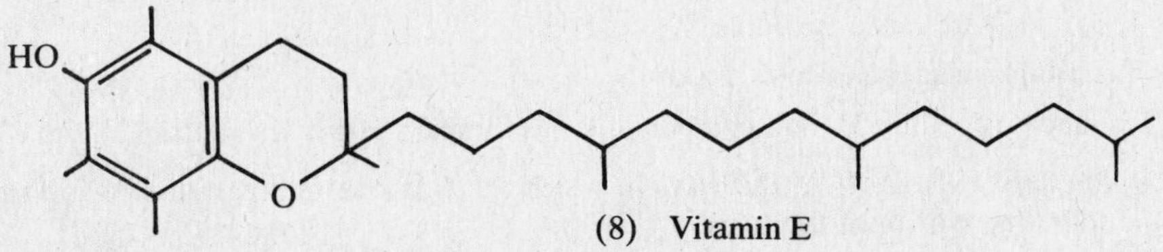

HO

O

(8) Vitamin E

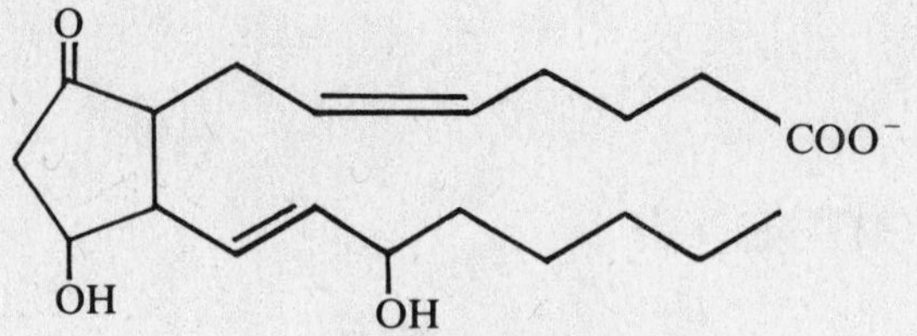

(9) Prostaglandin E_2

Compounds 1, 3, and 5 to 9 are lipids because they are biological in origin and are soluble in organic solvents. The latter property arises because they contain a *high proportion of carbon and hydrogen* and are therefore insoluble in water. Compound 4 is not a lipid because it does not occur free in living organisms. Compound 2 is water-soluble, but because it is a member of the same series of compounds as compound 1, it is usually considered to be a lipid.

6.2 CLASSES OF LIPIDS

A common feature of all lipids is that, biologically, their hydrocarbon content is derived from the polymerization of acetate followed by *reduction* of the chain so formed. (However, this process also occurs for the synthesis of some compounds that are not lipids and therefore cannot be used as a definition of lipids.) For example, polymerization of acetate can give rise to the following:

1. Long, linear hydrocarbon chains:

$$n\mathrm{CH_3COO^-} \longrightarrow\longrightarrow \mathrm{CH_3COCH_2CO}\cdots \longrightarrow\longrightarrow \mathrm{CH_3CH_2CH_2CH_2}\cdots$$

   The products are *fatty acids*, $CH_3(CH_2)_nCOOH$, which in turn can give rise to amines and alcohols. Lipids containing fatty acids include the *glycerolipids*, the *sphingolipids*, and *waxes*.

2. Branched-chain hydrocarbons via a five-carbon intermediate, isopentene (*isoprene*):

$$3\mathrm{CH_3COO^-} \longrightarrow\longrightarrow (\mathrm{CH_3{-}CH{=}\overset{\overset{\displaystyle CH_3}{|}}{C}{-}CH_2{-}}) \longrightarrow\longrightarrow \textit{terpenes}$$

   (including *steroids* and *carotenoids*)

3. Linear or cyclic structures that are only partially reduced:

$$n\mathrm{CH_3COO^-} \longrightarrow\longrightarrow \mathrm{CH_3CO{-}[CH_2CO]_{n-3}{-}CH_2CO{-}CH_2COO^-}$$

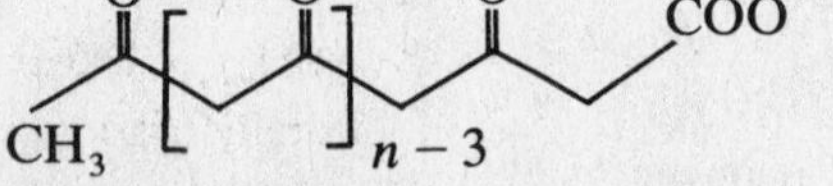

   These are called *acetogenins* (or sometimes *polyketides*). Many of these compounds are aromatic, and their pathway of formation is the principal means of synthesis of the benzene ring in nature. Not all are lipids, because partial reduction often leaves oxygen-containing groups, which render the product soluble in water.

EXAMPLE 6.2

For the lipids in Example 6.1, the routes of synthesis (just discussed) are:

Compounds 1, 3, and 9 are made via route (1).

Compounds 5 and 6 are made via route (2).

Compound 7 is made by route (3), and compound 8 has a mixed origin via routes (2) and (3).

For those lipids synthesized by route (2), the ways in which the isoprene units are linked are shown by the broken lines in the following structural formulas:

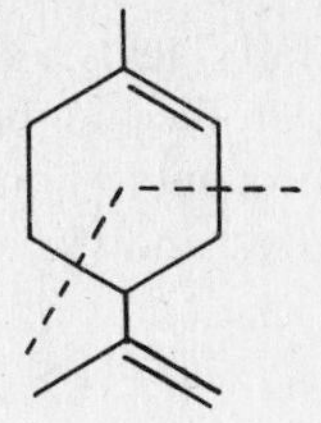

Limonene

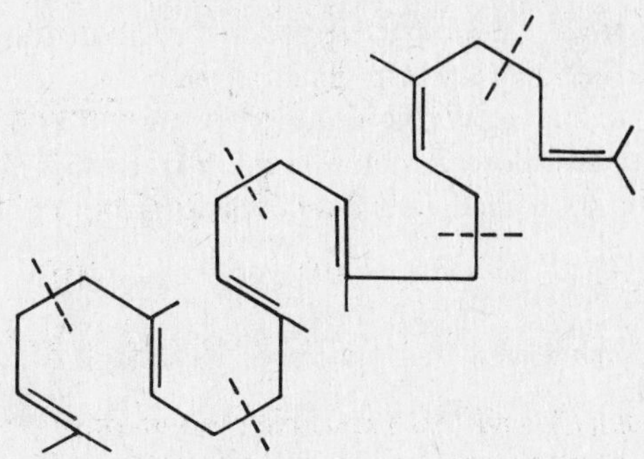

Squalene

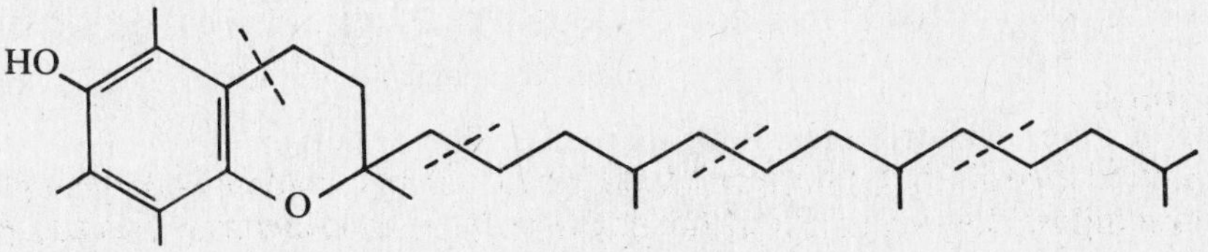

Vitamin E

6.3 FATTY ACIDS

Over 100 fatty acids are known to occur naturally. They vary in chain length and degree of unsaturation. Nearly all have an even number of carbon atoms. Most consist of linear chains of carbon atoms, but a few have branched chains. Fatty acids occur in very low quantities in the free state and are found mostly in an *esterified* state as components of other lipids. The pK_a of the carboxylic acid group is about 5, and under physiological conditions, this group will exist in an ionized state called an *acylate* ion; e.g., the ion of palmitic acid is *palmitate*, $CH_3(CH_2)_{14}COO^-$.

EXAMPLE 6.3

The following are some biologically important fatty acids.

(*a*) Saturated:

$CH_3(CH_2)_{14}COOH$	$CH_3(CH_2)_{16}COOH$
Palmitic acid	Stearic acid
(hexadecanoic acid)	(octadecanoic acid)
16:0	18:0

(*b*) In *unsaturated* fatty acids, the double bond nearly always has the cis conformation:

$$CH_3(CH_2)_5CH{=}CH(CH_2)_7COOH$$

Palmitoleic acid

(*cis*-9-hexadecenoic acid)

$16{:}1^{\Delta 9}$

(*c*) In *polyunsaturated* fatty acids, the double bonds are rarely conjugated:

$$CH_3(CH_2)_4CH{=}CH{-}CH_2{-}CH{=}CH(CH_2)_7COOH$$

Linoleic acid

(*cis*,*cis*-9,12-octadecadienoic acid)

$18{:}2^{\Delta 9,12}$

A *number notation* used widely for indicating the structure of a fatty acid is shown under the names of the fatty acids in Example 6.3. To the left of the colon is shown the number of C atoms in the acid; to the right, the number of double bonds. The position of the double bond is shown by a superscript Δ followed by the number of carbons between the double bond and the end of the chain, with the carbon of the carboxylic acid group being called 1.

EXAMPLE 6.4

The number notations, systematic name, and trivial name, respectively, for three more fatty acids are:

$$CH_3(CH_2)_7CH{=}CH(CH_2)_7COOH$$

$18{:}1^{\Delta 9}$ *cis*-9-octadecenoic acid (oleic acid)

$$CH_3CH_2CH{=}CH{-}CH_2{-}CH{=}CH{-}CH_2{-}CH{=}CH(CH_2)_7COOH$$

$18{:}3^{\Delta 9,12,15}$ all *cis*-9,12,15-octadecatrienoic acid (α-linolenic acid)

$$CH_3(CH_2)_4CH{=}CH{-}CH_2{-}CH{=}CH{-}CH_2{-}CH{=}CH{-}CH_2{-}CH{=}CH(CH_2)_3COOH$$

$20{:}4^{\Delta 5,8,11,14}$ all *cis*-5,8,11,14-eicosatetraenoic acid (arachidonic acid)

The melting points of different fatty acids differ markedly. For example:

Palmitic acid, 63°C; stearic acid, 70°C; oleic acid, 13°C

Elaidic acid (*trans*-9-octadecanoic acid), 44°C

Linoleic acid, −5°C; α-linolenic acid, −11°C

Question: Why do differences in melting point exist between fatty acids containing the same number of carbon atoms?

The preferred conformation of a chain of saturated C atoms is a long, straight structure. A cis double bond will cause a bend in the structure, making it less likely to pack into a crystal than will a saturated molecule of the same length. A trans double bond does not cause a bend in the chain. For example:

1. Saturated chain

COOH

2. Chain with one trans double bond

COOH

3. Chain with one cis double bond

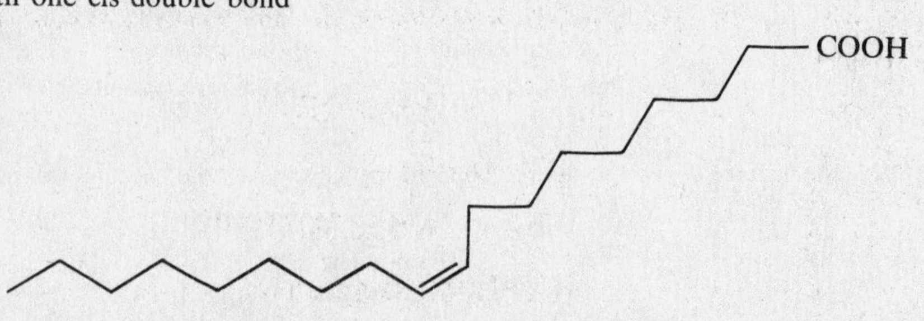

Straight molecules can pack together more densely and give crystals of higher melting point than the melting points of bent molecules of the same size; in other words, more energy is required to separate the molecules when they are heated.

The presence of cis rather than trans double bonds in unsaturated fatty acids ensures that lipids containing fatty acids have low melting points and are therefore fluid at physiological temperatures.

Question: Apart from unsaturation, what other structural feature of a fatty acid could affect its melting point?

Branching. For example, stearic acid (18 carbons) and arachidic acid (20 carbons) are both saturated and linear. They melt at 70°C and 75°C, respectively, whereas 10-methylstearic acid melts at 10°C. However, branched fatty acids are rare in living systems and the evolution of synthetic pathways for them has not occurred as a major device for keeping lipids fluid.

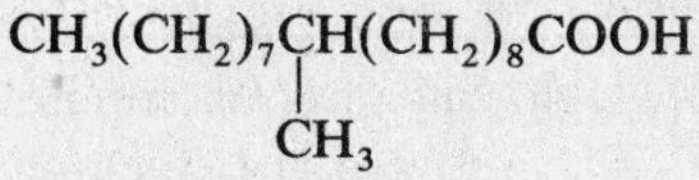

$$\begin{array}{c} CH_3(CH_2)_7\underset{\displaystyle CH_3}{\underset{|}{C}}H(CH_2)_8COOH \end{array}$$

10-Methylstearic acid

6.4 GLYCEROLIPIDS

Glycerolipids are lipids containing *glycerol* in which the hydroxyl groups are substituted in some way. In terms of quantity, these are by far the most abundant lipids in animals. Somewhat similar in structure, but occurring at levels of less than 1 percent of the glycerolipids, are lipids containing *diols*, e.g., ethylene glycol and 1,2- and 1,3-propanediol. Because of their rarity, lipids based on diols will not be discussed further.

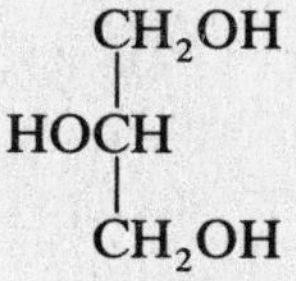

$$\begin{array}{ccc} \begin{array}{c} CH_2OH \\ | \\ HOCH \\ | \\ CH_2OH \end{array} & \begin{array}{c} CH_2OH \\ | \\ CH_2OH \\ \\ \\ \end{array} & \begin{array}{c} CH_2OH \\ | \\ HOCH \\ | \\ CH_3 \end{array} \\ \text{Glycerol} & \text{Ethylene glycol} & \text{1,2-Propanediol} \end{array}$$

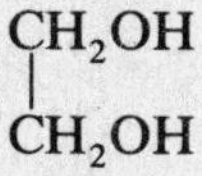

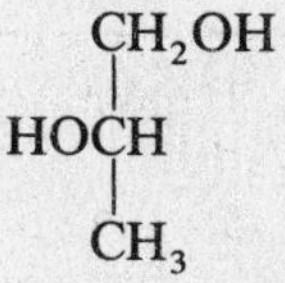

Triacylglycerols

Triacylglycerols (TAGs) are *neutral* glycerolipids and are also known as *triglycerides*. In the TAGs the three hydroxyl groups of glycerol are each esterified, usually by *different* fatty acids. This makes the second carbon of the glycerol moiety chiral. A special convention is used for dealing with the naming of TAGs and other glycerol derivatives. The derivative is drawn as a Fischer projection (Chap. 2) with the *secondary hydroxyl to the left*, and the carbon atoms are numbered 1, 2, and 3 from the top. The prefix *sn*- (for *stereospecifically numbered*) precedes the name *glycerol*. For example,

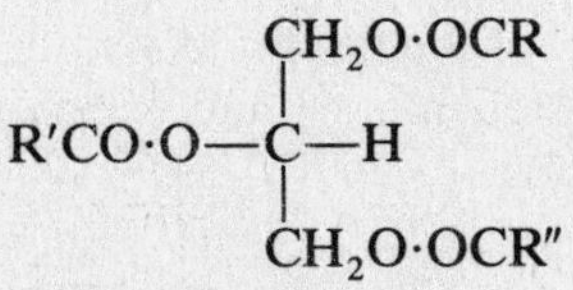

$$\begin{array}{r} CH_2O\cdot OCR \\ | \quad\quad\;\; \\ R'CO\cdot O\text{—}C\text{—}H \\ | \quad\quad\;\; \\ CH_2O\cdot OCR'' \end{array}$$

1,2,3-Triacyl-*sn*-glycerol

EXAMPLE 6.5

The structure of 1-oleoyl-2-palmitoyl-*sn*-glycerol is

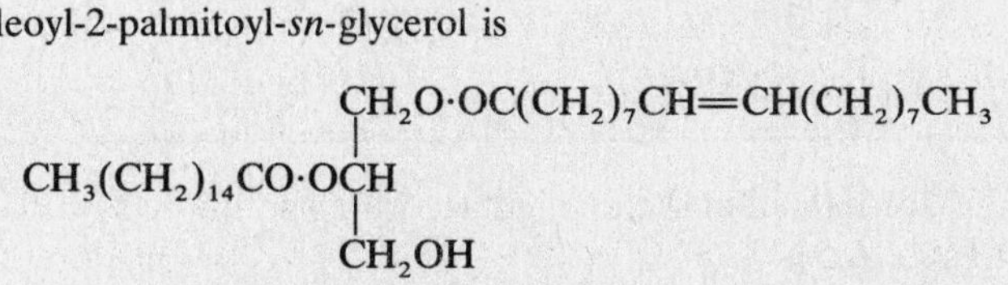

$$\begin{array}{l} \qquad\qquad\quad\; CH_2O\cdot OC(CH_2)_7CH{=}CH(CH_2)_7CH_3 \\ \qquad\qquad\quad\; | \\ CH_3(CH_2)_{14}CO\cdot OCH \\ \qquad\qquad\quad\; | \\ \qquad\qquad\quad\; CH_2OH \end{array}$$

This is a *diacylglycerol*. Diacyl- and monoacylglycerols are found in cells, but only in small amounts; they are metabolites of TAGs and phospholipids.

TAGs are the most abundant lipids found in animals. This is because they function as a food store. Although found in most cells, TAGs are particularly present in the cells of adipose tissue where they form *depot fat*. The hydrolysis of the ester bonds of TAGs and the release of glycerol and fatty acids from adipose tissue is called *fat mobilization*. Depot fat is a water-free mixture of TAGs differing from each other in the nature of the three fatty acyl groups from which they are built.

Question: Depot fat has a relatively high content of unsaturated fatty acids. What advantage does this have for the cell?

The fat exists in a liquid state. Solid fat would present only a small surface area for enzymes in the cytoplasmic water that mobilize it. Also, solid fat would render the adipose tissue rigid and unyielding during mechanical stress.

Usually the melting point of depot fat is only a few degrees below body temperature. The fatty acid composition of depot fat thus is a compromise between the requirement for keeping the fat liquid and the ability to store as much energy as possible. (Unsaturated fatty acids yield less energy than saturated ones of the same size, when oxidized.)

Phosphoglycerides

Phosphoglycerides are *polar* glycerolipids and are often referred to as *phospholipids*. However, some other lipids, not containing glycerol, also contain phosphorus, and the term phospholipid also describes these. The term *phospholipid* will be used here to describe *any* lipid containing phosphorus.

All phosphoglycerides are derived from *sn*-glycerol-3-phosphoric acid in which the phosphoric acid moiety is esterified with certain alcohols and the hydroxyl groups on C-1 and C-2 are esterified with fatty acids.

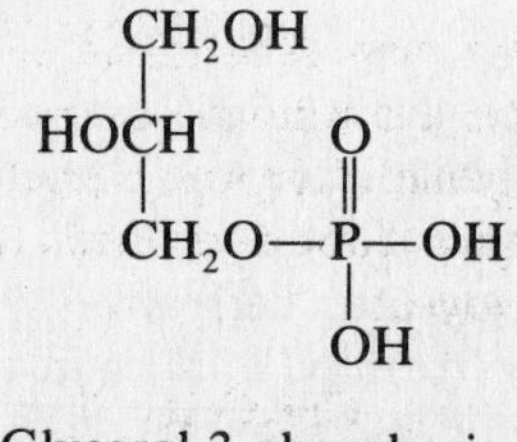

$$\begin{array}{l} \mathrm{CH_2OH} \\ \mathrm{HOCH} \qquad \mathrm{O} \\ \qquad\qquad\quad \| \\ \mathrm{CH_2O{-}P{-}OH} \\ \qquad\qquad\quad | \\ \qquad\qquad\ \mathrm{OH} \end{array}$$

sn-Glycerol-3-phosphoric acid

Question: In what form will *sn*-glycerol-3-phosphoric acid exist at physiological pH?

The pK_a values of phosphomonoesters are ~1 and ~6; therefore, there will be two ionic species at pH 7, with the dianion predominating:

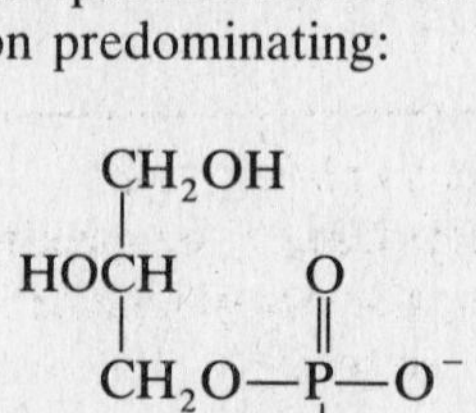

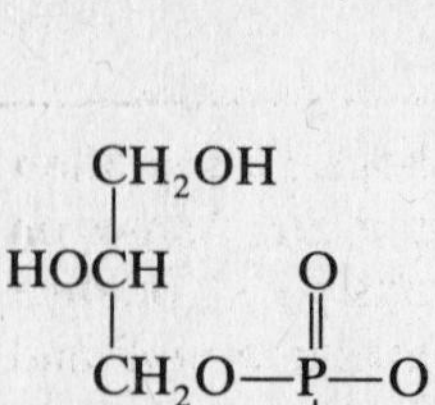

$$\begin{array}{l} \mathrm{CH_2OH} \\ \mathrm{HOCH} \qquad \mathrm{O} \\ \qquad\qquad\quad \| \\ \mathrm{CH_2O{-}P{-}O^-} \\ \qquad\qquad\quad | \\ \qquad\qquad\ \mathrm{OH} \end{array} \qquad\qquad \begin{array}{l} \mathrm{CH_2OH} \\ \mathrm{HOCH} \qquad \mathrm{O} \\ \qquad\qquad\quad \| \\ \mathrm{CH_2O{-}P{-}O^-} \\ \qquad\qquad\quad | \\ \qquad\qquad\ \mathrm{O^-} \end{array}$$

The phosphoglycerides are named and classified according to the nature of the alcohol esterifying the glycerol phosphate (Table 6.1).

Table 6.1. Some Major Classes of Phosphoglyceride

$$\begin{array}{l} \mathrm{CH_2O \cdot OCR} \\ | \\ \mathrm{R'CO \cdot OCH} \quad \overset{\displaystyle O}{\|} \\ | \\ \mathrm{CH_2O{-}\underset{\displaystyle O^-}{\overset{}{P}}{-}OX} \end{array}$$

Name of X—OH	Structure of X	Name of Phosphoglyceride (Symbol)
Water	$-\mathrm{H}$	Phosphatidate [ion of phosphatidic acid] (PA)
Ethanolamine	$-\mathrm{CH_2CH_2\overset{+}{N}H_3}$	Phosphatidylethanolamine (PE)
Choline	$-\mathrm{CH_2CH_2\overset{+}{N}(CH_3)_3}$	Phosphatidylcholine (PC)
Serine	$-\mathrm{CH_2\underset{COO^-}{CH}\overset{+}{N}H_3}$	Phosphatidylserine (PS)
Glycerol	$-\mathrm{CH_2CH(OH)CH_2OH}$	Phosphatidylglycerol (PG)
Phosphatidylglycerol	$-\mathrm{CH_2CH(OH)CH_2O\overset{O}{\overset{\|}{P}}OCH_2}$, with $^-\mathrm{O}$ on P; $\mathrm{CH_2}$—$\mathrm{HCO \cdot OCR}$—$\mathrm{R'CO \cdot OCH_2}$	Diphosphatidylglycerol [cardiolipin] (DPG)
Inositol	(inositol ring: OH, OH, OH HO, OH)	Phosphatidylinositol (PI)

In most phosphoglycerides, the fatty acid substituted on C-1 is saturated and that on C-2 is unsaturated. Although phosphoglycerides are referred to in the singular (as in Table 6.1), they are mixtures in which compounds with the same X group are esterified with a variety of different fatty acids. In some instances C-1 is *etherified* (not esterified) with a long-chain fatty *alcohol*. This phosphoglyceride is known as a *plasmalogen*:

$$\begin{array}{l} \mathrm{CH_2O \cdot R} \\ | \\ \mathrm{R'CO \cdot OCH} \quad \overset{\displaystyle O}{\|} \\ | \\ \mathrm{CH_2O{-}\underset{\displaystyle O^-}{P}{-}OX} \end{array}$$

Question: Why are phosphoglycerides often described as *polar* lipids?

The term arises because of the charges on the moiety bearing the esterified phosphate. The term *polar* is used in a relative sense; i.e., relative to TAGs, the phosphoglycerides are polar. But in an *absolute sense*, they are nonpolar overall and insoluble in water.

About 1 percent of the total phosphoglycerides that occur in animal cells are in the form of *lysophosphoglycerides*, in which one of the acyl substituents, usually from C-2, is missing. The lysophosphoglycerides are named by adding the prefix *lyso-* to the name of the parent phosphoglyceride.

Question: What is the structural formula for *lysophosphatidylcholine*?

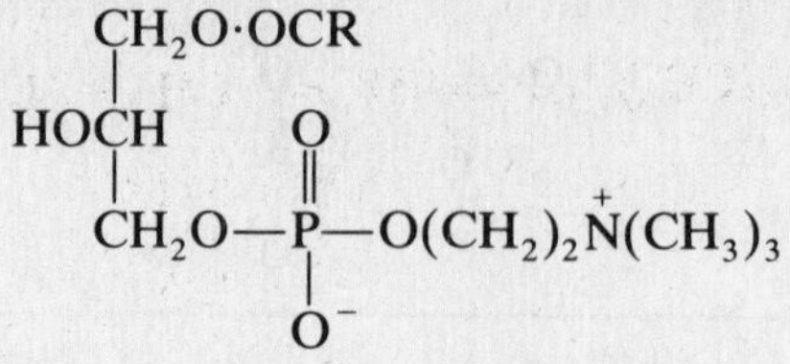

$$
\begin{array}{l}
CH_2O{\cdot}OCR \\
\,| \\
HOCH \qquad O \\
\,| \qquad\qquad \| \\
CH_2O{-}P{-}O(CH_2)_2\overset{+}{N}(CH_3)_3 \\
\qquad\quad\;\, | \\
\qquad\quad\; O^-
\end{array}
$$

Glycoglycerolipids

Glycoglycerolipids are similar in some respects to the phosphoglycerides; namely, they have hydrophobic and polar (hydrophilic) parts—the latter being provided by a carbohydrate moiety rather than by an esterified phosphate.

EXAMPLE 6.6

A typical glycoglycerolipid is 6-α-D-galactopyranosyl-β-D-galactopyranosyldiglyceride. The method for writing the carbohydrate portion is explained in Chap. 2, and the structure of the lipid is

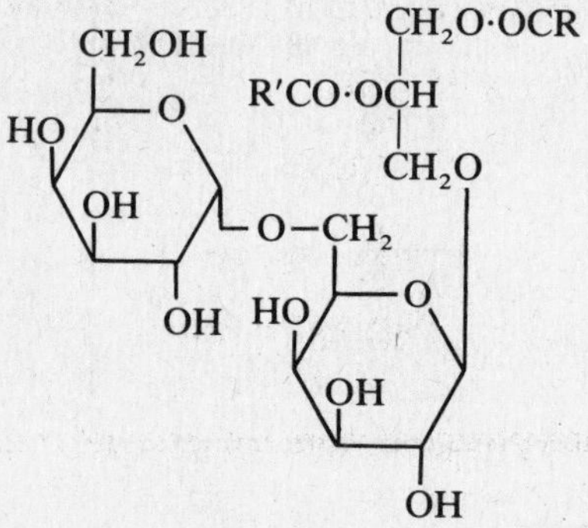

CH₂OH, O, HO, OH, OH, O—CH₂, HO, O, OH, OH; $CH_2O{\cdot}OCR$, $R'CO{\cdot}OCH$, CH_2O

6.5 SPHINGOLIPIDS

Sphingolipids are built from long-chain, hydroxylated bases rather than from glycerol. Two such bases are found in animals: *sphingosine* and dihydrosphingosine (*sphinganine*), with the former being much more common.

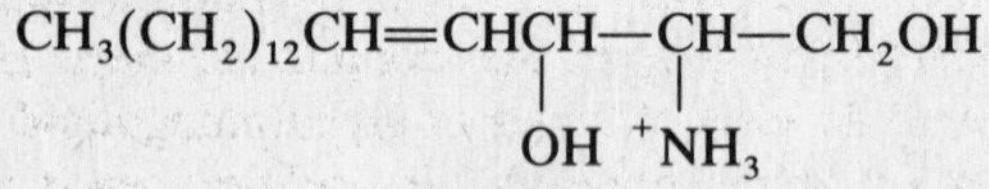

$$
\begin{array}{c}
CH_3(CH_2)_{12}CH{=}CHCH{-}CH{-}CH_2OH \\
\qquad\qquad | \qquad\; | \\
\qquad\qquad OH \;\; {}^+NH_3 \\
\text{Sphingosine}
\end{array}
\qquad
\begin{array}{c}
CH_3(CH_2)_{12}CH_2CH_2CH{-}CH{-}CH_2OH \\
\qquad\qquad | \qquad\; | \\
\qquad\qquad OH \;\; {}^+NH_3 \\
\text{Sphinganine}
\end{array}
$$

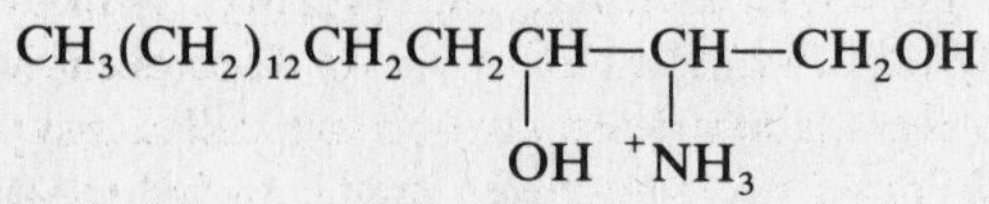

Question: From which compounds is sphinganine synthesized in living systems?

Notice the polar part is related to the amino acid *serine* (*a*) and the nonpolar part resembles *palmitate* (*b*). A reaction between these two compounds, with the elimination of CO_2, is followed by a reduction reaction which yields sphinganine.

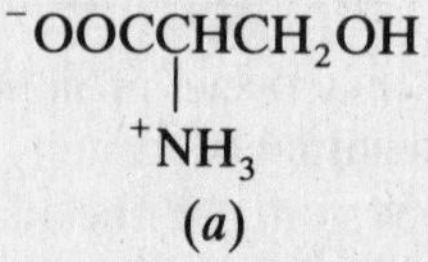

$$
\begin{array}{c}
{}^-OOCCHCH_2OH \\
| \\
{}^+NH_3 \\
(a)
\end{array}
\qquad
\begin{array}{c}
CH_3(CH_2)_{14}COO^- \\
\\
\\
(b)
\end{array}
$$

When the amino group of sphingosine or sphinganine is acylated with a fatty acid, the product is a *ceramide*,

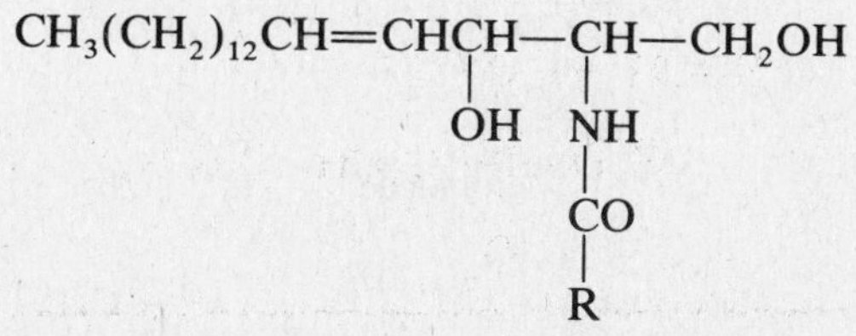

$$
\begin{array}{lcccc}
CH_3(CH_2)_{12}CH{=}CH & CH & - & CH & -CH_2OH \\
 & | & & | & \\
 & OH & & NH & \\
 & & & | & \\
 & & & CO & \\
 & & & | & \\
 & & & R &
\end{array}
$$

The primary hydroxyl group is substituted in one of two ways to give two classes of sphingolipid; these are the phosphosphingolipids and glycosphingolipids.

Phosphosphingolipids

In *phosphosphingolipids*, the primary hydroxyl group is esterified with choline phosphate. The lipid is known as *sphingomyelin*. It has the structure:

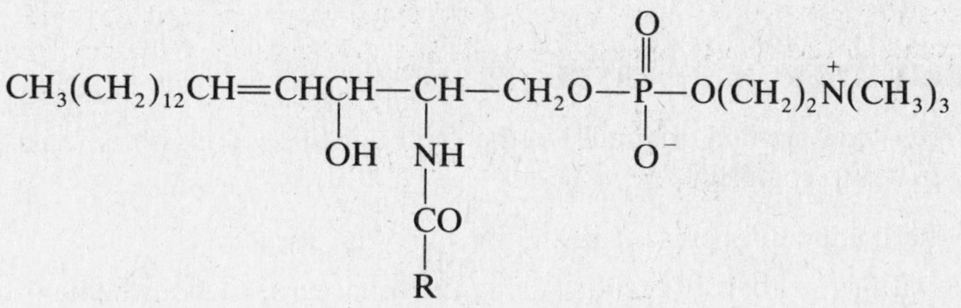

$$
\begin{array}{lcccccc}
 & & & & & O & \\
 & & & & & \| & \\
CH_3(CH_2)_{12}CH{=}CH & CH & - & CH & -CH_2O- & P & -O(CH_2)_2\overset{+}{N}(CH_3)_3 \\
 & | & & | & & | & \\
 & OH & & NH & & O^- & \\
 & & & | & & & \\
 & & & CO & & & \\
 & & & | & & & \\
 & & & R & & &
\end{array}
$$

Sphingomyelin

The fatty acyl (R) groups found in sphingomyelin are unusual; e.g.:

Lignoceric 24:0
Nervonic $24{:}1^{\Delta 15}$
Cerebronic 24:0, C-2 hydroxylated

Glycosphingolipids

In *glycosphingolipids*, the primary hydroxyl group is glycosylated, i.e., substituted with a carbohydrate, either a mono- or an oligosaccharide. Glycosphingolipids that contain the sugar *sialic acid* in the carbohydrate portion are called *gangliosides*. At least 50 types of glycosphingolipid are known, based on differences in the carbohydrate portion of the molecule. In addition, each type displays variation in the types of fatty acid found in the ceramide portion.

EXAMPLE 6.7

Some common glycosphingolipids are:

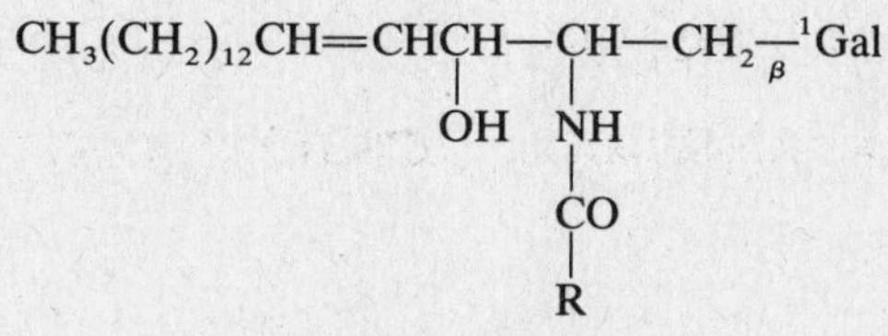

$$
\begin{array}{lcccc}
CH_3(CH_2)_{12}CH{=}CH & CH & - & CH & -CH_2 \overset{}{\underset{\beta}{-}}{}^1Gal \\
 & | & & | & \\
 & OH & & NH & \\
 & & & | & \\
 & & & CO & \\
 & & & | & \\
 & & & R &
\end{array}
$$

Galactosylceramide

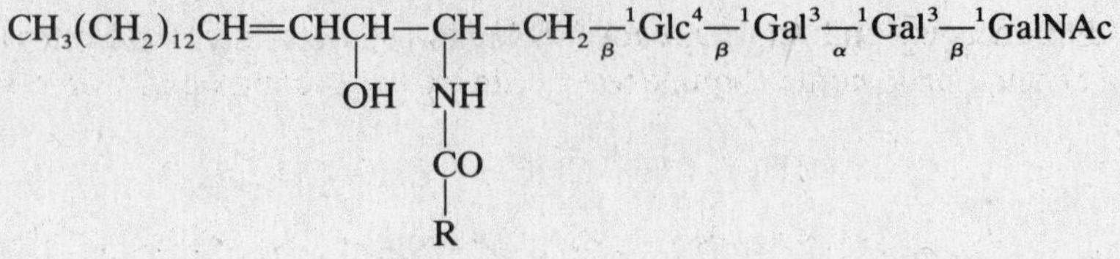

$$\mathrm{CH_3(CH_2)_{12}CH{=}CHCH{-}CH{-}CH_2 \underset{\beta}{-} {}^1Glc^4 \underset{\beta}{-} {}^1Gal^3 \underset{\alpha}{-} {}^1Gal^3 \underset{\beta}{-} {}^1GalNAc}$$

(OH on the CH carbon; NH—CO—R on the adjacent CH carbon)

Globoside

$$\mathrm{CH_3(CH_2)_{12}CH{=}CHCH{-}CH{-}CH_2 \underset{\beta}{-} {}^1Glc^4 \underset{\beta}{-} {}^1Gal^4 \underset{\beta}{-} {}^1GalNAc}$$

(OH on the CH carbon; NH—CO—R on the adjacent CH carbon; NeuNAc linked α 2→3 to Gal)

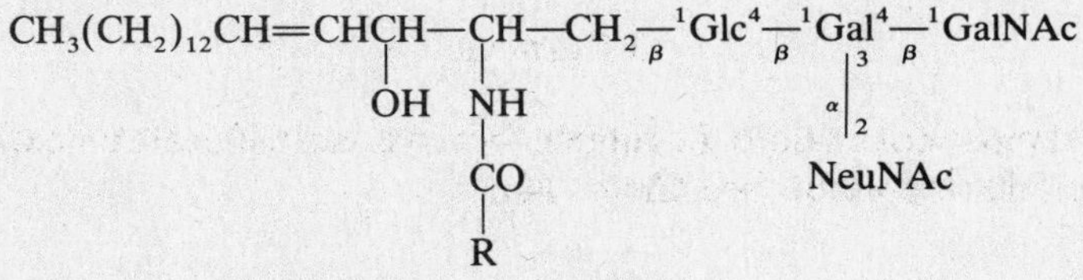

Tay-Sachs ganglioside

Note: Glc ≡ glucosyl; Gal ≡ galactosyl; GalNAc ≡ *N*-acetylgalactosaminyl; NeuNAc ≡ sialyl.

6.6 LIPIDS DERIVED FROM ISOPRENE (TERPENES)

The name *terpene* was applied originally to the steam-distillable oils obtained from turpentine (an extract of pine). It was recognized that:

1. Most of the compounds present in the oil have the formula $C_{10}H_{15}$.
2. Terpenes with more than 10 carbons exist, the number of carbons being usually a multiple of five. The structures are extraordinarily diverse.
3. Many similar water-insoluble compounds are distributed very widely; particularly large quantities are found in many plants, but they exist also in most other living organisms.

Terpenes

Terpenes with 10 C atoms are known as *monoterpenes*. Two examples are:

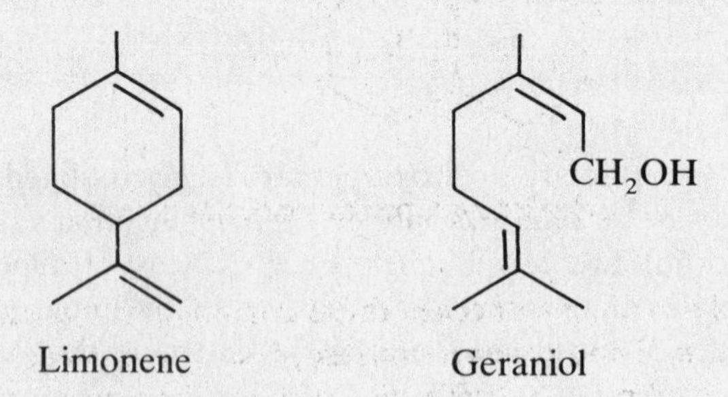

Limonene Geraniol

Sesquiterpenes and *diterpenes* have 15 and 20 C atoms, respectively. Two examples are:

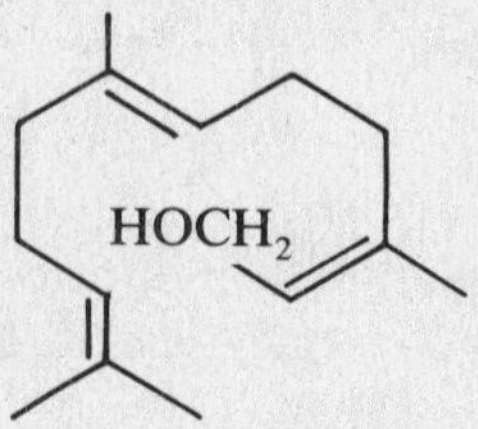

HOCH₂ CH₂OH

Farnesol 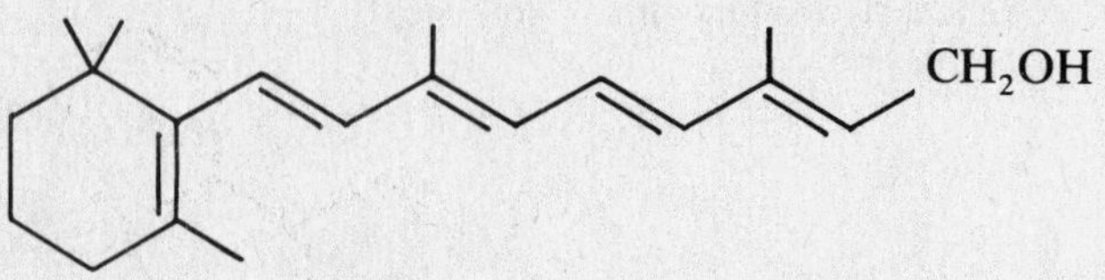 Vitamin A

Triterpenes (30 C atoms) and *tetraterpenes* (40 C atoms) require special attention because they give rise to *steroids* and *carotenoids*. Squalene (a triterpene) is compound 6 in Example 6.1.

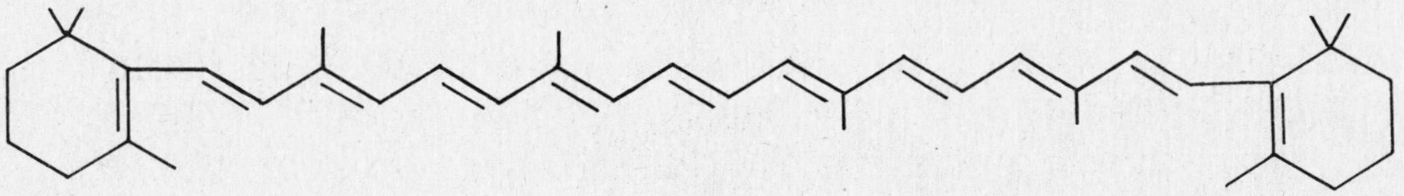

β-Carotene

Polyisoprenoid compounds exist, e.g., rubber, but in a biochemical context, the *ubiquinones* and *dolichols* are particularly important (see Chap. 14).

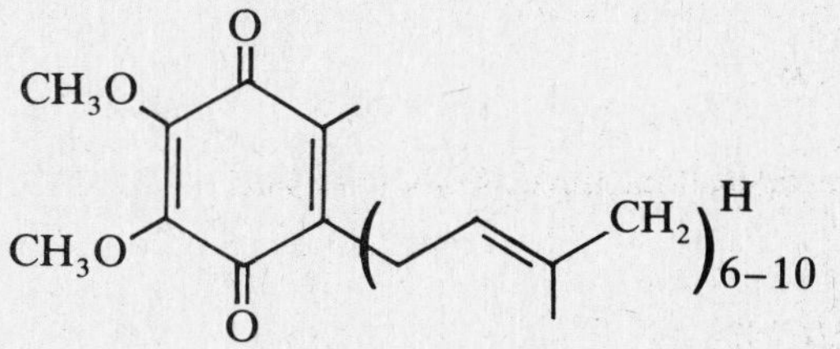

Ubiquinone

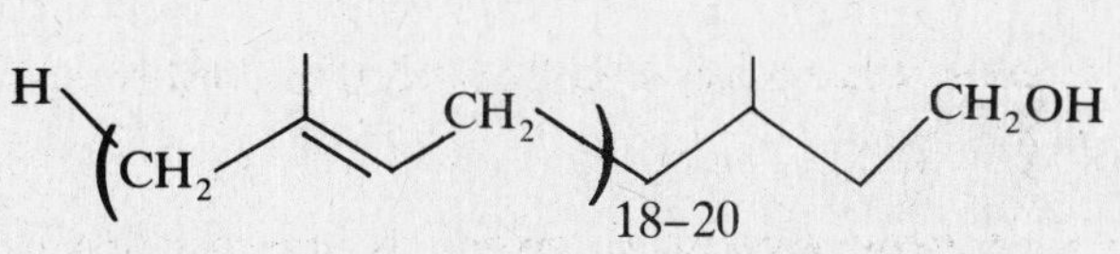

Dolichol

Steroids

Structurally, steroids are derivatives of the reduced aromatic hydrocarbon *perhydrocyclopentanophenanthrene*.

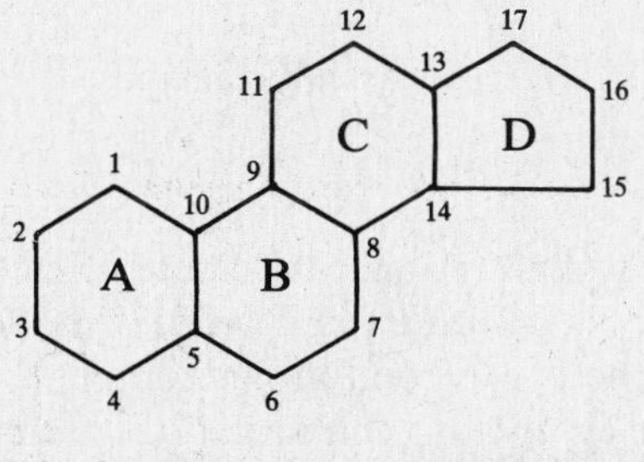

Perhydrocyclopentanophenanthrene

Although related structurally to phenanthrene, these compounds are *not acetogenins* but are true terpenes that are synthesized in living systems from isoprene via squalene. (Compare the structure of squalene—compound 6 in Example 6.1—with that of perhydrocyclopentanophenanthrene.) *Sterols*

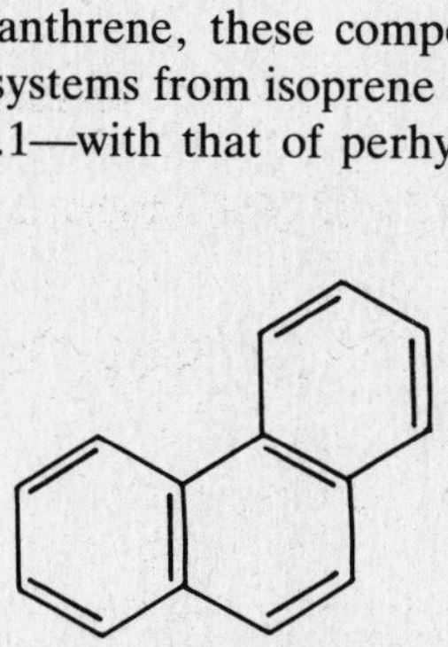

Phenanthrene

are steroids containing one or more hydroxyl groups. Some examples are *cholesterol*, a component of the cytoplasmic membrane of animal cells, *testosterone*, a hormone, and *cholic acid*, a constituent of bile.

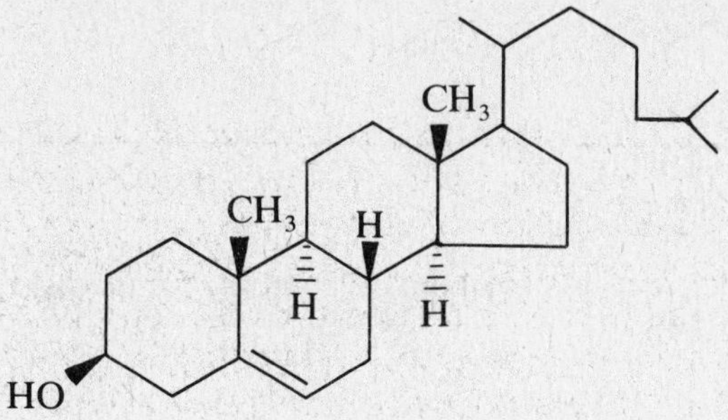

Cholesterol

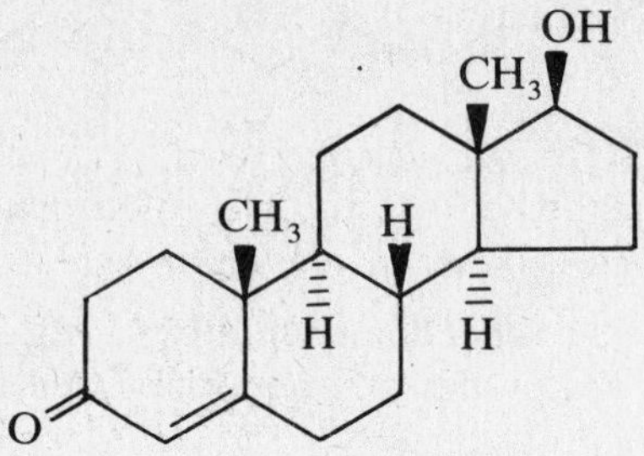

Testosterone

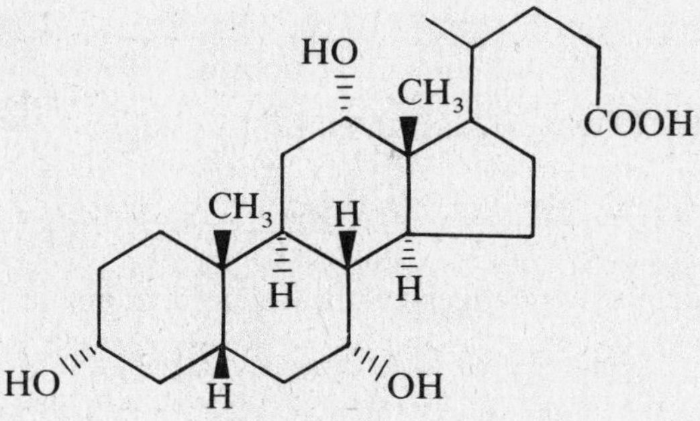

Cholic acid

The fused-ring system is an essentially planar structure. Bonds shown ▶ and ııı indicate the substituents that are, respectively, in *front* (β) and *behind* (α) the general plane of the sterol. The configuration of substituents of the two carbon atoms shared between two fused rings determines whether the rings are fused in the cis or trans conformation. Trans-fused rings are flat structures, but cis-fused rings are bent structures. Cholesterol and testosterone have trans-fused rings and are essentially planar structures.

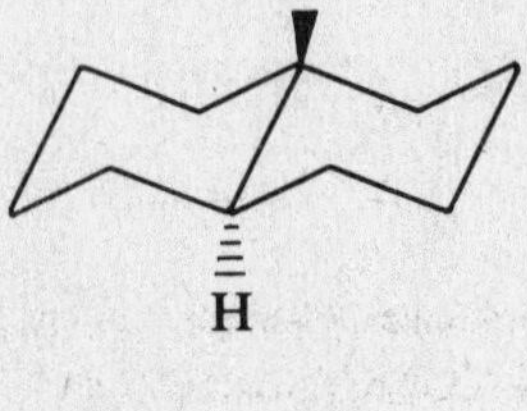

Trans-fused rings

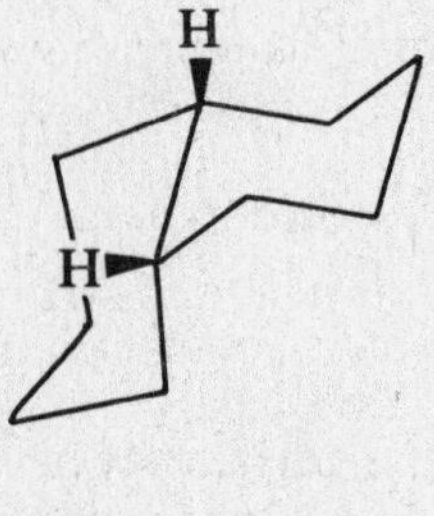

Cis-fused rings

Question: What is the three-dimensional arrangement of the rings of cholic acid?

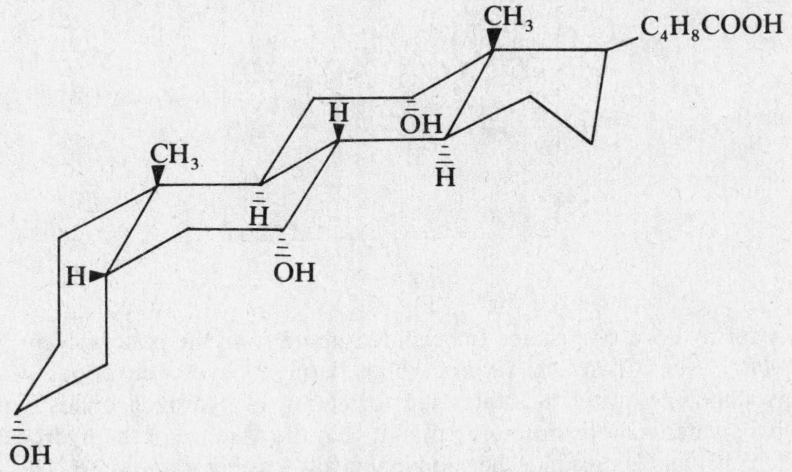

Carotenoids

These are hydroxylated derivatives of the 40-carbon hydrocarbons called *carotenes*. Because these compounds are highly conjugated, they absorb visible light; most of the yellow and red pigments occurring naturally are carotenes and carotenoids. These pigments are often involved with the interaction of living systems with light. Thus in animals, *β-carotene* (a tetraterpene) is metabolically converted to *vitamin A* (both structures are shown earlier in this section), which in turn is necessary for visual activity.

6.7 BEHAVIOR OF LIPIDS IN WATER

By definition, lipids are insoluble in water. Yet they exist in an aqueous environment, and their behavior toward water is therefore of critical importance biologically.

Many types of lipid are said to be *amphiphilic*, meaning they consist of two parts—a nonpolar hydrocarbon region and a region that is polar, ionic, or both. (The term *amphiphilic* has tended to replace *amphipathic*, used formerly.)

Question: Which of the following lipids are amphiphilic: fatty acids; acylate ions; TAGs; cholesterol; phosphoglycerides; phosphosphingolipids; glycosphingolipids?

Fatty acids, TAGs, and cholesterol are not amphiphilic; what polarity they have is extremely weak. All the others possess at least one *formal charge* or an abundance of hydroxyl groups in one part of the molecule.

When amphiphilic molecules are dispersed in water, their hydrophobic parts (i.e., hydrocarbon chains) segregate from the solvent by self-aggregation. The aggregated products are known as *micelles* (for those aggregates dispersed in water) and *monolayers* (for those aggregates at the water-air boundary). A diagram may be drawn (see Fig. 6-1) showing how an amphiphile (symbolized O—, where O represents the polar head and — represents the hydrocarbon tail) will form a monolayer on the surface of water. The polar heads are in contact with the polar water, thus ensuring that the nonpolar tails are as remote as possible from water.

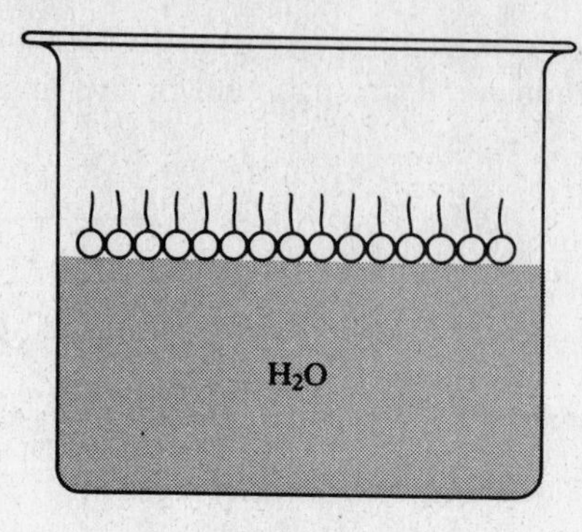

Fig. 6-1

The tendency for hydrocarbon chains to become remote from the polar solvent, water, is known as the *hydrophobic effect* (Chap. 4). Hydrocarbons form no hydrogen bonds with water, and a hydrocarbon surrounded by water facilitates the formation of hydrogen bonds between the water molecules. The bulk water is more structured than it is in the absence of the hydrocarbon; i.e., it has lost entropy (Chap. 10) and is thus in a thermodynamically less favorable state. This state is obviated by the hydrocarbon being organized so that it is remote from water, thus rendering the water molecules less ordered. Thus the hydrophobic effect is said to be *entropically driven*.

Only a small quantity of an amphiphilic lipid dispersed in water can form a monolayer (unless the water is spread as a very thin film), in which case the bulk of the lipid will form soluble micelles. Micelles can take a variety of forms, each satisfying the hydrophobic effect. Fig. 6-2 shows one such form, representing a spherical micelle, although ellipsoidal, diskoidal, and cylindrical variations are possible.

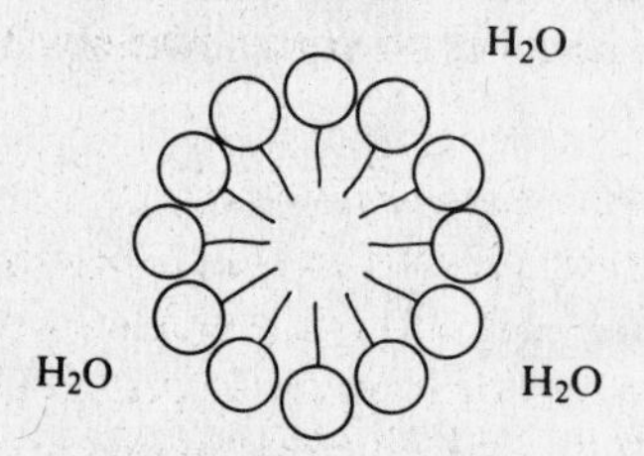

Fig. 6-2

Question: What is another form of micelle?

A bilayered structure in the form of a closed, hollow sphere is also possible [Fig. 6-3(*a*)]. This type of micelle is called a *vesicle*. The primary concept of a vesicle is two sheets of lipid with their hydrocarbon chains opposed [a *bilayer*, Fig. 6-3(*b*)]. An isolated bilayer cannot exist as such in water, because exposed hydrocarbon tails would exist at the edges of the sheet; however this situation is obviated by the sheet's curving to form a self-sealed, hollow sphere.

Thus, there are two types of micelle formed from amphiphilic lipids, small and large. Both types arise through the operation of two opposing forces: (1) attractive forces between hydrocarbon chains (van der Waals forces) caused by the hydrophobic effect forcing such chains together and (2) repulsive forces between the polar head groups.

Question: What determines the lower limit of micellar size?

The hydrophobic effect does; a minimum number of hydrocarbon chains must associate before the water-hydrocarbon interface is eliminated. This association process is a cooperative one, and the micelles therefore have a minimum size.

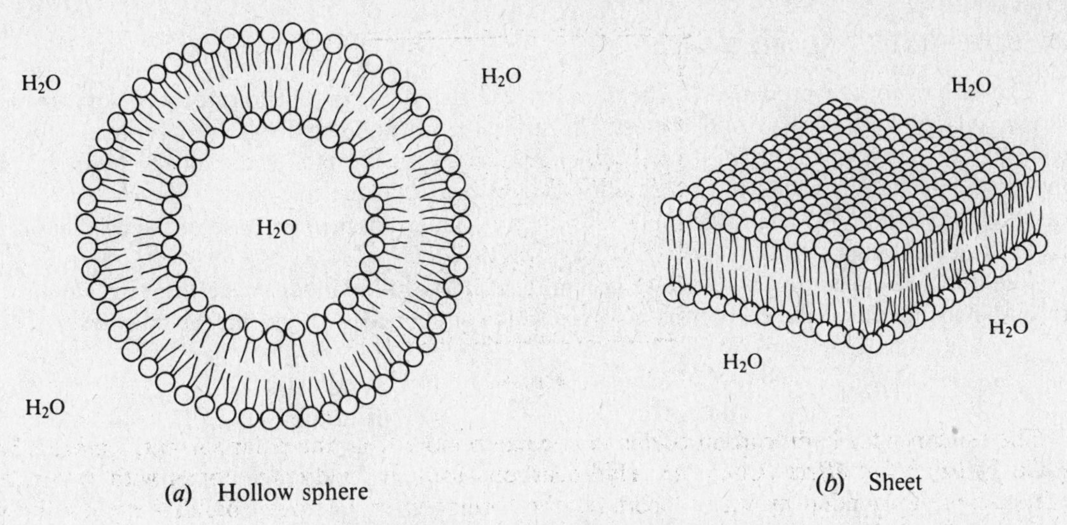

Fig. 6-3 Forms of lipid bilayers.

Question: What determines the upper limit of micellar size?

The repulsion of the polar heads does. If there are *two* hydrocarbon chains per polar head group, the nonpolar volume per head group is twice that of an amphiphilic lipid with one hydrocarbon chain. The greater repulsive force in the latter prevents the lipid molecules from coming too close and thus keeps the micellar size small. The weaker repulsive force and larger hydrocarbon volume in the former allows very much larger micelles to form; namely, bilayers and vesicles.

Question: Which lipids form small micelles and which form vesicles and bilayers?

The length of the hydrocarbon chain relative to the size of the polar head group of an amphiphile influences whether it forms micelles or vesicles in polar solvents such as water. The answers are acylate ions and lysophosphoglycerides form small micelles; phosphoglycerides, phosphosphingolipids, plasmalogens, glycoglycerolipids, and glycosphingolipids form vesicles and bilayers.

Question: How will (1) length of hydrocarbon chain, (2) ionic strength of the aqueous medium, and (3) concentration of amphiphile affect the size of small micelles?

1. Micellar size is larger with lipids having longer hydrocarbon chains, since the hydrocarbon-volume-to-head ratio is higher and the attractive forces between the tails (van der Waals forces) are greater than the ionic repulsive forces between the head groups.
2. Increasing ionic strength of the aqueous medium will permit formation of larger micelles since the higher ionic strength will decrease the ionic repulsive forces between the head groups and thus more lipid molecules will pack together.
3. At very low concentrations of amphiphile, micelles will not form; the transition from the unaggregated to the micellar state occurs over a narrow range of concentration known as the *critical micellar concentration*.

Question: What role can cholesterol play in micellar formation?

Cholesterol does not form micelles because (1) it is not amphiphilic and (2) its flat, rigid, fused-ring structure gives a solid rather than a liquid, mobile hydrocarbon phase necessary for micellar formation. Cholesterol forms *mixed micelles* with amphiphilic lipids and will enter monolayers.

6.8 BILE ACIDS AND BILE SALTS

The *bile acids* are produced in the liver by the degradation of cholesterol. They are di- and trihydroxylated steroids with 24 C atoms. The structure of cholic acid was seen earlier. *Deoxycholic acid* and *chenodeoxycholic acid* are two other bile acids. In the bile acids, all the hydroxyl groups have an α orientation, while the two methyl groups are β. Thus, one side of the molecule is more polar than the other. However, the molecules are not planar but *bent* because of the cis conformation of the A and B rings.

The bile acids produced by the liver accumulate in the gall bladder in the form of *bile salts*; they are bile acids in which the carboxylic acid group is conjugated with glycine or taurine.

$$^{+}NH_3CH_2COO^{-} \qquad\qquad ^{+}NH_3CH_2CH_2SO_3^{-}$$

Glycine Taurine

EXAMPLE 6.8

The structures for the salts of glycocholic acid and taurocholic acid are:

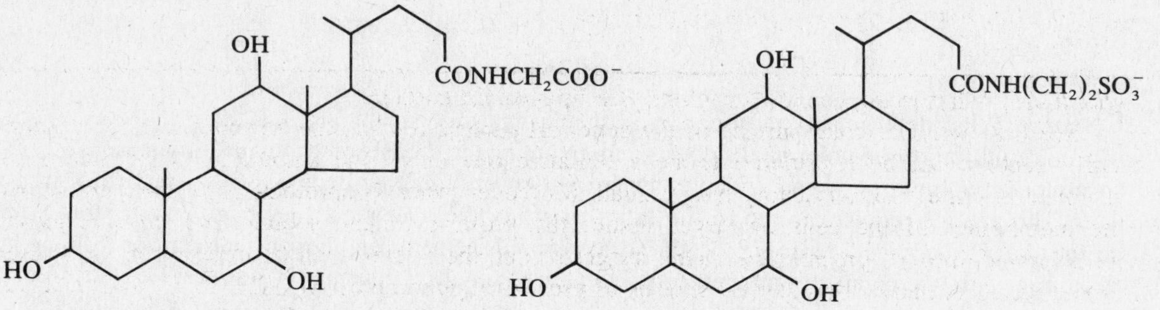

OH $CONHCH_2COO^-$ HO OH

Glycocholate

OH $CONH(CH_2)_2SO_3^-$ HO OH

Taurocholate

The bile salts have *detergent properties*, but they do not form typical micelles.

Question: Why don't the bile salts form typical micelles?

Although bile salts possess a polar head, the hydrocarbon tail is not pure hydrocarbon since there are two or three hydroxyl groups on one side of the molecule. Moreover, like cholesterol (which also does not form micelles), the rigid ring system would give a tightly packed, almost solid nonpolar phase, rather than a liquid nonpolar one.

The precise structure of aggregated bile salts is unknown, but it almost certainly involves hydrogen bonding between the hydroxyl groups in adjacent molecules. Nevertheless, the bile salts are potent detergents, and they are able to emulsify dietary lipid in the intestine, thereby making the lipid more accessible to attack by digestive enzymes. They are also required for the absorption of digested dietary lipids into the cells of the intestinal mucosa.

6.9 PLASMA LIPOPROTEINS

Blood plasma contains a number of soluble *lipoproteins*, which are classified, according to their densities, into four major types. These lipid-protein complexes function as a lipid transport system. Isolated lipids are insoluble in blood, but they are rendered soluble, and therefore transportable, by combination with specific proteins, the so-called lipoproteins. There are four basic types in human blood: (1) *chylomicrons*, (2) *very low density lipoproteins* (VLDL), (3) *low-density lipoproteins* (LDL), and (4) *high-density lipoproteins* (HDL). Their properties are summarized in Table 6.2.

Table 6.2. Plasma Lipoproteins

	Chylomicrons	VLDL	LDL	HDL
Density ($g\,mL^{-1}$)	<0.95	0.95–1.006	1.006–1.063	1.063–1.21
Max. diameter (nm)	500	70	25	15
% composition:*				
Protein	2	10	22	33
TAG	83	50	10	8
Phospholipid	7	18	22	29
Cholesterol and cholesterol esters	8	22	46	30

*Based on dry weight of the whole lipoprotein

The different compositions of the plasma lipoproteins give a clue to their function. Essentially, those lipoproteins rich in TAGs are synthesized by the liver and intestines and deliver the neutral fat to the extrahepatic tissues (particularly adipose tissue). The fat-depleted lipoproteins have a higher density, but their further metabolism is obscure.

Question: What is the general structure of a lipoprotein particle?

See Fig. 6-4. The polar surface of the spherical particle renders the assembly soluble in water. This structure can be considered to be a tentative one only. The amount of polar material in chylomicrons and VLDL is astonishingly small. Moreover, when lipoproteins come into contact with the membranes of the cells of target tissue, the proteins remain soluble and do not become incorporated into the membrane. This suggests that the proteins of lipoproteins have unusual properties. It is known that several species of proteins (*apoproteins*: AI, AII, B, CI, CII, CIII, D, and E) occur. The amino acid sequences of some of them have been determined, and they show hydrophobic regions; i.e., they have properties suggesting that parts of their structure are compatible with hydrocarbons (e.g., TAGs and the tails of phospholipids).

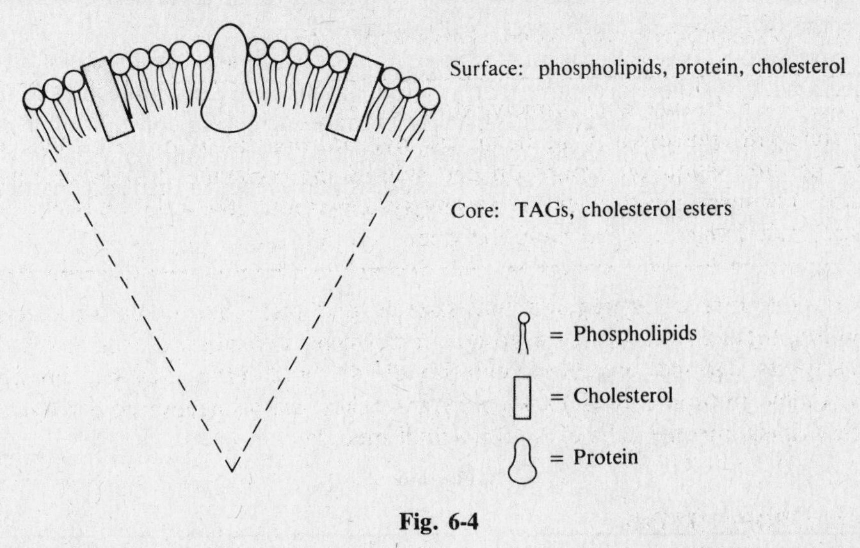

Fig. 6-4

6.10 VESICLES

When water is added to certain dry phospholipids with long hydrocarbon chains, the phospholipids swell, and when they are dispersed in more water, structures known as *liposomes* are

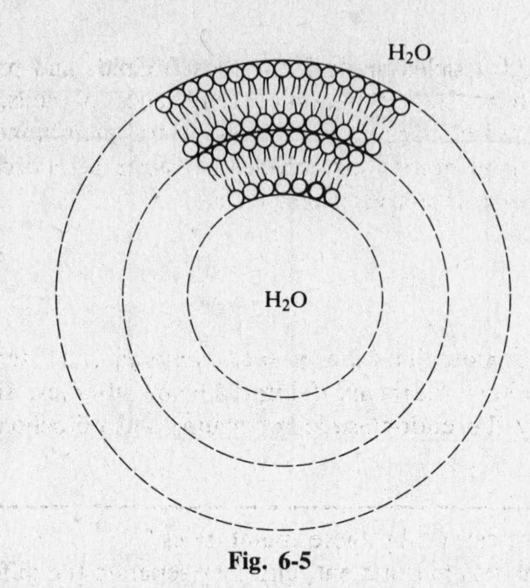

Fig. 6-5

formed. Liposomes are vesicles with multilayers of phospholipid. See Fig. 6-5. When subjected to ultrasonic vibration (*sonication*), liposomes are transformed into vesicles that have only a single bilayer of phospholipid.

Question: Vesicles can also be made by dialyzing (Chap. 10) a solution of phospholipid in detergent. How do vesicles form under these circumstances?

The phospholipid and detergent form mixed micelles, dominated by the detergent with its single chain of hydrocarbon; the micelles are therefore small. Dialysis lowers the concentration of the water-soluble detergent, so that the micelles become dominated by the phospholipid, which, having two hydrocarbon chains per molecule, is more bulky. This leads to the formation of bilayers by the coalescing of micelles.

Question: Vesicles made from a mixture of phospholipids have an asymmetric distribution of the lipids between the two leaves of the bilayer. Why is this so?

The high curvature of a vesicle can make the external surface area up to about three times the internal surface area. Thus there are more than twice the number of lipid molecules in the outer *leaf* (also called *leaflet* or *layer*). The outer leaf is thicker because increasing constriction at the center of the bilayer forces the hydrocarbon chains to be fully extended. Phospholipids with large heads (e.g., PC) tend to partition into the outer leaf, where repulsions are weaker in the less constricted space on the outside of the vesicle. See Fig. 6-6.

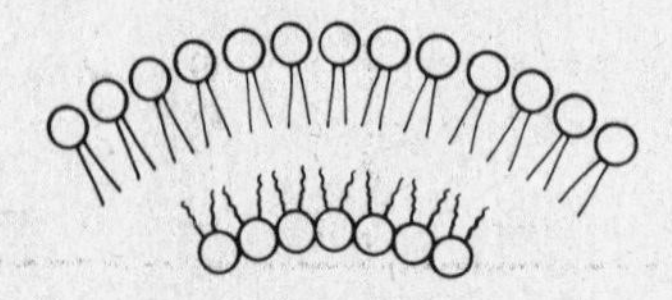

Fig. 6-6

Vesicles made by the methods described in the preceding questions are virtually impermeable to small cations and to most large polar molecules. They are slightly permeable to Cl^-, and the permeability of water is high because the solubility of water in liquid hydrocarbon is appreciable. When proteins are present during vesicle formation, they may be incorporated into the phospholipid bilayer. Such vesicles are known as *proteoliposomes*.

Apart from purely artificial vesicles made from phospholipids and proteins of choice, it is also possible to make vesicles from the cytoplasmic membranes of cells, usually by a sonication procedure. Artificial vesicles and vesicles derived from natural membranes have proved very useful in studying *transport* phenomena across membranes. Vesicles also occur naturally, e.g., by the budding of the Golgi apparatus in eukaryotic cells (Chap. 1).

6.11 MEMBRANES

The cytoplasm of cells is surrounded by a *plasma membrane*, and subcellular structures such as the nucleus, lysosomes, and mitochondria are delimited by membranes. The cytoplasm of eukaryotic cells is compartmentalized by the endoplasmic reticulum, and mitochondria have a highly folded internal membrane.

Question: What functions are served by these membranes?

They separate the cell from its environment, and they separate the different parts of the cell from each other, thus allowing certain activities to occur independently. Thus, a membrane is a physical barrier that, given the appropriate selective permeabilities, will allow the space enclosed by it to acquire and exclude useful and harmful substances, respectively, and to effect the efflux of selected compounds. Membranes also provide an environment in which chemical reactions that require nonaqueous conditions can occur.

Membranes contain lipids, proteins, and small amounts of carbohydrate. The mass ratios of these vary considerably according to the type of membrane. The carbohydrate is present as glycoglycerolipid, glycosphingolipid, and glycoprotein. The most common types of lipid found in all membranes are phosphoglycerides and phosphosphingolipid (sphingomyelin). Cholesterol is found in plasma membranes of animals but seldom in plants. Glycosphingolipids are found in the membranes of nerve and muscle tissue. Although the type and amounts of polar heads vary widely, the hydrocarbon chains found in all membrane lipids are similar.

The composition of some membranes is shown in Table 6.3.

Table 6.3. Membrane Composition

	Component (percent of weight of membrane)			Lipid Composition* (percent of total lipid)†									
Membrane	Protein	Lipid	Carbohydrate	PA	PC	PE	PI	PS	PG	DPG	SM	GS	Chol
Human erythrocyte	49	43	8	2	19	18	1	8	—	—	18	10	25
Human myelin	18	79	3	1	10	20	1	9	—	—	8	26	26
Rat liver:													
Plasma	55	40	5	1	19	12	4	9	—	—	14	—	30
Outer mitochondria	50	47	3	1	48	22	12	2	2	3	5	—	5
Inner mitochondria	75	23	2	1	43	24	6	1	2	18	2	—	3
Escherichia coli plasma	75	25	—	tr‡	—	65	—	tr	18	12	—	—	—

* SM ≡ sphingomyelin
GS ≡ glycosphingolipid
Chol ≡ cholesterol

†Some lipid analyses do not give a total of 100% because not all lipids present are recorded here.

‡tr ≡ trace.

Question: Considering (1) the nature of the lipids found in membranes and (2) the fact that membranes are two-dimensional structures, what is the most likely arrangement of the lipid molecules in a membrane?

It is a closed bilayer as occurs with artificial vesicles (Fig. 6-3). Naturally occurring membranes are sometimes called *biomembranes* to distinguish them from the membranes present in artificial vesicles and the vesicles made experimentally from natural membranes.

Some of the proteins can be removed from membranes by agents that disrupt ionic and polar bonding (e.g., urea or high concentrations of salt solution). These are known as *extrinsic* or *peripheral* proteins. Other proteins, called *intrinsic* or *integral* proteins, can be removed only by treating the membranes with detergents or with organic solvents.

Question: What do the observations concerning extrinsic and intrinsic proteins suggest regarding the location of proteins in membranes?

If a membrane is essentially a bilayer of polar lipids, then the peripheral proteins exist on the surfaces of the bilayer and are attached via ionic and polar bonds to the polar heads of the lipids or the integral proteins. The integral proteins are deeply embedded in the bilayer and are anchored in the membrane by van der Waals bonds and hydrophobic interactions.

Question: Some of the integral proteins isolated from membranes have high molecular weights. How might these fit into the lipid bilayer?

They do this by completely spanning the bilayer. (See Fig. 6-7.) The parts that project on either side are polar, while the parts embedded in the bilayer consist of amino acid residues with hydrophobic side chains or are folded in such a way that the hydrophilic side chains project into the center of the structure away from the lipid. Such integral proteins are called *transmembrane proteins*.

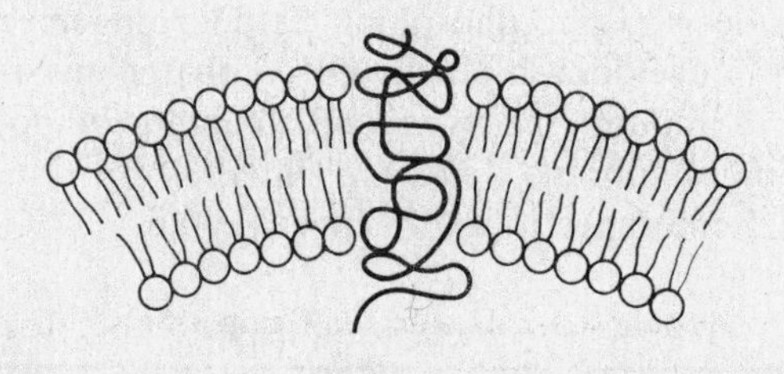

Fig. 6-7

Few integral proteins have been purified. Although some have an unusually high content of hydrophobic amino acids, this feature is not a universal one. Even fewer integral proteins have been sequenced, but those that have do possess long sequences (up to 20 and more) of residues that are hydrophobic.

So far, no protein has been found as a common constituent of all membranes (compare the almost universal existence of the lipids PC and PE), even from the same species. Thus, it seems unlikely that there is a universal structural protein in membranes. The numbers of different proteins in a membrane vary widely according to membrane type. The plasma membrane of the bacterium *Halobacterium halobium* contains only 1 protein (bacteriorhodopsin), whereas the membrane of another bacterium, *Escherichia coli*, contains about 100. The plasma membrane of the human red blood cell contains at least 17 different proteins.

The arguments developed from the beginning of this section concerning the occurrence and relationship between the lipids and proteins in membranes led S.J. Singer and G.L. Nicholson in 1972 to propose the so-called *fluid mosaic model* as a universal scheme for membrane structure (Fig. 6-8).

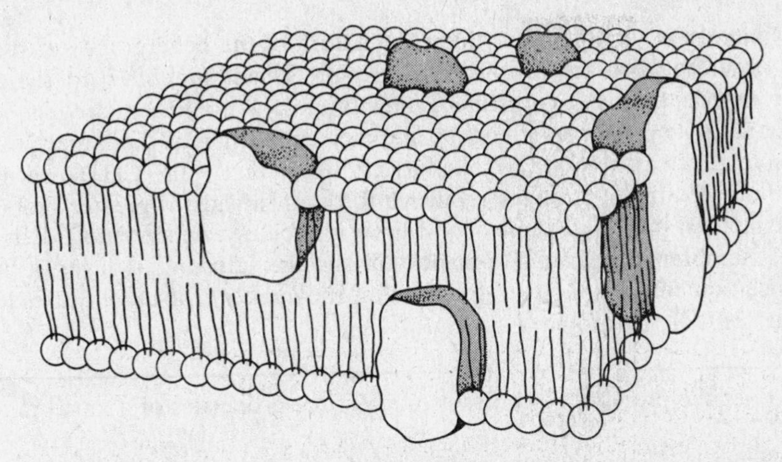

Fig. 6-8 The fluid mosaic model for membrane structure. The mosaic bilayer of polar lipids is about 5 nm thick. The integral proteins, including a transmembrane protein, are shown as irregular lumps.

Question: The model in Fig. 6-8 envisages a fluid, dynamic structure. What is meant by this?

The structure is not rigid. Because the hydrocarbon region is liquid, there is rapid *lateral diffusion* and *rotational motion* (about an axis perpendicular to the bilayer) of both lipid and protein components.

Question: What sort of motion is restricted in the model?

Movement of components from one leaf to the other (*flip-flop*). For this to occur, the polar regions of a lipid or a protein must pass, at some stage, through the hydrophobic core of the bilayer, and this is a thermodynamically unfavorable occurrence.

Question: What consequences does the restriction on flip-flop of lipids have for the structure of the membrane?

Membranes often show an *asymmetric distribution* of lipid and protein components between the two leaves of the bilayer. There is ample evidence for this. For example, PC and sphingomyelin, where it occurs, are found essentially in the outer leaf, whereas PE and PS are found predominantly in the inner leaf. The existence of unidirectional membrane processes, such as certain transport phenomena, indicates that the proteins involved must be asymmetrically distributed across the membrane. Moreover, the outer leaf of the plasma membrane contains the molecules (glycoproteins and glycolipids) that identify the cell as a particular type so that it can be recognized by components in the circulation. These glycoproteins and glycolipids do not occur in the inner leaf.

Question: How will temperature affect the physical properties of membranes?

At temperatures well below that at which a membrane occurs naturally, the lipid bilayer will be nonliquid. A reversible transition, usually over a range of ~10°C, occurs as the temperature is raised, and the hydrocarbon chains become disordered as the membrane becomes liquid. The midpoint of this transition is known as the *melting point* or *transition temperature*. Its value in a particular organism is maintained a few degrees below the ambient temperature and depends on (1) the lengths of the hydrocarbon chains and their degree of unsaturation and (2) the nature of the head groups of the lipids.

The dependence of the melting point upon the nature of the head groups of the lipids suggests that interactions occur between the various head groups. This is supported by the observation that PE is much less mobile than PC in artificial membranes. With PE, hydrogen bonds can occur between $—NH_3^+$ and a phosphate on a neighboring molecule, whereas the larger group $—^+N(CH_3)_3$ in the PC molecule relative to $—NH_3^+$ makes it less likely that $—^+N(CH_3)_3$ will fit into a tightly packed array without creating a stronger repulsive force; this leads to greater mobility.

A number of integral proteins retain bound lipid molecules when isolated. It is possible, then, that, in the intact membrane, the mobility of protein and the surrounding layer of lipid is restricted by a physical association between the two. Removal of the *lipids of solvation* causes loss of structural and functional integrity of the protein.

Question: How does the presence of cholesterol affect the properties of a membrane at physiological temperature?

Cholesterol decreases the freedom of motion of the lipids surrounding it, since cholesterol is a rigid structure having a high affinity for the hydrocarbon chains of its neighbors. Moreover, it is deeply embedded, so that its hydroxyl group is level with the ester bonds of neighboring lipids and there is hydrogen bonding between the hydroxyl group and the polar heads. On the other hand, the alkyl tail (isopropyl group) of the cholesterol increases the fluidity at the center of the bilayer.

6.12 TRANSPORT

A membrane is primarily a mechanical barrier separating two aqueous phases and is able to function as such by virtue of the two-dimensional hydrophobic bilayer of polar lipids that inhibits the free movement of solutes from one phase to the other. Free movement across a membrane is referred to as *simple diffusion*. The rate of simple diffusion of different substances varies considerably. For example, the following substances are arranged in order of decreasing rate of simple diffusion across a typical membrane: oxygen, benzene, glycerol, glucose, aspartate, hemoglobin. The rate of simple diffusion depends on the state of the substance (gases diffuse fastest), its size (the smaller, the faster), and its polarity (the more polar, the slower). In quantitative terms and for nongases, simple diffusion is of importance only in the transport of molecules with a large hydrophobic character. A few membranes, however, permit rapid simple diffusion of a wide range of substances, including polar compounds.

Question: By what mechanism do some membranes allow nonselective entry of certain solutes?

Some membranes contain *pores*. Examples of membranes that contain pores and allow quite hydrophilic compounds, generally of M_r less than 600, to pass freely include the outer membranes of mitochondria and those of gram-negative bacteria. The pores of the membranes are lined with proteins called *porins*.

Most membranes do not possess nonspecific pores allowing rapid diffusion of solutes, and for them, simple diffusion of solutes is a very slow process. Solutes can cross such membranes much more rapidly by *carrier-mediated transport*. The membrane components that mediate transport are proteins. They are known as *carriers* or *permeases* or *transport proteins*. The term *translocase* has sometimes been used but should be avoided since this term is also used to describe enzymes involved in protein biosynthesis.

Carriers are specific for the solute transported. A set of carriers will allow a cell to take up desirable substances rapidly and get rid of unwanted substances rapidly. By the same token, undesirable compounds can be kept out of cells, and useful ones retained, by the *absence* of suitable carriers, in which case simple diffusion would be the only way such compounds could cross membranes.

Although it is a simple matter, experimentally, to determine the *characteristics* of the movement of a particular compound across a membrane (e.g., the effect of pH and temperature on the rate), it is difficult to examine *how* a carrier operates at a molecular level.

Question: Why is determining the mechanisms of carrier transport so difficult?

For a carrier to be studied effectively it must usually be purified. But purifying it involves removing it from the membrane in which it exists naturally; i.e., the membrane must be disrupted, and as a result, the very quality by which the biological activity of the carrier can be measured is lost. This problem can be overcome by incorporating the purified carrier into artificial vesicles.

EXAMPLE 6.9

In order to measure the rate of transport of a solute into a cell (or artificial vesicle), radioactively labeled solute is used. A suspension of cells is incubated in a medium containing radioactive solute; portions of the suspension are removed at different times, and the radioactivity of the washed, filtered cells is measured. Fig. 6-9 shows a plot of the radioactivity of the washed cells at different times. The slope of the initial part of the plot gives the rate of uptake of the solute (after solute concentrations are first derived from measurements of radioactivity) corresponding to the solute concentration at time = 0, $[S]_o$.

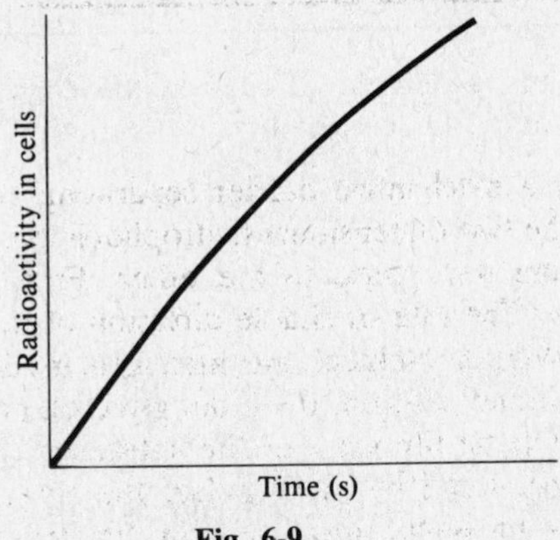

Fig. 6-9

Question: Are there any experimental artifacts likely in the method described?

Yes. Once in the cells, the solute may be metabolized. Thus, its actual concentration inside the cell will be less than that indicated by the measured radioactivity of the cells. Also, if there is solute already inside the cell, the measured rate of uptake is a *net* value since there must be some movement (by simple diffusion and carrier transport) of the solute out of the cell. (These objections do not usually apply to transport by carriers incorporated into artificial vesicles.) These problems are overcome partly by using a nonmetabolizable analog; e.g., in studying the transport of glucose, radioactive 2-deoxyglucose or α-methylglucoside could be used.

In studies on transport in bacteria, another method for overcoming the problem of solute metabolism is to use bacterial mutants which will transport the solute but which lack one or more enzymes for metabolizing the solute.

Question: What would be an objection to the use of analogs?

The precise rate of transport of the compound of interest, e.g., *glucose*, is not determined, although it can be *presumed* to be similar to that of the analog.

Question: Given that it is possible to measure v, the rate of transport of a solute into a cell, how will v vary with the concentration of solute $[S]_o$ in which the cells are suspended for (*a*) carrier-mediated transport and (*b*) simple diffusion?

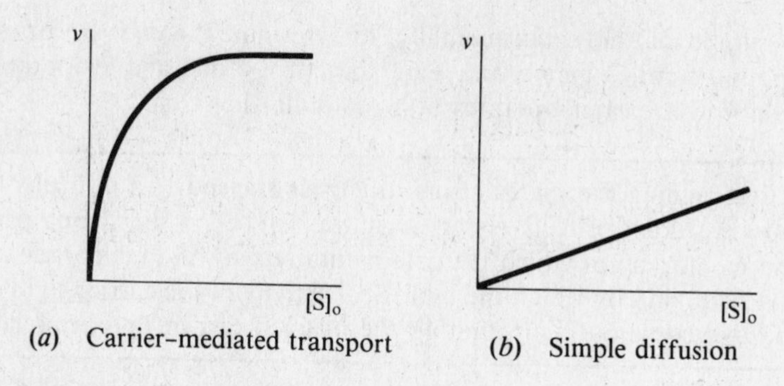

(*a*) Carrier–mediated transport (*b*) Simple diffusion

Fig. 6-10 Dependence of rate of transport on solute concentration.

For carrier-mediated transport [Fig. 6-10(*a*)] there must be a finite number of carrier molecules in the membrane. At low $[S]_o$, only some of these molecules will be bound to the solute, but at high $[S]_o$, *most* of the carrier molecules will be occupied, and there is therefore a maximal value for v (V_{max}). In simple diffusion [Fig. 6-10(*b*)], there is no carrier to *saturate*, and v is higher at high $[S]_o$ because the concentration gradient of solute across the membrane, which determines the rate of diffusion, is greater than it is at low $[S]_o$.

Mathematical equations can be derived to explain the curves demonstrated in Fig. 6-10 (see Chap. 9). The curve in Fig. 6-10(*a*) is described by the equation for a rectangular hyperbola.

$$v = \frac{V_{max}[S]_o}{[S]_o + \text{a constant}} \tag{6.1}$$

The equation for the graph in Fig. 6-10(*b*) is

$$v = \frac{\text{a constant}}{l}[S]_o \tag{6.2}$$

where l is the thickness of the membrane. (In the derivation of these equations it is assumed there is no solute on one side of the membrane when a concentration of $[S]_o$ is introduced on the other side.)

Question: Suppose carrier-mediated transport *and* appreciable simple diffusion occurred simultaneously in an experiment on transport. How could this be demonstrated?

A plot of v against $[S]_o$ would not show saturation, except when simple diffusion was trivial compared with carrier-mediated transport. See Fig. 6-11. However, it is often impossible to determine v at high $[S]_o$; e.g., the solute may not be soluble enough to achieve high $[S]_o$. Mixed processes of movement of solute across a membrane can be demonstrated using a *double-reciprocal plot*. (See also Chap. 9.)

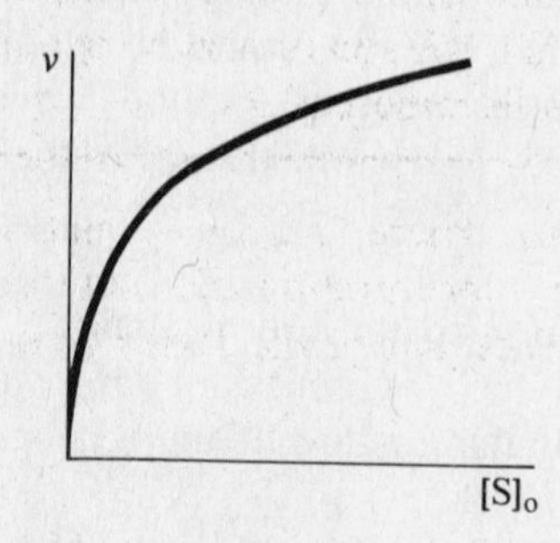

Fig. 6-11

EXAMPLE 6.10

The inverted forms of Eqs. (*6.1*) and (*6.2*) are

$$\frac{1}{v} = \frac{1}{V_{\max}} + \frac{\text{a constant}}{V_{\max}}\frac{1}{[\mathrm{S}]_{\mathrm{o}}} \quad \text{and} \quad \frac{1}{v} = \frac{l}{\text{a constant}}\frac{1}{[\mathrm{S}]_{\mathrm{o}}}$$

Plots of $1/v$ against $1/[\mathrm{S}]_{\mathrm{o}}$ for both of these are straight lines [see Fig. 6-12(*a*)]. For simultaneous simple diffusion and carrier-mediated transport, the double-reciprocal plot is shown in Fig. 6-12(*b*):

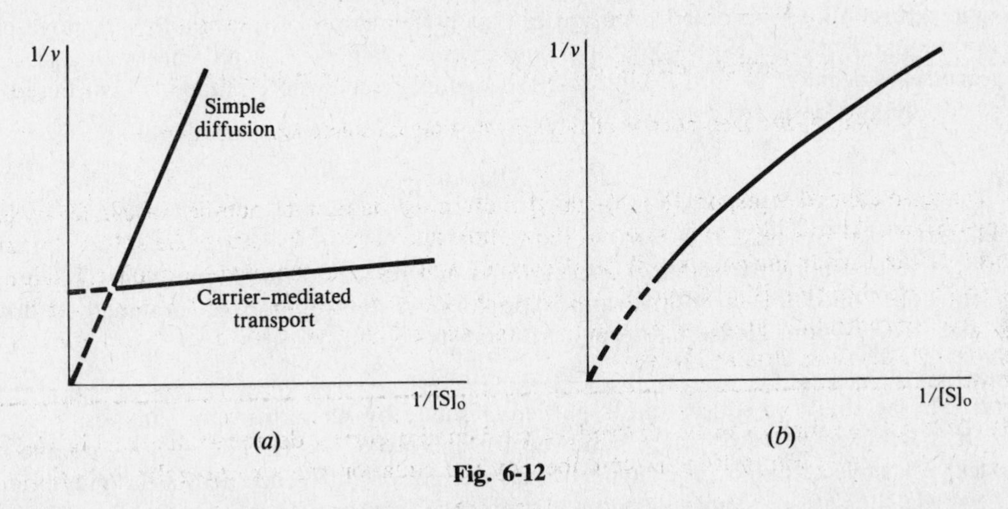

Fig. 6-12

An alternative method for distinguishing simple diffusion from carrier-mediated transport can be used if the solute contains a chiral atom. Simple diffusion will occur for both enantiomers and at the same rate, but carrier-mediated transport is invariably stereospecific; i.e., the carrier will recognize only one of the enantiomers.

Question: Which would be *recognized* (bound) by a carrier located in a cell membrane, D-glucose or L-glucose?

D-Glucose would be recognized by a carrier because this isomer occurs naturally. The other does not, and there would be little point in living systems evolving such a carrier.

Question: Under what conditions would the process of (*a*) simple diffusion and (*b*) carrier-mediated transport cease?

(*a*) Simple *net* diffusion would cease when the concentrations of solute on both sides of the membrane were the same.

(*b*) With some carrier-mediated transport systems it has been found that, as in simple diffusion, net transport ceases when the solute concentration is the same on both sides of the membrane. With other systems, the solute continues to be transported even after the solute concentrations on both sides become equal.

As the preceding question shows, there are *two* types of carrier-mediated transport: (1) *facilitated diffusion* (which allows the concentration of solute on both sides of a membrane to be equalized) and (2) *active transport* (which allows the solute to move *up*, or *against*, a concentration gradient).

EXAMPLE 6.11

Figure 6-13 shows the ways in which solutes can move across a membrane. [S] and [s] represent high and low concentrations of solute, respectively.

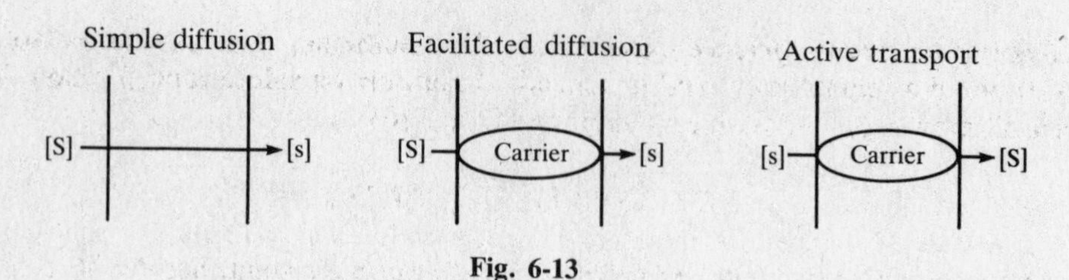

Fig. 6-13

Simple diffusion and facilitated diffusion are spontaneous processes; solute will move down a concentration gradient (i.e., [S]→[s]) until equilibrium is attained. The free energy change, ΔG, for these processes is negative (Chap. 10) because the solute becomes more randomly distributed. Thus, no input of energy is required for the process.

$$\Delta G = 2.3RT \log \frac{[s]}{[S]} \tag{6.3}$$

At equilibrium [s] = [S] and ΔG is zero.

With active transport, the solute is concentrated on one side of the membrane. This process is not spontaneous, since ΔG is positive and some form of energy must be provided to drive the carrier-mediated reaction. Hence the term *active* transport.

Question: Is the ability to concentrate a particular solute by active transport limitless?

No. If the concentration of solute builds up to high levels on one side of a membrane, solute leaks back by simple diffusion. The higher the concentration difference across the membrane, the greater will be the rate of simple diffusion. In practice, concentration ratios greater than several hundred to one never occur, and the ratios are usually very much smaller.

The solute will also diffuse back across a membrane if the membrane contains a carrier for facilitated diffusion. That is, simple diffusion and facilitated diffusion are *bidirectional*; active transport is *unidirectional*.

Question: The movement of an *ionized* solute across certain types of membranes does not always give rise to equal concentrations of the ion on both sides of the membrane, even when only simple diffusion, facilitated diffusion, or both are available. How can this be explained?

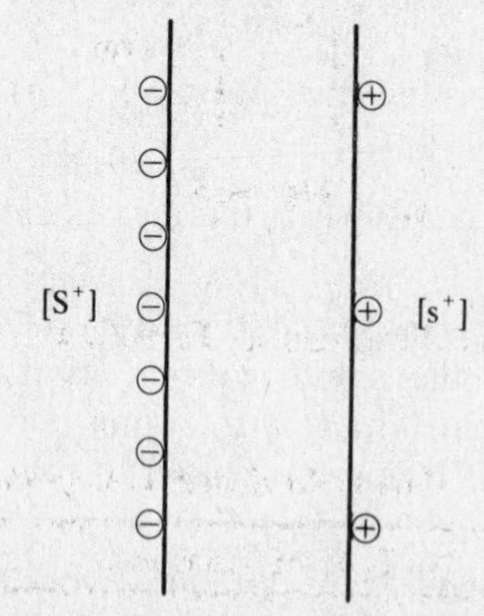

Fig. 6-14 $[S^+]$ and $[s^+]$ = equilibrium concentrations. Note that there are more negative than positive charges on the membrane.

If an *electrical potential difference* exists across the membrane, then the ionized solute will be unevenly distributed at equilibrium. ΔG is dependent not only on solute concentration, but also on the potential difference $\Delta\Psi$ across the membrane (Chap. 10).

$$\Delta G = 2.3RT \log \frac{[s^+]}{[S^+]} + ZF\,\Delta\Psi$$

where Z is the net charge on the solute, and F is the Faraday constant. See Fig. 6-14.

For active transport to occur, some form of energy must be provided. In *primary active transport*, energy is provided either by the hydrolysis of ATP or by utilization of electron flow down an electron-transport chain (see Chap. 10). In *secondary active transport*, the energy is provided by ions moving down a concentration gradient that runs in the same direction as the solute's concentration gradient. The ion gradient is set up by primary active transport. A special type of active transport is *group translocation*. In this, the substance transported is modified chemically while crossing the membrane, e.g., by accepting a chemical group; energy is expended since the chemical group is transferred as a result of cleavage of a *high-energy* compound.

EXAMPLE 6.12

Figure 6-15 shows the three types of active transport. The source of energy is assumed to be ATP. The symbols [s], [i$^+$] and [S], [I$^+$] represent low and high concentrations of solute and ion, respectively. AX is a compound from which a chemical group X (e.g., a phosphate) is transferred to the solute to form SX.

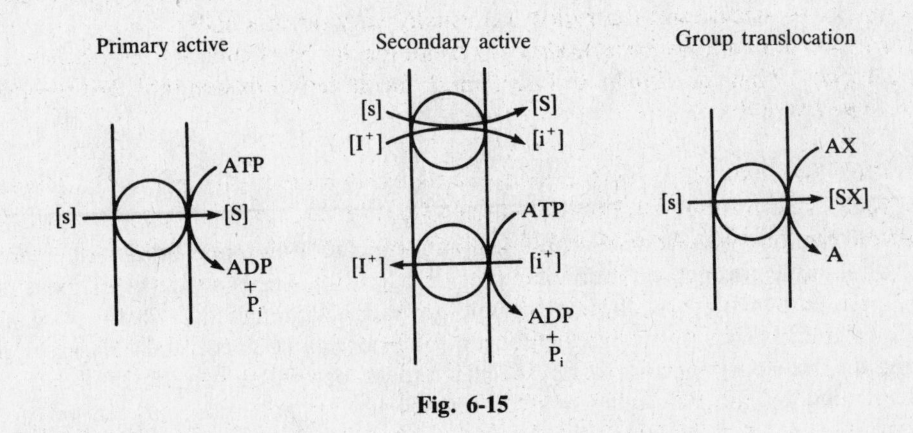

Fig. 6-15

Some examples of carrier systems are given in Table 6.4.

Table 6.4. Carrier-Mediated Transport Systems

Solute	Type of Transport	Occurrence
Glucose	Facilitated diffusion	Most animal cells
ATP out—ADP in	Facilitated diffusion	Mitochondria
H^+	Primary active (uses ATP)	Stomach epithelia
Na^+ out—K^+ in	Primary active (uses ATP)	Animal cells
Glucose	Secondary active (Na^+ cotransport)	Some animal cells
Glucose	Group translocation	Many bacteria

6.13 MOLECULAR MECHANISMS OF TRANSPORT ACROSS MEMBRANES

A protein engaged in the transport of a specific solute across a membrane must have two properties: (1) the ability to bind the solute and (2) the ability to carry out a *vectorial* process, i.e., a *directional* process which delivers the solute from one side of the membrane to the other. In no instance have the details of the vectorial process been worked out.

Question: What are some simple means by which the vectorial process might occur?

1. The transport protein could behave as a *mobile carrier*; e.g., a large, transmembrane protein could rotate, as in Fig. 6-16(*a*), or a small protein could traverse the membrane as in Fig. 6-16(*b*).
2. The transport protein could constitute a pore or channel as in Fig. 6-16(*c*).

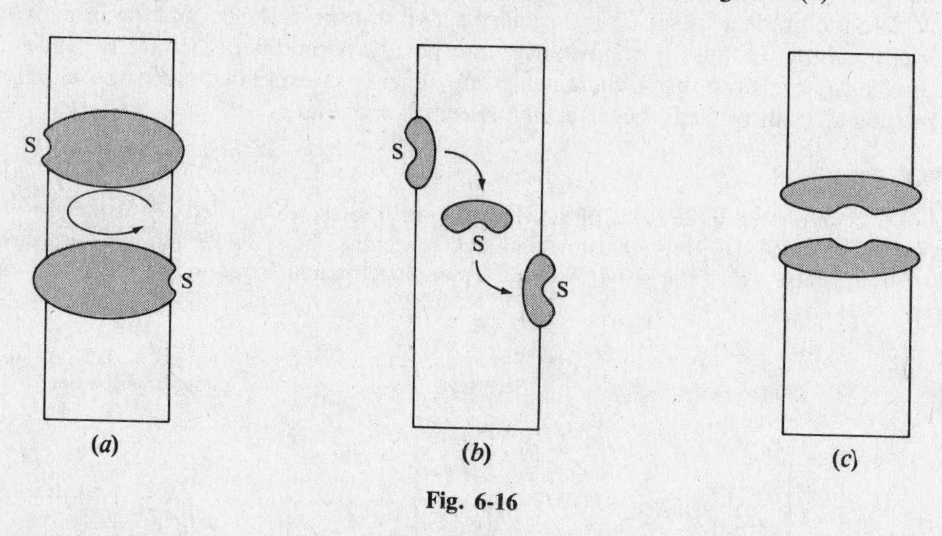

Fig. 6-16

There are difficulties with both concepts presented in the above question. Movement of polar segments of a molecule through the nonpolar part of a lipid bilayer is a rare event, and the mobile carrier concept is not favored on thermodynamic grounds. Although the existence of *nonspecific* channels in some membranes is known, the concept of a pore can be accepted only if the pore has a *gate* enabling the transport protein to be specific and to permit movement of the solute in the direction demanded by the particular type of transport. The pore concept can be modified by proposing that the transport protein is an oligomeric protein in which the spaces between subunits constitute a water-filled channel closed by contact of the subunits (Fig. 6-17). Binding of solute to the protein triggers a conformational change altering the relative positions of the subunits and so opens the channel. This would be *facilitated* diffusion. If metabolic energy is expended to cause the conformational change, rather than simply binding of the solute, then this would be *active transport*. So far, experimental studies on the structures of transport systems have provided more support for the pore concept than for the mobile carrier hypothesis.

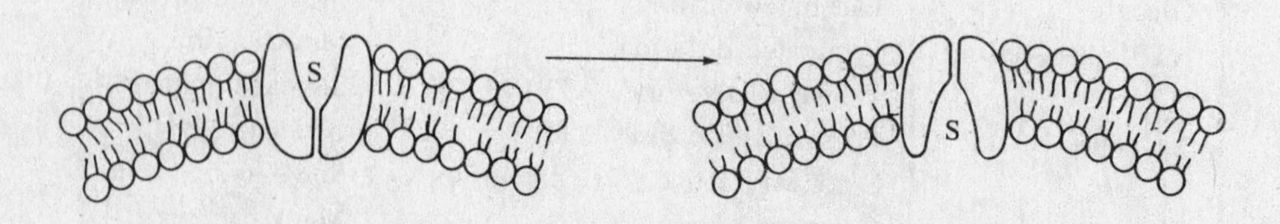

Fig. 6-17 Conformational change in a pore protein allowing transport of the solute.

Artificial systems permitting the transport of cations have been inserted into natural membranes and into artificial vesicles. These are known as *ionophores*, and there are two types, exemplified by *gramicidin A* and *valinomycin*.

1. Gramicidin A is an antibiotic composed of 15 amino acids:

   CHO—NH—L-Val—Gly—L-Ala—D-Leu—L-Ala—D-Val—L-Val—D-Val—L-Trp—D-Leu—L-Trp—D-Leu—L-Trp—D-Leu—L-Trp—CONH(CH_2)OH

   Because of its hydrophobic nature, gramicidin A penetrates a lipid bilayer: two molecules form a head-to-tail dimer, which can twist to form a hollow helix with six turns and a total length of 3 nm, about the distance across the hydrophobic region of a lipid bilayer. The helix is stabilized by intramolecular hydrogen bonds between peptide links in the polypeptide chain. The hydrophobic side chains are on the outside of the helix, leaving a hydrophilic aqueous channel down the center of the helix through which cations can pass. This type of ionophore is nonspecific, and the cations do not bind strongly to the gramicidin A.

2. Valinomycin is an ionophore but with different properties from gramicidin A: (*a*) it specifically transports K^+, and no other ion, when inserted into membranes or vesicles; (*b*) it can transport K^+ only above the phase transition temperature of the membrane, whereas cation transport by gramicidin A is insensitive to temperature.

   Valinomycin has a cyclic structure resembling a peptide chain containing 12 monomers; basically it is a four-unit structure that is repeated three times:

   NH—(L-Val—D-HyV—D-Val—L-Lac)$_3$CO (cyclic)

   where D-HyV is D-hydroxyisovalerate and L-Lac is L-lactate. D-HyV and L-Lac are bonded in the chain via their hydroxyl groups that form an ester bond with the carboxyl group of a neighboring residue in the same way that an amino acid binds via its amino group to form a peptide bond.

Question: What would be a possible mechanism of K^+ transport by valinomycin?

Since valinomycin transports K^+ specifically, it must bind the ion rather than provide a channel, as does gramicidin. Since transport can occur only when the membrane is fluid and since valinomycin is a small molecule, this suggests that the valinomycin-K^+ complex must move through the membrane; i.e., it behaves as a *mobile carrier*. The hydrocarbon side chains of the cyclic structure can be imagined on the outside of the structure, thus making the latter compatible with the hydrocarbon portion of the lipid bilayer. The inside contains the 12 carbonyl groups of the ester and amide bonds, six of which coordinate the K^+ in the space at the center of the structure (Fig. 6-18).

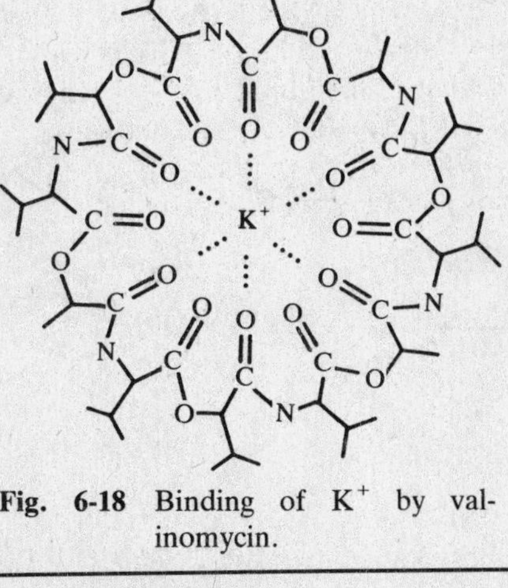

Fig. 6-18 Binding of K^+ by valinomycin.

Solved Problems

CLASSES OF LIPIDS

6.1. Will the melting point of lactobacillic acid ($C_{19}H_{36}O_2$, a *cyclopropane fatty acid*) be higher or lower than that of the linear, saturated fatty acid of the same chain length?

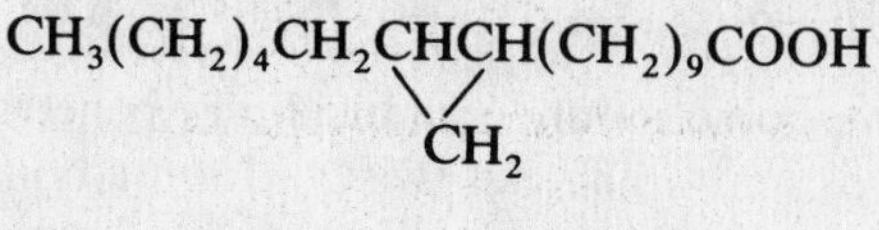

$$CH_3(CH_2)_4CH_2\underset{\diagdown\!/\atop CH_2}{CHCH}(CH_2)_9COOH$$

SOLUTION

The melting point will be lower; the cyclopropane group makes it difficult for an array of molecules to pack regularly. The melting points of lactobacillic acid and the 19:0 fatty acid are 28°C and 69°C, respectively.

6.2. Mevalonic acid, radioactively labeled at the α-carbon atom, was fed to an organism that synthesizes cholesterol. Which atoms in the cholesterol will be labeled?

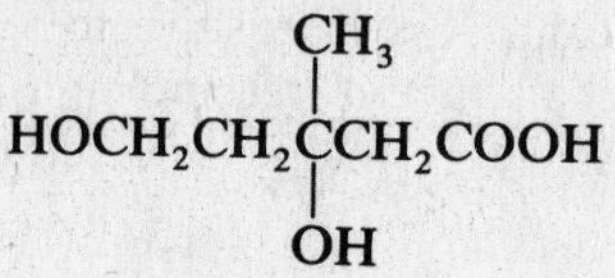

$$HOCH_2CH_2\overset{CH_3}{\underset{OH}{C}}CH_2COOH$$

Mevalonic acid

SOLUTION

Sterols are synthesized in nature from squalene and, therefore, ultimately from isoprene. Mevalonic acid is the immediate precursor of the isoprene unit, and the carboxylic acid group is lost as carbon dioxide when two mevalonic acid molecules combine head to tail. Thus, if the α carbon of mevalonic acid is labeled, then this carbon is always adjacent to the carbon bearing a side-chain methyl group. Examination of the way in which six isoprene units are linked in squalene (Example 6.2) shows that they are not all linked head to tail; there is a point of symmetry in the structure of squalene (marked * in the structure below). At this point a set of three isoprene units, linked head to tail, is joined *head-to-head* to a similar set of three isoprene units, to give the labeling pattern shown.

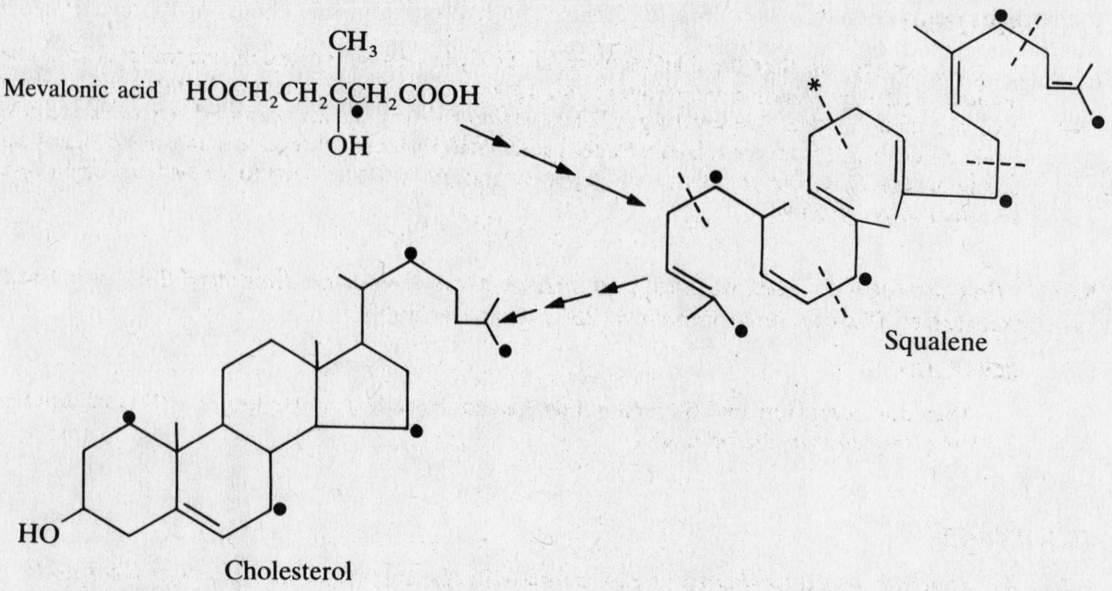

Mevalonic acid $HOCH_2CH_2\overset{CH_3}{\underset{OH}{C}}\overset{\bullet}{C}H_2COOH$

Squalene

HO

Cholesterol

BEHAVIOR OF LIPIDS IN WATER

6.3. Will phospholipids with short hydrocarbon chains form bilayers?

SOLUTION

No. They form small micelles (see the fifth question in Sec. 6.7). The hydrocarbon chains need to contain at least six carbon atoms for bilayers to form.

6.4. Explain why acylate ions have detergent properties.

SOLUTION

Acylate ions are amphiphilic, and the hydrocarbon chains are able to penetrate fatty (hydrophobic) particles, leaving the surface of the particle ionic. (See Fig. 6-19.) Thus, the particle behaves as a micelle and is readily soluble in water. The sodium and potassium salts of fatty acids are *soaps*. Soaps have poor detergent properties in hard water because the calcium present in such water causes micelles to aggregate and precipitate. The divalent calcium ion can act as a bridge between two micelles, but since a micelle is polyvalent, a small amount of calcium relative to the amount of the soap can cause all the micelles to aggregate.

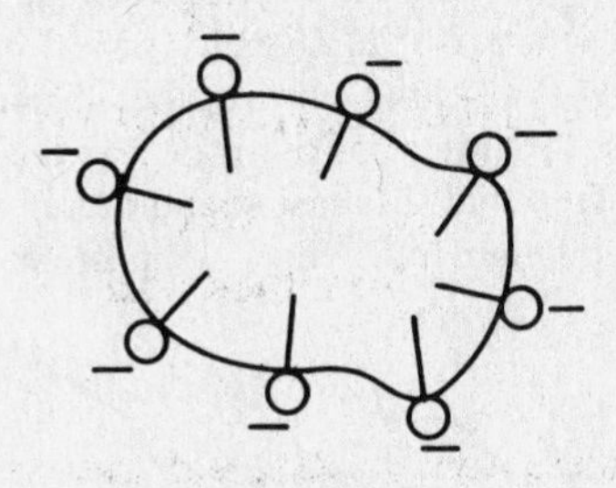

Fig. 6-19

MEMBRANES

6.5. Which would be better for solubilizing the integral proteins present in membranes, ionic or nonionic detergents?

SOLUTION

Ionic detergents alter the conformation of the hydrophobic portions of integral proteins, whereas nonionic detergents dissolve membranes and form mixed micelles of detergent-lipid-protein in which the proteins' conformations are unchanged. Thus nonionic detergents are preferred. However, removal of lipid and detergent from the mixed micelles causes conformational changes and loss of biological activity of the protein. This loss of activity arises because integral proteins need to be surrounded by *lipids of solvation* to be effective.

6.6. What strategy do bacteria adopt to restore their membrane fluidity if they are suddenly transferred from an environment at 25°C to one at 35°C?

SOLUTION

They can incorporate into their phospholipids fatty acids that are (1) longer, (2) more saturated, or (3) less branched than the originals.

TRANSPORT

6.7. To what use could L-glucose be put in studying the transport of D-glucose into cells?

SOLUTION

L-Glucose can be used for distinguishing between simple diffusion and carrier-mediated transport. L-Glucose would be transported by simple diffusion only. The difference in the rate of uptake of D-glucose and L-glucose would represent the true rate of carrier-mediated transport of D-glucose because the glucose transporter is stereo-selective for D-glucose.

6.8. To each of six tubes containing buffer was added 50 μL of a suspension of rat liver mitochondria that contained 20 mg mL^{-1} of protein. Radioactively labeled pyruvate (0.07 mCi mmol^{-1}) was added to each tube. α-Cyano-3-hydroxycinnamate (an inhibitor of transport) was added to each tube in turn at intervals of 5 s. Each solution was filtered immediately after the addition of inhibitor, and the radioactivities of the filters (which retained mitochondria) were found to be 0.17, 0.35, 0.51, 0.66, 0.82, and 0.96 nCi for the filters obtained at 5, 10, 15, 20, 25, and 30 s, respectively. What is the rate of uptake of pyruvate in nmol min^{-1} mg^{-1} of protein?

SOLUTION

The radioactivity of the filters is a measure of the pyruvate present in the mitochondria at the time of their isolation. A graph (Fig. 6-20) of the radioactivity of the filters versus the length of time of incubation of the mitochondria with pyruvate gives a curve, but the *initial* rate of uptake of pyruvate, obtained from the slope of the tangent to the curve at the origin, is 2.1 nCi min^{-1}. This represents a rate of 30 nmol of pyruvate min^{-1} since 1 mmol of pyruvate has 0.07 mCi of radioactivity. Each incubation mixture contains 50 μL of mitochondrial suspension and therefore contains 1 mg of protein. The rate of uptake of pyruvate is therefore 30 nmol min^{-1} mg^{-1} of protein.

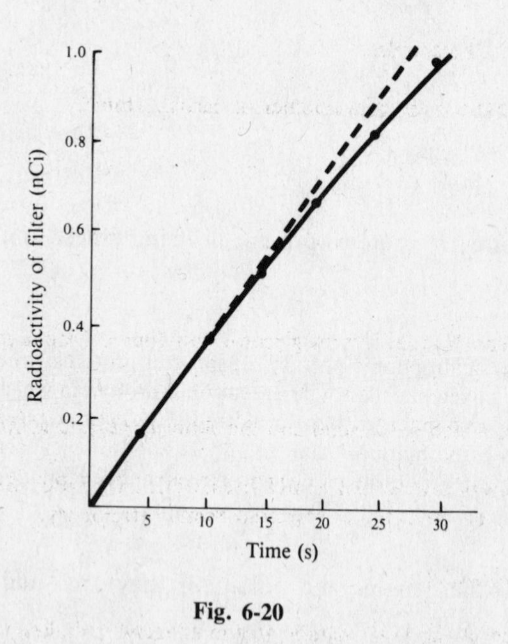

Fig. 6-20

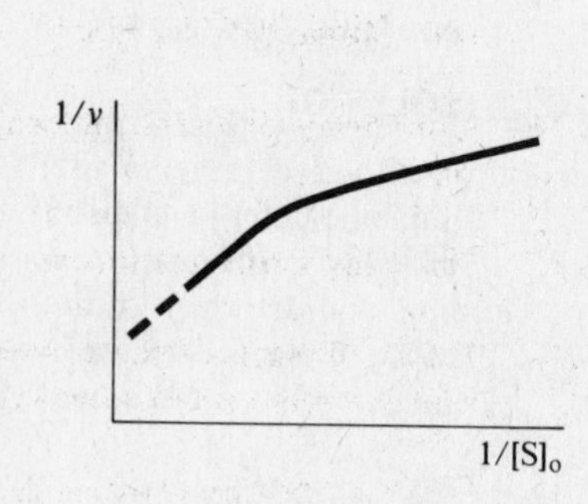

Fig. 6-21

6.9. Suppose two independent carriers existed for the transport of a solute into a cell and simple diffusion was negligible. How could the existence of *two* carriers be demonstrated?

SOLUTION

The existence of two carriers can be demonstrated by measuring the rate of uptake (v) of the solute at different concentrations of solute ($[S]_o$). A double-reciprocal plot would have the appearance shown in Fig. 6-21.

Supplementary Problems

6.10. Indicate the ways in which isoprene units are linked together in forming the following compounds:

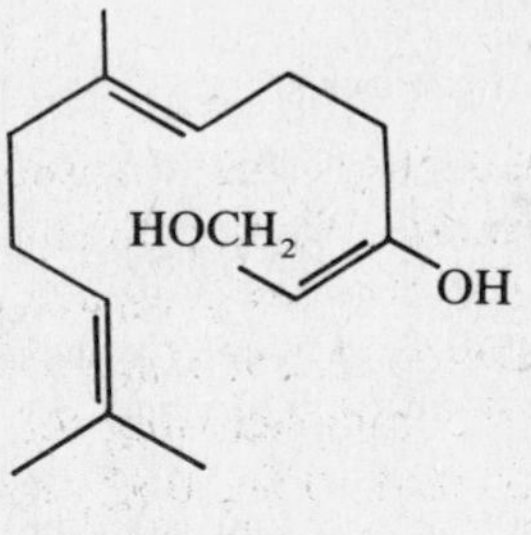

HOCH$_2$ OH

(*a*) Farnesol

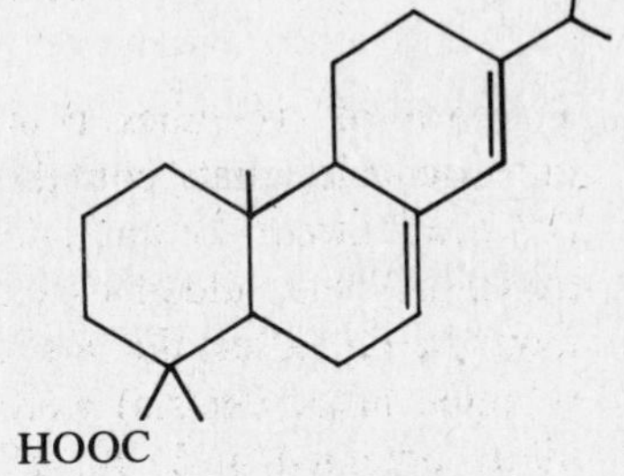

HOOC

(*b*) Abietic acid

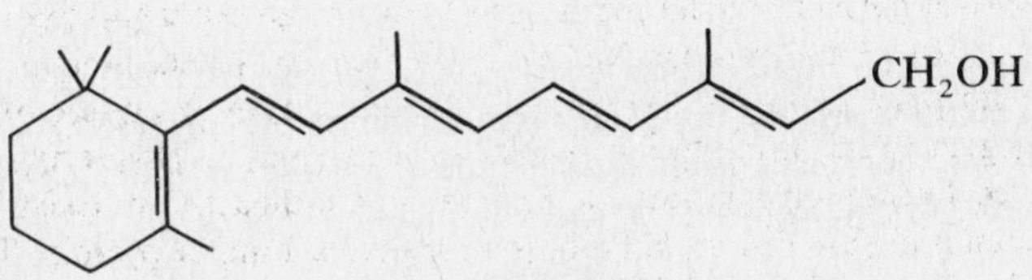

CH$_2$OH

(*c*) Vitamin A

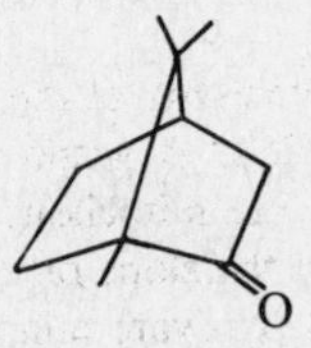

O

(*d*) Camphor

6.11. (*a*) How would you define a *glycolipid*? (*b*) Give two examples.

6.12. Why do almost all naturally occurring fatty acids contain an even number of carbon atoms?

6.13. Write the structures for the following fatty acids:

(*a*) Myristic acid (14:0)
(*b*) Myristoleic acid ($14{:}1^{\Delta 9}$)
(*c*) Ricinoleic acid ($18{:}1^{\Delta 9}$ C-12 hydroxylated)

6.14. How many different molecules of triacylglycerol can be made from glycerol and four different fatty acids?

6.15. Write the structures for two triacylglycerols, one of which is (*a*) solid and the other (*b*) liquid at 37°C.

6.16. Electrophoresis of the following mixture of lipids was carried out at pH 7: PE, PS, PG, DPG. State whether these lipids would move toward the anode or the cathode or would remain stationary.

6.17. Draw a structure showing the conformation of cholesterol.

6.18. A solution of 1-palmitoyl-2-stearoyl-3-myristoylglycerol and phosphatidic acid in benzene is shaken with an equal volume of water. After the two phases separate, which lipid will be in the higher concentration in the aqueous phase?

6.19. How many phospholipid molecules are there in a 1-μm^2 region of a phospholipid bilayer? Assume that a phospholipid molecule occupies 0.7 nm^2 of the surface area.

6.20. The lipids isolated from the membranes of 4.74×10^9 human red blood cells were spread as a monolayer with an area of 0.89 m^2. Assuming the red blood cell approximates a disk 7 μm in diameter and 1 μm thick, show that the membrane covering the red blood cell must be two lipid molecules thick.

6.21. Why does phosphatidylethanolamine partition preferentially into the inner leaf of artificial vesicles composed of PE and PC?

6.22. Predict the effects of the following operations on the phase-transition temperature and on phospholipid mobility in vesicles made from dipalmitoylphosphatidylcholine:

(*a*) Introducing dipalmitoleoylphosphatidylcholine into the vesicles

(*b*) Introducing a high concentration of cholesterol into the vesicles

(*c*) Introducing integral membrane proteins into the vesicles

6.23. Calculate the average density of a membrane composed of 30 percent (by weight) protein (density, 1.33 g cm^{-3}) and 70 percent by weight phosphoglyceride (density, 0.92 g cm^{-3}).

6.24. For the membrane referred to in Prob. 6.23, how many molecules of lipid are there for each molecule of protein? Assume an average M_r of 800 for the phosphoglyceride and 40,000 for the protein.

6.25. What will be the order of simple diffusion of the following compounds through a biological membrane: propionic acid, 1,3-propanediol, propionamide, 1-propanol, alanine?

6.26. In the bacterium *E. coli*, glucose is taken up by group translocation, lactose is taken up by secondary active transport (using H^+), and maltose is taken up by means of a binding-protein system. Outline how it would be possible to determine whether melibiose (a disaccharide of glucose and galactose) is taken up by *E. coli* and, if it is, whether one of the mechanisms described earlier applies.

6.27. A number of identical cell suspensions are treated with different amounts of radioactively labeled leucine. The initial rates of leucine uptake are measured for each suspension (see the table below). What is the maximum possible rate of uptake of leucine by cells using the same transport system?

Leucine Concentration (μm)	Initial Rate of Uptake (cpm)*
0.5	55
1	110
5	480
10	830
20	1,300
30	1,700
50	2,100
100	2,600

*cpm represents counts per minute, a measure of radioactivity.

6.28. The pH of gastric juice is 1. The cells that produce gastric juice have an internal pH of 7. Calculate the ΔG for transport of protons from these cells into the stomach at 37°C. (*R*, the gas constant, is 8.3 J mol^{-1} K^{-1}.)

6.29. Predict the effects of (*a*) valinomycin and (*b*) gramicidin A on the initial rate of glucose transport into vesicles derived from cells that accumulate glucose by cotransport with Na^+. Assume that the outside medium contains 0.2 *M* Na^+ and that the interior of the vesicle contains an equivalent concentration of K^+.

Chapter 7

Nucleic Acids

7.1 INTRODUCTION

In 1868 Friederich Miescher isolated a substance from the nucleus of pus cells. It was considered to be characteristic of the nucleus, and he called it *nuclein*. A similar substance was subsequently isolated from salmon sperm heads. Nuclein was later shown to be a mixture of a basic protein and a phosphorus-containing organic acid, now called *nucleic acid*.

7.2 NUCLEIC ACIDS AND THEIR CHEMICAL CONSTITUENTS

The major nucleic acid in the nucleus of cells is *deoxyribose nucleic acid* (or *DNA*). It contains the pentose sugar *deoxyribose* as one of its chemical constituents. DNA is now known to be the genetic material. Another type of nucleic acid, *ribonucleic acid* (or *RNA*), contains *ribose* instead of deoxyribose. Its main role is in the transmission of the genetic information from DNA into protein.

DNA molecules are very large—much larger than proteins. RNA is more comparable to proteins in size. Complete hydrolysis of DNA (or RNA) by acid cleaves it into a mixture of nitrogenous bases, 2-deoxy-D-ribose (or D-ribose for RNA), and orthophosphate. There are two general types of nitrogenous bases in both DNA and RNA, *pyrimidines* and *purines*.

Pyrimidines are derivatives of the heterocyclic compound *pyrimidine*:

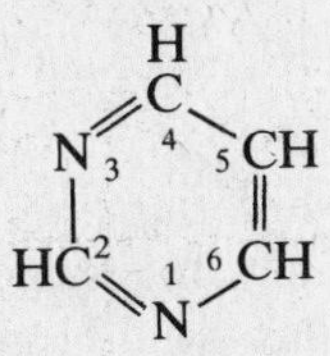

Pyrimidine

The numbering of the positions in the ring has been established by convention (IUPAC).

Purines are derivatives of the fused-ring compound *purine*:

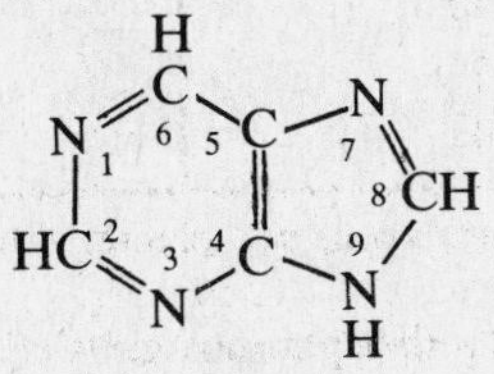

Purine

Pyrimidines

The major pyrimidines found in DNA are *thymine* and *cytosine*; in RNA, they are *uracil* and *cytosine*. These three pyrimidines differ in the types and positions of chemical groups attached to the ring.

Thymine is 5-methyl-2,4-dioxypyrimidine.

Cytosine is 2-oxy-4-aminopyrimidine.

Uracil is 2,4-dioxypyrimidine.

EXAMPLE 7.1

Write the structures of (*a*) thymine, (*b*) cytosine, and (*c*) uracil.

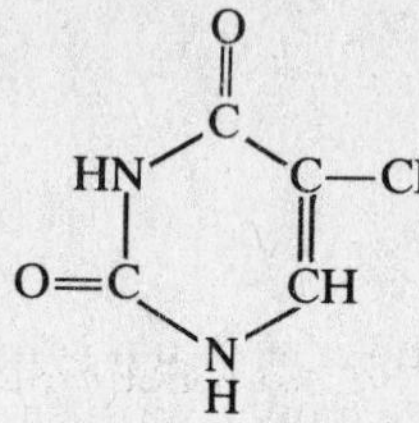

(*a*) Thymine 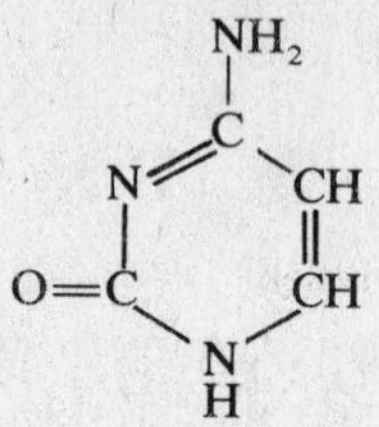 (*b*) Cytosine 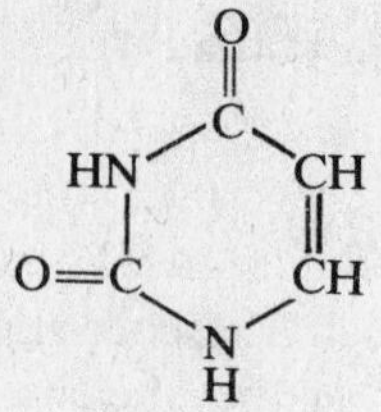 (*c*) Uracil

Thymine can also be described as 5-methyluracil. Other methylated pyrimidines are found in some nucleic acids.

EXAMPLE 7.2

Write the structure of 5-methylcytosine.

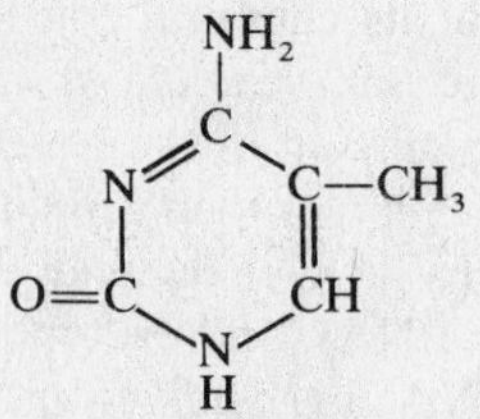

5-Methylcytosine

Methylation of cytosine in both DNA and RNA has important biological implications with respect to protection of the genetic material and its expression.

5-Bromouracil is an analog of thymine, differing only in the substituent on C-5 (Br versus CH_3). These two substituents occupy approximately the same space, and the enzyme responsible for making DNA can accommodate either, allowing 5-bromouracil to be incorporated into DNA in certain types of cells and viruses. This has been of considerable value in studies of DNA synthesis.

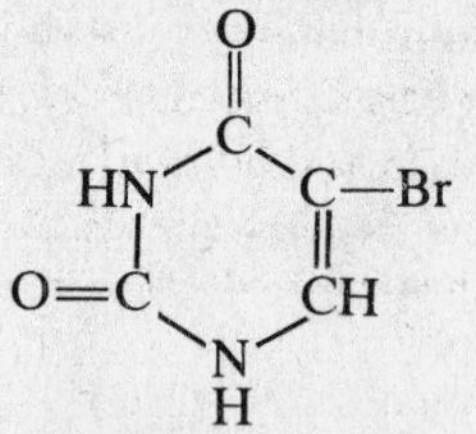

5-Bromouracil

Purines

The major purines found in DNA and RNA are *adenine* and *guanine*. They differ in the types and positions of chemical groups attached to the purine ring, as shown below:

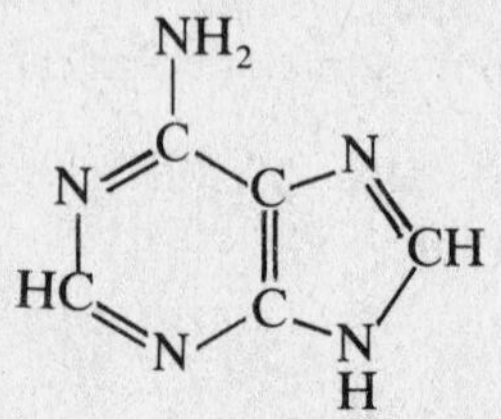

Adenine 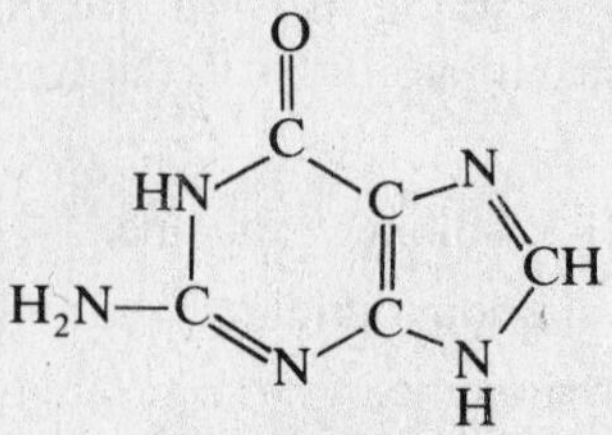 Guanine

EXAMPLE 7.3

Describe adenine and guanine in terms that indicate the nature and positions of substituent chemical groups on the purine ring.

Adenine is 6-aminopurine.

Guanine is 6-oxy-2-aminopurine.

Tautomeric Forms of Pyrimidines and Purines

All pyrimidines and purines can exist in alternative isomeric forms called *tautomers*. Thus, uracil can exist in *keto* and *enol* forms.

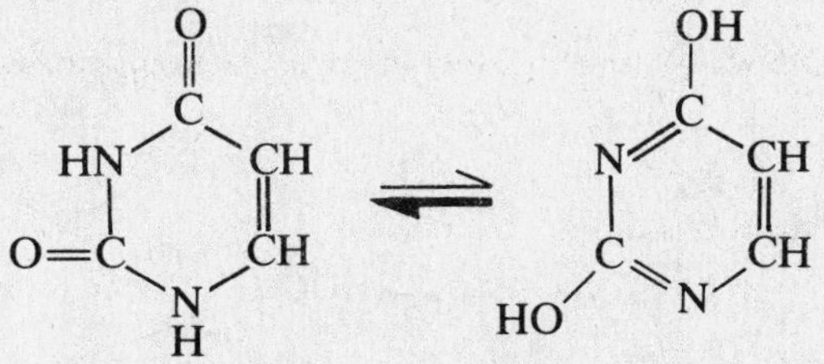

Keto ⇌ Enol

The heavy arrow indicates that the keto form is strongly preferred at neutral pH.

EXAMPLE 7.4

Write the enol form of guanine.

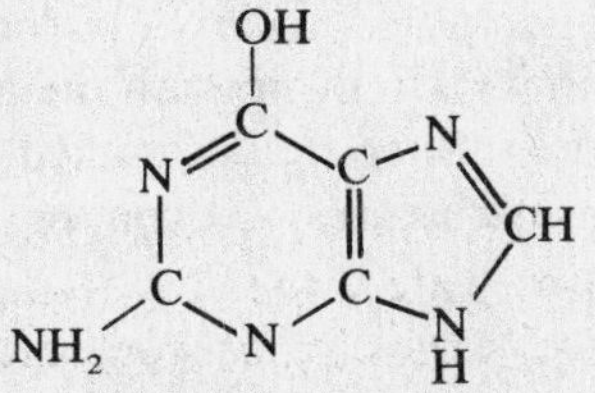

Enol form of guanine

Question: Is it possible to write an enol form for adenine?

No, because it does not contain keto groups. It can, however, isomerize to the tautomeric *imino* form, but the *amino* form shown earlier in this section predominates.

Sugars

The sugar in DNA is 2-deoxy-D-ribose; in RNA it is D-ribose.

EXAMPLE 7.5

Write the forms of these sugars as they occur in DNA and RNA.

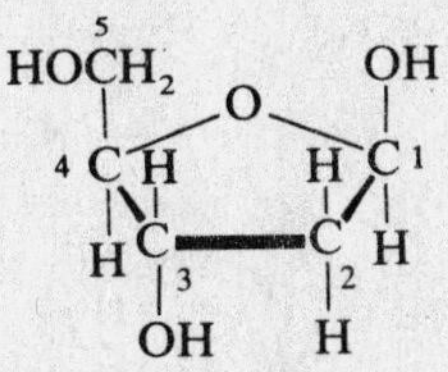

2-Deoxy-β-D-ribose
(2-deoxy-β-D-ribofuranose)

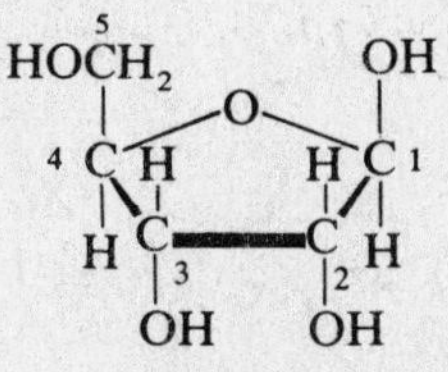

β-D-Ribose
(β-D-ribofuranose)

Note that it is the β anomer in each case that is present in the nucleic acid (Chap. 2).

7.3 NUCLEOSIDES

Within the structure of the nucleic acids, a pyrimidine or purine is linked to the sugar (2-deoxy-D-ribose or D-ribose) to give a *nucleoside*. The nucleosides are referred to as *deoxyribonucleosides* if they contain deoxyribose, and *ribonucleosides* if they contain ribose. The purine nucleosides have a *β-glycosidic linkage* from N-9 of the base to C-1 of the sugar. In pyrimidine nucleosides, the linkage is from N-1 of the base to C-1 of the sugar.

EXAMPLE 7.6

Write the structures of (*a*) the ribonucleoside containing adenine and (*b*) the deoxyribonucleoside containing cytosine.

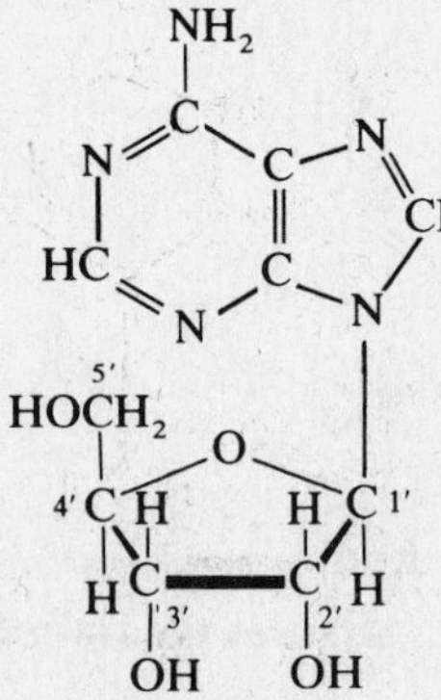

(*a*) 9-β-D-Ribofuranosyl adenine (adenosine)

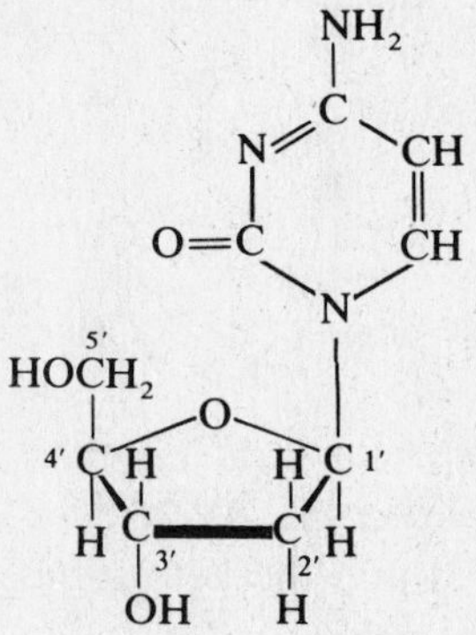

(*b*) 1,2′-Deoxy-β-D-ribofuranosyl cytosine (deoxycytidine)

Because the glycosidic linkage in Example 7.6 is to a nitrogen in the pyrimidine or purine, these nucleosides are referred to as N-*glycosides*. To distinguish the atoms in the furanose ring of the sugar from those in the rings of the bases, the former are designated 1′, 2′, . . . , 5′, as shown. The chemical names written immediately below the structures are concise but awkward to use, and it is more convenient to use simpler terms. Thus:

Adenine linked to ribose ≡ adenosine
Uracil linked to ribose ≡ uridine
Guanine linked to ribose ≡ guanosine
Guanine linked to deoxyribose ≡ deoxyguanosine
Cytosine linked to deoxyribose ≡ deoxycytidine
Thymine linked to deoxyribose ≡ deoxythymidine

EXAMPLE 7.7

Write the structures of (*a*) deoxyguanosine and (*b*) cytidine.

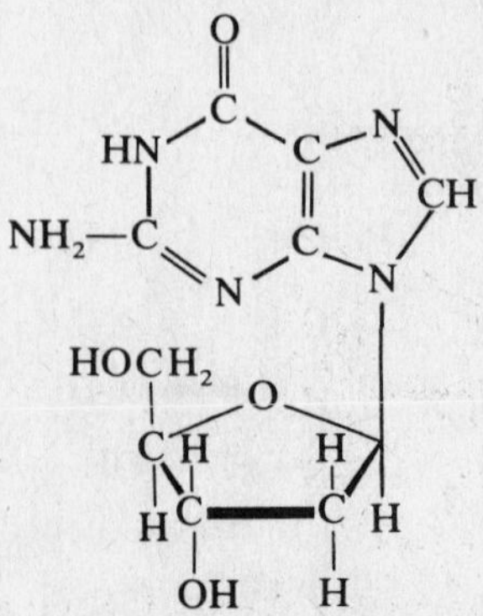

(*a*) Deoxyguanosine

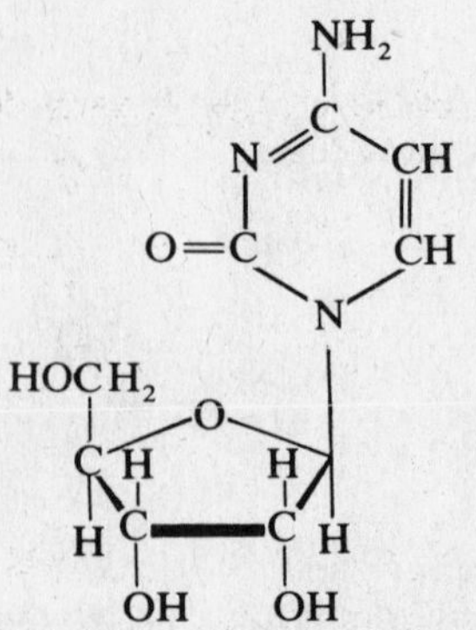

(*b*) Cytidine

7.4 NUCLEOTIDES

The *nucleotides* are phosphoric acid esters of nucleosides, with phosphate at position C-5′. Nucleotides with phosphorylation at other positions are known, but they are not components of the nucleic acids. Nucleotides containing deoxyribose are called *deoxyribonucleotides*; those containing ribose are known as *ribonucleotides*.

EXAMPLE 7.8

Write the structures of (*a*) the ribonucleotide containing adenine and (*b*) the deoxyribonucleotide containing thymine.

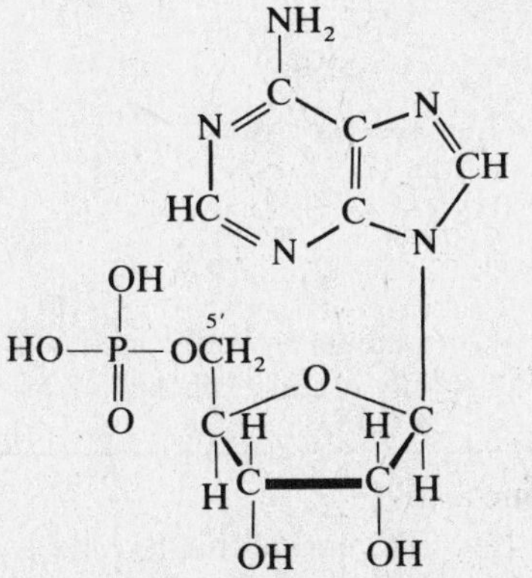

(*a*) Adenosine 5′-phosphate (AMP) 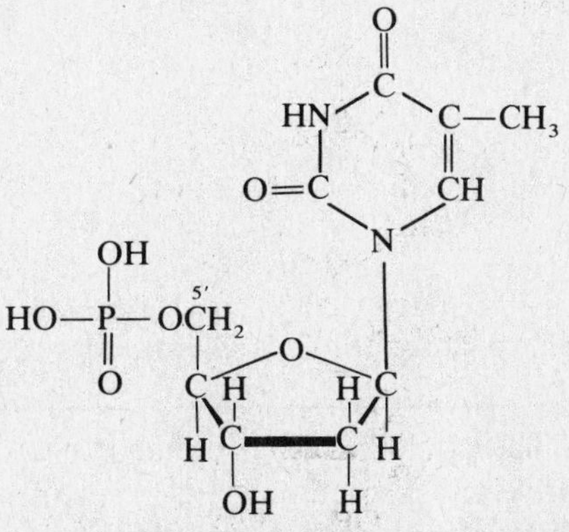 (*b*) Deoxythymidine 5′-phosphate (dTMP)

Adenosine 5′-phosphate is also known as *AMP* (for adenosine monophosphate) or *adenylic acid*. If deoxyribose replaces ribose in adenosine 5′-phosphate, the terminology is *dAMP* or deoxyadenylic acid. The abbreviated names for some ribonucleotides and deoxyribonucleotides are listed below.

Base	Ribonucleotide	Deoxyribonucleotide
Adenine, A	Adenylic acid, AMP	Deoxyadenylic acid, dAMP
Guanine, G	Guanylic acid, GMP	Deoxyguanylic acid, dGMP
Cytosine, C	Cytidylic acid, CMP	Deoxycytidylic acid, dCMP
Uracil, U	Uridylic acid, UMP	Deoxyuridylic acid, dUMP
Thymine, T	Thymidylic acid, TMP	Deoxythymidylic acid, dTMP

The terminology tells us that the nucleotides are acids. This results from the primary phosphate ionization, which has a pK_a value of approximately 1 (Chap. 10). The nucleotides are thus negatively charged at neutral pH; also contributing to this negative charge is the ionization of the secondary phosphate, which has a pK_a value of approximately 6. At neutral pH, there is no charge on any of the bases.

EXAMPLE 7.9

Write the structure of the charged form of AMP at pH 7.

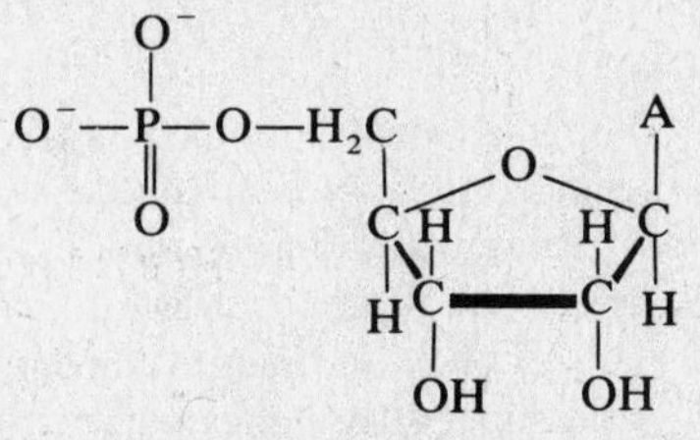

AMP (charged form)

Note that two negative charges reside on the phosphate. In this structure, adenine is represented by A; it carries no charge.

All the common 5′-nucleotides exist also as 5′-diphosphates and 5′-triphosphates. These contain two and three phosphates, respectively. The corresponding adenosine 5′-nucleotides are referred to as *ADP* and *ATP*.

EXAMPLE 7.10

Write the structure of ATP.

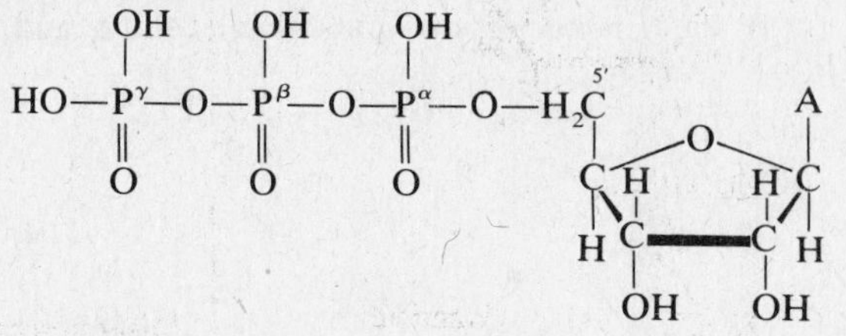

```
       OH      OH      OH
       |       |       |                5'
HO—Pγ—O—Pβ—O—Pα—O—H2C                    A
       ||      ||      ||        |    O      |
       O       O       O         CH      HC
                                 | \    / |
                                 H C————C H
                                   |    |
                                   OH   OH
```

ATP

Note that this has been written in the uncharged form. The phosphorus atoms are designated α, β, and γ, the α phosphorus being attached to the 5′ C of the ribose. *ADP* contains only α and β phosphates.

Question: What would be the net charge on ATP at neutral pH?

It is negative and in the range -2 to -4. This is because ATP, and all nucleoside triphosphates, can dissociate four protons from the phosphate groups. The first has a pK_a of ~ 1, and the second, third, and fourth in the range 6 to 7, approximately.

EXAMPLE 7.11

Arrange ATP, dAMP, and CDP (cytidine *di*phosphate) in order of increasing net negative charge at pH 7.

The order is dAMP, CDP, ATP, because the phosphates have the dominating effect at neutral pH, and the more phosphates there are in a molecule, the greater will be its negative charge.

The ribonucleoside di- and triphosphates (NDPs, NTPs) and deoxyribonucleoside di- and triphosphates (dNDPs, dNTPs) have important functions in the cell. They operate as *energy carriers* in various reactions and as precursors for the synthesis of nucleic acids (Chaps. 10 and 16).

7.5 POLYNUCLEOTIDES

The nucleic acids, both DNA and RNA, are *polynucleotides*; that is, they are polymers containing nucleotides (various types) as the repeating subunits. The nucleotides are joined to one another through *phosphodiester linkages* between the 3′ C of one nucleotide and the 5′ C of the adjacent one. This linkage is repeated many times to build up large structures (chains or strands) containing hundreds to millions of nucleotides within a single giant molecule.

EXAMPLE 7.12

Write the structure (top of page 181) of a section of a poly*ribo*nucleotide (RNA) chain containing adenine, guanine, and cytosine as a sequence of nucleotides. Note that the structure is written in its charged form. Because the phosphodiester linkages join different carbons, 3′ and 5′, in adjacent nucleotides, the chain has a chemical direction, or *polarity*. By convention, the structure shown here has the direction $5' \rightarrow 3'$ downward (or $3' \rightarrow 5'$ upward). Also, the order (or sequence) of nucleotides is conventionally written in the $5' \rightarrow 3'$ direction. Thus, the section shown has the sequence adenine, guanine, cytosine; or AGC, for short. To indicate that there are phosphates attached at the 5′ and 3′ ends of the structure shown, it is more accurately referred to as pApGpCp. This shorthand form does not indicate that the sequence of three nucleotides is just a portion of a much longer structure; pApGpCp could also refer to a molecule containing only three nucleotide units (a trinucleotide) in which the 5′ and 3′ ends are phosphorylated. If the sequence AGC had been present in a poly*deoxy*nucleotide, the structure would be written in the shorthand form as d-pApGpCp, or just dAGC.

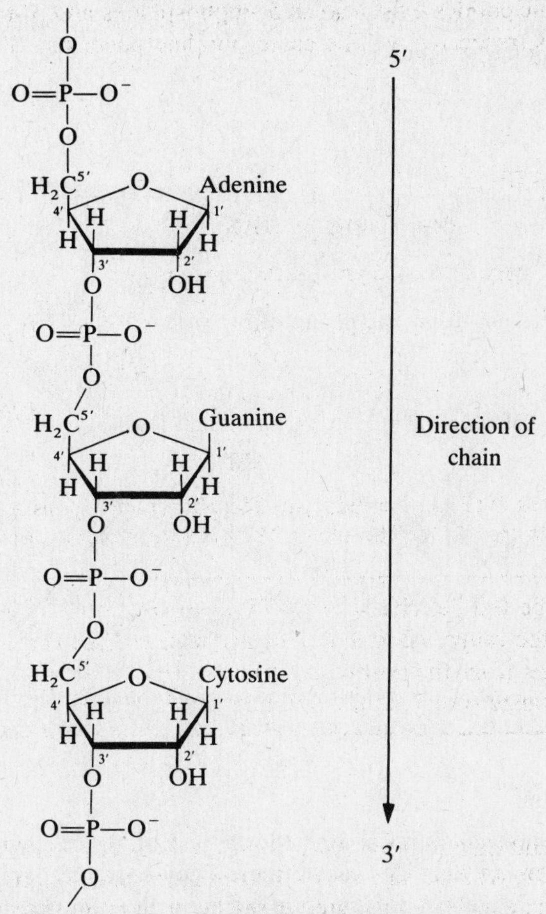

EXAMPLE 7.13

Write the structure of the dinucleotide d-ApTp.

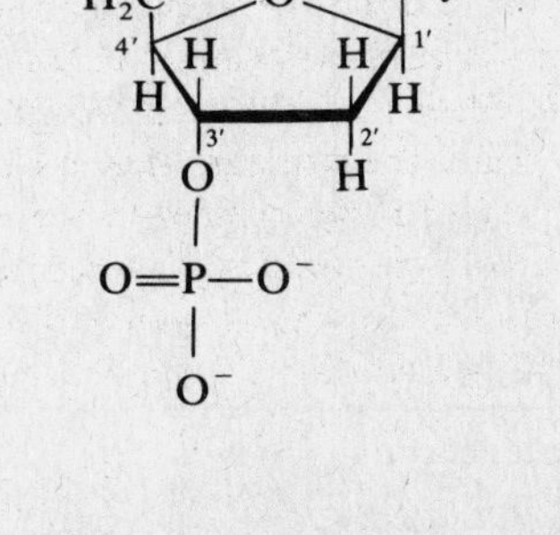

Note the absence of a phosphate at the 5′ end and the presence of deoxyribose. Another commonly used shorthand form for describing this structure is

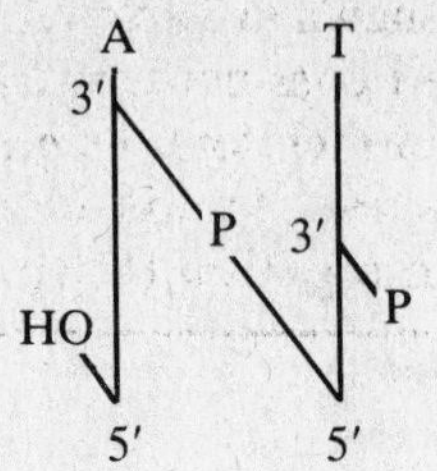

A T
3′ P 3′ P
HO
5′ 5′

Note that this latter form does not define the nature of the sugar.

EXAMPLE 7.14

Write the structure of pApUpGpCpApCp in the shorthand form.

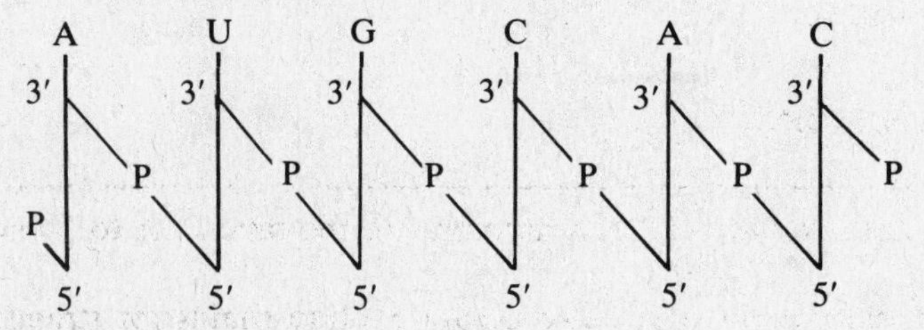

This structure contains six nucleotide units and is referred to as a *hexanucleotide*. The general term for structures containing a few nucleotides (10 or less) is *oligonucleotide*.

7.6 STRUCTURE OF DNA

DNA is a *polydeoxynucleotide* and among the largest of the biological macromolecules; some DNA molecules comprise more than 10^8 nucleotides. They contain adenine, thymine, guanine, and cytosine as the bases, and the genetic information is encoded within the nucleotide sequence, which is precisely defined over the entire length of the molecule. One of the simplest methods for determining the nucleotide sequence of DNA makes use of an enzyme, *DNA polymerase*, which catalyzes the synthesis of DNA. The properties of this enzyme are discussed in Chap. 16.

Base Composition of DNA

The base composition of DNA from many different species has been determined. It varies from one to another (see Table 7.1).

Table 7.1. Base Composition of DNA in Various Species

	Base Composition (mol %)			
Species	G	A	C	T
Sarcina lutea	37.1	13.4	37.1	12.4
Alcaligenes faecalis	33.9	16.5	32.8	16.8
E. coli K12	24.9	26.0	25.2	23.9
Wheat germ	22.7	27.3	22.8*	27.1
Bovine thymus	21.5	28.2	22.5*	27.8
Human liver	19.5	30.3	19.9	30.3
Saccharomyces cerevisiae	18.3	31.7	17.4	32.6
Clostridium perfringens	14.0	36.9	12.8	36.3

*Cytosine + methylcytosine.

Question: What is the base composition for DNA from human kidney?

It is the same as for human liver, as shown in Table 7.1, because the base composition of DNA is a characteristic of a particular species and does not vary from one cell type to another. This reflects the fact that the nucleotide sequence, and therefore the genetic information present, in each type of cell within an organism is exactly the same. As will be seen later, this information is expressed differently in the various cell types of an organism (Chap. 17).

EXAMPLE 7.15

Are there any features common to DNA from various species with respect to the ratio of one base (or type of base) to another?

The ratio of purines (A + G) to pyrimidines (T + C) is close to unity in all cases. Perhaps more remarkable is that the ratios of both A to T and G to C are each close to unity. These two facts reflect an important structural feature of most DNAs.

Double-Helical Structure of DNA

Question: What structural feature of DNA accounts for the ratio of A to T and G to C being close to unity?

DNA is a duplex molecule in which two polynucleotide chains (or strands) are linked to one another through specific *base pairing* (Fig. 7-1). Adenine in one strand is paired to thymine in the other, and guanine is paired to cytosine. The two chains are said to be *complementary*. This was one of the essential features of Watson and Crick's proposal regarding the structure of DNA. Hydrogen bonds form between the opposing bases within a pair. In the structure proposed by Watson and Crick, A:T and G:C base pairs are roughly planar, with H bonds (dotted lines), as shown in Fig. 7-1. Note that two H bonds form in an A:T pair and three in a G:C pair.

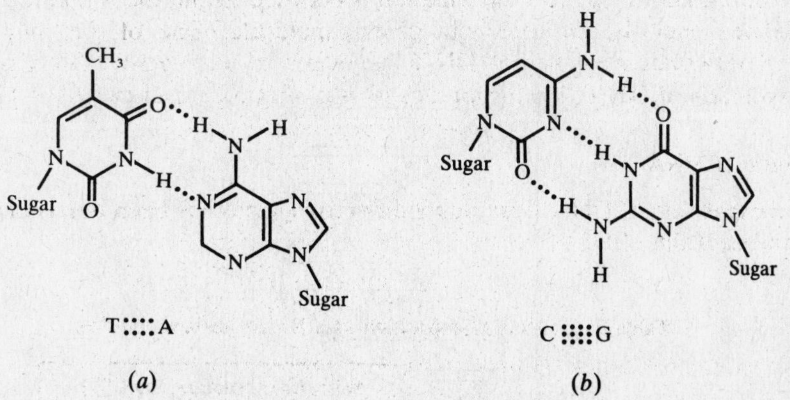

Fig. 7-1 Base pairing in DNA.

The base pairs are stacked on top of one another, with the plane of the base pairs being perpendicular to the length of the duplex. This is shown diagramatically in the ladder-type structure in Fig. 7-2.

Question: In the duplex, ladder-type structure shown in Fig. 7-2, why are the two chains orientated in opposite directions?

A model for DNA incorporating base pairing between complementary strands and consistent

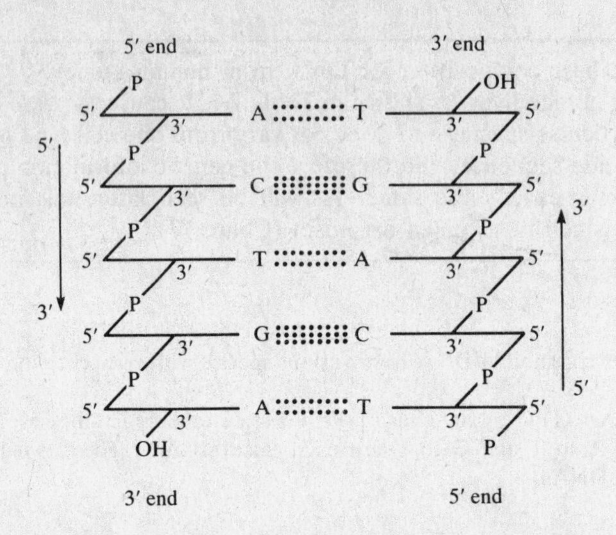

Fig. 7-2 Stacking of base pairs within the DNA duplex. Each group of dotted lines, representing the H bonds between the base pairs, is in a plane perpendicular to the surface of the page.

with x-ray diffraction data was developed by J. Watson and F. Crick in 1953. Basic to the structure was the twisting of the two strands around one another to give a *right-handed* helix (the *double helix*), and to achieve a structure consistent with data available at the time, it was necessary to orient the complementary chains in opposite directions (Fig. 7-3). Direct proof for this *opposite polarity* in chain direction was achieved about 10 years later.

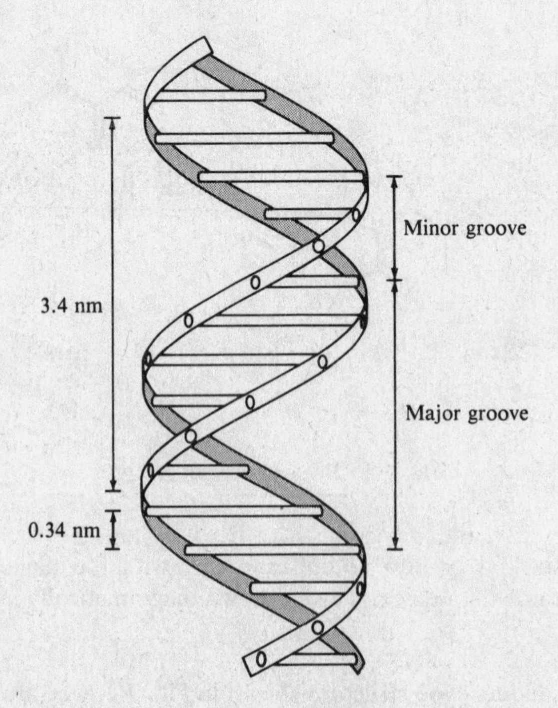

Fig. 7-3 Diagram of the DNA double helix (the so-called B form).

Question: How does the twisting into a helix contribute to the stability of the overall structure of DNA?

One of the most significant effects of twisting into a helix is to bring the stacked base pairs very close to one another (0.34 nm). Consequently, water is excluded from what is now a hydrophobic core; the charged phosphates are on the surface. The hydrophobic interactions within the core contribute, along with the H bonds between the base pairs, to the overall stability of the helix. It should also be noted that there is one complete twist of the helix every 10 base pairs or 3.4 nm. This distance is referred to as the *pitch* of the helix. The surface of the helix shows alternating *major* and *minor grooves*, which follow the twist of the double-stranded molecule along its length. The major groove is now known to accommodate interactions with proteins that recognize and bind to specific nucleotide sequences.

Question: The structure of the double helix shown in Fig. 7-3 is called the *B form*. What are the other forms, and do they have a biological role?

The B form of DNA appears to be the main one that occurs in vivo and in purified DNA in solution. It exists in fibers of DNA at high humidity. At lower humidities ($<$75 percent), a fiber of DNA will shorten. This is a result of a change to the *A form*, in which the base pairs are not perpendicular to the helix axis; they are tilted about 20°, and the pitch is reduced to 2.8 nm with 11 base pairs per turn.

A dramatically different form of the double helix has been observed in DNA containing the alternating $d(GC)_n$ structure along each strand. It is a left-handed, rather than a right-handed, helix and is known as the *Z form* of DNA. Space-filling models of the Z form and B form of DNA are compared in Fig. 7-4.

In the Z form, the repeating unit is a *dinucleotide*, and the resultant structure gives a staggered zigzag, instead of a smooth twist, for the sugar-phosphate backbone. It is possible that the Z form of DNA has an important biological role, although at present this is uncertain.

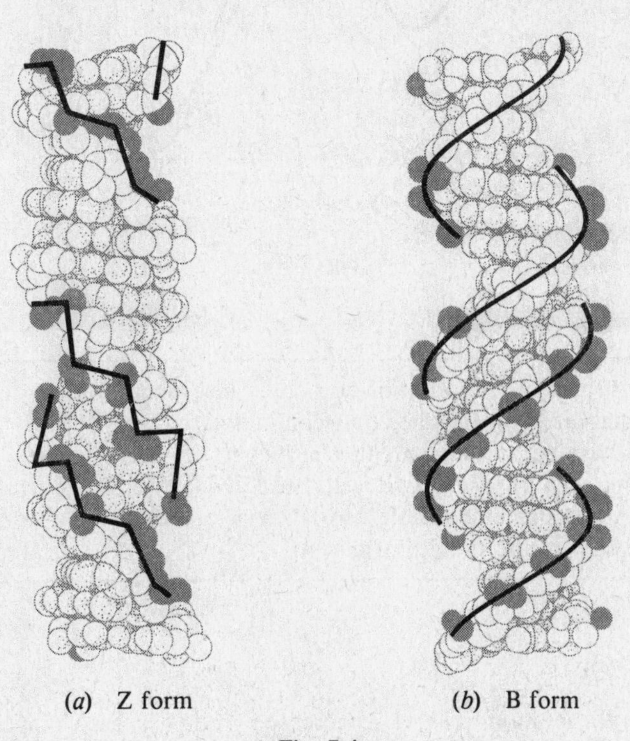

(*a*) Z form (*b*) B form

Fig. 7-4

7.7 DENATURATION OF DNA

The double helix is a relatively stiff and elongated molecule. Consequently, a solution of DNA has a high viscosity. If such a solution is heated to ~95°C, the viscosity drops markedly, reflecting a collapse of the double-helical structure. This is known as *denaturation* and is accompanied by separation of the duplex into its single strands, which are fairly flexible. Denaturation and renaturation provide valuable information on important properties of the DNA obtained from various sources. Denaturation also provides the basis for very precise and sensitive approaches to the identification of specific sequences in both DNA and RNA. This has been central to the rapid developments in molecular genetics.

While denaturation can be detected readily through changes in viscosity, a much more convenient way to detect it is by ultraviolet (uv) absorption measurement. The difference in the uv absorption spectra of the native (double-helical) and denatured (single-stranded) forms of DNA is shown in Fig. 7-5. At the wavelength of maximum absorption (260 nm), absorption by single-stranded DNA is approximately 40 percent higher than by double-stranded DNA. This is referred to as the *hyperchromic effect* and results from the *unstacking* of the base pairs in the helix.

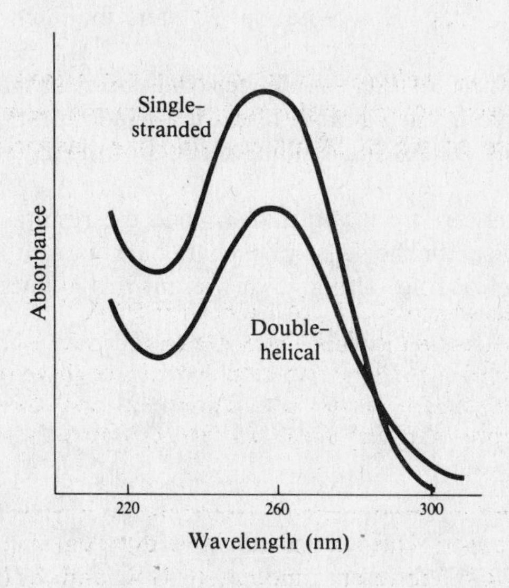

Fig. 7-5

Question: Will other treatments, in addition to heat, cause denaturation of DNA?

The DNA helix is stabilized by H bonds between individual base pairs as well as by hydrophobic forces between stacked base pairs. Reagents that reduce the H bonding and decrease the polarity of the surrounding medium, such as *formamide*, will cause denaturation. *Extremes of pH*, which endow the bases with a charge, are also effective. Thus, DNA at pH 12 shows absorption at 260 nm that is 40 percent higher than that of the native form.

Heat Denaturation of DNA

If the temperature of a solution of DNA is increased gradually, the change to the denatured form can be monitored by the change in absorbance at 260 nm. Typical results for several types of DNA are shown in Fig. 7-6.

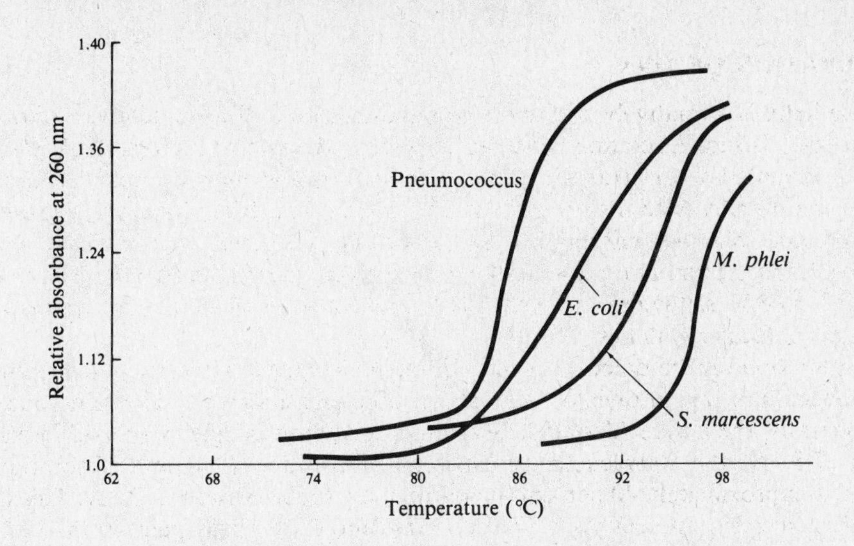

Fig. 7-6 Melting curves for DNA from different species.

The curves are referred to as *melting curves*, because the region over which the absorbance increases reflects the collapse (or *melting*) of the highly organized, semicrystalline state of double-helical DNA. The temperature at which 50 percent melting has occurred is called the *melting temperature*, or T_m.

EXAMPLE 7.16

What other factor besides heat affects the T_m of a particular DNA?

The T_m at neutral pH is dependent on the salt concentration (or *ionic strength*; Chap. 10) of the medium. The curves shown in Fig. 7-6 were obtained at an ionic strength of just above 0.15. If this were reduced by 90 percent, all T_m values would be lowered by about 20°C. This results from the additional negative charge and consequent greater electrostatic repulsion (which aids in disruption of the helix) within the DNA structure at the lower ionic strength.

Question: Why do the DNAs from various sources have different T_m values?

This is because the DNAs have different amounts of G:C and A:T base pairs, and the former confer the greater stability to the helix, perhaps through the presence of three H bonds per base pair rather than two (Fig. 7-1). Thus, the higher the GC content, the higher is the T_m. The value of T_m under standard conditions can be used, therefore, to obtain an estimate of the G + C content of an unknown DNA. This is obvious from Fig. 7-7, which shows a plot of T_m versus G + C content for a number of DNAs.

Renaturation of DNA

The complementary strands of DNA, separated by heat, spontaneously reassociate when the temperature is lowered below the T_m. This renaturation is also referred to as *annealing*.

Question: Is the rate of annealing the same for all types of DNA?

The rate of renaturation depends on the concentration of complementary sequences. Viral DNA has a smaller variety of sequences than does bacterial DNA; this reflects the higher level of genetic complexity in bacteria. Thus, for viral and bacterial DNA fragments of the same average size and at

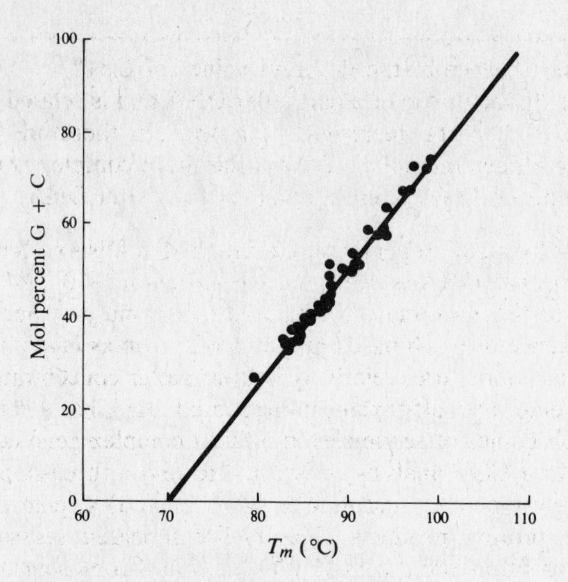

Fig. 7-7 Melting temperature of DNA as a function of the G + C content.

the same molar concentration, there would be a higher concentration of complementary sequences in the former. Viral DNA therefore would renature faster than bacterial DNA. In other words, bacterial DNA has greater *sequence heterogeneity*.

Rates of renaturation and sequence heterogeneity (or complexity) can be examined quantitatively through *COT analysis*. If C_0 is the initial concentration of DNA (moles per liter DNA phosphate) and k is the rate constant for association of the complementary strands, it can be shown that the fraction f of single-stranded molecules decreases with time t (s) according to the expression.

$$f = \frac{1}{1 + kC_0 t} \tag{7.1}$$

It is usual to plot the results of a COT analysis as f versus $C_0 t$. The behavior of several DNAs for a fixed set of conditions (size of DNA fragments, temperature, pH, ionic strength) is shown in Fig. 7-8. The value of $C_0 t$ when $f = 0.5$ is known as $C_0 t_{\frac{1}{2}}$.

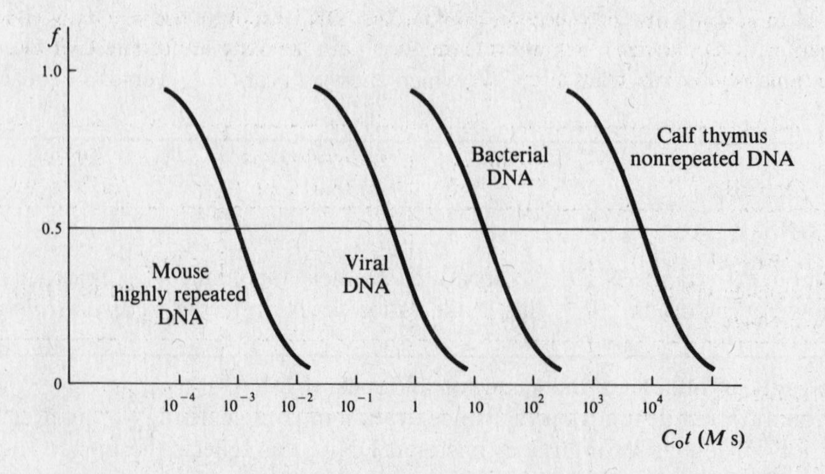

Fig. 7-8 COT analysis of various DNAs.

Question: What is the significance of the different values of $C_0t_{\frac{1}{2}}$?

The rate constant k is characteristic of a particular DNA and is related to its complexity in terms of sequence composition. $C_0t_{\frac{1}{2}}$ is the reciprocal of k and can therefore be used as a measure of sequence complexity. The higher the value of $C_0t_{\frac{1}{2}}$, the more complex is the DNA. Thus from the analysis shown above, bacterial DNA is more complex than viral DNA.

Question: What is meant by highly repeated and nonrepeated DNAs, shown in the above analysis?

The mouse highly repeated DNA is seen to have a $C_0t_{\frac{1}{2}}$ of $\sim 10^{-3}$ M s. It represents the most rapidly renaturing DNA of those examined. The mouse genome (see Sec. 7.8) contains about 10^6 copies of a repeating sequence of ~300 base pairs; this is known as *highly repeated DNA*. Thus, this fraction of DNA is simple in structure, relatively high in molar concentration, and able to renature rapidly. At the other extreme, the calf thymus nonrepeated DNA has a very high $C_0t_{\frac{1}{2}}$. This reflects the reassociation of unique copies of sequences in a fairly complex genome. When total DNA from animal cells is examined in a COT analysis, it is usual to find a three-stepped curve resulting from highly repeated, moderately repeated, and unique (nonrepeated) sequences. The unique sequences are those that code for protein products. The highly repeated sequences are located in the centromeric region of chromosomes and could be involved in chromosome-chromosome recognition. Little is known about the moderately repeated sequences. Viral and bacterial DNA do not show multiple steps in a COT analysis and contain no highly or moderately repeated sequences.

7.8 SIZE, ORGANIZATION, AND TOPOLOGY OF DNA

DNA molecules are very long. For example the DNA in a bacterial cell is contained within a single double-helical molecule, which, when spread out, is about 1,000 times as long as the diameter of the rod-shaped cell. This molecule carries all the genetic information of the cell and thus describes the *genome* (a single complement of the genetic material).

The term *chromosome* refers to a physical or organizational unit within which part of or all the genome is contained. Thus, the *Escherichia coli* genome is contained within just one chromosome, comprising a single DNA molecule. It has a size of 2.5×10^9 daltons and contains approximately 4×10^6 base pairs. The size of DNA molecules is more commonly expressed in *kilobase pairs* (kb, 1,000 base pairs). The *E. coli* chromosome is 4,000 kb in size. Another feature of this particular molecule is that it is a closed, or "circular," structure, i.e., there are no free ends.

EXAMPLE 7.17

What variation in genome size, chromosome number, and DNA topology occurs among various organisms?

This information is summarized below for a number of commonly investigated viruses, bacteria, and eukaryotes (organisms whose cells contain nuclei).

Organism	Genome Size (kilobases)	Chromosomes per Genome	DNA Topology
Viruses			
Simian virus 40 (SV40)	5.1	1	Circular
Bacteriophage ϕX174	5.4	1	Circular, single-stranded
Bacteriophage λ	48.6	1	Linear
Bacteria			
Escherichia coli	4,000	1	Circular
Eukaryotes			
Yeast	13,500	17	Linear
Human	2,900,000	23	Linear

Note that in Example 7.17, the DNA of bacteriophage ϕX174 is *single-stranded*, not double-stranded. In this case, the genome size given in kb refers to the number of base pairs in an equivalent duplex form. In progressing from the simple viruses to eukaryotes, the amount of information in the genome increases through many orders of magnitude. As a rough approximation, the number of kilobases in the genome of viruses and bacteria can be considered equivalent to the number of genes, each coding for a protein product. This equivalency of base number to genes does not exist in eukaryotes because of the presence of unexpressed sequences of bases (see Chap. 17). In viruses and bacteria (prokaryotes), circular DNAs are common. In eukaryotes, it is generally considered that each chromosome contains a single linear molecule of double-helical DNA.

Question: How are the very long DNA molecules condensed into more compact structures within the chromosomes?

In eukaryotes, DNA does not exist free. It is complexed with an approximately equal mass of basic proteins called *histones*. For a long time it was thought that bacterial DNA did not form such complexes. While histones are absent from bacteria, there is increasing evidence for the presence of histonelike proteins in them that enable condensation of their DNA into its compact *nucleoid* form.

There are five types of histones, designated H1, H2A, H2B, H3, and H4. They are of fairly low molecular weight (M_r = 11,000–23,000) and contain a large portion of the basic amino acids arginine and lysine. In each case there is an unusual distribution of amino acids along the single polypeptide chain; the basic amino acids tend to be clustered in one half, with the other half being relatively hydrophobic. In addition, the histones contain many modified amino acid side chains, e.g., methylated arginines and acetylated lysines.

The nucleoprotein complex that is formed is called *chromatin*. Chromatin can be isolated as fibers from nuclei. When the fibers are spread and examined under the electron microscope, they appear as "beads on a string." Most of the histone is contained in the beads, called *nucleosomes*. The nucleosome bead consists of a set of eight histones, two each of H2A, H2B, H3, and H4, around which approximately 200 base pairs of DNA are wrapped. Digestion of chromatin with *nucleases* (see below) yields a core particle still containing the eight histones but only 140 base pairs of DNA. The rest of the DNA, which is accessible to digestion, functions as a linker between the cores, and it is likely that histone H1 is associated with this linker. The structure of the core particles has been examined by x-ray diffraction. Figure 7-9 shows the arrangement of the DNA with the histone octamer. To compact the DNA further within the nucleus, the nucleosomes are organized into condensed higher-order structures.

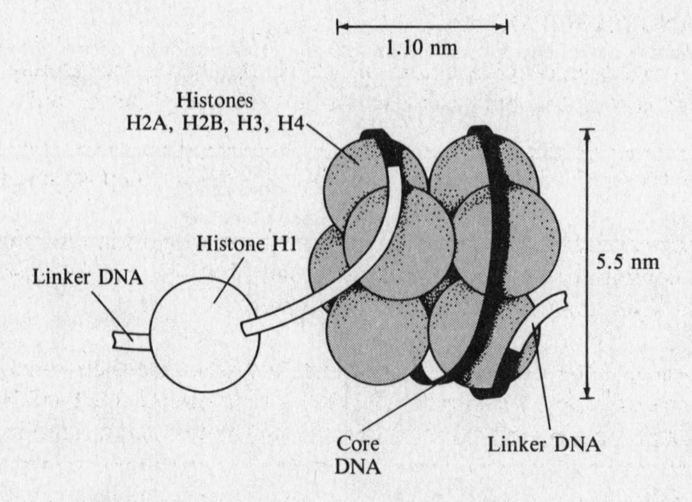

Fig. 7-9 The arrangement of histone and DNA in a nucleosome.

Question: What are plasmids?

Bacterial cells frequently contain additional DNA molecules called *plasmids*. These are relatively small (up to 200 kb) and are present in the form of circular duplexes. They can replicate independently of the bacterial chromosome and multiple copies can be present in a cell. Normally, they exist within a cell in a negatively supercoiled conformation, as does all DNA. Eukaryotic cells can also contain additional DNA to that present in the nucleus. Such *extrachromosomal* DNA is present in mitochondria and chloroplasts.

Question: What is meant by supercoiled DNA?

Supercoiling represents a twisting of the DNA double helix upon itself. It occurs in circular DNAs and in DNAs that are *topologically constrained* by being complexed to proteins. If a circular DNA molecule in a *relaxed* conformation (no supercoiling) is broken across both strands and one or more additional right-handed helical turns are inserted before rejoining the ends, the molecule would twist on itself to form a *positive supercoil*. If, on the other hand, the helix is unwound (given left-handed turns) before rejoining, the result is a *negative supercoil*. These forms, *topoisomers*, are shown in Fig. 7-10. Each supercoil depicted here contains a single *supertwist*. The number of supertwists in a molecule can be very large. Supercoiled DNA is readily converted into its relaxed form by the introduction of a break (or *nick*) between adjacent nucleotides in one of the two strands; this form is no longer topologically constrained.

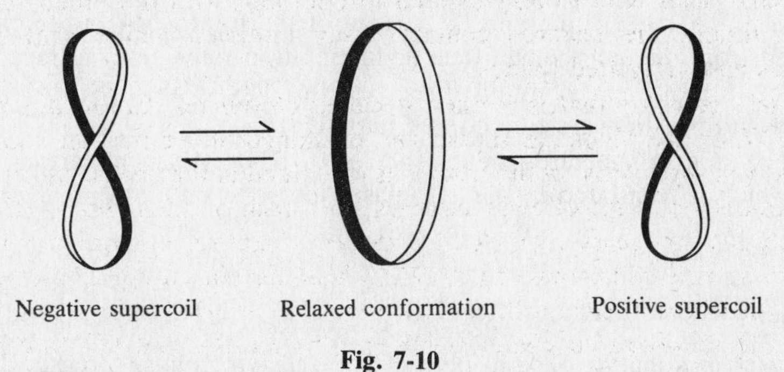

Fig. 7-10

7.9 STRUCTURE AND TYPES OF RNA

RNA comprises *polyribonucleotide* chains in which the bases are usually adenine, guanine, uracil, and cytosine. It is found in both the nucleus and the cytoplasm of cells.

Question: What are the main differences, other than chemical composition, between RNA and DNA?

There are a greater variety of RNA forms, with molecular weights in the range 25,000 to several million. Most RNAs contain a *single* polynucleotide chain, but this can fold back on itself to form double-helical regions containing A:U and G:C base pairs.

Question: What types of RNA occur in a typical cell and what are their functions?

There are three major types, *transfer RNA* (tRNA), *ribosomal RNA* (rRNA), and *messenger* RNA (mRNA); their roles in the expression of genetic information are treated in detail in Chap. 17.

Transfer RNA ($M_r \cong 25{,}000$) functions as an *adapter* in polypeptide chain synthesis. It comprises 10–20 percent of the total RNA in a cell, and there is at least one type of tRNA for each type of amino acid. Transfer RNAs are unique in that they contain a relatively high proportion of nucleosides of unusual structure (e.g., pseudouridine, inosine, and 2′-*O*-methylnucleosides) and many types of modified bases (e.g., methylated or acetylated adenine, cytosine, guanine, and uracil). As examples, the structures of pseudouridine and inosine are shown below. Inosine has an important role in *codon-anticodon* pairing (Chap. 17).

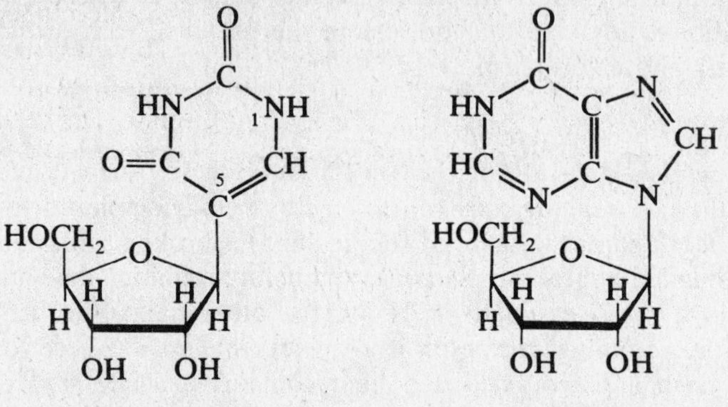

Pseudouridine (5-ribosyluracil) **Inosine**

Ribosomal RNA is present in the ribosomes, which contain approximately an equal mass of protein. Ribosomal RNA makes up about 80 percent of the total RNA in the cell and is of several types, distinguished from one another by their sedimentation rates in an ultracentrifuge (Chap. 4). Bacterial ribosomes, for example, contain three types of RNA: 5S, 16S, and 23S. The details of ribosome structure and function are treated in Chap. 17.

Messenger RNA is a very heterogeneous species of RNA. Each molecule carries a copy of a DNA sequence, which is translated in the cytoplasm into one or more polypeptide chains (Chap. 17).

7.10 NUCLEASES

Nucleases are enzymes that degrade nucleic acids by cleaving phosphodiester linkages. They may be specific for DNA or RNA, or they may act on both. A nuclease specific for DNA is called a *deoxyribonuclease* (DNase) and for RNA, *ribonuclease* (RNase).

EXAMPLE 7.18

Is there any specificity with respect to the location of the bond to be cleaved by a nuclease within the polynucleotide chain?

Nucleases are of two general types: (1) *exonucleases* and (2) *endonucleases*. Exonucleases bind to a terminus (5′ or 3′) and remove nucleotides either one or a few at a time. Some require a 5′ terminus and operate in the 5′→3′ direction (5′→3′ exonucleases); others (3′→5′ exonucleases) start at a 3′ terminus and degrade in the opposite direction. There are some exonucleases that will work from either terminus. Exonucleases show no base or sequence specificity.

Endonucleases do not require a terminus and will catalyze the cleavage of a polynucleotide chain at one or more sites. Frequently, they are specific for certain sites (specific base sequences) within the polynucleotide.

Question: Will nucleases cleave both single-stranded and double-stranded nucleic acids?

Many exo- and endonucleases discriminate between these two forms, although some will catalyze the hydrolysis of both. For example, DNase I (bovine pancreas) and DNase II (calf thymus) hydrolyze both forms. The difference between the two enzymes is that DNase I yields 5′-P-terminated oligonucleotides and DNase II yields 3′-P-terminated oligonucleotides. On the other hand, exonuclease III (*E. coli*), a 3′→5′ exonuclease, requires double-stranded DNA as substrate.

Question: What are restriction endonucleases?

Restriction endonucleases are part of a "DNA immunity system" in bacteria. They protect the cell against entry by foreign DNA by catalyzing double-strand cleavages; the cell's own DNA is protected. There are three types of restriction endonucleases. *Type II* restriction endonucleases have been very useful in the analysis and construction of DNA molecules. These enzymes cleave double-stranded DNA at *specific sites* defined by a four- to eight-nucleotide sequence. An important feature of these sequences is their *twofold rotational symmetry*. The sites of cleavage by nearly 300 type II restriction endonucleases have now been determined, and, in many cases, the cleavage is staggered to yield overlapping 3′-hydroxyl or 5′-phosphoryl termini. The recognition sequence and sites of cleavage by the commonly used *Eco*RI endonuclease are shown below. The arrows show the site of cleavage in each strand.

```
   ↓
5′ G A A T T C

   C T T A A G 5′
           ↑
```

Solved Problems

NUCLEIC ACIDS AND THEIR CHEMICAL CONSTITUENTS; NUCLEOSIDES; NUCLEOTIDES

7.1. Write the structure of 5-fluorouracil.

SOLUTION

Uracil is 2,4-dioxypyrimidine, and 5-fluorouracil is uracil containing a fluorine atom attached to C-5. Thus, its structure is

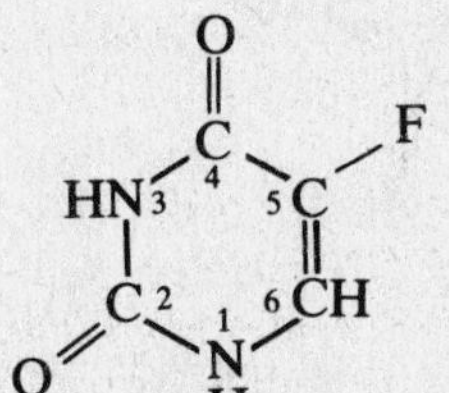

```
          O
          ‖
          C       F
       ╱  4  ╲  ╱
   HN3        5C
    |          ‖
    C2    1   6CH
  ╱   ╲   ╱
O       N
        H
```

7.2. 5-Bromouracil is an analog of thymine, but 5-fluorouracil is not. Why?

SOLUTION

Thymine is 5-methyluracil. The fluorine atom is much smaller in size than the methyl group, but the bromine atom is similar in size to it. 5-Bromouracil can be readily incorporated into DNA through the action of the enzyme DNA polymerase (Chap. 16), while 5-fluorouracil cannot.

7.3. Write the tautomeric forms of adenine.

SOLUTION

Adenine undergoes amino $\rightleftharpoons$ imino tautomerism. Thus,

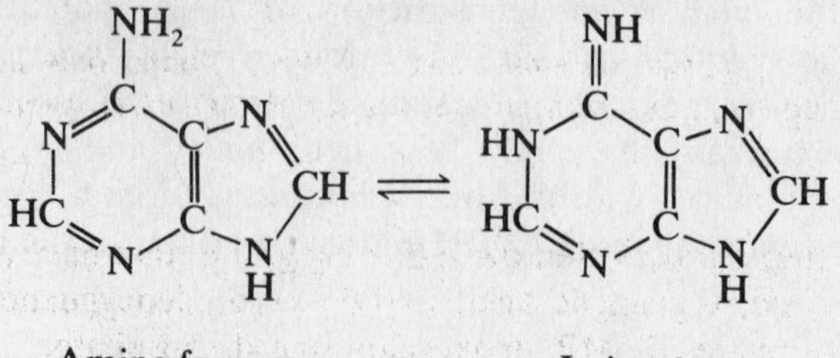

Amino form $\rightleftharpoons$ Imino form

7.4. Write the structure of 2,3-dideoxy-β-D-ribose.

SOLUTION

This compound is β-D-ribose, in which the oxygen atoms are missing from the C-2 and C-3 positions. Thus, its structure is

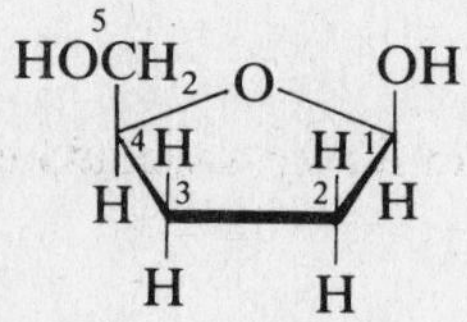

7.5. If DNA were hydrolyzed so that 2-deoxy-D-ribose were produced, in what anomeric forms would this sugar exist?

SOLUTION

Both 2-deoxy-β-D-ribose and 2-deoxy-α-D-ribose would be present because the former isomer, which is stabilized within the DNA structure, can now convert to the α form (Chap. 2).

7.6. Write the structure of deoxyadenosine.

SOLUTION

Deoxyadenosine is 9-2′-deoxy-β-D-ribofuranosyladenine, or 2-deoxy-D-ribose linked through a β-glycosidic bond from N-9 of the base to C-1 of the sugar. Thus, its structure is

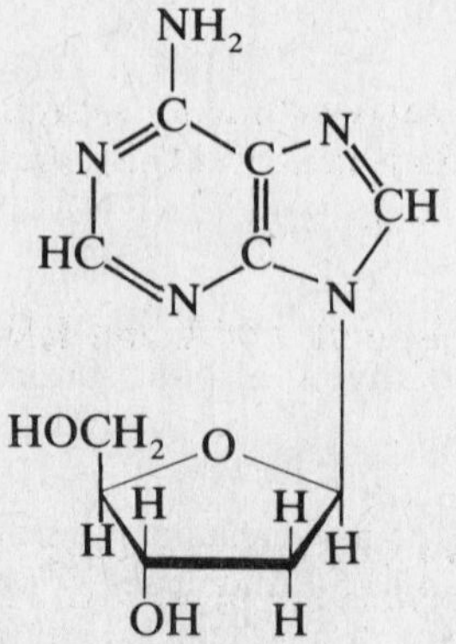

7.7. Of the nucleosides and nucleoside components adenosine, uridine, D-ribose, deoxyguanosine, cytidine, and thymidine, which would not be expected to occur in significant amounts in a partial hydrolysate of RNA?

SOLUTION

Deoxyguanosine, which contains 2-deoxy-D-ribose (found only in DNA), and thymidine, which contains thymine (present in significant quantities only in DNA), would be absent.

7.8. Match up the compounds, within the following group, that have the same structure: guanosine monophosphate, deoxyguanylic acid, dGDP, GTP, deoxyguanosine monophosphate, GMP, guanosine triphosphate, dGMP, deoxyguanosine diphosphate.

SOLUTION

$$\text{guanosine monophosphate} \equiv \text{GMP}$$
$$\text{deoxyguanylic acid} \equiv \text{dGMP} \equiv \text{deoxyguanosine monophosphate}$$
$$\text{dGDP} \equiv \text{deoxyguanosine diphosphate}$$
$$\text{GTP} \equiv \text{guanosine triphosphate}$$

POLYNUCLEOTIDES

7.9. How many 3′,5′ phosphodiester linkages would be present in a linear polynucleotide containing 20 nucleotide units?

SOLUTION

A phosphodiester linkage joins each nucleotide to the adjacent one, so the total number within a polynucleotide is always one less than the number of nucleotide units. Phosphates present at the 5′ or 3′ end of the chain do not constitute phosphodiesters. Thus, the answer is 19.

7.10. Write the chemical structure of the tetranucleotide ApGpUpCp.

SOLUTION

Remember that sequences are always written in the 5′→3′ direction, from left to right. The 3′ terminus is therefore phosphorylated. The sugar present is D-ribose because there is no *d* prefix to indicate that it is a *deoxy* tetranucleotide. Thus, the structure is as shown on top of page 196.

7.11. What would be the charge carried by ApGpUpC at neutral pH?

SOLUTION

This tetranucleotide would contain three phosphates, each dissociating one proton. The rest of the molecule would be uncharged at neutral pH. Thus, the charge would be −3.

7.12. Do polynucleotides containing both DNA and RNA within a single covalent structure occur naturally?

SOLUTION

Yes, but in small quantities and only transiently. The nascent (or Okazaki) fragments formed through discontinuous DNA replication contain a short stretch of RNA which serves as a primer for DNA chain growth (Chap. 16).

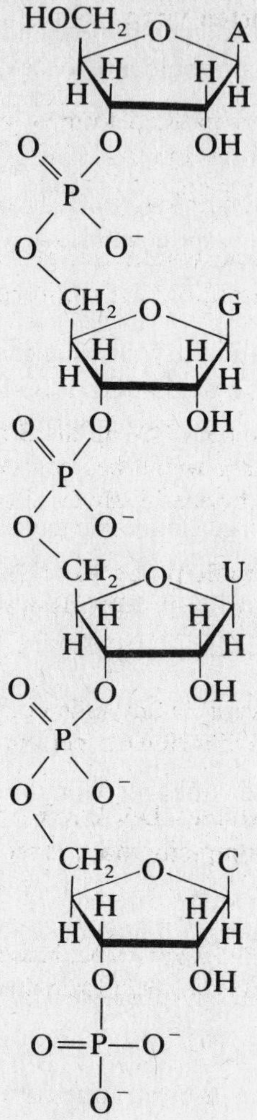

HOCH$_2$ O A
H H
H H
O O OH
P
O O$^-$
CH$_2$ O G
H H
H H
O O OH
P
O O$^-$
CH$_2$ O U
H H
H H
O O OH
P
O O$^-$
CH$_2$ O C
H H
H H
O OH
O=P—O$^-$
O$^-$

STRUCTURE OF DNA

7.13. Write the complementary DNA sequence for the following: GCTTAGTA.

SOLUTION

In complementary base pairing within DNA, G pairs with C and A with T. Thus the answer is CGAATCAT.

7.14. Write the complementary RNA sequence for the sequence shown in Prob. 7.13.

SOLUTION

In pairing between DNA and RNA, A and G in DNA link, respectively, with U and C in RNA. Thus the answer is CGAAUCAU.

7.15. With respect to base pairing, 5-methylcytosine behaves in the same way as does cytosine. How is this possible?

SOLUTION

In 5-methylcytosine, the methyl group is positioned away from the region of H bonding with guanine and does not interfere with the formation of the three H bonds occurring in a G : C base pair.

7.16. Why does a fiber of DNA shorten upon drying?

SOLUTION

The DNA duplex exists largely in the B form at high humidities. The pitch (or repeat distance) is 3.4 nm, which covers 10 base pairs. At low humidities, DNA converts to the A form, in which the pitch is less, 2.8 nm, and covers 11 base pairs.

7.17. By approximately what percent would a fiber of DNA shorten upon transfer from an environment of high to low humidity, i.e., following conversion from the B to the A form of DNA?

SOLUTION

Consider 100 base pairs of duplex DNA. In the B form this would have a length of 34 nm (10 base pairs per 3.4 nm). In the A form it would be reduced to approximately 25.5 nm (11 base pairs per 2.8 nm). Thus, the length would shorten by about 25 percent.

7.18. The A and G composition (in mole percent) of one of the strands of a duplex DNA is $A = 27$ and $G = 30$. What would be the T and C contents of the complementary strand?

SOLUTION

In duplex DNA, T (mole percent) in one strand equals A in the other, and C equals G. Thus, in the complementary strand $T = 27$ and $C = 30$ mole percent.

7.19. With respect to the strand of duplex DNA referred to in Prob. 7.18, what can be said about the A and G contents of its complementary strand?

SOLUTION

From the data provided, knowledge of the individual A and G contents of the complementary strand is not possible. However, $A + G$ would together comprise $100 - (27 + 30)$, or 43, mole percent of the complementary strand.

DENATURATION OF DNA

7.20. Why does circular duplex DNA renature more rapidly than linear duplex DNA?

SOLUTION

When denatured, the component single strands of circular DNA remain interlocked (assuming neither is broken) and, upon renaturation, can find one another more readily than the completely separated strands of linear DNA.

7.21. Why does DNA denature when it is put into pure water, i.e., at an ionic strength of ~0?

SOLUTION

The melting (or denaturation) temperature of DNA is dependent on ionic strength; lowering the ionic strength lowers the melting temperature. In the extreme situation of zero ionic strength, the electrostatic repulsion within the DNA, whose anionic groups are not shielded by counterions, is sufficient to lower the melting temperature to less than 20°C.

7.22. Why does viral DNA have less sequence heterogeneity than does bacterial DNA?

SOLUTION

The genome of viruses carries fewer genes than does that of bacteria. Thus, in a fixed amount of DNA (say several viral genome equivalents), viral genes would be present in more copies than would individual bacterial genes. Different genes have different nucleotide sequences, but various sequences would be repeated more frequently in the viral DNA; i.e., it would have less sequence heterogeneity.

SIZE, ORGANIZATION, AND TOPOLOGY OF DNA

7.23. The *E. coli* chromosome has a size of approximately 4,000 kb. What length of DNA (B form) would be contained in it?

SOLUTION

The B form of DNA has a length of 3.4 nm per 10 base pairs, or 340 nm per kb. Thus the total length would be 1.36×10^6 nm, or approximately 1.4 mm.

7.24. What length of B-form DNA would be present in a human cell (diploid)?

SOLUTION

The size of the human genome is 2.9×10^6 kb. A diploid cell would contain 5.8×10^6 kb of DNA. This is equivalent to approximately 2×10^9 nm or 2 m of B-form DNA.

7.25. The nucleus of a human cell is of the order of 10 μm in diameter. How can it accommodate 2×10^6 μm of duplex DNA?

SOLUTION

The DNA is condensed, by complexing with histones, into chromatin. The basic unit of condensation is the nucleosome, and these repeating structures (200 base pairs of DNA each) are further organized into more compact structures that make up the chromosomes.

7.26. What features of DNA structure are essential for supercoiling?

SOLUTION

Supercoiling results from the introduction (or removal) of extra double-helical twists that are maintained within the DNA structure. Thus, for supercoiling, the DNA must be double-helical, and it must be topologically constrained, e.g., by being circular or by being complexed to protein in such a way as to "tie" two strands at certain points.

7.27. If the two strands of a circular DNA molecule were pulled apart (unwound) at a certain position, what type of supercoiling would be introduced into the rest of the molecule?

SOLUTION

In the double-helical section of the molecule that remains, the DNA would be overwound. Thus, extra right-hand turns would be introduced into this section of DNA. This would cause positive supercoiling.

STRUCTURE AND TYPES OF RNA

7.28 What are the main chemical differences between RNA and DNA?

SOLUTION

The sugar present in RNA is ribose, while DNA contains deoxyribose. Also, uracil is present in RNA; this is replaced by thymine in DNA.

7.29. What is the most abundant species of RNA in a typical cell?

SOLUTION

Ribosomal RNA (rRNA) comprises about 80 percent of all RNA present. It is by far the most abundant.

7.30. Write the structure of inosine 5′-phosphate (IMP).

SOLUTION

IMP consists of inosine to which a phosphate group is attached at C-5 of the ribose residue. Thus, its structure is

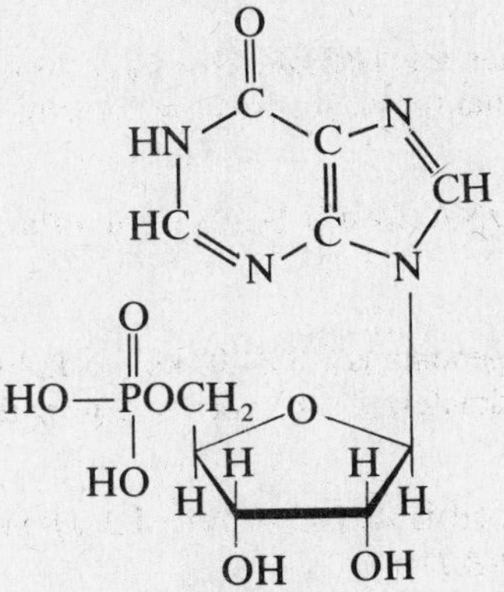

```
               O
               ‖
           HN ⁄ C ⁄ C ⁄ N
            |        ‖     CH
           HC ⁄ N ⁄ C ⁄ N
  O                    |
  ‖                    |
HO—POCH2   O
   |    H      H
  HO   H        H
       OH    OH
```

7.31. How would the melting curve for RNA compare with that for DNA?

SOLUTION

DNA is nearly always double-helical. Upon heating through the melting temperature, it undergoes a transition from the native (double-helical) to denatured (random-coil) state over a relatively narrow range of temperature. This is accompanied by a 40 percent increase in A_{260} (absorbance at wavelength 260). RNA, on the other hand, is nearly always single-stranded, and the extent of intrastrand base pairing is generally low and variable. Furthermore, the short base-paired (or double-helical) regions vary in stability. Thus, increasing the temperature of a solution of a typical RNA would result in a *gradual* increase in A_{260}, reflecting the successive melting of short helical regions, and the extent of increase would be significantly less than 40 percent.

NUCLEASES

7.32. What type of bond is cleaved by a nuclease?

SOLUTION

Nucleases cleave 3′,5′ phosphodiester linkages. They always catalyze the hydrolysis of the ester bond where it is connected to either the C-3 or the C-5 of the sugar moiety to give a 5′ or 3′ phosphate plus a 3′ or 5′ OH.

7.33. Why wouldn't an exonuclease degrade the DNA from bacteriophage ϕX174?

SOLUTION

Exonucleases require a 5′ or 3′ terminus of DNA or RNA as substrate. ϕX174 DNA is a single-stranded circle and has no 5′ or 3′ terminus.

7.34. Approximately how many *Eco*RI sites would you expect to find in the *E. coli* chromosome?

SOLUTION

*Eco*RI recognizes and cleaves at a sequence-specific six-base-pair site on DNA. Considering that each site can be occupied by one of four correctly orientated base pairs, the enzyme would cleave a random sequence once every 4^6 (or 4,096) base pairs. The *E. coli* chromosome has a size of 4×10^6 base pairs. Thus, there would be about $4 \times 10^6/4{,}096$, or approximately 1,000, *Eco*RI sites on the chromosome.

Supplementary Problems

7.35. Which of the following compounds are pyrimidines or purines?

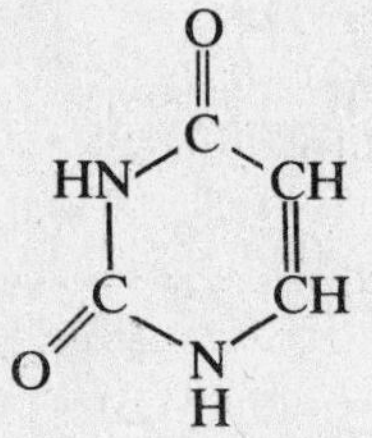

(a) 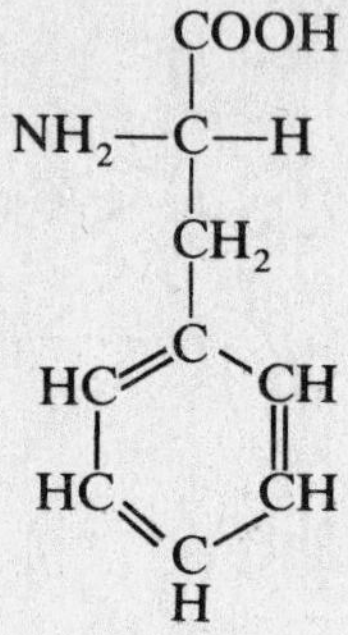 (b) 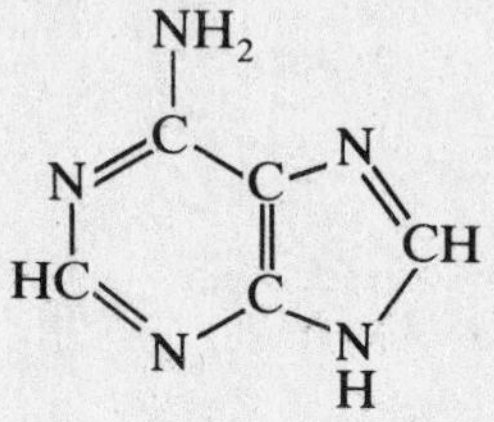 (c)

7.36. List the common nitrogenous bases found in (*a*) RNA and (*b*) DNA.

7.37. (*a*) Write the structure of 5-methylcytosine. (*b*) Is it a commonly occurring constituent of DNA?

7.38. Write the imino form of cytosine.

7.39. Write the structures of (*a*) deoxythymidine and (*b*) ribothymidine (occurs as a minor constituent in some forms of RNA).

7.40. Give an example of an *N*-glycoside.

7.41. Distinguish between a *nucleoside* and a *nucleotide*.

7.42. Write the structure of 3′,5′-AMP (cyclic AMP).

7.43. Which of the following compounds is different from the others: (*a*) GMP, (*b*) deoxyguanosine monophosphate, (*c*) guanylic acid, (*d*) guanosine 5′-phosphate?

7.44. The tetranucleotide d-ApGpUpCp is unlikely to be formed by partial hydrolysis of DNA. Why?

7.45. Would DNA from *Clostridium perfringens* be more or less resistant to denaturation than that from *Alcaligenes faecalis* (see Table 7.1)?

7.46. What is meant by the term *sequence complementarity*?

7.47. List three of the major structural differences between the B and Z forms of DNA.

7.48. Why is the double-helical structure of DNA more stable in solution at higher rather than lower ionic strength?

7.49. Why do nonrepeated sequences of bacterial DNA renature more rapidly than those of eukaryotic DNA?

7.50. What is meant by the term *DNA sequence complexity*?

7.51. Distinguish between the terms *chromosome* and *genome*.

7.52. Distinguish between *core* and *linker* histones.

7.53 What feature of the amino acid composition of histones enables them to interact strongly with DNA?

7.54. (*a*) Distinguish between *positive* and *negative* supercoiling of DNA. (*b*) What is meant by *relaxation* of supercoiled DNA, and (*c*) how can it be achieved?

7.55. List the major differences between the chemical composition of DNA and RNA.

7.56. Why does messenger RNA (mRNA) have the most heterogeneous base sequences of the major forms of RNA present in the cytoplasm of a cell?

7.57. Which of the following sequences would not be cut by a restriction endonuclease and why? (*a*) GAATTC; (*b*) GTATAC; (*c*) GTAATC; (*d*) CAATTG.

Chapter 8

Enzyme Catalysis

8.1 BASIC CONCEPTS

Question: What are enzymes?

Enzymes are proteins that are catalysts of biochemical reactions. They usually exist in very low concentrations in cells, where they increase the *rate* of a reaction without altering its equilibrium position; i.e., both forward *and* reverse reaction rates are enhanced by the same factor. This factor is usually around 10^3–10^{12}.

EXAMPLE 8.1

Carbon dioxide gas dissolves readily in water and is spontaneously hydrated to form carbonic acid, which rapidly dissociates to a proton and a bicarbonate ion:

$$CO_2 + H_2O \longrightarrow H^+ + HCO_3^-$$

The forward hydration reaction rate for 20 mmol L^{-1} CO_2 at 25°C and pH 7.2 is ~0.6 mmol L^{-1} s^{-1}.

In mammalian red blood cells, the enzyme *carbonic anhydrase* is present at a concentration of 1–2 g per liter of cells; its M_r is 30,000; thus its *molar concentration* is $\sim 50 \times 10^{-6}$. The flux through the forward reaction in the presence of this concentration of enzyme, under the above-mentioned conditions, is ~50 mol L^{-1} s^{-1}, a *rate enhancement* over the noncatalyzed process of 8×10^4.

There are over 2,500 different biochemical reactions with specific enzymes adapted for their rate enhancement. Since different species of organism produce different structural variants of enzymes, the number of different enzyme proteins in all of biology is well in excess of 10^6. Each enzyme is characterized by *specificity* for a narrow range of chemically similar *substrates* (reactants) and also other molecules that modulate their activities; these are called *effectors* and can be *activators*, *inhibitors*, or both; in more complex enzymes, one compound may have either effect, depending on other physical or chemical conditions. Enzymes range in size from large multiple-subunit complexes (called *multimeric* enzymes; $M_r \simeq 10^6$) to small single-subunit forms.

EXAMPLE 8.2

Aspartate carbamoyltransferase, $M_r = 310{,}000$, catalyzes the formation of carbamoyl aspartate from carbamoyl phosphate and aspartate in the first committed step of pyrimidine biosynthesis (Chap. 15). The enzyme from the bacterium *E. coli* consists of twelve subunits, six regulatory and six catalytic. CTP is a negative effector; i.e., it inhibits the enzyme, and does so through binding to the regulatory subunits. ATP is a positive effector that acts through the regulatory subunits, while succinate inhibits the reaction by direct competition with aspartate at the active site (see Chap. 10 for more on effectors).

The surface area of even the smallest enzymes (such as ribonuclease, $M_r = 12{,}000$) that is occupied by the chemical groups to which the reactants bind is less than 5 percent of the total area; this region is called the *active site*.

Question: What part of an enzyme is responsible for its substrate specificity?

The particular arrangement of an enzyme's amino acid side chains in the active site determines the type of molecules that can bind *and* react there; there are usually about five such side chains in any particular enzyme. In addition, many enzymes have small nonprotein molecules associated with or near the active site that determine substrate specificity. These molecules are called *cofactors* if they are noncovalently linked to the protein; they are called *prosthetic groups* if covalently bound. In some enzymes a specific metal ion is required for activity.

EXAMPLE 8.3

Carbonic anhydrase has one Zn^{2+} ion per molecule of enzyme, and the metal ion resides in the active site. *Aspartate carbamoyltransferase* has six Zn^{2+} ions per dodecamer; these are required for the stabilization of the complex, since without Zn^{2+}, the hexamer dissociates.

8.2 CLASSIFICATION OF ENZYMES

Question: On what basis are enzymes given their particular names?

All enzymes are named according to a classification system designed by the Enzyme Commission (EC) of the IUPAC and based on the type of reaction they catalyze. Each enzyme type has a specific, four-integer *EC number* and a complex, but unambiguous, name that obviates confusion about enzymes catalyzing similar but not identical reactions. In practice, many enzymes are known by a *common name*, which is derived from the name of its principal, specific reactant, with the suffix *-ase* added. Some common names do not even have *-ase* appended, but these tend to be enzymes studied and named before systematic classification of enzymes was undertaken.

EXAMPLE 8.4

Examples of typical enzyme names are *arginase*, which acts on *arginine*, and *urease*, which acts on urea (Chap. 15). Two atypical common names are *pepsin*, a digestive tract *proteolytic* enzyme (EC number 3.4.23.1), and, more exotically, *rhodanese* (thiosulfate : cyanide sulfurtransferase, EC 2.8.1.1), which is in mammalian liver and kidney and catalyzes the removal of cyanide and thiosulfate from the body. In the latter case, it is understandable why the old name has remained in common use.

The first integer in the EC number designates to which of the six major classes an enzyme belongs (see Table 8.1 for details).

The second integer in an EC enzyme number indicates the *type of bond* acted upon by the enzyme. For the hydrolase enzyme class, the second integers are listed in Table 8.2.

EXAMPLE 8.5

Arginase is a *hydrolase* that is in the liver of urea-producing organisms (*ureoteles*). It catalyzes the reaction:

$$
\begin{array}{c}
NH_2 \\
| \\
\overset{+}{C}{=}NH_2 \\
| \\
NH \\
| \\
(CH_2)_3 \\
| \\
H{-}C{-}\overset{+}{N}H_3 \\
| \\
COO^-
\end{array}
\xrightarrow{H_2O}
\begin{array}{c}
NH_2 \\
| \\
C{=}O \\
| \\
NH_2
\end{array}
+
\begin{array}{c}
\overset{+}{N}H_3 \\
| \\
(CH_2)_3 \\
| \\
H{-}C{-}\overset{+}{N}H_3 \\
| \\
COO^-
\end{array}
$$

Arginine Urea Ornithine

Table 8.1. Major Enzyme Classes

First EC Number	Enzyme Class	Type of Reaction Catalyzed
1.	Oxidoreductase	Oxidation-reduction. A hydrogen or electron donor is one of the substrates.
2.	Transferase	Chemical group transfer of the general form $A{-}X + B \longrightarrow A + B{-}X$.
3.	Hydrolase	Hydrolytic cleavage of C—C, C—N, C—O, and other bonds.
4.	Lyase	Cleavage (*not* hydrolytic) of C—C, C—N, C—O, and other bonds, leaving double bonds; alternatively, addition of groups to a double bond.
5.	Isomerase	Change of geometrical (spatial) arrangement of a molecule.
6.	Ligase	Ligating (joining together) of two molecules, *with* the accompanying hydrolysis of a high-energy bond.

Table 8.2. Hydrolase Subclassification

First Two EC Integers	Type of Bond Acted Upon
3.1	Ester, $-\overset{\overset{\displaystyle O}{\|}}{C}-O\overset{\downarrow}{-}R$, or with S or P in place of C, or $-\overset{\overset{\displaystyle O}{\|}}{C}-S-R$
3.2	Glycosyl, sugar$-C-O\overset{\downarrow}{-}R$, or with N or S in place of O
3.3	Ether, $R-O\overset{\downarrow}{-}R'$, or with S in place of O
3.4	Peptide, $C\overset{\downarrow}{-}N$
3.5	Nonpeptide, $C\overset{\downarrow}{-}N$
3.6	Acid anhydride, $R-\overset{\overset{\displaystyle O}{\|}}{C}-O\overset{\downarrow}{-}\overset{\overset{\displaystyle O}{\|}}{C}-R'$
3.7	$C\overset{\downarrow}{-}C$
3.8	Halide (X), $C\overset{\downarrow}{-}X$, or with P in place of C
3.9	$P\overset{\downarrow}{-}N$
3.10	$S\overset{\downarrow}{-}N$
3.11	$C\overset{\downarrow}{-}P$

The official EC name of this enzyme is *L-arginine amidinohydrolase*; the last word refers to the fact that the *amidino group* (dotted circle in the equation) is cleaved from arginine by introduction of a molecule of water across the C—N bond. In the reaction, a nonpeptide C—N bond is cleaved; thus, the second EC number for arginase is 5; its whole classification number is 3.5.3.1.

The third number is a subclassification of the bond type acted upon or the group transferred in the reaction or both, and the categories vary from one main EC class to the next. The fourth number is simply a serial number.

8.3 MODES OF BOND CLEAVAGE BY ENZYMES

Question: Among all the different types of bonds cleaved during the various enzymatic reactions, are there any common features in electron behavior?

Much of the experimental study of enzymes concerns the quest for an understanding of the basic chemistry of the catalytic process and how covalent chemical bonds are broken and others are formed. A covalent bond between two atoms consists of a shared pair of electrons, and in bond *fission*, one or both atoms combine with new partners with which to share an electron pair. Of the hundreds of enzymatic reactions that involve C—H bond fission, the basic process can occur in only two ways: (1) *homolytic cleavage* in which one electron remains with the carbon and one with the hydrogen to produce two radicals; i.e., distributed through the molecular orbitals is an odd, *unpaired* electron; (2) *heterolytic cleavage*, which leaves *both* electrons on one atom; if they remain on the carbon, a *carbanion* intermediate species is formed plus an H^+; if they depart on the hydrogen, an electron-deficient *carbocation* and a *hydride ion* result.

EXAMPLE 8.6

Free radicals generally have extremely short lifetimes, in the region of 1 ns. The previously mentioned homolytic cleavage reaction is represented simply as

$$-\overset{|}{\underset{|}{C}}\!:\!H \longrightarrow -\overset{|}{\underset{|}{C}}\cdot + H\cdot$$

where the dots designate electrons and the bonds are a pair of electrons, shared by the adjacent atoms.

A completely different free-radical reaction involves molecular oxygen and Fe^{2+} in the heme of hemoglobin in red blood cells. There is a finite but small probability of the following *side reaction* occurring, with the production of the *superoxide* free radical

$$-Fe^{2+}\div O{=}O \longrightarrow -Fe^{3+} + :O{=}O^{\cdot -}$$

Approximately 1 percent of all the heme in human red blood cell hemoglobin undergoes this conversion per day. The resulting hemoglobin is called *methemoglobin*, and a metabolic process exists to convert it back to ferrous hemoglobin; another metabolic process eliminates the highly reactive superoxide radicals.

EXAMPLE 8.7

Carbanions form during heterolytic cleavage of bonds, as follows for the case of a C—H bond:

$$-\overset{|}{\underset{|}{C}}\div H \longrightarrow -\overset{|}{\underset{|}{C}}\!:^{-} + H^+$$

Carbocations and hydride ions are intermediate species in many enzyme-catalyzed reactions; the most notable examples are the dehydrogenases. The basic reaction is as follows:

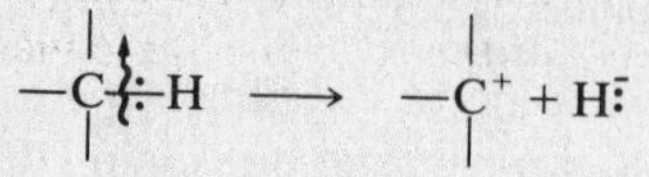

$$-\overset{|}{\underset{|}{C}}-H \longrightarrow -\overset{|}{\underset{|}{C^+}} + H^{:-}$$

Since carbon is more electronegative than hydrogen, carbanion formation is generally favored in enzymatic reactions, but the particular mechanism is determined by the constituents in the active site. *Both* heterolytic paths of cleavage of C—H bonds involve ionic intermediates that are formed in the conversion of substrate(s) to product(s). These reactions predominate in organic chemistry and biochemistry, and this has led to the broad categorization of reactants as either electron-rich (*nucleophiles*) or electron-deficient (*electrophiles*). Because atomic nuclei are positively charged, atoms that readily combine with positively charged, or electron-deficient, species are called nucleophilic. Conversely, the recipient atoms are electrophilic. Generally, nucleophiles are anionic or contain a lone pair of electrons that can *attack* other molecules.

Question: What are the common *nucleophiles* in biology?

They are those molecules containing oxygen, sulfur, and nitrogen, e.g., H—O—H, R—O—H, R—O$^-$, R—S—H, R—S$^-$, R—N—H. There are only a few biologically important *electrophiles* available to the various nucleophiles on substrates or in active sites of enzymes. Electrophiles are electron-deficient, often cationic (e.g., metal cations) or having an unfilled valence electron shell.

EXAMPLE 8.8

Biologically important *electrophiles* are H^+ and metal cations such as Cu^+, Fe^{2+}, Fe^{3+}, Mo^{6+}, Zn^{2+}, and some cofactors, such as derivatives of vitamins B_1 and B_6.

Writing the chemical mechanism of a reaction involves describing the rearrangement of electrons as the substrate is converted to the product via some sort of *transition state(s)*. The best way of depicting the pathway of rearrangement of bonds is by use of curved arrows that indicate the directions of electron flow.

EXAMPLE 8.9

The electron flow diagram for the hydrolysis of a peptide bond is as follows:

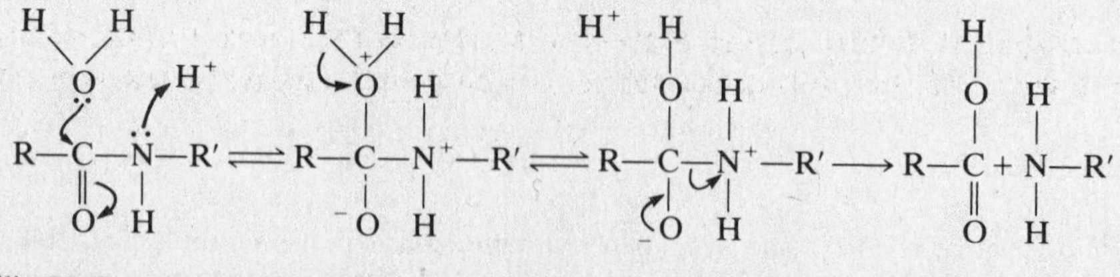

$$\underset{\substack{\| \\ O}}{\overset{\substack{H\;\;H \\ \diagdown\diagup \\ O \\ }}{R-C}}-\underset{\substack{| \\ H}}{\ddot{N}}-R' \;\rightleftharpoons\; R-\underset{\substack{| \\ {}^-O}}{\overset{\substack{H\;\;H \\ \diagdown\diagup \\ {}^+O \\ |}}{C}}-\underset{\substack{| \\ H}}{\overset{\substack{H \\ |}}{N^+}}-R' \;\rightleftharpoons\; R-\underset{\substack{| \\ {}^-O}}{\overset{\substack{H^+\;H \\ | \\ O \\ |}}{C}}-\underset{\substack{| \\ H}}{\overset{\substack{H \\ |}}{N^+}}-R' \longrightarrow R-\underset{\substack{\| \\ O}}{\overset{\substack{H \\ | \\ O \\ |}}{C}}+\underset{\substack{| \\ H}}{\overset{\substack{H \\ |}}{N}}-R'$$

The nucleophilic H_2O attacks the electrophilic carbonyl carbon, which is rendered *electron-depleted* by the electron withdrawing attached carbonyl oxygen. Note the tetrahedral intermediates in the second and third structures; in these, the carbon has the "usual" tetrahedral arrangement of four bonds.

8.4 MODES OF ENHANCEMENT OF RATES OF BOND CLEAVAGE

The basic mechanisms by which enzymes increase the rates of chemical reactions can be classified into four groups.

Facilitation of Proximity

This effect is also called the *propinquity effect* and means that the rate of a reaction between two molecules is enhanced if they are abstracted from dilute solution and held in close proximity to each other in the enzyme's active site; this raises the *effective* concentration of the reactants.

Covalent Catalysis

The side chains of amino acids present a number of nucleophilic groups for catalysis; these include $RCOO^-$, $R—NH_2$, aromatic—OH, histidyl, R—OH, and RS^-. These groups attack electrophilic (electron-deficient) parts of substrates to form a covalent bond between the substrate and the enzyme, thus forming a *reaction intermediate*. This type of process is particularly evident in the group-transfer enzymes (EC class 2; see Table 8.1). In the formation of a covalently bonded intermediate, attack by the enzyme nucleophile (Enz-X in Example 8.10) on the substrate can result in acylation, phosphorylation, or glycosylation of the nucleophile. About 100 different enzymatic reactions occur via this mechanism.

EXAMPLE 8.10

A phosphoenzyme intermediate is formed in one type of covalent catalysis in enzymes:

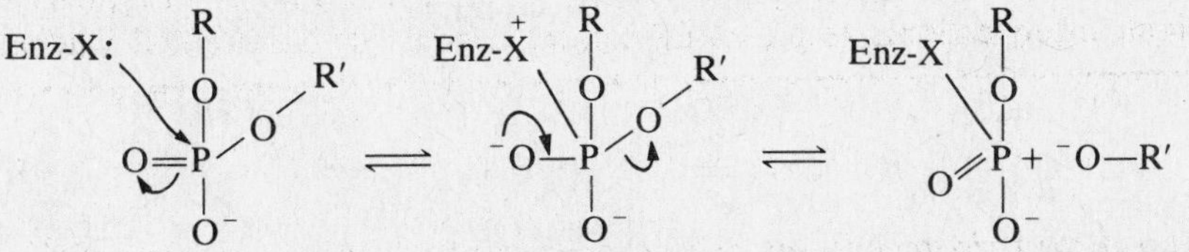

$$\text{Enz-X:} + R{-}O{-}P(=O)(O^-){-}O{-}R' \rightleftharpoons \text{Enz-}\overset{+}{X}{-}P(O^-)(O^-)(OR)(OR') \rightleftharpoons \text{Enz-X}{-}P(=O)(O^-){-}O{-}R + {}^-O{-}R'$$

Numerous examples of this basic mechanism of catalysis can be found among the EC class 2 enzymes. One example is *hexokinase*.

The covalent intermediates can be attacked by a second nucleophile to cause the release of the product. When the second nucleophile is water, the overall reaction is called *hydrolysis*. Also, in many cases the nucleophile is not simply an amino acid side chain of the enzyme but a prosthetic group; an example is *pyridoxal phosphate* in the *transaminases* (Chap. 15).

General Acid-Base Catalysis

Acid-base catalysis is defined as the process of *transferral of a proton* in the transition state. It does not involve covalent bond formation per se, but an overall enzymatic reaction can involve this as well.

EXAMPLE 8.11

An example of general acid-base catalysis from organic chemistry illustrates the above-mentioned point, but note that hemiacetals also form in some enzymatic reactions.

Overall reaction:

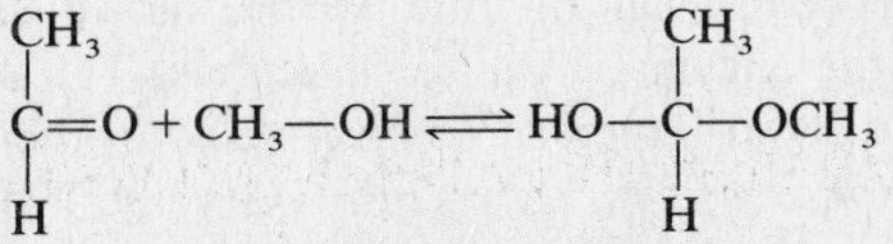

$$\underset{\text{Acetaldehyde}}{CH_3{-}\overset{}{C}H{=}O} + \underset{\text{Methanol}}{CH_3{-}OH} \rightleftharpoons \underset{\text{Hemiacetal}}{HO{-}C(CH_3)(H){-}OCH_3}$$

Reaction mechanism A: A *base* (OH^-) accelerates hemiacetal formation as follows:

$$CH_3—OH + OH^- \rightleftharpoons CH_3—\ddot{O}^{-} + H_2O$$

Nucleophile

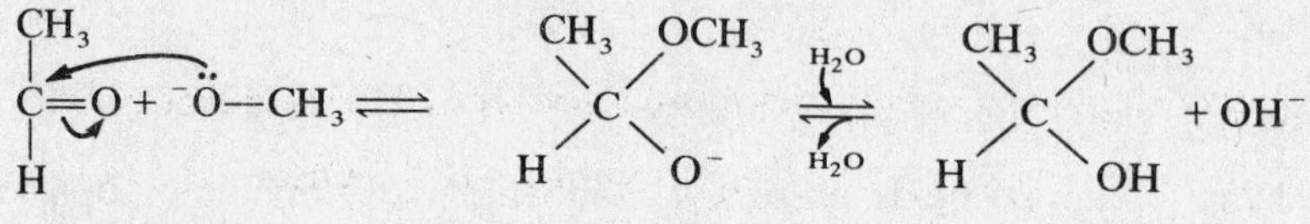

$$\underset{\text{Nucleophilic attack}}{\overset{CH_3}{\underset{H}{C}}=O} + {}^{-}\ddot{O}—CH_3 \rightleftharpoons \overset{CH_3\quad OCH_3}{\underset{H\qquad O^-}{C}} \underset{H_2O}{\overset{H_2O}{\rightleftharpoons}} \overset{CH_3\quad OCH_3}{\underset{H\qquad OH}{C}} + OH^-$$

Note: The OH^- is recycled in the reaction; thus, it can be considered to be a catalyst in the true sense of the word.

Reaction mechanism B: Acid catalysis also occurs in the reaction, and it involves the formation of the *oxonium* salt, followed by reaction with the alcohol as follows:

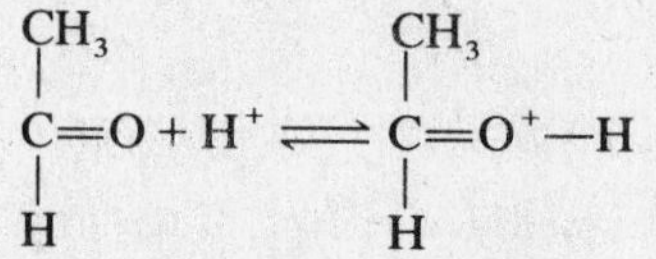

$$\overset{CH_3}{\underset{H}{C}}=O + H^+ \rightleftharpoons \underset{\text{Oxonium}}{\overset{CH_3}{\underset{H}{C}}=O^+—H}$$

$$\overset{CH_3}{\underset{H}{C}}=O^+—H + :\underset{H}{O}—CH_3 \rightleftharpoons \overset{CH_3\quad {}^{+}O(CH_3)H}{\underset{H\qquad OH}{C}} \rightleftharpoons \overset{CH_3\quad O—CH_3}{\underset{H\qquad OH}{C}} + H^+$$

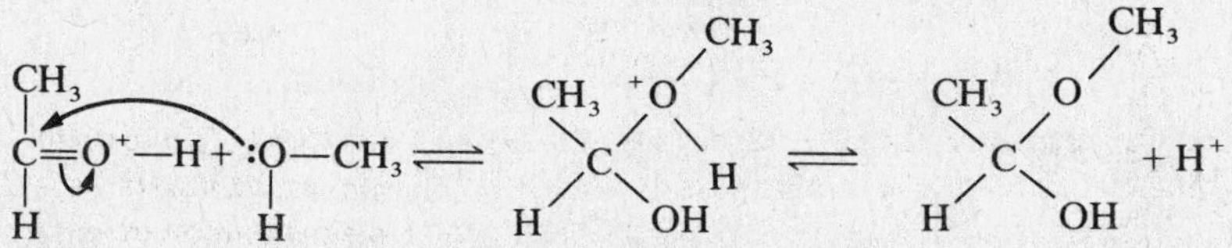

In the preceding example, the rate of hemiacetal formation is enhanced in strong acid *or* strong base. In other cases, only one—either base or acid—might be a catalyst.

EXAMPLE 8.12

The hydrolysis of *nitramide* is susceptible to base, but not acid, catalysis. An elevation of pH leads to an increased rate of reaction with no net consumption of base, as is shown in the following reaction:

$$NH_2NO_2 + OH^- \longrightarrow H_2O + NHNO_2^-$$
$$NHNO_2^- \longrightarrow N_2O + OH^-$$

OH^- is not the only base that will catalyze the hydrolysis; other bases such as acetate also react; e.g.,

$$NH_2NO_2 + CH_3COO^- \longrightarrow CH_3COOH + NHNO_2^-$$
$$NHNO_2^- \longrightarrow N_2O + OH^-$$
$$OH^- + CH_3COOH \longrightarrow H_2O + CH_3COO^-$$

According to the Bronsted-Lowry definitions, and as implied in the previous example, an acid is any moiety that will donate a proton while a base is one that will accept a proton from another moiety.

Acid-base catalysis does not contribute to rate enhancement by a factor greater than ~100, but together with other mechanisms that operate in the active site of an enzyme, it contributes considerably to increasing the enzymatic rate of reactions. The amino acid side chains of glutamic acid, histidine, aspartic acid, lysine, tyrosine, and cysteine in their *protonated* forms can act as *acid catalysts* and in their *unprotonated* forms as base catalysts (see Prob. 8.11). Clearly, the effectiveness of the side chain as a catalyst will depend on the pK_a (Chap. 3) in the environment of the active site and on the pH at which the enzyme operates.

Strain, Molecular Distortion, and Shape Change

Strain in the bond system of reactants and the release of the strain as the transition state converts into the products (like cutting a wound clock spring) can provide rate enhancement of chemical reactions.

EXAMPLE 8.13

The following two chemical reactions involve hydrolysis of a phosphate ester bond.

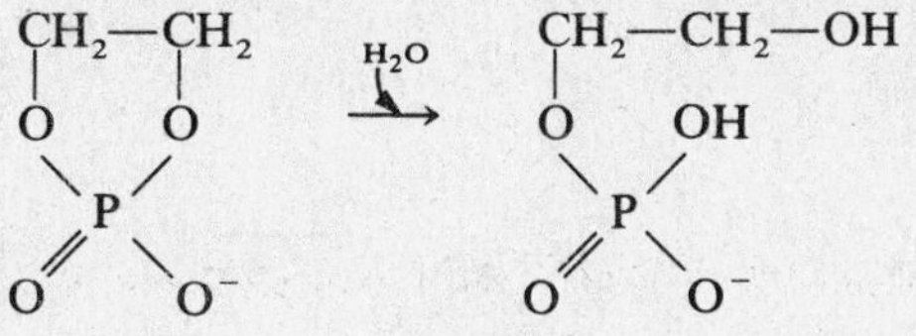

$$\begin{array}{c}CH_2\text{—}CH_2\\ |\quad\quad |\\ O\quad\quad O\\ \diagdown\ \diagup\\ P\\ \diagup\!\!\diagup\ \diagdown\\ O\quad\quad O^-\end{array} \xrightarrow{H_2O} \begin{array}{c}CH_2\text{—}CH_2\text{—}OH\\ |\quad\quad\quad\\ O\quad\quad OH\\ \diagdown\ \diagup\\ P\\ \diagup\!\!\diagup\ \diagdown\\ O\quad\quad O^-\end{array}$$

(*a*)

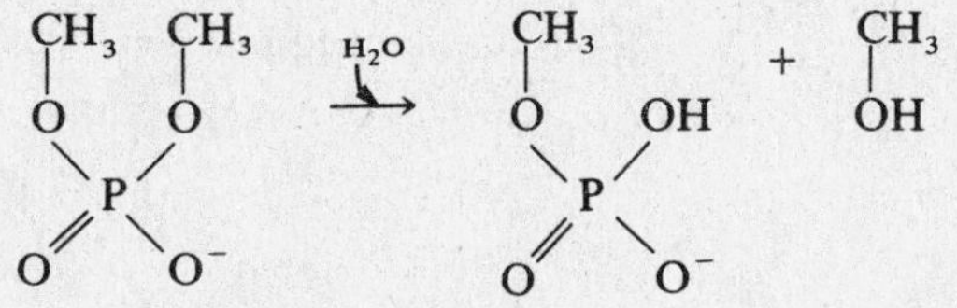

$$\begin{array}{c}CH_3\quad CH_3\\ |\quad\quad |\\ O\quad\quad O\\ \diagdown\ \diagup\\ P\\ \diagup\!\!\diagup\ \diagdown\\ O\quad\quad O^-\end{array} \xrightarrow{H_2O} \begin{array}{c}CH_3\quad\quad\\ |\quad\quad\quad\\ O\quad\quad OH\\ \diagdown\ \diagup\\ P\\ \diagup\!\!\diagup\ \diagdown\\ O\quad\quad O^-\end{array} + \begin{array}{c}CH_3\\ |\\ OH\end{array}$$

(*b*)

Under standard conditions, reaction (*a*) is 10^8 times faster than reaction (*b*). The explanation is that the cyclic compound in (*a*) has considerable *bond strain* (potential energy in this configuration is high), which is released on ring opening during hydrolysis. This type of strain is not present in the diester in (*b*).

In the case of enzymes, not only may the substrate be distorted (have strain) but an extra degree of freedom is introduced, namely, the enzyme with all its amino acid side chains. The binding of a substrate to an enzyme involves *interaction energy*, which may facilitate catalysis. Also for an increase in catalytic rate, there must be an overall *destabilization* of the enzyme-substrate complex and an increase in the stability of the transition state. This idea is illustrated in Fig. 8-1.

In the *uncatalyzed* reaction [Fig. 8-1(*a*)], the reactant has a relatively low probability of assuming the *strained* conformation necessary for interaction between the two reactive groups. In order for the reaction to take place, the molecule must cross this so-called *activation-energy barrier*. In the *catalyzed* reaction [Fig. 8-1(*b*)], the *binding* of the reactant to the enzyme leads to the formation of a combined structure (enzyme-substrate complex) in which the tendency for the substrate to form into the *transition state* is greater; i.e., less energy is involved in bringing the reactive groups together. Therefore, the reaction proceeds faster.

The destabilization of the enzyme-substrate complex can be imagined to be due to distortion of bond angles and lengths from their previously more stable configuration; this may be achieved by electrostatic attraction or repulsion by groups on the substrate and enzyme. Or, it could involve *desolvation* (removal of water) of a charged group in a hydrophobic active site. A further consideration is that of *entropy* change in the reaction; this is discussed in the next section.

Question: Does tight binding between an enzyme and its substrate imply rapid catalysis?

If a substrate were to bind *without* significant transformation of binding energy into distortion strain, then binding would be stronger. But this does not necessarily mean that $\Delta G^{\ddagger}$ (see Fig. 8-1) is altered by the binding interaction. Hence, to dispel a common belief, tight binding is *not* necessarily useful in the rate enhancement of enzyme catalysis.

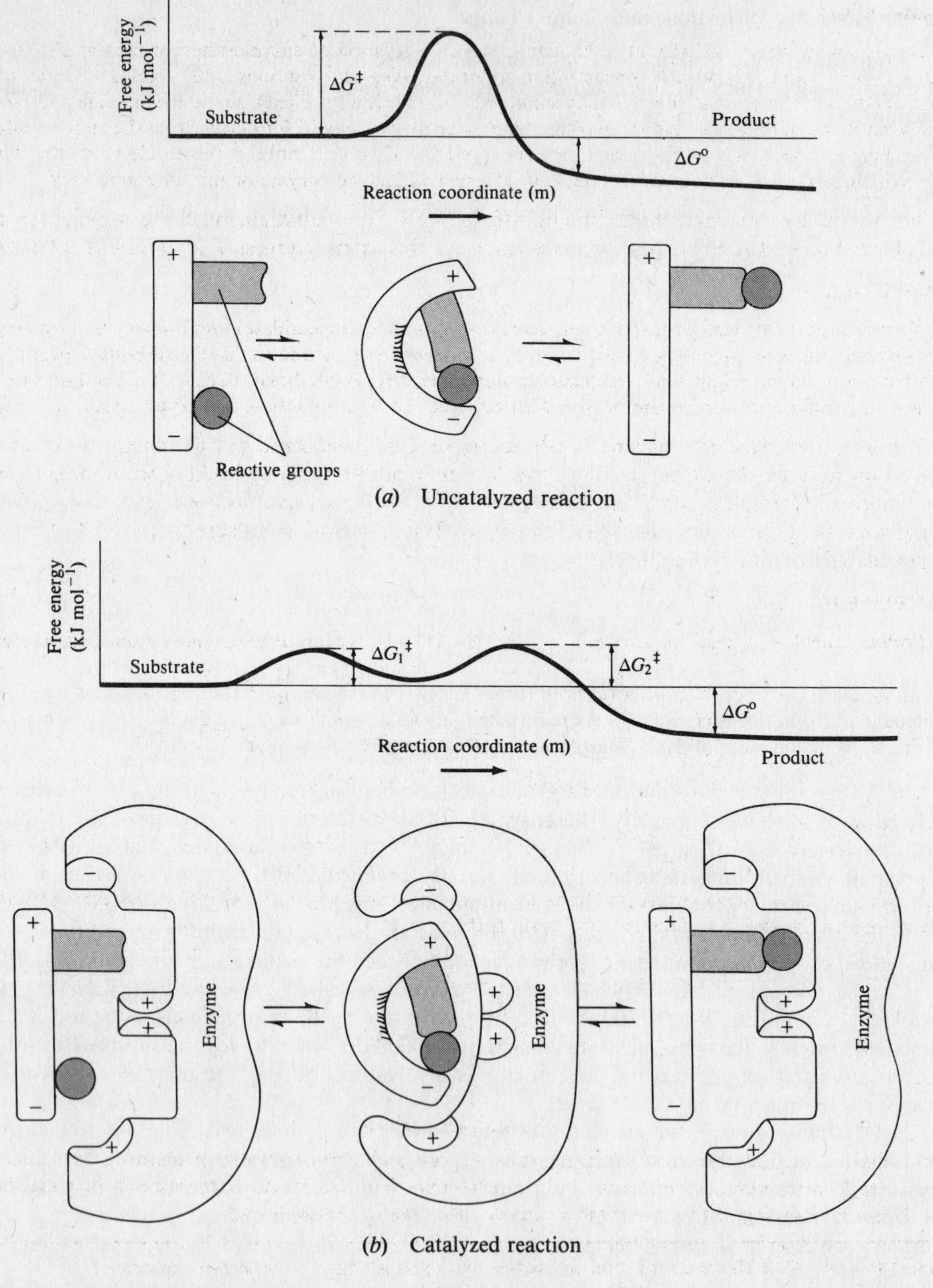

Fig. 8-1 Activation energy is lowered in catalyzed reactions. The graphs above each reaction scheme indicate the energy of the substrate (depicted here as *potential* energy of the *bent* substrate) at each stage of the reaction. The arrows indicate, according to their length and boldness, the probability and, in this case, the rate of the reactions. $\Delta G^\ddagger$ is the activation energy of the transition-state(s) molecule, and ΔG° is the *overall* free energy of the reaction (Chap. 10). N.B. Although the reaction coordinate may be viewed in terms of *distance* (as depicted here), an alternative approach is to consider the coordinate as some sort of *extent of reaction*, expressed in arbitrary units.

EXAMPLE 8.14

Suppose a substrate half-saturates the active sites of a solution of enzyme when present at 10^{-7} mol L^{-1} (i.e., $K_d = 10^{-7}$ mol L^{-1}), but the concentration under *physiological* conditions is 10^{-3} mol L^{-1}. Under physiological conditions, the enzyme sites are fully saturated (i.e., all sites are filled), so the enzyme rate enhancement is not what it could be if a large portion of the binding energy were used for *destabilizing* the enzyme-substrate (ES) complex. If the binding affinity were reduced, such that $K_d = 10^{-3}$ mol L^{-1}, this would imply that some of the binding energy was *used* in the introduction of strain within the enzyme or substrate molecules.

Many enzymes, in fact, have binding affinities for their substrates that are around the mean physiological concentrations, possibly as a result of evolutionary pressure for efficient catalysis.

EXAMPLE 8.15

X-ray analysis of crystals of *carboxypeptidase A* (a pancreatic exopeptidase) with bound *pseudo substrate* (a false substrate that is not degraded by the enzyme, i.e., an inhibitor), indicates that the *susceptible* peptide bond is twisted out of the *normal* planar configuration usually seen in peptide bonds (Chap. 4). This distortion leads to a loss of resonance energy in the bond, which enhances its susceptibility to hydrolytic attack.

Because in catalysis the enzyme-substrate complex is destabilized and the energy so involved is released on forming the transition state, the enzyme binds very tightly in the transition state. Some enzymes can be dramatically inhibited by so-called *transition-state analogs*. The transition state normally has only a fleeting existence ($<10^{-13}$ s), but the analogs are stable structures that *resemble* the postulated transition-state complex.

EXAMPLE 8.16

Proline racemase, a bacterial enzyme, catalyzes the interconvention of the D and L isomers of proline:

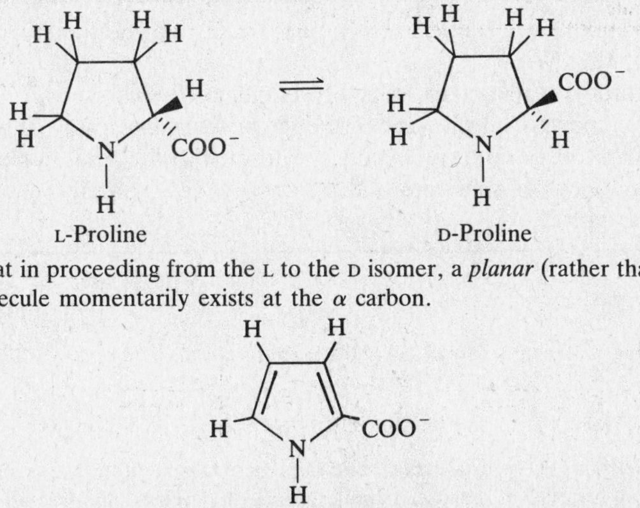

L-Proline ⇌ D-Proline

It was postulated that in proceeding from the L to the D isomer, a *planar* (rather than the usual tetrahedral) configuration of the molecule momentarily exists at the α carbon.

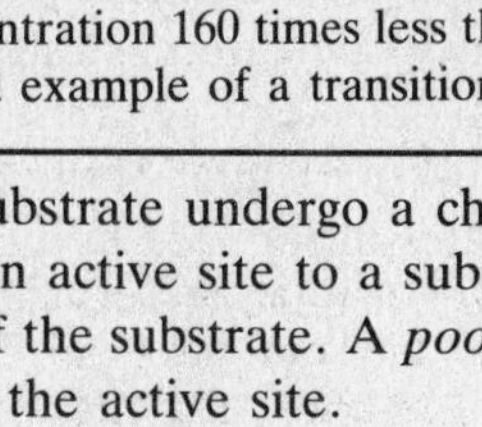

Pyrrole 2-carboxylate

A planar analog of proline is *pyrrole 2-carboxylate*, and this proves to be a potent inhibitor of the racemase; it gives rise to 50 percent inhibition at a concentration 160 times less than the concentration of D- or L-proline that gives 50 percent binding. Thus, it is a good example of a transition-state analog.

Question: Do *both* the enzyme and substrate undergo a change when they interact?

Yes, the concept of *induced fit* of an active site to a substrate emphasizes the *adaptation* of the active site to fit the functional groups of the substrate. A *poor* substrate or inhibitor does not induce the correct conformational response in the active site.

EXAMPLE 8.17

Hexokinase demonstrates the *induced fit* phenomenon; it catalyzes phosphoryl transfer from ATP to C-6 of glucose as follows:

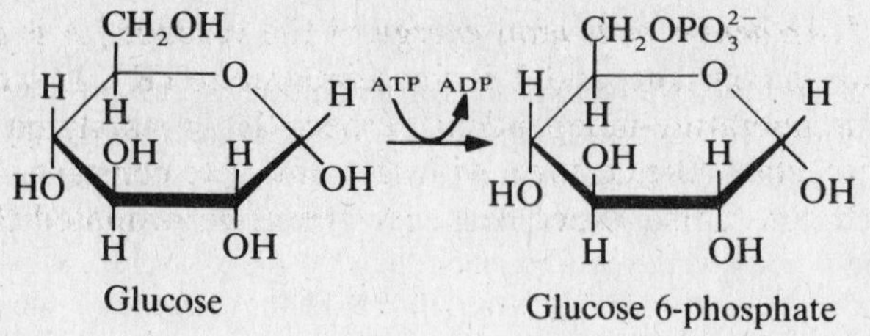

$$\text{Glucose} \xrightarrow{\text{ATP} \to \text{ADP}} \text{Glucose 6-phosphate}$$

Glucose Glucose 6-phosphate

The enzyme can also catalyze the transfer of the terminal phosphoryl of ATP to water; i.e., it acts as an ATPase but at a rate 5×10^6 times slower than the above reaction. The basic and nucleophilic properties of water versus the C-6 hydroxyl of glucose are sufficiently similar to suggest no marked differences in rate. Therefore, the explanation of the rate difference is that glucose induces a conformational change that *establishes* the correct active-site geometry in the enzyme, whereas a water molecule is too small to do so.

The induction of the correct geometry in the active site of an enzyme is *paid for* by a *good substrate*, with binding energy. An alternative explanation to that of induced fit is that some small molecules (e.g., H_2O in the hexokinase example) bind *nonproductively*, i.e., their small size allows them to assume many orientations with respect to the other substrate (ATP in the case of hexokinase) that do not lead to reaction. Large substrates are restricted in motion and are held in a catalytically *correct* orientation millions of times more often during molecular vibrations than is, say, water.

8.5 RATE ENHANCEMENT AND ACTIVATION ENERGY

Question: Some biochemicals are stable when in pure form on a shelf and yet in the presence of an enzyme break down rapidly. Why?

There is an important distinction to be made between *thermodynamic stability* (expressed in terms of the equilibrium constant of the reaction) and the *kinetic stability* of a substance; the latter merely refers to how fast the reaction proceeds, the former to the final position of the reaction in terms of the relative amounts of substrate and product. (See Example 8.18.) Enzymes affect the kinetic stability of a substance.

EXAMPLE 8.18

Most reduced organic molecules, such as glucose, are *thermodynamically unstable* in our oxidizing atmosphere.

$$\text{glucose} + 6O_2 \longrightarrow 6CO_2 + 6H_2O \qquad -\Delta G^0 = 2{,}872 \text{ kJ mol}^{-1}$$

Thus, oxidation is very *exergonic* (heat-producing), and the reaction is favored by the large $-\Delta G^0$ (Chap. 10) of the reaction. But we are all aware that glucose is stable *on the shelf*. Thus, it is *thermodynamically unstable* but *kinetically stable*.

The distinction between kinetic and thermodynamic stability is important and is explained by the concept of the *free energy* necessary to activate the substrate to its transition state. In order for the substrate to form products, its internal free energy must exceed a certain value; i.e., it must *surmount an energy barrier*. The energy barrier is that of the free energy of the transition state, $\Delta G^{\ddagger}$. The transition-state theory of reaction rates introduced by H. Eyring relates the rate of the reaction to the magnitude of $\Delta G^{\ddagger}$.

Question: Is there a simple mathematical relationship between reaction rate and $\Delta G^{\ddagger}$?

Yes. In the 1880s, Arrhenius observed that the rate constant k for a simple chemical reaction varies with the temperature according to

$$k = Ae^{-E_a/RT} \tag{8.1}$$

where E_a is the so-called *Arrhenius activation energy* of the reaction, A is called the *preexponential factor*, R is the universal gas constant, and T is the temperature (K). However, it became apparent that A was *not quite* temperature-independent, especially in catalyzed reactions; thus Eyring proposed that *all* transition states break down with the same rate constant, $\kappa T/h$; where κ and h are Boltzmann's and Planck's constants, respectively. He therefore proposed that for any reaction,

$$k = \frac{\kappa T}{h} e^{-\Delta G^{\ddagger}/RT} \tag{8.2}$$

where, again, $\Delta G^{\ddagger}$ is the activation energy of the *transition-state* complex.

In Chap. 10 it will be shown that the Gibbs free energy of a system is made up of two components such that

$$\Delta G = \Delta H - T\,\Delta S \tag{8.3}$$

where ΔS is the *entropy* change and ΔH the *enthalpy* change in the reaction system. Therefore, Eq. (*8.2*) can be written as

$$k = \left(\frac{\kappa T}{h} e^{\Delta S^{\ddagger}/R}\right) e^{-\Delta H^{\ddagger}/RT} \tag{8.4}$$

Entropy is an equilibrium thermodynamic entity that is *interpreted* mechanically as the *degree of disorder* in a system. From Eq. (*8.4*), it is therefore seen that (1) the preexponential factor A [Eq. (*8.1*)] can be interpreted as being related to the *organization* of a reactant in an enzyme as the transition-state complex is formed and (2) the exponential factor relates to the enthalpy (heat) of the reaction.

Any molecular factors that tend to stabilize the transition state decrease $\Delta G^{\ddagger}$ and thus increase the rate of the reaction. Thus, this rate enhancement can result from either entropy or enthalpy effects, or from both.

Solved Problems

BASIC CONCEPTS

8.1. When and by whom was (*a*) the notion of enzymes first introduced; (*b*) the *-ase* nomenclature introduced; (*c*) the first enzyme crystallized, and which enzyme was it?

SOLUTION

(*a*) Although phenomena of fermentation and digestion had long been known, the first clear recognition of an enzyme was made by Payen and Persoz (*Ann. Chim.* (*Phys*), 53, 73, 1833) when they found that an alcohol precipitate of malt extract contained a thermolabile substance that converted starch into sugar.

(*b*) The above-mentioned substance was called *diastase* (Greek: "separation") because of its ability to separate soluble dextrin from insoluble envelopes of starch grains. Diastase became a generally applied term for these enzyme mixtures until 1898, when Duclaux suggested the use of *-ase* in the name of an enzyme; this classification procedure still holds today.

(*c*) Many enzymes were purified from a large number of sources, but it was J. B. Sumner who was the first to *crystallize* one. The enzyme was urease from jack beans. For his travail, which took over six years (1924–1930), he was awarded the 1946 Nobel prize. The work demonstrated once and for all that enzymes are distinct chemical entities.

8.2. Justify the claim that the surface area of the active site of an enzyme is less than 5 percent of its total surface area. Consider the specific case of a 27,000-dalton globular enzyme with five amino acids in its active site.

SOLUTION

The ratio of the volume of the active site to that of the whole protein is $\sim 5{:}27{,}000/110 = 0.02$; this is so because the mean amino acid residue weight is ~ 110. If we assume that the enzyme is spherical, the above volume ratio corresponds to a surface-area ratio of $(0.02)^{2/3} \times 0.5$. The factor 0.5 accounts for the fact that half of the active-site residues face outward and contribute also to the total surface area. The answer is 0.04, or 4 percent of the surface area.

8.3. The nerve gas *diisopropyl fluorophosphate* (DFP) reacts with the serine —OH in some enzymes to form HF and the *O*-phosphoryl ester as follows:

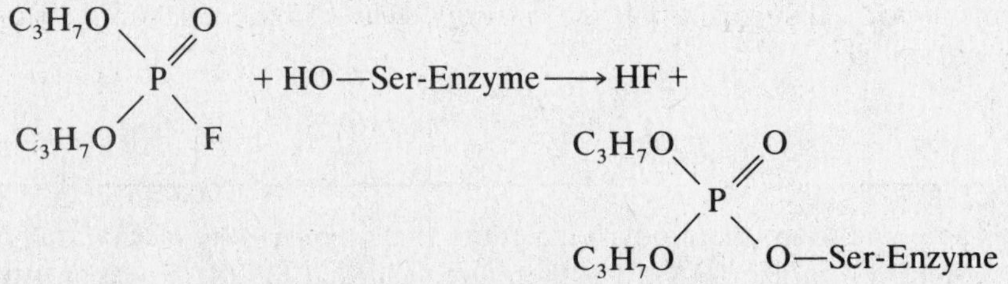

$$(C_3H_7O)_2P(=O)F + HO\text{—Ser-Enzyme} \longrightarrow HF + (C_3H_7O)_2P(=O)\text{O—Ser-Enzyme}$$

Since DFP is a potent inhibitor of the enzyme *chymotrypsin*, what might we infer about the amino acid side-chain composition of the active site?

SOLUTION

In a chemical enzyme-modification experiment conducted by E. F. Jansen and colleagues in 1949, the enzyme was incubated with ^{32}P-labeled DFP and then hydrolyzed with a strong acid. Separation of the constituent amino acids revealed 1 mol of *labeled O*-phosphorylserine per 25,000 g of chymotrypsin. Since $M_r = 25{,}000$, only a single serine had reacted out of a total of 27. This indicated that a particular serine is an important component of the active site. This experiment is the archetypal form of many enzyme-modification procedures that are now used frequently to identify active-site constituents.

CLASSIFICATION OF ENZYMES

8.4. What is the order of abundance of enzymes in the six EC groups?

SOLUTION

Of the ~2,500 different, named enzymes, the most abundant group is the oxidoreductases, group 1. The order is $1 > 2 > 4 > 3 > 6 > 5$.

8.5. Classify the following enzyme-catalyzed reactions into their major EC groups and suggest possible common names for each.

(*a*) D-Glyceraldehyde 3-phosphate + P_i + $NAD^+ \rightleftharpoons$ 1,3-diphosphoglycerate + NADH

(*b*)

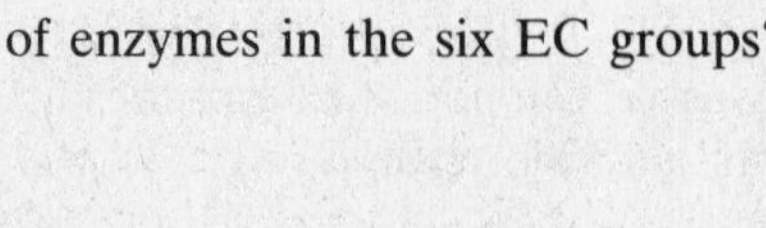

$$H_2N\text{—}C(=O)\text{—}NH_2 + H_2O \longrightarrow NH_3 + CO_2$$

(c)

$$\underset{\text{2-Oxoglutarate}}{{}^-OOC-CH_2-CH_2-C(=O)-COO^-} + \underset{\text{L-Alanine}}{CH_3-CH(\overset{+}{N}H_3)-COO^-} \rightleftharpoons \underset{\text{L-Glutamate}}{{}^-OOC-CH_2-CH_2-CH(\overset{+}{N}H_3)-COO^-} + \underset{\text{Pyruvate}}{CH_3-C(=O)-COO^-}$$

(d)

$$\underset{\text{L-Histidine}}{\text{(imidazolium)}-CH_2-CH(\overset{+}{N}H_3)-COO^-} + H_2O \longrightarrow \underset{\text{Histamine}}{\text{(imidazolium)}-CH_2-CH_2-\overset{+}{N}H_3} + HCO_3^-$$

(e) L-Alanine $\rightleftharpoons$ D-Alanine

(f) L-Ribulose 5-phosphate $\rightleftharpoons$ D-Xylulose 5-phosphate

(g) The enzyme that catalyzes the rearrangement of S—S bonds in proteins.

(h) ATP + L-Tyrosine + $tRNA^{Tyr}A \longrightarrow$ AMP + PP_i + L-Tyrosyl-$tRNA^{Tyr}A$

(i) ATP + L-Asparagine + $tRNA^{Asn} \longrightarrow$ AMP + PP_i + L-Asparaginyl-$tRNA^{Asn}$

(j) ATP + γ-L-Glutamyl-L-cysteine + Glycine $\longrightarrow$ ADP + P_i + Glutathione

SOLUTION

(a) Glyceraldehyde 3-phosphate dehydrogenase, EC 1.2.1.12. The systematic name is *D-glyceraldehyde-3-phosphate:NAD^+ oxidoreductase* (phosphorylating); it is an important glycolytic enzyme.

(b) Urease, EC 3.5.1.5. Its systematic name is *urea amidohydrolase*. Interestingly, it is a nickel-containing enzyme.

(c) Alanine transaminase, EC 2.6.1.12. The systematic name is *L-alanine:2-oxoglutarate aminotransferase*. Note also that aminotransferases almost invariably have *pyridoxal phosphate* as a cofactor.

(d) Histidine decarboxylase, EC 4.1.1.22. The systematic name is *L-histidine carboxy-lyase*; it, too, requires pyridoxal phosphate in animals, but the bacterial enzyme does not.

(e) Alanine racemase, EC 5.1.1.1. The systematic name is the same. It has the honor of being the first enzyme in group 5; it also requires pyridoxal phosphate as a cofactor. (A *racemic* mixture is a mixture of optical isomers of one chemical species.)

(f) Ribulose phosphate epimerase, EC 5.1.3.4. Its official name is *L-ribulose-5-phosphate 4-epimerase*. This is a key enzyme in the pentose phosphate pathway. (An *epimer* is a stereoisomer of a sugar differing in the configuration on only *one* carbon.)

(g) Disulfide bond (S—S) rearrangease, EC 5.3.4.1. The official name is *protein disulfide-isomerase*. The common name is an example of naming the enzyme after the phenomenon with which it is associated; enzymes in this category are called *phenomenases*, and the EC of IUPAC and the Nomenclature Committee of the International Union of Biochemistry (IUB) have generally disapproved of such naming. Another common example is the use of the name translocase for an enzyme or protein carrier that catalyzes the movement of a moiety between biological structural compartments, e.g., *ATP translocase* in mitochondria (Chap. 14).

(*h*) Tyrosyl-tRNA synthetase, EC 6.1.1.1. The official name is *L-tyrosine:tRNA^{Tyr} ligase* (*AMP-forming*). It is the first of the group 6 enzymes but, more importantly, is essential to life because of its role in protein synthesis (Chap. 17).

(*i*) Asparaginyl-tRNA synthetase, EC 6.1.1.22. The official name is *L-asparagine:tRNA^{Asn} ligase* (*AMP-forming*). It is the last in the series of tRNA ligases (Chap. 17).

(*j*) Glutathione synthase, EC 6.3.2.3. The official name is *γ-L-glutamyl-L-cysteine:glycine ligase* (*ADP-forming*). This is the key enzyme in glutathione production and is found in many tissues and organisms.

MODES OF ENHANCEMENT OF RATES OF BOND CLEAVAGE

8.6. Covalent enzyme catalysis involves the formation of a transient covalent bond between an enzyme and its substrate. Below are the general structures of commonly encountered so-called *acyl-enzyme intermediates*.

1. $\text{Enz—OH} + {}^{-}\text{O—C(=O)—R} \longrightarrow \text{OH}^{-} + \text{Enz—O—C(=O)—R}$

2. $\text{Enz—SH} + {}^{-}\text{O—C(=O)—R} \longrightarrow \text{OH}^{-} + \text{Enz—S—C(=O)—R}$

3. $\text{Enz—(imidazole)NH} + {}^{-}\text{O—C(=O)—R} \longrightarrow \text{OH}^{-} + \text{Enz—(imidazole)N—C(=O)—R}$ (imidazolium ring, +, N—H)

4. $\text{Enz—NH}_2 + \text{O=C}(R_1)(R_2) \longrightarrow \text{H}_2\text{O} + \text{Enz—N=C}(R_1)(R_2)$

5. $\text{Enz—C(=O)—OH} + \text{O=C}(R_1)(R_2) \longrightarrow R_1\text{H} + \text{Enz—C(=O)—O—C(=O)—}R_2$

(*a*) Give examples of amino acid residues that have side chains with the reactive groups indicated above. (*b*) Give chemical names to the acyl-enzyme intermediate compounds.

SOLUTION

(*a*) (1) Serine, (2) cysteine, (3) histidine, (4) lysine, (5) aspartate or glutamate.

(*b*) (1) Ester, (2) thioester, (3) acylimidazole, (4) Schiff base, (5) anhydride.

8.7. The glycolytic pathway enzyme *fructose 1,6-diphosphate aldolase* forms an acyl-enzyme intermediate with its ketone substrate *fructose 1,6-diphosphate*. Given that the enzyme contains a lysine residue that is essential for its activity, what type of covalent intermediate is likely to be formed?

SOLUTION

The general structure is that of (4) in Prob. 8.6(*b*), namely, a Schiff base (also called a *ketimine*).

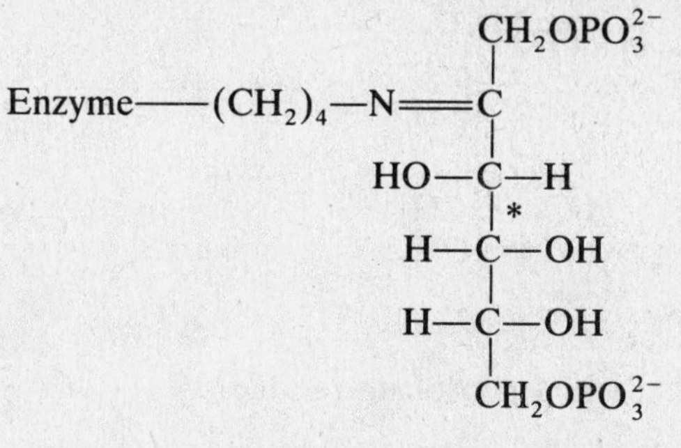

$$
\begin{array}{rcl}
 & CH_2OPO_3^{2-} & \\
 & | & \\
\text{Enzyme}—(CH_2)_4—N═ & C & \\
 & | & \\
HO— & C & —H \\
 & |^{*} & \\
H— & C & —OH \\
 & | & \\
H— & C & —OH \\
 & | & \\
 & CH_2OPO_3^{2-} &
\end{array}
$$

The formation of the ketimine labilizes the bond marked with the asterisk, leading to its cleavage.

8.8. *Propinquity* (*proximity*) *effects* are important in reaction rate enhancement. In the case of the following compounds, *anhydrides* (products formed on removal of water) form at different rates. Arrange the compounds in order of their rates of anhydride formation and explain the reasons for the ordering.

(*a*) $H_2C(COOR)$—$CH_2(COO^-)$: $\begin{array}{c} H \\ H—C—C(=O)—O—R \\ H—C—C(=O)—O^- \\ H \end{array}$

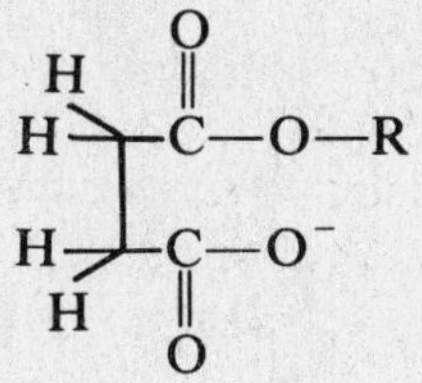

(*b*) $\begin{array}{c} H\ \ C(=O)—O—R \\ \diagdown\!\!\diagup \\ \| \\ \diagup\!\!\diagdown \\ H\ \ C(=O)—O^- \end{array}$

(*c*) Bicyclic ring compound bearing C(=O)—O—R and C(=O)—O⁻ on adjacent ring carbons, with H substituents.

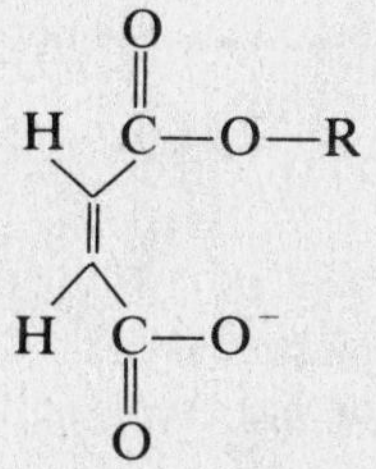

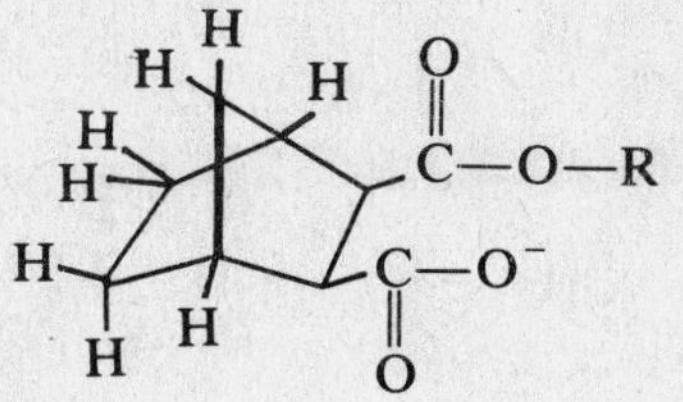

(*d*) $\begin{array}{c} H—CH—C(=O)—O—R \\ | \\ H—CH \\ | \\ H—CH—C(=O)—O^- \end{array}$

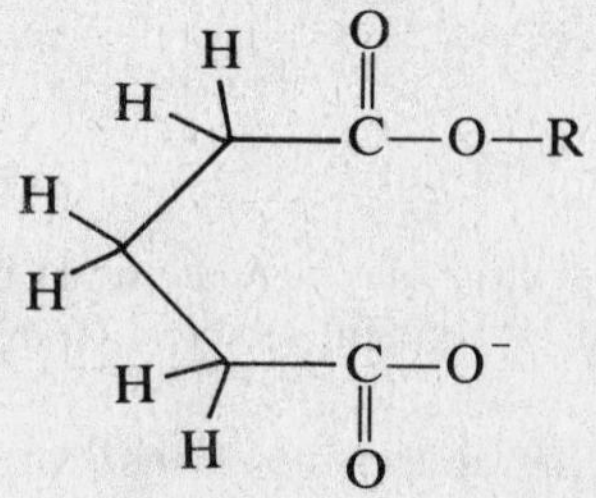

SOLUTION

The relative rates of anhydride formation are as follows: (*d*) 1; (*a*) 230; (*b*) 10,100; (*c*) 53,000. A greater rate enhancement occurs in the compounds in which the reacting carboxyl groups are held more rigidly; this increases the time during which the transition state can form and therefore the time in which the products can be formed.

8.9. Transition-state analogs are potent inhibitors of enzymes. In the enzyme *cytidine deaminase* from the bacterium *E. coli*, the following chemical transformation takes place:

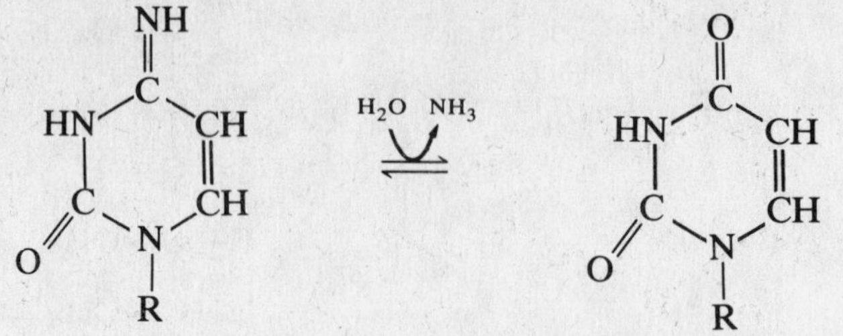

NH
C
HN CH
C CH
O N
R

H_2O NH_3

O
C
HN CH
C CH
O N
R

where R denotes a ribose 5-phosphate residue.

(*a*) Draw a possible transition-state compound.

(*b*) The two following compounds have different effects on the reaction rate; one is a transition-state analog, while the other is a substrate. Give reasons for your proposal for which is the analog.

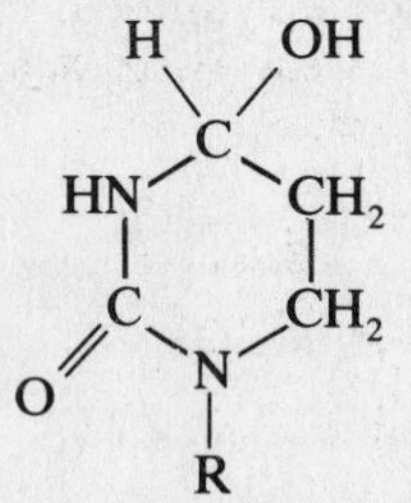

H OH
C
HN CH_2
C CH_2
O N
R

3,4,5,6-Tetrahydrouridine
(1)

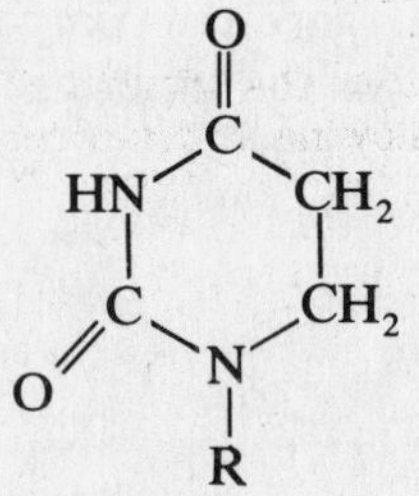

O
C
HN CH_2
C CH_2
O N
R

5,6-Dihydrouridine
(2)

SOLUTION

(*a*) The likely transition state is the so-called *tetrahedral intermediate*; namely, an intermediate species in which the carbon has the usual four-bond arrangement for carbon:

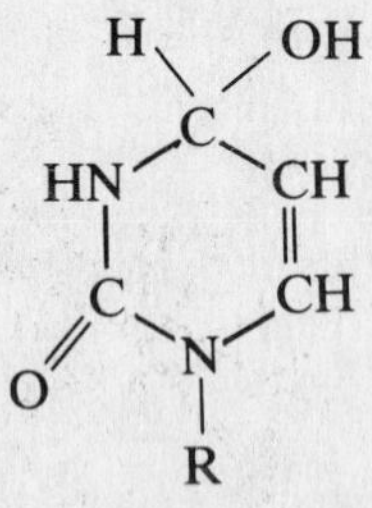

H OH
C
HN CH
C CH
O N
R

(*b*) It is clear that compound (1) has a structure very similar to that of the intermediate in (*a*); it is indeed a potent inhibitor (transition-state analog) of cytidine deaminase.

8.10. The bacterial enzyme *chorismate mutase–prephenate dehydrogenase* is peculiar because it is a single protein unit with *two* catalytic activities. It catalyzes the sequential reactions of mutation of chorismate to prephenate and then the reactions that lead to the formation of *phenylalanine* and *tyrosine*, through oxidation of prephenate. The first of these reactions is interesting because it is one of the few strictly *single-substrate* enzymatic reactions; it entails the migration of a side chain from one part of the ring to another, as shown in the scheme below.

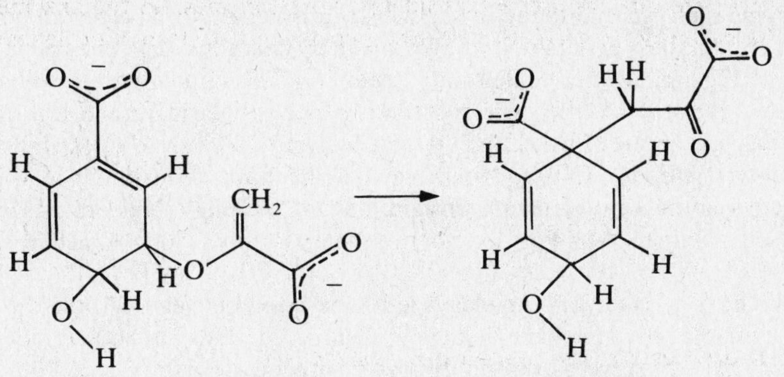

Chorismate Prephenate

(*a*) Predict a likely transition-state structure.

(*b*) Suggest a likely transition-state analog that might be a potent inhibitor of the enzyme.

SOLUTION

(*a*) By using molecular orbital calculations, P. R. Andrews and G. D. Smith in 1973 suggested the following structure as that of the transition-state molecule:

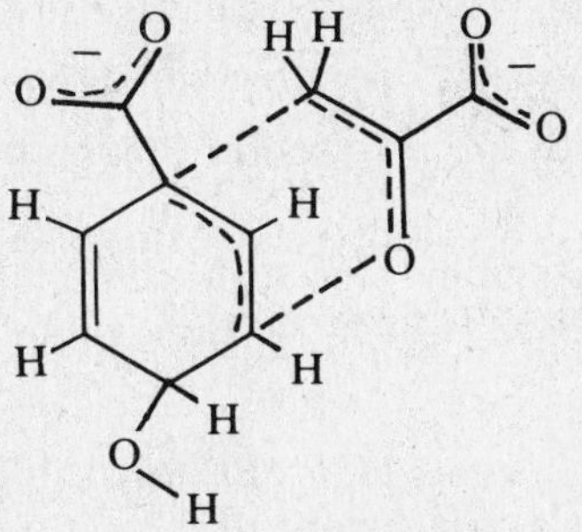

Transition state

(*b*) Andrews and Smith recognized the similarity between the structure of the transition state and *adamantane*.

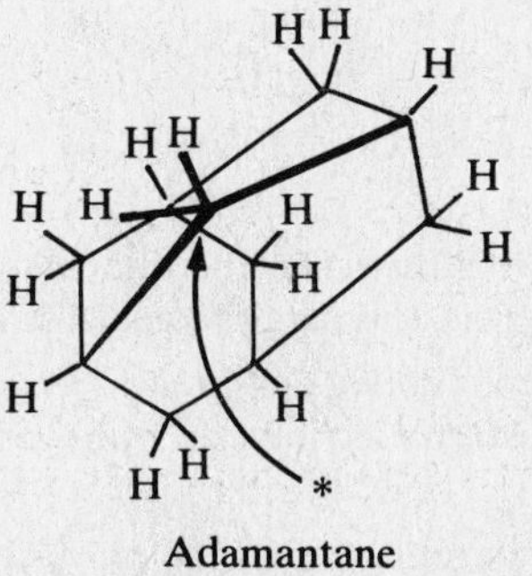

Adamantane

Adamantane has an *extra* methylene bridge (the asterisk in the diagram) linked to the six-membered ring, thus stabilizing a cagelike structure. The authors indeed subsequently showed that some adamantane derivatives are potent inhibitors of chorismate mutase; thus, these are examples of transition-state analogs.

Note: Since the enzyme is not found in mammals, inhibitors of this enzyme *may* be an effective means of controlling bacterial infection. Certainly, *species-selective toxicity* is an important consideration in the development of new antimicrobial agents.

8.11. *Lysozyme* is an enzyme found in tears. It hydrolyzes bacterial cell wall polysaccharides, and it has one of the best understood of all enzyme mechanisms. The enzyme is a single polypeptide chain of 129 amino acids folded into a shape like a grain of puffed wheat, with a cleft along one side. Into the cleft fits the substrate, a polysaccharide made up of alternating units of *N*-acetylglucosamine (NAG) and *N*-acetylmuramic acid (NAM). Details of the binding of a competitive inhibitor $(NAG)_3$ to the active site have been obtained using x-ray crystallography. Using the x-ray structure, insight into the binding of substrates such as $(NAG\text{-}NAM)_3$ has been obtained (see Fig. 8-2 for a schematic model of the active site and the binding groups).

The enzyme catalyzes the cleavage of the bond between carbon 1 of residue [4] and the oxygen atom of the glycosidic linkage of residue [5]. Two amino acid side chains in the region of this bond can serve as proton donors or acceptors: Asp 52 and Glu 35, each of which is about 0.3 nm from the bond. Asp 52 is in a polar environment and is ionized at the pH optimum of lysozyme (pH 5), whereas Glu 35 is in a nonpolar region and is not ionized. The proposed catalytic mechanism is given in Fig. 8-3.

(*a*) To which EC group does lysozyme belong, and what are the first two integers in its EC number?

(*b*) Describe in words the various basic chemical processes that take place in the cleavage of $(NAM\text{-}NAG)_3$, as shown in Fig. 8-3.

(*c*) Is homolytic or heterolytic bond cleavage involved?

(*d*) What type of catalysis is operating here, covalent or noncovalent?

(*e*) What type of bonds are involved in the binding of the substrate $(NAM\text{-}NAG)_3$ to the enzyme?

(*f*) On binding to the enzyme, the sugar residue [4] is distorted from a *chair* conformation to that of a *half-chair* (Chap. 2). How might this aid catalysis?

SOLUTION

(*a*) Lysozyme is a *hydrolase*; thus its first EC number (Table 8.1) is 3. Since it catalyzes the hydrolysis of a C—O bond, its second number is 2 (Table 8.2).

(*b*) The carboxyl of Glu 35 donates a proton, cleaving the C-1—O bond and releasing the disaccharide [5]-[6]. The resulting carbocation, C-1 of ring [4], is stabilized by the negatively charged Asp 52. The carbocation then reacts with the OH^- from the solvent water to release the tetrasaccharide [1]-[2]-[3]-[4]. Glu 35 is then reprotonated in readiness for the *next* round of reactions. The glutamic acid acts as a proton donor for the reaction, which is thus classified as general acid catalysis.

(*c*) Since both electrons leave C-1 of residue [4] to form the carbocation, this is heterolytic bond cleavage.

(*d*) Since proton donation is *central* to the catalytic process, this is an example of general acid-base catalysis; specifically, it is acid catalysis and it is noncovalent.

(*e*) In Fig. 8-3, the numerous dotted lines drawn between O's and NH's of amino acid residues and O's and NH's of the oligosaccharide indicate hydrogen bonding; however, van der Waals, noncovalent bonding also occurs.

(*f*) The binding of the substrate distorts the previous chair conformation of residue [4]; this *reduces* the tendency for binding; i.e., ΔG for binding will be elevated. However, this energy of distortion (strain) contributes to the total activation energy required for subsequent bond cleavage via formation of the carbonium ion.

8.12. *Carboxypeptidase A* (EC 3.4.17.1) is a pancreatic digestive enzyme consisting of a single polypeptide chain of 307 amino acids with a total M_r of 36,000. It catalyzes the cleavage of amino acid residues from C termini of polypeptides. Importantly, for its mechanisms of action, it contains one Zn^{2+} in its active site. The amino acid side chains that form its active site and the catalytic sequence are shown in Fig. 8-4.

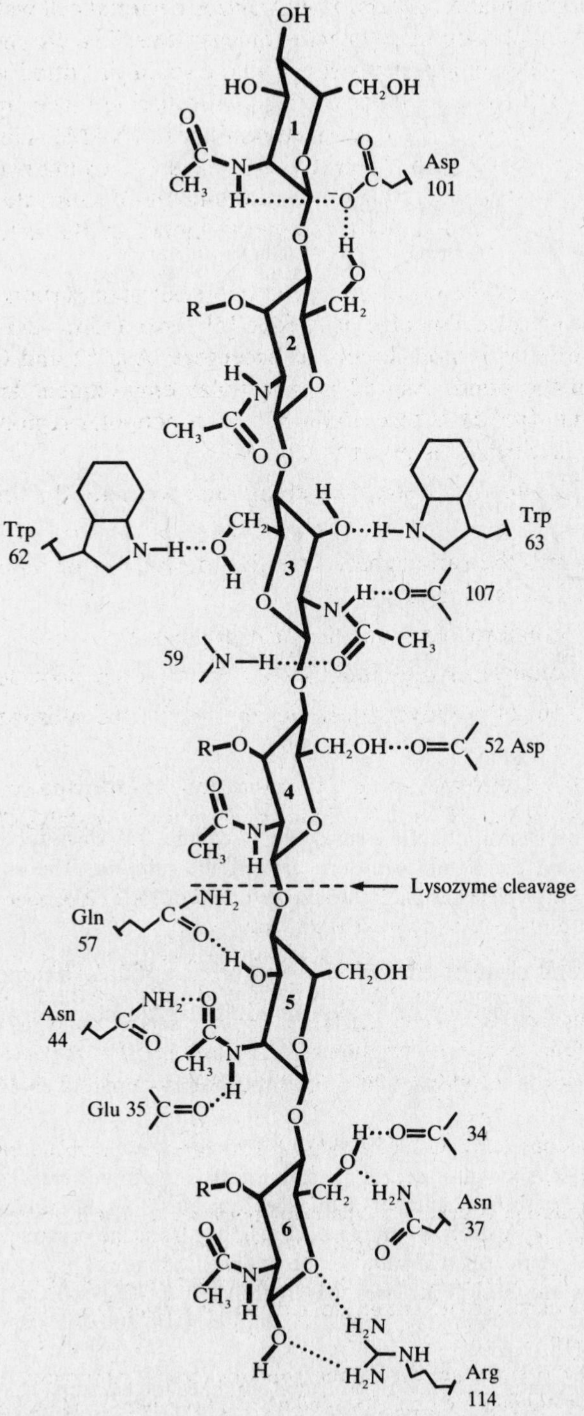

Fig. 8-2 Binding of the substrate $(NAG\text{-}NAM)_3$ to the active site of lysozyme. The substrate is drawn with bold bonds, the enzyme groups with light-face bonds. H bonds are indicated by dotted lines.

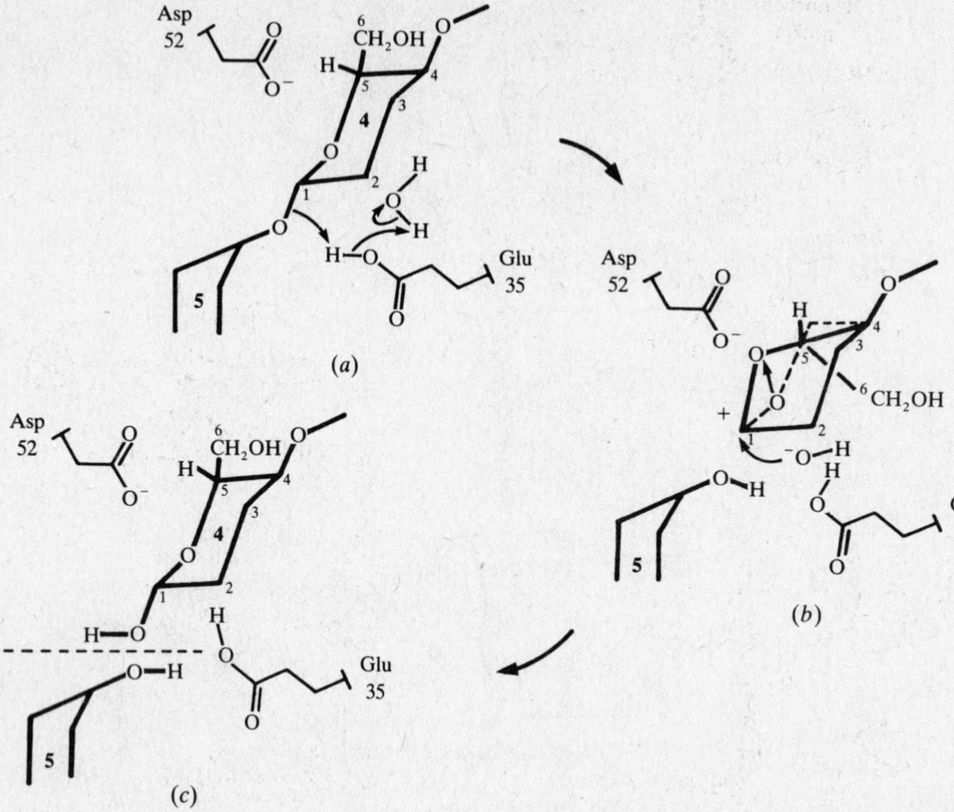

Fig. 8-3 Probable mechanism of bond cleavage by lysozyme. The thick solid line structure is the substrate, and the lightface groups are on the enzyme. The small arrows indicate displacement of electron pairs during the reaction.

(*a*) To what general class of enzyme does carboxypeptidase belong?

(*b*) What basic role does the Zn^{2+} play in the catalytic mechanism? What general type of catalysis occurs?

(*c*) Describe in words the sequence of events that is depicted in Fig. 8-4.

SOLUTION

(*a*) Carboxypeptidase is a hydrolase; i.e., it catalyzes the hydrolytic cleavage of a (peptide) bond. The second integer in its EC number sequence indicates that it cleaves a C—N bond. It is an *exopeptidase*; i.e., it hydrolyzes amino acid residues from the carboxyl termini of peptides. There exist also *amino*peptidases that catalyze the hydrolytic removal of N-terminal amino acid residues. *Endo*peptidases are those hydrolases that hydrolyze peptide bonds, not at the C or N termini, but *within* the chain; examples are pepsin in the stomach and the pancreatic peptidases, such as chymotrypsin and trypsin.

(*b*) The Zn^{2+} acts as an electrophile that further polarizes the carbonyl oxygen [Fig. 8-4(*c*)] before the formation of an ester linkage with the γ carboxyl of Glu 270 of the enzyme. This linkage is covalent, so this reaction is an example of covalent catalysis.

(*c*) (1) The peptide substrate binds to the active site, which contains Arg 145, the Zn^{2+} ion, and the so-called *hydrophobic pocket*, which contains aromatic and aliphatic amino acid side chains. (2) Nucleophilic attack by Glu 270 on the peptide bond is accompanied by the uptake of an H^+ from Tyr 248. (3) This results in cleavage of the peptide bond and diffusion away of the C-terminal free amino acid. (4) The covalently bound polypeptide is then released to regenerate free enzyme by nucleophilic attack of an H_2O molecule on the anhydride bond of Glu 270; this is followed by reprotonation of Tyr 248.

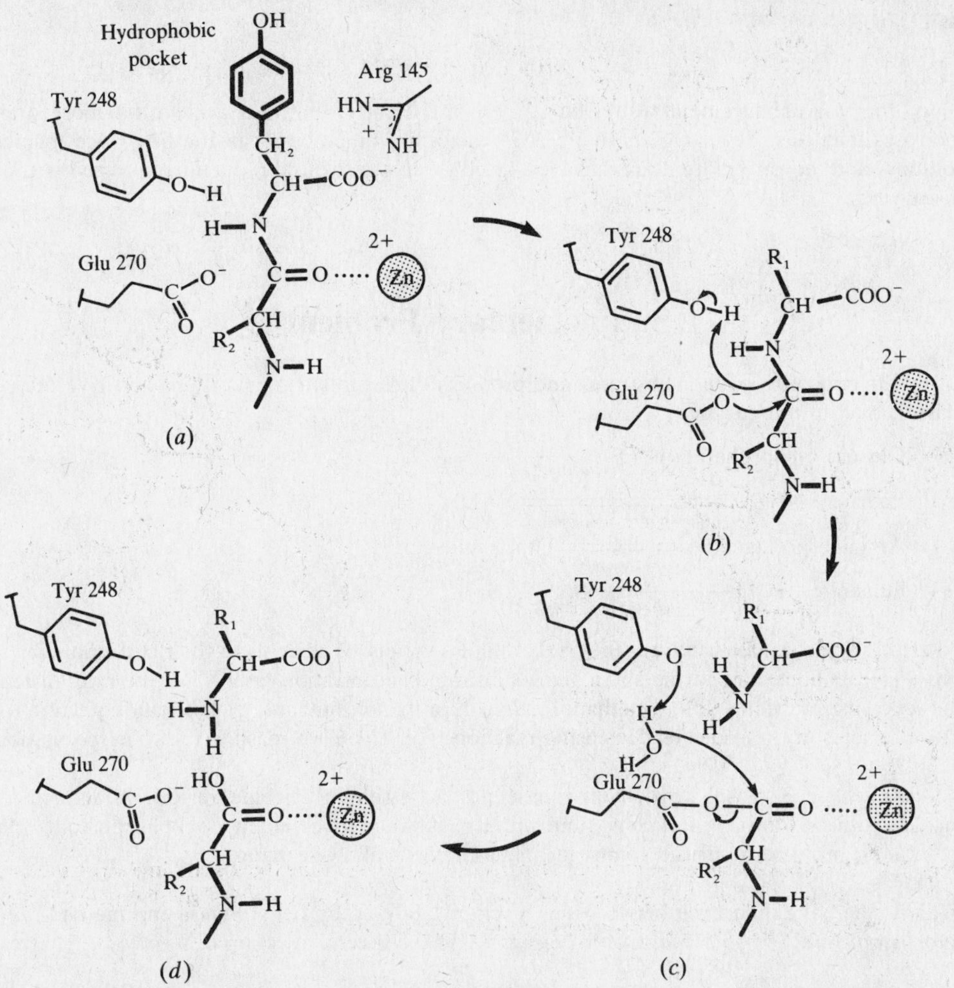

Fig. 8-4 The mechanism of covalent catalysis of the hydrolysis of a C-terminal amino acid residue from a peptide by carboxypeptidase A. The reaction is $(a) \rightarrow (d)$, and the bold line structure is the peptide substrate. The C-terminal tyrosine side chain of the substrate shown in (a) is indicated by R_1 in (b), (c), and (d).

RATE ENHANCEMENT AND ACTIVATION ENERGY

8.13. Calculate the rate enhancement that would be achieved if the activation energy of the transition-state complex of an enzyme with its substrate were halved.

SOLUTION

From Eq. (*8.2*) k is proportional to $e^{-\Delta G^{\ddagger}/RT}$. Thus,

$$k_{\mathrm{NEW}} = k_{\mathrm{OLD}} \exp(-\Delta G^{\ddagger}/2RT)$$

It is thus clear that the rate enhancement is dependent on both the original $\Delta G^{\ddagger}$ *and* the temperature. So, unless these values are given, the enhancement cannot be calculated.

8.14. If $\Delta G^{\ddagger}$ and T in the previous example were -1 kJ mol^{-1} and 300 K, respectively, what would be the rate enhancement factor?

SOLUTION

$$k_{NEW} = k_{OLD} \exp[1{,}000/(2 \times 8.314 \times 300)] = k_{OLD}\, 1.22$$

Thus, the rate enhancement would be 22 percent. Clearly, enzymes achieve far more dramatic rate increases than this. This suggests that $-\Delta G^{\ddagger}$ values are much larger in the first place (namely, in free solution) and they are more dramatically reduced (than the twofold reduction discussed in this problem) in enzymes.

Supplementary Problems

8.15. Draw the structures of the substrates and products of the following reactions and give the general EC classification of the enzyme involved:

(*a*) Glucose 6-phosphate + $H_2O \xrightarrow[\text{Glucose 6-phosphatase}]{}$ . . . .

(*b*) Lactate + $NAD^+ \underset{\text{Lactate dehydrogenase}}{\rightleftharpoons}$ NADH + H^+ + ...

(*c*) Argininosuccinate $\underset{\text{Argininosuccinase}}{\rightleftharpoons}$ Fumarate + ...

(*d*) Fumarate + $H_2O \xrightarrow[\text{Fumarase}]{}$ . . . .

8.16. Alcohol dehydrogenase catalyzes the oxidation of a variety of alcohols to their corresponding aldehydes. For a given amount of enzyme and a given substrate concentration (mol L^{-1}), the rates of reaction with the following substrates differ: methanol, ethanol, propanol, butanol, cyclohexanol, phenol. (*a*) Arrange the substrates in the order of decreasing reaction rate. (*b*) Give reasons for your speculations.

8.17. Glyceraldehyde 3-phosphate dehydrogenase has an essential cysteine residue in its active site. The enzyme forms a transient acyl compound with its substrate, glyceraldehyde 3-phosphate. (*a*) What is the general chemical name of the compound? (*b*) Draw its likely structure.

8.18. What is the basic difference between the reaction catalyzed by (*a*) a *mutase* enzyme and that catalyzed by an *isomerase*; (*b*) an *oxidase*, an *oxygenase*, and the reverse reaction of a *reductase*? Give examples.

8.19. Given that the spontaneous hydration of CO_2 is reasonably fast (Example 8.1), what might be a physiological rationalization for the need for the enzyme carbonic anhydrase?

8.20. What is a *suicide* substrate of an enzyme?

8.21. If the enzyme concentration in a reaction mixture at equilibrium is comparable with that of the reactants, is the ratio of product concentrations to substrate concentrations the same as if no enzyme were present?

8.22. Why are most enzymes so large relative to their substrates?

8.23. Give some examples of enzymes that are smaller than their substrates.

Chapter 9

Enzyme Kinetics

9.1 INTRODUCTION AND DEFINITIONS

Question: What determines the rate of an enzyme-catalyzed reaction and how can the physical and chemical effects be quantified?

Enzyme kinetics is concerned with measuring the rates of enzymatic reactions and with factors that affect the rates, such as pH, temperature, presence of cofactors, and metal ions. It is essentially an experimental subject with two main aspects to it. First, there is the design of experiments, including the means of determining the progress of reactions. Second, there is the interpretation of the data; this usually depends on writing mathematical expressions for model reaction schemes, which are then tested for consistency with the experimental data. This analysis is then used to design subsequent experiments, and in the end an interpretation is made of the mechanism of the reaction. Thus, the process is very much an iterative one; new data suggest new models of the mechanism, and new models suggest new experiments to check for flaws in the models.

The basic principles and definitions used frequently in enzyme kinetic analysis are discussed below.

Principle of Mass Action

The rate of a chemical reaction is proportional to the concentrations of the reactants involved in the *elementary* chemical process. The constant of proportionality is called the *rate constant*, or the *unitary rate constant* to highlight the fact that it applies to an elementary process. A subtlety that may be introduced into rate expressions is to use *chemical activities* (see Chap. 10) and not simply concentrations, but activity coefficients in biological systems are generally taken to be near 1.

EXAMPLE 9.1

Application of the principle of mass action to the reaction scheme

$$A + B \underset{k_{-1}}{\overset{k_1}{\rightleftharpoons}} P + Q$$

with forward and reverse rate constants k_1 and k_{-1}, leads to the following expressions for the forward and reverse reaction rates:

$$\begin{aligned} \text{forward rate} &= k_1[A][B] \\ \text{reverse rate} &= k_{-1}[P][Q] \end{aligned} \tag{9.1}$$

where the square brackets denote concentration in mol L^{-1}. At *chemical equilibrium*, the forward and reverse rates are equal, so there is no *net* production of any of the reactants with time. Thus,

$$\frac{k_1}{k_{-1}} = \frac{[P]_e[Q]_e}{[A]_e[B]_e} = K_e \tag{9.2}$$

where K_e is termed the *equilibrium constant* and the subscript e denotes the equilibrium value of the concentrations.

Reaction rates are simply concentration changes of a species per unit of time and therefore can be written mathematically as *derivatives*. Note, however, that the mathematical expression for the rate of change of, say, [A] must include forward and reverse *fluxes* (Greek: "to flow"); for example,

$$\frac{d[A]}{dt} = -k_1[A][B] + k_{-1}[P][Q] \tag{9.3}$$

Molecularity

Molecularity refers to the *number of molecules* involved in an elementary reaction. Usually, only two molecules collide in one instant to give product(s) (molecularity = 2) or a single molecule undergoes *fission* (also called *scission*; molecularity = 1). Example 9.1 is of a reaction in which the forward and reverse processes have a molecularity of two.

Order of a Reaction

This is the *sum of the powers* to which the concentration (or chemical activity) terms are raised in a rate expression.

EXAMPLE 9.2

If the forward flux rate of a reaction between A and B is given by:

$$\text{rate} = k[\text{A}]^{1/2}[\text{B}]^{1/3}$$

then the order of the reaction is 5/6.

Units of Rate Constants

Unitary rate constants obey the dimensional relationship $(\text{mol L}^{-1})^{-(n-1)}\ \text{s}^{-1}$, where n is the *order* of the reaction. As shown in Example 9.1, a rate constant is denoted by a lowercase k with a subscripted integer (e.g., k_{-1}), where the sign of the subscript indicates the direction of the reaction to which it applies; − refers to the reverse direction.

EXAMPLE 9.3

In the first-order reaction

$$\text{A} \xrightarrow{k} \text{P}$$

the expression for the *rate of change* of [A] is

$$\frac{d[\text{A}]}{dt} = -k[\text{A}] \tag{9.4}$$

Since the left-hand side of the expression has the units-of-reaction rate ($\text{mol L}^{-1}\ \text{s}^{-1}$), then these units must also apply to the right-hand side (balance of dimensions). Therefore, the units of $k[\text{A}]$ must be $\text{mol L}^{-1}\ \text{s}^{-1}$, implying that k has units of s^{-1}. Thus, simple *dimensional analysis* leads directly to the general expression for the units of a particular constant in a particular reaction scheme.

Extent of Reaction

The extent of reaction is the *fraction* of progress of a reaction from its commencement and is thus a *dimensionless* ratio. In a reaction where the product P *increases* from zero with time, the extent of reaction at time t is given by $[\text{P}]_t/[\text{P}]_\infty$, where the subscript t denotes time and ∞ denotes a very long time. For a substrate S which *declines* with time, the expression for the extent of reaction is $([\text{S}]_0 - [\text{S}]_t)/([\text{S}]_0 - [\text{S}]_\infty)$, where the subscript 0 indicates concentration at time zero and t indicates any subsequent time.

9.2 DEPENDENCE OF ENZYME REACTION RATE ON SUBSTRATE CONCENTRATION

Experimentally, the effect of substrate concentration on enzyme reaction rate is studied by recording the progress of an enzyme-catalyzed reaction, using a fixed amount of enzyme and a series of different substrate concentrations. The *initial velocity*, v_0, is measured as the slope of the tangent of the progress curve at time = 0. The initial velocity is used because enzyme degradation during the reaction or inhibition by reaction products may occur, thus yielding results that may be difficult to interpret.

When $[S]_0 \gg$ the enzyme concentration, v_0 is usually directly proportional to the enzyme concentration in the reaction mixture, and for most enzymes v_0 is a rectangular hyperbolic function of $[S]_0$ (Fig. 9-1). If there are other (co-) substrates, then these are usually held constant during the series of experiments in which $[S]_0$ is varied.

The equation describing the rectangular hyperbola that successfully represents enzyme-reaction data (e.g., Fig. 9-1) is called the *Michaelis-Menten equation*:

$$v_0 = -\left(\frac{d[S]}{dt}\right)_{t=0} = \frac{V_{max}[S]_0}{K_m + [S]_0} \tag{9.5}$$

The equation has the property that when $[S]_0$ is very large, $v_0 = V_{max}$ (the so-called *maximal velocity*); also when $v_0 = V_{max}/2$, the value of $[S]_0$ is K_m, the so-called *Michaelis constant.*

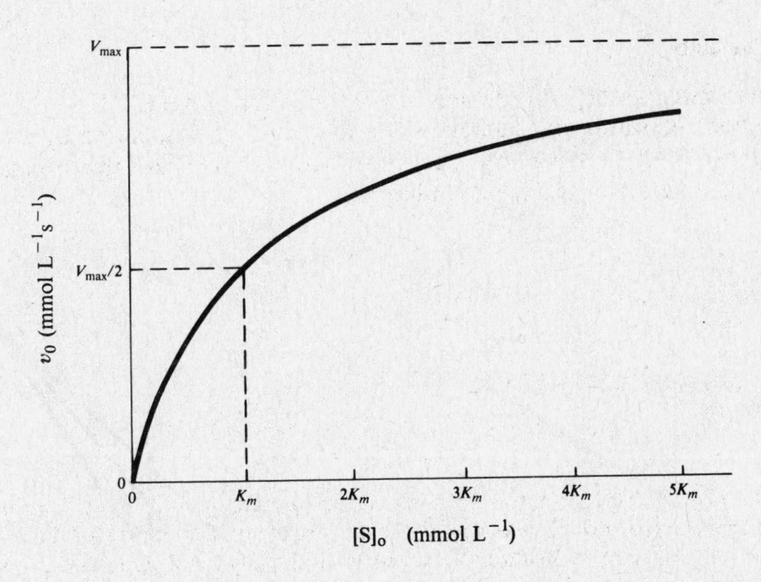

Fig. 9-1 The hyperbolic relationship between initial velocity (v_0) and initial substrate concentration ($[S]_0$) of an enzyme-catalyzed reaction.

9.3 GRAPHICAL EVALUATION OF K_m AND V_{max}

Eq. (*9.5*) can be rearranged into several new forms that yield straight lines when one *new* variable is plotted against the other. The advantages of this mathematical manipulation are that (1) V_{max} and K_m can be determined readily by fitting a straight line to the transformed data; (2) departures of the data from a straight line are more easily detected than is nonconformity to a hyperbola (these departures may indicate an inappropriateness of the simple enzyme model); (3) the effects of inhibitors on the reaction can be analyzed more easily.

The four commonly used transformations of the Michaelis-Menten equation are discussed below.

Lineweaver-Burk Equation

The Lineweaver-Burk equation was first introduced in 1935. By taking reciprocals of both sides of Eq. (*9.5*) we get

$$\frac{1}{v_0} = \frac{K_m}{V_{\max}} \frac{1}{[\mathrm{S}]_0} + \frac{1}{V_{\max}} \tag{9.6}$$

A plot of data pairs $(1/[\mathrm{S}]_{0,i}, 1/v_{0,i})$, for $i = 1, \ldots, n$, where n is the number of data pairs, gives a straight line with ordinate and abscissal intercepts $1/V_{\max}$ and $-1/K_m$, respectively [Fig. 9-2(*a*)].

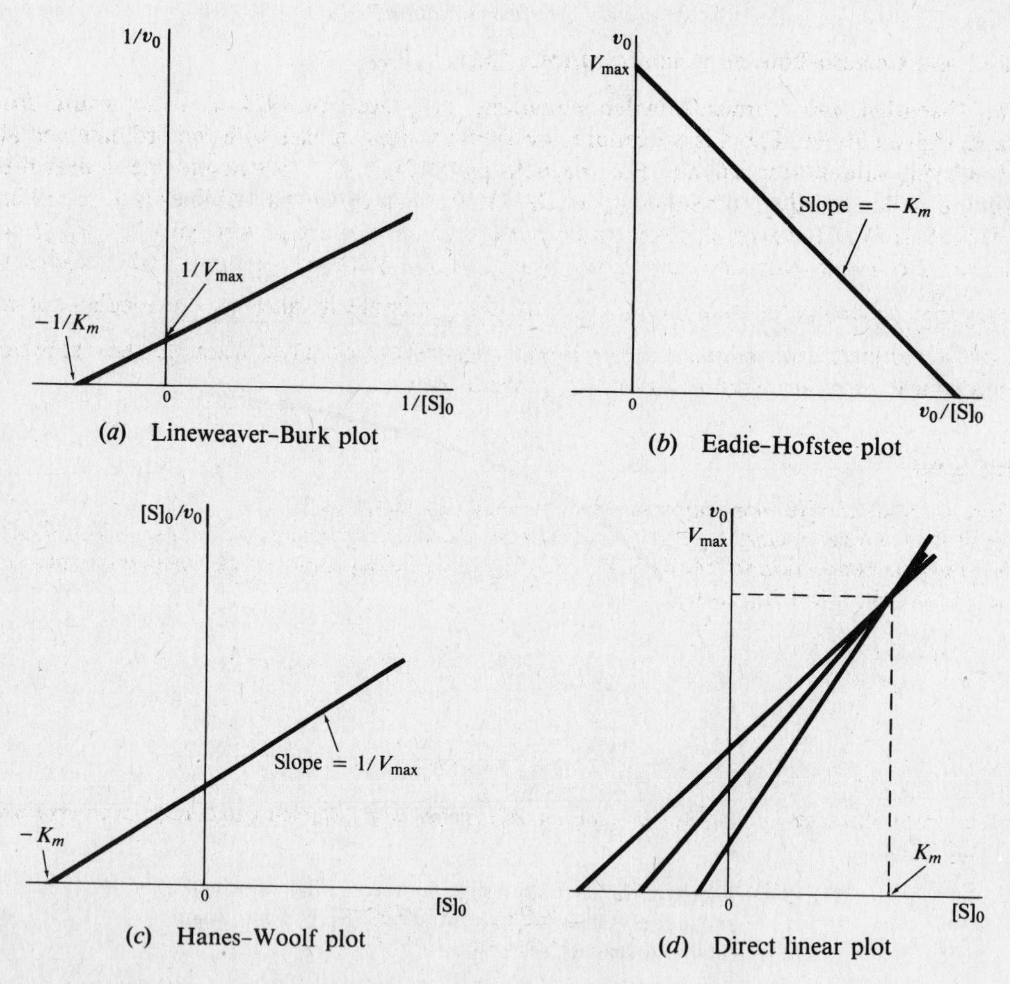

Fig. 9-2 Graphical procedures for determining the two steady-state kinetic parameters in the Michaelis-Menten equation.

Eadie-Hofstee Equation

The Eadie-Hofstee equation takes the form

$$v_0 = -K_m \frac{v_0}{[\mathrm{S}]_0} + V_{\max} \tag{9.7}$$

A plot of the data pairs consisting of $(v_{0,i}/[\mathrm{S}]_{0,i}, v_{0,i})$ gives a straight line with a slope of $-K_m$ and ordinate intercept $V_{\max}$ [Fig. 9-2(*b*)].

Hanes-Woolf Equation

The Hanes-Woolf equation is

$$\frac{[S]_0}{v_0} = \frac{[S]_0}{V_{max}} + \frac{K_m}{V_{max}} \tag{9.8}$$

A plot of the data pairs $([S]_{0,i}, [S]_{0,i}/v_{0,i})$ gives a straight line with a slope of $1/V_{max}$ and abscissal intercept of $-K_m$ [Fig. 9-2(c)].

Eisenthal and Cornish-Bowden Equation (Direct Linear Plot)

The Eisenthal and Cornish-Bowden equation, introduced in 1974, is a departure from the approaches given above. For this procedure, we set up Cartesian axes with the ordinate and abscissa for v_0 and $[S]_0$ values, respectively. For the data pair $([S]_{0,i}, v_{0,i})$, a straight line is drawn to pass through the points $(-[S]_{0,i}, 0)$ and $(0, v_{0,i})$. The intersection of any pair of lines is at the point (K_m, V_{max}) [Fig. 9-2(d)]. However, experimental data are never perfect, so a set of (K_m, V_{max}) points is determined for *every* pair of lines in the graph. If there are n data pairs, then there will be $\binom{n}{2} = \frac{n!}{2!(n-2)!}$ estimates of K_m and V_{max}; the *best* estimate is taken as the median (or middle) value of the sequence of estimates. The median values are obtained after all the estimates of a parameter have been arranged in ascending order of value.

EXAMPLE 9.4

There is an alternative, nongraphical way of viewing this analysis.

Recall that two parameters K_m and V_{max} are to be evaluated. Therefore, we require two pairs of data to solve for the two unknowns in the Michaelis-Menten equation. The equation is rearranged to give two equations that are linear in V_{max} and K_m, as follows:

$$v_{0,i} = V_{max} - \frac{v_{0,i}}{[S]_{0,i}} K_m \tag{9.9}$$

$$v_{0,j} = V_{max} - \frac{v_{0,j}}{[S]_{0,j}} K_m$$

where the i and j subscripts specify the two different data pairs. Straightforward algebra leads to expressions for $V_{max,ij}$, and $K_{m,ij}$:

$$V_{max,ij} = \frac{([S]_{0,i} - [S]_{0,j})v_{0,i}v_{0,j}}{([S]_{0,i}v_{0,j} - [S]_{0,j}v_{0,i})} \tag{9.10}$$

$$K_{m,ij} = \frac{(v_{0,j} - v_{0,i})[S]_{0,i}[S]_{0,j}}{([S]_{0,j}v_{0,i} - [S]_{0,i}v_{0,j})} \tag{9.11}$$

This procedure is readily programmed in a computer, as is the sorting required to give the medians of the estimates of the two parameters.

Question: Which graphical procedure should be used to obtain the best estimates of steady-state parameters?

The direct linear plot has been shown to yield the *least biased* estimates. However, with the general availability of computers, *all* the graphical methods have to a large extent been replaced by the procedure of *nonlinear least squares regression* of the kinetic equation *directly* onto the *untransformed* data. Nevertheless, the graphical procedures are still valuable for obtaining a quick guide to the parameter values or for obtaining the initial estimates that are required by the numerical methods.

9.4 ENZYME INHIBITION—DEFINITIONS

Often, rates of enzymatic reactions are affected by substances that are not reactants; when there is a reduction in rate caused by a compound, then the compound is said to be an *inhibitor*. Increased reaction rate by an *activator* is the opposite of this effect. In dealing with inhibitors it is important to distinguish between the *effects* that are observed experimentally and the *mechanisms* (or models) proposed to explain them.

There are three basic types of inhibition. They are defined in terms of the *degree of inhibition* i, which itself is defined as

$$i = \frac{v_0 - v_i}{v_0} \tag{9.12}$$

where v_0 and v_i are the uninhibited and inhibited initial reaction rates, respectively.

1. *Pure noncompetitive inhibition* is said to exist if i is unaffected by the concentration of substrate.
2. *Competitive inhibition* exists if i decreases as the substrate concentration is increased.
3. *Anti- or uncompetitive inhibition* exists if i increases as the substrate concentration is increased.

In addition to the above, there is *mixed inhibition*, which exists if i increases or decreases as substrate concentration increases, but *not to the same extent* as for the pure competitive or anticompetitive cases, respectively. In fact, it can be shown that *mechanistically*, noncompetitive inhibition is a special case of mixed inhibition; however, *operationally*, as defined here, mixed inhibition is a combination of two of the three basic types.

9.5 ENZYME INHIBITION—EQUATIONS

The mathematical expressions relating reaction rate and inhibitor concentration are often rather complicated, but there are four simple equations that are extensions of the Michaelis-Menten formula. These merit special consideration because the kinetics of many enzymes can be satisfactorily described by them. In the equations in Table 9.1, [I] denotes the inhibitor concentration and K_I and K'_I are inhibition constants, the units of which are those of a dissociation equilibrium constant (mmol L^{-1}).

9.6 MECHANISTIC BASIS OF THE MICHAELIS-MENTEN EQUATION

Equilibrium Analysis

To explain their results on the conversion of sucrose to glucose and fructose by the enzyme *invertase*, Michaelis and Menten proposed in 1913 the following scheme of reactions:

$$\mathrm{E} + \mathrm{S} \underset{k_{-1}}{\overset{k_1}{\rightleftharpoons}} \mathrm{ES} \xrightarrow{k_2} \mathrm{E} + \mathrm{P}$$

Table 9.1. Rate Equations for the Four Types of Enzyme Inhibition

Pure Noncompetitive	Pure Competitive
$v = \dfrac{V_{max}[S]}{(K_m + [S])\left(1 + \dfrac{[I]}{K_I}\right)}$ $i = \dfrac{[I]}{K_I + [I]}$	$v = \dfrac{V_{max}[S]}{K_m\left(1 + \dfrac{[I]}{K_I}\right) + [S]}$ $i = \dfrac{K_m \dfrac{[I]}{K_I}}{K_m\left(1 + \dfrac{[I]}{K_I}\right) + [S]}$
Anticompetitive	**Mixed**
$v = \dfrac{V_{max}[S]}{K_m + \left(1 + \dfrac{[I]}{K_I}\right)[S]}$ $i = \dfrac{[S]\dfrac{[I]}{K_I}}{K_m + \left(1 + \dfrac{[I]}{K_I}\right)[S]}$	$v = \dfrac{V_{max}}{K_m\left(1 + \dfrac{[I]}{K_I}\right) + \left(1 + \dfrac{[I]}{K_I'}\right)[S]}$ $i = \dfrac{[I]\left\{\dfrac{K_m}{K_I} + \dfrac{[S]}{K_I'}\right\}}{K_m\left(1 + \dfrac{[I]}{K_I}\right) + \left(1 + \dfrac{[I]}{K_I'}\right)[S]}$

They assumed that both k_1 and k_{-1} were large, compared with k_2; thus, the first part of the reaction could be described by the equilibrium constant for the dissociation of the enzyme-substrate complex: $K_s = [E][S]/[ES]$. The concentrations of S and E at any time are derived from the known *initial conditions*:

$$[S]_0 = [S] + [ES] + [P] \tag{9.13}$$

$$[E]_0 = [E] + [ES] \tag{9.14}$$

Since experimentally, $[S]_0 \gg [E]_0$, then $[S] \simeq [S]_0$ early in the reaction, and by using the expression for K_s and Eq. (*9.14*), we obtain:

$$K_s = \frac{([E]_0 - [ES])[S]_0}{[ES]} \tag{9.15}$$

This equation can be rearranged to give

$$[ES] = \frac{[E]_0[S]_0}{K_s + [S]_0} \tag{9.16}$$

The second step of the reaction is a simple first-order one, and thus

$$v = k_2[ES] \tag{9.17}$$

Hence the overall rate of decline of substrate is described by

$$v_0 = \frac{k_2[E]_0[S]_0}{K_s + [S]_0} \tag{9.18}$$

which is exactly the form of Eq. (*9.5*) if $V_{max} = k_2[E]_0$ and $K_m = K_s$.

Steady-State Analysis

Briggs and Haldane in 1925 examined the earlier Michaelis-Menten analysis and made an important development. Instead of assuming that the first stage of the reaction was at equilibrium, they merely assumed for all intents and purposes that the concentration of the enzyme-substrate complex scarcely changed with time; i.e., it was in a *steady state*. Written mathematically, this amounts to

$$\frac{d[\mathrm{ES}]}{dt} = 0 \tag{9.19}$$

Now, the flux equation for [ES] is

$$\frac{d[\mathrm{ES}]}{dt} = k_1[\mathrm{E}][\mathrm{S}] - (k_{-1} + k_2)[\mathrm{ES}] \tag{9.20}$$

and by using Eqs. (*9.13*) and (*9.14*) with $[\mathrm{S}] \simeq [\mathrm{S}_0]$, then

$$0 = k_1[\mathrm{E}]_0[\mathrm{S}]_0 - (k_1[\mathrm{S}]_0 + k_{-1} + k_2)[\mathrm{ES}] \tag{9.21}$$

By rearranging this equation and using the fact that $v = k_2[\mathrm{ES}]$, we obtain

$$v_0 = \frac{k_2[\mathrm{E}]_0[\mathrm{S}]_0}{\dfrac{(k_{-1} + k_2)}{k_1} + [\mathrm{S}]_0} \tag{9.22}$$

Again, this equation has the same form as Eq. (*9.5*), provided K_m is *identified with* $(k_{-1} + k_2)/k_1$ and V_{max} is identified with $k_2[\mathrm{E}]_0$.

Several features of Eq. (*9.22*) are worth noting:

1. Because k_2 describes the number of molecules of substrate converted to product per second per molecule of enzyme, it is called the *turnover number* of the enzyme. Generally, in more complex enzyme mechanisms, the expression for V_{max} is complicated by k_2 being replaced by an expression with sums of products of unitary rate constants; this grouped expression is called k_{cat}.
2. If an enzyme is not pure, it may not be possible to accurately determine the concentration of the active form, $[\mathrm{E}]_0$. Nevertheless, V_{max} can still be obtained by steady-state kinetic analysis. So, to standardize experimental results, we refer to *one enzyme unit* (or *katal*) as the amount of enzyme solution required to transform 1 μmol of substrate into product(s) in 1 min under standard conditions of pH, ionic strength, and temperature.
3. When $[\mathrm{S}]_0$ is very large compared with K_m, virtually all of E is in the form of ES, so the enzyme is said to be *saturated*, i.e., it is then operating at its maximum velocity.

9.7 DERIVATION OF COMPLICATED STEADY-STATE EQUATIONS

In principle, the steady-state rate expression for any enzyme with any number of reactants can be derived using the methods of the previous section. In practice, the procedure is very laborious, so use is made of an *algorithmic method*, introduced by King and Altman in 1956; it is *not* applicable to (1) nonenzymatic reactions (each reactant concentration must be $\gg[\mathrm{E}]_0$), (2) mixtures of enzymes, or (3) reactions with nonenzymatic steps. However, these are not severe restrictions. It is applied as follows:

1. Draw the reaction scheme (the *master pattern*) with the required reaction arrows interconnecting all relevant enzyme species (free and complexed forms).
2. Annotate all reaction arrows with the corresponding unitary rate constant. For forward reactions where a substrate is involved, place its letter of designation next to the rate constant in the scheme; do the same for reverse reactions involving a product.
3. For the n enzyme forms (one of free enzyme and the rest complexes or isomeric forms of the enzyme), draw *reaction patterns* that have $n - 1$ arrows and yield a continuous path or paths that lead to each enzyme form. In addition, *no closed loops* of steps are allowed in the pattern.

4. The expression for $[ES_i] \cdot [E]_T$, where $[ES_i]$ is any enzyme form and $[E]_T$ is the *total* enzyme concentration, is given by the summed products of concentrations and rate constants from each pattern.
5. The expression for the overall rate of *product formation* alone is given by multiplying the concentration of the relevant enzyme-substrate complex (say, $[ES_k]$) and the first-order rate constant of its breakdown (say k_j):

$$v = \frac{k_j[E]_T[ES_k]}{\text{(sum of all expressions for [E] and the different } [ES_i]\text{, where } i = 1, \ldots, n-1)} \tag{9.23}$$

In applying the algorithmic method, note that:

(*a*) In all mechanisms not involving alternative reaction sequences, the numerator of the rate expression will have only two terms: one, the product of all *forward* rate constants and substrate concentrations [as in Eq. (*9.23*)] and the other, the corresponding product for the *reverse* reaction.

(*b*) With complicated mechanisms, it is easy to overlook some patterns at step 3, above. A useful formula that yields the total number of $(n-1)$-line patterns for m steps, where n is the number of enzyme forms, is

$$\text{Total } (n-1)\text{-line patterns} = \binom{m}{n-1} = \frac{m!}{(n-1)!(m-n+1)!} \tag{9.24}$$

(*c*) Eq. (*9.24*) predicts the number of $(n-1)$-line patterns, of which some may have closed loops; the closed-loop patterns must be eliminated. The total number of $(n-1)$-line patterns with r-sided closed loops, Z, is

$$Z = \binom{m-r}{n-1-r} = \frac{(m-r)!}{(n-1-r)!(m-n+1)!} \tag{9.25}$$

Hence, Z is determined for a range from 1 to $n-1$, and the sum of all these Z values is the total number of excluded patterns.

EXAMPLE 9.5

By using the King-Altman procedure, derive the steady-state rate equation for the following enzyme mechanism:

$$A + E \underset{k_{-1}}{\overset{k_1}{\rightleftharpoons}} EA \underset{k_{-2}}{\overset{k_2}{\rightleftharpoons}} EP \underset{k_{-3}}{\overset{k_3}{\rightleftharpoons}} E + P$$

Steps (1) and (2): The master pattern is:

$$\begin{array}{ccc} E & \underset{k_{-1}}{\overset{k_1 A}{\rightleftharpoons}} & EA \\ k_3 \nwarrow\searrow k_{-3}P & & k_{-2} \nearrow\swarrow k_2 \\ & EP & \end{array}$$

Step (3): The reaction patterns for EP are:

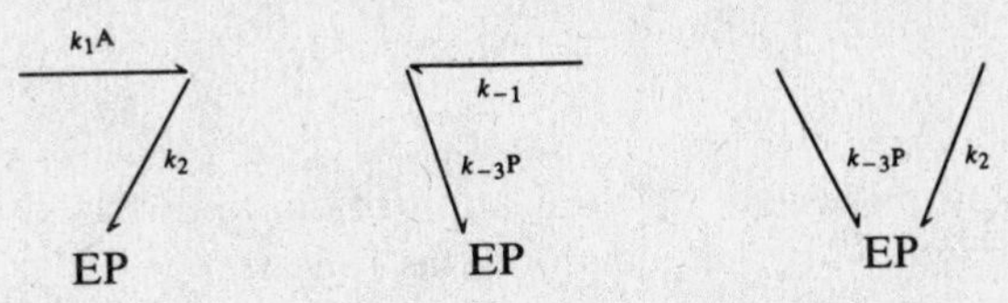

Step (4): Therefore,

$$[EP] = k_1k_2[A] + k_{-1}k_{-3}[P] + k_2k_{-3}[P] \tag{9.26}$$

Similarly, the expressions for the other two enzyme species are

$$[\mathrm{EA}] = k_1k_{-2}[\mathrm{A}] + k_{-2}k_{-3}[\mathrm{P}] + k_1k_3[\mathrm{A}]$$

$$[\mathrm{E}] = k_{-1}k_3 + k_2k_3 + k_{-1}k_{-2}$$

Since

$$v = \frac{d[\mathrm{P}]}{dt} = k_3[\mathrm{EP}] - k_{-3}[\mathrm{E}][\mathrm{P}] \tag{9.27}$$

and

$$[\mathrm{E}]_T = [\mathrm{E}] + [\mathrm{EA}] + [\mathrm{EP}] \tag{9.28}$$

then

$$v = \frac{[\mathrm{E}]_T\{k_3(k_1k_2[\mathrm{A}] + k_{-1}k_{-3}[\mathrm{P}] + k_2k_{-3}[\mathrm{P}]) - k_{-3}[\mathrm{P}](k_{-1}k_3 + k_2k_3 + k_{-1}k_{-2})\}}{k_1k_2[\mathrm{A}] + k_{-1}k_{-3}[\mathrm{P}] + k_2k_{-3}[\mathrm{P}] + k_1k_{-2}[\mathrm{A}] + k_{-2}k_{-3}[\mathrm{P}] + k_1k_3[\mathrm{A}] + k_{-1}k_3 + k_2k_3 + k_{-1}k_{-2}} \tag{9.29}$$

which is simplified to

$$v = \frac{[\mathrm{E}]_T(k_1k_2k_3[\mathrm{A}] - k_{-1}k_{-2}k_{-3}[\mathrm{P}])}{(k_{-1}k_3 + k_2k_3 + k_{-1}k_{-2}) + k_1(k_2 + k_3 + k_{-2})[\mathrm{A}] + k_{-3}(k_{-1} + k_2 + k_{-2})[\mathrm{P}]} \tag{9.30}$$

If $[\mathrm{P}] = 0$, then the expression of Eq. (*9.30*) is of the form

$$v = \frac{[\mathrm{E}]_T(\mathrm{num1})[\mathrm{A}]}{\mathrm{coef} + \mathrm{coefA}[\mathrm{A}]} \quad \text{or} \quad v = \frac{[\mathrm{E}]_T\left(\frac{\mathrm{num1}}{\mathrm{coefA}}\right)[\mathrm{A}]}{\left(\frac{\mathrm{coef}}{\mathrm{coefA}}\right) + [\mathrm{A}]} \tag{9.31}$$

where the numerator coefficient is designated num1 and the two denominator coefficients are designated coef and coefA, respectively. When the numerator and denominator are divided by coefA, the form of Eq. (*9.31*) is identical to the Michaelis-Menten Eq. (*9.5*). Thus, the analysis illustrates a very important rule of steady-state enzyme kinetics: The introduction to a mechanism of steps that involve only isomerization between enzyme complexes (EP in this case) does not change the *form* of the rate equation.

9.8 MULTIREACTANT ENZYMES

General

The most common enzymatic reactions are those with two or more substrates and as many products. But many of the simpler single-substrate schemes are valuable for the development of kinetic ideas concerning effects of pH, temperature, etc., on enzyme reaction rates. Although the mechanisms of multisubstrate reactions are complicated, their kinetics can often be described by an equation of the form:

$$v = \frac{V_{\max}^{\mathrm{app}}[\mathrm{A}]}{K_m^{\mathrm{app}} + [\mathrm{A}]} \tag{9.32}$$

This is so if the concentrations of all substrates other than A are held constant during the experiments; the values of $V_{\max}^{\mathrm{app}}$ and K_m^{app} are functions of the concentrations of the other reactants (Example 9.8).

Nomenclature

In 1963 W. W. Cleland published a classification of enzymes based on the number of substrates and products in the reaction. This classification is as follows:

1. The *reactancy* is the number of kinetically significant substrates or products and is designated by the syllables *Uni*, *Bi*, *Ter*, *Quad*.

EXAMPLE 9.6

The associated Cleland designations for the reactions below are

$A \rightleftharpoons P$	Uni Uni
$A \rightleftharpoons P + Q$	Uni Bi
$A + B \rightleftharpoons P + Q$	Bi Bi
$A + B + C \rightleftharpoons P + Q + R + S$	Ter Quad

2. If all substrates add to the enzyme *before* any products are released, the mechanism is defined as *sequential*. If substrates add in an *obligatory* order, the mechanism is called *sequential ordered*. If there is no obligatory order of addition of substrates or release of products, it is called *sequential random*. When one or more products are released before all the substrates have been added, the enzyme will exist in two or more stable forms between which it oscillates during the reaction; this type of mechanism is therefore called *Ping Pong*. If *isomerization* of stable, as distinct from transitory (e.g., $EAB \rightleftharpoons EPQ$ in Example 9.7), enzyme forms occurs, the term *Iso* is added to the designation of the mechanism.

EXAMPLE 9.7

The ordered Bi Bi mechanism can be written as

$$E + A \underset{k_{-1}}{\overset{k_1}{\rightleftharpoons}} EA \qquad EA + B \underset{k_{-2}}{\overset{k_2}{\rightleftharpoons}} EAB \qquad EAB \underset{k_{-3}}{\overset{k_3}{\rightleftharpoons}} EPQ$$

$$EPQ \underset{k_{-4}}{\overset{k_4}{\rightleftharpoons}} EQ + P \qquad EQ \underset{k_{-5}}{\overset{k_5}{\rightleftharpoons}} E + Q$$

or simply, using Cleland's diagrammatical convention,* as

$$\begin{array}{ccccc} A & B & & P & Q \\ k_1 \downarrow\uparrow k_{-1} & k_2 \downarrow\uparrow k_{-2} & & k_4 \uparrow\downarrow k_{-4} & k_5 \uparrow\downarrow k_{-5} \\ \hline E \quad EA & & (EAB \underset{k_{-3}}{\overset{k_3}{\rightleftharpoons}} EPQ) & EQ & E \end{array}$$

In Cleland's convention the letters A, B, C, D denote substrates and P, Q, R, S denote products, in the order in which they add and leave the enzyme, respectively.

EXAMPLE 9.8

By using the King-Altman procedure, the rate expression for the ordered Bi Bi mechanism in the previous example can be shown to be:

$$\begin{aligned} v = {} & (k_1k_2k_3k_4k_5[A][B] - k_{-1}k_{-2}k_{-3}k_{-4}k_{-5}[P][Q]) \div \{k_5k_{-1}(k_3k_4 + k_4k_{-2} + k_{-2}k_{-3}) \\ & + [A]k_1k_5(k_3k_4 + k_4k_{-2} + k_{-2}k_{-3}) + [B]k_2k_3k_4k_5 + [P]k_{-1}k_{-2}k_{-3}k_{-4} \\ & + [Q]k_{-1}k_{-5}(k_3k_4 + k_4k_{-2} + k_{-2}k_{-3}) + [A][B]k_1k_2(k_3k_4 + k_3k_5 + k_4k_5 + k_5k_{-3}) \\ & + [P][Q]k_{-4}k_{-5}(k_3k_{-1} + k_{-1}k_{-2} + k_{-1}k_{-3} + k_{-2}k_{-3}) + [A][P]k_1k_{-2}k_{-3}k_{-4} \\ & + [B][Q]k_2k_3k_4k_{-5} + [A][B][P]k_1k_2k_{-4}(k_3 + k_{-3}) + [B][P][Q]k_2k_{-4}k_{-5}(k_3 + k_{-3})\} \end{aligned} \tag{9.33}$$

The expression is simplified dramatically if we assume that at the start of the reaction [P] and [Q] = 0. Furthermore, if [B] is *saturating*, we can divide both the numerator and denominator by [B] and take the limit as $[B] \rightarrow \infty$, and the equation reduces to

$$v_0 = \frac{\text{num1}[A]_0}{\text{coefB} + \text{coefAB}[A]_0} \tag{9.34}$$

*In its original form, Cleland's notation had only *single* arrows; thus this mechanism would be written as

$$\begin{array}{ccccc} & A & B & P & Q \\ & \downarrow & \downarrow & \uparrow & \uparrow \\ \hline E & EA & EAB & EQ & E \end{array}$$

where the terms num1 and coefAB and coefB relate to Eq. (*9.33*). Therefore, the expressions for V_{max} and K_m, respectively, are given by

$$V_{max} = \frac{\text{num1}}{\text{coefAB}} \tag{9.35}$$

$$K_m = \frac{\text{coefB}}{\text{coefAB}} \tag{9.36}$$

Expressions for the parameters relating to B may be similarly derived.

9.9 pH EFFECTS ON ENZYME REACTION RATES

General

To describe completely the effects of pH changes on enzyme catalysis is an impossible task. Many of the amino acid side chains in an enzyme are ionizable, but in environments with polarities different from that of the free solution, the pK_a's (Chap. 3) will probably be significantly altered. However, experimentally, it is a simple matter to determine values of steady-state kinetic parameters (K_m, V_{max}) of an enzyme for various pH conditions.

The possible effects of pH are to change the ionization state of (1) groups involved in catalysis, (2) groups involved in the binding of substrate, (3) groups involved in binding at sites other than the active site, defined as *allosteric effector sites*, and (4) groups on the substrates. These altered charge states will affect the affinity of the enzyme for its substrates and the rate of catalysis (Chap. 8).

Simple Models of pH Effects

Ionization of Free Enzyme

The simplest Michaelis-Menten-type scheme for showing analytically the effect of enzyme ionization on enzyme kinetic parameters is

$$\begin{array}{l} E^- \\ \quad \Updownarrow K_{2e} \\ EH + S \underset{k_{-1}}{\overset{\kappa_1}{\rightleftharpoons}} EHS \xrightarrow{k_2} EH + P \\ \quad \Updownarrow K_{1e} \\ EH_2^+ \end{array}$$

where

$$K_{2e} = \frac{[E^-][H^+]}{[EH]} \qquad K_{1e} = \frac{[EH][H^+]}{[EH_2^+]} \qquad K_s = \frac{[EH][S]}{[EHS]} \tag{9.37}$$

The rate expression corresponding to this scheme is derived most readily by using the equilibrium analysis of Michaelis and Menten. The *conservation of mass equation* for the enzyme is

$$[E]_0 = [E^-] + [EH] + [EHS] + [EH_2^+] \tag{9.38}$$

By using the relationship in Eq. (*9.38*), we obtain

$$[EHS] = \frac{[E]_0}{1 + \dfrac{K_s}{[S]_0}\left(1 + \dfrac{K_{2e}}{[H^+]} + \dfrac{[H^+]}{K_{1e}}\right)} \tag{9.39}$$

Since $v_0 = k_2[EHS]$, then

$$v_0 = \frac{k_2[E]_0[S]_0}{K_s\left(1 + \dfrac{K_{2e}}{[H^+]} + \dfrac{[H^+]}{K_{1e}}\right) + [S]_0} \tag{9.40}$$

Eq. (*9.40*) corresponds to the Michaelis-Menten equation with the same expression for V_{max}, $k_2[\mathrm{E}]_0$, but

$$K_m = K_s\left(1 + \frac{K_{2e}}{[\mathrm{H}^+]} + \frac{[\mathrm{H}^+]}{K_{1e}}\right) \tag{9.41}$$

The following facts about these equations should be noted:

(*a*) Since at any pH the term in parentheses in Eq. (*9.41*) is always >1, then $K_m > K_s$.

(*b*) The concentration of H^+ has no effect on V_{max}.

(*c*) If $[\mathrm{H}^+] \gg K_{2e}$, that is, when the pH is low (i.e., $[\mathrm{H}^+]$ is high), then Eq. (*9.40*) becomes

$$v_0 = \frac{V_{max}[\mathrm{S}]_0}{K_s\left[\left(1 + \frac{[\mathrm{H}^+]}{K_{1e}}\right)\right] + [\mathrm{S}]_0} \tag{9.42}$$

which is the same form as the competitive inhibition equation (Table 9.1) with H^+ as the *inhibitor*.

(*d*) A plot of $-\log$ (apparent K_m) $\equiv \mathrm{p}K_m$ versus pH has the form shown in Fig. 9-3, with a maximum value of $\mathrm{p}K_m$ at $\mathrm{pH} = -1/2 \log (K_{1e} \cdot K_{2e})$; i.e., the optimum pH for the reaction is $(\mathrm{p}K_{1e} + \mathrm{p}K_{2e})/2$.

(*e*) At low pH, $K_m \simeq K_s[\mathrm{H}^+]/K_{1e}$; hence, $\mathrm{p}K_m = \mathrm{p}K_s + \mathrm{pH} - \mathrm{p}K_{1e}$, and a plot of $\mathrm{p}K_m$ versus pH is a straight line with a slope of 1 (Fig. 9-3). Similarly, at high pH, where $K_{1e} \gg K_{2e} \gg [\mathrm{H}^+]$, $K_m = K_sK_{2e}/[\mathrm{H}^+]$ and $\mathrm{p}K_m = \mathrm{p}K_s + \mathrm{p}K_{2e} - \mathrm{pH}$, and a plot of $\mathrm{p}K_m$ versus pH has a slope of -1. In the intermediate pH range, when $K_{1e} \gg [\mathrm{H}^+] \gg K_{2e}$, $K_m \simeq K_s$ and a plot of $\mathrm{p}K_m$ versus pH is a horizontal line. Hence the intersection of the horizontal line with the two straight-line segments yields estimates of $\mathrm{p}K_{1e}$ and $\mathrm{p}K_{2e}$ (Fig. 9-3).

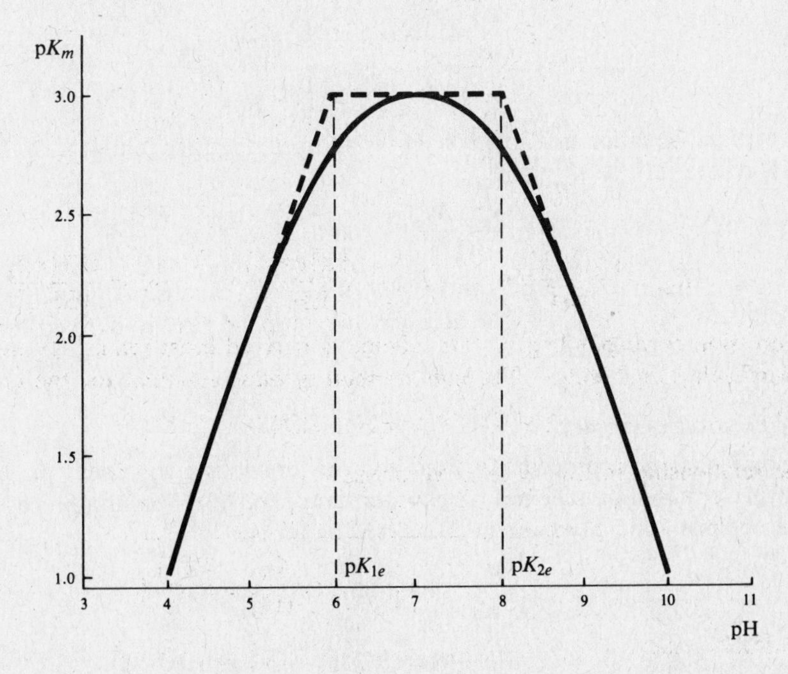

Fig. 9-3 A plot of $-\log$ (apparent K_m) $\equiv \mathrm{p}K_m$ versus pH for the enzyme scheme shown on page 236. The parameters used in Eq. (*9.40*) were $K_S = 10^{-3}$ mol L^{-1}, $K_{1e} = 10^{-6}$ mol L^{-1}, $K_{2e} = 10^{-8}$ mol L^{-1}.

Ionization of ES Complex

The ionization of the groups in the enzyme-substrate complex and not the free enzyme may produce pH effects on certain enzyme reaction rates. Consider the scheme

$$\begin{array}{c} \mathrm{ES^-} \\ \Updownarrow K_{2es} \\ \mathrm{EH + S} \underset{k_{-1}}{\overset{k_1}{\rightleftharpoons}} \mathrm{EHS} \xrightarrow{k_2} \mathrm{EH + P} \\ \Updownarrow K_{1es} \\ \mathrm{EH_2S^+} \end{array}$$

By using the analytical procedures of the previous section, the expression for v_0 is obtained:

$$v_0 = \frac{V_{max}[\mathrm{S}]_0}{K_s + [\mathrm{S}]_0\left(1 + \dfrac{[\mathrm{H^+}]}{K_{1es}} + \dfrac{K_{2es}}{[\mathrm{H^+}]}\right)} \tag{9.43}$$

This equation has the same form as the uncompetitive inhibition case (Table 9.1), and furthermore, both V_{max}^{app} and K_m^{app} are functions of $[\mathrm{H^+}]$:

$$V_{max}^{app} = \frac{V_{max}}{1 + \dfrac{[\mathrm{H^+}]}{K_{1es}} + \dfrac{K_{2es}}{[\mathrm{H^+}]}} \tag{9.44}$$

$$K_m^{app} = \frac{K_s}{1 + \dfrac{[\mathrm{H^+}]}{K_{1es}} + \dfrac{K_{2es}}{[\mathrm{H^+}]}} \tag{9.45}$$

So, in the particular case that $[\mathrm{H^+}]$ is high (low pH) and $\gg K_{2es}$, Eq. (*9.43*) reduces to

$$v_0 = \frac{V_{max}[\mathrm{S}]_0}{K_s + [\mathrm{S}]_0\left(1 + \dfrac{[\mathrm{H^+}]}{K_{1es}}\right)} \tag{9.46}$$

By analogy with the equation in Table 9.1, $\mathrm{H^+}$ is an uncompetitive inhibitor of the reaction. Thus, at very low pH, where $[\mathrm{H^+}] \gg K_{1es} \gg K_{2es}$

$$V_{max}^{app} = \frac{V_{max}K_{1es}}{[\mathrm{H^+}]} \quad \text{and} \quad K_m^{app} = \frac{K_s K_{1es}}{[\mathrm{H^+}]} \tag{9.47}$$

So, $\log V_{max}^{app} = \log V_{max} - \mathrm{p}K_{1es} + \mathrm{pH}$, and a plot of $\log V_{max}^{app}$ versus pH gives the curve shown in Fig. 9-4. Other plots for K_m^{app} and V_{max}^{app} at high pH may also be readily derived, enabling estimates of $\mathrm{p}K_{1es}$ and $\mathrm{p}K_{2es}$ to be obtained.

Ionization of E and ES Complex

The general analytical procedures used for the previous two cases can be applied to more complex enzyme-protonation schemes. A counterpart of the mixed inhibition case was introduced by von Euler, Josephson, and Myrbäck in 1924; it is as follows:

$$\begin{array}{ccc} \mathrm{E^-} & & \mathrm{ES^-} \\ K_{2e}\Updownarrow & & K_{2es}\Updownarrow \\ \mathrm{EH + S} & \underset{k_{-1}}{\overset{k_1}{\rightleftharpoons}} & \mathrm{EHS} \xrightarrow{k_2} \mathrm{EH + P} \\ K_{1e}\Updownarrow & & K_{1es}\Updownarrow \\ \mathrm{EH_2^+} & & \mathrm{EH_2S^+} \end{array}$$

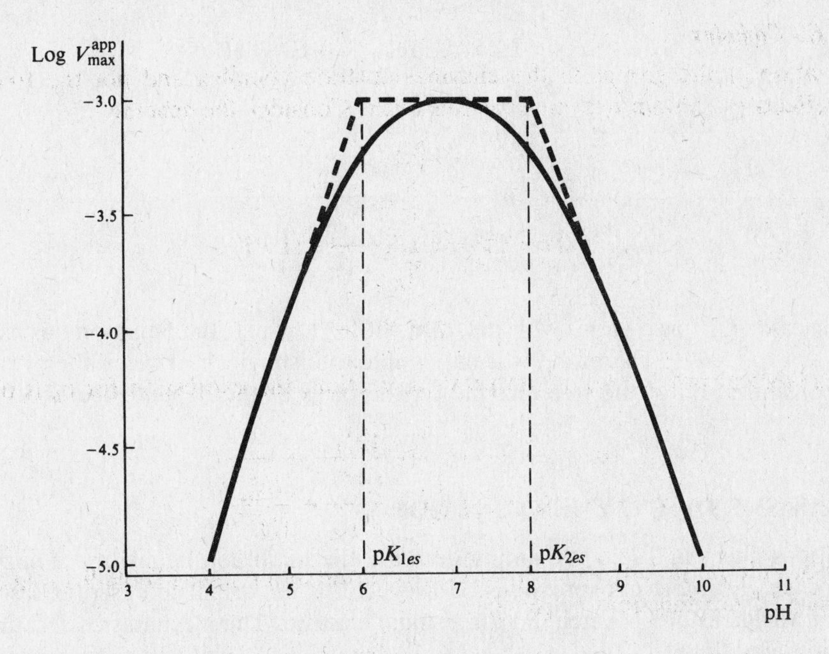

Fig. 9-4 Diagrammatic plot of $\log V_{max}^{app}$ versus pH for the reaction scheme shown on page 238, with $V_{max} = 10^{-3}$ mol L^{-1} min^{-1}, p$K_{1es} = 6.0$, and p$K_{2es} = 8.0$.

The corresponding steady-state rate equation is

$$v_0 = \frac{k_2[\mathrm{E}]_0[\mathrm{S}]_0}{K_s\left(1 + \frac{[\mathrm{H}^+]}{K_{1e}} + \frac{K_{2e}}{[\mathrm{H}^+]}\right) + [\mathrm{S}]_0\left(1 + \frac{[\mathrm{H}^+]}{K_{1es}} + \frac{K_{2es}}{[\mathrm{H}^+]}\right)} \tag{9.48}$$

It is left to the reader to analyze the dependence of V_{max} and K_m on $[\mathrm{H}^+]$ and thus to study the *form* of plots of log (V_{max}^{app}) versus pH, etc. From these plots, in the case of real enzyme systems, estimates of the values of K_{1e}, K_{2e}, K_{1es}, and K_{2es} will be obtained.

Ionization of Substrate

For the pH effects on the ionization of substrate, there are two very simple models that can be studied:

1.

$$\begin{array}{l} \mathrm{S} + \mathrm{E} \underset{k_{-1}}{\overset{k_1}{\rightleftharpoons}} \mathrm{ES} \xrightarrow{k_2} \mathrm{E} + \mathrm{P} \\ K_e \Big\updownarrow \\ \mathrm{SH}^+ \end{array}$$

$$v_0 = \frac{V_{max}[\mathrm{S}]_0}{K_m\left(1 + \frac{[\mathrm{H}^+]}{K_e}\right) + [\mathrm{S}]_0} \tag{9.49}$$

Thus, K_m^{app} varies with pH, but V_{max} does *not*.

2.
$$\begin{array}{c} SH^+ + E \underset{k_{-1}}{\overset{k_1}{\rightleftharpoons}} ESH^+ \xrightarrow{k_2} E + PH^+ \\ K_e \Big\updownarrow \\ S \end{array}$$

$$v_0 = \frac{V_{max}[S]_0}{K_m\left(1 + \frac{K_e}{[H^+]}\right) + [S]_0} \tag{9.50}$$

In this case, K_m^{app} also varies with pH, and yet for low pH, the equation has exactly the same form as Eq. (*9.40*). Therefore, it is impossible to distinguish between the present model and that for ionization of the free enzyme on the basis of the dependence of K_m on pH.

9.10 MECHANISMS OF ENZYME INHIBITION

The equations given in Table 9.1 simply describe the inhibition behaviors of enzymes and thus can be called *phenomenological* expressions. However, it is important to describe basic mechanisms, in the way that Michaelis and Menten did for a single enzyme. The mechanisms account for the *form* of the inhibition equations.

Competitive Inhibition

The simplest scheme for competitive inhibition is

$$\begin{array}{c} I \\ + \\ E + S \underset{k_{-1}}{\overset{k_1}{\rightleftharpoons}} ES \xrightarrow{k_2} E + P \\ k_I \Big\updownarrow k_{-I} \\ EI \end{array}$$

where $K_I = \dfrac{k_{-I}}{k_I}$ and is the dissociation constant of the enzyme-inhibitor complex.

By using the steady-state analysis of Briggs and Haldane, it is easy to show that

$$v_0 = \frac{k_2[E]_0[S]_0}{\left(\frac{k_{-1} + k_2}{k_1}\right)\left(1 + \frac{[I]}{k_{-I}/k_I}\right) + [S]_0} \tag{9.51}$$

$$= \frac{V_{max}[S]_0}{K_m\left(1 + \frac{[I]}{K_I}\right) + [S]_0} \tag{9.52}$$

The reciprocal is

$$\frac{1}{v_0} = \frac{1}{V_{max}}\left(1 + \frac{K_m}{[S]_0}\right) + \frac{K_m}{V_{max}} \frac{[I]}{K_I[S]_0} \tag{9.53}$$

Thus a plot of $1/v_0$ versus [I] (called a *Dixon plot*) gives a straight line with a slope of $K_m/(V_{max}K_I[S]_0)$ [Fig. 9-5(*a*)], and if several experiments are performed with a range of substrate concentrations, the above-mentioned lines intersect at $[I] = -K_I$.

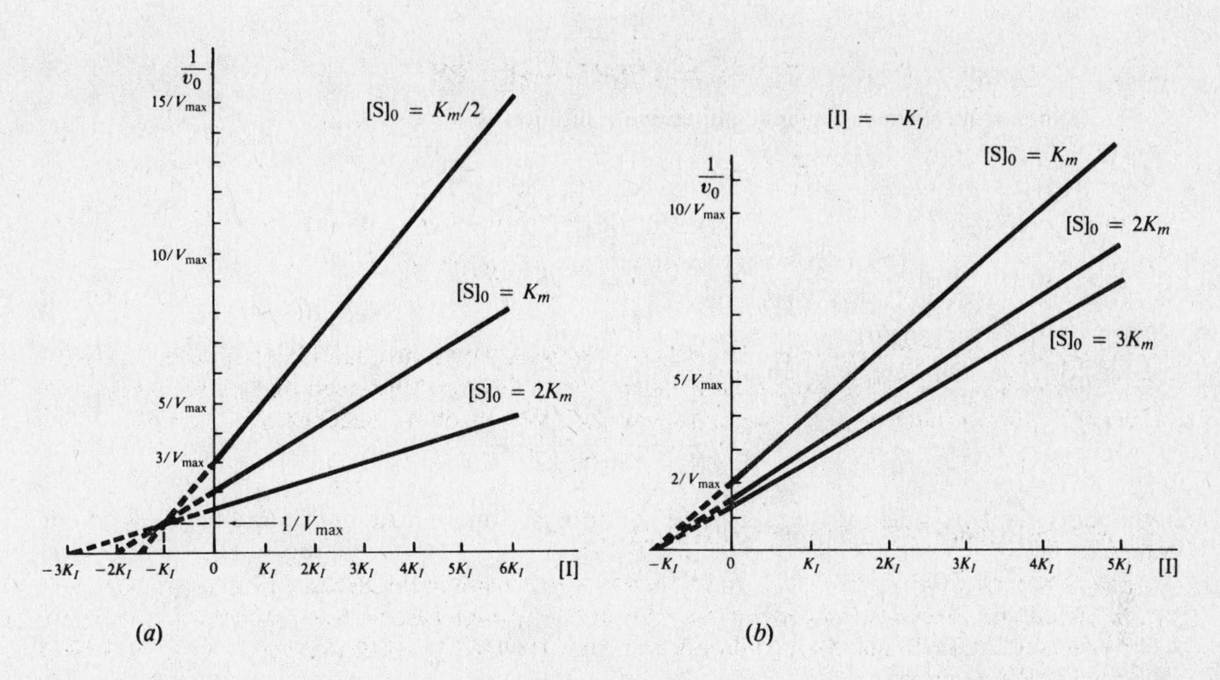

Fig. 9-5 Dixon plot: $1/v_0$ versus [I] for (*a*) simple competitive inhibition and (*b*) simple noncompetitive inhibition.

Noncompetitive Inhibition and Mixed Inhibition

The simplest scheme for noncompetitive and mixed inhibitions is

$$\begin{array}{ccccc} \mathrm{I} & & & & \\ + & & & & \\ \mathrm{E} + \mathrm{S} & \underset{k_{-1}}{\overset{k_1}{\rightleftharpoons}} & \mathrm{ES} & \xrightarrow{k_2} & \mathrm{E} + \mathrm{P} \\ K_I \Big\Updownarrow & & \Big\Updownarrow K_I' & & \\ \mathrm{EI} & & \mathrm{ESI} & & \end{array}$$

The corresponding rate equation is

$$v_0 = \frac{V_{\max}[\mathrm{S}]_0}{K_m\left(1 + \dfrac{[\mathrm{I}]}{K_I}\right) + \left(1 + \dfrac{[\mathrm{I}]}{K_I'}\right)[\mathrm{S}]_0} \tag{9.54}$$

This equation corresponds to that for mixed inhibition (Table 9.1) if $K_I \neq K_I'$. However, if $K_I = K_I'$, then pure noncompetitive inhibition is the result. Thus, it can be seen that, mechanistically speaking, pure noncompetitive inhibition is a special case of mixed inhibition.

For pure noncompetitive inhibition, the Dixon plot, $1/v_0$ versus [I], relies on the expression

$$\frac{1}{v_0} = \left(\frac{K_m + [\mathrm{S}]_0}{V_{\max}[\mathrm{S}]_0}\right) + \left(\frac{K_m + [\mathrm{S}]_0}{V_{\max}[\mathrm{S}]_0}\right)\frac{[\mathrm{I}]}{K_I} \tag{9.55}$$

In experiments, a range of inhibitor concentrations is used with a range of substrate concentrations; the Dixon plots are straight lines for each fixed $[\mathrm{S}]_0$, and the lines intersect on the [I] axis at $-K_I$ [Fig. 9-5(*b*)].

Un- or Anticompetitive Inhibition

The simplest mechanism for anticompetitive inhibition is

$$\begin{array}{c} \mathrm{E} + \mathrm{S} \underset{k_{-1}}{\overset{k_1}{\rightleftharpoons}} \overset{\mathrm{I} \\ +}{\mathrm{ES}} \xrightarrow{k_2} \mathrm{E} + \mathrm{P} \\ K_I \Big\Updownarrow \\ \mathrm{ESI} \end{array}$$

and the corresponding rate equation is

$$v_0 = \frac{V_{\max}[\mathrm{S}_0]}{K_m + [\mathrm{S}]_0(1 + [\mathrm{I}]/K_I)} \tag{9.56}$$

In this case, the Dixon plot of $1/v_0$ versus [I] does *not* give an estimate of K_I since the lines do not intersect.

9.11 FAST REACTIONS

The time period in a reaction before the concentration(s) of ES complex(es) have reached a steady state is called the *transient* or *pre-steady-state* phase. Usually in laboratory bench experiments, this stage of the reaction lasts less than 1 s. Therefore, it should be clear that the experimental methods for studying reactions in the early phase must be special because the usual simple kinetic methods take several seconds for adequate mixing of reactants. Also, the kinetic equations that are used to describe and analyze the data are quite different from the steady-state equations described hitherto. Steady-state kinetic analyses do not, in general, provide estimates of unitary rate constants of an enzymatic reaction; pre-steady-state kinetic studies do. The procedures fall into two main classes.

Rapid-Flow Techniques

In 1923 Hartidge and Roughton described an apparatus for the study of the fast reaction between hemoglobin and oxygen. The device drove two solutions, under constant pressure, through a mixing chamber into a transparent tube, along which was placed a movable detector; today, the detector is usually a spectrophotometer (Fig. 9-6). The time of reaction is proportional to the distance of the spectrophotometer observation site from the mixing chamber and may be calculated from the known constant-volume flow rate and cross section of the flow tube.

The main disadvantage of this apparatus is the large quantity of reactants that is needed. In 1951 Britton Chance described the stopped-flow apparatus (Fig. 9-7), in which the continuous flow of reactants is stopped automatically after an observation cell has been filled; this action triggers a spectrophotometric recording device. The recording of time courses of reactions typically commences at ~0.2 ms after the mixing of reactants.

Relaxation Methods

The development of one of these methods earned Eigen the 1967 Nobel prize. Instead of rapidly mixing reagents, a reaction mixture is allowed to reach equilibrium. Then its equilibrium is abruptly perturbed, and the progress of the reaction toward a new equilibrium position is monitored spectroscopically. Reactions with half-lives as short as 10^{-10} s have been studied with this technique. The basis of the method is that chemical equilibrium constants, K_e, generally depend on one or more

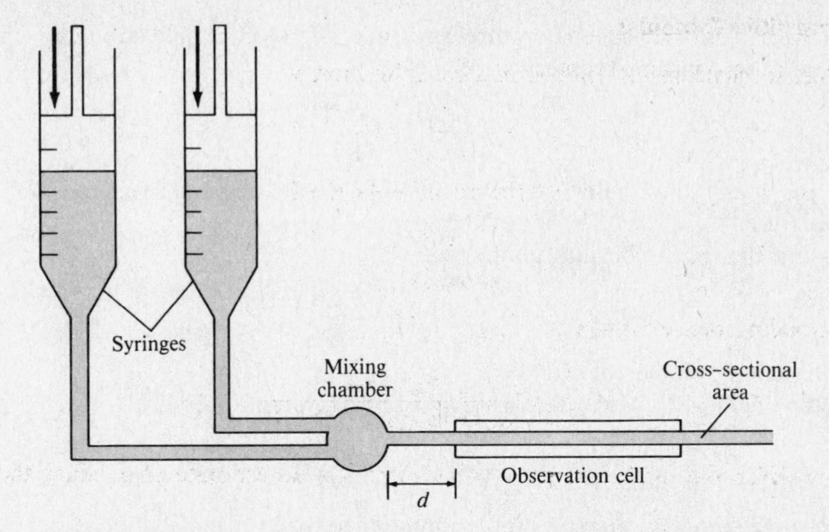

Fig. 9-6 Schematic representation of a continuous-flow apparatus. The reactants are driven from the syringes at a constant velocity and combined in the mixing chamber; the products are monitored along the observation cell. The distance (d) of the observation cell from the mixing chamber can be varied.

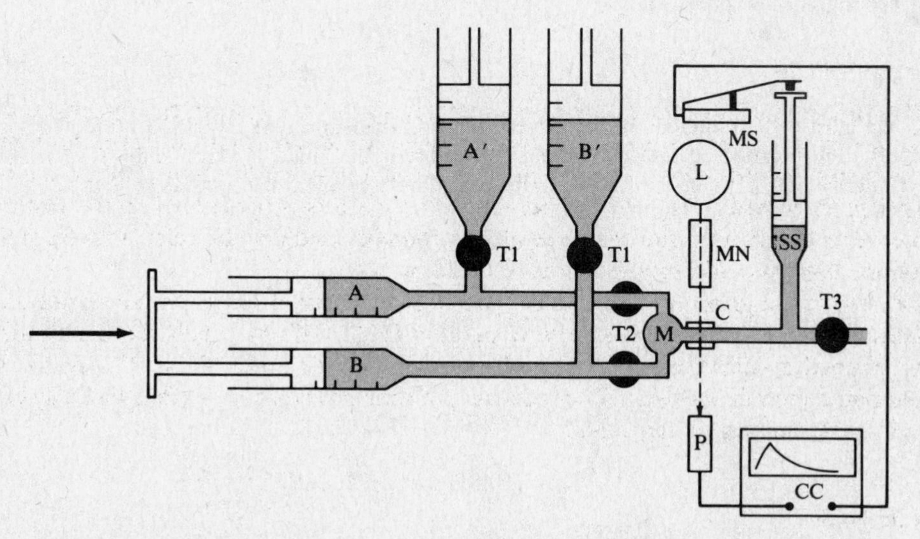

Fig. 9-7 Schematic representation of a stopped-flow apparatus. A and B are the hydraulically driven syringes, and A′ and B′ are reservoirs. T1–T3 are taps, M is the mixing chamber, and C is the observation cell. L is the light source, MN is the monochromator, and P is the light detector. SS is the stopping syringe; MS is a microswitch to trigger data acquisition in the computer, CC.

thermodynamic variables, such as temperature, pressure, and electric field strength (Chap. 10). For a temperature change at constant pressure P (Chap. 10)

$$\left(\frac{\partial \ln K_e}{\partial T}\right)_P = \left(\frac{\Delta H^\circ}{RT^2}\right) \tag{9.57}$$

where ΔH° is the standard enthalpy of the reaction, R is the universal gas constant, and T is the absolute temperature.

For a pressure change at constant temperature

$$\left(\frac{\partial \ln K_e}{\partial P}\right)_T = -\left(\frac{\Delta V}{RT}\right) \tag{9.58}$$

where ΔV is the molar volume change.

For an electric field (E) change at constant temperature and pressure

$$\left(\frac{\partial \ln K_e}{\partial E}\right)_{T,P} = \left(\frac{\Delta M}{RT}\right) \tag{9.59}$$

where ΔM is the difference in electric dipole moment between a mole of substrate and a mole of product.

Temperature jump is the most frequently used perturbation method for enzyme kinetic studies. The change in temperature is brought about in a few microseconds by the discharge of a high-voltage capacitor through platinum electrodes in a specially constructed spectrophotometer cell, or by microwave or laser-light irradiation.

Pre-steady-state Rate Equations

Pre-steady-state rate equations are differential flux equations written using the principle of mass action and conservation of mass equations. An important simplifying assumption, leading to linear differential equations that are easily solved, is to let $[S]_0$ be $\gg [E]_0$ or $[E]_0 \gg [S]_0$. The solutions for $[S]_t$ or $[P]_t$ are sums of exponentials.

EXAMPLE 9.9

Among the most widely studied enzymatic reactions is that of α-chymotrypsin, a proteolytic enzyme, that also hydrolyzes p-nitrophenyl acetate to liberate the colored compound p-nitrophenol. The enzyme forms a stable acetyl intermediate (EQ in the following reaction) in its covalent catalysis as follows:

$$E + S \underset{k_{-1}}{\overset{k_1}{\rightleftharpoons}} ES \xrightarrow[\searrow P]{k_2} EQ \xrightarrow{k_3} E + Q$$

If S is the ester, then P is the alcohol (p-nitrophenol) and Q, the second product, is the acid (acetate); EQ is the acetyl- (acyl)-enzyme intermediate. Rate and conservation equations for each of the species can be written as previously described (Sec. 9.1):

$$[E]_0 = [E] + [ES] + [EQ]$$

$$\frac{d[ES]}{dt} = k_1[E][S] - (k_{-1} + k_2)[ES] \tag{9.60}$$

$$\frac{d[EQ]}{dt} = k_2[ES] - k_3[EQ] \tag{9.61}$$

$$\frac{d[P]}{dt} = k_2[ES] \tag{9.62}$$

$$\frac{d[Q]}{dt} = k_3[EQ] \tag{9.63}$$

If we ensure experimentally that $[S]_0 \gg [E]_0$, then Eqs. (*9.60*)–(*9.63*) can be solved using the methods applicable to simultaneous linear differential equations. The expression for [P] at any time after initiation of the reaction is

$$[\mathrm{P}]_t = \frac{k_{\mathrm{cat}}[\mathrm{E}]_0[\mathrm{S}]_0 t}{[\mathrm{S}]_0 + K_m} + \frac{k_2 \mathrm{A}(1-e^{-\lambda_1 t})}{\lambda_1} + \frac{k_2 \mathrm{B}(1-e^{-\lambda_2 t})}{\lambda_2} \tag{9.64}$$

where

$$\mathrm{A}+\mathrm{B} = \frac{k_{\mathrm{cat}}[\mathrm{E}]_0[\mathrm{S}]_0}{[\mathrm{S}]_0 + K_m}, \qquad k_{\mathrm{cat}} = \frac{k_2 k_3}{k_2 + k_3}, \qquad K_m = \frac{k_3(k_1 + k_2)}{k_1(k_2 + k_3)} \tag{9.65}$$

$$\lambda_1 \lambda_2 = k_1(k_2 + k_3)([\mathrm{S}]_0 + K_m) \tag{9.66}$$

and

$$\lambda_1 + \lambda_2 = k_1[\mathrm{S}]_0 + k_{-1} + k_2 + k_3 \tag{9.67}$$

Typical data obtained from a stopped-flow experiment are shown in Fig. 9-8.

Data analysis involves nonlinear regression of the double exponential expression [Eq. (*9.64*)] onto the data to determine values of λ_1 and λ_2. The identities, Eqs. (*9.66*) and (*9.67*), enable estimates of the rate constants to be obtained.

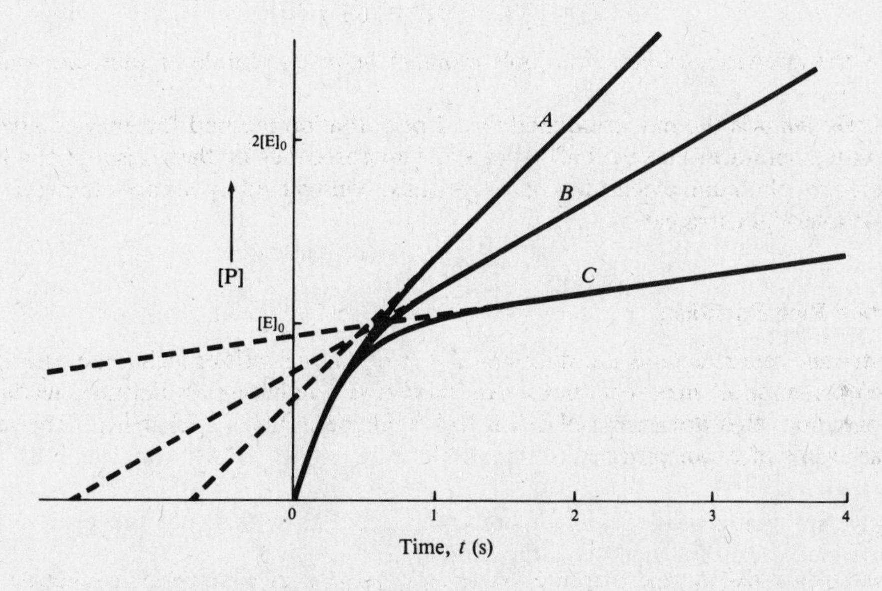

Fig. 9-8 The progress of the release of P in the reaction scheme for chymotrypsin. The concentration of P is given relative to the total enzyme concentration, and $[\mathrm{S}]_0 \gg [\mathrm{E}]_0$. The curves show the effect of changing the value of k_3, the rate constant for the breakdown of the acyl-enzyme complex. Note that for small values of k_3, the intercept on the ordinate tends to $[\mathrm{E}]_0$. Thus, for some substrates that form a stable acyl-enzyme intermediate, it is possible to titrate the enzyme with the substrate to determine the number of active sites in the solution. The parameter values used in Eq. (*9.64*) for this diagram were $[\mathrm{E}]_0 = 10^6$ mol L^{-1}, $[\mathrm{S}]_0 = 10^{-3}$ mol L^{-1}, $k_1 = 10^4$ mol^{-1} L s^{-1}, $k_{-1} = 10^2$ s^{-1}, $k_2 = 50$ s^{-1}. (*A*) $k_3 = 1$ s^{-1}; (*B*) $k_3 = 0.5$ s^{-1}; (*C*) $k_3 = 0.1$ s^{-1}.

Relaxation Kinetic Equations

The simplest reaction scheme that adequately illustrates the principle of the relaxation method is

$$\mathrm{A} \underset{k_{-1}}{\overset{k_1}{\rightleftharpoons}} \mathrm{B}$$

Suppose A is more stable than B at low temperature; then if the system is rapidly heated (*temperature jump*), a certain quantity of A will be converted to B. Let $[\mathrm{A}]_e^{\mathrm{L}}$ and $[\mathrm{B}]_e^{\mathrm{L}}$ be the equilibrium concentrations of the corresponding species at the low (starting) temperature and $[\mathrm{A}]_e^{\mathrm{H}}$ and $[\mathrm{B}]_e^{\mathrm{H}}$ be the concentrations at the high temperature. We also denote the concentrations of A and B at any time t after the perturbation by $[\mathrm{A}]_t$ and $[\mathrm{B}]_t$. The appropriate flux equations are

$$\frac{d[\mathrm{A}]_t}{dt} = -k_1[\mathrm{A}]_t + k_{-1}[\mathrm{B}]_t \tag{9.68}$$

$$\frac{d[\mathrm{B}]_t}{dt} = k_1[\mathrm{A}]_t - k_{-1}[\mathrm{B}]_t \tag{9.69}$$

The equations that describe $[\mathrm{A}]_t$ and $[\mathrm{B}]_t$ after the perturbation are obtained by integrating Eqs. (*9.68*) and (*9.69*).

The mathematical procedures for solving simultaneous linear differential equations such as these are well known, and only the expression for [B] is given here.

Denote

$$[\mathrm{S}]_0 = [\mathrm{A}]_e^{\mathrm{L}} + [\mathrm{B}]_e^{\mathrm{L}} = [\mathrm{A}]_t + [\mathrm{B}]_t \tag{9.70}$$

$$[\mathrm{B}]_t = \frac{k_1[\mathrm{S}]_0}{k_1 + k_{-1}}\left(1 - e^{-(k_1+k_{-1})t}\right) + [\mathrm{B}]_e^{\mathrm{L}} e^{-(k_1+k_{-1})t} \tag{9.71}$$

But

$$[\mathrm{B}]_e^{\mathrm{H}} = \frac{k_1[\mathrm{S}]_0}{k_1 + k_{-1}} \tag{9.72}$$

so Eq. (*9.71*) becomes

$$\frac{[\mathrm{B}]_t - [\mathrm{B}]_e^{\mathrm{H}}}{[\mathrm{B}]_e^{\mathrm{L}} - [\mathrm{B}]_e^{\mathrm{H}}} = e^{-(k_1+k_{-1})t} \tag{9.73}$$

This can be written as

$$\ln\left([\mathrm{B}]_t - [\mathrm{B}]_e^{\mathrm{H}}\right) = -(k_1 + k_{-1})t + \ln\left([\mathrm{B}]_e^{\mathrm{L}} - [\mathrm{B}]_e^{\mathrm{H}}\right) \tag{9.74}$$

Therefore, a plot of the left-hand side of Eq. (*9.74*) versus t gives a straight line with a slope of $-(k_1 + k_{-1})$; using the equilibrium constant $K_e = k_1/k_{-1} = [\mathrm{B}]_e^{\mathrm{H}}/[\mathrm{A}]_e^{\mathrm{H}}$ we can thus evaluate k_1 and k_{-1}.

Obviously, enzymatic processes are more complicated than this, but the general principle of setting up the mechanism for analysis is the same.

9.12 REGULATORY ENZYMES

General

The enzyme kinetics discussed thus far can be called *Michaelis-Menten kinetics*, since a plot of reaction rate versus substrate concentration is a pure rectangular hyperbola; alternatively, a plot of reciprocal initial velocity versus reciprocal substrate concentration is linear. Many reactions involving two or more substrates also give such linear plots with respect to one substrate, with the initial concentration of the other substrate(s) fixed. While many enzymes obey Michaelis-Menten kinetics, a significant number do not. These enzymes typically give velocity versus substrate curves that are *sigmoidal* rather than hyperbolic; they are called *control* or *regulatory* enzymes and are usually situated at the *beginning* or at *branch points* of a metabolic pathway.

EXAMPLE 9.10

The simplest form of regulation of a metabolic pathway is the *inhibition* of an enzyme by the product of the pathway. In Fig. 9-9, the E_i's denote enzymes, A and B are metabolites, and the circled minus sign indicates inhibition. If there were no inhibitor of the enzyme (E_1) acting on A, the concentration of B would depend entirely on its rate of synthesis or utilization. If the rate of utilization of B decreased or B was supplied from an outside source, its concentration would rise, perhaps even to toxic levels. However, if B is an *inhibitor* of the

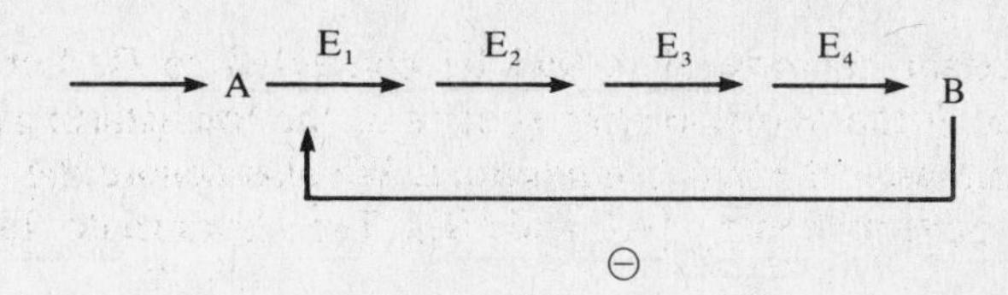

Fig. 9-9

first enzyme, then as its concentration rises, the extent of inhibition will increase and its rate of synthesis will decrease. This effect is called *feedback inhibition* or negative feedback control; it is a concept also used in describing electronic circuits.

EXAMPLE 9.11

Control in *branched* metabolic pathways is more complex. Consider the metabolic scheme in Fig. 9-10. Here, B reacts with C, and D is produced further along the pathway; for most effective control of D, B should inhibit the first enzyme (E_1) and C should activate it. In this case, if B is supplied from an external source so that $B \gg C$, then B would inhibit its own synthesis from A and the concentration of B and C would tend to become equal. Alternatively, if $C \gg B$, then C would activate the production of B and this again would tend to equalize the concentrations of B and C. This activation by C is usually the result of C competing for the same binding site as B on E_1 and thus reducing the inhibition by B. The first enzyme of pyrimidine synthesis, aspartate carbamoyltransferase, in *E. coli* is subject to this type of control (Chap. 15); in this case B is CTP, C is ATP, and D is the nucleic acids.

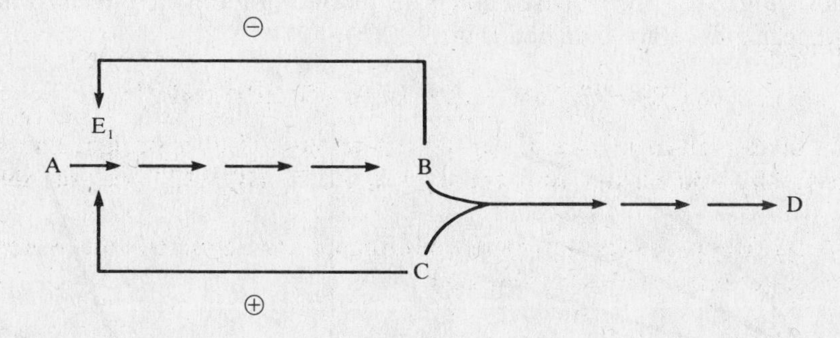

Fig. 9-10

EXAMPLE 9.12

The previous example was of a *convergent* pathway; it is more usual for biosynthetic pathways to *diverge*, for example, see Fig. 9-11. For optimal control of this pathway, the levels of C and D as well as of B should be regulated. There are several ways this could be achieved, but the most common is called *isoenzymic control*; in this case the first enzyme (E_1) exists as two isoenzymes, one of which is inhibited by C and the other by D (hence the two arrows under E_1 in the diagram). However, if this were the only control, then if C were at a high level so that the C isoenzyme was completely inhibited, C would still be formed from B via the D isoenzyme. More effective control would be achieved if C and D also inhibit their respective enzymes at the branch point B.

Metabolic pathways are also controlled at the enzyme-synthesis level via enzyme *repression* and *derepression* (Chap. 17), which occur in response to variations of metabolite levels; however, the time scale of this control is hours to days, in contrast to less than a few seconds for direct binding effects. Finally, *molecular conversion*, namely, reversible covalent modification, of enzymes is another method of enzyme control. The best example of this is the *enzyme phosphorylation* and *dephosphorylation* that occurs in the control of glycogen synthesis and degradation (Chap. 11).

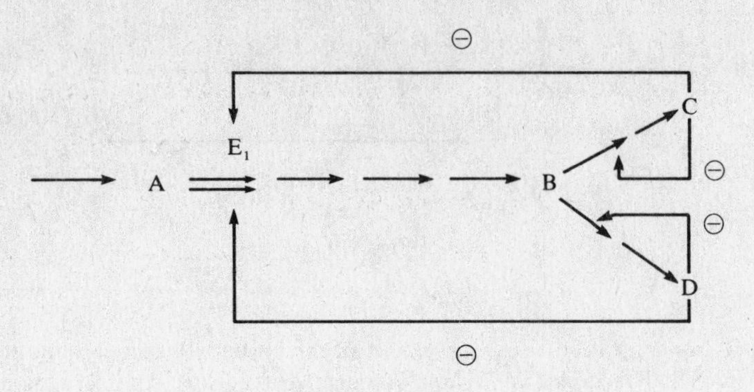

Fig. 9-11

Kinetic Behavior of Regulatory Enzymes

Usually effector molecules bear little structural resemblance to the substrates of the enzymes they control. The control is therefore not likely to be due to binding at the active site but at an alternative site, the *allosteric site*. The effect on the reaction at the active site is mediated by conformational changes in the protein.

If an effector of an enzyme is also the substrate, it is called a *homotropic* effector; if it is a nonsubstrate, it is called *heterotropic*.

Regulatory enzymes are usually identified by the deviation of their kinetics from Michaelis-Menten kinetics; plots of velocity versus substrate concentration can be a sigmoidal curve or a modified hyperbola [Fig. 9-12(*a*)]. If these curves are plotted in the double-reciprocal (Lineweaver-Burk) form, nonlinear graphs are obtained [Fig. 9-12(*b*)].

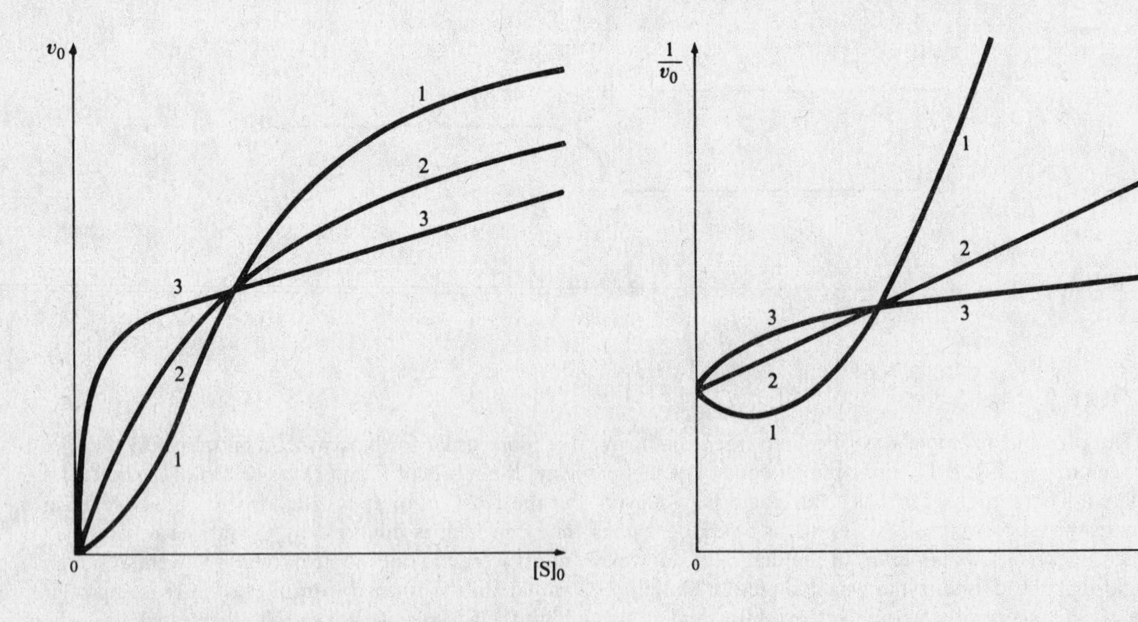

(*a*) Plots of v_0 versus $[S]_0$: (1) sigmoidal; (2) hyperbolic; (3) apparent hyperbolic

(*b*) The same data is plotted in double-reciprocal form. Note that only (2) is linear and therefore indicative of hyperbolic kinetics, while (1) is concave and (3) convex

Fig. 9-12 Possible kinetic behavior of regulatory enzymes.

A useful parameter for comparing regulatory enzymes is the ratio R_s:

$$R_s = \frac{\text{substrate concentration at 0.9 maximal velocity}}{\text{substrate concentration at 0.1 maximal velocity}} \tag{9.75}$$

For a Michaelis-Menten enzyme $R_s = 81$. For a sigmoidal curve, $R_s < 81$ and the enzyme is said to exhibit *positive cooperativity* with respect to the substrate. Positive cooperativity implies that the substrate binding or catalytic rate, or both, increases with increasing substrate concentration more than would be expected for a simple Michaelis-Menten enzyme. If $R_s > 81$, the enzyme is said to display *negative cooperativity* with respect to the substrate; substrate binding or catalysis, while increasing, becomes progressively less than would be found with a simple enzyme as substrate concentration is increased.

EXAMPLE 9.13

Hetero- and homotropic effects can also operate with binding proteins. The best example is the binding of O_2 to hemoglobin; the binding of one molecule of O_2 to hemoglobin *cooperates* in a *positive* way with the binding of the next molecule, so that the apparent affinity of hemoglobin for O_2 increases as the *degree of saturation* by O_2 increases. When the fractional saturation is plotted against the partial pressure of O_2 (equivalent to O_2 concentration), the curve is not hyperbolic but *sigmoidal* (Fig. 9-13).

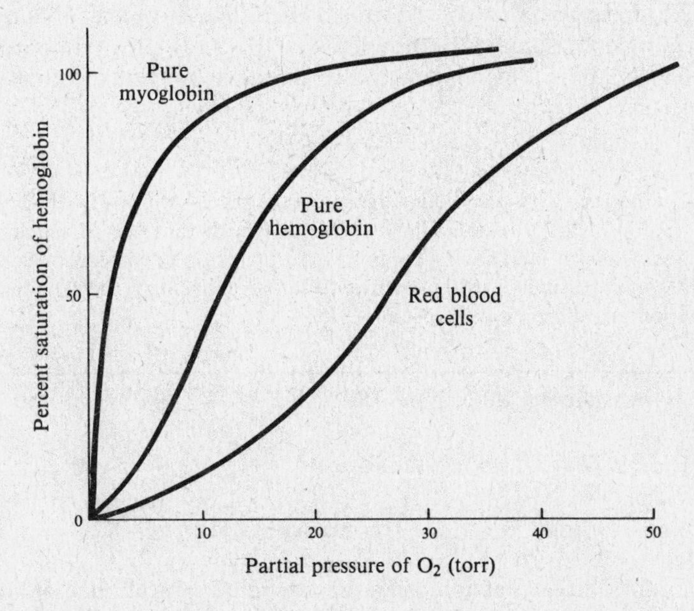

Fig. 9-13 Oxygen binding curves for hemoglobin and myoglobin.

In human red blood cells, the *heterotropic effector* 2,3-diphosphoglycerate reduces the affinity of all four binding sites of the hemoglobin tetramer for O_2 (Chap. 5). On the other hand, myoglobin, the O_2-binding protein of muscle, is a single subunit and does *not* display sigmoidal O_2-binding behavior, nor is it affected by 2,3-diphosphoglycerate.

Mathematical Models of Cooperativity

Although hemoglobin is not an enzyme, but an O_2-binding protein, the study of it has contributed a great deal to our understanding of molecular cooperativity. The cooperative O_2-binding behavior of hemoglobin was first recognized by Bohr (1903), long before the effect was seen with enzymes; much effort was expended in developing theories to explain the effect. Therefore, it is worth considering some of the earlier theories, which still have considerable relevance today.

The Hill Equation

In 1909 A. V. Hill proposed that the binding reaction between hemoglobin (Hb) and oxygen could be described by a reaction of molecularity $n + 1$:

$$\mathrm{Hb} + n\mathrm{O}_2 \rightleftharpoons \mathrm{Hb(O_2)}_n$$

where n is the number of O_2-binding sites on a hemoglobin molecule. The equation for the fractional saturation Y (the fraction of binding sites occupied by O_2 at any instant) is

$$Y = \frac{K_b(p\mathrm{O}_2)^n}{1 + K_b(p\mathrm{O}_2)^n} \tag{9.76}$$

where pO_2 is the partial pressure of O_2 (torr) and K_b is the *association binding constant*, defined as

$$K_b = \frac{[\mathrm{Hb(O_2)}_n]}{[\mathrm{Hb}](p\mathrm{O}_2)^n} \tag{9.77}$$

Note that Y is a dimensionless quantity that has the range 0–1; n is called the *Hill coefficient*. For values of n around 2.5, a sigmoidal curve akin to those in Fig. 9-13 was obtained by Hill. The fact that n is not an integer raised some early problems in the mechanistic interpretations of the data analyzed with the Hill equation. This enigma led to the development of other mechanistic models.

EXAMPLE 9.14

If $n = 1$ in the Hill equation, what form does the equation take?

When $n = 1$, Eq. (*9.76*) yields a rectangular hyperbola. Recall that the Michaelis-Menten expression is written with K_m being equivalent to (and having the units of) a *dissociation* constant, whereas binding equations, such as the Hill equation, are usually written with *association* constants (their numerical values being the reciprocal of the corresponding dissociation constants).

Question: How is it possible to determine the value of K_b and n in the Hill equation from experimental data?

By rearranging Eq. (*9.77*) and taking the logarithm of both sides, we get

$$\log\left[\frac{[\mathrm{Hb(O_2)}_n]}{[\mathrm{Hb}]}\right] = \log K_b + n\log(p\mathrm{O}_2) \tag{9.78}$$

Or, a more general form of the expression using the fractional saturation Y of a binding protein with substrate S is

$$\log\left(\frac{Y}{1-Y}\right) = \log K_b + n\log[\mathrm{S}] \tag{9.79}$$

A plot of the left-hand side of Eq. (*9.79*) versus log [S] is called a *Hill plot*; it yields an estimate of n from the slope, and K_b from the ordinate intercept.

EXAMPLE 9.15

For an *enzymatic* reaction (in contrast to a *binding* reaction), the initial velocity v_0 is determined by the concentration of the enzyme-substrate complex; therefore, what form does equation (*9.79*) take?

A fractional saturation of unity corresponds to V_{max}, and thus the Hill plot consists of $\log\left(\frac{v_0}{V_{max} - v_0}\right)$ versus $\log[\mathrm{S}]_0$. Clearly, an initial estimate of V_{max} must be made before analyzing the data this way.

The Adair Equation

G. S. Adair in 1925 determined that the molecular weight of hemoglobin was about four times as great as previously thought. He postulated that hemoglobin had four O_2-binding sites that were filled in a four-step process as follows:

$$\begin{aligned}
Hb + O_2 &\rightleftharpoons HbO_2 \\
HbO_2 + O_2 &\rightleftharpoons Hb(O_2)_2 \\
Hb(O_2)_2 + O_2 &\rightleftharpoons Hb(O_2)_3 \\
Hb(O_2)_3 + O_2 &\rightleftharpoons Hb(O_2)_4
\end{aligned}$$

Consider *one* binding site on a totally empty (no sites filled) hemoglobin molecule; the binding reaction for this site can be characterized by the *association* equilibrium constant $K_1 = k_1/k_{-1}$, where k_1 and k_{-1} are the unitary rate constants for the forward and reverse reactions, respectively. This equilibrium constant is called an *intrinsic* constant since it refers to one site only. The *overall*, or *extrinsic*, binding constant relates to all four of the binding sites and thus has the value $4K_1$. The extrinsic binding constants for the other three sites are also expressed in terms of *intrinsic* binding constants and their so-called *statistical factors*: $\frac{3}{2}K_2$, $\frac{2}{3}K_3$, and $\frac{1}{4}K_4$. Thus, the Adair binding function Y for four sites is expressed as follows:

$$Y = \frac{\text{number (or moles) of binding sites occupied}}{\text{total number (or moles) of binding sites}} \tag{9.80}$$

$$= \frac{K_1[X] + 3K_1K_2[X]^2 + 3K_1K_2K_3[X]^3 + K_1K_2K_3K_4[X]^4}{1 + 4K_1[X] + 6K_1K_2[X]^2 + 4K_1K_2K_3[X]^3 + K_1K_2K_3K_4[X]^4} \tag{9.81}$$

where $[X] = pO_2$ in the particular case of hemoglobin binding.

EXAMPLE 9.16

An alternative way of viewing the concept of extrinsic binding constants is as follows. The experimentally *measured* binding constant for each site of a polymeric protein, or enzyme, will depend on the number of available sites on each molecule; e.g., for the hemoglobin tetramer the first binding reaction is

$$Hb + O_2 \underset{k_{-1}}{\overset{k_1}{\rightleftharpoons}} HbO_2$$

There are four sites available for binding O_2, but only one from which bound O_2 can dissociate. Thus, from the law of mass action, the (overall) extrinsic equilibrium constant is equal to $4k_1/k_{-1} = 4K_1$. Similarly, the *extrinsic* constant denoted by K_2^e is given by $\frac{3}{2}K_2$, where K_2 is the intrinsic constant, and by the same reasoning $K_3^e = \frac{2}{3}K_3$ and $K_4^e = \frac{1}{4}K_4$.

Question: If the hemoglobin tetramer had four *identical noninteracting* binding sites, what relative values would the intrinsic binding constants have?

They would all be equal, i.e., $K_1 = K_2 = K_3 = K_4$. There would thus be no cooperativity in this case (see Example 9.17). If each binding step facilitated the next, i.e., $K_1 < K_2 < K_3 < K_4$, then positive cooperativity would exist. Negative cooperativity at each stage in the binding requires that $K_1 > K_2 > K_3 > K_4$. Clearly, more complex relationships between the values of the K's *could* exist; e.g., $K_1 > K_2 < K_3 < K_4$.

EXAMPLE 9.17

Show that if $K_1 = K_2 = K_3 = K_4$, the four-site Adair equation becomes that of a rectangular hyperbola.

Note that the statistical factors in the numerator and denominator of the four-site Adair equation, and in fact in the general n-site equation, are the coefficients of the *binomial expansion*, i.e., the coefficients of the expansion of $(1-x)^n$. In the present case with $n = 4$, Eq. (*9.81*) applies and

$$Y = \frac{K_1[\mathrm{X}] + 3K_1^2[\mathrm{X}]^2 + 3K_1^3[\mathrm{X}]^3 + K_1^4[\mathrm{X}]^4}{1 + 4K_1[\mathrm{X}] + 6K_1^2[\mathrm{X}]^2 + 4K_1^3[\mathrm{X}]^3 + K_1^4[\mathrm{X}]^4} \tag{9.82}$$

$$= \frac{K_1[\mathrm{X}](1 + K_1[\mathrm{X}])^3}{(1 + K_1[\mathrm{X}])^4} = \frac{K_1[\mathrm{X}]}{1 + K_1[\mathrm{X}]} \tag{9.83}$$

The final expression is that of a hyperbola; it is evident that it is equivalent to the Michaelis-Menten expression when it is recalled that K_1 is an *association* constant.

The Model of Monod, Wyman, and Changeux (MWC Model) for Homotropic Effects

The cooperative binding of O_2 by hemoglobin and the allosteric effects in many enzymes require interaction between sites that are widely separated in space. The MWC model was proposed in 1965 to incorporate allosteric and conformational effects in an explanation of enzyme cooperativity. The seminal observation was that most cooperative proteins have several identical subunits (*protomers*) in each molecule (*oligomer*); this situation is imperative for binding cooperativity but *not* for *kinetic cooperativity* (see "Other Models of Cooperative Behavior" later in this section and Prob. 9.33). The MWC model is defined as follows:

1. Each protomer can exist in either of two conformations, designated R and T; these originally referred to *relaxed* and *tense*, with the latter not being catalytically active.
2. All subunits of one oligomer *must* occupy the *same* conformation; hence, for a tetrameric protein, the conformational states R_4 and T_4 are the only ones permitted. Mixed states such as R_3T are forbidden. In other words, a *concerted* (concert: "all together") transition takes place in the conversion of R_4 to T_4 and vice versa.
3. The two states of the protein are in equilibrium *independent* of whether any ligand (substrate) is bound; for a tetramer the equilibrium constant is

$$L = [\mathrm{T}_4]/[\mathrm{R}_4] \tag{9.84}$$

4. A ligand molecule can bind to a subunit in either conformation, but the association constants are different. For each R subunit,

$$K_\mathrm{R} = [\mathrm{RX}]/[\mathrm{R}][\mathrm{X}] \tag{9.85}$$

and for each T subunit,

$$K_\mathrm{T} = [\mathrm{TX}]/[\mathrm{T}][\mathrm{X}] \tag{9.86}$$

These postulates imply the following multiple equilibrium scheme for the protein and ligand (X):

$$\begin{array}{rcccl}
 & \mathrm{R}_4 & \overset{L}{\rightleftharpoons} & \mathrm{T}_4 & \\
4K_\mathrm{R}[\mathrm{X}] & \downarrow\uparrow & & \downarrow\uparrow & 4K_\mathrm{T}[\mathrm{X}] \\
 & \mathrm{R}_4\mathrm{X} & \rightleftharpoons & \mathrm{T}_4\mathrm{X} & \\
\frac{3}{2}K_\mathrm{R}[\mathrm{X}] & \downarrow\uparrow & & \downarrow\uparrow & \frac{3}{2}K_\mathrm{T}[\mathrm{X}] \\
 & \mathrm{R}_4\mathrm{X}_2 & \rightleftharpoons & \mathrm{T}_4\mathrm{X}_2 & \\
\frac{2}{3}K_\mathrm{R}[\mathrm{X}] & \downarrow\uparrow & & \downarrow\uparrow & \frac{2}{3}K_\mathrm{T}[\mathrm{X}] \\
 & \mathrm{R}_4\mathrm{X}_3 & \rightleftharpoons & \mathrm{T}_4\mathrm{X}_3 & \\
\frac{1}{4}K_\mathrm{R}[\mathrm{X}] & \downarrow\uparrow & & \downarrow\uparrow & \frac{1}{4}K_\mathrm{T}[\mathrm{X}] \\
 & \mathrm{R}_4\mathrm{X}_4 & \rightleftharpoons & \mathrm{T}_4\mathrm{X}_4 &
\end{array}$$

The concentrations of the 10 forms of the protein are related by the following equilibrium expressions, in which c is the ratio of $K_\mathrm{T}/K_\mathrm{R}$.

$$
\begin{aligned}
&[R_4X] = 4K_R[R_4][X] \\
&[R_4X_2] = \tfrac{3}{2}K_R[R_4X][X] = 6K_R^2[R_4][X]^2 \\
&[R_4X_3] = \tfrac{2}{3}K_R[R_4X_2][X] = 4K_R^3[R_4][X]^3 \\
&[R_4X_4] = \tfrac{1}{4}K_R[R_4X_3][X] = K_R^4[R_4][X]^4 \\
&[T_4] = L[R_4] \\
&[T_4X] = 4K_T[T_4][X] = 4LcK_R[R_4][X] \\
&[T_4X_2] = 6K_T^2[T_4][X]^2 = 6Lc^2K_R^2[R_4][X]^2 \\
&[T_4X_3] = 4K_T^3[T_4][X]^3 = 4Lc^3K_R^3[R_4][X]^3 \\
&[T_4X_4] = K_T^4[T_4][X]^4 = Lc^4K_R^4[R_4][X]^4
\end{aligned}
\tag{9.87}
$$

Again, the statistical factors 4, $\frac{3}{2}$, etc., arise because K_R and K_T are *intrinsic* association binding constants, yet the overall expressions for the complexes require *extrinsic* parameters; e.g., $K_R = [RX]/[R][X] = \frac{3}{2}[R_4X_2]/[R_4X][X]$ because there are three unfilled sites in the R_4X molecule and two in each R_4X_2 molecule (see also Example 9.16). The fractional saturation Y is again defined by Eq. (*9.80*):

$$
Y = \frac{[R_4X] + 2[R_4X_2] + 3[R_4X_3] + 4[R_4X_4] + [T_4X] + 2[T_4X_2] + 3[T_4X_3] + 4[T_4X_4]}{4([R_4] + [R_4X] + [R_4X_2] + [R_4X_3] + [R_4X_4] + [T_4] + [T_4X] + [T_4X_2] + [T_4X_3] + [T_4X_4])} \tag{9.88}
$$

$$
= \frac{(1 + K_R[X])^3K_R[X] + Lc(1 + cK_R[X])^3K_R[X]}{(1 + K_R[X])^4 + L(1 + cK_R[X])^4} \tag{9.89}
$$

The general n-site MWC model with $K_R[X] = \alpha$ is

$$
Y = \frac{\alpha(1+\alpha)^{n-1} + Lc\alpha(1 + c\alpha)^{n-1}}{(1+\alpha)^n + L(1 + c\alpha)^n} \tag{9.90}
$$

The behavior of this expression for various ranges of parameter values is shown in Fig. 9-14.

EXAMPLE 9.18

The shape of the saturation curve defined by Eq. (*9.90*) depends on the values of L and c. If $L = 0$, then the T form of the protein does not exist and $Y = K_R[X]/(1 + K_R[X])$. This defines a hyperbolic binding function. Similarly if $L = \infty$, $Y = K_T[X]/(1 + K_T[X])$. Thus, deviations from hyperbolic binding occur only if both R and T forms exist; otherwise the situation described for the Adair equation in Example 9.17 applies since binding is *independent* and *identical* at each site.

EXAMPLE 9.19

The fact that Eq. (*9.90*), or (*9.89*), defines sigmoidal curves is not obvious unless $c = 0$. In this case, for $n = 4$,

$$
Y = \frac{(1 + K_R[X])^3K_R[X]}{L + (1 + K_R[X])^4} \tag{9.91}
$$

Hence, when [X] is large and $K_R[X] \gg L$, the term in L is negligible and $Y \simeq \dfrac{K_R[X]}{1 + K_R[X]}$; i.e., the hyperbolic binding equation is the result. At *low* values of [X], the term in L dominates the denominator and the slope of the binding curve as $[X] \to 0$ approaches $K_RL/(L+1)^2$, in contrast to the slope of the hyperbola, which approaches K_R. In symbols,

$$
\lim_{[X]\to 0} \frac{dY}{d[X]} = \frac{K_RL}{(L+1)^2} \tag{9.92}
$$

and the curve *must* be sigmoidal if L is large compared with 1; in the limit of infinite L, the initial slope is zero.

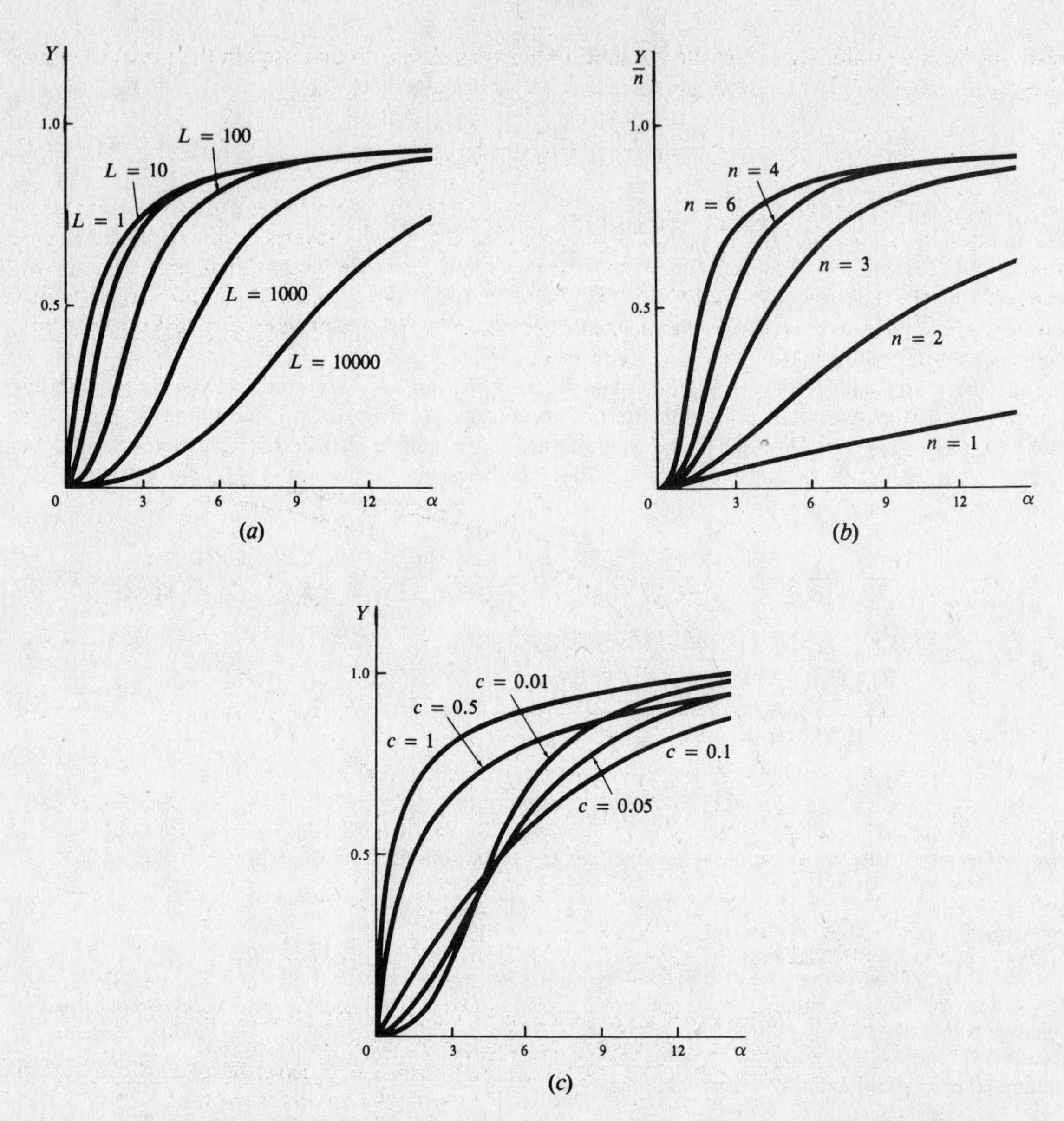

Fig. 9-14 Binding curves described by the MWC equation (*9.90*). (*a*) The effect of varying the value of the isomerization constant (L) with $n = 4$ and $c = 0.01$. (*b*) The effect of varying the number of binding sites (n) with $L = 100$ and $c = 0.01$. (*c*) The effect of varying c, the ratio of K_T/K_R, with $L = 1{,}000$ and $n = 4$.

The MWC Model for Heterotropic Effects

The MWC model of the previous section accounts only for *homotropic* effects, as occur, for example, with O_2 binding to hemoglobin, but *heterotropic* effects are frequently observed with regulatory enzymes. To incorporate these effects into the MWC model, we must introduce the binding of a second ligand to either the R or the T state, or to both. Let us consider a simplified model and thus assume that the substrate X (ligand) binds only to the R state. For the binding of another ligand, A, to bring about activation of the enzyme, it must promote a shift of the equilibrium between the R and T states in favor of the R state. An activator must therefore bind to the R state. By the same reasoning an inhibitor, I, must bind to the T state and thus cause a shift of the enzyme into this inactive state. By a multiple equilibrium analysis similar to that used in the previous section the expression for Y is obtained:

$$Y = \frac{\alpha(1+\alpha)^{n-1}}{(1+\alpha)^n + [L(1+\beta)^n/(1+\gamma)^n]} \tag{9.93}$$

where $\alpha = K_R[X]$, $\beta = K_I[I]$, $\gamma = K_A[A]$, and $L = [T_n]/[R_n]$. The term $L(1+\beta)^n/(1+\gamma)^n$ is regarded as an *allosteric coefficient*; if this term is zero, then Eq. (*9.93*) reverts to a hyperbolic binding function. Clearly, an increase in the activator term γ *decreases* the value of the allosteric coefficient, thus making the function more hyperbolic (see Example 9.19). An increase of inhibitor concentration increases the inhibition term (β), thus increasing the allosteric coefficient and making the curve more sigmoidal. Both these effects are seen in Fig. 9-15.

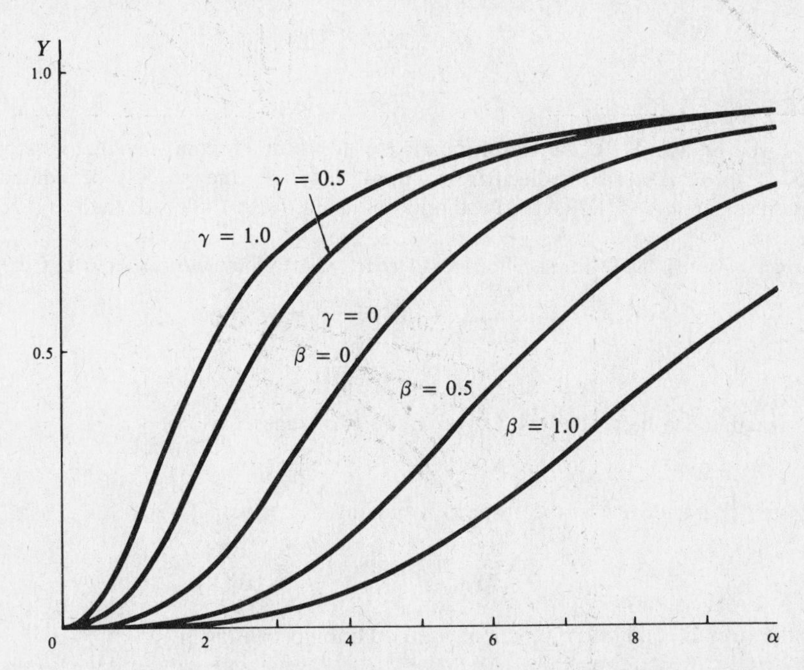

Fig. 9-15 Behavior of an MWC allosteric enzyme in the presence of positive and negative heterotropic effectors. The activator term (γ) in Eq. (*9.93*) causes the curve to become more hyperbolic, whereas the inhibitor term (β) renders it more sigmoidal. The curves were constructed using Eq. (*9.93*) with $L = 1{,}000$ and $n = 4$.

The Model of Koshland, Nemethy, and Filmer (KNF Model)

The failure of the MWC model to describe negatively cooperative hetero- and homotropic effects in the binding function of an oligomeric protein led Koshland et al. (1966) to develop a more general model of cooperativity. Koshland's 1958 *induced-fit hypothesis* for enzyme specificity extended Fischer's *lock-and-key* concept. Koshland claimed that the binding of substrate to an enzyme *creates* the correct three-dimensional arrangement of reactive groups for catalysis to occur. This abstract notion was extended in the KNF model of oligomeric cooperative enzymes; in this case, the change in conformation is *induced* (in contrast to the MWC model, where equilibrium between states of the subunits exists whether substrate is bound or not) by the binding of the substrate to a protomer (subunit). A change in the conformation in the protomer is transmitted to neighboring protomers to affect their binding and catalytic properties. This basic idea can explain both negative *and* positive cooperativity.

The KNF binding function is of the same form as the Adair equation, namely, a *ratio of polynomials* in [X]; but the coefficients of each term in $[X]^i$, where $i = 1, \ldots, n$, have a new mechanistic interpretation that depends on the following model:

1. The enzyme or binding protein is composed of *identical protomers*, which exist in either of two states, A or B; this interaction (isomerization) $A \rightleftharpoons B$ is described by the parameter

$$K_t = [B]/[A] \tag{9.94}$$

2. In the absence of substrate binding, the protomers remain in the A conformation; i.e., generally it is assumed that $K_t \ll 1$.
3. Any protomer that binds the substrate changes to the B state; the association binding constant for the interaction $B + X \rightleftharpoons BX$ is

$$K_x = [BX]/[B][X] \tag{9.95}$$

EXAMPLE 9.20

It may well be asked, "If the transformation of a protomer from state A to state B requires the binding of X to A (i.e., a bimolecular reaction), how can the process be characterized by the isomerization constant $K_t = [B]/[A]$?" Furthermore, in Eq. (*9.95*), K_x is defined as if X binds to B and not to A.

The explanation is as follows: The overall reaction of X with A to yield BX can be written as

$$A + X \rightleftharpoons BX$$

with

$$K = \frac{[BX]}{[A][X]}$$

Also, this reaction can be written as if it occurs in two stages

$$A + X \rightleftharpoons B + X \rightleftharpoons BX$$

with each step characterized by the respective binding constants

$$K_t = \frac{[B]}{[A]} \quad \text{and} \quad K_x = \frac{[BX]}{[B][X]}$$

And, for the *overall* equilibrium reaction, the equilibrium constant is given by $K = K_t K_x = \dfrac{[BX]}{[A][X]}$.

In words, the Koshland analysis merely highlights the presence of binding *and* conformational changes by assigning separate energy (binding) terms to each. Also, note, as will be evident later, that in the KNF equation, K_t and K_x always occur together, so in fact they really constitute only *one* binding constant.

4. Between any two protomers there are three possible strengths of interaction characterized by *intrinsic* binding (interaction) constants:

$$A + B \rightleftharpoons AB \qquad K_{AB} = \frac{[AB]}{[A][B]}$$

$$A + A \rightleftharpoons AA \qquad K_{AA} = \frac{[AA]}{[A]^2}$$

$$B + B \rightleftharpoons BB \qquad K_{BB} = \frac{[BB]}{[B]^2}$$

The strength of interaction between two protomers thus depends on whether zero, one, or two substrate molecules are bound to the pair. Koshland et al. assigned K_{AA} to unity, so that if $K_{AB} > 1$, the interaction between A and B is more stable than between A and A. In this case, the *interaction* will stabilize the complex ABX, and thus enhance binding of the substrate.

EXAMPLE 9.21

Describe the involvement of the three classes of KNF interaction parameters (binding constants) K_t, K_x, and K_{AA}, etc., in the binding of X to a dimeric protein:

$$\bigcirc\bigcirc + \mathrm{X} \rightleftharpoons \bigcirc\boxed{\mathrm{X}}$$

$$\mathrm{AA} \qquad\qquad \mathrm{ABX}$$

The equilibrium constant for this reaction (K_1) is the product of K_t, K_x, and the ratio of the subunit interaction constants K_{AB}/K_{AA}. Thus,

$$K_1 = \frac{2K_tK_xK_{AB}}{K_{AA}}$$

The statistical factor 2 is included because there are *two* forms of complex (A · BX and BX · A) and only *one* form of AA. Similarly, for the binding of the second molecule of substrate:

$$\bigcirc\boxed{\mathrm{X}} + \mathrm{X} \rightleftharpoons \boxed{\mathrm{X}}\boxed{\mathrm{X}} \qquad \text{and} \qquad K_2 = \frac{K_tK_xK_{BB}}{2K_{AB}}$$

$$\mathrm{ABX} \qquad\qquad \mathrm{BBX_2}$$

Since $K_{AA} = 1$ (by definition), the binding of the second molecule of X will be more avid than the first ($K_2 > K_1$) if the $2K_{AB} < \frac{1}{2}K_{BB}/K_{AB}$ (positive cooperativity). If $2K_{AB} > \frac{1}{2}K_{BB}/K_{AB}$, negative cooperativity pertains.

5. **The value of the parameter that defines the interaction between a particular protomer and another one depends on the spatial *arrangement* of the protomers.**

EXAMPLE 9.22

Koshland et al. showed that for a tetrameric protein, the interaction between protomers can be described *as if* the protomers are arranged as (1) tetrahedral, (2) square, or (3) linear configurations. A special case, equivalent to the MWC model, is where there is a concerted transition of A to B as soon as one molecule of X binds to A_4. See Fig. 9-16.

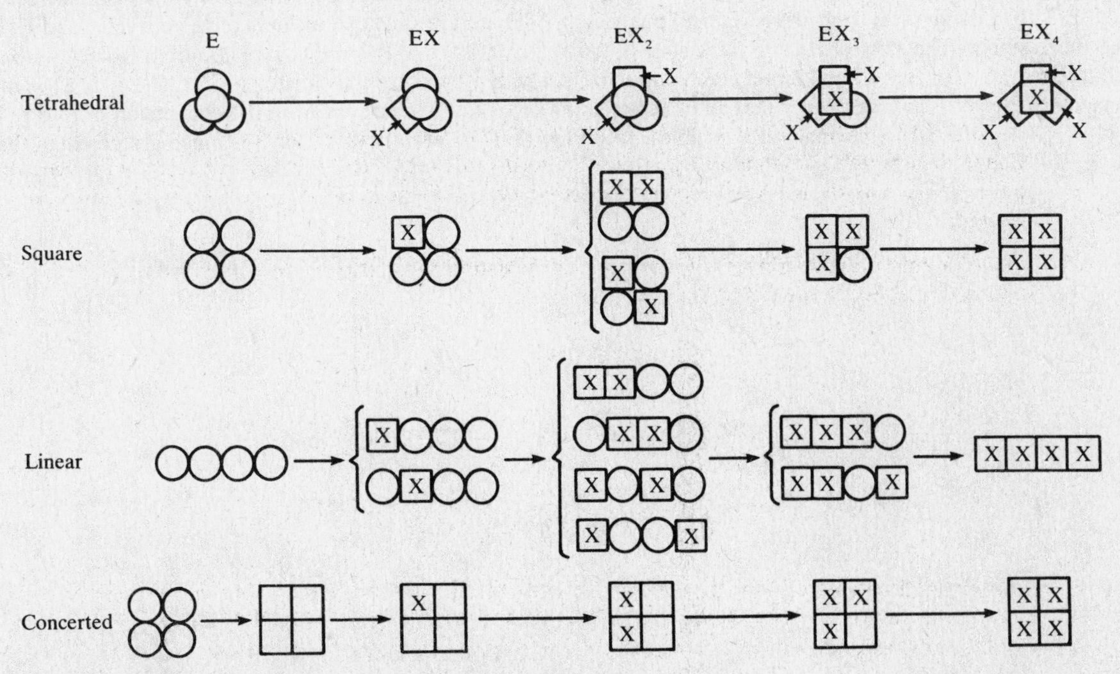

Fig. 9-16 The three types of KNF interaction models and the special concerted case. The circles denote the A state and the squares the B state of the protomers.

The KNF binding equation appears to be extremely complex (see Probs. 9.32 and 9.50). However, using an analysis like that in Example 9.21, one can show that positive or negative cooperativity in an n-protomer protein can arise from this model. But it is interesting, and important, to note that only *pure positive* or *pure negative* cooperativity (i.e., $K_1 > K_2 > K_3 > K_4$ or $K_1 < K_2 < K_3 < K_4$) can be described by the model as presented here. More complex relationships with *mixed cooperativity* require even more complex models.

Other Models of Cooperative Behavior

(*a*) *Protein polymerization*. An extension of the MWC model, made by Frieden and by Nichol, Jackson, and Winzor in 1967, was to propose that the protomers can actually dissociate from the oligomers; and in the two forms, the protomers have different affinities for the substrate.

EXAMPLE 9.23

It appears that the polymerization model provides a complete explanation of the cooperativity of binding various nucleotides to glutamate dehydrogenase.

Protein polymerization may also *contribute* to the cooperativity of hemoglobin, which dissociates to a dimer in high salt concentrations, but the model does not describe the cooperativity completely since cooperativity is observed under conditions when no dissociation occurs.

(*b*) *Kinetic cooperativity*. All the previous models in this section have been essentially *equilibrium models*. They can be applied to kinetic experiments by assuming that $v_0 = V_{max} Y$. However, cooperativity can arise for purely *kinetic* reasons in mechanisms that would show no cooperativity if binding could be measured at equilibrium; the cooperativity can even, in principle, arise in single-subunit enzymes, but few real examples exist.

EXAMPLE 9.24

A simple enzyme model that displays kinetic cooperativity was proposed by Rabin (1967); it is as shown in Fig. 9-17. The initial reaction entails the formation of the enzyme-substrate complex E′S, which isomerizes to give E″S. E″S then reacts to give free E″ and product(s) P. There are thus two isomers of the enzyme, and E′ is the more stable. The slowest, *rate-limiting* step in the reaction is k_2. Also, the conversion of E″ to E′ is *slow*. Thus, generation of free E″ and direct combination of this with S bypasses the slowest step (E′S ⟶ E″S) in the reaction pathway. The free energy that allows conversion of E′ to E″ is derived from the conversion of S to P. If $k_2 \gg k_{-2}$ (i.e., this step is virtually irreversible), it follows that the affinity of E″ for S is much greater than that of E′ for S.

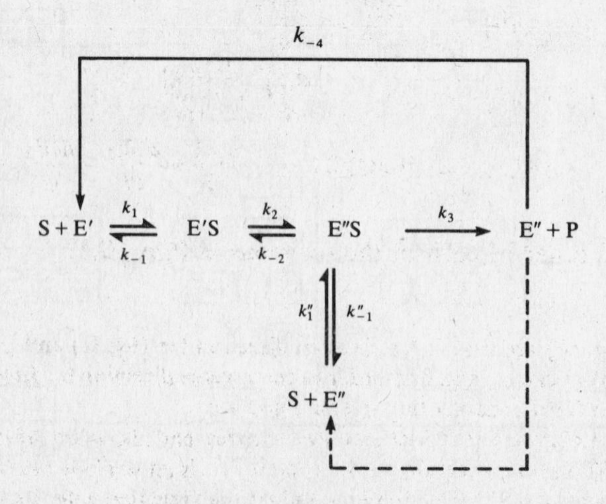

Fig. 9-17

Question: How does kinetic cooperativity arise in the Rabin enzyme model?

At low [S], [E′S] is low and the rate of conversion of E′S to E″S (which is proportional to [E′S]) will be slow compared with the rate of conversion of E″ to E′ (i.e., k_{-4}); thus, the concentration of the high-affinity form of the enzyme, E″, is low, and hence the reaction rate is low. At higher [S] values, [E′S] will be higher, and the conversion rate of E′S to E″S will be greater, resulting in an increase in [E″]. Thus the likelihood of E″ combining with S before being converted back to E′ is increased, and so the rate of catalysis is increased to an extent greater than expected for a simple enzyme.

Another way of viewing the model is to consider that there is cooperation between one S molecule and the next to bind to the enzyme; this does not occur spatially but by virtue of the fact that the enzyme "remembers" the first molecule. It is an interaction transmitted *through time* rather than *space*.

Solved Problems

INTRODUCTION AND DEFINITIONS

9.1. Use the principle of mass action to write expressions for the rate equations for the following reactions:

(*a*) $A \xrightarrow{k_1} P$ (*d*) $A + B \underset{k_{-1}}{\overset{k_1}{\rightleftharpoons}} P$

(*b*) $A \underset{k_{-1}}{\overset{k_1}{\rightleftharpoons}} P$ (*e*) $A + B \underset{k_{-1}}{\overset{k_1}{\rightleftharpoons}} P + Q + R$

(*c*) $A + B \xrightarrow{k_1} P$

SOLUTION

(*a*) $\dfrac{d[A]}{dt} = -k_1[A] \qquad \dfrac{d[P]}{dt} = -\dfrac{d[A]}{dt}$

(*b*) $\dfrac{d[A]}{dt} = -k_1[A] + k_{-1}[P] \qquad \dfrac{d[P]}{dt} = -\dfrac{d[A]}{dt}$

(*c*) $\dfrac{d[A]}{dt} = \dfrac{d[B]}{dt} = -k_1[A][B] \qquad \dfrac{d[P]}{dt} = -\dfrac{d[A]}{dt}$

(*d*) $\dfrac{d[A]}{dt} = \dfrac{d[B]}{dt} = -k_1[A][B] + k_{-1}[P] \qquad \dfrac{d[P]}{dt} = -\dfrac{d[A]}{dt}$

(*e*) $\dfrac{d[A]}{dt} = \dfrac{d[B]}{dt} = -k_1[A][B] + k_{-1}[P][Q][R] \qquad \dfrac{d[P]}{dt} = \dfrac{d[Q]}{dt} = \dfrac{d[R]}{dt} = -\dfrac{d[A]}{dt}$

9.2. What is the kinetic order of each of the reactions in Prob. 9.1?

SOLUTION

Reactions (*a*) and (*b*) are first-order, in each direction for (*b*); (*c*) and (*d*) are second-order in the *forward* direction (left to right), and first-order in the reverse direction for (*d*); (*e*) is second-order in the forward direction and third-order in the reverse direction.

Note: For the uncatalyzed reactions, the *molecularity* and the *order* have been given as being the same. However, a trimolecular reaction, as depicted in (*e*), is an unlikely occurrence; what is more likely is the interaction between two of the reactants, *followed* by reaction with the third. This amounts to two second-order reactions in sequence.

9.3. What are the *units* of the *rate constants* in the reaction schemes of Prob. 9.1?

SOLUTION

(*a*) k_1: s^{-1}; (*b*) k_1 and k_{-1}: s^{-1}; (*c*) k_1: $mol^{-1}\,L\,s^{-1}$; (*d*) k_1: $mol^{-1}\,L\,s^{-1}$, and k_{-1}: s^{-1}; (*e*) k_1: $mol^{-1}\,L\,s^{-1}$, and k_{-1}: $mol^{-2}\,L^2\,s^{-1}$.

9.4. The data in Table 9.2 were obtained for the *rate* of a reaction that has the stoichiometry $aA + bB \longrightarrow pP$; the rate was determined with various concentrations of A and B. Determine the order of the reaction; i.e., estimate the values of a and b.

Table 9.2. Reaction Order Data: v_0 (μmol L^{-1} s^{-1})

		$[A]_0$ (mmol L^{-1})			
		10	20	50	100
$[B]_0$ (mmol L^{-1})	10	1.2	2.0	2.8	3.9
	20	2.6	4.0	5.9	7.9
	50	6.5	8.8	14.5	19.8
	100	12.5	17.9	29.0	40.5

SOLUTION

The order of a reaction is an *experimentally* determined parameter, and the simplest way of obtaining it is to measure the initial reaction rate for a range of concentrations of reactants. Then a plot of log(rate) versus log(concentration) of one of the reactants, while holding the rest constant, gives a line with a slope equal to the order. If the concentrations of all reactants are varied in a constant ratio, then the slope of the graph is the *overall* reaction order; if just one is varied, the order refers only to *that* reactant. Hence, from the data in Table 9.2, a plot of log(diagonal elements) versus log(concentration) yields Fig. 9-18(*a*).

The slope in Fig. 9-18(*a*) is $1\frac{1}{2}$, which is therefore the overall order of the reaction. A family of four order plots is obtained for variable $[A]_0$ and constant $[B]_0$ *and* vice versa [Fig. 9-18(*b*) and (*c*)]. From the slopes of these plots, we see that a—the order of A—is $\frac{1}{2}$ and b—the order of B—is 1.

DEPENDENCE OF ENZYME REACTION RATE ON SUBSTRATE CONCENTRATION

9.5. (*a*) Use the Michaelis-Menten equation to complete the enzyme kinetic data set; the K_m is known to be 1 mmol L^{-1}. (*b*) Draw a graph of your results and determine the reaction order.

$[S]_0$ (mmol L^{-1})	v_0 (μmol L^{-1} min^{-1})
0.5	50
1.0	—
2.0	—
3.0	—
10.0	—

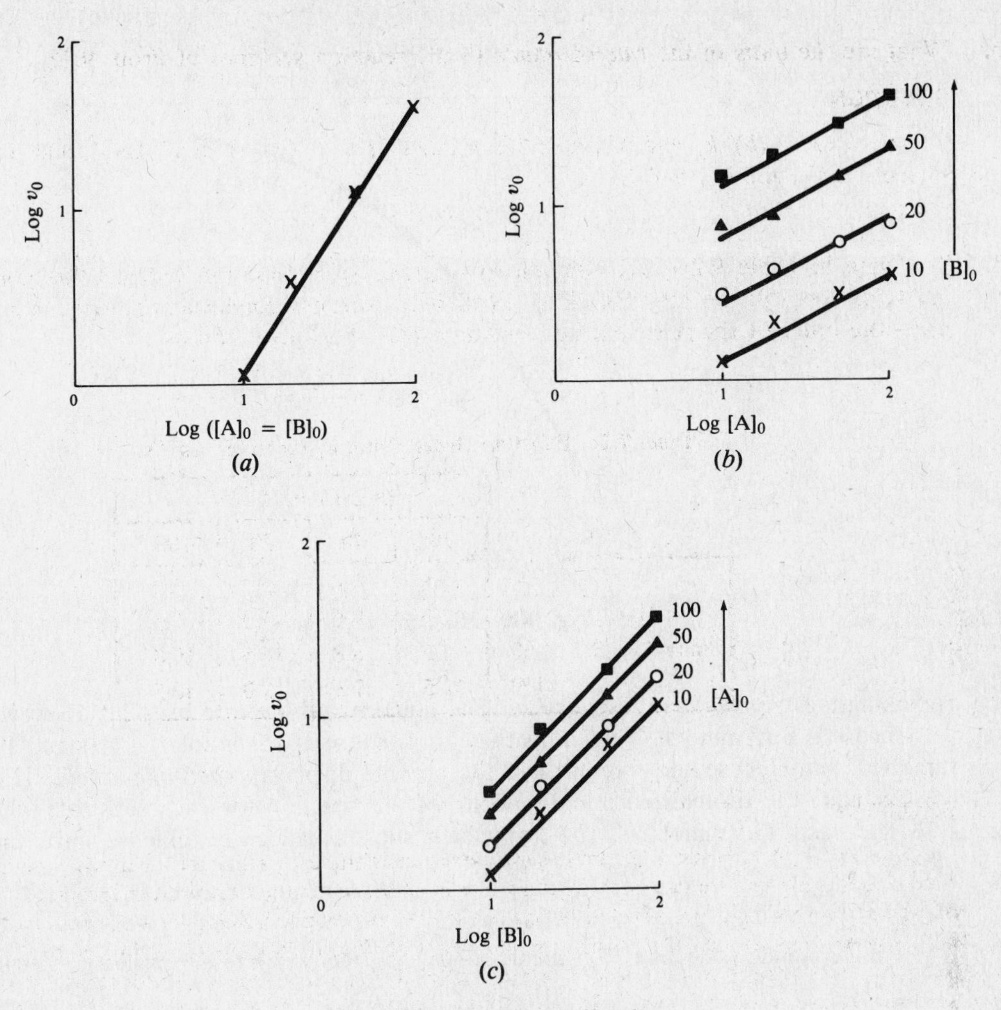

Fig. 9-18 Reaction order plots.

SOLUTION

(*a*) Using Eq. (*9.5*) and the first entry in the table gives $V_{max} = 150\ \mu\text{mol L}^{-1}\ \text{min}^{-1}$. The other entries simply follow by substituting the $[S]_0$ values into Eq. (*9.5*); the results are shown below.

$[S]_0$ (mmol L^{-1})	v_0 (μmol L^{-1} min^{-1})
0.5	50
1.0	75
2.0	100
3.0	112.5
10.0	136.4

(*b*) The order of the reaction is determined by plotting the log of normalized values to $[S]_0$ and v_0, that is, $\log([S]_0/K_m)$ versus $\log(v_0/V_{max})$; see Fig. 9-19. Enzyme-catalyzed reactions have variable order; in this case, at low values of $[S]_0/K_m$, the order is 1; it is 0 at very high $[S]_0/K_m$ values and is intermediate between 0 and 1 for intermediate values of $[S]_0/K_m$.

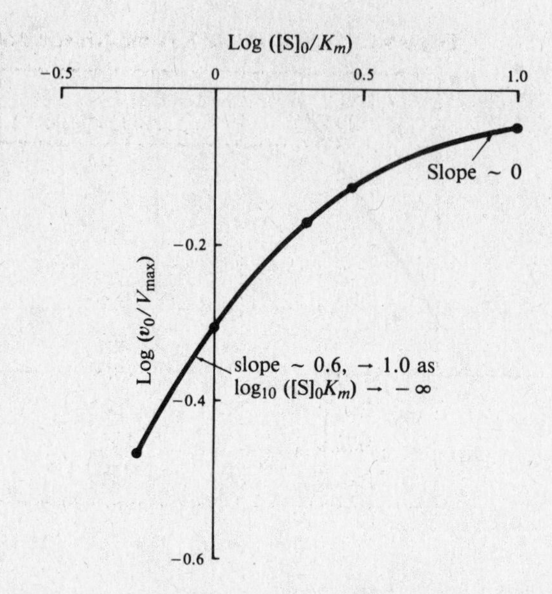

Fig. 9-19 Reaction order plot.

9.6. Hexokinase catalyzes the phosphorylation of glucose and fructose by ATP. However the K_m for glucose is 0.13 mmol L^{-1}, whereas that for fructose is 1.3 mmol L^{-1}. Assume V_{max} is the same for both glucose and fructose and the enzyme displays *hyperbolic kinetics* [Eq. (*9.5*)]. (*a*) Calculate the normalized initial velocity of the reaction for each substrate when $[S]_0 =$ 0.13, 1.3, and 13.0 mmol L^{-1}. (*b*) For which substrate does hexokinase have the greater affinity?

SOLUTION

(*a*) For glucose the values of v_0/V_{max} are 0.5, 0.91, and 0.99; for fructose the values are 0.091, 0.5, and 0.91.

(*b*) Glucose; at lower concentrations, the reaction rate is a greater fraction of V_{max} than it is with fructose.

GRAPHICAL EVALUATION OF K_m AND V_{max}

9.7. A constant amount of enzyme solution was added to a series of reaction mixtures containing different substrate concentrations. The initial reaction rates were obtained by measuring the initial slope of the progress curve of product formation. The data in Table 9.3 were obtained:

(*a*) What is the V_{max} for this enzyme-reaction mixture?

(*b*) What is the K_m of the enzyme for the substrate?

SOLUTION

We can use the Lineweaver-Burk equation; for this, the reciprocals of the variables in Table 9.3 must be calculated and then plotted as shown in Fig. 9-20. (Ignore the asterisks in Fig. 9-20; they refer to Prob. 9.11.) (*a*) From the reciprocal of the ordinate intercept, $V_{max} = 160\ \mu$mol L^{-1} min^{-1}, and (*b*) from the reciprocal of the abscissal intercept, $K_m = 60\ \mu$mol L^{-1}.

9.8. One of the critical factors in experimental design for enzyme kinetics is the correct choice of substrate concentrations. On the basis of the data in Fig. 9-20, what is the possible optimal substrate concentration range?

Table 9.3. Steady-State Enzyme Kinetic Data

$[S]_0$ (μmol L^{-1})	v_0 (μmol L^{-1} min^{-1})
0.1	0.27
2.0	5.0
10.0	20
20.0	40
40.0	64
60.0	80
100.0	100
200.0	120
1,000.0	150
2,000.0	155

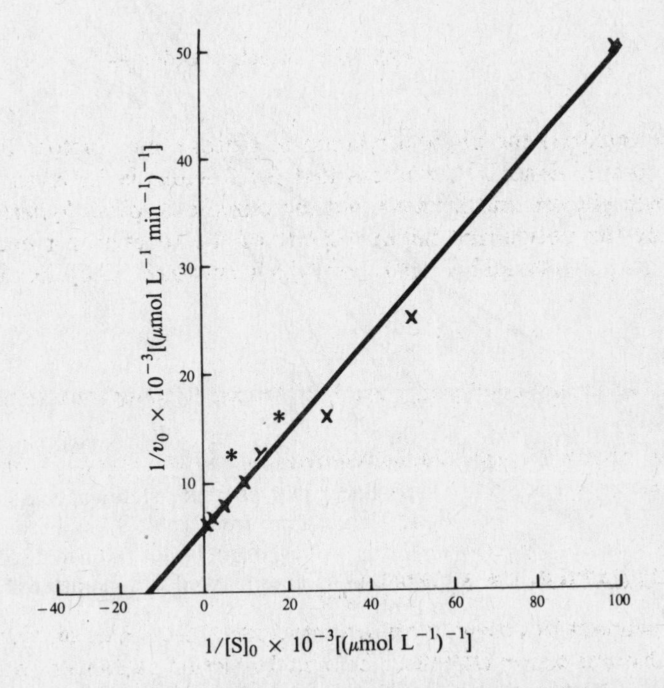

Fig. 9-20 Lineweaver-Burk plot of all but the first two data pairs in Table 9.3. (The asterisks refer to Prob. 9.11.)

SOLUTION

The most useful concentrations are those around K_m. In fact, a useful rule of thumb is 1/5 to 5 K_m; this way, v_0 will range from 0.17 to 0.83 times V_{max}. In other words, v_0 will be ~5 times greater at the highest concentration than at the lowest. If initial concentrations are too low, then velocity estimates become less accurate because of the nonlinearity of the reaction-progress curves near time zero.

9.9. (*a*) Use the *direct linear plot* to analyze the data in Table 9.3. (*b*) What advantages does this procedure have over the Lineweaver-Burk plot?

SOLUTION

(*a*) See Fig. 9-21. (*b*) The advantages of this method are as follows. First, the data can be used directly without the need to calculate reciprocals. Second, K_m and V_{max} are read off the graph directly or are readily determined from the median of the estimates arranged in echelon form (i.e., in ascending or descending order). Third, the results are much less sensitive to erroneous data pairs (outliers).

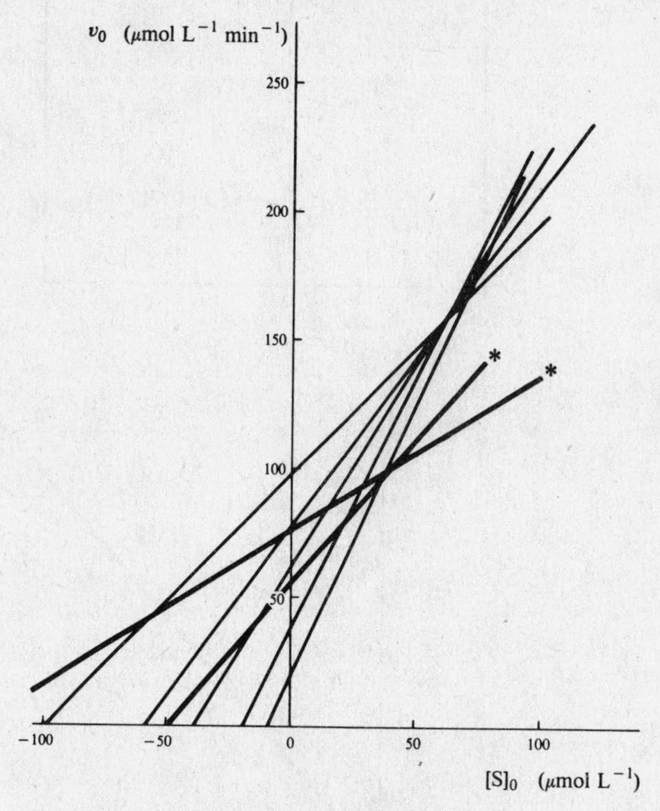

Fig. 9-21 Direct linear plot of steady-state enzyme kinetic data. The data are from Table 9.3, and the pairs with $[S]_0$ values from 10 to 100 μmol L^{-1} are plotted. (The lines marked with an asterisk refer to Prob. 9.11.)

9.10. Plot the data of Table 9.3 using the (*a*) Eadie-Hofstee and (*b*) Hanes-Woolf equations; include error bars of ±20 percent of the v_0 estimate. (*c*) What advantage, if any, do these methods have over the previous two procedures?

SOLUTION

(*a*) See Fig. 9-22.

(*b*) See Fig. 9-23.

(*c*) The Eadie-Hofstee plot has the advantage of being able to handle a much greater range of $[S]_0$ values than the Lineweaver-Burk plot. With the Lineweaver-Burk plot, the points at high $[S]_0$ are all lumped together. Thus, the Eadie-Hofstee plot is useful for indicating whether the enzyme conforms to Michaelis-Menten kinetics over a very large substrate concentration range. On the other hand, the Eadie-Hofstee plot is statistically not recommended because v_0 has much greater error than $[S]_0$ and appears on both sides of the equation. The Hanes-Woolf plot also handles a large $[S]_0$ domain. Furthermore, since errors in $[S]_0/v_0$ are a reasonably faithful reflection of those in v_0 (they are usually small in $[S]_0$), this is, statistically, the best of the straight-line plots.

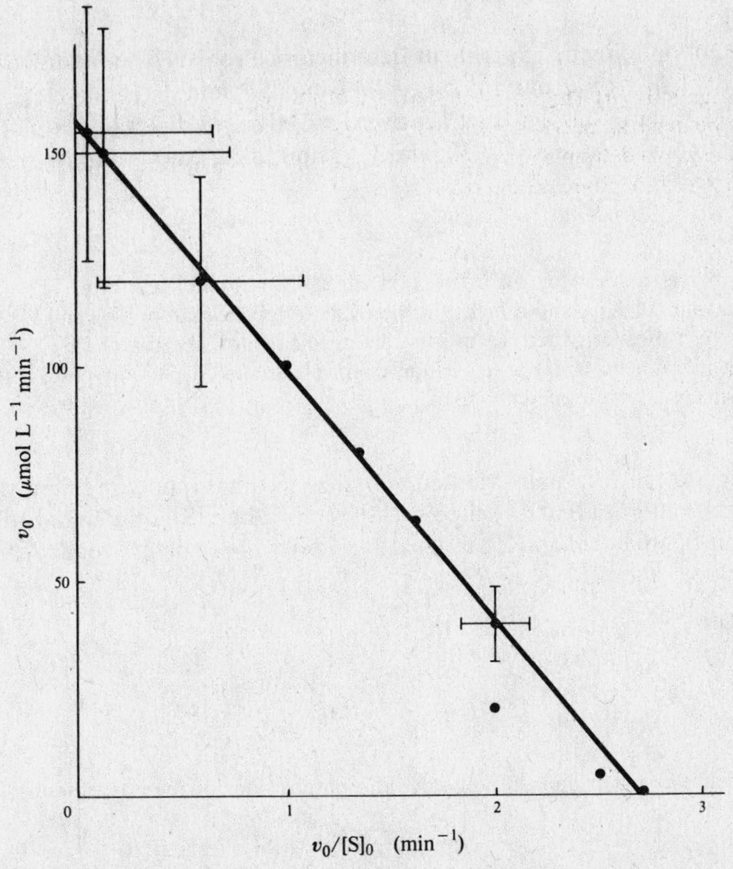

Fig. 9-22 Eadie-Hofstee plot of the data in Table 9.3. The error bars indicate a coefficient of variation in v_0 of ± 20 percent.

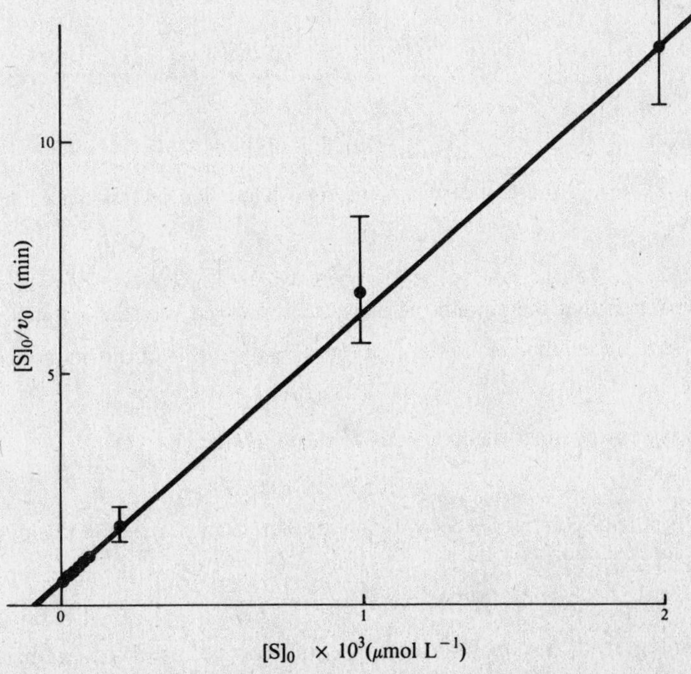

Fig. 9-23 Hanes-Woolf plot of data in Table 9.3. The error bars indicate a coefficient of variation in v_0 of ± 20 percent.

9.11. For the enzyme kinetic experiment described in Prob. 9.7, an additional two data pairs were obtained: ($[S]_0 = 50\ \mu$mol L^{-1}, $v_0 = 60\ \mu$mol L^{-1} min^{-1}) and ($[S]_0 = 150\ \mu$mol L^{-1}, $v_0 = 80\ \mu$mol L^{-1} min^{-1}). Construct Lineweaver-Burk and direct linear plots from the whole data set. What are the estimates of K_m and V_{max} now, as distinct from the results obtained in Probs. 9.7 and 9.9?

SOLUTION

The points marked by an asterisk in Fig. 9-20 are the new ones. They lead to a much lower estimated value of V_{max} and a higher value of K_m—approximately 50 μmol L^{-1} min^{-1} and 120 mol L^{-1}, respectively. However, when we use the direct linear plot, the lines in Fig. 9-21 are so clearly outliers that we would not be tempted to include them in the analysis; in any case, if they were included, they would scarcely alter the median value.

9.12. (*a*) Integrate the Michaelis-Menten equation to obtain an expression that relates time (t) to the concentration of S at any time t, ($[S]_t$). (*b*) When $[S]_0$, the initial substrate concentration, is very much greater than K_m, to what form does the equation reduce? (*c*) What is the form of the equation when $[S]_0 \ll K_m$?

SOLUTION

(*a*) Eq. (*9.5*) is

$$\frac{d[S]_t}{dt} = -\frac{V_{max}[S]_t}{K_m + [S]_t}$$

Collect like terms and differentials on separate sides of the equation:

$$\left(\frac{K_m}{[S]_t} + 1\right) d[S]_t = -V_{max}\, dt$$

Integrate [S] from $[S]_0$ to $[S]_t$ and t from $t = 0$ to t:

$$\int_{[S]_0}^{[S]_t} \left(\frac{K_m}{[S]_t} + 1\right) d[S]_t = -V_{max} \int_0^t dt$$

$$K_m \ln [S]_t \Big|_{[S]_0}^{[S]_t} + [S]_t \Big|_{[S]_0}^{[S]_t} = -V_{max} \cdot t$$

Therefore,

$$V_{max} \cdot t = ([S]_0 - [S]_t) + K_m \ln ([S]_0/[S]_t)$$

(*b*) When $[S]_0 \gg K_m$, the first term on the right-hand side of this equation is much greater than the second. So,

$$V_{max} \cdot t \simeq [S]_0 - [S]_t$$

This indicates that the amount of substrate consumed is a linear function of time.

(*c*) If $[S]_0 \ll K_m$, then the second term on the right-hand side of the integrated equation dominates. So,

$$V_{max} \cdot t \simeq K_m \ln ([S]_0/[S]_t)$$

By taking exponentials of both sides of this equation, we get

$$[S]_t \simeq [S]_0 \exp -(V_{max} \cdot t/K_m)$$

which is a simple first-order decay function; the extent of reaction at any time is independent of the initial substrate concentration.

9.13. Can the integrated Michaelis-Menten equation be used to estimate K_m and V_{max} from experimental enzyme kinetics data?

SOLUTION

Yes. For a simple graphical procedure, the equation is rearranged to give

$$\frac{1}{t}\ln([\mathrm{S}]_0/[\mathrm{S}]_t) = -\frac{1}{K_m}\left(\frac{[\mathrm{S}]_0-[\mathrm{S}]_t}{t}\right)+\frac{V_{max}}{K_m}$$

When the left-hand term is plotted versus $([\mathrm{S}]_0-[\mathrm{S}]_t)/t$, the slope of the line is $-1/K_m$, the ordinate intercept is V_{max}/K_m, and the abscissal intercept is V_{max} (Fig. 9-24).

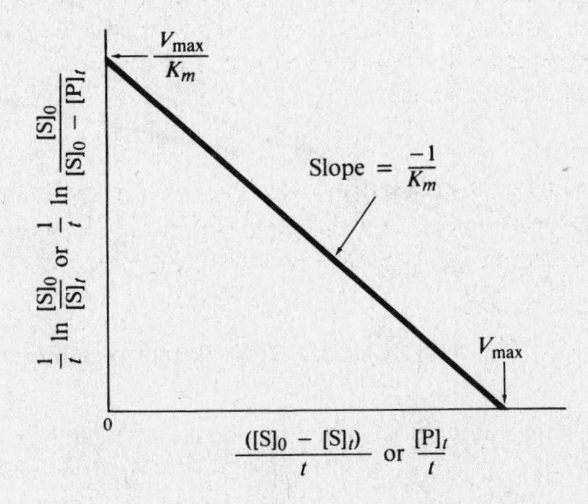

Fig. 9-24 Data analysis using the integrated Michaelis-Menten equation.

ENZYME INHIBITION

9.14. The effect of an inhibitor I on the rate of a single-substrate enzyme-catalyzed reaction was investigated and gave the following results:

Substrate Concentration $[\mathrm{S}]_0$ (mmol L^{-1})	Inhibitor Concentration (mmol L^{-1}) 0	0.5	1.0
	Rate of Reaction v_0 (μmol L^{-1} min^{-1})		
0.05	0.33	0.20	0.14
0.10	0.50	0.33	0.25
0.20	0.67	0.50	0.40
0.40	0.80	0.67	0.57
0.50	0.83	0.71	0.63

(*a*) What is the mode of action of the inhibitor? (*b*) Estimate values for V_{max}, K_m, and the inhibition constant(s) for the reaction.

SOLUTION

(*a*) The simplest way to deduce the mode of action of the inhibitor is to plot $1/v_0$ as a function of $1/[\mathrm{S}]_0$ for each inhibitor concentration (Fig. 9-25).

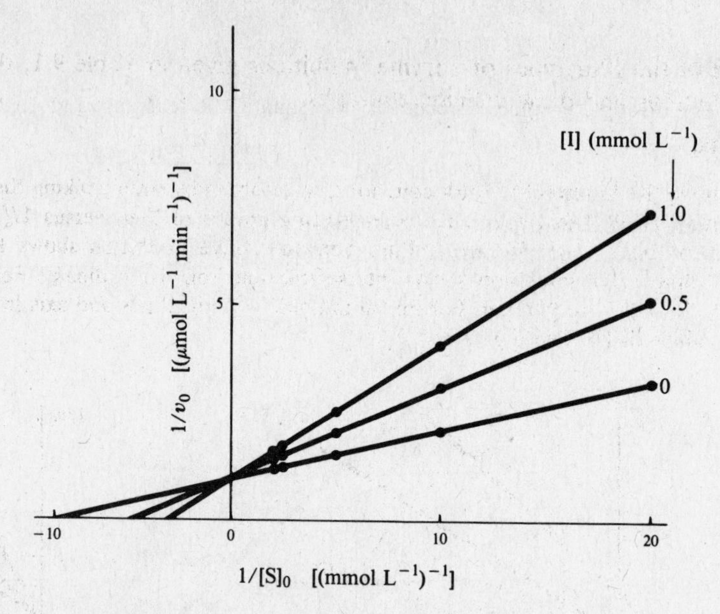

Fig. 9-25 Lineweaver-Burk plot of inhibitor effect.

When the pure competitive inhibition equation (Table 9.1) is transformed into reciprocal form, we obtain:

$$\frac{1}{v_0} = \frac{1}{V_{\max}} + \frac{K_m}{V_{\max}[\mathrm{S}]_0}\left(1 + \frac{[\mathrm{I}]}{K_I}\right)$$

This equation predicts that the slope of the Lineweaver-Burk plot will change with changes in inhibitor concentration, but the intercept on the $1/v_0$ axis ($1/V_{\max}$) will not. Thus, the above data are consistent with the inhibitor (I) acting as a *pure competitive* inhibitor.

(*b*) From the intercept on the $1/v_0$ axis, $V_{\max}$ may be estimated. We note from the equation above that the slope (Sl) of the plot is:

$$\mathrm{Sl} = \frac{K_m}{V_{\max}}\left(1 + \frac{[\mathrm{I}]}{K_I}\right)$$

Thus, when the slope is plotted as a function of [I], we get the results shown in Fig. 9-26. The intercept of this *secondary* plot on the abscissa is $-K_I$; hence K_I can be estimated as 0.5 mmol L^{-1}. The intercept on the ordinate is $K_m/V_{\max}$. Since we know $V_{\max}$, then K_m can be estimated as 0.1 mmol L^{-1}.

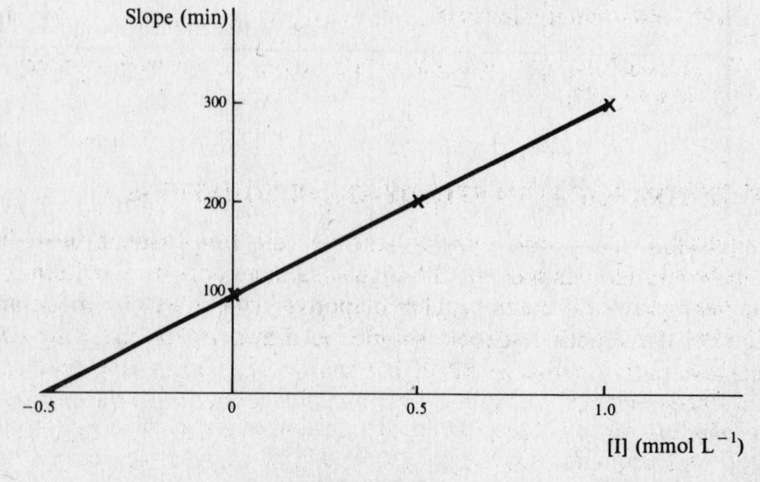

Fig. 9-26

9.15. For each of the four types of enzyme inhibitions given in Table 9.1, derive the Lineweaver-Burk equations and draw *archetypal* graphs.

SOLUTION

To derive the Lineweaver-Burk equations, we proceed by simply taking the reciprocals of each side of the equations in Table 9-1. The corresponding graphs of $1/v_0$ versus $1/[S]_0$ have varying slopes, intercepts, or both as [I] is varied. Pure noncompetitive inhibition shows lines intersecting on the abscissa. Competitive inhibition shows intersecting lines on the ordinate. For anti- or uncompetitive inhibition, the lines are parallel. For mixed inhibition, both slopes and axis intercepts are different for different values of [I]. See Fig. 9-27.

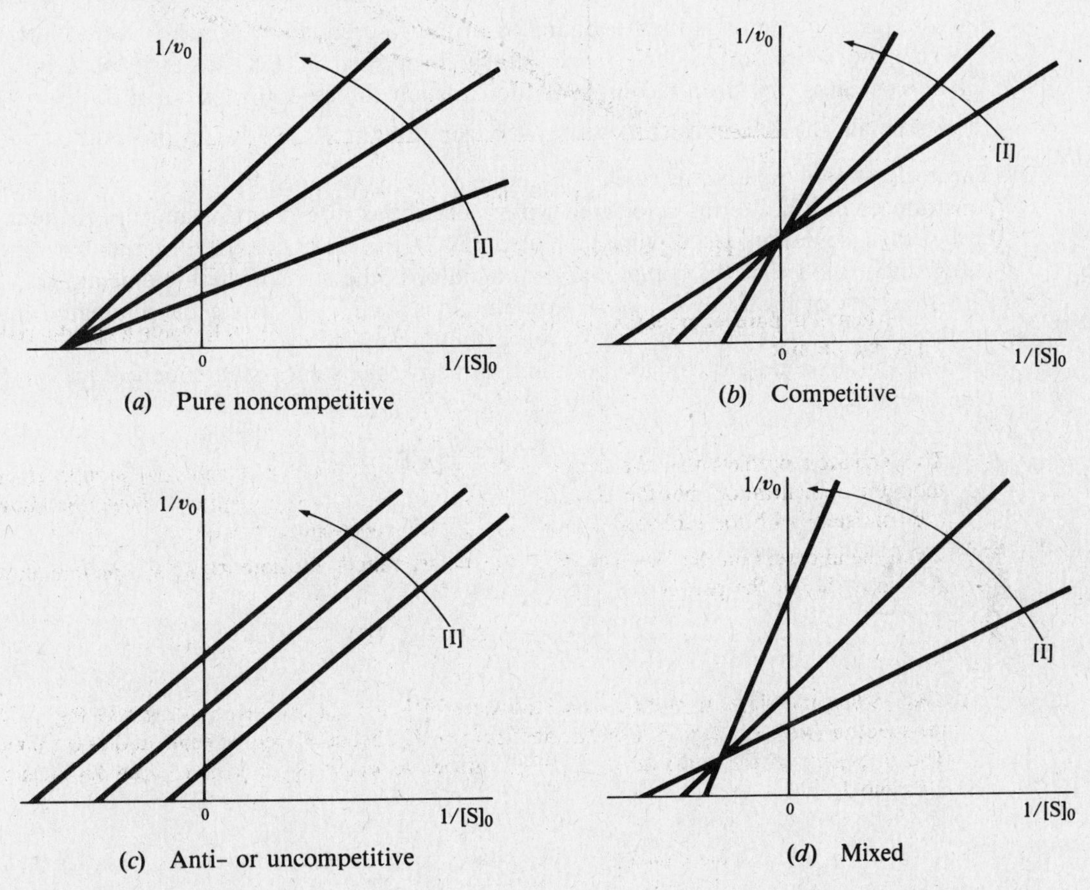

Fig. 9-27 Lineweaver-Burk plots for various types of enzyme inhibitors. The arrows indicate increasing [I].

DERIVATION OF COMPLICATED STEADY-STATE EQUATIONS

9.16. For complicated steady-state enzyme schemes, the King-Altman procedure for deriving rate equations is very tedious indeed, although several authors have written computer programs to perform the task. A different, algebraic approach was taken by Indge and Childs in 1976. It requires that the reaction scheme should be drawn as for the King-Altman procedure (in mathematical parlance this is called a *nonorientated linear connected graph*). Additionally, so-called Wang algebra is applied. This algebraic system has the properties that the *sum* or *product* of identical elements (variables) is *zero*, but all other algebraic operations apply as usual; i.e., for any numbers a and b, $a \cdot a = 0$, $a + a = 0$, but $a + b \neq 0$, unless a and b are zero.

The procedure is as follows:

1. Carry out step 1 of the King-Altman procedure; i.e., draw the scheme.
2. Circle $n-1$ *nodes* (i.e., enzyme forms) of the total n in the scheme.
3. List separately the rate constants of reaction arrows cut by each of the $n-1$ circles.
4. List all combinations of *pairs* of the rate constants that *leave* each individual node of the graph. These are called *forbidden combinations*.
5. Use the rules of Wang algebra to eliminate the forbidden combinations listed at step 4 for each of the nodes. This entails multiplying the terms in step 3 and then eliminating the forbidden combinations listed in step 4 and the combinations that represent cycles.
6. The result of step 5 is the denominator of the rate equation, which is sorted into nodes. The nodes are sorted; e.g., to isolate a form such as EX_i select those terms in the denominator that do *not* contain a rate constant directed *away* from EX_i.
7. Assemble the rate constants in the way used for the King-Altman procedure.

Three additional operations assist in deriving the equations: (*a*) in steps 3–5, omit any substrate or product terms associated with a particular rate constant and insert them at the end of the calculation; (*b*) separate the elements of the array of rate constants by commas, it being understood that the completed denominator is the sum of these elements; (*c*) express rate constants of the form k_{-1} as i'. Combinations such as ii' would be eliminated from the solution by the rules of Wang algebra since they represent cycles.

Use this procedure of Indge and Childs to derive the steady-state kinetic equation for the *Uni Uni* scheme:

$$E + A \underset{k_{-1}}{\overset{k_1}{\rightleftharpoons}} EA \underset{k_{-2}}{\overset{k_2}{\rightleftharpoons}} EP \underset{k_{-3}}{\overset{k_3}{\rightleftharpoons}} E + P$$

SOLUTION

Step 1. Draw the scheme.

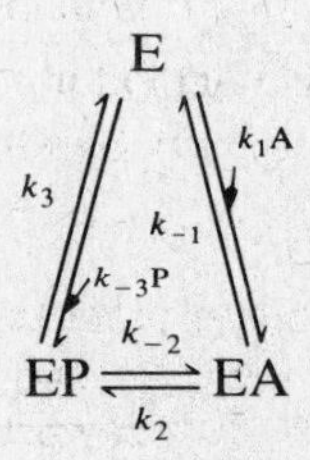

Step 2. Circle *only* $n-1$ nodes.

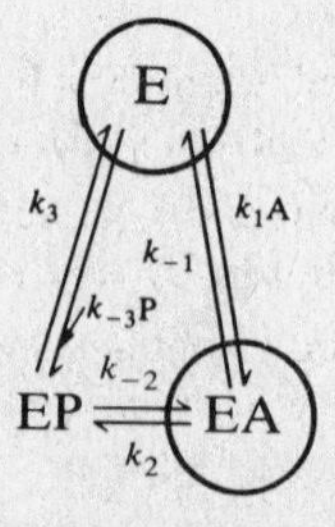

Step 3. Arrange the $n-1$ lists in readiness for multiplying the terms in step 5.

$$(1,1',3,3')(1,1',2,2')$$

Step 4. List forbidden combinations.

E:	13′
EA:	1′2
EP:	2′3

Step 5. Multiply the terms in step 3 and eliminate the forbidden combinations (step 4), as well as combinations that represent cycles.

~~11~~	~~1′1~~	(31)	~~3′1~~
~~11′~~	~~1′1′~~	31′	[3′1′]
(12)	~~1′2~~	32	[3′2]
(12′)	1′2′	~~32′~~	[3′2′]

Step 6. Sort the nodes.

Node E (those terms *not* containing 1 or 3′ belong here):	1′2′	32	31′
Node EA (*not* 1′ or 2):	12′	31	3′2′
Node EP (*not* 2′ or 3):	12	3′1′	3′2

Step 7. Assemble the rate constants: The circled terms that contain 1 are associated with [A], and the boxed terms that contain 3′ are associated with [P].

The expression given by the denominator of Eq. (*9.30*) is the result:

$$(k_{-1}k_3 + k_2k_3 + k_{-1}k_{-2}) + k_1(k_2 + k_3 + k_{-2})[\mathrm{A}] + k_{-3}(k_{-1} + k_2 + k_{-2})[\mathrm{P}]$$

Thus, at least for this case, we have confirmed that the method works.

9.17. Consider the ordered Bi Bi reaction (Example 9.8):

$$\mathrm{A + B \rightleftharpoons P + Q}$$

Predict the inhibition patterns that will be obtained when product P or Q is used as an inhibitor, with varied concentrations of A.

SOLUTION

We have four cases to consider—the effects of different Q concentrations with:

(1) Varied [A]; fixed, nonsaturating [B]; P inhibitor

(2) Varied [A]; fixed, saturating [B]; P inhibitor

(3) Varied [A]; fixed, nonsaturating [A]; Q inhibitor

(4) Varied [A]; fixed, saturating [B]; Q inhibitor

It is convenient to consider the inhibition patterns in terms of plots of $1/v_0$ versus $1/[\text{substrate}]_0$ at various levels of inhibitor. Varying inhibitor may affect only the slopes of the plots (competitive inhibition), both slopes and intercepts (noncompetitive or mixed inhibition), or only the intercepts (uncompetitive inhibition). Inhibition patterns can be predicted from the following so-called Cleland's rules:

(*a*) Competitive inhibition is obtained when the inhibitor and varied substrate bind to the *same* form of the enzyme, or different forms that are in *equilibrium* with each other.

(*b*) Noncompetitive or mixed inhibition is obtained when the inhibitor and varied substrate bind to *different* forms of the enzyme, linked by *reversible* interconversions of enzyme forms.

(*c*) Anti- or uncompetitive inhibition is obtained when the inhibitor and substrate bind to *different* forms of the enzyme, linked by *irreversible* interconversions of enzyme forms.

The reaction above can be written in full as:

$$\mathrm{E} \underset{k_{-1}}{\overset{k_1\mathrm{A}}{\rightleftharpoons}} \mathrm{EA} \underset{k_{-2}}{\overset{k_2\mathrm{B}}{\rightleftharpoons}} \begin{matrix}\mathrm{EAB}\\ \mathrm{EPQ}\end{matrix} \underset{k_{-3}\mathrm{P}}{\overset{k_3}{\rightleftharpoons}} \mathrm{EQ} \underset{k_{-4}\mathrm{Q}}{\overset{k_4}{\rightleftharpoons}} \mathrm{E}$$

Case 1 (varied [A], nonsaturating [B], P inhibitor): From rule (*b*) above, noncompetitive or mixed inhibition will be obtained, as A and P bind to different forms of the enzyme (E and EQ, respectively) linked by reversible interconversions.

Case 2 (varied [A], saturating [B], P inhibitor): In this case, the link between E and EQ is irreversible (because B is present at saturating levels); hence, from rule (*c*) above, uncompetitive inhibition will be obtained.

Case 3 (varied [A], nonsaturating [B], Q inhibitor): Both A and Q bind to the *same* form (E) of the enzyme; hence, from rule (*a*) above, competitive inhibition will be obtained.

Case 4 (varied [A], saturating [B], Q inhibitor): Although the interconversion of EA and EAB is irreversible in the direction of EAB formation at saturating [B], A and Q still bind to the *same* form of the enzyme (E); hence, from rule (*a*) above, competitive inhibition will be obtained.

Establishing the inhibition patterns in an enzyme-catalyzed reaction is usually an important step in elucidating the reaction mechanism. One complication in the interpretation of such data is the possible formation of *dead-end* complexes (i.e., a complex of the form EAP in the above scheme). This is especially important in *rapid-equilibrium* reactions [ones in which all steps except the rate constants for the central isomerization step ($EAB \rightleftharpoons EPQ$ in the above example) are very large].

pH EFFECTS ON ENZYME REACTION RATES

9.18. A hypothetical enzyme has activity, with respect to the hydrolysis of a neutral substrate, that is unaffected by pH over a very broad range, pH 6 to pH 8. However, it shows a much narrower pH-activity range when an alternative substrate, which contains an imidazole group, is used. A total of 50 assays were carried out by using 10 different substrate concentrations with a fixed amount of enzyme, at 5 different pH values. For each pH, the enzyme gave linear Lineweaver-Burk plots, and the estimates of K_m and V_{max} were as follows:

pH	K_m (mmol L^{-1})	V_{max} (μmol L^{-1} min^{-1})
8.0	1.1	75
7.5	1.3	80
7.0	2.0	77
6.5	4.2	83
6.0	11.0	75

(*a*) Describe the effect of varying pH on K_m and V_{max}. (*b*) What is the significance of the fact that the enzyme activity with an uncharged substrate scarcely alters with pH in the range 6–8? (*c*) Propose a reaction scheme that is consistent with the results. (*d*) Calculate the relevant dissociation constant of the H^+-reactant complex.

SOLUTION

(*a*) From the data it is evident that within experimental error V_{max} does not change with pH. However, K_m is strongly pH-dependent, with the enzyme having the highest affinity for the substrate at high pH.

(*b*) If the rate of an enzyme-catalyzed reaction is insensitive to pH, this suggests a lack of ionizable groups involved in binding, catalysis, or both. This is the case with the present enzyme. The fact that with the ionizable substrate the reaction rate is pH-dependent suggests that the ionization of the substrate affects the rate.

(*c*) In view of the above statements, a possible reaction scheme is

$$\begin{array}{l} S + E \underset{k_{-1}}{\overset{k_1}{\rightleftharpoons}} ES \overset{k_2}{\longrightarrow} E + P \\ K_e \updownarrow \\ SH^+ \end{array}$$

(*d*) This is the first model presented under "Ionization of Substrate," in Sec. 9.9, and the corresponding rate expression is Eq. (*9.49*). We see that V_{max} is independent of pH, but $K_m^{app} = K_m(1 + [H^+]/K_e)$. A plot of K_m^{app} versus $[H^+]$ for the data in the table yields K_m/K_e from the slope and K_m from the intercept; thus, $K_e = 10^{-7}$ mol L^{-1} and $K_m = 1.0$ mmol L^{-1}. See Fig. 9-28.

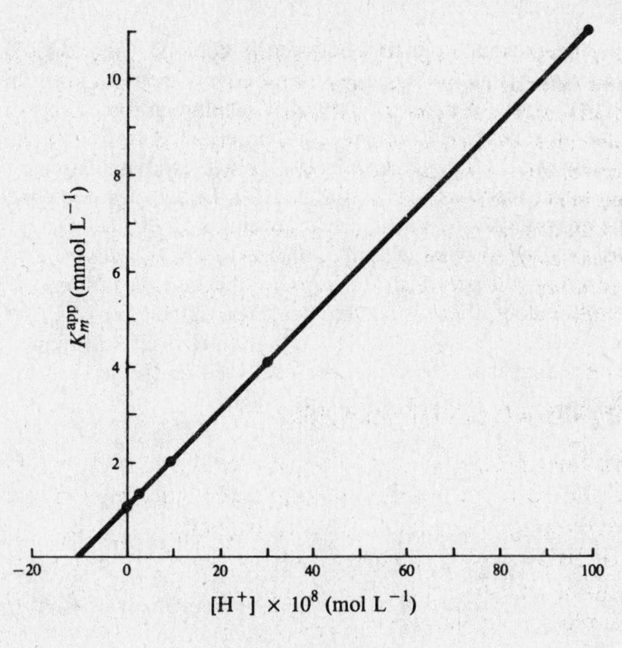

Fig. 9-28

FAST REACTIONS

9.19. The mathematical procedures used to analyze pre-steady-state kinetics are used also in the so-called *isotope-exchange methods*. In this technique, an enzyme-catalyzed reaction is allowed to come to *chemical* equilibrium, and then a small amount of radioactive substrate (too little to significantly alter the chemical equilibrium) is added. However, the reaction is not at *isotopic* equilibrium, and the radioactivity will exchange from the substrate into the corresponding product until isotopic equilibrium is reached. The rate at which isotopic equilibrium is approached can be measured as a function of the concentrations of substrates and products. Note that these concentrations must be manipulated so that *chemical* equilibrium is maintained.

The enzyme *lactate dehydrogenase* catalyzes the reaction:

$$\underset{\text{Lactate}}{\text{A}} + \underset{\text{NAD}^+}{\text{B}} \rightleftharpoons \underset{\text{Pyruvate}}{\text{P}} + \underset{(\text{NADH} + \text{H}^+)}{\text{Q}}$$

Isotope-exchange reactions can be monitored between lactate and pyruvate ^{14}C labeled at carbon 3, and between NAD$^+$ and NADH^{14} (labeled in the nicotinamide ring). Given that the reaction is of the ordered Bi Bi type, with NAD$^+$ binding first (see Example 9.7), what would be the effects on both exchange reactions as the concentrations of either lactate and pyruvate, or NAD$^+$ and NADH, are increased?

SOLUTION

The above reaction may be written in full as:

$$\text{E} \underset{k_{-1}}{\overset{k_1\text{NAD}^+}{\rightleftharpoons}} \text{ENAD}^+ \underset{k_{-2}}{\overset{k_2\,\text{lactate}}{\rightleftharpoons}} \begin{matrix}\text{E-NAD}^+\text{-Lactate} \\ \text{E-NADH-Pyruvate}\end{matrix} \underset{k_{-3}\,\text{pyruvate}}{\overset{k_3}{\rightleftharpoons}} \text{ENADH} \underset{k_{-4}\,\text{NADH}}{\overset{k_4}{\rightleftharpoons}} \text{E}$$

Consider the exchange reaction between [^{14}C]lactate and pyruvate first. As the concentrations of lactate and pyruvate are increased (note that *both* must be increased in order to maintain chemical equilibrium), the rate of the isotope-exchange reaction will increase until the enzyme is saturated with these two compounds (as determined by rate constants k_2, k_{-2}, k_3, and k_{-3}). Similarly, as the concentrations of NAD$^+$ and NADH are increased, the rate of the isotope-exchange reaction will increase until the enzyme is saturated with NAD$^+$ and NADH.

Turning now to the exchange reaction between [^{14}C]NAD$^+$ and NADH, as the concentrations of NAD$^+$ and NADH increase, the exchange reaction rate will increase until the enzyme is saturated with NAD$^+$ and NADH. However, a different pattern is seen when the effects of varying the pyruvate and lactate concentrations are examined. Initially, as these concentrations are increased, the rate of the exchange reaction between [^{14}C]NAD$^+$ and NADH will increase, because the enzyme has to bind pyruvate and lactate to interconvert NAD$^+$ and NADH. But at elevated concentrations of pyruvate and lactate, more of the enzyme will be in the central complex (E-NAD$^+$-lactate/E-NADH-pyruvate), and the concentration of the free enzyme will fall. Hence, as NAD$^+$ *must* bind to the free enzyme, the rate of the exchange reaction will decrease to zero as the concentrations of pyruvate and lactate are increased, because, ultimately, there is no free enzyme available for [^{14}C]NAD$^+$ to bind to, and hence no possible isotopic exchange between NAD$^+$ and NADH. This technique is probably the method of choice in establishing the mechanisms of enzyme-catalyzed reactions, as it avoids many of the problems inherent in initial velocity and product inhibition studies.

REGULATORY ENZYMES

9.20. The reaction scheme in Fig. 9-29 depicts isoleucine (E) synthesis from aspartate (A) by the bacterium *Rhodopseudomonas spheroides*. The control is not isoenzymic (Example 9.12) but is called *sequential feedback control*. Describe the operation of this metabolic control system.

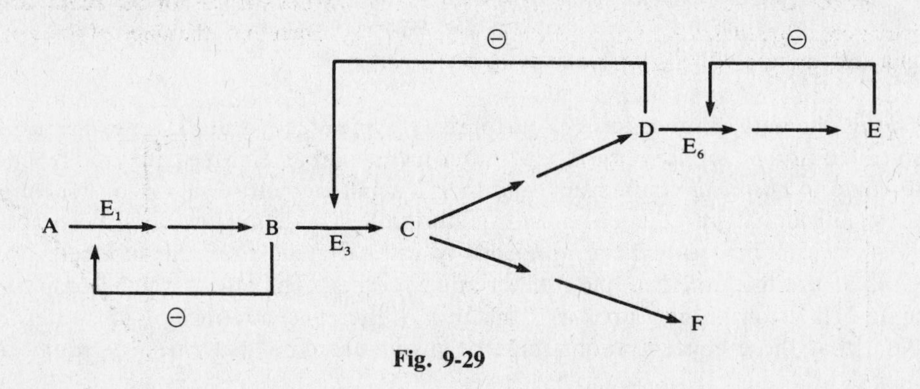

Fig. 9-29

SOLUTION

Overproduction of E (isoleucine) inhibits enzyme E_6 (threonine deaminase), and the consequent rise of D (threonine) reduces the rate of production of C (homoserine) via enzyme E_3 (homoserine dehydrogenase). The concentration of B (aspartate semialdehyde) rises, and this in turn inhibits E_1 (aspartokinase). It is therefore obvious why the control system is called a *negative feedback network*, or *sequential feedback system*.

9.21. In liver phosphofructokinase, ATP, ADP, and citrate are *effectors* of the reaction rate (see Fig. 9-30). Define what type of effectors they are.

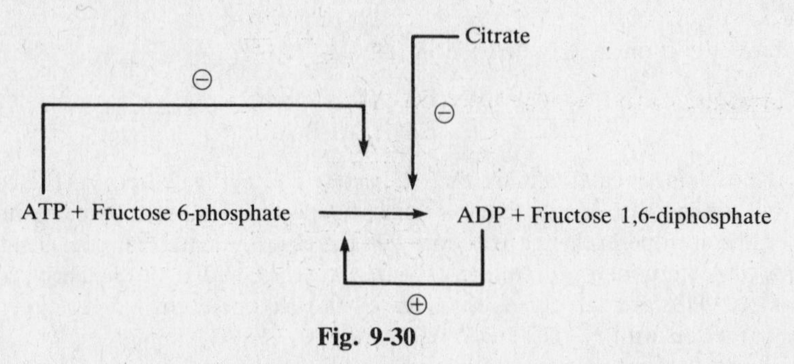

Fig. 9-30

SOLUTION

1. ATP: negative homotropic effector
2. Citrate: negative heterotropic effector
3. ADP: positive homotropic effector

ATP exerts *negative feedforward* control (contrast with Example 9.10), while ADP exerts *positive feedback* control.

9.22. Describe the shapes of $1/v_0$ versus $1/[S]_0$ curves for enzymes exhibiting positive cooperativity, no cooperativity, and negative cooperativity of binding of the substrate (S).

SOLUTION

In plots of $1/v_0$ versus $1/[S]_0$, the slope of the line is directly proportional to K_m. In a positively cooperative case, the affinity of the enzyme for the substrate *increases*, and hence K_m *decreases* with increasing $[S]_0$ (or decreasing $1/[S]_0$). Therefore, the slope of the plot of $1/v_0$ versus $1/[S]_0$ will decrease as $1/[S]_0$ decreases [Fig. 9-31(*a*)].

In the case of *no cooperativity*, the affinity of the enzyme for the substrate, and hence K_m, is *constant* with changing $[S]_0$; therefore, the slope of the plot of $1/v_0$ versus $1/[S]_0$ will be constant with $1/[S]_0$ [Fig. 9-31(*b*)].

In the case of *negative* cooperativity, the affinity of the enzyme for the substrate *decreases* with increasing $[S]_0$, hence K_m *increases* with increasing $[S]_0$. Therefore, the slope of the $1/v_0$ versus $1/[S]_0$ plot will *increase* with *decreasing* $1/[S]_0$ [Fig. 9-31(*c*)].

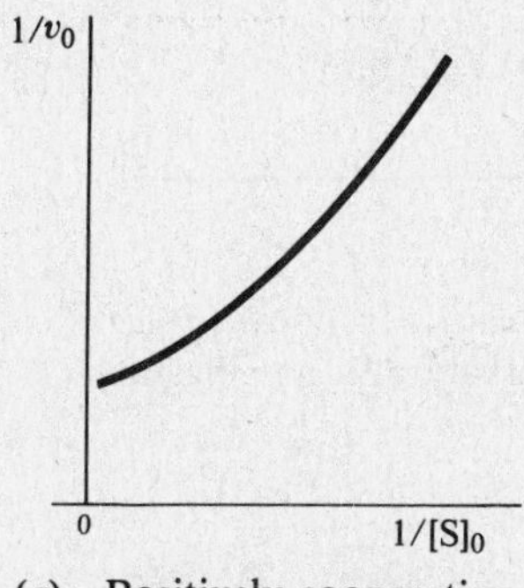

(*a*) Positively cooperative enzyme

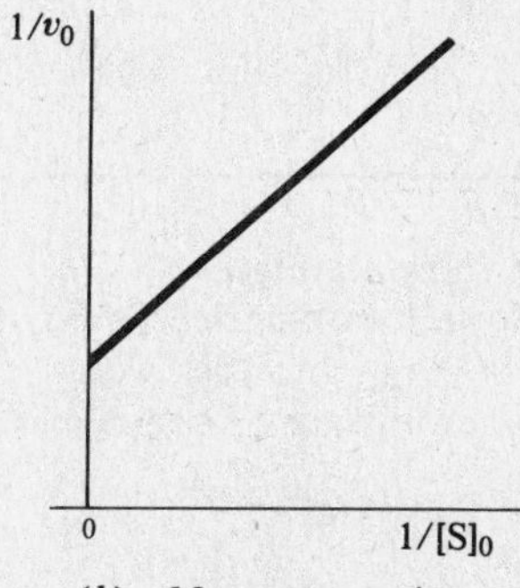

(*b*) Noncooperative enzyme

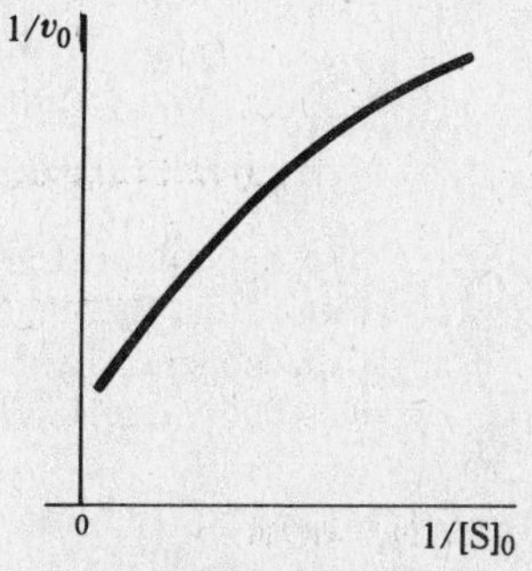

(*c*) Negatively cooperative enzyme

Fig. 9-31

9.23. Prove that for a Michaelis-Menten enzyme, R_s of Eq. (*9.75*) is equal to 81.

SOLUTION

The rate equation is

$$v_0 = \frac{V_{max}[S]_0}{K_m + [S]_0}$$

and it is arranged to give an expression for $[S]_0$:

$$[S]_0 = \frac{v_0 K_m}{V_{max} - v_0}$$

$$R_s = \frac{0.9V_{max}K_m(V_{max} - 0.1V_{max})}{0.1V_{max}K_m(V_{max} - 0.9V_{max})}$$

$$= \frac{(0.9)(0.9)}{(0.1)(0.1)} = 81$$

9.24. If a solution contains two isozymes of an enzyme in equal amounts (as determined by gel electrophoresis) and it is known that the k_{cat} of each is the same but that the K_m for the substrate in one is 1 mmol L^{-1} and the other is 0.1 mmol L^{-1}, comment on the appearance of the Lineweaver-Burk plot of initial velocity data obtained for nine substrate concentrations ranging from 0.02 to 5 mmol L^{-1}.

SOLUTION

An *ideal* data set is generated from this expression:

$$v_0 = \frac{1[\mathrm{S}]_0}{1 + [\mathrm{S}]_0} + \frac{1[\mathrm{S}]_0}{0.1 + [\mathrm{S}]_0}$$

where $[\mathrm{S}]_0 = [0.02, \ldots, 5]$ mmol L^{-1} $V_{max} = 1$ (arbitrary units)

$[\mathrm{S}]_0$ (mmol L^{-1})	$1/[\mathrm{S}]_0$ [(mmol L^{-1})$^{-1}$]	v_0 (arbitrary units)	$1/v_0$ [(arbitrary units)$^{-1}$]
0.020	50	0.19	5.3
0.025	40.0	0.22	4.5
0.04	25.0	0.32	3.1
0.10	10.0	0.59	1.7
0.20	5.0	0.83	1.2
0.50	2.0	1.17	0.85
1.00	1.0	1.41	0.71
2.00	0.5	1.63	0.61
5.0	0.2	1.82	0.55

The Lineweaver-Burk plot is shown in Fig. 9-32.

Notice that the line curves down from the right in Fig. 9-32. This is identical to the effect seen in negative cooperativity [see Fig. 9-31(*c*)]. In other words, this curvature is not diagnostic of *control* enzymes but can arise simply from a mixture of isoenzymes.

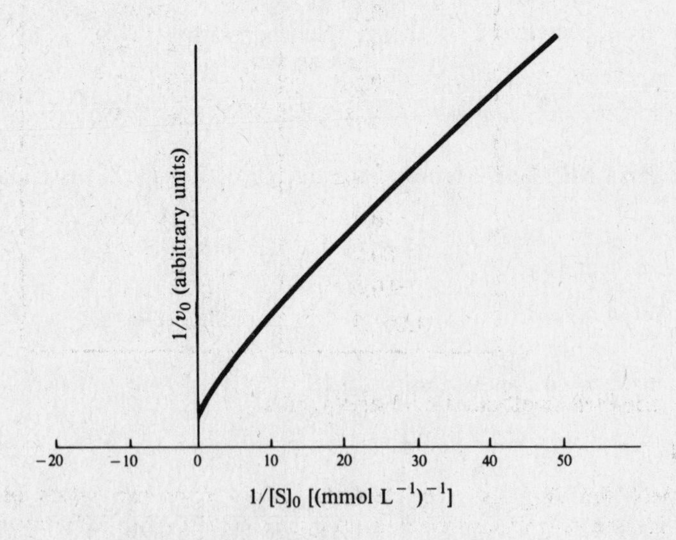

Fig. 9-32 Lineweaver-Burk plot for the sum of two rectangular hyperbolas: $V_{max1} = V_{max2} = 1$ and $K_{m1} = 1.0$ mmol L^{-1}, $K_{m2} = 0.1$ mmol L^{-1}.

9.25. Derive the Hill equation, Eq. (9.76).

SOLUTION

From the definition of K_b, Eq. (9.77), the concentration of the hemoglobin-O_2 complex is given by

$$[Hb(O_2)_n] = K_b[Hb](pO_2)^n$$

Y, the fractional saturation, is

$$Y = \frac{\text{concentration of } Hb(O_2)_n}{\text{total concentration of hemoglobin}}$$

$$= \frac{K_b[Hb](pO_2)^n}{[Hb] + K_b[Hb](pO_2)^n}$$

Cancellation of [Hb] from the numerator and denominator yields the Hill equation:

$$Y = \frac{K_b(pO_2)^n}{1 + K_b(pO_2)^n}$$

9.26. Prove that for $n > 1$, the Hill equation conforms to a positively cooperative binding or enzyme system.

SOLUTION

For an enzyme the Hill equation, rearranged, gives

$$[S]_0 = \left(\frac{v_0 K_m}{V_{max} - v_0}\right)^{1/n}$$

Therefore, from Prob. 9.23, $R_s = (81)^{1/n}$. So if $n > 1$, then $R_s < 81$, and according to the definition given in Sec. 9.12, under "Kinetic Behavior of Regulatory Enzymes," the enzyme is positively cooperative.

9.27. The initial velocity of a reaction catalyzed by a regulatory enzyme was determined over the following range of initial substrate concentrations:

$[S]_0$ (mmol L^{-1})	v_0 (mmol L^{-1} min^{-1})
0.1	1.6×10^{-4}
0.3	1.4×10^{-3}
0.5	4.0×10^{-3}
1.0	1.6×10^{-2}
3.0	0.14
5.0	0.39
10.0	1.45
100.0	14.6
500.0	15.9
1,000.0	16.0

Determine the Hill coefficient and apparent K_m.

SOLUTION

It is clear that V_{max} is ~16.0 mmol L^{-1} min^{-1} since the values of v_0 for $[S]_0$ of 500 and 1,000 mmol L^{-1} are very similar; thus, the enzyme is almost saturated at these high substrate concentrations. If we assume V_{max} has this value and use the expression in Example 9.15, we can construct the following table expressing S as molar concentration. From Prob. 9.26 a plot of these data give Fig. 9.33 yielding a slope of n and an ordinate intercept of $\log_{10}(1/K_m)$; with these data $n = 2$ and $K_m = 1$ mmol L^{-1}.

$\log_{10}\left(\frac{v_0}{V_{max} - v_0}\right)$	$\log_{10}[S]_0$
−5.00	−4.00
−4.06	−3.52
−3.60	−3.30
−3.00	−3.00
−2.05	−2.52
−1.60	−2.30
−1.00	−2.00
+1.02	−1.00
+2.20	−0.30
—	0.00

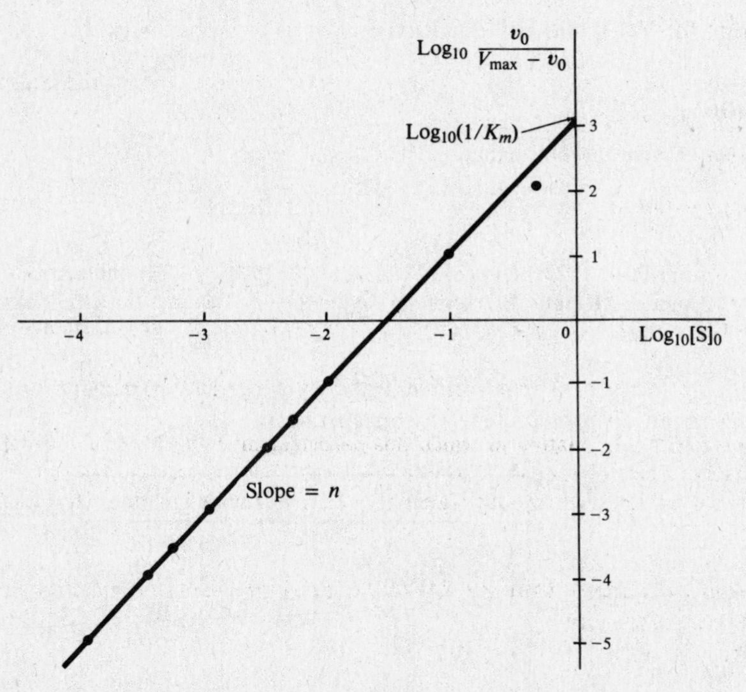

Fig. 9-33

9.28. Derive the four-site Adair equation for O_2 binding to hemoglobin.

SOLUTION

The relevant binding reaction scheme is given under "Mathematical Models of Cooperativity" in Sec. 9.12, and from the expressions for the *extrinsic* equilibrium constants, the concentration of each of the four complexes is given by:

$$[\text{HbX}] = 4K_1[\text{Hb}][\text{X}]$$
$$[\text{HbX}_2] = \tfrac{3}{2}K_2[\text{HbX}][\text{X}] = 6K_1K_2[\text{Hb}][\text{X}]^2$$
$$[\text{HbX}_3] = \tfrac{2}{3}K_3[\text{HbX}_2][\text{X}] = 4K_1K_2K_3[\text{Hb}][\text{X}]^3$$
$$[\text{HbX}_4] = \tfrac{1}{4}K_4[\text{HbX}_3][\text{X}] = K_1K_2K_3K_4[\text{Hb}][\text{X}]^4$$

where [X] denotes pO_2, and HbX denotes HbO_2, etc.

The expressions for the concentrations of the complexes are substituted into Eq. (*9.80*) to give

$$Y = \frac{[\mathrm{HbX}] + 2[\mathrm{HbX_2}] + 3[\mathrm{HbX_3}] + 4[\mathrm{HbX_4}]}{4([\mathrm{Hb}] + [\mathrm{HbX}] + [\mathrm{HbX_2}] + [\mathrm{HbX_3}] + [\mathrm{HbX_4}])}$$

$= \text{Eq. } (9.81)$, as required.

9.29. Under what special conditions of the Adair model does the corresponding binding equation, (*9.81*), become equivalent to the Hill equation, (*9.76*)?

SOLUTION

If K_4 of the Adair equation is very large compared with K_1, K_2, and K_3 and $[\mathrm{X}] \gg K_1, K_2, K_3$, Eq. (*9.81*) simplifies to

$$Y = \frac{K_1K_2K_3K_4[\mathrm{X}]^4}{1 + K_1K_2K_3K_4[\mathrm{X}]^4}$$

If $K_1K_2K_3K_4$ is equated with K_b, we obtain the Hill equation. Physically, the condition that $K_4 \gg K_1, K_2, K_3$ implies insignificant amounts of the complexes other than $\mathrm{HbX_4}$, which was indeed a starting premise of the Hill derivation.

9.30. What form does the MWC equation, (*9.90*), take when the ratio of *intrinsic* binding constants, $K_\mathrm{T}/K_\mathrm{R} = c$, is equal to 1?

SOLUTION

The equation becomes

$$Y = \frac{\alpha(1+\alpha)^{n-1} + L\alpha(1+\alpha)^{n-1}}{(1+\alpha)^n + L(1+\alpha)^n}$$

By factoring out $(1+\alpha)^{n-1}$ in the numerator and $(1+\alpha)^n$ in the denominator, we get

$$Y = \frac{\alpha(1+L)(1+\alpha)^{n-1}}{(1+L)(1+\alpha)^n} = \frac{\alpha}{1+\alpha} = \frac{K_\mathrm{R}[\mathrm{X}]}{1 + K_\mathrm{R}[\mathrm{X}]}\,.$$

Thus, this is a third situation in which this particular model reduces to a simple hyperbolic expression; the other two cases are discussed in Example 9.18. The outcome is physically "reasonable," since if the ligand binds equally well to both states, the relative amounts of the states are irrelevant to the binding behavior.

9.31. Prove that the general four-site MWC equation predicts only positive cooperativity and never negative cooperativity.

SOLUTION

Examine Eq. (*9.89*) in relation to the Adair equation. In the latter, positive cooperativity requires that $K_1 < K_2 < K_3 < K_4$. If Eq. (*9.89*) is expanded and the terms in [X], $[\mathrm{X}]^2$, etc., are collected, the equation assumes the form of the Adair equation, if

$$K_1 = \frac{K_\mathrm{R}(1+Lc)}{1+L} \qquad K_2 = \frac{K_\mathrm{R}(1+Lc^2)}{1+Lc}$$

$$K_3 = \frac{K_\mathrm{R}(1+Lc^3)}{1+Lc^2} \qquad K_4 = \frac{K_\mathrm{R}(1+Lc^4)}{1+Lc^3}$$

Now examine the ratio of any of these four Adair constants, e.g., K_4/K_3:

$$\frac{K_4}{K_3} = \frac{(1+Lc^4)(1+Lc^2)}{(1+Lc^3)^2} = \frac{1 + Lc^2(1+c^2) + L^2c^6}{1 + 2Lc^3 + L^2c^6}\,.$$

Since L and c are both positive, $Lc^2(1+c^2) > 2Lc^3$, and thus $K_4 > K_3$. The same applies to the other pairs of Adair constants, and, in fact, the result may be generalized to the n-site model; i.e., only positive cooperativity or hyperbolic binding can be described by this model.

9.32. Derive the saturation function for the KNF tetrahedral model.

SOLUTION

The binding reactions are as follows:

$$X + A_4 \rightleftharpoons A_3BX$$
$$X + A_3BX \rightleftharpoons A_2B_2X_2$$
$$X + A_2B_2X_2 \rightleftharpoons AB_3X_3$$
$$X + AB_3X_3 \rightleftharpoons B_4X_4$$

The concentrations of the protein-ligand complexes are obtained by using Example 9.21 as a guide:

$$[A_3BX] = 4K_{AB}^3 K_x K_t\ [A_4]$$
$$[A_2B_2X_2] = 6K_{AB}^4 K_{BB}(K_x K_t)^2\ [A_4]$$
$$[AB_3X_3] = 4K_{AB}^3 K_{BB}^3(K_x K_t)^3\ [A_4]$$
$$[B_4X_4] = K_{BB}^6(K_x K_t)^4\ [A_4]$$

The statistical factors arise because there are four equivalent ways to bind one molecule of X to the tetramer, six equivalent ways to bind two molecules, four equivalent ways to bind three, and one way to bind four. Also, the number of interacting subunits is three A-B and three A-A for the A_3BX species; therefore, the constants K_{AA} and K_{AB} are raised to the power 3. For the $A_2B_2X_2$ species, the number of interactions is one B-B, four A-B, and one A-A. For AB_3X_3, the number is three B-B and three A-B interactions. Finally, for B_4X_4, there are six B-B interactions. Thus, by substituting the KNF terms above into the general saturation function (Eq. *9-80*), the following expression is obtained:

$$Y = \frac{K_{AB}^3(K_x K_t[X]) + 3K_{AB}^4 K_{BB}(K_x K_t[X])^2 + 3K_{AB}^3 K_{BB}^3(K_x K_t[X])^3 + K_{BB}^6(K_x K_t[X])^4}{1 + 4K_{AB}^3(K_x K_t[X]) + 6K_{AB}^4 K_{BB}(K_x K_t[X])^2 + 4K_{AB}^3 K_{BB}^3(K_x K_t[X])^3 + K_{BB}^6(K_x K_t[X])^4}$$

Similar Y expressions can be derived for the *square* and *linear* models; furthermore, these are *simpler* than the saturation function for the tetrahedral model.

9.33. The Rabin model (Example 9.24) is one example of an enzyme system that describes *kinetic cooperativity*. Another is the random Bi Bi mechanism depicted in Fig. 9-34.

By using the King-Altman or Indge and Childs method, it can be shown that the *initial rate* expression for such a system, where $[B]_0$ is held constant and $[A]_0$ is allowed to vary, is

$$v_0 = \frac{C_1[A]_0 + C_2[A]_0^2}{C_3 + C_4[A]_0 + C_5[A]_0^2}$$

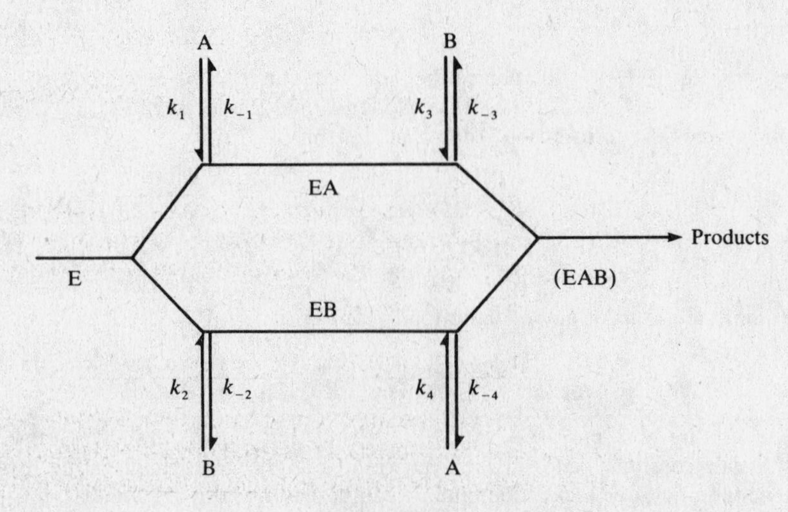

Fig. 9-34

where the C_i's are sums of products rate constants also involving the initial concentration of the second substrate B.

It can be shown that when $C_3C_2 > C_1C_4$, sigmoidal kinetics results. Assuming that $k_1k_3 > k_2k_4$ and that the affinity of E for B is less than EA for B, give a qualitative interpretation of the operation of this kinetic model and state how cooperativity may arise.

SOLUTION

From the assumptions above, the kinetic pathway

$$\text{E} \longrightarrow \text{EA} \longrightarrow \text{EAB, etc., is preferred over E} \longrightarrow \text{EB} \longrightarrow \text{EAB, etc.}$$

Let $[\text{B}]_0$ be high and constant. If $[\text{A}]_0$ is low, A will react mostly with EB, and thus the reaction will proceed via the *slower* pathway. At higher $[\text{A}]_0$ values, A will react with free E and also will *recruit* free E from the complex EB, and thus a greater proportion of the reaction flux will be via the *fast* pathway. Therefore, a sigmoidal v_0 versus [A] plot will result.

MISCELLANEOUS

9.34. The technique of *affinity labeling* is widely used to establish the nature of the amino acid residues in the active site of an enzyme. In this technique, a substrate or product analog (S′), able to form covalent bonds with suitable amino acid residues, is used to *label* residues in the active site:

$$\text{E} + \text{S}'\text{—X} \overset{K_d}{\rightleftharpoons} \text{E} \cdots \text{S}'\text{—X} \xrightarrow{k_{\text{inact}}} \text{E—S}' + \text{X}'$$

The affinity label is specific, because, in principle, the covalent reaction, via reactive group X, occurs only when the label is bound in the active site, by virtue of the similarity of the group S′ to the true substrate S. Derive an expression for the rate of inactivation of the enzyme as a function of [S′—X].

SOLUTION

The rate of inactivation of the enzyme is given by the rate of formation of E—S′. Thus, we have

$$\frac{d[\text{E—S}']}{dt} = k_{\text{inact}}[\text{E} \cdots \text{S}'\text{—X}]$$

Using the equilibrium expression

$$K_d = \frac{[\text{E}][\text{S}'\text{—X}]}{[\text{E} \cdots \text{S}'\text{—X}]}$$

and the conservation equation

$$[\text{E}]_T = [\text{E}] + [\text{E} \cdots \text{S}'\text{—X}] + [\text{E—S}']$$

we obtain,

$$[\text{E} \cdots \text{S}'\text{—X}] = \frac{[\text{E}]_T - [\text{E—S}']}{1 + K_d/[\text{S}'\text{—X}]}\,.$$

Hence,

$$\frac{d[\text{E—S}']}{dt} = k_{\text{inact}} \frac{[\text{E}]_T - [\text{E—S}']}{1 + K_d/[\text{S—X}']}$$

or

$$\frac{d[\text{E—S}']}{[\text{E}]_T - [\text{E—S}']} = \frac{k_{\text{inact}}\, dt}{1 + K_d/[\text{S—X}']}$$

Integrating gives

$$\ln([\text{E}]_T - [\text{E—S}']) = \{-k_{\text{inact}}t/(1 + K_d/[\text{S—X}'])\} + \ln[\text{E}]_T$$

Hence, if the natural logarithm of the enzyme activity $(\ln([\text{E}]_T - [\text{E—S}'])$ is plotted as a function of time t, the observed rate constant, k_{obs} (the negative slope of the plot), is given by:

$$k_{\text{obs}} = \frac{k_{\text{inact}}[\text{S}'\text{—X}]}{[\text{S}'\text{—X}] + K_d}$$

Supplementary Problems

9.35. Derive the Eadie-Hofstee equation from the Michaelis-Menten equation.

9.36. Derive the Hanes-Woolf equation from the Michaelis-Menten equation.

9.37. (*a*) Draw the 5 percent coefficient of variation envelope around the data of Fig. 9-20. (*b*) Do this plot and analysis yield unbiased parameter estimates?

9.38. (*a*) Integrate the competitive inhibition Michaelis-Menten equation (Table 9.1) and (*b*) derive a graphical procedure that enables the estimation of K_I.

9.39. (*a*) Integrate the noncompetitive inhibition Michaelis-Menten equation (Table 9.1) and (*b*) derive a graphical procedure that enables the estimation of K_I.

9.40. Prove that if an enzyme is inhibited competitively by its product and the numerical value of K_I is the same as that of K_m, the equation from Prob. 9.38 may be written as

$$[\mathrm{S}]_t = [\mathrm{S}]_0 \exp\left[-V_{\max} t/(K_m + [\mathrm{S}]_0)\right]$$

which is a pure single exponential decay.

9.41. Draw the direct linear plots that correspond to those in Fig. 9-21 for the various types of enzyme inhibition given in Table 9.1.

9.42. An ATPase from yak saliva was isolated, and the following ATP hydrolysis-rate data obtained, at an enzyme concentration of 10^{-8} mol L^{-1}:

[ATP] (μmol L^{-1})	5.0	1.7	1.0	0.7	0.56
v_0 (μmol L^{-1} min^{-1})	2.6	1.95	1.7	1.4	1.24

Determine K_m, V_m, and k_{cat} for the enzyme.

9.43. A *Theorell-Chance* enzyme mechanism is one in which the steady-state concentration of the central complex is effectively zero. In the case of a Bi Bi reaction, it may be represented as:

$$\mathrm{E} \underset{k_{-1}}{\overset{k_1\mathrm{A}}{\rightleftharpoons}} \mathrm{EA} \underset{k_{-2}\mathrm{P}}{\overset{k_2\mathrm{B}}{\rightleftharpoons}} \mathrm{EQ} \underset{k_{-3}\mathrm{Q}}{\overset{k_3}{\rightleftharpoons}} \mathrm{E}$$

Horse-liver *alcohol dehydrogenase* conforms to this basic scheme. Note that the central EAB/EPQ complex is not present. (*a*) Draw the reaction scheme by using the Cleland convention. (*b*) Show that the inhibition patterns for P and Q, with varied substrate A, both at saturating and at nonsaturating levels of B, are as follows:

P inhibition, nonsaturating B:	noncompetitive inhibition
P inhibition, saturating B:	no inhibition
Q inhibition, nonsaturating B:	competitive inhibition
Q inhibition, saturating B:	competitive inhibition

9.44. How does varying the substrate concentration relative to the enzyme concentration affect the validity of the steady-state assumption and the applicability of the Michaelis-Menten expression?

9.45. In a Michaelis-Menten enzyme mechanism, what substrate concentrations (relative to K_m) are needed for the reaction rate to be: (*a*) 0.1 $V_{\max}$; (*b*) 0.25 $V_{\max}$; (*c*) 0.5 $V_{\max}$; (*d*) 0.9 $V_{\max}$?

9.46. For an ionizable enzyme with a nonionizable substrate, the plot of pK_m versus pH has a maximum at pH $= -0.5 \log(K_{1e} \cdot K_{2e})$; prove this.

9.47. Using Fig. 9-13, calculate R_S for (*a*) hemoglobin in red blood cells, (*b*) pure hemoglobin, and (*c*) pure myoglobin.

9.48. Can the value of n in the Hill equation, when it is fitted to real data, ever be less than 1?

9.49. Derive the two-site Adair equation.

9.50. Derive the binding function Y for the KNF square model of a tetrameric protein.

Chapter 10

Metabolism: Underlying Theoretical Principles

10.1 INTRODUCTION

All living organisms transform energy taken from their surroundings. This energy is required for the synthesis of macromolecules to be used in growth and differentiation of the organism. These transformations are achieved via the action of a large number of enzymes (Chap. 8), catalyzing a complex network of chemical reactions collectively known as *metabolism*. This chapter reviews the theoretical principles underlying the study of this network of reactions; the detailed chemistry of metabolic processes is discussed in succeeding chapters.

10.2 THERMODYNAMICS

Living organisms can be considered as *physicochemical systems* interacting with their surroundings. *Thermodynamics* is the science of the energetics of such systems. It is a *macroscopic theory*, being concerned with the *bulk* properties of matter; the link between thermodynamics and molecular processes is provided by the theory of *statistical mechanics*.

Basic Concepts

(*a*) *System*: that part of the universe with which we are concerned (e.g., an organism or a glass vessel in which a chemical reaction is occurring). The universe apart from the system is the system's *surroundings*.

(*b*) *Open system*: a system in which both matter and heat can exchange with the surroundings.

(*c*) *Closed system*: a system in which only *heat* can exchange with the surroundings.

(*d*) *Isolated* (*adiabatic*) *system*: a system in which neither matter nor heat can exchange with the surroundings.

EXAMPLE 10.1

Should living organisms be considered open, closed, or isolated systems?

Living organisms are *open* systems because they exchange both heat and matter (nutrients, excreta) with their surroundings.

(*e*) *State functions*: properties relating to changes in a system that are dependent *only* on its initial and final states. Many of the important properties of systems discussed below (e.g., internal energy, enthalpy, entropy, and Gibbs free energy) are state functions.

First Law of Thermodynamics

The first law of thermodynamics states that if a system exchanges heat with, or does work on, its surroundings, then a *change* in its *internal energy* (ΔU) occurs. This is expressed mathematically as

$$\Delta U = \Delta q - \Delta w \qquad (10.1)$$

where Δq is the heat exchanged with the surroundings and Δw is the work done on the surroundings.

EXAMPLE 10.2

What determines whether Δq and Δw are positive or negative?

If Δq is positive, heat has been transferred *to* the *system*, giving an *increase* in internal energy. When Δq is negative, heat has been transferred *to* the *surroundings*, giving a *decrease* in internal energy. When Δw is positive, work has been done by the system, giving a *decrease* in internal energy. When Δw is negative, work has been done by the surroundings, giving an *increase* in internal energy.

Question: How can ΔU values be measured experimentally?

For chemical reactions, ΔU values can be obtained by allowing the reaction to proceed in an *adiabatic calorimeter*, in which temperature changes during the reaction can be measured and converted to values for Δq. The most common source of work in chemical reactions is changes in volume (V) during the reaction. The work done against a pressure P is given by:

$$\Delta w = \int_{v_1}^{v_2} P\, dV \tag{10.2}$$

where v_1 and v_2 are the initial and final volumes of the system. Hence, from Eq. (*10.1*)

$$\Delta U = \Delta q - \int_{v_1}^{v_2} P\, dV \tag{10.3}$$

In a *constant volume* adiabatic calorimeter, dV is zero, and therefore $\Delta U = \Delta q_v$, while in a *constant pressure* adiabatic calorimeter, the above integral is $P\,\Delta V$, and hence:

$$\Delta U = \Delta q_p - P\,\Delta V \tag{10.4}$$

Question: How do calorimeters work?

In a *constant volume* adiabatic calorimeter (Fig. 10-1), the reaction of interest is initiated in a constant volume vessel, which is surrounded by a water bath. The calorimeter is well insulated, ensuring that there is no loss of heat from the system. The temperature change (ΔT) arising from the completed reaction is converted to Δq_v using the *heat capacity* (C) of the calorimeter: $\Delta q_v = C\,\Delta T$. A *constant pressure* adiabatic calorimeter is based on the same principles, except the vessel in which the reaction occurs is designed to operate at constant pressure, rather than constant volume.

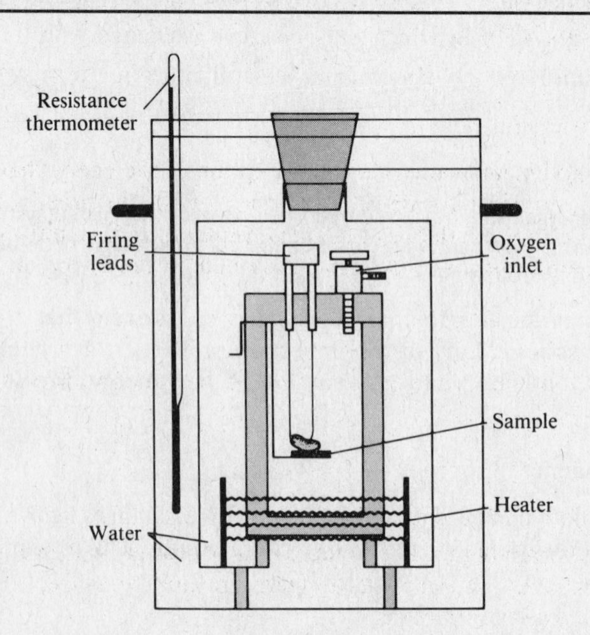

Fig. 10-1 Constant volume bomb calorimeter.

Enthalpy Changes

Many chemical reactions are conveniently studied at constant pressure, e.g., atmospheric pressure. From Eq. (*10.4*) above, the heat change during such a reaction is:

$$\Delta q_p = \Delta U + P\,\Delta V \qquad (10.5)$$

This heat change is called the *enthalpy change* for a reaction and is symbolized by ΔH. Most biochemical reactions occur in solution; in these cases volume changes are negligible, and thus $\Delta H = \Delta U$.

EXAMPLE 10.3

The ΔU values for oxidation of glucose ($C_6H_{12}O_6$) and stearic acid ($C_{18}H_{36}O_2$) are -2.9×10^3 kJ mol^{-1} and -11.36×10^3 kJ mol^{-1}, respectively, at 310 K. (*a*) Calculate ΔH for these processes; (*b*) which of the two substances is more useful as an *energy store* in the body?

(*a*) For glucose the reaction is

$$C_6H_{12}O_6 + 6O_2 \longrightarrow 6CO_2 + 6H_2O$$

In this case, there is no volume change (both O_2 and CO_2 are gases at 310 K). Hence, from Eq. (*10.5*), $\Delta U = \Delta H = -2.9 \times 10^3$ kJ mol^{-1}.

For stearic acid the reaction is

$$C_{18}H_{36}O_2 + 26O_2 \longrightarrow 18CO_2 + 18H_2O$$

In this case, 8 moles of gas are consumed per mole of stearic acid ($26\,O_2 - 18\,CO_2$). Assuming that the ideal gas law applies, i.e.,

$$P\,\Delta V = \Delta n\,RT \qquad (10.6)$$

where n is the number of moles of gas, and R is the gas constant (8.314 J mol^{-1} K^{-1}), and T is the absolute temperature, then

$$\Delta H = \Delta U + \Delta n\,RT = -\left(11.36 \times 10^3 + \frac{8 \times 8.31 \times 310}{1{,}000}\right) \text{kJ mol}^{-1} = -11{,}381 \text{ kJ mol}^{-1}$$

(*b*) Clearly, a fat such as stearic acid is a far more energy-rich storage substance for a given number of carbon atoms than is glucose.

Second Law of Thermodynamics

The first law of thermodynamics provides a description of the energy balance for a given process; the second law provides a criterion for deciding whether or not the process will occur *spontaneously*. The second law of thermodynamics defines the *entropy change* (ΔS, in units of J K^{-1}) associated with a change in a closed system in terms of the heat absorbed by the system at constant temperature T;

$$\Delta S \geq \Delta q / T \qquad (10.7)$$

The equality in this expression refers to a *reversible* process, while the inequality refers to an *irreversible* process.

Gibbs Free Energy

The second law of thermodynamics provides that a process will occur spontaneously if $\Delta q < T\,\Delta S$. For processes occurring at constant temperature and pressure, $\Delta q = \Delta H$ (see Eq. (*10.5*) above). Hence, for a spontaneous process

$$\Delta H - T\,\Delta S < 0 \qquad (10.8)$$

The quantity $(\Delta H - T\,\Delta S)$ is symbolized by ΔG. Thus, if a process occurs spontaneously, $\Delta G < 0$. The symbol G refers to the *Gibbs free energy*, defined as

$$G = H - TS \tag{10.9}$$

and for a reaction at constant temperature and pressure

$$\Delta G = \Delta H - T\,\Delta S \tag{10.10}$$

The value of ΔG gives the maximum *work* available from the process, apart from work associated with changes in pressure or volume. ΔG may be expressed in joules or calories per mole ($4.186\ \text{J} = 1\ \text{cal}$).

EXAMPLE 10.4

Most of the energy transformed by higher animals derives from the oxidation of glucose:

$$C_6H_{12}O_6 + 6O_2 \longrightarrow 6CO_2 + 6H_2O$$

Given $\Delta H = -2{,}808\ \text{kJ mol}^{-1}$ and $\Delta S = 182.4\ \text{J K}^{-1}\ \text{mol}^{-1}$ for this reaction, how much energy is available from oxidation of 1 mole of glucose at 310 K?

We use:

$$\Delta G = \Delta H - T\,\Delta S = -(2{,}808 \times 10^3 + 310 \times 182.4)\ \text{J mol}^{-1} = -2{,}865\ \text{kJ mol}^{-1}$$

Hence, digestion of 1 mole (180.2 g) of glucose at 310 K provides an animal with 2,865 kJ of non-*PV* work.

Standard States

The *standard state* of a pure substance is defined as that form, at a specified temperature, that is stable at 1 atmosphere pressure (101.325 kPa). For *solutes*, the standard state is more conveniently defined as a $1\ \text{mol L}^{-1}$ solution of the solute. For chemical reactions in solution, the *standard free energy change* (ΔG°) is that for converting $1\ \text{mol L}^{-1}$ of reactants into $1\ \text{mol L}^{-1}$ of products:

$$\Delta G^\circ = \Delta H^\circ - T\,\Delta S^\circ \tag{10.11}$$

Biochemical Standard States

Many biological processes involve hydrogen ions; the standard state of an H^+ solution is (by definition) a $1\ \text{mol L}^{-1}$ solution, which would have a pH of nearly 0—a condition incompatible with most forms of life. Hence, it is convenient to define the *biochemical* standard state for solutes, in which all components except H^+ are at $1\ \text{mol L}^{-1}$, and H^+ is present at $10^{-7}\ \text{mol L}^{-1}$ (i.e., pH 7). Biochemical standard-state free energy changes are symbolized by $\Delta G^{\circ\prime}$, and the other thermodynamic parameters are indicated analogously ($\Delta H^{\circ\prime}$, $\Delta S^{\circ\prime}$, etc.).

Equilibrium and Free Energy

The free energy change for the reaction

$$a\text{A} + b\text{B} \rightleftharpoons c\text{C} + d\text{D}$$

is given by

$$\Delta G = \Delta G^\circ + RT \ln\left(\frac{([\text{C}]/[\text{C}]^\circ)^c([\text{D}]/[\text{D}]^\circ)^d}{([\text{A}]/[\text{A}]^\circ)^a([\text{B}]/[\text{B}]^\circ)^b}\right) \tag{10.12}$$

where $[\text{A}]^\circ$, $[\text{B}]^\circ$, $[\text{C}]^\circ$, and $[\text{D}]^\circ$ are the concentrations of the reactants in their standard states.

However, for reactions in solution, these concentrations are all 1 mol L^{-1}, hence Eq. (*10.12*) can be simplified to

$$\Delta G = \Delta G^\circ + RT \ln\left(\frac{[C]^c}{[A]^a}\frac{[D]^d}{[B]^b}\right) \tag{10.13}$$

If the above chemical reaction is at equilibrium (i.e., a reversible process), then $\Delta G = 0$. Hence, from Eq. (*10.13*)

$$\Delta G^\circ = -RT \ln\left(\frac{[C]_e^c}{[A]_e^a}\frac{[D]_e^d}{[B]_e^b}\right) \tag{10.14}$$

where the subscript e indicates the equilibrium concentrations of the reactants. However, the expression in parentheses is the *equilibrium constant* (K_e) for the reaction. Hence

$$\Delta G^\circ = -RT \ln K_e \tag{10.15}$$

EXAMPLE 10.5

The enzyme phosphoglucomutase catalyzes the reaction

$$\text{Glucose 1-phosphate} \rightleftharpoons \text{Glucose 6-phosphate}$$

which has ΔG° of -7.3 kJ mol^{-1}. If the enzyme is added to a 2×10^{-4} mol L^{-1} solution of glucose 1-phosphate at 310 K, what will be the equilibrium composition of the solution?

From Eq. (*10.15*)

$$K_e = \exp(-\Delta G^\circ/RT) = \exp(7.3 \times 10^3/8.31 \times 310) = 17.00 \tag{10.16}$$

Hence

$$\frac{[\text{Glucose 6-phosphate}]_e}{[\text{Glucose 1-phosphate}]_e} = 17.00$$

and

$$[\text{Glucose 1-phosphate}] + [\text{Glucose 6-phosphate}] = 2 \times 10^{-4} \text{ mol L}^{-1}$$

Therefore

$$[\text{Glucose 6-phosphate}]_e = 1.89 \times 10^{-4} \text{ mol L}^{-1}$$
$$[\text{Glucose 1-phosphate}]_e = 1.11 \times 10^{-5} \text{ mol L}^{-1}$$

Variation of Equilibrium Constant with Temperature

From Eq. (*10.15*), we have

$$\ln K_e = -\Delta G^\circ/RT \tag{10.17}$$

Hence

$$\frac{d \ln K_e}{dT} = -\frac{1}{R}\,\frac{d\left(\frac{\Delta G^\circ}{T}\right)}{dT} \tag{10.18}$$

Using Eq. (*10.11*)

$$\Delta G^\circ = \Delta H^\circ - T\,\Delta S^\circ$$

we have

$$\frac{\Delta G^\circ}{T} = \frac{\Delta H^\circ}{T} - \Delta S^\circ \tag{10.19}$$

Assuming that ΔH° and ΔS° are *independent* of temperature

$$\frac{d\left(\frac{\Delta G^\circ}{T}\right)}{dT} = -\frac{\Delta H^\circ}{T^2} \tag{10.20}$$

Hence, from (*10.18*) and (*10.20*)

$$\frac{d \ln K_e}{dT} = \frac{\Delta H^\circ}{RT^2} \tag{10.21}$$

This equation, the *van't Hoff isochore*, is often more useful in the integrated form:

$$\int_{K_1}^{K_2} d \ln K_e = \int_{T_1}^{T_2} \frac{\Delta H^\circ}{RT^2}\, dT \tag{10.22}$$

EXAMPLE 10.6

The enzyme phosphorylase *b* binds AMP:

$$\text{Phosphorylase } b \cdot \text{AMP} \rightleftharpoons \text{Phosphorylase } b + \text{AMP}$$

The equilibrium constant for this reaction is 2.75×10^{-5} mol L^{-1} at 286 K and 5.9×10^{-5} mol L^{-1} at 313 K. What are $\Delta H°$, $\Delta S°$, and $\Delta G°$ for the reaction at 303 K?

From Eq. (*10.22*), assuming that $\Delta H°$ is independent of temperature,

$$\ln\left(\frac{5.9 \times 10^{-5}}{2.75 \times 10^{-5}}\right) = \frac{-\Delta H°}{8.31}\left(\frac{1}{313} - \frac{1}{286}\right)$$

Hence $$\Delta H° = 21.03 \text{ kJ mol}^{-1}$$

Using this value, we can calculate the value of the equilibrium constant at 303 K

$$\ln\left(\frac{K_{303}}{5.9 \times 10^{-5}}\right) = \frac{-21.03 \times 10^3}{8.31}\left(\frac{1}{303} - \frac{1}{313}\right) = -0.2668$$

Therefore $$K_{303} = 4.5 \times 10^{-5} \text{ mol L}^{-1}$$

Now, substituting this value into the equation

$$\Delta G° = -RT \ln K_e$$

we have $$\Delta G°_{303\text{ K}} = 25.2 \text{ kJ mol}^{-1}$$

Hence, using $$\Delta G° = \Delta H° - T\,\Delta S°$$

$\Delta S°$ can be calculated to be -13.8 J K^{-1} mol^{-1}.

10.3 REDOX REACTIONS

Chemical reactions involving oxidation and reduction processes (*redox* reactions) are central to metabolism. The energy derived from the oxidation of carbohydrates is coupled to the synthesis of ATP via a series of redox reactions, the mitochondrial electron-transport chain (see Chap. 14). Moreover, most life on earth is dependent on a series of redox reactions in *photosynthesis*, the process in which solar energy is used to produce ATP and O_2 and to synthesize carbohydrates from CO_2.

Basic Concepts

(*a*) Oxidation reaction: reaction in which a substance *loses* electrons.

(*b*) Reduction reaction: reaction in which a substance *gains* electrons.

(*c*) Half-cell reaction: the oxidation *or* reduction step in a *redox* reaction.

EXAMPLE 10.7

What are the half-cell reactions in the following redox reaction?

$$Zn + Cu^{2+} \rightleftharpoons Zn^{2+} + Cu$$

The *oxidation* half-cell reaction is

$$Zn \longrightarrow Zn^{2+} + 2e^-$$

The *reduction* half-cell reaction is

$$Cu^{2+} + 2e^- \longrightarrow Cu$$

Free Energy Changes in Redox Reactions

The free energy change for a redox reaction is given by

$$\Delta G = -nFE \qquad (10.23)$$

where n is the number of electrons transferred from each molecule of the substance being oxidized to that being reduced, E is the *electromotive force* (in volts) required to prevent the electron transfer, and F is the *Faraday constant*, a conversion factor, approximately equal to 96.5 kJ V^{-1} mol^{-1}. When the components of a redox reaction are in their standard states, Eq. (*10.24*) applies:

$$\Delta G^\circ = -nFE^\circ \tag{10.24}$$

EXAMPLE 10.8

ΔG° for the reaction

$$Zn + Cu^{2+} \rightleftharpoons Zn^{2+} + Cu$$

is -213 kJ mol^{-1}. What is E° for this reaction?

From Eq. (*10.24*), we have

$$E^\circ = -\frac{\Delta G^\circ}{nF} \tag{10.25}$$

$$= \frac{213}{2 \times 96.5} = 1.1 \text{ V}$$

(because *two* electrons are transferred).

Electrochemical Cells

Redox reactions can be studied using electrochemical cells. An electrochemical cell for the chemical reaction in Example 10.8 is shown in Fig. 10-2. The Cu and Zn electrodes dip into solutions of their respective ions and the *salt bridge* (containing concentrated KCl) maintains electrical contact between the two solutions. Electrons will flow from the Zn half-cell to the Cu half-cell if Zn is oxidized to Zn^{2+}, with concomitant reduction of Cu^{2+} to Cu in the Cu half-cell. The value of E for this reaction may be determined by measuring the potential difference (in volts) that has to be applied to the cell to prevent the electron flow.

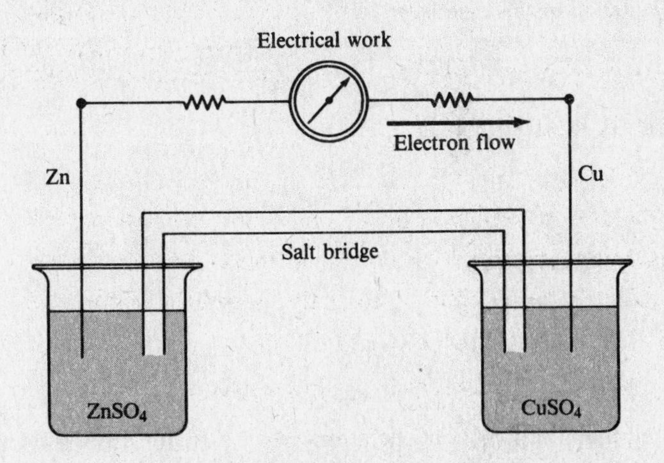

Fig. 10-2 Electrochemical cell.

Half-Cell Electrode Potentials

Half-cell reactions cannot be studied in isolation; all that can be measured is the difference in potential (ΔE) when two half-cells are linked to form an electrochemical cell. *Relative* electrode potentials for half-cells are obtained by reference to a standard half-cell, the *hydrogen electrode*, which is assigned an E° of zero. The half-cell reaction for hydrogen is

$$H^+ + e^- \rightleftharpoons \tfrac{1}{2} H_2 \text{ (Pt)}$$

Hydrogen gas at 1 atm pressure is bubbled over a Pt electrode in a 1 mol L^{-1} H^+ solution. $E°$ values for other half-cell reactions (with their components in their standard states) may be measured using electrochemical cells in which the hydrogen electrode is linked to an electrode at which the reaction of interest occurs.

The Relationship between $E°$ and E

The free energy change for the general redox reaction occurring in solution

$$A_{ox} + B_{red} \longrightarrow A_{red} + B_{ox}$$

is

$$\Delta G = \Delta G° + RT \ln \left(\frac{[A_{red}][B_{ox}]}{[A_{red}]°[B_{ox}]°} \bigg/ \frac{[A_{ox}][B_{red}]}{[A_{ox}]°[B_{red}]°} \right) \tag{10.26}$$

Using Eq. (*10.23*)

$$\Delta G = -nFE$$

and recalling that, for solutes, $[A_{red}]° = [B_{ox}]° = 1$ mol L^{-1}, etc. [cf. Eq. (*10.13*)], we have

$$E = E° - \frac{RT}{nF} \ln \left(\frac{[A_{red}][B_{ox}]}{[B_{red}][A_{ox}]} \right) \tag{10.27}$$

This equation, the *Nernst redox equation*, provides a way of relating E and $E°$ for any redox reaction or half-cell reaction.

EXAMPLE 10.9

What is E for the half-cell reaction

$$NAD^+ + H^+ + 2e^- \longrightarrow NADH$$

at pH 7, 298 K, for 1 mol L^{-1} solutions of NAD^+ and NADH, given $E°$ is -0.11 V?

The Nernst redox equation in this case is

$$E = E° - \frac{RT}{2F} \ln \left(\frac{[NADH]}{[NAD^+][H^+]} \right)$$

Substituting in values for $E°$, [NADH], $[NAD^+]$, etc., we get

$$E = -0.11 - \frac{8.31 \times 298}{2 \times 9.65 \times 10^4} \ln (10^{+7}) = -0.32 \text{ V}$$

Since the components of the reaction are in their *biochemical standard states* (see Sec. 10.2), the above value for E is, in fact, $E°'$ for this half-cell reaction.

10.4 ATP AND ITS ROLE IN BIOENERGETICS

Properties of ATP

ATP is present in all living cells at concentrations of 10^{-3} to 10^{-2} mol L^{-1} of cell water. The ATP molecule is composed of three parts: adenine, D-ribose, and three phosphate groups in ester linkages (Chap. 7). The analogous compounds containing one and two phosphate groups are designated AMP and ADP, respectively.

ATP has a strong tendency to hydrolyze to ADP and phosphate; this is predicted from thermodynamics.

$$ATP^{4-} + H_2O \longrightarrow ADP^{3-} + HPO_4^{2-} + H^+$$

Under biochemical standard-state conditions ($[ATP^{4-}] = [ADP^{3-}] = [HPO_4^{2-}] = 1$ mol L^{-1}, $[H^+] = 10^{-7}$ mol L^{-1}, and $T = 310$ K), the thermodynamic parameters for this reaction are $\Delta G°' = -30.5$ kJ mol^{-1}, $\Delta H°' = -20$ kJ mol^{-1}, and $\Delta S°' = 34$ J K^{-1} mol^{-1}.

Question: Why is the hydrolysis of ATP thermodynamically favorable?

The *enthalpy* change for the reaction is favorable because (1) electrostatic repulsion between the negative charges in ATP exceeds that in the reaction products, (2) the reaction products are resonance stabilized, and (3) the enthalpies of solvation of the products are larger than that for ATP. The *entropy* change for the reaction is favorable because of the release of a phosphate group. Note that this implies that ATP hydrolysis is strongly temperature-dependent (cf. Eq. (*10.7*)).

ATP is often referred to as being *energy-rich* or containing *high-energy phosphate ester bonds*. This nomenclature should be thought of as implying that the overall process of ATP hydrolysis has large negative values of $\Delta G^{\circ\prime}$ and $\Delta H^{\circ\prime}$.

ATP in Bioenergetics

ATP is *not* a high-energy compound in comparison with other biological compounds. The functions of ATP depend on its having a ΔG value for hydrolysis that is *intermediate* in value compared with ΔG values for hydrolysis of other phosphate esters. Thus, ATP and ADP can act as a *donor-acceptor* pair for *phosphoryl-group transfer*. In many cases the free energy of ATP hydrolysis is used to *support* reactions that would otherwise be thermodynamically unfavorable. This usually occurs via phosphorylation of one of the reactants in an *unfavorable* reaction.

EXAMPLE 10.10

The synthesis of glutamine from glutamate and NH_4^+ is thermodynamically unfavorable:

$$\text{Glutamate}^- + NH_4^+ \longrightarrow \text{Glutamine} \qquad (\Delta G^{\circ\prime} = 14.2\ \text{kJ mol}^{-1})$$

How could the hydrolysis of ATP be used to render this reaction usable for the synthesis of glutamine?

The original reaction is altered by using ATP. The new reaction, which is thermodynamically favorable, consists of two *partial reactions* linked through a common intermediate via ATP hydrolysis:

$$\text{Glutamate}^- + \text{ATP}^{4-} \longrightarrow \text{5-Phosphoglutamate}^{2-} + \text{ADP}^{3-}$$

$$\text{5-Phosphoglutamate}^{2-} + NH_4^+ \longrightarrow \text{Glutamine} + \text{Phosphate}^-$$

The sum of these two reactions is:

$$\text{Glutamate}^- + \text{ATP}^{4-} + NH_4^+ \longrightarrow \text{Glutamine} + \text{ADP}^{3-} + \text{Phosphate}^-$$

$\Delta G^{\circ\prime}$ for this reaction is -16.3 kJ mol^{-1}, and hence it is thermodynamically favorable. The enzyme *glutamine synthetase* catalyzes this reaction in animal cells (Chap. 15).

10.5 CONTROL POINTS IN METABOLIC PATHWAYS

Enzymes that are subject to *control signals* generally fulfill two criteria: they are present at low enzymatic activities and catalyze reactions that are not at equilibrium under cellular conditions. Both criteria arise because control enzymes are likely to be those catalyzing the slowest (*rate-determining*) step in a metabolic pathway. This is likely to be the case if an enzyme is present at low activity. If this is the case, the enzyme-catalyzed reaction is unlikely to be at equilibrium in vivo because there is insufficient enzyme present to allow equilibration of its reactants before they react with other compounds.

Identification of Rate-Determining Steps in Metabolic Pathways

Identifying rate-determining steps can be done by measuring the activities of the enzymes in a metabolic pathway in vitro, using a homogenate of the tissue of interest and assay conditions [pH, temperature, ionic strength (Prob. 10.3), and substrate concentrations] similar to those in the cell. In some cases it is possible to measure enzyme activities in vivo using nuclear magnetic resonance spectroscopy (NMR).

Detection of Nonequilibrium Reactions

Nonequilibrium reactions can be detected by determining metabolite concentrations in the tissue of interest. Conventionally, a tissue sample is rapidly frozen by compression between metal plates that have been cooled to 77 K by immersion in liquid nitrogen (*freeze-clamping*). This procedure rapidly halts any enzymatic processes that might alter the metabolite concentrations; the concentrations can then be determined by enzymatic or chemical assays. Recently, ^{31}P-nuclear magnetic resonance spectroscopy has shown considerable promise for measuring the concentrations of such metabolites as ATP, ADP, AMP, phosphate, and phosphocreatine in living cells or tissues.

Once the metabolite concentrations are known, the *mass-action ratio* (Γ) can be calculated. For the reaction

$$A + B + C + \cdots \rightleftharpoons P + Q + R + \cdots$$

$$\Gamma = \frac{[P][Q][R]\ldots}{[A][B][C]\ldots} \tag{10.28}$$

A comparison of Γ with the equilibrium constant for the reaction will establish whether or not the reaction is near equilibrium in vivo.

EXAMPLE 10.11

The following data were obtained for metabolite concentrations in a rat heart perfused with glucose:

Metabolite	Intracellular Concentration (mmol L^{-1})
ATP	11.5
ADP	1.3
AMP	0.17
Fructose 6-phosphate	0.09
Fructose 1,6-diphosphate	0.02

Are the enzymes (*a*) phosphofructokinase and (*b*) adenylate kinase likely to be control enzymes in heart metabolism?

(*a*) For the phosphofructokinase reaction

$$\text{Fructose 6-phosphate} + \text{ATP} \xrightarrow{Mg^{2+}} \text{Fructose 1,6-diphosphate} + \text{ADP}$$

$$\Gamma = \frac{[\text{fructose 1,6-diphosphate}][\text{ADP}]}{[\text{fructose 6-phosphate}][\text{ATP}]}$$

$$= (0.02 \times 1.3)/(0.09 \times 11.5) = 0.025$$

This value is far smaller than that for the equilibrium constant for the reaction (~1,200); therefore, the reaction is *far* from equilibrium in rat heart, and hence phosphofructokinase is likely to be a control enzyme.

(*b*) For the adenylate kinase reaction

$$\text{ATP} + \text{AMP} \overset{Mg^{2+}}{\rightleftharpoons} 2\text{ADP}$$

$$\Gamma = \frac{[\text{ADP}]^2}{[\text{ATP}][\text{AMP}]} = (1.3)^2/(11.5 \times 0.17) = 0.86$$

This value is similar to that for the equilibrium constant for the reaction (~0.4); hence, this reaction is unlikely to be the control point of its metabolic pathway.

Note that we have assumed that all the intracellular metabolites are available to the enzymes mentioned; no account has been taken of possible effects of *metabolic compartmentation* (see Sec. 10.7). Hence, deductions about possible control enzymes, based on mass-action ratios, should be viewed with some caution.

The Crossover Theorem

This principle relates to experiments in which a metabolic pathway is perturbed, for example, by adding an inhibitor or activator of one of the enzymes in the pathway. Following such a perturbation, the metabolite concentrations before and after a control enzyme in the pathway will change in opposite directions. This arises because such an enzyme catalyzes a slow, often rate-determining, step in the pathway. For example, if the perturbation *decreases* the activity of the enzyme, there will be an *increase* in the concentration of its substrate(s) and a *decrease* in the concentration of its product(s). This theorem is strictly valid only for nonbranching linear segments of metabolic pathways.

10.6 AMPLIFICATION OF CONTROL SIGNALS

The response of the *flux* (mass flow per unit time) through a metabolic pathway, but not the flux through a single enzyme-catalyzed reaction, to a regulatory signal can be subject to *amplification mechanisms*. Regulatory signals are usually changes in the concentrations of substrates or other effector molecules. These chemical signals can be amplified by *substrate cycles* or by cycles of *interconvertible enzymes*.

Substrate Cycles

In the following segment of a metabolic pathway

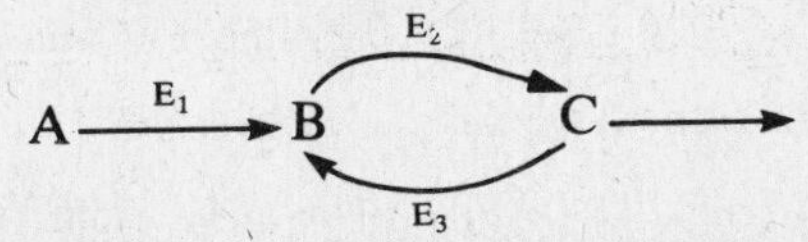

a *substrate cycle* occurs between B and C: separate enzymes catalyze the interconversion of these two metabolites. The net flux through this step is the rate of E_2 minus the rate of E_3. If the rates of *both* enzymes are separately controlled, then the flux through the pathway is far more sensitive to changes in regulator concentration than if only one enzyme is involved.

EXAMPLE 10.12

In glycolysis, the enzymes phosphofructokinase (PFK) and fructose 1,6-diphosphatase (FDP) form a substrate cycle:

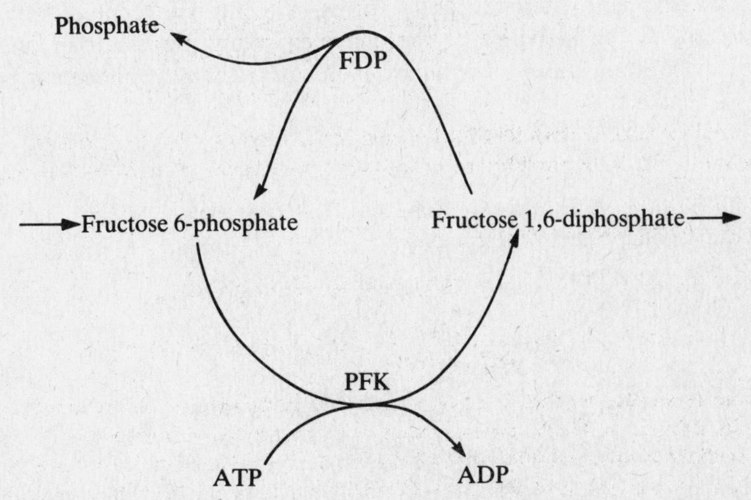

The enzymatic activity of PFK is increased and that of FDP is decreased by elevated concentrations of AMP. The following data were obtained for the fractional saturation (Chap. 9) of the two enzymes with AMP:

AMP Concentration (arbitrary units)	Fractional Saturation
0	0
2.5	0.093
12.5	0.89

Calculate the change in flux through the pathway as [AMP] changes from 2.5 to 12.5 concentration units, assuming: that the maximum enzymatic activities for PFK and FDP are 100 and 10 mmol L^{-1} min^{-1}, respectively; that PFK is inactive in the absence of bound AMP; and that FDP is inactive when it has bound AMP.

The activity of PFK is given by

$$\text{PFK activity} = \text{maximum activity} \times \text{proportion of enzyme with bound AMP}$$

while that for FDP is given by

$$\text{FDP activity} = \text{maximum activity} \times \text{proportion of enzyme without bound AMP}$$

At an AMP level of 2.5 concentration units, we have

$$\text{PFK activity} = 100 \times 0.093 = 9.3 \text{ mmol L}^{-1} \text{ min}^{-1}$$

$$\text{FDP activity} = 10 \times (1.0 - 0.093) = 9.1 \text{ mmol L}^{-1} \text{ min}^{-1}$$

Hence, the relative flux through this section of glycolysis is

$$\text{Flux} = \text{PFK activity} - \text{FDP activity}$$
$$= 9.3 - 9.1 = 0.2 \text{ mmol L}^{-1} \text{ min}^{-1}$$

At an AMP level of 12.5 concentration units, we have analogously

$$\text{PFK activity} = 100 \times 0.89 = 89 \text{ mmol L}^{-1} \text{ min}^{-1}$$

$$\text{FDP activity} = 10 \times (1.0 - 0.89) = 1.1 \text{ mmol L}^{-1} \text{ min}^{-1}$$

$$\text{Flux} = 87.9 \text{ mmol L}^{-1} \text{ min}^{-1}$$

The substrate cycle has allowed a nearly 440-fold increase (87.9/0.2) in the flux through the pathway for only a 5-fold (12.5/2.5) change in the concentration of AMP.

Cycles of Interconvertible Enzymes

Some regulatory enzymes exist in different forms, the interconversions between these forms being enzyme-catalyzed. If the activities of the enzymes catalyzing the interconversions are controlled by changes in concentrations of effector molecules, then *amplification* of these chemical signals can occur.

EXAMPLE 10.13

The activity of the *pyruvate dehydrogenase complex* (PDHC) in mammals can be altered by an enzyme-catalyzed phosphorylation (Chap. 12) of the complex:

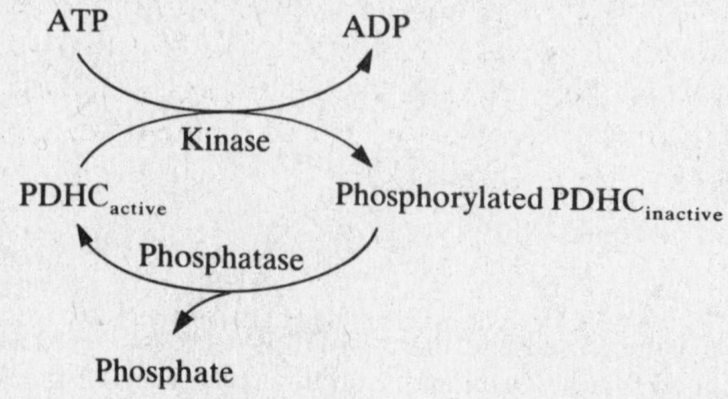

How could the activities of the kinase and phosphorylase be regulated so as to control the entry of pyruvate into the tricarboxylic acid cycle?

Pyruvate enters the tricarboxylic acid cycle after conversion into acetyl-CoA via the PDHC. The tricarboxylic acid cycle generates NADH from NAD^+, and the NADH then enters the mitochondrial electron-transport chain. Hence, if either the ratios of acetyl-CoA/CoA or NADH/NAD^+ were high, it would be advantageous to render the PDHC inactive. There are three ways in which this is achieved: the kinase is active (1) when the NADH/NAD^+ ratio is high and (2) when the acetyl-CoA/CoA ratio is high; the phosphatase is inhibited (3) when the ratio of NADH/NAD^+ is high. Thus, the portion of the pyruvate dehydrogenase complex that is active is sensitive to the metabolic requirements of the cell.

10.7 INTRACELLULAR COMPARTMENTATION AND METABOLISM

Prokaryotic cells have volumes in the range 0.03–500 μm^3 and appear to contain no subcellular structures, whereas eukaryotic cells are larger, with volumes mostly in the range 50–15,000 μm^3, and contain many discrete subcellular organelles (Chap. 1). These organelles can be isolated, following disruption of the cells or tissue of interest, by differential centrifugation; organelles of different sizes and densities will sediment at different rates when placed in a centrifugal field.

Compartmentation of Metabolic Pathways

There are many connections between the main metabolic pathways. Many substrates and regulatory molecules, and some enzymes, are common to several pathways. An understanding of these connections requires knowledge of (1) the subcellular locations and concentrations of the enzymes involved, (2) the concentrations of metabolites within different subcellular organelles, and (3) the nature of permeability barriers for metabolites between the organelles; these barriers divide the cell into a number of compartments for each metabolite.

Location of Enzymes

The distribution of enzymes among the various subcellular organelles can be established by measurement of enzyme activities in the various fractions isolated by differential centrifugation. In this way, for example, the enzymes of glycolysis have been localized to the cytosol, while the distribution of the enzymes of the reverse process, gluconeogenesis (Chap. 11), has been found to be more complex. In this pathway, fructose diphosphatase, together with the enzymes common to glycolysis, is cytosolic, while pyruvate carboxylase is found in the mitochondrial matrix, and phosphoenolpyruvate carboxykinase is cytosolic in rats, but is found in the mitochondria of pigeons.

Location of Metabolites

Establishing the concentrations of metabolites within different subcellular organelles or compartments is a complex task, because metabolite concentrations may change during the time required to isolate the organelles and low-molecular-weight metabolites may diffuse across organelle membranes during the isolation process. Most attention has been given to the concentrations of adenine and nicotinamide nucleotides and to intermediates of the tricarboxylic acid cycle in the cytosol and mitochondria. In some cases these concentrations can be estimated indirectly from measurements of enzyme-catalyzed reactions that are at equilibrium.

EXAMPLE 10.14

Lactate dehydrogenase is a cytosolic enzyme that is present at high enough concentrations for its reaction to be at equilibrium.

$$\text{Lactate} + NAD^+ \rightleftharpoons \text{Pyruvate} + NADH + H^+$$

The equilibrium constant for this reaction is known from in vitro studies. How could this be used to calculate the cytosolic $NADH/NAD^+$ ratio?

The cytosolic concentrations of pyruvate and lactate can be obtained from the results of freeze-clamping experiments. Hence, from the known $[H^+]$ of the cytosol and the known equilibrium constant (K_e)

$$K_e = \frac{[\text{Pyruvate}][\text{NADH}][\text{H}^+]}{[\text{Lactate}][\text{NAD}^+]}$$

the $NADH/NAD^+$ ratio can be estimated as ~1:1,000.

Solved Problems

THERMODYNAMICS

10.1. A sample of acetic acid (1 mol) was completely oxidized to CO_2 and H_2O in a constant volume adiabatic calorimeter, at 298 K. The heat released in the oxidation was 874 kJ. Calculate ΔH for the oxidation of acetic acid.

SOLUTION

At constant volume, the heat absorbed by the system (Δq_v) is equal to ΔU. Hence, because heat is released (transferred *from* the system *to* the surroundings)

$$\Delta U = -\Delta q_v = -874 \text{ kJ mol}^{-1}$$

ΔU is related to ΔH by the equation

$$\Delta H = \Delta U + P\,\Delta V$$

However, given the reaction for oxidation of acetic acid

$$CH_3COOH + 2O_2 \longrightarrow 2CO_2 + 2H_2O$$

there is no volume change (as both O_2 and CO_2 are gases at 298 K), and hence ΔH for the oxidation is -874 kJ mol^{-1}.

10.2. The $\Delta H°$ value for oxidation of fumaric acid (HOOC—CH=CH—COOH) is $-1{,}336$ kJ mol^{-1}. Given the following data:

$$C + O_2 \longrightarrow CO_2 \qquad \Delta H° = -393 \text{ kJ mol}^{-1}$$

$$H_2 + \tfrac{1}{2}O_2 \longrightarrow H_2O \qquad \Delta H° = -285.5 \text{ kJ mol}^{-1}$$

calculate the enthalpy of formation ($\Delta H°$) of fumaric acid from its elements.

SOLUTION

Because enthalpy is a state function, we can calculate $\Delta H°$ for fumaric acid by summation of the enthalpy changes for the following reactions:

	$\Delta H°$
$4C + 4O_2 \longrightarrow 4CO_2$	$-1{,}572$ kJ
$2H_2 + O_2 \longrightarrow 2H_2O$	-571 kJ
$4CO_2 + 2H_2O \longrightarrow C_4H_4O_4 + 3O_2$	$+1{,}336$ kJ

Net equation:

$$4C + 2H_2 + 2O_2 \longrightarrow C_4H_4O_4 \qquad \Delta H° = -807 \text{ kJ mol}^{-1}$$

This procedure, which is an example of *Hess's law of constant heat summation*, relies upon the fact that ΔH for any reaction depends only on the final and initial states, and illustrates a convenient method for calculating ΔH values that would be difficult to measure experimentally.

10.3. The respective concentrations of Na^+ and K^+ ions in tissues are approximately 10 mmol L^{-1} and 90 mmol L^{-1} inside cells and 140 mmol L^{-1} and 4 mmol L^{-1} outside cells. Calculate the free-energy requirements for maintenance of these ion gradients.

SOLUTION

We can represent the ion gradients by the equilibria:

$$Na^+_{in} \rightleftharpoons Na^+_{out}$$

$$K^+_{in} \rightleftharpoons K^+_{out}$$

In both cases, the ΔG values are given by the expression:

$$\Delta G = \Delta G^\circ + RT \ln\left(\frac{[\text{ion}]_{out}}{[\text{ion}]_{in}}\right)$$

$$\Delta G^\circ = -RT \ln\left(\frac{[\text{ion}]_{out_e}}{[\text{ion}]_{in_e}}\right)$$

where the subscript e indicates the equilibrium concentrations of the ions. However, at equilibrium one would expect equal concentrations of the ions inside and outside the cells. Hence, $\Delta G^\circ = 0$, and therefore the free energy required to maintain the ion gradients is:

$$\Delta G = RT \ln\left(\frac{[\text{ion}]_{out}}{[\text{ion}]_{in}}\right)$$

Substituting the Na^+ and K^+ concentrations into this equation, we have

$$\Delta G_{Na^+} = 6.8 \text{ kJ mol}^{-1}$$

$$\Delta G_{K^+} = 8.0 \text{ kJ mol}^{-1}$$

We have made a number of assumptions in this calculation, the most notable being that the ionic solutions are *ideal*, in that there are no interactions (attractive or repulsive) between solute molecules. It is most unlikely that this is the case, especially in moderately concentrated solutions of ions. In order to correct for nonideality (interactions between solute molecules), we need to substitute the *activities* of solute molecules for their *concentrations* in all thermodynamic calculations. The activity (a) of a solute molecule is related to its concentration (C) by an *activity coefficient* (γ).

$$a = \gamma C \tag{10.29}$$

For solutions of electrolytes, γ can be calculated using *Debye-Hückel theory*. The mean activity coefficient ($\gamma_\pm$) of positively and negatively charged ions in solution is given by:

$$\log \gamma_\pm = -0.5 z_+ z_- I^{1/2} \tag{10.30}$$

where z_+ and z_- are the charges carried by the ions and I is the *ionic strength* of the solution:

$$I = \tfrac{1}{2} \Sigma\, C_i z_i^2 \tag{10.31}$$

where the sum is carried out over all ion types (i) in the solution.

10.4. Calculate the mean activity coefficient in a solution of 0.25 mol L^{-1} Na_3PO_4.

SOLUTION

This salt is completely dissociated in water. First, we calculate the ionic strength:

$$I = \tfrac{1}{2}(C_{Na^+} z^2_{Na^+} + C_{PO_4^{3-}} z^2_{PO_4^{3-}}) = \tfrac{1}{2}(0.75 \times 1^2 + 0.25 \times 3^2) = 1.5 \text{ mol L}^{-1}$$

Now, using

$$\log \gamma_\pm = -0.5 z_+ z_- I^{1/2}$$

with $z_+ = 1$, and $z_- = 3$

$$\log \gamma_\pm = -1.837$$

$$\gamma_\pm = 0.0146$$

Strictly, this value is valid only at 298 K, as the constant (0.5) in the equation for $\log \gamma_\pm$ is slightly different at other temperatures.

10.5. Consider the general dissociation of an acid HA

$$HA \rightleftharpoons H^+ + A^-$$

Derive an expression for the dependence of the concentrations of dissociated and undissociated acid ($[A^-]$ and [HA]) on pH.

SOLUTION

The equilibrium constant for the dissociation is

$$K_a = \frac{[H^+][A^-]}{[HA]}$$

Hence,

$$\log K_a = \log [H^+] + \log\left(\frac{[A^-]}{[HA]}\right)$$

and since $pH = -\log [H^+]$

$$-\log K_a = pH - \log\left(\frac{[A^-]}{[HA]}\right)$$

Substituting pK_a for $-\log K_a$

$$pK_a = pH - \log\left(\frac{[A^-]}{[HA]}\right)$$

This is the Henderson-Hasselbalch equation (Chap. 3).

REDOX REACTIONS

10.6. Calculate $\Delta G^{\circ\prime}$ for the following reaction:

$$\text{Pyruvate} + \text{NADH} + H^+ \rightleftharpoons \text{Lactate} + \text{NAD}^+$$

The half-cell reactions are:

$$\text{Pyruvate} + 2H^+ + 2e^- \longrightarrow \text{Lactate} \qquad E^{\circ\prime} = -0.19\text{ V}$$

$$\text{NAD}^+ + H^+ + 2e^- \longrightarrow \text{NADH} \qquad E^{\circ\prime} = -0.32\text{ V}$$

SOLUTION

From the half-cell reactions, $E^{\circ\prime}$ for the reaction is $-0.19 - (-0.32) = 0.13$ V. Now, using the equation

$$\Delta G^{\circ\prime} = -nFE^{\circ\prime}$$

with $n = 2$ in this case,

$$\Delta G^{\circ\prime} = -2 \times 9.65 \times 10^4 \times 0.13\text{ J mol}^{-1} = -25.1\text{ kJ mol}^{-1}$$

Note that a *positive* $E^{\circ\prime}$ value for a redox reaction implies that it is thermodynamically *favorable*, having a *negative* value of $\Delta G^{\circ\prime}$.

10.7. The *electrochemical potential* μ of an ion of charge z in an *electrostatic field* ψ is (assuming ideal behavior) defined by the equation

$$\mu = zF\psi + \mu^\circ + RT \ln X \tag{10.32}$$

where μ° is the *chemical potential* (free energy per mole) of the ion in its standard state and X is the *mole fraction* of the ion, given by the expression

$$X = \frac{(\text{moles of ion})}{(\text{moles of ion}) + (\text{moles of solvent})} \tag{10.33}$$

For dilute solutions (moles of solvent) $\gg$ (moles of ion); hence

$$X = \frac{\text{moles of ion}}{\text{moles of solvent}} \tag{10.34}$$

In the *chemiosmotic theory* for oxidative phosphorylation (Chap. 14), electron flow in the electron-transport chain is coupled to generation of a proton concentration gradient across the inner mitochondrial membrane. Derive an expression for the difference in electrochemical potential for a proton across the membrane.

SOLUTION

The difference in chemical potential is given by the expression

$$\Delta\mu = \mu_{\text{in}} - \mu_{\text{out}}$$

From Eq. (*10.32*),

$$\Delta\mu = zF(\psi_{\text{in}} - \psi_{\text{out}}) + RT \ln\left(\frac{[\text{H}^+]_{\text{in}}}{[\text{H}^+]_{\text{out}}}\right)$$

Hence

$$\Delta\mu = zF\,\Delta\psi - 2.3\,RT\,\Delta\text{pH} \tag{10.35}$$

where $\Delta\psi$ is the potential difference across the membrane, and ΔpH the pH difference across the membrane. This is the *Nernst potential equation*.

ATP AND ITS ROLE IN BIOENERGETICS

10.8. The negatively charged phosphate groups of ATP provide an effective means of chelating divalent cations such as Mg^{2+}. Given that the following reaction

$$\text{Mg}^{2+} + \text{ATP}^{4-} \rightleftharpoons \text{MgATP}^{2-}$$

has an association equilibrium constant of 10^4 L mol^{-1}, calculate the proportion of ATP in a cell that is present as the Mg^{2+} chelate if the free Mg^{2+} concentration is 2×10^{-2} mol L^{-1}.

SOLUTION

$$\frac{[\text{MgATP}^{2-}]}{[\text{Mg}^{2+}][\text{ATP}^{4-}]} = 10^4\ \text{L mol}^{-1}$$

Hence,

$$\frac{[\text{MgATP}^{2-}]}{[\text{ATP}^{4-}]} = 2 \times 10^2$$

Therefore, more than 99 percent of the ATP in a cell is present as the Mg^{2+} chelate. It is thus not surprising that most ATP-utilizing enzymes use $MgATP^{2-}$, rather than ATP^{4-}, as a substrate.

CONTROL POINTS IN METABOLIC PATHWAYS

10.9. Given the following hypothetical scheme for the enzyme-catalyzed interconversions of metabolites A_1–A_6:

$$A_1 \xrightarrow{E_1} A_2 \xrightarrow{E_3} A_4 \xrightarrow{E_4} A_5 \xrightarrow{E_5} A_6$$
$$A_2 \xrightarrow{E_2} A_3$$

which enzyme in the scheme would be most suitable for being subject to metabolic control in terms of regulating the levels of the metabolite A_6?

SOLUTION

If enzyme E_1 were subject to feedback inhibition at high levels of metabolite A_6, then high levels of A_6 would lead to decreased production of metabolite A_2. This could have the undesirable effect of decreasing the production of metabolite A_3, via the action of enzyme E_2. This problem is overcome if enzyme E_3, which catalyzes the *first committed step* in the production of A_6, is subject to metabolic control.

AMPLIFICATION OF CONTROL SIGNALS

10.10. Does the existence of the phosphofructokinase–fructose 1,6-diphosphatase *substrate cycle* have any energetic disadvantages?

SOLUTION

This substrate cycle has the disadvantage of acting, in effect, as an *ATPase*; ATP is converted to ADP and phosphate without net production of fructose 1,6-diphosphate. This may be outweighed by the advantage conferred on an animal of having an efficient control mechanism in the form of a cycle.

INTRACELLULAR COMPARTMENTATION AND METABOLISM

10.11. The enzyme succinate dehydrogenase (Chaps. 12 and 14) catalyzes the reaction:

$$\text{Succinate} + \text{FAD} \longrightarrow \text{Fumarate} + \text{FADH}_2$$

A sample of rat liver was homogenized, the various cell organelles were isolated by differential centrifugation, and the succinate dehydrogenase enzymatic activity in each organelle preparation was determined:

Organelle Preparation	Succinate Dehydrogenase Activity (% of total)
Nuclei	11
Mitochondria	75
Lysosomes	6
Plasma membrane	5
Cytosol	3

How could these data be useful in investigating intracellular compartmentation?

SOLUTION

The data show that succinate dehydrogenase is found primarily in the mitochondria; it can thus act as a *marker enzyme* for these organelles. Marker enzymes can be used to determine the extent of contamination of a particular subcellular fraction by other organelles; in this case, for example, it would appear that the preparation of nuclei is contaminated with mitochondria.

Supplementary Problems

10.12. The *osmotic pressure* (π) of a solution is given by the equation

$$\pi = RTa \tag{10.36}$$

where a is the activity of the solute. Sketch the curve you would expect to see for the relationship between the osmotic pressure of a solution of NaCl and the concentration of NaCl.

10.13. Given the half-cell potentials

$$\tfrac{1}{2}O_2 + 2H^+ + 2e^- \rightleftharpoons H_2O \qquad E^{\circ\prime} = 0.82\ \text{V}$$

$$NAD^+ + H^+ + 2e^- \rightleftharpoons NADH \qquad E^{\circ\prime} = -0.32\ \text{V}$$

calculate $\Delta G^{\circ\prime}$ across the electron-transport chain.

10.14. Given that $\Delta G^{\circ\prime}$ for the synthesis of ATP from ADP and phosphate is 30.5 kJ mol^{-1}, and assuming that two protons are translocated by the mitochondrial ATPase per molecule of ATP synthesized, use Eq. (*10.35*) in Prob. 10.7 to calculate the *minimum* value of $\Delta\mu$ necessary for synthesis of ATP. (*Hint*: $\Delta\mu$ has units of kJ mol^{-1}; how can this be converted to units of mV to give $\Delta\mu$ as proton-motive force Δp?)

10.15. Calculate the ionic strengths of solutions of Na_2SO_4 at (*a*) 0.01 mol L^{-1}; (*b*) 0.05 mol L^{-1}; (*c*) 0.1 mol L^{-1}; and (*d*) 1.0 mol L^{-1}.

10.16. The enzyme pyruvate carboxylase (Chap. 12) from chicken liver is an oligomer composed of four identical subunits. The enzyme loses its catalytic activity when cooled below 277 K. Assuming that this loss of activity reflects dissociation of the tetrameric enzyme into its subunits, what can you deduce about the relative importance of enthalpic and entropic effects in the association of the subunits in the tetrameric enzyme?

10.17. (*a*) For the following alcohol dehydrogenase–catalyzed reaction

$$CH_3CH_2OH + NAD^+ \rightleftharpoons CH_3CHO + NADH + H^+$$

derive an expression for the variation of ΔG° with pH. (*b*) Given that $\Delta G^{\circ\prime}$ for this reaction is 18.5 kJ mol^{-1}, what is ΔG° at a pH of 6 ($T = 298$ K)?

10.18. For the ionization of acetic acid

$$CH_3COOH \rightleftharpoons CH_3COO^- + H^+$$

ΔG° and ΔH° values at 298 K are 27.1 kJ mol^{-1} and -0.39 kJ mol^{-1}. (*a*) Calculate ΔS° for the reaction at 298 K. (*b*) Evaluate the equilibrium constant for the reaction at 323 K, noting any assumptions necessary for the calculation.

10.19. The hydrolysis of acetyl phosphate

$$\text{Acetyl phosphate} \rightleftharpoons \text{Acetate} + \text{Phosphate}$$

has a ΔG° value of -42.3 kJ mol^{-1} at 298 K. In principle, which of the following phosphate ester hydrolysis reactions could be used to synthesize acetyl phosphate by a mechanism analogous to that in Example 10.10?

(*a*) 1,3 Diphosphoglycerate $\longrightarrow$ 3-Phosphoglycerate + Phosphate
($\Delta G^{\circ\prime} = -57$ kJ mol^{-1})

(*b*) Creatine phosphate $\longrightarrow$ Creatine + Phosphate
($\Delta G^{\circ\prime} = -42.7$ kJ mol^{-1})

(*c*) ADP $\longrightarrow$ AMP + Phosphate
($\Delta G^{\circ\prime} = -27.6$ kJ mol^{-1})

10.20. The *control strength* (C) provides a way of describing relative changes in the flux through a metabolic pathway arising from changes in the activity of an enzyme in the pathway. The control strength is defined as:

$$C_i = \frac{\partial \ln v_g}{\partial \ln v_i} = \frac{\partial v_g}{\delta v_i} \cdot \frac{v_i}{v_g} \tag{10.37}$$

where v_g is the net flux through the pathway, and v_i that through the step catalyzed by enzyme E_i. If the substrates of the enzyme are present in large excess, show that

$$C_i = \frac{[E_i]}{v_g} \cdot \frac{\partial v_g}{\partial [E_i]} \tag{10.38}$$

(*Hint*: The control strength can also be written as

$$C_i = v_i(\partial v_g/\partial p)/v_g(\partial v_i/\partial p)$$

where p is an arbitrary parameter. What is the relationship between v_i and $[E_i]$ at high substrate concentrations?)

10.21. Prove that the sum of the control strengths of all enzymes in a linear sequence of reactions is 1.

Chapter 11

Carbohydrate Metabolism

11.1 GLYCOLYSIS

Glycolysis is a process that results in the conversion of a molecule of glucose to two molecules of pyruvate. It is a primitive metabolic pathway since it operates in even the simplest and archaic cells and does not require oxygen. The pathway of glycolysis performs *five* functions in the cell:

1. Glucose is converted to pyruvate, which can be oxidized in the citric acid cycle (Chap. 12).
2. Many compounds other than glucose can enter the pathway at intermediate stages.
3. In some cells the pathway of enzymes is modified to enable glucose to be synthesized.
4. The pathway contains intermediates that are involved in alternative metabolic reactions.
5. For each glucose that is consumed, two molecules of ADP are phosphorylated by *substrate-level* phosphorylation to produce two molecules of ATP.

Question: What is the overall, balanced chemical equation for glycolysis?

$$\underset{\text{Glucose}}{C_6H_{12}O_6} + 2ADP + 2NAD^+ + 2P_i \longrightarrow \underset{\text{Pyruvate}}{2C_3H_4O_3} + 2ATP + 2NADH + 2H^+ + 2H_2O$$

The apparent simplicity of this equation conceals the complexity of the glycolytic pathway, which involves 9 intermediate compounds and 10 enzymes; the enzymes are located in the cytoplasm of the cell.

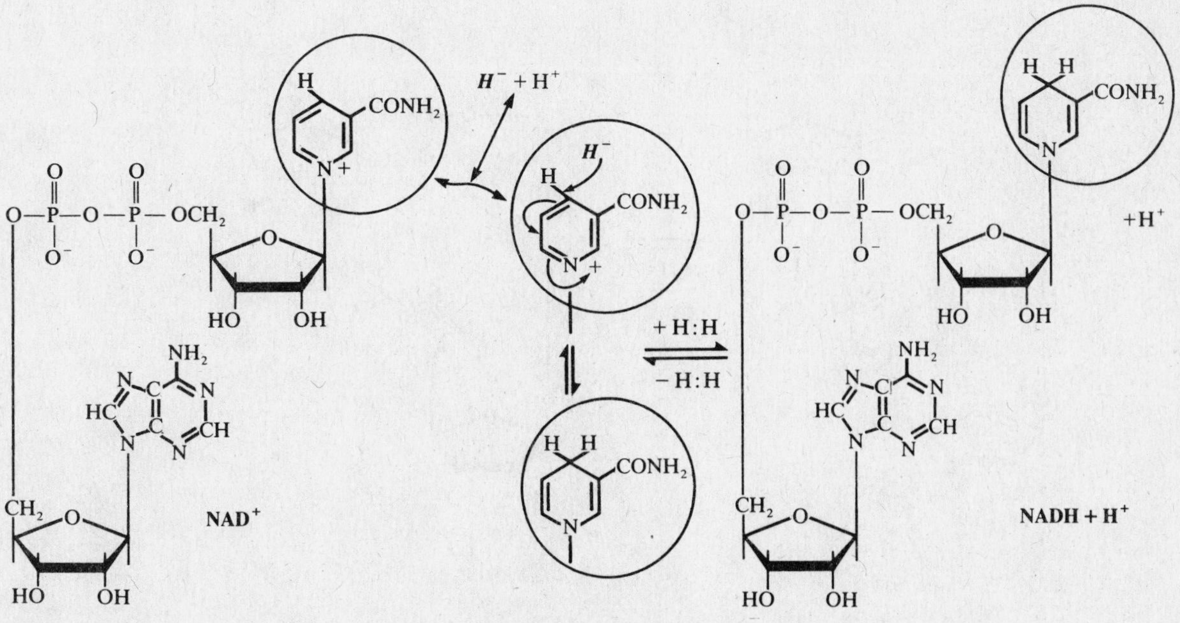

Fig. 11-1 The interconversion of NAD^+ and NADH.

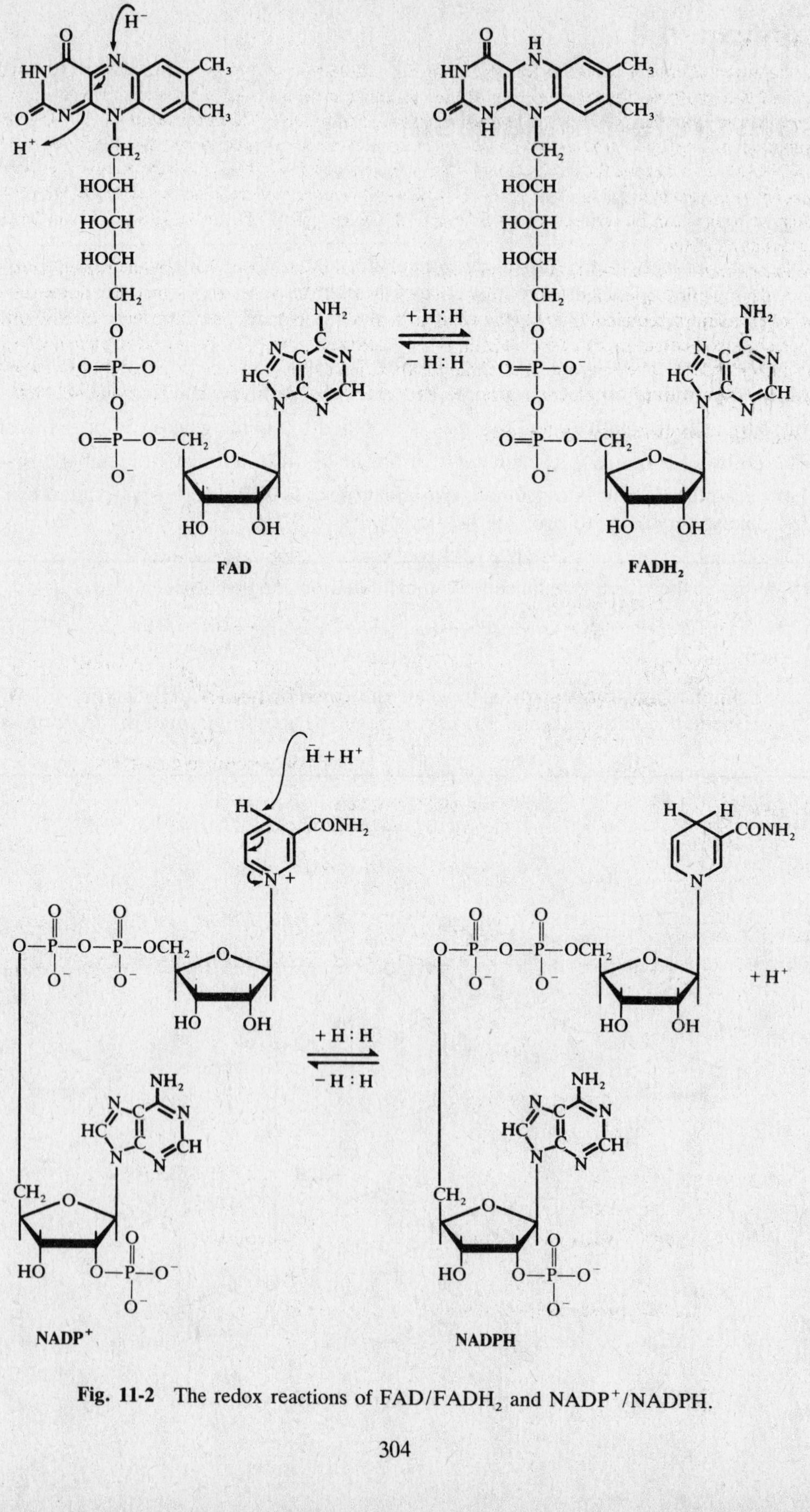

Fig. 11-2 The redox reactions of FAD/$FADH_2$ and $NADP^+$/NADPH.

EXAMPLE 11.1

In the balanced chemical equation for glycolysis, two molecules of NAD^+ are converted to two molecules of NADH and two protons. The structure of NAD^+ (*n*icotinamide *a*denine *d*inucleotide) is given in Fig. 11.1. This is a reduction, and the NAD^+, an enzyme cofactor, has accepted the equivalent of H^- (a *hydride ion*). When a substrate is oxidized, it loses a pair of electrons that are contained within the structure H:H, and the function of NAD^+ is to accept these electrons. The relevant portion of the molecule involved in accepting the electrons as H^- is the aromatic ring. When the reduction has been completed, the reaction has produced NADH and H^+. Just as NAD^+ can be *reduced* to NADH and H^+ by accepting 2H's, NADH and H^+ can be *oxidized* to NAD^+ by donating 2H's.

There are two other cofactors that can participate in *redox* processes; these are *f*lavin *a*denine *d*inucleotide (FAD) and *n*icotinamide *a*denine *d*inucleotide *p*hosphate ($NADP^+$), both of which are shown in Fig. 11-2. FAD accepts 2H's and is reduced to $FADH_2$, whereas $NADP^+$ accepts H^- and is reduced to NADPH and H^+. Both of these reduced cofactors can be oxidized, thereby donating their H's (or *reducing equivalents*), similar to the oxidation of NADH. The enzymes that catalyze those reactions involving an oxidation or a reduction are usually very selective toward a particular cofactor (NAD or NADP) in a particular oxidation state.

The steps in glycolysis are as follows:

Step 1

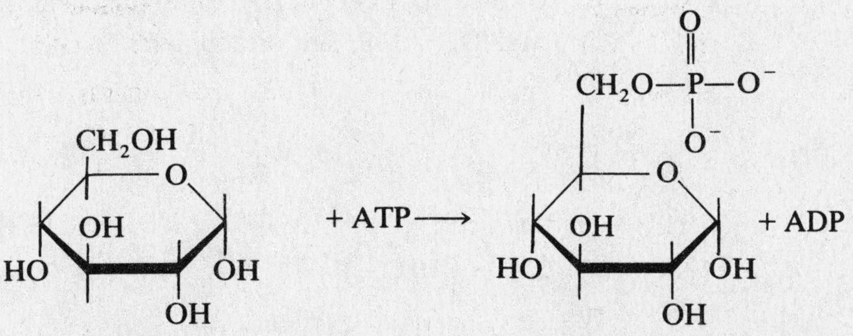

α-D-Glucose + ATP ⟶ α-D-Glucose 6-phosphate + ADP

Reaction type:	phosphorylation
Enzyme:	hexokinase
Coreactants:	Mg^{2+} or Mn^{2+}, ATP
Effectors:	
Inhibited by:	glucose 6-phosphate, MgATP, 2,3-diphosphoglycerate
Activated by:	orthophosphate (which competes with MgATP)

Step 2

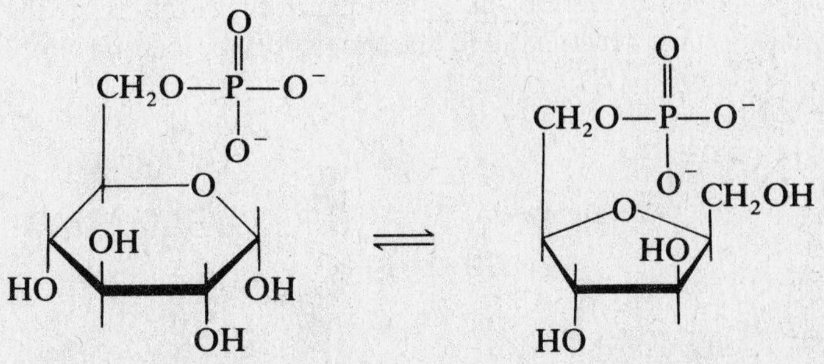

α-D-Glucose 6-phosphate ⇌ α-D-Fructose 6-phosphate

Reaction type:	isomerization
Enzyme:	glucose 6-phosphate isomerase
Coreactants:	none

Step 3

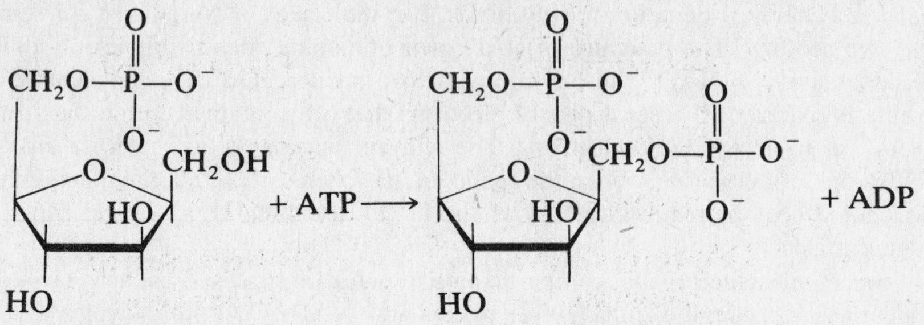

α-D-Fructose 6-phosphate + ATP ⟶ α-D-Fructose 1,6-diphosphate + ADP

Reaction type:	phosphorylation
Enzyme:	phosphofructokinase
Coreactants:	Mg^{2+}, ATP
Effectors:	
Inhibited by:	citrate, MgATP, NADH, and long-chain, fatty acids
Activated by:	MgADP, AMP, and orthophosphate

Step 4

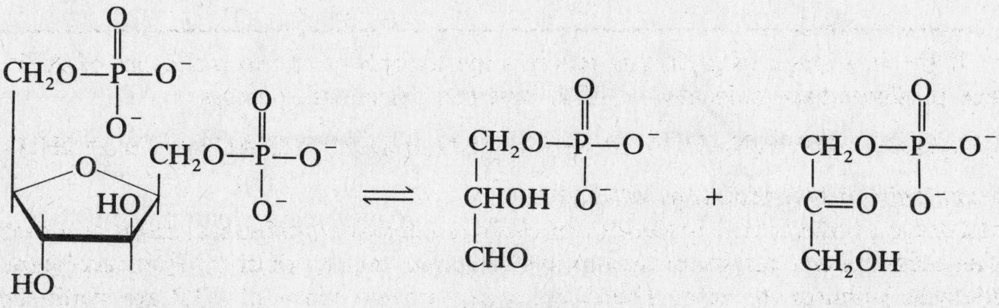

α-D-Fructose 1,6-diphosphate ⇌ D-Glyceraldehyde 3-phosphate + Dihydroxyacetone phosphate

Reaction type:	reverse- or retro-aldol condensation
Enzyme:	fructose 1,6-diphosphate, usually called aldolase
Coreactants:	none

The mechanism of Step 4 involves a retro-aldol condensation reaction on the open-chain form of D-fructose 1,6-diphosphate. This reaction, and the origin of the carbon atoms in the products, is shown below.

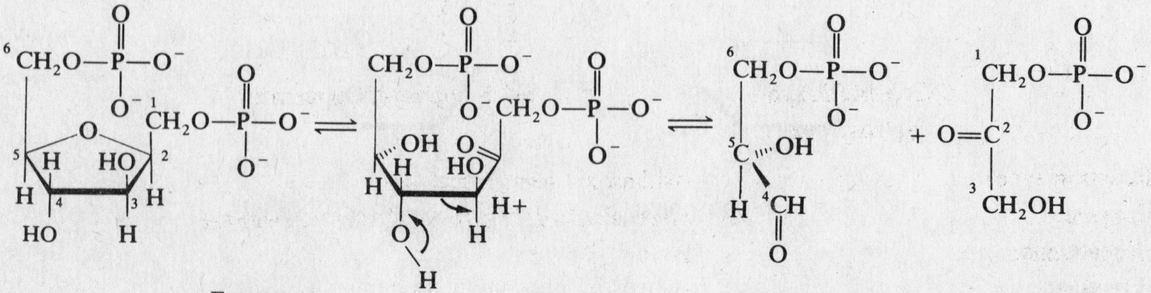

Fructose 1,6-diphosphate ⇌ ⇌ D-Glyceraldehyde 3-phosphate + Dihydroxyacetone phosphate

Step 5

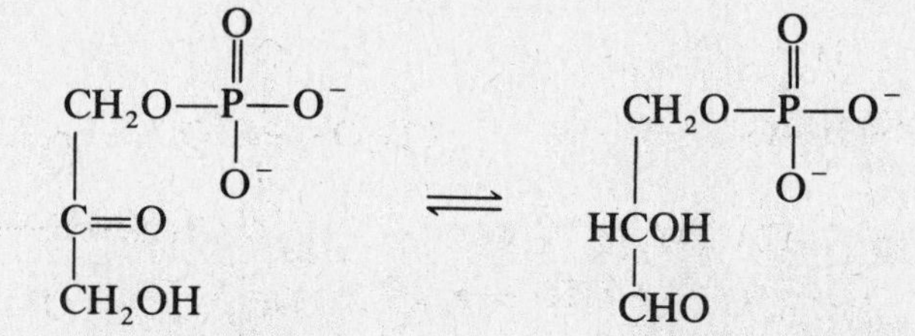

$$\begin{array}{c} CH_2O{-}\overset{\overset{\textstyle O}{\|}}{\underset{\underset{\textstyle O^-}{|}}{P}}{-}O^- \\ | \\ C{=}O \\ | \\ CH_2OH \end{array} \rightleftharpoons \begin{array}{c} CH_2O{-}\overset{\overset{\textstyle O}{\|}}{\underset{\underset{\textstyle O^-}{|}}{P}}{-}O^- \\ | \\ HCOH \\ | \\ CHO \end{array}$$

Dihydroxyacetone phosphate　　D-Glyceraldehyde 3-phosphate

Reaction type:	isomerization
Enzyme:	triosephosphate isomerase
Coreactants:	none

Steps 4 and 5 result in the production of two molecules of D-glyceraldehyde 3-phosphate from one molecule of α-D-fructose 1,6-diphosphate. The glycolytic pathway to this point (i.e., the conversion of glucose into two molecules of D-glyceraldehyde 3-phosphate) is called the *first stage* of glycolysis, and two molecules of ATP are hydrolyzed (at Steps 1 and 3) to provide the necessary energy. The remaining five steps compose the *second stage* of glycolysis and yield two molecules of ATP from ADP.

Question: If the first stage of glycolysis requires the hydrolysis of two molecules of ATP and the second stage produces two molecules of ATP, how can the overall process

$$C_6H_{12}O_6 + 2ADP + 2NAD^+ + 2P_i \longrightarrow 2C_3H_4O_3 + 2ATP + 2NADH + 2H^+ + 2H_2O$$

produce a net yield of two molecules of ATP?

The first stage produces *two* molecules of D-glyceraldehyde 3-phosphate from one molecule of glucose. The second stage results in the formation of *two* molecules of ATP for *each* molecule of D-glyceraldehyde 3-phosphate used. Therefore, a net two molecules of ATP are synthesized from each molecule of glucose.

Step 6

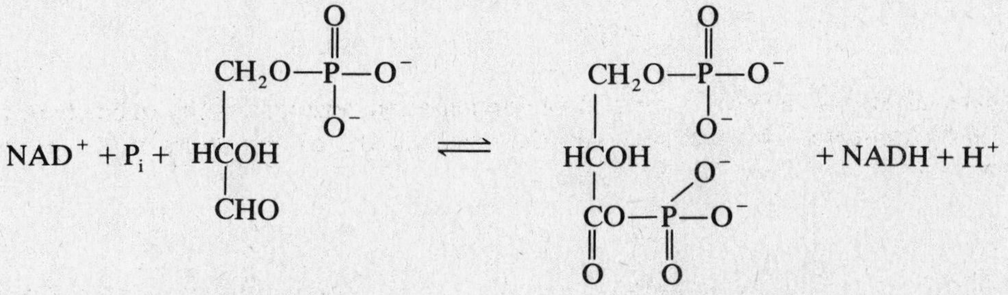

$$NAD^+ + P_i + \begin{array}{c} CH_2O{-}\overset{\overset{\textstyle O}{\|}}{\underset{\underset{\textstyle O^-}{|}}{P}}{-}O^- \\ | \\ HCOH \\ | \\ CHO \end{array} \rightleftharpoons \begin{array}{c} CH_2O{-}\overset{\overset{\textstyle O}{\|}}{\underset{\underset{\textstyle O^-}{|}}{P}}{-}O^- \\ | \\ HCOH \\ | \\ \underset{\underset{\textstyle O}{\|}}{C}O{-}\overset{\overset{\textstyle O^-}{/}}{\underset{\underset{\textstyle O}{\|}}{P}}{-}O^- \end{array} + NADH + H^+$$

D-Glyceraldehyde 3-phosphate　　D-1,3-Diphosphoglycerate

Reaction type:	oxidation + phosphorylation
Enzyme:	glyceraldehyde 3-phosphate dehydrogenase
Coreactants:	NAD^+, P_i
Effectors:	
Inhibited by:	NADH

Step 7

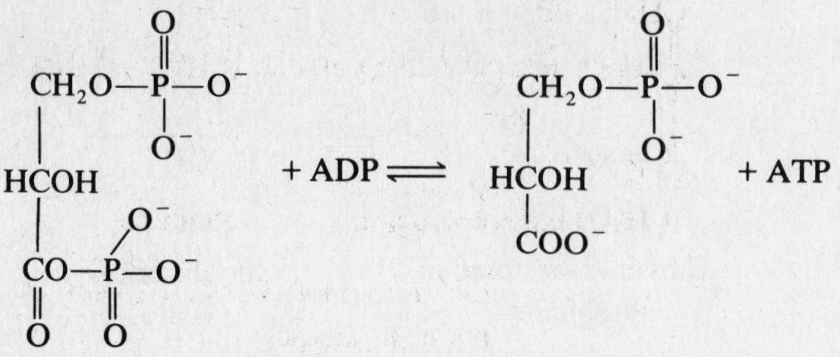

$$\text{D-1,3-Diphosphoglycerate} + \text{ADP} \rightleftharpoons \text{D-3-Phosphoglycerate} + \text{ATP}$$

D-1,3-Diphosphoglycerate: $CH_2O{-}PO_3^{2-}$ / HCOH / $CO{-}O{-}PO_3^{2-}$ (C=O)

D-3-Phosphoglycerate: $CH_2O{-}PO_3^{2-}$ / HCOH / COO^-

Reaction type:	substrate-level phosphorylation
Enzyme:	phosphoglycerate kinase
Coreactants:	ADP, Mg^{2+}

Step 8

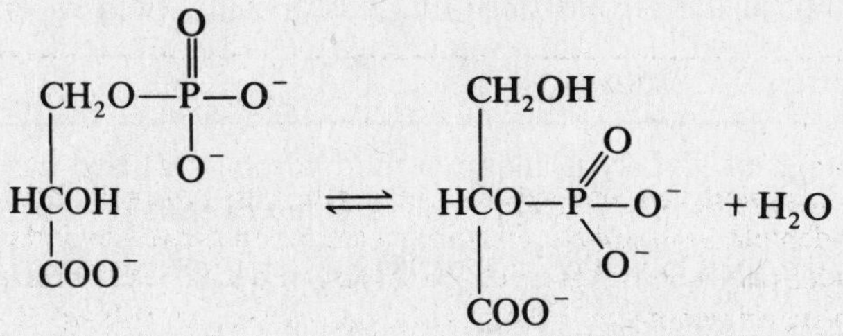

$$\text{D-3-Phosphoglycerate} \rightleftharpoons \text{D-2-Phosphoglycerate} + H_2O$$

D-3-Phosphoglycerate: $CH_2O{-}PO_3^{2-}$ / HCOH / COO^-

D-2-Phosphoglycerate: CH_2OH / $HCO{-}PO_3^{2-}$ / COO^-

Reaction type:	isomerization (phosphate transfer)
Enzyme:	phosphoglyceromutase
Coreactants:	Mg^{2+}

Step 9

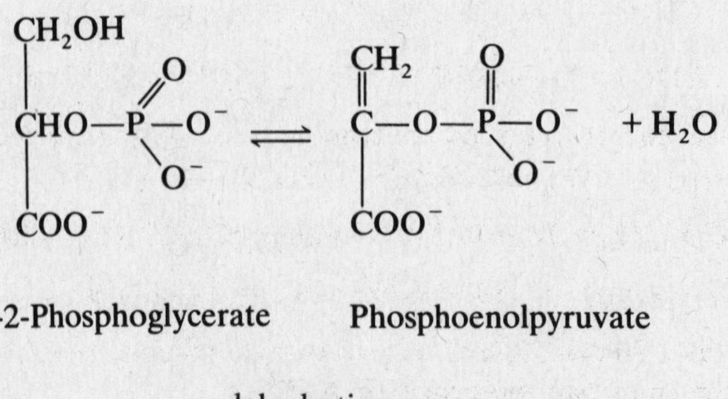

$$\text{D-2-Phosphoglycerate} \rightleftharpoons \text{Phosphoenolpyruvate} + H_2O$$

D-2-Phosphoglycerate: CH_2OH / $CHO{-}PO_3^{2-}$ / COO^-

Phosphoenolpyruvate: CH_2 (=) / $C{-}O{-}PO_3^{2-}$ / COO^-

Reaction type:	dehydration
Enzyme:	enolase
Coreactants:	Mg^{2+} or Mn^{2+}
Effectors:	
Inhibited by:	F^-

Step 10

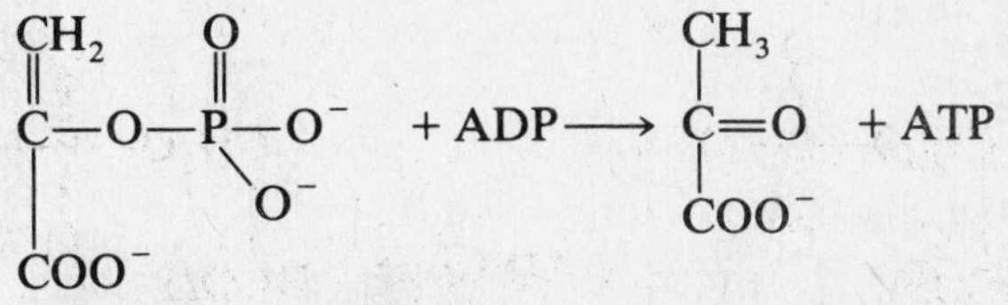

$$\begin{array}{c}CH_2 \quad O \\ \| \quad\quad \| \\ C{-}O{-}P{-}O^- \\ | \quad\quad \backslash \\ | \quad\quad O^- \\ COO^-\end{array} + ADP \longrightarrow \begin{array}{c}CH_3 \\ | \\ C{=}O \\ | \\ COO^-\end{array} + ATP$$

Phosphoenolpyruvate **Pyruvate**

Reaction type:	substrate-level phosphorylation
Enzyme:	pyruvate kinase
Coreactants:	Mg^{2+} or Mn^{2+}; K^+
Effectors:	
Inhibited by:	MgATP, citrate, long-chain fatty acids
Activated by:	fructose 1,6-diphosphate

Question: What do Steps 1, 3, and 10 of glycolysis have in common?

These are the only three steps that are not reversible. They are the main control points in the pathway.

EXAMPLE 11.2

Several compounds can inhibit various enzymes in the glycolytic pathway. Enolase is inhibited by fluoride ions, and in vitro, glyceraldehyde 3-phosphate dehydrogenase is inactivated by iodoacetamide. In the presence of arsenate ions, glyceraldehyde 3-phosphate dehydrogenase can catalyze the conversion of glyceraldehyde 3-phosphate to phosphoglycerol arsenate, which hydrolyzes spontaneously to 3-phosphoglycerate and the arsenate ion. While this is not an *inhibition* of the enzyme, no ATP is produced at the phosphoglycerate kinase catalyzed reaction.

The Energetics of Glycolysis

The conversion of glucose to two molecules of pyruvic acid is an *exergonic* process (Chap. 10).

$$C_6H_{12}O_6 \longrightarrow 2C_3H_4O_3 \qquad \Delta G^{\circ\prime} = -147\ \text{kJ mol}^{-1}$$

There is a sufficient decrease in the free energy of glucose to couple the process of glucose degradation with the substrate-level phosphorylation of two molecules of ADP, which is an *endergonic* reaction:

$$ADP + P_i \longrightarrow ATP + H_2O \qquad \Delta G^{\circ\prime} = +30\ \text{kJ mol}^{-1}$$

Since the change in free energy required to produce two ATP molecules is +60 kJ, the coupled reaction of glucose ⟶ 2 pyruvate and 2ADP ⟶ 2ATP is:

$$C_6H_{12}O_6 + 2P_i + 2ADP \longrightarrow 2C_3H_4O_3 + 2ATP + 2H_2O$$

$$\Delta G^{\circ\prime} = -147 + 60\ \text{kJ mol}^{-1} = -87\ \text{kJ mol}^{-1}$$

This means that the conversion of glucose to pyruvate is favored, even though it is coupled to the energy-requiring production of two molecules of ATP.

Four steps in glycolysis are exergonic and are, consequently, *favored* reactions (Fig. 11-3). All these steps involve either the hydrolysis of ATP (Steps 1 and 3) or the production of ATP (Steps 7 and 10). While the overall process favors the production of pyruvate, the individual changes in free energy are not particularly great. This suggests that it should be possible to reverse most of the steps without having to sacrifice large amounts of energy.

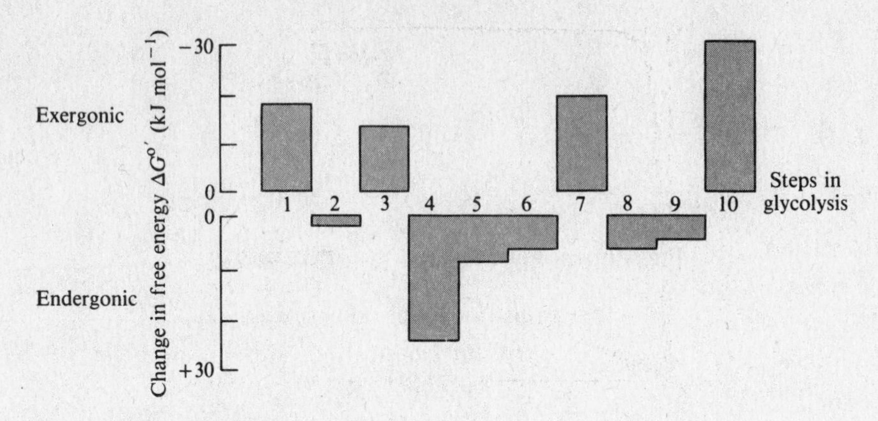

Fig. 11-3 Free energy of the glycolytic reactions.

In the complete oxidation of glucose to CO_2:

$$C_6H_{12}O_6 + 6O_2 \longrightarrow 6CO_2 + 6H_2O$$

The standard free energy change is $-2{,}870\ \text{kJ mol}^{-1}$. When compared with the conversion of glucose to pyruvate ($\Delta G^{\circ\prime} = -87\ \text{kJ mol}^{-1}$), only 3 percent of the energy available in the glucose molecule is released in producing pyruvate.

The Control Points of Glycolysis

The rate of glycolysis, as with all metabolic pathways, is under control; it is controlled at *three* stages.

The First Control Point

Step 1 is the first control point. In this step, glucose is converted to glucose 6-phosphate via hexokinase. This enzyme, which is present in all cells, is not specific for glucose but will catalyze the phosphorylation of many hexoses and hexose derivatives. However, the activity of the enzyme is regulated by the concentration of its principal product, *glucose 6-phosphate*; this product inhibits the activity of hexokinase in a process known as *product inhibition*. In mammalian cells, this regulatory function serves two purposes. First, it ensures that if a cell has sufficient glucose 6-phosphate to meet its energy demands, then subsequent phosphorylation of glucose will be reduced in that particular cell. Second, since the rate of the removal of glucose from the blood (by its conversion to glucose 6-phosphate within cells) is decreased, there will be an increase in the concentration of blood glucose if the glucose supply from elsewhere continues. The outcome of this is that the glucose will become more available to another phosphorylating enzyme, *glucokinase*. Glucokinase, which is *specific* for *D-glucose* and which is found only in the liver, also converts glucose to glucose 6-phosphate. Fig. 11-4 shows the relative activities of hexokinase (in all cells) and glucokinase (liver only).

Under normal conditions, glucose is available from the blood to all cells. The low K_m of hexokinase ($0.1\ \text{m}M$) implies that even at *low* concentrations, glucose entering a cell is rapidly converted to glucose 6-phosphate, which then enters the glycolytic pathway. As the energy requirements of the cell are met, the concentration of glucose 6-phosphate rises and thus reduces the activity of hexokinase. If the blood glucose concentration rises (e.g., after a carbohydrate-rich meal), then glucose flux through liver glucokinase increases. This occurs because hexokinase is fully saturated but glucokinase does not operate near its maximal rate until glucose levels rise beyond its K_m of $10\ \text{m}M$. Furthermore, glucokinase is not inhibited by glucose 6-phosphate. The interplay of these two enzymes ensures that if glucose is in excess of normal demands, it is converted into glucose 6-phosphate *specifically* within the liver.

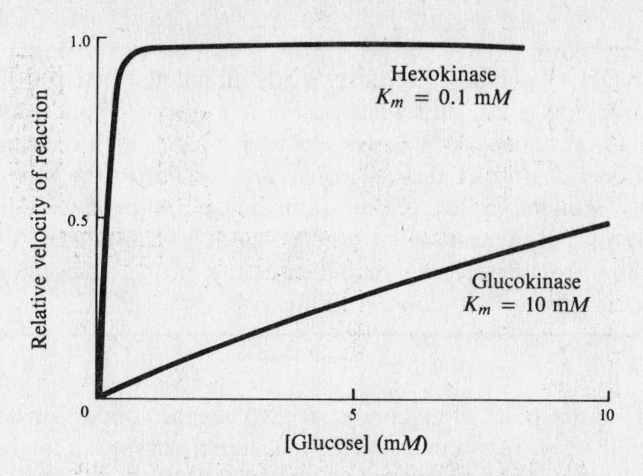

Fig. 11-4 The dependence of the rate of glucose phosphorylation on glucose concentration for hexokinase and glucokinase. The K_m for hexokinase is much lower than that for glucokinase.

The Second Control Point

Step 3 is the second control point of glycolysis and involves the conversion of fructose 6-phosphate into fructose 1,6-diphosphate, catalyzed by phosphofructokinase.

Question: Which enzyme is the main point of control of glycolysis?

It is phosphofructokinase, which is an allosteric enzyme; thus, its activity is regulated by a number of effectors, all of which are involved in energy transduction.

The activity of phosphofructokinase is enhanced by ADP or AMP and inhibited by ATP, NADH, citrate, or long-chain fatty acids. When a cell is in a *low*-energy state, the amounts of ADP and AMP are high relative to normal, whereas the amount of ATP is low. Under these conditions, the enzyme is fully activated and has a high affinity for its substrate, fructose 6-phosphate (Fig. 11-5). When the cell is in a high-energy state, the concentration of ATP is high, but the concentrations of AMP and ADP are low. In this circumstance, ATP binds to a regulatory site (Chap. 9) on the enzyme, causing its velocity curve to change from hyperbolic to sigmoidal. The enzyme now has a lower affinity for its substrate, and the rate of the reaction decreases.

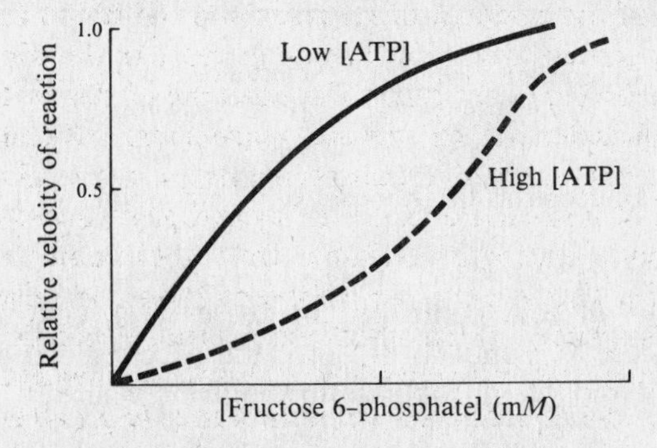

Fig. 11-5 The dependence of the rate of fructose 6-phosphate phosphorylation by phosphofructokinase on the presence of low and high concentrations of ATP.

Question: Citrate, NADH, and long-chain fatty acids inhibit the activity of phosphofructokinase. From where do these effectors arise, and what is the significance of their action?

Cytoplasmic NADH is produced in glycolysis at Step 6, so a high concentration of this *reduced* cofactor implies a high-energy state of the cell; thus, an increase in the rate of glucose degradation does not occur. High amounts of long-chain fatty acids are produced by the degradation of triglycerides (Chap. 13) and, in the case of citrate, by the degradation of certain amino acids (Chap. 15). Since these substrates are available for oxidation in the citric acid cycle (Chap. 12), their effect in inhibiting phosphofructokinase is to conserve glucose.

The Third Control Point

Step 10 is the third control point of glycolysis. It involves the conversion of phosphoenolpyruvate to pyruvate, catalyzed by pyruvate kinase. This allosteric enzyme is activated by fructose 1,6-diphosphate and phosphoenolpyruvate and is inhibited by ATP, citrate, and long-chain fatty acids. This means that the activity of pyruvate kinase is regulated in a manner similar to that of phosphofructokinase; both enzymes are inhibited when the cell is in a high-energy state or when alternative fuels to glucose are available. Furthermore, fructose 1,6-diphosphate (the product of the reaction catalyzed by phosphofructokinase) activates pyruvate kinase, as does phosphoenolpyruvate (the substrate of pyruvate kinase). These are examples of *positive feedforward control* akin to some electronic circuitry. So, when phosphofructokinase is activated by low levels of ATP, it produces one activator for pyruvate kinase (fructose 1,6-diphosphate) that is ultimately converted into a second activator of pyruvate kinase (phosphoenolpyruvate). This *cooperation* between the two enzymes to accelerate glycolysis also extends to their combined ability to retard the process. When the concentration of ATP is high, both enzymes are inhibited. Phosphofructokinase has reduced activity toward fructose 6-phosphate, so the concentration of this component rises, and because it interconverts to glucose 6-phosphate (via glucose 6-phosphate isomerase), the concentration of this substrate increases, thereby inhibiting hexokinase (Fig. 11-6).

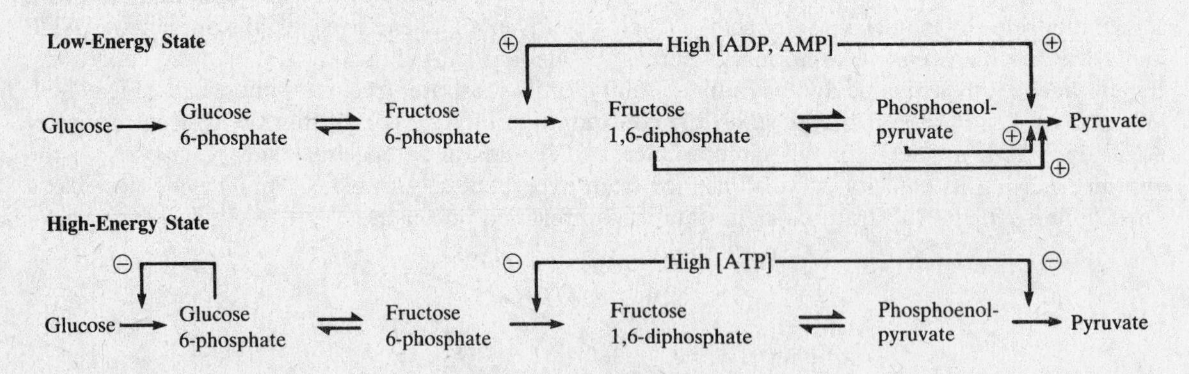

Fig. 11-6 The controls of glycolysis. ⊕ = activation; ⊖ = inhibition.

Not only is glycolysis in all cells controlled by these three enzymes, but if the cell is in a high-energy state, or if glucose is plentiful, or both, the excess glucose is not degraded by other tissues but is selectively captured via glucokinase in the liver, where it is stored (Sec. 11.5).

11.2 THE FATE OF PYRUVATE

The production of two molecules of pyruvate from one molecule of glucose occurs in virtually all cells. This process has three important characteristics: (1) no oxygen is required; (2) two molecules of ADP are phosphorylated by substrate-level phosphorylation; and (3) two molecules of NAD^+ are reduced. The subsequent fate of pyruvate in a particular cell depends on conditions related to these three criteria. The first is oxygen availability to the cell; the second is the energy status of the cell; and the third concerns the mechanisms available to the cell to oxidize the NADH to NAD^+.

Question: Why must the NADH produced in glycolysis be oxidized to NAD^+ and thus be recycled?

NAD^+ is required by glyceraldehyde 3-phosphate dehydrogenase (Step 6); therefore, NAD^+ is essential for this step and for glycolysis to proceed.

A further criterion governing the fate of pyruvate is the type of cell in which it is formed, since some cells (e.g., red blood cells) lack the metabolic capability to carry out the complete oxidation of pyruvate to CO_2.

Complete Oxidation of Pyruvate to CO_2

On complete oxidation, a mole of glucose liberates 2,870 kJ of free energy. In its conversion to two moles of pyruvate, only 3 percent of this available energy is released and only two moles of ATP are produced. The majority of the remaining energy is available to the cell only if it has the capacity to oxidize the pyruvate completely to CO_2.

The chemical equation for the complete oxidation of pyruvate is

$$C_3H_4O_3 + 2\tfrac{1}{2}O_2 \longrightarrow 3CO_2 + 2H_2O$$

This equation actually represents two oxidative processes, the first being the oxidation of pyruvate to CO_2 in the citric acid cycle (Chap. 12):

$$C_3H_4O_3 + 3H_2O \longrightarrow 3CO_2 + 10H$$

This process, in which pyruvate yields the equivalent of 10 H's, also results in the direct phosphorylation of 1 molecule of ADP. The second process is

$$10H + 2\tfrac{1}{2}O_2 \longrightarrow 5H_2O$$

It describes the oxidation of the 10 H's in the electron-transport chain (Chap. 14), with the concomitant phosphorylation of 14 molecules of ADP by oxidative phosphorylation. Consequently, by being completely oxidized to CO_2, 1 molecule of pyruvate leads to the phosphorylation of 15 molecules of ADP. When glucose is degraded to 2 molecules of pyruvate, only 2 molecules of ATP are formed, but the 2 pyruvate molecules have the potential to generate a further 30 molecules of ATP. The ability to produce this ATP depends on two factors: (1) the cell must have the capacity to perform both the citric acid cycle and electron transport, and (2) it must have a supply of oxygen. If either of these two criteria is lacking, then the 30 molecules of ATP cannot be produced. One example of a mammalian cell that lacks the ability to perform the citric acid cycle and electron transport is the red blood cell; it lacks this ability because it does not possess mitochondria. Skeletal muscle, when it is very active, has a limited capacity to oxidize pyruvate completely.

Conversion of Pyruvate to Lactate

If a cell lacks the ability to oxidize pyruvate, it is restricted to the glycolytic process for its production of ATP. If sufficient glucose is available to the cell, pyruvate is disposed of, so long as ADP, NAD^+, and P_i are present. All cells have adequate amounts of ADP and P_i, since these are hydrolysis products of ATP, but the amounts of NAD^+ are more limited. Step 6 is the only oxidative reaction in glycolysis; glyceraldehyde 3-phosphate is oxidized to 1,3-diphosphoglycerate, while NAD^+ is reduced to NADH. For continued glycolysis, this NADH must be reoxidized to NAD^+. This occurs in red blood cells and active muscle cells by the reduction of pyruvate to give lactate, which diffuses out of the cell via a specific membrane transport protein. The reaction is catalyzed by *lactate dehydrogenase*:

$$\begin{array}{c} CH_3 \\ | \\ C{=}O \\ | \\ COO^- \end{array} + NADH + H^+ \rightleftharpoons \begin{array}{c} CH_3 \\ | \\ HCOH \\ | \\ COO^- \end{array} + NAD^+$$

Pyruvate Lactate

Reaction type: redox
Enzyme: lactate dehydrogenase
Coreactants: $NAD^+/NADH$

All mammalian cells possess lactate dehydrogenase, but the activity of the enzyme varies from tissue to tissue. This variation is due to there being five forms of the enzyme, called *isozymes* or *isoenzymes*, each possessing a different "apparent" K_m for pyruvate. Each isozyme of lactate dehydrogenase consists of four subunits of either type M or H. So the five isozymes of lactate dehydrogenase are M_4, M_3H, M_2H_2, MH_3, and H_4. Isozyme M_4 has a relatively high K_m (low affinity) for pyruvate but has a turnover number approximately twice that of H_4. This means that at high concentrations of pyruvate, M_4 will convert pyruvate to lactate at a faster rate than will an equivalent amount of isozyme H_4. Even though H_4 has a low K_m (high affinity) for pyruvate, it has a *lower* turnover number so that at high concentrations of pyruvate it will convert pyruvate to lactate at a slower rate than will an equivalent amount of M_4. Furthermore, isozyme H_4 is inhibited by high concentrations of pyruvate. The other isozymes have degrees of inhibition intermediate to these two extremes. All cells possess varying amounts of the five isozymes; skeletal muscle, for example, has a predominance of the M_4 isozyme, while heart muscle has a predominance of H_4. The fact that heart muscle cells contain a predominance of the isozyme with the lowest turnover number for pyruvate and highest inhibition by pyruvate is consistent with the tendency of these cells to oxidize pyruvate to CO_2 and not, as is the case of active skeletal muscle, to convert it to lactate.

Although liver cells can oxidize pyruvate to CO_2, they contain a predominance of the M_4 isozyme, with its low affinity for pyruvate. Lactate enters the liver from the blood plasma and is rapidly converted to pyruvate.

This metabolic scheme, which is called *lactate fermentation*, is shown in Fig. 11-7. The coreactant cycle between the two dehydrogenase enzymes, glyceraldehyde 3-phosphate dehydrogenase (Step 6) and lactate dehydrogenase, ensures that there is regeneration of NAD^+ in this particular oxidation state so that glycolysis, lactate fermentation, and the production of ATP can continue.

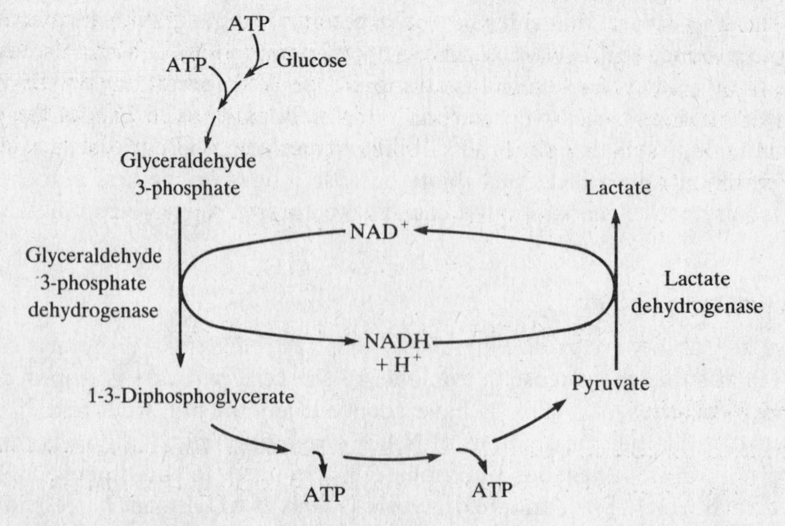

Fig. 11-7 The cooperation between glyceraldehyde 3-phosphate dehydrogenase and lactate dehydrogenase.

Conversion of Pyruvate to Ethanol

There is another fate of pyruvate that, while not occurring in mammalian tissues, is nevertheless very important. Some organisms can exist under aerobic or anaerobic conditions. These are called *facultative anaerobes*, and they can alter their metabolism to adapt to the presence or absence of oxygen. The most important facultative anaerobes are the yeasts. They convert glucose to pyruvate and then, if oxygen is present, oxidize the pyruvate to CO_2. If there is no oxygen available, then a pathway for the regeneration of NAD^+ comes into operation. Yeasts do not have lactate dehydrogenase but do possess *pyruvate decarboxylase*, which is not present in mammalian cells. This enzyme catalyzes the conversion of pyruvate to *acetaldehyde*:

$$\begin{array}{c} CH_3 \\ | \\ C{=}O \\ | \\ COO^- \end{array} \longrightarrow \begin{array}{c} CH_3 \\ | \\ CHO \end{array} + CO_2$$

Pyruvate — Acetaldehyde

Reaction type:	decarboxylation
Enzyme:	pyruvate decarboxylase
Coreactants:	thiamine pyrophosphate

Acetaldehyde is then reduced to ethanol (ethyl alcohol) via the Zn-containing enzyme *alcohol dehydrogenase*:

$$\begin{array}{c} CH_3 \\ | \\ CHO \end{array} + NADH + H^+ \rightleftharpoons \begin{array}{c} CH_3 \\ | \\ CH_2OH \end{array} + NAD^+$$

Acetaldehyde — Ethanol

Reaction type:	redox
Enzyme:	alcohol dehydrogenase
Coreactants:	$NAD^+/NADH$

This, the final step in alcohol fermentation, is analogous to lactate fermentation. Both reactions regenerate NAD^+ and produce low-molecular-weight, water-soluble, metabolic end products that diffuse out of the cells in which they were produced. In the case of alcoholic fermentation, the second reaction is reversible, so that if oxygen becomes available to previously anaerobic yeast cells, the ethanol is oxidized to acetaldehyde. Unlike lactate fermentation, in which the lactate is oxidized to pyruvate, alcoholic fermentation cannot form pyruvate from acetaldehyde. Instead, the acetaldehyde is converted to acetic acid, which is then completely oxidized to CO_2 in the citric acid cycle. The oxidation of acetic acid yields 8 H's, which generate ATP in the process of *oxidative phosphorylation* (Chap. 14):

$$CH_3COOH + 2H_2O \longrightarrow 2CO_2 + 8H$$

$$8H + 2O_2 \longrightarrow 4H_2O$$

The overall reaction is:

$$CH_3COOH + 2O_2 \longrightarrow 2CO_2 + 2H_2O$$

11.3 GLUCONEOGENESIS

Question: It has been stated above that tissues can break down glucose, but can they synthesize it too?

Yes, but only certain tissues have this ability.

In mammalian cells, glucose is the most abundant carbohydrate energy source. It is metabolized in all cells as a glycolytic fuel and is stored in liver and muscle as the polymer glycogen. But certain cells have the enzymes to catalyze the synthesis of glucose under certain conditions. The requirements are (1) the availability of *specific* carbon skeletons (carbon backbone structures of various types), (2) the energy, in the form of ATP, necessary to accomplish the sequence of reactions, and (3) the enzymes to catalyze reactions of the sequence.

The carbon skeletons used for the synthesis of glucose are *not* of *carbohydrate* origin but are derived from particular amino acids. One exception to this is the carbon skeleton of lactate, itself a product of carbohydrate metabolism, which can be incorporated into a new glucose molecule. This process, the synthesis of new glucose from essentially noncarbohydrate precursors, is called *gluconeogenesis*.

Question: Can the glycolytic pathway operate in the reverse direction; that is, can pyruvate be converted into glucose?

The pathway cannot operate *directly* in reverse because of the three irreversible steps, but pyruvate can be converted into glucose because of additional reactions.

The three steps in glycolysis that are irreversible are those catalyzed by hexokinase (glucokinase in the liver), phosphofructokinase, and pyruvate kinase. However, those tissues that carry out gluconeogenesis (e.g., liver and kidney) possess enzymes that allow these three steps to be reversed. When this occurs, the net flux of carbon atoms through the other reactions of glycolysis is reversed. A summary of the process is given in Fig. 11-8. Three nonglycolytic steps occur when pyruvate is converted to glucose via the *net reversal* of glycolysis.

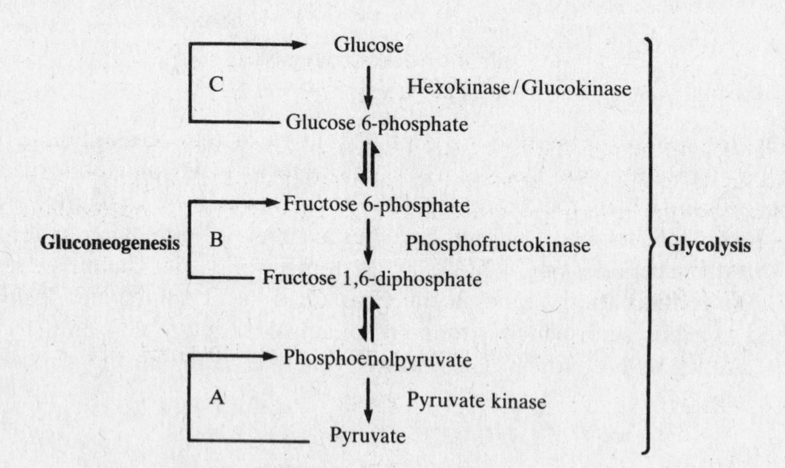

Fig. 11-8 Gluconeogenesis and glycolysis. A, B, and C denote steps in gluconeogenesis that bypass irreversible glycolytic reactions (in the reverse direction to that of net glycolytic flux).

Step A, the conversion of pyruvate to phosphoenolpyruvate, is accomplished by a circuitous process commencing with pyruvate entering the mitochondrion, which for gluconeogenesis to occur must be in a high-energy state. Under these conditions, the mitochondrial enzyme *pyruvate carboxylase* catalyzes the conversion of pyruvate to *oxaloacetate*:

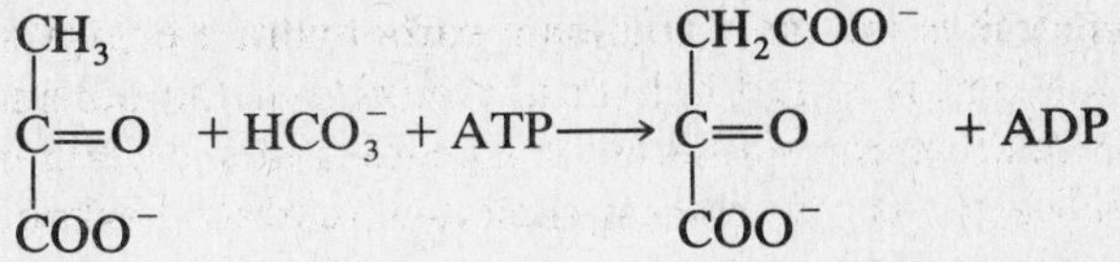

$$\begin{array}{l} CH_3 \\ | \\ C{=}O \\ | \\ COO^- \end{array} + HCO_3^- + ATP \longrightarrow \begin{array}{l} CH_2COO^- \\ | \\ C{=}O \\ | \\ COO^- \end{array} + ADP$$

Pyruvate Oxaloacetate

Reaction type: carboxylation
Enzyme: pyruvate carboxylase
Coreactants: HCO_3^-, ATP, Mg^{2+}, biotin
Effectors:
Activated by: acetyl-CoA

Pyruvate carboxylase is an allosteric enzyme being activated by its effector, acetyl-CoA. The enzyme contains tightly bound Mn^{2+} ions, and a covalently attached prosthetic group, biotin. When the mitochondria are in a high-bond-energy state, the concentrations of acetyl-CoA and ATP are relatively high, so the modulator of the enzyme and a source of energy are both available. The oxaloacetate is then converted within mitochondria to *malate*:

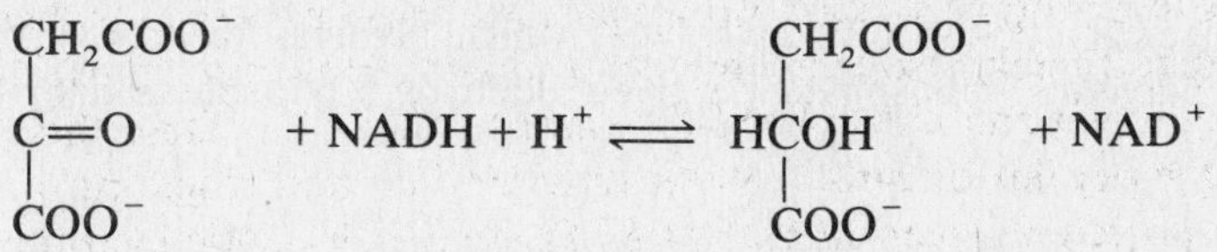

$$\begin{array}{l} CH_2COO^- \\ | \\ C{=}O \\ | \\ COO^- \end{array} + NADH + H^+ \rightleftharpoons \begin{array}{l} CH_2COO^- \\ | \\ HCOH \\ | \\ COO^- \end{array} + NAD^+$$

Oxaloacetate Malate

Reaction type: redox
Enzyme: malate dehydrogenase (mitochondrial)
Coreactants: $NADH/NAD^+$

The next reaction occurs in the cytoplasm. Malate is transported to the cytoplasm by a *dicarboxylate carrier* which is specific for malate, succinate, and fumarate and which requires the entry of P_i or one of these dicarboxylate anions. Cytoplasmic malate is then converted to oxaloacetate by *cytoplasmic malate dehydrogenase*:

$$\begin{array}{l} CH_2COO^- \\ | \\ HCOH \\ | \\ COO^- \end{array} + NAD^+ \rightleftharpoons \begin{array}{l} CH_2COO^- \\ | \\ C{=}O \\ | \\ COO^- \end{array} + NADH + H^+$$

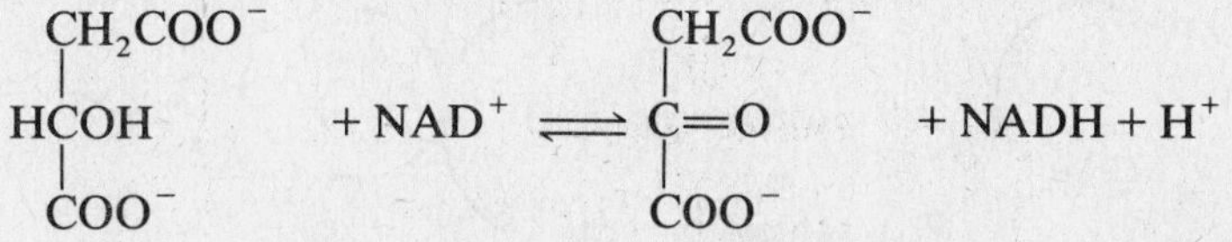

Malate Oxaloacetate

Reaction type: redox
Enzyme: cytoplasmic malate dehydrogenase
Coreactants: $NAD^+/NADH$

The preceding two reactions are necessary to achieve the overall result of transporting oxaloacetate to the cytoplasm from the mitochondria, as there is no direct mechanism for this. Cytoplasmic oxaloacetate is then converted irreversibly to phosphoenolpyruvate by way of *phosphoenolpyruvate carboxykinase*, a cytoplasmic enzyme that operates only when the ATP concentration is high:

$$\begin{array}{c} CH_2COO^- \\ | \\ C{=}O \\ | \\ COO^- \end{array} + GTP \longrightarrow \begin{array}{c} CH_2 \quad O \\ \| \quad\;\; \| \\ C{-}O{-}P{-}O^- \\ | \quad\;\; | \\ COO^- \;\; O^- \end{array} + CO_2 + GDP$$

Oxaloacetate **Phosphoenolpyruvate**

Reaction type: decarboxylation + phosphorylation
Enzyme: phosphoenolpyruvate carboxykinase
Coreactants: GTP

These four reactions convert pyruvate to phosphoenolpyruvate (Step A, p. 316) and bypass the irreversible step in glycolysis that is catalyzed by pyruvate kinase (Step 10).

Reversal of glycolytic Steps 9 through to 4 is achieved at the expense of supplying energy, to Step 7, in the form of ATP. In other words, net gluconeogenesis occurs only under conditions of relatively high cellular free energy. The high-energy state also implies that the NADH concentration is high; this enables the reversal of Step 6 in glycolysis, which involves the conversion of 1,3-diphosphoglycerate to glyceraldehyde 3-phosphate via glyceraldehyde 3-phosphate dehydrogenase (Fig. 11-9).

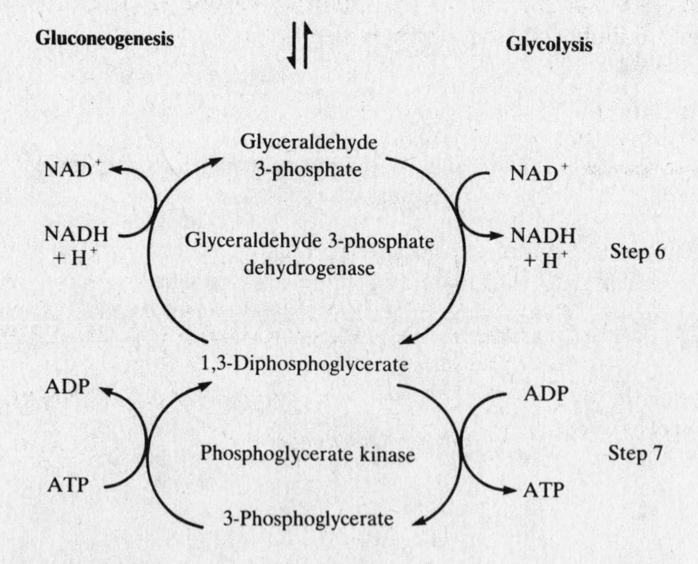

Fig. 11-9 The reversal of Steps 6 and 7 in glycolysis is favored by a high-energy state.

As a result of the reversal of Step 4 in glycolysis, the equivalent of two molecules of pyruvate is condensed to give one molecule of fructose 1,6-diphosphate. This compound is the product of the irreversible Step 3 in glycolysis. Gluconeogenic cells have the enzyme *fructose 1,6-diphosphatase*, which catalyzes the reverse reaction (Step 4, p. 306).

$$\text{Fructose 1,6-diphosphate} \longrightarrow \text{Fructose 6-phosphate} + PO_3^{2-}$$

Reaction type:	hydrolysis
Enzyme:	fructose 1,6-diphosphatase
Coreactants:	none
Effectors:	
Inhibited by:	AMP
Activated by:	citrate, 3-phosphoglycerate

Fructose 1,6-diphosphatase is inhibited by AMP but activated by citrate and 3-phosphoglycerate. Thus, in a high-bond-energy state, an increase of citrate and a decrease of AMP combine to activate fructose 1,6-diphosphatase and to inhibit phosphofructokinase (Fig. 11-10). This promotes the hydrolysis of fructose 1,6-diphosphate to fructose 6-phosphate.

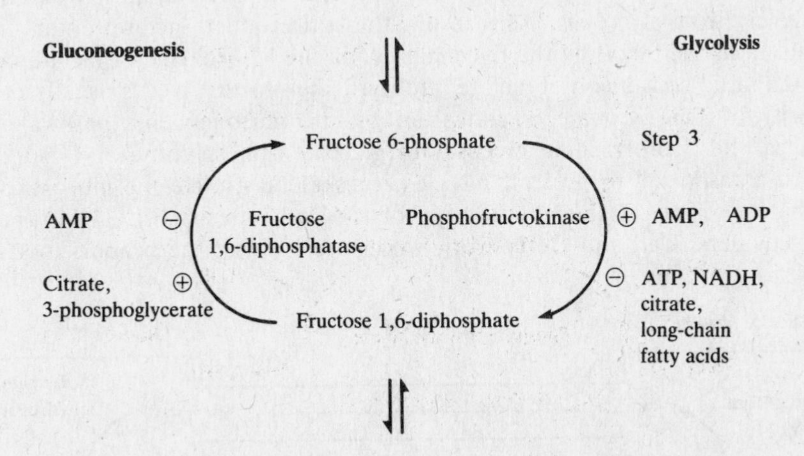

Fig. 11-10 The reversal of Step 3 in glycolysis is favored by a high-energy state. ⊕ = activation; ⊖ = inhibition.

Fructose 6-phosphate is converted to glucose 6-phosphate via glucose 6-phosphate isomerase. Glucokinase (liver only) and hexokinase, the enzymes that produce glucose 6-phosphate from glucose, are not able to catalyze the reverse reaction, but the liver has a specific enzyme, *glucose 6-phosphatase*, that catalyzes the hydrolysis of glucose 6-phosphate to glucose. The glucose that is produced can enter the blood.

$$\text{Glucose 6-phosphate} + H_2O \longrightarrow \text{Glucose} + P_i$$

Glucose 6-phosphate Glucose

Reaction type:	hydrolysis
Enzyme:	glucose 6-phosphatase
Coreactants:	none

An alternative metabolic route for glucose 6-phosphate is its storage as glycogen in the liver and muscles. (The production of glycogen is discussed in Sec. 11.5.)

In summary, glucose can be synthesized in the liver and kidney from lactate and noncarbohydrate precursors. The synthesis is accomplished essentially by a reversal of the glycolytic pathway, because the cells of these tissues possess the enzymes that are necessary to overcome the three irreversible glycolytic steps.

11.4 THE CORI CYCLE

The localization of particular enzymes in only certain cells means that some organs depend on others for the complete metabolism of certain substrates. So far as carbohydrates are concerned, the liver and skeletal muscle exhibit a special metabolic cooperation. Skeletal muscle obtains ATP during exercise almost solely from glycolysis. As a result, the end product, lactate, enters the blood. The lactate is removed from the blood by the liver mainly via the M_4 isozyme of lactate dehydrogenase, which catalyzes the rapid conversion of lactate into pyruvate. As the liver is usually in a high-energy state, the majority of this pyruvate is converted by the gluconeogenic pathway to glucose 6-phosphate. This can be hydrolyzed to glucose via glucose 6-phosphatase and can then enter the blood, where it is transported to skeletal muscles. In skeletal muscle the glucose is converted to glucose 6-phosphate via hexokinase and enters glycolysis. This process (Fig. 11-11) is called the *Cori cycle* after its discoverers, Carl and Gerty Cori, who were Nobel prize winners in 1937.

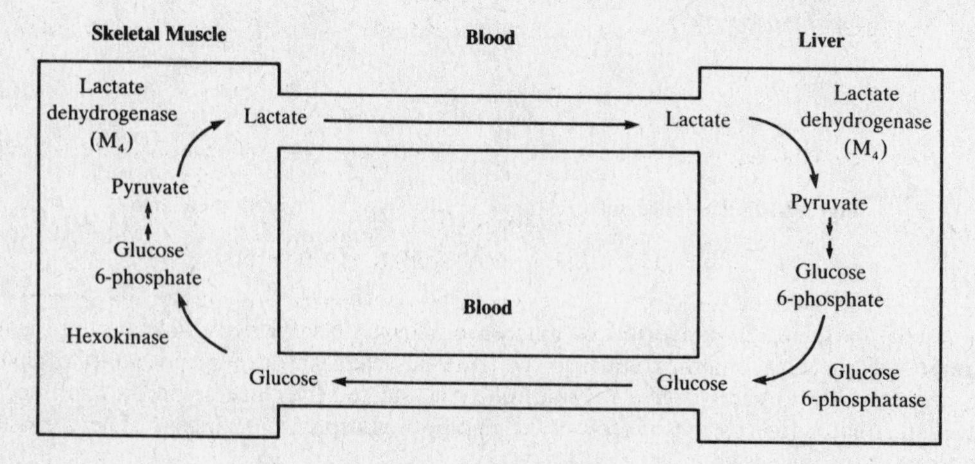

Fig. 11-11 The Cori cycle.

11.5 GLYCOGEN METABOLISM

The Synthesis of Glycogen

Glycogen is synthesized from glucose 6-phosphate in the liver and muscle and is stored within these tissues as *glycogen granules*. The glycogen, which is a polymer of glucose, is an energy store that can be rapidly broken down to glucose 6-phosphate, which then enters the glycolytic pathway. While the mechanism of synthesis of glycogen in all tissues is identical, the origin of the glucose 6-phosphate needs to be considered. In the liver, glucose 6-phosphate can arise either from blood glucose or through gluconeogenesis. In skeletal muscle, the glucose 6-phosphate originates *solely* from blood glucose.

The first step in glycogen synthesis is the formation of glucose 1-phosphate:

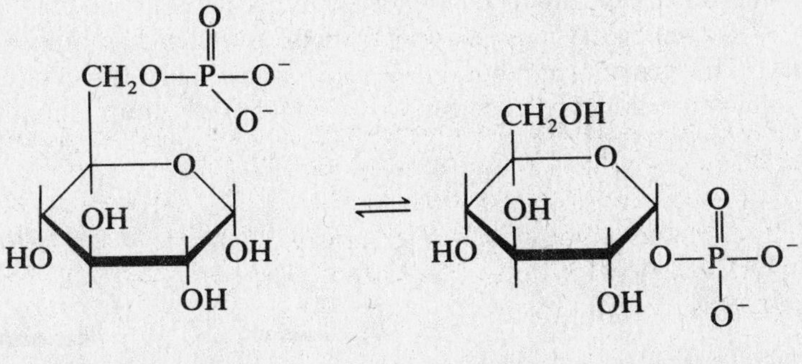

Glucose 6-phosphate ⇌ Glucose 1-phosphate

Reaction type:	isomerization (phosphate transfer)
Enzyme:	phosphoglucomutase
Coreactants:	glucose 1,6-diphosphate (intermediate)

The glucose 1-phosphate is then *activated* to enable its incorporation into glycogen. This activation involves the expenditure of energy derived from the hydrolysis of a molecule of *uridine triphosphate* (UTP).

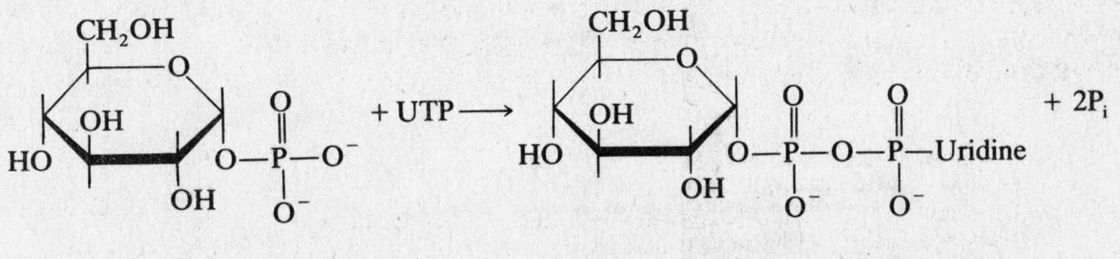

Glucose 1-phosphate + UTP ⟶ Uridine diphosphoglucose (UDP-glucose) + $2P_i$

Reaction type:	esterification
Enzyme:	UDP-glucose pyrophosphorylase

Question: What is the origin of the two phosphate groups in UDP-glucose?

One is from glucose 1-phosphate and the other is from *uridine monophosphate* (UMP). The pyrophosphate that is liberated from the terminal phosphates of UTP is hydrolyzed to inorganic phosphate by the enzyme *pyrophosphatase*. This hydrolysis, which is *irreversible*, drives the reaction in the direction of UDP-glucose synthesis.

The modified glucose molecule is a substrate for the enzyme *glycogen synthase*:

$$(\text{Glucose})_n + \text{UDP-glucose} \longrightarrow (\text{Glucose})_{n+1} + \text{UDP}$$

Reaction type: esterification, chain elongation
Enzyme: glycogen synthase
Coreactants: none
Effectors:
  Activated by: glucose 6-phosphate

The addition of a glucose unit to a polymeric chain of glucose is specific; i.e., the ester linkage forms only between the hydroxyl group on the C-4 of a terminal glucose unit of glycogen and the oxygen of the C-1 of the α isomer of the incoming UDP-glucose molecule (Fig. 11-12). The bond formed is, therefore, an $\alpha(1 \longrightarrow 4)$ linkage (Chap. 2). In addition, glycogen has branched chains joined in $\alpha(1 \longrightarrow 6)$ linkages. The enzyme *amylo-*(1,4$\longrightarrow$1,6)*-transglycosylase* catalyzes the transfer of fragments of 6 or 7 glucose residues in glycogen to a C-6 hydroxyl group on another glucose residue within the glycogen polymer (Fig. 11-13).

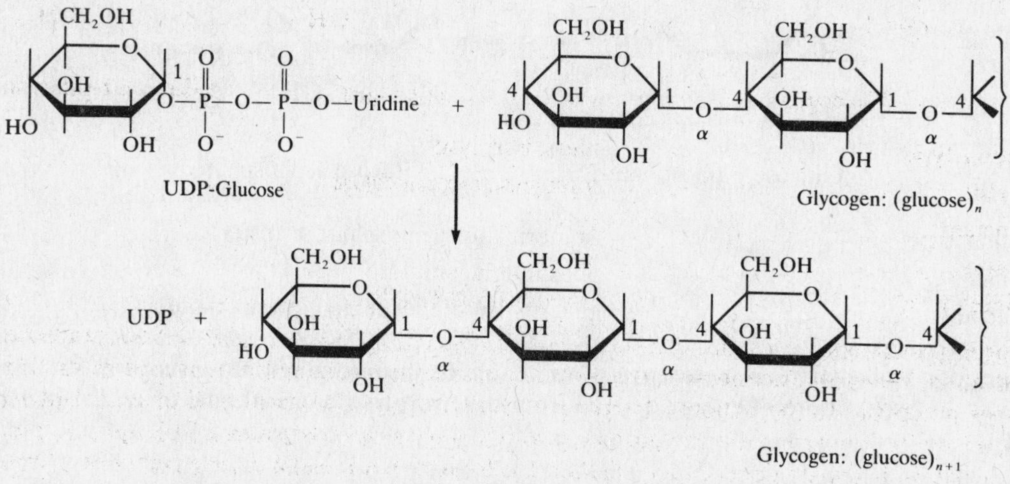

Fig. 11-12 The addition of a glucose unit to glycogen.

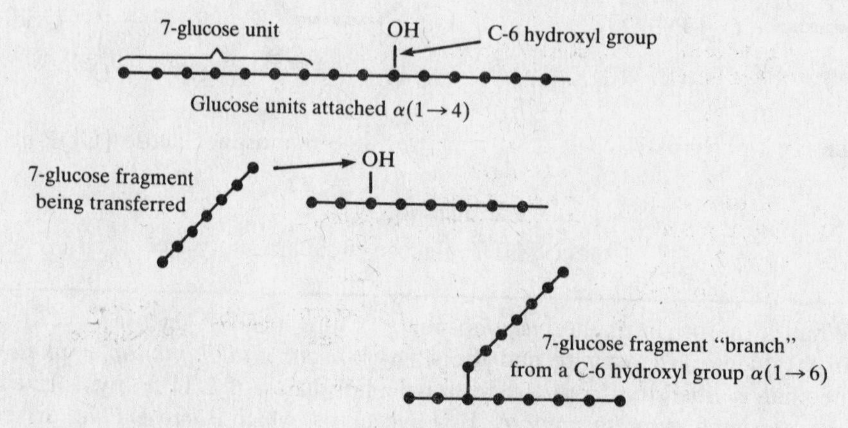

Fig. 11-13 The formation of branched chains of glucose residues within glycogen. ●– = glucose residue.

The branching within glycogen molecules makes the polymers more *compact* and more *soluble* and produces more *terminal glucose residues*. The increase in terminal residues is important when glycogen is to be degraded, because this occurs through a stepwise cleavage of terminal glucose residues from the polymer.

The Degradation of Glycogen (Glycogenolysis)

Glycogen is degraded to glucose 6-phosphate by a pathway that differs from its synthesis. The first step is via the enzyme *glycogen phosphorylase*, which, with inorganic phosphate, catalyzes the cleavage of a terminal $\alpha(1 \longrightarrow 4)$ bond, provided an $\alpha(1 \longrightarrow 6)$ linkage is not attached, to produce glycogen with one residue less and a molecule of glucose 1-phosphate.

$$\text{Glycogen} \xrightarrow{P_i} \text{Glycogen} + \text{Glucose 1-phosphate}$$

Reaction type: phosphorolysis
Enzyme: glycogen phosphorylase
Coreactants: P_i
Effectors:
  Inhibited by: glucose 6-phosphate

Glycogen phosphorylase cannot cleave $\alpha(1 \longrightarrow 6)$ linkages. This is carried out by another enzyme, called an $\alpha(1 \longrightarrow 6)$-glucosidase, which hydrolyzes these bonds and thus makes more $\alpha(1 \longrightarrow 4)$ linkages accessible to the actions of glycogen phosphorylase. Production of glucose 1-phosphate is followed by its conversion via the enzyme *phosphoglucomutase* to glucose 6-phosphate. The fate of the glucose 6-phosphate depends on whether it is formed in a skeletal muscle cell or in a liver cell. In skeletal muscle, the compound continues along the glycolytic pathway; while this also occurs in the liver, glucose 6-phosphatase can also convert glucose 6-phosphate to glucose.

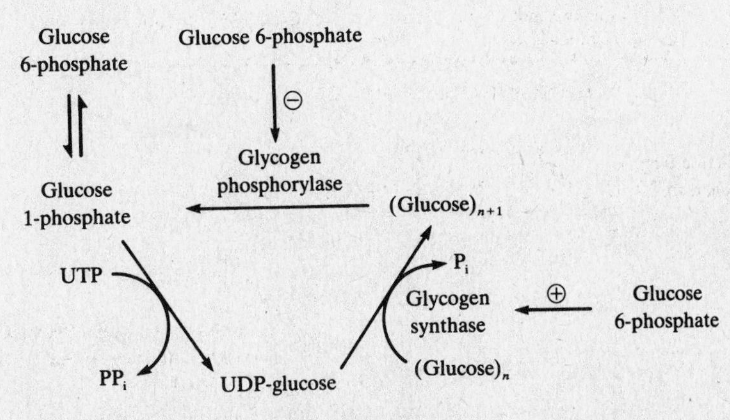

Fig. 11-14 The synthesis and degradation of glycogen. ⊕ = activation; ⊖ = inhibition.

Control of Glycogen Synthesis and Degradation

There are a number of factors controlling the synthesis and degradation of glycogen. The two enzymes concerned in these processes, *glycogen synthase* and *glycogen phosphorylase*, are allosterically controlled, and their activities are modulated by glucose 6-phosphate; when the concentration of glucose 6-phosphate is high, glycogen synthase is activated, thus favoring glycogen synthesis over glycogenolysis. A high concentration of glucose 6-phosphate inhibits the activity of glycogen phosphorylase, so that the degradation of glycogen is inhibited (Fig. 11-14). Another control in glycogenolysis is mediated by the hormones *epinephrine* and *glucagon* (Sec. 11.9).

11.6 THE ENTRY OF OTHER CARBOHYDRATES INTO GLYCOLYSIS

Glycolysis was defined as the process of converting one molecule of glucose into two molecules of pyruvate. However, many carbohydrates can contribute to the cell their carbon skeletons and the bond energy contained within their structures via the glycolytic pathway. In so doing, these other carbohydrates can ultimately be converted to pyruvate or to glucose and therefore be stored as glycogen.

EXAMPLE 11.3

The degradation of glycogen to glucose 1-phosphate is an example of the entry of a polysaccharide into glycolysis, since glucose 1-phosphate can be converted into glucose 6-phosphate. Another polysaccharide that can contribute its carbohydrate units to glycolysis is starch. *Starch* is the storage form of glucose in plants and has a structure similar to that of glycogen. It is a polymer of glucose units joined in $\alpha(1 \longrightarrow 4)$ and $\alpha(1 \longrightarrow 6)$ linkages but has fewer $\alpha(1 \longrightarrow 6)$ linkages than glycogen. Starch is really a mixture of two types of glucose polymers: an unbranched form called *amylose* and a branched form called *amylopectin*. During digestion, starch is hydrolyzed ultimately to maltose and glucose. Maltose is a *disaccharide* of two glucose units, joined by an $\alpha(1 \longrightarrow 4)$ link, and this is cleaved via the enzyme *maltase* to produce two glucose molecules. Consequently, the digestion of starch leads to the formation of glucose.

EXAMPLE 11.4

Other disaccharides commonly ingested by humans are *sucrose* (cane sugar) and *lactose* (milk sugar). Sucrose is cleaved into glucose and fructose by the action of *sucrase*, and lactose is cleaved into glucose and galactose by the action of *lactase*. All the monosaccharides can produce glycolytic intermediates. Glucose is converted via hexokinase into glucose 6-phosphate (Step 1 in glycolysis); fructose also reacts with hexokinase to give fructose 6-phosphate, as does another monosaccharide, *mannose*, which gives mannose 6-phosphate and then fructose 6-phosphate by way of the enzyme *phosphomannose isomerase* (Fig. 11-15).

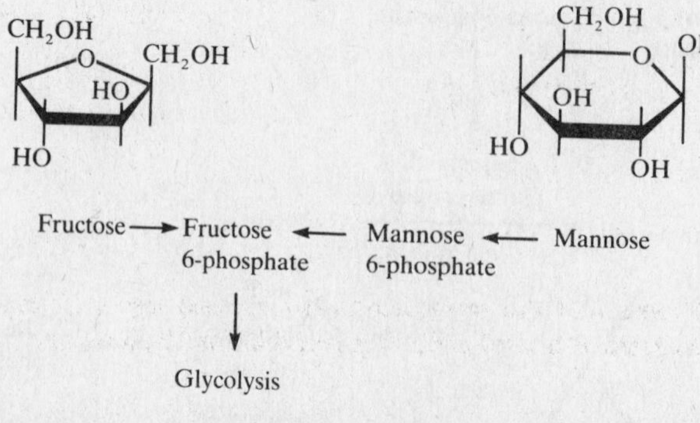

Fructose ⟶ Fructose 6-phosphate ⟵ Mannose 6-phosphate ⟵ Mannose

↓

Glycolysis

Fig. 11-15 The conversion of the monosaccharides fructose and mannose into the glycolytic intermediate fructose 6-phosphate.

Not all monosaccharides have such a simple or direct entry into the glycolytic pathway. Galactose, for example, is phosphorylated by *galactokinase* to galactose 1-phosphate, which then reacts with UTP via *galactose 1-phosphate uridylyltransferase* to yield UDP-galactose. This is converted into UDP-glucose by *UDP-glucose 4-epimerase*. The UDP-glucose can then be incorporated into glycogen and reappear as glucose 6-phosphate (Fig. 11-16).

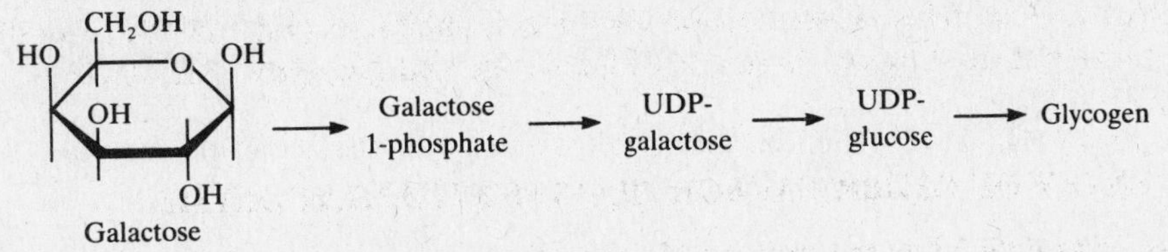

Fig. 11-16 The conversion of the monosaccharide galactose into glycogen.

While the entry of some monosaccharides (such as galactose) into the glycolytic pathway is circuitous, for others, alternative sequences of reactions are available. An alternate pathway exists for the metabolism of fructose, apart from its direct conversion into fructose 6-phosphate via hexokinase. This involves the conversion of fructose by way of *fructokinase* into fructose 1-phosphate, which is then cleaved by the action of *fructose 1-phosphate aldolase* into dihydroxyacetone phosphate and glyceraldehyde. Glyceraldehyde is then phosphorylated via *glyceraldehyde kinase* to give glyceraldehyde 3-phosphate. Fructose is, therefore, converted into dihydroxyacetone phosphate and glyceraldehyde 3-phosphate, two C_3 intermediates of the glycolytic pathway (Fig. 11-17).

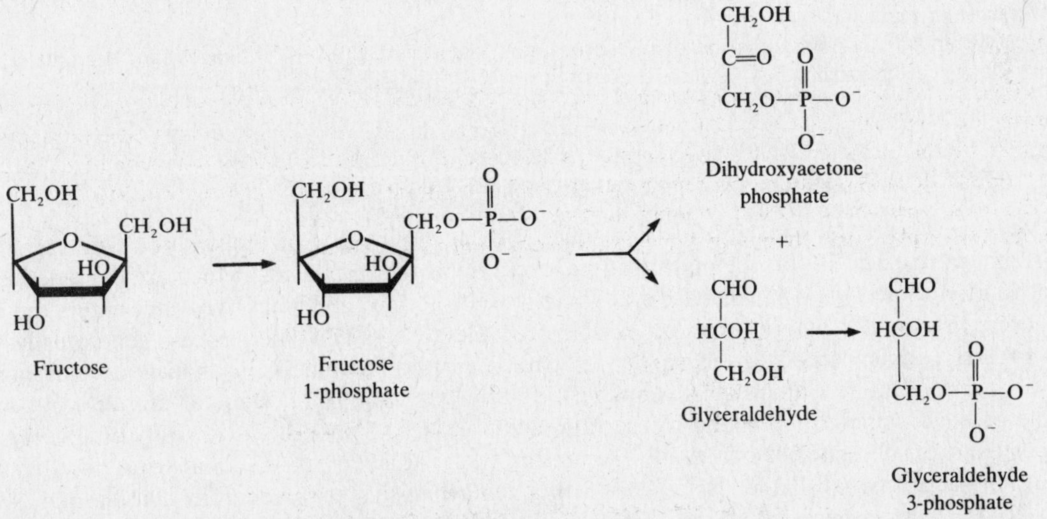

Fig. 11-17 The conversion of the monosaccharide fructose into the glycolytic intermediates dihydroxyacetone phosphate and glyceraldehyde 3-phosphate.

Another C_3 compound that can enter the glycolytic pathway is *glycerol*. Glycerol is converted via *glycerol kinase* to *glycerol 3-phosphate*, which, with glycerol 3-phosphate dehydrogenase, produces dihydroxyacetone phosphate (Fig. 11-18).

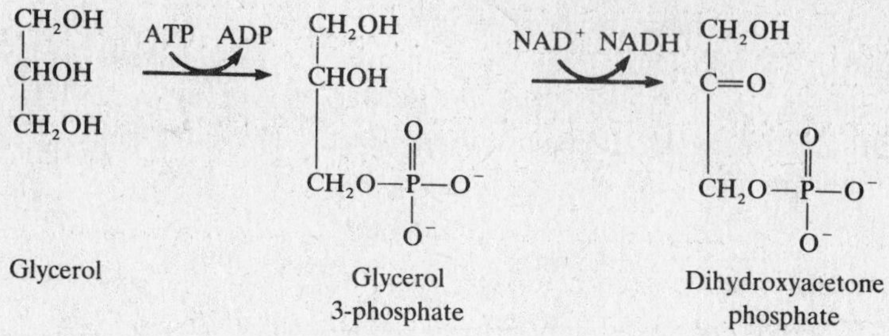

$$\underset{\text{Glycerol}}{CH_2OH-CHOH-CH_2OH} \xrightarrow{\text{ATP} \quad \text{ADP}} \underset{\text{Glycerol 3-phosphate}}{CH_2OH-CHOH-CH_2O-\overset{O}{\overset{\|}{P}}(O^-)-O^-} \xrightarrow{\text{NAD}^+ \quad \text{NADH}} \underset{\text{Dihydroxyacetone phosphate}}{CH_2OH-C{=}O-CH_2O-\overset{O}{\overset{\|}{P}}(O^-)-O^-}$$

Fig. 11-18 The conversion of glycerol into the glycolytic intermediate dihydroxyacetone phosphate.

Question: What is the metabolic origin of glycerol?

It is derived principally from the hydrolysis of triglycerides (Chap. 13).

With one exception all these carbohydrates yield the same amount of ATP when they are degraded to pyruvate; that is, one mole of ATP per mole of pyruvate formed. The exception is galactose, which requires the equivalent of three molecules of ATP to produce glucose 1-phosphate and a further molecule of ATP to give fructose 1,6-diphosphate, so the yield of ATP per molecule of pyruvate formed from galactose is 0.5.

EXAMPLE 11.5

One of the most abundant forms of D-glucose is cellulose. Can this be converted into a glycolytic intermediate in mammals?

No. Cellulose is a polysaccharide of D-glucose units joined by $\beta(1 \longrightarrow 4)$ linkages, and mammalian cells cannot cleave this particular link because they do not possess the enzyme *cellulase*.

11.7 REGENERATION OF CYTOPLASMIC NAD^+ LEVELS

Most cells have the capacity to derive energy from three types of fuels: carbohydrates, amino acids, and fatty acids. By far the major source of energy is the fatty acids, which are degraded in the mitochondria to acetyl-CoA. Acetyl-CoA, which can also be formed from pyruvate, enters the citric acid cycle (Chap. 12), where it can be oxidized completely to CO_2. This process generates NADH and $FADH_2$, which are oxidized in the electron-transport system. Cells which do not possess mitochondria (e.g., red blood cells) or which contain very few (e.g., skeletal muscle), either are unable or have a diminished ability to use fatty acids as fuels. Thus, these cells rely on glycolysis as their means of phosphorylating ADP. To ensure that sufficient NAD^+—so vital to Step 6 in glycolysis—is always available, these cells contain mechanisms to regenerate cytoplasmic NAD^+ from NADH: these mechanisms result in the conversion of pyruvate to lactate in mammalian cells (lactate fermentation; Sec. 11.2).

EXAMPLE 11.6

What mechanisms do nonmammalian cells use to regenerate NAD^+ for continued glycolysis?

Yeast, a facultative anaerobe, uses alcoholic fermentation (Sec. 11.2); pyruvate decarboxylase catalyzes the conversion of pyruvate to acetaldehyde, and then alcohol dehydrogenase converts the acetaldehyde to ethyl alcohol and oxidizes NADH to NAD^+.

Question: Cells that possess mitochondria can use pyruvate when oxygen is available. How do these cells cope with the problem of regenerating NAD^+?

The fact that mitochondria are present means that these cells can convert cytoplasmic NADH into NAD^+. The H's, or reducing equivalents, are transported into the mitochondria via metabolite *shuttles*.

Although mitochondria contain both NAD^+ and NADH, as does the cytoplasm, the mitochondrial and cytoplasmic pools are unable to exchange their contents directly, as the mitochondrial membranes are impermeable to the cytoplasmic compounds. The shuttle mechanisms allow the H on cytoplasmic NADH to be transported into the mitochondria, where it is donated to NAD^+ (to form mitochondrial NADH) or to FAD (to form mitochondrial $FADH_2$). There are several shuttle mechanisms that are used by mammalian cells; two of the most important are the *malate-aspartate shuttle* and the *glycerol 3-phosphate shuttle*.

The Malate-Aspartate Shuttle

Cytoplasmic malate dehydrogenase reduces oxaloacetate to malate, thereby oxidizing cytoplasmic NADH to NAD^+ (see Example 11.1). The malate enters the mitochondria via a specific protein carrier (a dicarboxylate carrier) in the inner membrane and is oxidized to oxaloacetate by *mitochondrial* malate dehydrogenase, a process that is coupled to the reduction of mitochondrial NAD^+ to NADH (see Fig. 11-19). While this results in the shuttling of cytoplasmic H into the mitochondria, the cytoplasmic oxaloacetate must be replaced so that the process can continue. This is accomplished by a *transaminase* enzyme, which concomitantly converts oxaloacetate to *aspartate* and glutamate to *2-oxoglutarate* (Chap. 12). The aspartate and 2-oxoglutarate are transported via specific carriers into the cytoplasm, where another transaminase converts them back to oxaloacetate and glutamate, respectively.

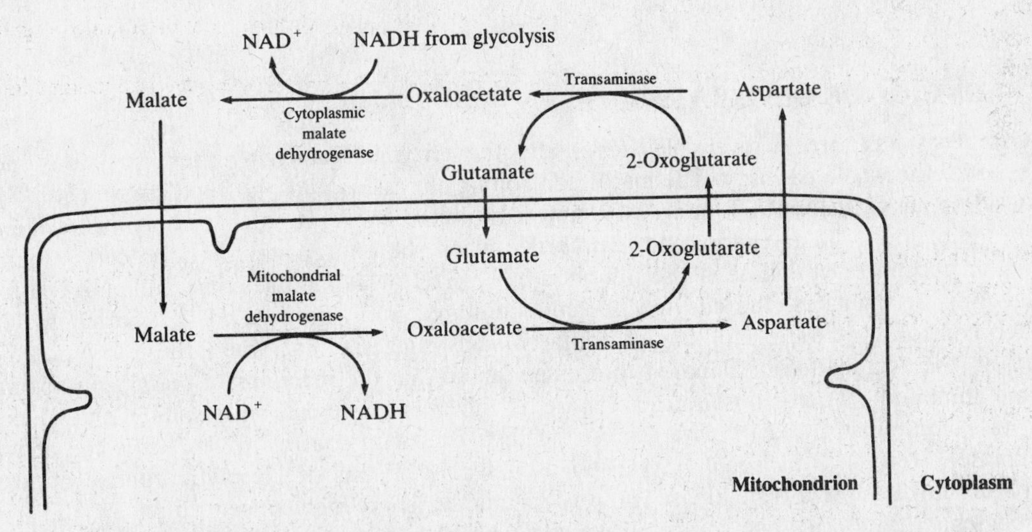

Fig. 11-19 The malate-aspartate shuttle, a mechanism for the transferral of reducing equivalents (H's) between cytoplasm and mitochondria.

The Glycerol 3-Phosphate Shuttle

Cytoplasmic *glycerol 3-phosphate dehydrogenase* reduces dihydroxyacetone phosphate to *glycerol 3-phosphate*, with the accompanying oxidation of NADH to NAD^+. The glycerol 3-phosphate passes through the outer mitochondrial membrane via a specific transport protein and is reoxidized to dihydroxyacetone phosphate by way of *mitochondrial* glycerol 3-phosphate dehydrogenase that is located within the inner mitochondrial membrane (Fig. 11-20). The redox coreactant for this enzyme is not NAD^+, but FAD (see Example 11.1, Fig. 11-2), and this is reduced to $FADH_2$. The dihydroxyacetone phosphate diffuses into the cytoplasm, where it is available for the shuttle to continue.

The net outcome of the two mechanisms just described is that cytoplasmic NADH is oxidized to cytoplasmic NAD^+. The H's, or reducing equivalents, are transferred to the mitochondria as NADH or $FADH_2$, according to the particular shuttle that is used.

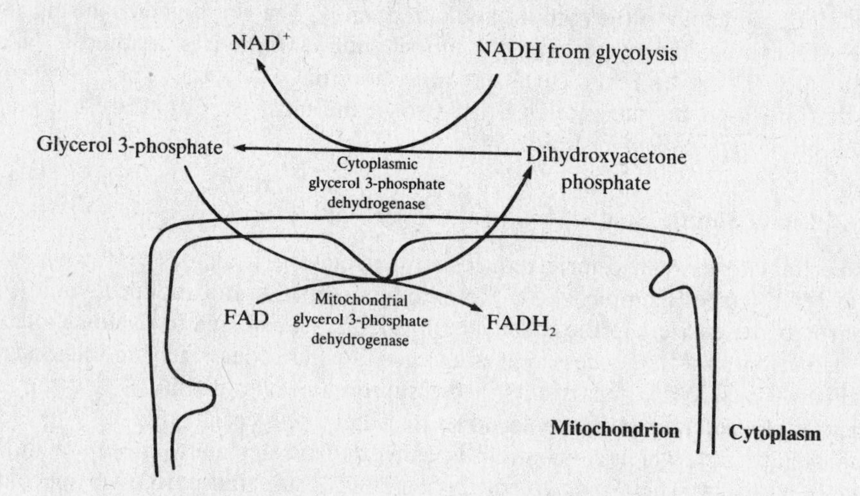

Fig. 11-20 The glycerol 3-phosphate shuttle, a mechanism for the transferral of reducing equivalents (H's) between the cytoplasm and mitochondria.

11.8 CONTROL OF GLYCOLYSIS

Glycolysis has three irreversible steps, and the enzymes catalyzing these reactions mediate control over the whole pathway. The major controlling enzyme is phosphofructokinase (Step 3). It is allosteric, is activated by ADP and AMP, and is inhibited by ATP, so that it is *most active* under conditions of low cellular energy and *least active* when the energy status of the cell is high. The activity of phosphofructokinase is also inhibited by NADH, which implies that if the NADH formed in glycolysis accumulates, the enzyme becomes inhibited until the NADH is oxidized to NAD^+. Further control by citrate and long-chain fatty acids suggests that when these compounds are abundant, the degradation of glucose is not essential, so that the reactions of the glycolytic pathway become inhibited. When phosphofructokinase is inhibited, fructose 6-phosphate and glucose 6-phosphate accumulate, and the latter substrate then exerts its control by *inhibiting* the activity of hexokinase (Step 1). This could be viewed as a second means of glycolytic control effected by phosphofructokinase. Pyruvate kinase (Step 10) is inactivated by high concentrations of ATP, citrate, and long-chain fatty acids and so acts in concert with phosphofructokinase.

Another mode of control is via glyceraldehyde 3-phosphate dehydrogenase (Step 6); this enzyme is activated by its oxidized coreactant (NAD^+) and inhibited by its reduced coreactant (NADH). Again, if the energy level of the cell is high, all the controlling enzymes cooperate to reduce the rate of glycolysis and, consequently, to conserve glucose.

The rate of glycolysis changes in a particular cell from moment to moment as its energy requirements change. This rate is referred to as the *glycolytic flux*. If, for example, a cell requires energy and can obtain this only from glucose, then the glycolytic flux would be high. But if a cell has sufficient energy for its immediate requirements as a result of degrading glucose or any other energy-producing compound, then the controls would operate and the glycolytic flux would be low.

EXAMPLE 11.7

A good illustration of how glycolytic flux can rapidly change is seen when yeast are grown under aerobic and anaerobic conditions with glucose as the carbon source. This effect, first observed by Louis Pasteur, is called the *Pasteur effect* and is depicted in Fig. 11-21.

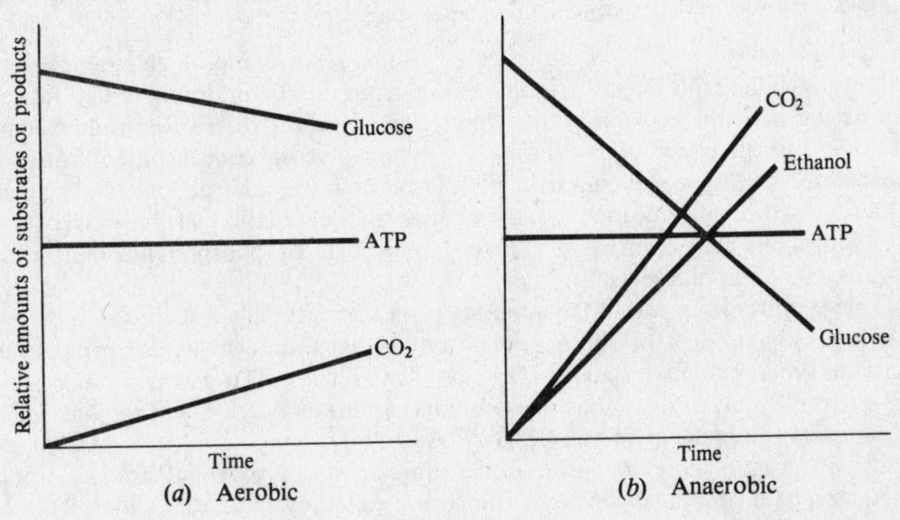

Fig. 11-21 The Pasteur effect: metabolic activity of yeast grown under aerobic and anaerobic conditions.

When grown under *aerobic* conditions, the yeast produces two ATP molecules from one molecule of glucose by substrate-level phosphorylation in glycolysis. The two molecules of pyruvate produced can then be completely oxidized to CO_2, and each yields a further 15 molecules of ATP. This leads to a slow decrease in the concentration of glucose, a steady production of CO_2, and relatively little change in the amount of ATP. Also, the two molecules of NADH can be reoxidized to NAD^+ by the electron-transport system. (This produces yet more ATP, as discussed in Chap. 14.)

When yeast are grown in an *anaerobic* environment, the utilization of glucose is markedly increased and the production of CO_2 and ethanol rises dramatically with very little change in the concentration of ATP. While this phenomenon has been explained in many ways, the simplest requires the assumption that the yeast needs a constant amount of ATP for its energy requirements, irrespective of the conditions under which it is grown.

When yeast are deprived of oxygen, the generation of ATP is possible only by using the reactions of the glycolytic pathway. Furthermore, the ability of the yeast to regenerate NAD^+ by the electron-transport system is denied in the absence of oxygen. If this situation is not modified, the level of ATP will decline and the cells will head toward a low-energy state. Controls in glycolysis are now lifted and the glycolytic flux increases. To cope with this, alcoholic fermentation comes into operation to remove the pyruvate as ethanol and to regenerate NAD^+ so that glycolysis can continue. As this permits only two molecules of ATP to be generated per molecule of glucose consumed, the rate of glycolysis increases; ethanol is produced and glucose is consumed more quickly than under aerobic conditions. The production of CO_2 also increases even though the maximum theoretical yield is two molecules per molecule of glucose (anaerobic) as against a possible six molecules per molecule of glucose under aerobic conditions. So, to maintain a constant concentration of ATP, yeast grown in an aerobic environment require the consumption of less glucose than do those grown under anaerobic conditions. Also, aerobic growth will result in the production of less CO_2 than will growth under anaerobic conditions.

11.9 EFFECTS OF HORMONES ON GLYCOLYSIS

Hormones do not exert any direct control on the rate of glycolysis, but three hormones have some indirect influence. These three are *insulin*, *glucagon*, and *epinephrine*. Basically, insulin is involved in the transport of glucose into all cells (except liver and red blood cells), whereas glucagon and epinephrine are both concerned with the degradation of glycogen in the liver; epinephrine is also involved with the degradation of glycogen in muscle.

EXAMPLE 11.8

Where are insulin and glucagon produced, and what controls their secretion into the blood?

Insulin and glucagon are polypeptide hormones synthesized in, and secreted by, the *pancreas*. Insulin is produced by the β cells of the pancreas, and glucagon by the α cells. The secretion of either of these hormones depends on the blood glucose concentration: above 4.5–5.5 mM (80–100 mg/100 mL) of glucose, insulin is secreted, but below 4.5 mM (80 mg/100 mL), glucagon is secreted.

Insulin and glucagon have opposing effects; insulin *reduces* blood glucose concentration by several methods, including (1) the enhancement of glucose uptake into muscle and fat cells, (2) the activation of glycogen synthase, and (3) the stimulation of the activities of phosphofructokinase and glucokinase. The overall effect of insulin is that blood glucose concentration *decreases* and the glucose is converted into glycogen, used in glycolysis, or both. The action of glucagon is one of *increasing* blood glucose concentration by stimulating the degradation of liver glycogen to glucose 1-phosphate, which can be converted to glucose 6-phosphate by phosphoglucomutase and then to glucose by glucose 6-phosphatase.

Epinephrine is synthesized within the *adrenal cortex* and is rapidly released into the bloodstream in response to a sudden fright or shock, which may signal the necessity for rapid evasive action requiring instantaneous muscular activity (fight or flight reflex). This hormone also stimulates the degradation of liver glycogen, as well as the degradation of muscle glycogen to glucose 6-phosphate in anticipation of its requirement in muscle glycolysis.

The action of glucagon and epinephrine in the stimulation of glycogen degradation in their target tissues is mediated by a process involving a compound called *cyclic AMP* (Chap. 7).

EXAMPLE 11.9

Cyclic AMP, or *cAMP* for short, is produced by the intramolecular cyclization of ATP, a reaction catalyzed by the enzyme *adenylate* (or *adenyl*) *cyclase*.

$$\text{ATP} \longrightarrow PP_i + \text{cAMP}$$

ATP cAMP

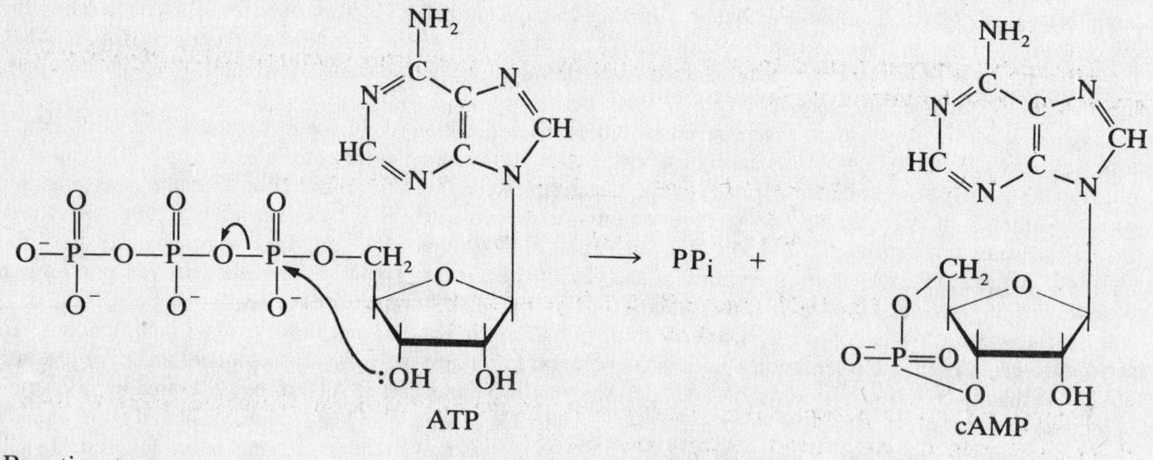

Reaction type:	intramolecular cyclization
Enzyme:	adenylate (or adenyl) cyclase
Coreactants:	Mg^{2+}

Being hormones derived from many amino acids (glucagon) or from a single amino acid (epinephrine), these compounds cannot enter their target cells directly. Consequently, liver cells have separate *receptors* on their outer membrane for glucagon and epinephrine, and muscle cells have only the receptor for epinephrine. When occupied by the hormone(s), these receptors undergo a conformational change. This change is transmitted through the membrane to the inner surface and results in the activation of adenyl cyclase. This enzyme, which lies under the receptors, is dormant or inactive when the receptors are unoccupied but is activated when the receptors are occupied. Once activated, adenyl cyclase catalyzes the cyclization of ATP to cyclic AMP (Fig. 11-22).

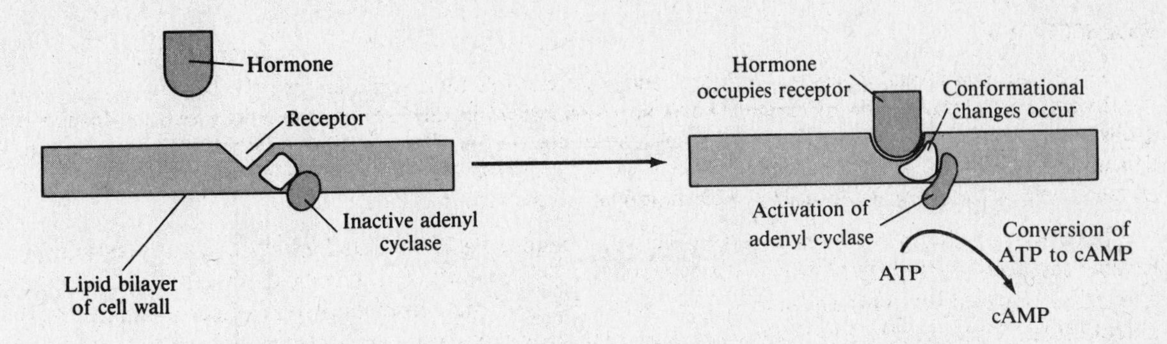

Fig. 11-22 The activation of adenyl cyclase by glucagon or epinephrine.

The cAMP is called the *second messenger*. It is produced within the liver or muscle cell when the specific hormone (the *first messenger*) occupies its particular receptor. The effect of cAMP is the activation of an enzyme, a *protein kinase*, by binding to and removing the enzyme's regulatory subunit (Fig. 11-23).

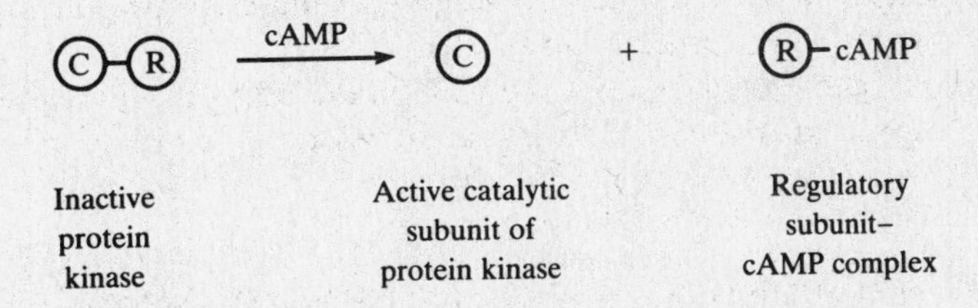

Fig. 11-23 The activation of protein kinase by cAMP.

The active protein kinase then uses ATP to activate another enzyme, called *phosphorylase kinase*, by phosphorylating it (Fig. 11-24).

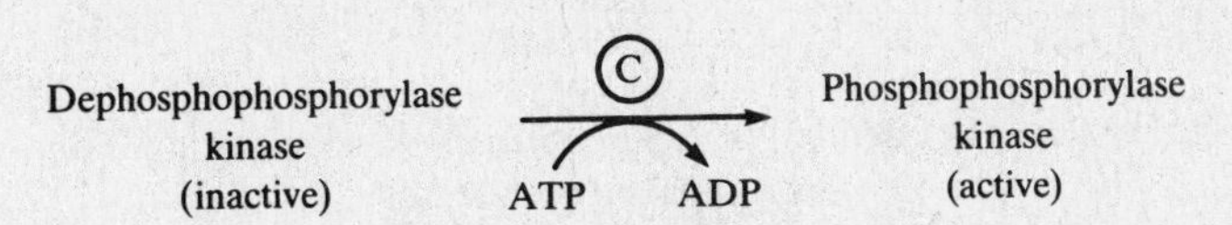

Fig. 11-24 The activation of phosphorylase kinase by protein kinase.

Finally, the active phosphorylase kinase converts the inactive form of another enzyme, *phosphorylase* b, into its active form, *phosphorylase* a (Fig. 11-25).

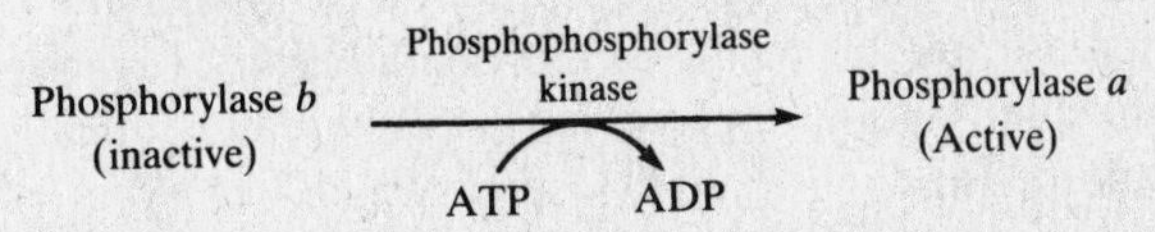

Fig. 11-25 The conversion of phosphorylase *b* to phosphorylase *a* via phosphorylase kinase.

Phosphorylase *a* is the active form of glycogen phosphorylase, and this can now catalyze the conversion of glycogen to glucose 1-phosphate. The entire process is shown in Fig. 11-26, which depicts the action of epinephrine on a liver or muscle cell.

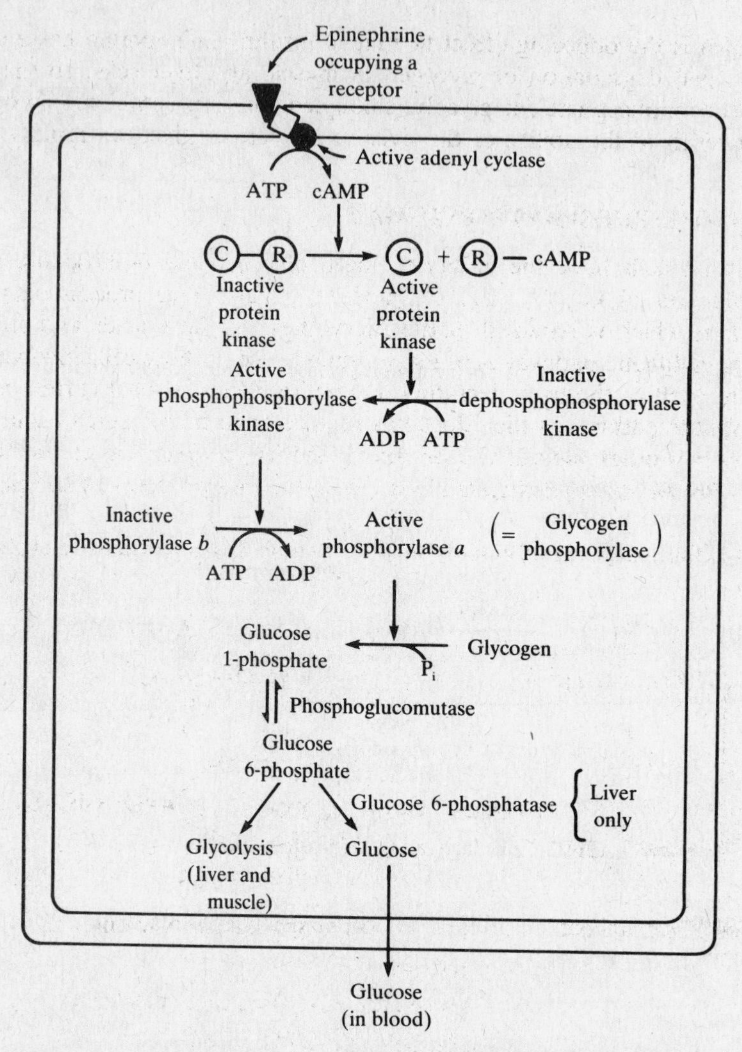

Fig. 11-26 The action of epinephrine on the degradation of glycogen in a liver or muscle cell.

Question: What is the purpose of this multistep process and what stops it?

Each step requires activation of an enzyme, and each enzyme can perform the next step many times. This gives a *cascade* effect, in which each step is an *amplification* of the one preceding it. There are a number of ways in which the process can slow down or stop. The epinephrine can leave its receptor; this would cause adenyl cyclase to assume its inactive or dormant conformation. The cAMP can be hydrolyzed to AMP; this process is catalyzed by an enzyme called a *phosphodiesterase*. The level of ATP will increase (as the result of the degradation of glycogen and the production of ATP by substrate-level phosphorylation in glycolysis); this will allosterically inhibit the conversion of phosphorylase *b* to phosphorylase *a*.

Under normal conditions the activity of glycogen phosphorylase (phosphorylase *a*) is regulated by the cellular concentration of glucose 6-phosphate, ATP, and AMP; the first two effectors inhibit its activation, and the third promotes its activation.

The mechanism of action of glucagon on the mobilization of glycogen in the liver is similar to that of epinephrine. The function of glucagon is to increase the concentration of blood glucose to normal levels, which is the opposing effect to that of insulin. Epinephrine has an immediate action that leads to the rapid degradation of glycogen in muscle and liver cells, so that muscles have an abundance of glucose 6-phosphate for glycolysis coupled with a plentiful supply of glucose available in the blood as a result of the ability of the liver to convert its glycogen rapidly to glucose.

11.10 THE PENTOSE PHOSPHATE PATHWAY

Some mammalian cells have the ability to metabolize glucose 6-phosphate in a pathway that involves the production of C_3, C_4, C_5, C_6, and C_7 sugars. This process also yields a reduced coenzyme, NADPH, which is oxidized in the biosynthesis of fatty acids and steroids (Chap. 13). Consequently, this metabolic pathway is of major importance in those cells involved in fatty acid and steroid production, such as the liver, lactating mammary gland, adrenal cortex, and adipose tissue. The pentose phosphate pathway, which does not require oxygen and which occurs in the cytoplasm of these cells, has two other names: the *phosphogluconate pathway* (after the first product in the pathway) and the *hexose monophosphate shunt* (since the end products of the pathway can reenter glycolysis).

The first step is the oxidation of glucose 6-phosphate at C-1 to produce a cyclic ester (also called a *lactone*):

$$\text{Glucose 6-phosphate} + NADP^+ \rightleftharpoons \text{6-Phosphogluconolactone} + NADPH + H^+$$

Reaction type:	redox
Enzyme:	glucose 6-phosphate dehydrogenase
Coreactants:	$NADP^+/NADPH$

The 6-phosphogluconolactone is unstable, and the ester spontaneously hydrolyzes, although a specific enzyme exists that catalyzes this reaction:

$$\text{6-Phosphogluconolactone} + H_2O \longrightarrow \text{6-Phosphogluconate}$$

Reaction type:	hydrolysis
Enzyme:	gluconolactonase
Coreactants:	none

The third step in the pathway yields a C_5 sugar, D-ribulose 6-phosphate, and concomitantly reduces another molecule of $NADP^+$:

```
 COO⁻                                   CH₂OH
  |                                      |
 HCOH                                   C=O
  |                                      |
HOCH                                    HCOH
  |                                      |
 HCOH           + NADP⁺ ⇌               HCOH                 + CO₂ + NADPH + H⁺
  |                                      |       O
 HCOH                                    |       ||
  |       O                             CH₂O—P—O⁻
  |       ||                                     |
 CH₂O—P—O⁻                                       O⁻
          |
          O⁻
```

6-Phosphogluconate D-Ribulose 6-phosphate

Reaction type:	oxidative decarboxylation
Enzyme:	6-phosphogluconate dehydrogenase
Coreactants:	$NADP^+$/NADPH

D-Ribulose 5-phosphate then undergoes an isomerization to D-ribose 5-phosphate:

```
 CH₂OH                     CHO
  |                         |
 C=O                       HCOH
  |                         |
 HCOH                      HCOH
  |             ⇌           |
 HCOH                      HCOH
  |       O                 |       O
  |       ||                |       ||
 CH₂O—P—O⁻                 CH₂O—P—O⁻
          |                         |
          O⁻                        O⁻
```

D-Ribulose 5-phosphate D-Ribose 5-phosphate

Reaction type:	isomerization
Enzyme:	ribosephosphate isomerase
Coreactants:	none

These four steps constitute the *first phase* of the pentose phosphate pathway and result in the consumption of one molecule of glucose 6-phosphate and the formation of one molecule of ribose 5-phosphate, two of NADPH, and one of CO_2.

The requirement for NADPH far exceeds an equal requirement for ribose 5-phosphate (necessary for the production of nucleic acids and nucleotides), and so the *second phase* of the pentose phosphate pathway converts the C_5 sugar, by a series of reversible reactions, into the glycolytic intermediates fructose 6-phosphate and glyceraldehyde 3-phosphate. This interconversion is shown in Fig. 11-27. Not only does the second phase of the pathway conserve all the carbon atoms of the C_5 sugar, but it produces erythrose 4-phosphate (C_4), xylulose 5-phosphate (C_5), and sedoheptulose 7-phosphate (C_7), which are available to other metabolic processes.

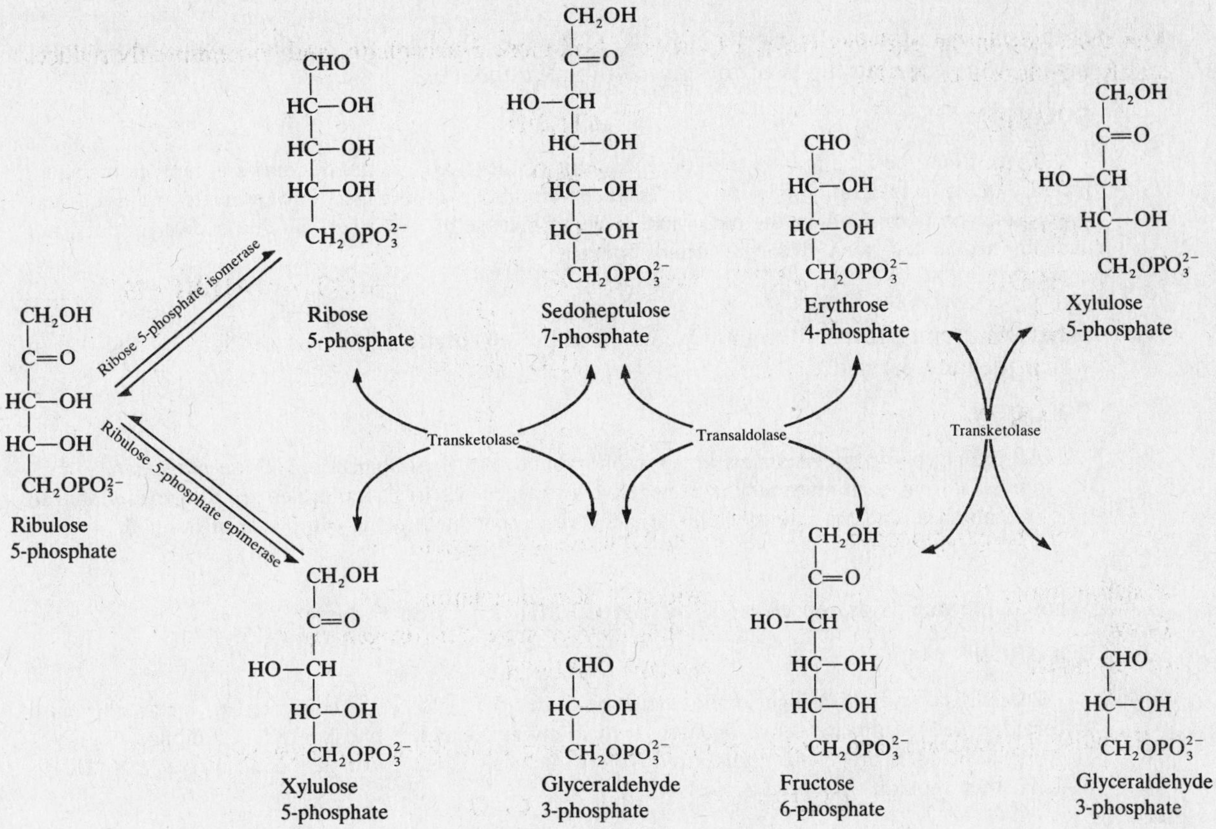

CH$_2$OH
C=O
HC—OH
HC—OH
CH$_2$OPO$_3^{2-}$
Ribulose 5-phosphate

Ribose 5-phosphate isomerase

Ribulose 5-phosphate epimerase

CHO
HC—OH
HC—OH
HC—OH
CH$_2$OPO$_3^{2-}$
Ribose 5-phosphate

CH$_2$OH
C=O
HO—CH
HC—OH
CH$_2$OPO$_3^{2-}$
Xylulose 5-phosphate

Transketolase

CH$_2$OH
C=O
HO—CH
HC—OH
HC—OH
HC—OH
CH$_2$OPO$_3^{2-}$
Sedoheptulose 7-phosphate

CHO
HC—OH
CH$_2$OPO$_3^{2-}$
Glyceraldehyde 3-phosphate

Transaldolase

CHO
HC—OH
HC—OH
CH$_2$OPO$_3^{2-}$
Erythrose 4-phosphate

CH$_2$OH
C=O
HO—CH
HC—OH
HC—OH
CH$_2$OPO$_3^{2-}$
Fructose 6-phosphate

Transketolase

CH$_2$OH
C=O
HO—CH
HC—OH
CH$_2$OPO$_3^{2-}$
Xylulose 5-phosphate

CHO
HC—OH
CH$_2$OPO$_3^{2-}$
Glyceraldehyde 3-phosphate

Fig. 11-27 The second phase of the pentose phosphate pathway.

EXAMPLE 11.10

What is the overall reaction of the pentose phosphate pathway?

The balanced sequence is

$$\text{3 Glucose 6-phosphate} + 6\text{NADP}^+ \longrightarrow 3\text{CO}_2 + 6\text{NADPH} + \text{2 Fructose 6-phosphate} + \text{Glyceraldehyde 3-phosphate}$$

This shows the production of NADPH in the pathway and its return of all but one of the carbon atoms of glucose 6-phosphate to the glycolytic pathway. However, the overall scheme belies its adaptability to produce and to deal with a variety of sugar phosphates.

Solved Problems

GLYCOLYSIS

11.1. A compound is an inhibitor of glyceraldehyde 3-phosphate dehydrogenase. If this compound were added to liver cells where D-glucose was the only substrate, what effect would it have on the concentrations of the glycolytic intermediates?

SOLUTION

There would be an accumulation of those intermediates from glucose 6-phosphate to glyceraldehyde 3-phosphate and a depletion of those from 1,3-diphosphoglycerate to pyruvate.

11.2. If the substrate for the liver cells in Prob. 11.1 were L-lactate, what effect would the inhibitor have on the concentrations of the glycolytic intermediates?

SOLUTION

There would be no effect on the concentrations of the glycolytic intermediates except, perhaps, an increase in the concentration of pyruvate. The cells would convert the lactate to pyruvate and use this as a precursor of acetyl-CoA in the citric acid cycle. If glucose or any other suitable carbohydrate is not available to the cell, then glycolysis cannot operate.

11.3. What constraint prevents the intermediates of the glycolytic pathway from leaving the cell in which they are formed?

SOLUTION

All the glycolytic intermediates are phosphorylated. At physiological pH, these phosphate groups are ionized so that each intermediate is negatively charged. Charged molecules are not readily able to cross membranes, and the intermediates are, therefore, confined to the cytoplasm of the cell.

11.4. What is the net oxidation change when glucose is converted to lactate?

SOLUTION

There is no overall change in the oxidation state when glucose is converted to lactate, because glyceraldehyde 3-phosphate dehydrogenase *oxidizes* glyceraldehyde 3-phosphate to 1,3-diphosphoglycerate, but lactate dehydrogenase *reduces* pyruvate to lactate. These two reactions also *reduce* NAD^+ to NADH, then reoxidize the NADH to NAD^+.

11.5. If all the glycolytic enzymes, ATP, ADP, NAD^+, and glucose were incubated together under ideal conditions, would pyruvate be produced?

SOLUTION

No, because an important omission is inorganic phosphate. Even if phosphate were added to the incubation mixture, pyruvate would be produced only in an amount equivalent to that of the NAD^+ present. Glycolysis requires NAD^+ for Step 6, the reaction catalyzed by glyceraldehyde 3-phosphate dehydrogenase.

THE FATE OF PYRUVATE

11.6. What is the metabolic fate of lactate in mammalian cells?

SOLUTION

Lactate can undergo only one reaction: it is oxidized to pyruvate via lactate dehydrogenase.

THE CORI CYCLE

11.7. From where do skeletal muscles obtain their carbohydrate for glycolysis?

SOLUTION

One source is blood glucose. Glucose entering a muscle cell is rapidly converted to glucose 6-phosphate, which enters glycolysis. Another source is muscle glycogen, which is degraded to glucose 1-phosphate. The latter is then converted to glucose 6-phosphate, which proceeds through glycolysis.

ENTRY OF OTHER CARBOHYDRATES INTO GLYCOLYSIS

11.8. Which of the following could theoretically yield the maximum net number of molecules of ATP by substrate-level phosphorylation in glycolysis: a molecule of sucrose, two molecules of glucose, or two molecules of fructose?

SOLUTION

All can yield four molecules of ATP. Sucrose is converted into one molecule each of glucose and fructose; each of these sugars requires two molecules of ATP to reach the stage of two molecules of glyceraldehyde 3-phosphate. From here to pyruvate, each glyceraldehyde 3-phosphate molecule yields two molecules of ATP by substrate-level phosphorylation of ADP. Thus, one molecule of glucose or of fructose generates two molecules of ATP.

11.9. Which reactions are catalyzed by the enzymes sucrase and lactase?

SOLUTION

Sucrase catalyzes the hydrolysis of sucrose to glucose and fructose. Lactase catalyzes the hydrolysis of lactose to glucose and galactose.

CONTROL OF GLYCOLYSIS

11.10. If a molecule of glucose produces 2 molecules of ATP by substrate-level phosphorylation of ADP in glycolysis and the resulting 2 molecules of pyruvate can each yield 15 molecules of ATP when oxygen is available, how many glucose molecules will be necessary to produce 160 molecules of ATP by yeast grown under (*a*) aerobic and (*b*) anaerobic conditions?

SOLUTION

(*a*) Growth under aerobic conditions can produce $2 + 30 = 32$ molecules of ATP per molecule of glucose. To produce 160 molecules of ATP, $160 \div 32 = 5$ molecules of glucose are required.

(*b*) Growth under anaerobic conditions can produce only two molecules of ATP per molecule of glucose. To produce 160 molecules of ATP, $160 \div 2 = 80$ molecules of glucose are required.

11.11. In Prob. 11.10, how many molecules of CO_2 would be evolved in producing 160 molecules of ATP during growth under (*a*) aerobic and (*b*) anaerobic conditions?

SOLUTION

(*a*) Aerobic:

$$1\ \text{Glucose} \longrightarrow 6CO_2$$
$$\therefore\ 5\ \text{Glucose} \longrightarrow 30CO_2$$

(*b*) Anaerobic:

$$1\ \text{Glucose} \longrightarrow 2CO_2$$
$$\therefore\ 80\ \text{Glucose} \longrightarrow 160CO_2$$

Supplementary Problems

11.12. In glycolysis, the conversion of dihydroxyacetone phosphate to glyceraldehyde 3-phosphate, catalyzed by triosephosphate isomerase, is reversible. Given that at equilibrium the reaction strongly favors the formation of dihydroxyacetone phosphate, how does glycolysis proceed?

11.13. In red blood cells, 2,3-diphosphoglycerate decreases the affinity of oxygen for hemoglobin by stabilizing the hemoglobin in its deoxygenated form (Chap. 5). This effector is synthesized in one step from a glycolytic intermediate and is converted to another glycolytic intermediate also in one step. What are the two glycolytic intermediates, and which enzymes catalyze the two reactions?

11.14. What is the fate of the six carbon atoms of glucose when it is metabolized by yeast grown under anaerobic conditions?

11.15. Calculate the number of high-energy phosphate bonds that are required for the conversion of two molecules of pyruvate into a glucose unit within glycogen.

11.16. A liver cell has a high concentration of glucose 6-phosphate, which inhibits the activity of hexokinase. What is the fate of the glucose 6-phosphate?

11.17. Termites exist almost entirely on a diet of cellulose. How is the cellulose degraded by the termites to carbohydrates, which can be used as an energy source?

11.18. When yeast are incubated in an oxygenated medium containing all necessary nutrients and glucose as the sole carbon source, the concentration of glucose decreases with time, carbon dioxide is evolved, and the levels of ADP and ATP remain fairly constant. Explain the metabolic processes that occur when the incubation is performed (*a*) in the absence of oxygen or (*b*) in the presence of both oxygen and an inhibitor of glyceraldehyde 3-phosphate dehydrogenase.

11.19. How is the concentration of glucose in the blood maintained (*a*) during rest after a carbohydrate-rich meal and (*b*) during prolonged exercise?

11.20. Diabetes mellitus is a condition in which the pancreas fails to produce sufficient active insulin or produces an ineffective form of insulin. What effects on carbohydrate metabolism would this condition impart to an otherwise healthy person?

11.21. For a sequence of reactions a cell requires NADPH far in excess of its requirements for ribose 5-phosphate. (*a*) How can the cell achieve this, and (*b*) what is the fate of the excess ribose 5-phosphate?

11.22. What are the four reactions that occur in liver cells that involve glucose 6-phosphate as a substrate?

Chapter 12

The Citric Acid Cycle

12.1 INTRODUCTION

The citric acid cycle is a sequence of reactions in which the two carbon atoms of acetyl-CoA are completely oxidized to CO_2. It is a central pathway for the release of energy from acetyl-CoA, which is produced from the catabolism of carbohydrates (Chap. 11), fatty acids (Chap. 13), and some amino acids (Chap. 15) and is closely involved with two other processes, namely, electron transport and oxidative phosphorylation (Chap. 14).

Question: What is acetyl-CoA?

The name "acetyl-CoA" is an abbreviation for the compound acetyl coenzyme A, which has the structure shown in Fig. 12-1. Coenzyme A has three components: ADP with an additional 3′ phosphate group, pantothenic acid, and β-mercaptoethylamine. Coenzyme A is a *carrier* of *acyl*

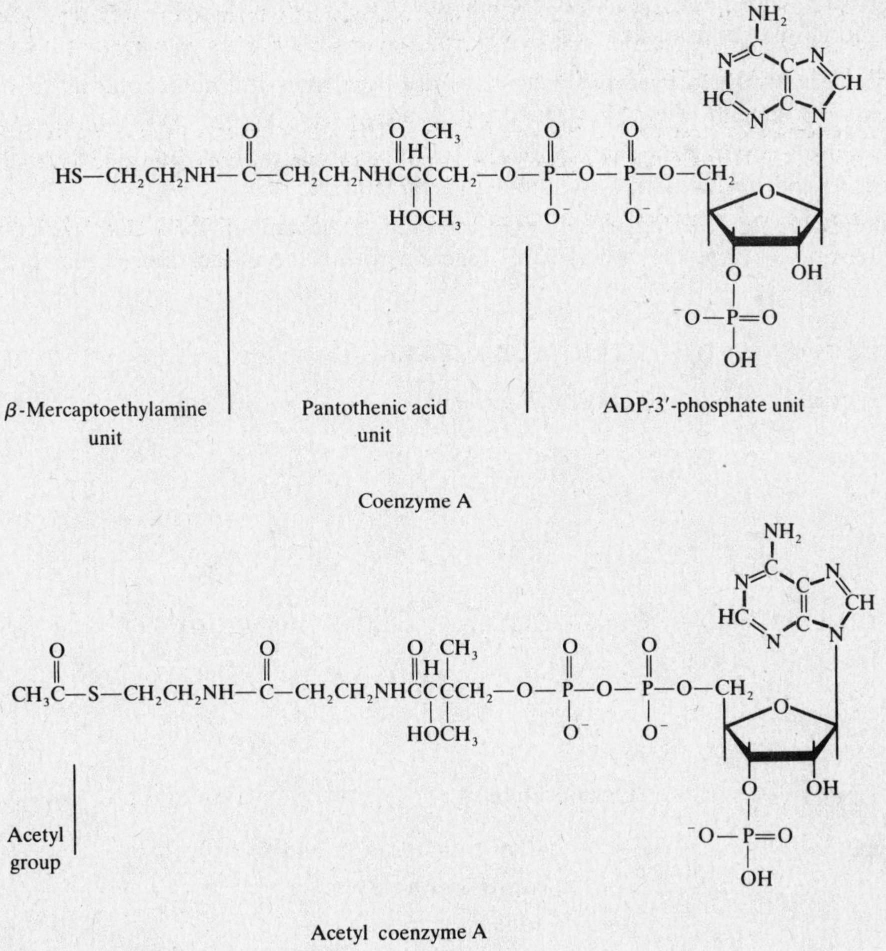

β-Mercaptoethylamine unit | Pantothenic acid unit | ADP-3′-phosphate unit

Coenzyme A

Acetyl group

Acetyl coenzyme A

Fig. 12-1 The structures of coenzyme A and acetyl coenzyme A.

groups, especially acetyl groups, which are attached to the thiol of β-mercaptoethylamine. Thus, the abbreviation CoA is used for coenzyme A and acetyl-CoA for acetyl coenzyme A. In some cases the thiol group is emphasized by incorporating SH into the abbreviations; i.e., CoASH (coenzyme A) and $CoASCOCH_3$ or $CH_3COSCoA$ (acetyl coenzyme A).

EXAMPLE 12.1

The degradation of carbohydrates (such as glucose) in the glycolytic pathway produces pyruvate, but how does the acetyl-CoA originate in carbohydrate metabolism?

Pyruvate is converted into acetyl-CoA by a group of enzymes known as the *pyruvate dehydrogenase complex* (see Example 12.3 and Chap. 5). Acetyl-CoA and the enzymes that catalyze the steps of the citric acid cycle are situated within the matrix of the mitochondria, except for one enzyme that is located in the inner mitochondrial membrane.

Question: Why is the citric acid cycle so named?

The name stems from the first step in the cycle, which is a condensation of oxaloacetate with acetyl-CoA to form *citric acid*. However, as this product is a tricarboxylic acid, the cycle has an alternative name, the *tricarboxylic acid cycle*.

Acetyl-CoA cannot permeate the two mitochondrial membranes, and it is formed exclusively in the mitochondrial matrix from three basic sources:

1. Glycolysis produces pyruvate, which readily permeates the mitochondrial membranes and enters the matrix, where it is converted to acetyl-CoA via pyruvate dehydrogenase.
2. Fatty acids enter from the cytoplasm (as their CoA derivatives) via a specific transport protein, and in the matrix, they undergo oxidation to acetyl-CoA.
3. Proteins are hydrolyzed to amino acids, and those amino acids that are converted into acetyl-CoA (Chap. 15) undergo this reaction within the mitochondrial matrix.

12.2 REACTIONS OF THE CITRIC ACID CYCLE

There are eight steps in the citric acid cycle.

Step 1

$$\underset{\text{Acetyl-CoA}}{CH_3COSCoA} + \underset{\text{Oxaloacetate}}{O{=}C(COO^-)-CH_2-COO^-} \overset{H_2O}{\rightleftharpoons} \underset{\text{Citrate}}{HO-C(COO^-)(CH_2COO^-)_2} + \underset{\text{Coenzyme A}}{CoASH}$$

Reaction type:	aldol condensation and hydrolysis
Enzyme:	citrate synthase
Coreactants:	none
Effectors:	
Inhibited by:	acetyl-CoA, oxaloacetate, NADH, succinyl-CoA
Activated by:	NAD^+

Step 2

$$
\begin{array}{c}
\mathrm{COO^-} \\
| \\
\mathrm{CH_2} \\
| \\
\mathrm{HO{-}C{-}COO^-} \\
| \\
\mathrm{CH_2} \\
| \\
\mathrm{COO^-} \\
\text{Citrate}
\end{array}
\underset{}{\overset{-\mathrm{H_2O}}{\rightleftharpoons}}
\left[
\begin{array}{c}
\mathrm{COO^-} \\
| \\
\mathrm{CH_2} \\
| \\
\mathrm{C{-}COO^-} \\
\| \\
\mathrm{CH} \\
| \\
\mathrm{COO^-}
\end{array}
\right]
\overset{+\mathrm{H_2O}}{\rightleftharpoons}
\begin{array}{c}
\mathrm{COO^-} \\
| \\
\mathrm{CH_2} \\
| \\
\mathrm{CH{-}COO^-} \\
| \\
\mathrm{HO{-}CH} \\
| \\
\mathrm{COO^-} \\
\text{Isocitrate}
\end{array}
$$

Citrate [*cis*-Aconitate] Isocitrate

Reaction type: dehydration and hydration; overall an isomerization
Enzyme: aconitase
Coreactants: none

The intermediate in this reaction, cis-*aconitate*, is bound to aconitase and is not usually classed as a discrete intermediate of the citric acid cycle.

Step 3

$$
\mathrm{NAD^+} +
\begin{array}{c}
\mathrm{COO^-} \\
| \\
\mathrm{CH_2} \\
| \\
\mathrm{CH{-}COO^-} \\
| \\
\mathrm{HO{-}CH} \\
| \\
\mathrm{COO^-}
\end{array}
\longrightarrow
\left[
\begin{array}{c}
\mathrm{COO^-} \\
| \\
\mathrm{CH_2} \\
| \\
\mathrm{CH{-}COO^-} \\
| \\
\mathrm{C{=}O} \\
| \\
\mathrm{COO^-}
\end{array}
\right]
\longrightarrow
\begin{array}{c}
\mathrm{COO^-} \\
| \\
\mathrm{CH_2} \\
| \\
\mathrm{CH_2} \\
| \\
\mathrm{C{=}O} \\
| \\
\mathrm{COO^-}
\end{array}
+ \mathrm{NADH} + \mathrm{H^+} + \mathrm{CO_2}
$$

Isocitrate [Oxalosuccinate] 2-Oxoglutarate (α-ketoglutarate)

Reaction type: oxidative decarboxylation
Enzyme: isocitrate dehydrogenase
Coreactants: Mg^{2+} or Mn^{2+}
Effectors:
  Inhibited by: ATP, NADH
  Activated by: ADP, NAD^+

The intermediate in this reaction, *oxalosuccinate*, does not dissociate from the enzyme and is not usually classed as a discrete intermediate of the citric acid cycle.

Step 4

$$
\begin{array}{c}
\mathrm{COO^-} \\
| \\
\mathrm{CH_2} \\
| \\
\mathrm{CH_2} \\
| \\
\mathrm{C{=}O} \\
| \\
\mathrm{COO^-}
\end{array}
+ \mathrm{CoASH} \rightleftharpoons
\begin{array}{c}
\mathrm{COO^-} \\
| \\
\mathrm{CH_2} \\
| \\
\mathrm{CH_2} \\
| \\
\mathrm{COSCoA}
\end{array}
+ \mathrm{CO_2}
$$

2-Oxoglutarate Coenzyme A Succinyl-CoA

Reaction type: thioesterification, decarboxylation, and oxidation
Enzyme: 2-oxoglutarate dehydrogenase complex
Coreactants: The coreactants and the mechanism of action are similar to those of the pyruvate dehydrogenase complex (Sec. 12.5).
Effectors:
Inhibited by: NADH, succinyl-CoA

Step 5

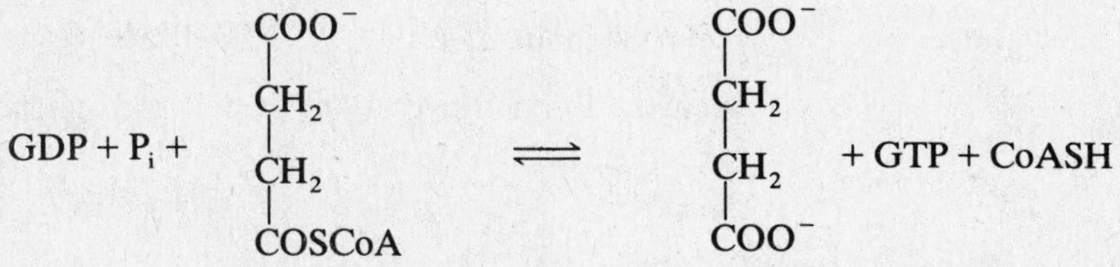

$$\mathrm{GDP + P_i +} \begin{array}{c} \mathrm{COO^-} \\ | \\ \mathrm{CH_2} \\ | \\ \mathrm{CH_2} \\ | \\ \mathrm{COSCoA} \end{array} \rightleftharpoons \begin{array}{c} \mathrm{COO^-} \\ | \\ \mathrm{CH_2} \\ | \\ \mathrm{CH_2} \\ | \\ \mathrm{COO^-} \end{array} \mathrm{+ GTP + CoASH}$$

Succinyl-CoA **Succinate** **Coenzyme A**

Reaction type: phosphorylation and hydrolysis
Enzyme: succinyl-CoA synthetase
Coreactants: Mg^{2+}

The phosphorylation of GDP in this reaction is an example of *substrate-level phosphorylation*, and this is the only reaction in the citric acid cycle to produce a high-energy phosphate bond directly. The energy for this phosphorylation is derived from the hydrolysis of the *thioester bond* of succinyl-CoA. Subsequently, GTP phosphorylates ADP, catalyzed by *nucleoside diphosphokinase*, but this reaction

$$\mathrm{GTP + ADP} \rightleftharpoons \mathrm{GDP + ATP}$$

is not metabolically an essential part of the citric acid cycle.

Step 6

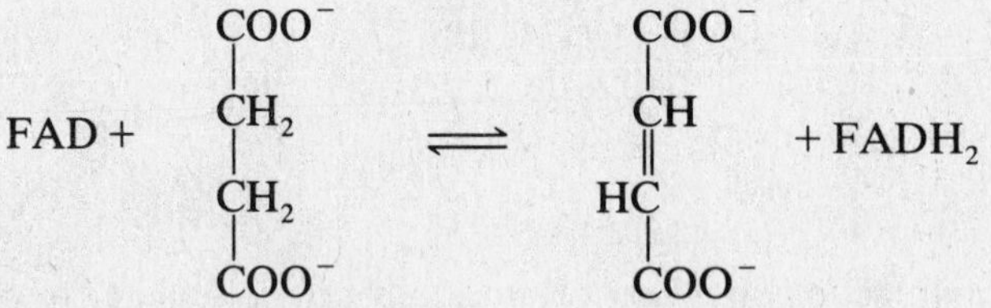

$$\mathrm{FAD +} \begin{array}{c} \mathrm{COO^-} \\ | \\ \mathrm{CH_2} \\ | \\ \mathrm{CH_2} \\ | \\ \mathrm{COO^-} \end{array} \rightleftharpoons \begin{array}{c} \mathrm{COO^-} \\ | \\ \mathrm{CH} \\ \| \\ \mathrm{HC} \\ | \\ \mathrm{COO^-} \end{array} \mathrm{+ FADH_2}$$

Succinate **Fumarate**

Reaction type: redox
Enzyme: succinate dehydrogenase
Coreactants: none
Effectors:
Inhibited by: oxaloacetate

Succinate dehydrogenase catalyzes the so-called *trans elimination* of two H's. This is the only reaction in the citric acid cycle involving FAD, and succinate dehydrogenase is the only enzyme in the cycle that is *membrane-bound*. The importance of this will be discussed in Chap. 14.

Step 7

$$H_2O + \begin{array}{c} COO^- \\ | \\ CH \\ \| \\ HC \\ | \\ COO^- \end{array} \rightleftharpoons \begin{array}{c} COO^- \\ | \\ HOCH \\ | \\ CH_2 \\ | \\ COO^- \end{array}$$

Fumarate L-Malate

Reaction type: hydration

Enzyme: fumarate hydratase (usually called *fumarase*)

Coreactants: none

Step 8

$$NAD^+ + \begin{array}{c} COO^- \\ | \\ HOCH \\ | \\ CH_2 \\ | \\ COO^- \end{array} \rightleftharpoons \begin{array}{c} COO^- \\ | \\ C{=}O \\ | \\ CH_2 \\ | \\ COO^- \end{array} + NADH + H^+$$

L-Malate Oxaloacetate

Reaction type: redox

Enzyme: malate dehydrogenase (mitochondrial)

Coreactants: none

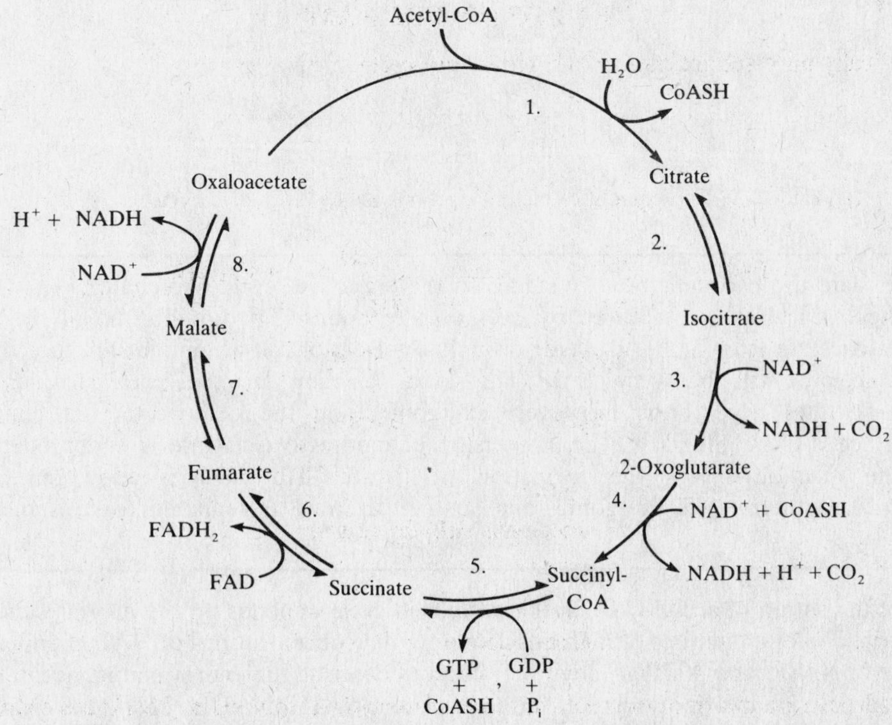

Fig. 12-2 The citric acid cycle. The numbered reactions refer to the steps given in the text.

With the regeneration of oxaloacetate in step 8, the cycle of reactions is completed, and the oxaloacetate condenses with another molecule of acetyl-CoA to commence another turn of the cycle. However, note that there is actually no beginning or end to the citric acid cycle (Fig. 12-2); if any of the intermediates are produced within the mitochondria or gain access to the mitochondria, they can participate in the cycle of reactions but they will always be regenerated. The only actual *fuel* for the cycle is acetyl-CoA and this is not regenerated; its carbon atoms are liberated at steps 3 and 4 as CO_2.

Question: The citric acid cycle is part of the process called *respiration*. Is oxygen directly involved in the reactions of the cycle?

No. While oxygen is not involved in any of the steps of the citric acid cycle, CO_2 is liberated at two of them, giving rise to part of the stoichiometry of respiration.

12.3 THE ENERGETICS OF THE CITRIC ACID CYCLE

The overall consumption of one molecule of acetyl-CoA in the citric acid cycle is an *exergonic* process; $\Delta G^{\circ\prime} = -60\ \text{kJ mol}^{-1}$. All but two of the individual reactions are exergonic. Step 2 (citrate $\longrightarrow$ isocitrate) and step 8 (malate $\longrightarrow$ oxaloacetate) are endergonic (Fig. 12-3).

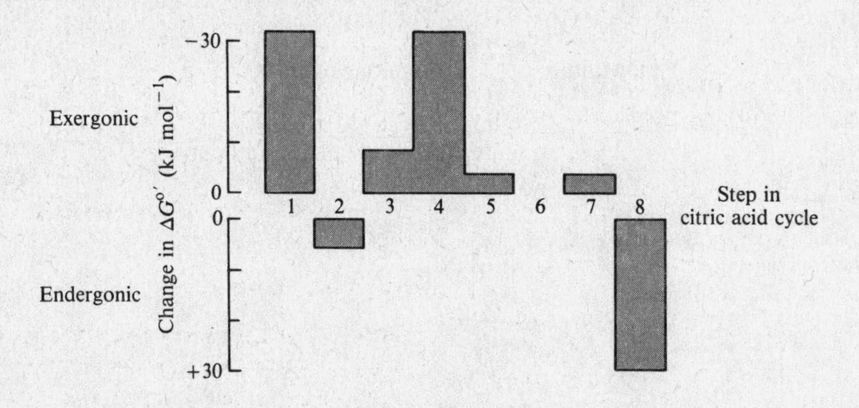

Fig. 12-3 The energetics of the reactions in the citric acid cycle.

Question: How are the two unfavored reactions in the citric acid cycle overcome?

Both of these reactions are followed by exergonic reactions. The equilibrium of the reaction malate $\rightleftharpoons$ oxaloacetate (step 8) lies in favor of malate formation, so at equilibrium the concentration of oxaloacetate will be very low. The next reaction in the cycle (oxaloacetate + acetyl-CoA $\longrightarrow$ citrate) (step 1) is, however, exergonic, and the oxaloacetate is removed to condense with acetyl-CoA. Similarly, the conversion of citrate to isocitrate is endergonic, and at equilibrium the reaction favors the formation of citrate. The next reaction in the cycle (isocitrate $\longrightarrow$ 2-oxoglutarate) is exergonic, and so the isocitrate is removed for this reaction to proceed.

The rate of utilization of acetyl-CoA in the citric acid cycle depends on the *energy status* within the mitochondria. Under conditions of high bond energy, the concentrations of NADH and ATP are high, and those of NAD^+ and AMP are low. It is usual to describe the energy status, not in terms of absolute concentrations, but in terms of ratios such as NAD^+/NADH and AMP/ATP. So a high-energy status would mean a low NAD^+/NADH ratio or a high NADH/NAD^+ ratio, and a low AMP/ATP ratio or a high ATP/AMP ratio.

EXAMPLE 12.2

What would be the effect on the reactions of the citric acid cycle if the NADH and $FADH_2$ were not reoxidized?

The reactions of the cycle would cease once all the NAD^+ and FAD were reduced to NADH and $FADH_2$, respectively. The reoxidation of NADH and $FADH_2$ occurs in the electron-transport system and is slow in comparison with the *potential* rate at which the citric acid cycle could function. Thus, because the reoxidation of NADH and $FADH_2$ results in energy transduction, the cycle is controlled by the energy requirements of the mitochondria.

12.4 REGULATION OF THE CITRIC ACID CYCLE

There are four major regulatory enzymes in the citric acid cycle. These are *citrate synthase* (step 1), *isocitrate dehydrogenase* (step 3), *2-oxoglutarate dehydrogenase* (step 4), and *succinate dehydrogenase* (step 6).

In the first control point, citrate synthase catalyzes the condensation of acetyl-CoA with oxaloacetate to produce citrate ($\Delta G^{\circ\prime} = -32.2\ \text{kJ mol}^{-1}$). Although the reaction is reversible, the equilibrium lies very much in favor of citrate formation because of the hydrolysis of a bond in the intermediate compound, citroyl-CoA (Fig. 12-4). Citroyl-CoA is bound to citrate synthase, and the hydrolysis of the thioester bond, to produce citrate and coenzyme A, is an exergonic process. Citrate synthase is inhibited by its substrates (acetyl-CoA and oxaloacetate), and its activity is affected by the energy status of the mitochondria (low $NAD^+/NADH$ inhibits) and by succinyl-CoA, which competes with acetyl-CoA for the active site.

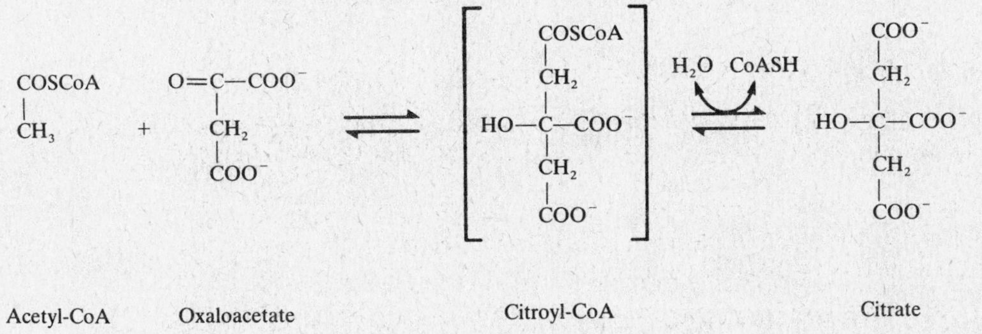

COSCoA—CH_3 (Acetyl-CoA) + O=C(—COO^-)—CH_2—COO^- (Oxaloacetate) ⇌ [$COSCoA$—CH_2—C(OH)(COO^-)—CH_2—COO^-] (Citroyl-CoA) ⇌ (H_2O in, CoASH out) COO^-—CH_2—C(OH)(COO^-)—CH_2—COO^- (Citrate)

Acetyl-CoA Oxaloacetate Citroyl-CoA Citrate

Fig. 12-4 The citroyl-CoA intermediate produced by citrate synthase.

The second point of regulation in the cycle is at the conversion of isocitrate to 2-oxoglutarate, catalyzed by isocitrate dehydrogenase. The decarboxylation in this step drives the reaction irreversibly toward the formation of 2-oxoglutarate. Isocitrate dehydrogenase is an allosteric enzyme (Chap. 8) that is activated by ADP and NAD^+ and inhibited by ATP and NADH. When the concentration of compounds with high-energy bonds (or those with the ability to produce compounds with high-energy bonds) rises, the enzyme is inhibited; thus, the remaining reactions of the cycle are impeded, and an accumulation of isocitrate occurs. The production of isocitrate from citrate is a reversible reaction, and if the isocitrate is not removed by other reactions, the citrate $\rightleftharpoons$ isocitrate reaction tends toward equilibrium. At equilibrium, the relative amounts of these two components are ~93 percent and ~7 percent, respectively. So the inhibition of isocitrate dehydrogenase results in an increase in the concentration of citrate, which can leave the mitochondria and enter the cytoplasm. This is the source of the citrate that is the inhibitor of the two glycolytic enzymes, phosphofructokinase and pyruvate kinase (Fig. 12-5).

1. Isocitrate dehydrogenase is inhibited by relatively high concentrations of high-energy compounds.
2. Isocitrate accumulates.
3. Equilibrium tends to become established.
4. Citrate accumulates.
5. Citrate enters the cytoplasm and inhibits two enzymes in glycolysis.

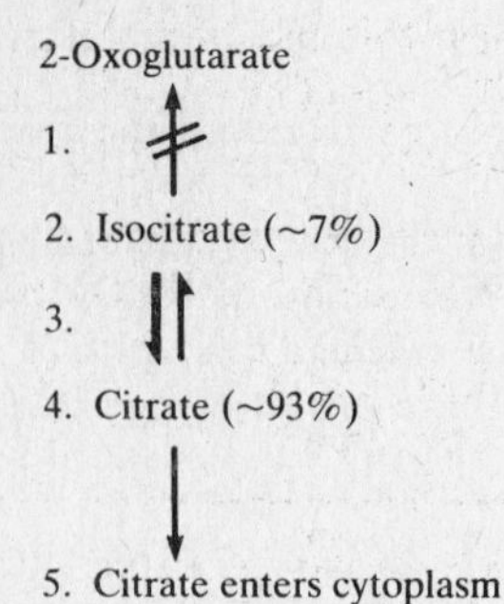

Fig. 12-5 The control of glycolysis by elevated concentrations of high-energy compounds within the mitochondria.

The third control step in the citric acid cycle is catalyzed by 2-oxoglutarate dehydrogenase. This multienzyme complex is subject to *product inhibition* by both NADH and succinyl-CoA. Yet again, the presence of a high-energy state (high $NADH/NAD^+$) acts to slow the cycle at this reaction.

The fourth control in the cycle is at the conversion of succinate to fumarate via succinate dehydrogenase. This enzyme is inhibited by oxaloacetate, so that if for any reason oxaloacetate accumulates, the enzymes will be inhibited; thus, oxaloacetate *feeds back* and inhibits a reaction that is required for its synthesis. This phenomenon is called *negative feedback* (See Chap. 9.)

The major regulatory sites in the citric acid cycle are shown in Fig. 12-6.

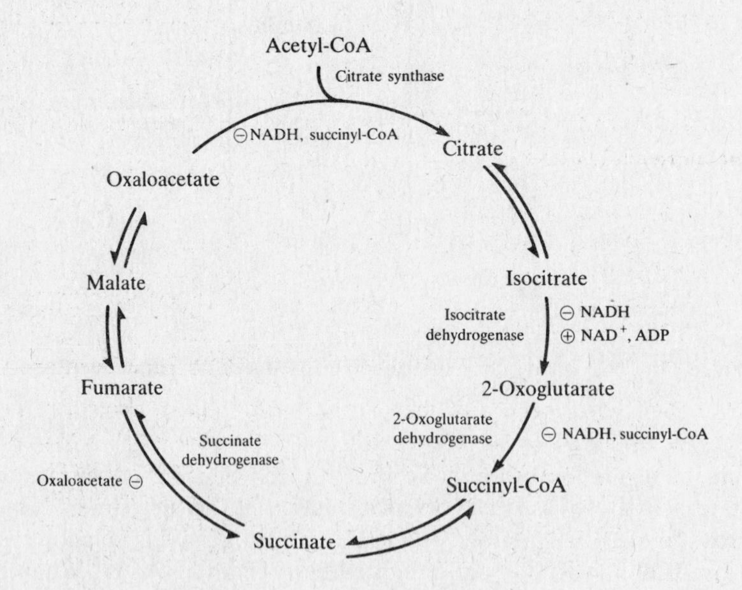

Fig. 12-6 The four regulatory steps in the citric acid cycle. ⊕ = activation; ⊖ = inhibition.

12.5 THE PYRUVATE DEHYDROGENASE COMPLEX

Acetyl-CoA is produced from fatty acids, proteins, and carbohydrates and is a central and major compound in intermediary metabolism. The mechanism of its formation from the degradation of fatty acids and proteins is discussed in Chaps. 13 and 15, respectively; here, the means whereby *carbohydrates* form this most important molecule will be presented. The glycolytic pathway can yield pyruvate from all degradable sugars, and this can be converted to acetyl-CoA. Pyruvate enters the mitochondrial matrix and is the substrate for the multienzyme complex pyruvate dehydrogenase.

Question: What is the overall reaction catalyzed by the pyruvate dehydrogenase complex?

$$\underset{\text{Pyruvate}}{CH_3COCOO^-} + NAD^+ + CoASH \longrightarrow \underset{\text{Acetyl-CoA}}{CH_3COSCoA} + CO_2 + NADH$$

The complex consists of three enzymes: *pyruvate dehydrogenase*, *dihydrolipoyl transacetylase*, and *dihydrolipoyl dehydrogenase*. The first reaction is a decarboxylation of pyruvate to produce a 2-hydroxyethyl residue bonded to pyruvate dehydrogenase via its prosthetic group *thiamine pyrophosphate* (TPP). The second enzyme of the complex (dihydrolipoyl transacetylase) catalyzes two steps, namely, the transfer of an acetyl group (derived from the oxidation of the 2-hydroxyethyl residue) to a lipoic acid *arm* of the enzyme and then transfer of this acetyl group to CoA to form acetyl-CoA. At this step the lipoic acid is converted to its fully reduced state; it then reduces the FAD prosthetic group of dihydrolipoyl dehydrogenase to $FADH_2$. Finally, the reduced prosthetic group, $FADH_2$, is reoxidized to FAD by converting NAD^+ to NADH. The overall sequence of reactions is operationally irreversible because of the decarboxylation step, which is irreversible by virtue of the relatively low CO_2 concentration in tissues. However, all the *other* reactions are reversible. The various reactions of the complex are given in Fig. 12-7.

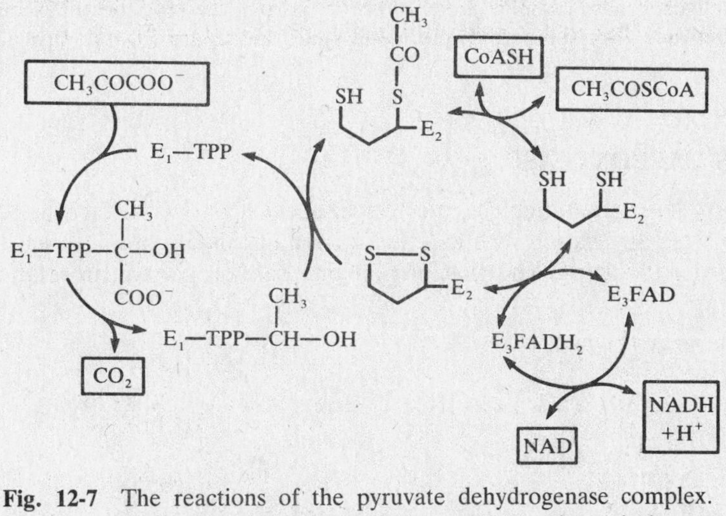

Fig. 12-7 The reactions of the pyruvate dehydrogenase complex. The reactants in the overall reaction are shown in boxes. E_1 = pyruvate dehydrogenase (TPP = thiamine pyrophosphate as prosthetic group), E_2 = dihydrolipoyl transacetylase (oxidized lipoic acid as prosthetic group), E_3 = dihydrolipoyl dehydrogenase (FAD as prosthetic group).

Since a major function of the citric acid cycle is to oxidize acetyl-CoA, with the subsequent generation of ATP, the rate at which the cycle of reactions operates depends on the availability of acetyl-CoA. Thus, the *energy status* of the cell exerts control over the activity of the pyruvate dehydrogenase complex. If the concentration of acetyl-CoA or ATP is low, then the complex is activated to produce acetyl-CoA. If the concentration of acetyl-CoA is high, because of either the breakdown of fatty acids or protein or the existence of a high-energy state within the mitochondria, then the complex is inhibited. This latter control reduces the rate of pyruvate degradation which in turn reduces the rate of glucose degradation.

EXAMPLE 12.3

Control of the activity of the pyruvate dehydrogenase complex is exerted by the phosphorylation of pyruvate dehydrogenase (E_1), which renders it *inactive*. This process is catalyzed by *pyruvate dehydrogenase kinase*, which is always tightly bound to E_1. The kinase is activated by high-energy conditions, and it requires ATP to accomplish the phosphorylation step. Another enzyme, *phosphoprotein phosphatase*, is weakly bound to E_1 and is capable of reactivating the system by removing the inhibitory phosphate group (Fig. 12-8).

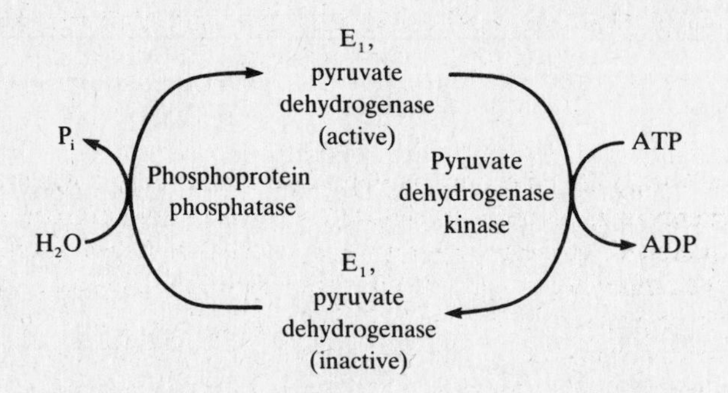

Fig. 12-8 The control of the pyruvate dehydrogenase complex.

The pyruvate dehydrogenase complex is *not directly* a part of the reactions that constitute the citric acid cycle. It is the *link* between glycolysis and the citric acid cycle, and its activity is controlled by the energy status of the mitochondria.

12.6 PYRUVATE CARBOXYLASE

Pyruvate carboxylase is another enzyme which is *not* a part of the citric acid cycle per se but which functions in close association with it. The function of this enzyme is described in Chap. 11, but it is useful to consider its action, and that of pyruvate dehydrogenase, in relation to the citric acid cycle.

$$\text{ATP} + \underset{\text{Pyruvate}}{\begin{array}{c}\text{CH}_3\\|\\\text{C=O}\\|\\\text{COO}^-\end{array}} + \text{HCO}_3^- \overset{\text{H}_2\text{O}}{\rightleftharpoons} \underset{\text{Oxaloacetate}}{\begin{array}{c}\text{CH}_2\text{COO}^-\\|\\\text{C=O}\\|\\\text{COO}^-\end{array}} + \text{ADP} + \text{P}_i$$

Reaction type: carboxylation
Enzyme: pyruvate carboxylase
Coreactants: Mg^{2+}
Effectors:
 Activated by: acetyl-CoA

The product of this reaction, oxaloacetate, can either enter the gluconeogenic pathway (Chap. 11) by way of malate or condense with acetyl-CoA to yield citrate. Pyruvate carboxylase is an allosteric enzyme, and it is activated by the heterotropic effector, acetyl-CoA. Thus, pyruvate in the mitochondria is the substrate for either pyruvate dehydrogenase or pyruvate carboxylase, the activities of which, in turn, are controlled by reactants associated with the citric acid cycle. The interplay among pyruvate dehydrogenase, pyruvate carboxylase, pyruvate, and the citric acid cycle is shown in Fig. 12-9.

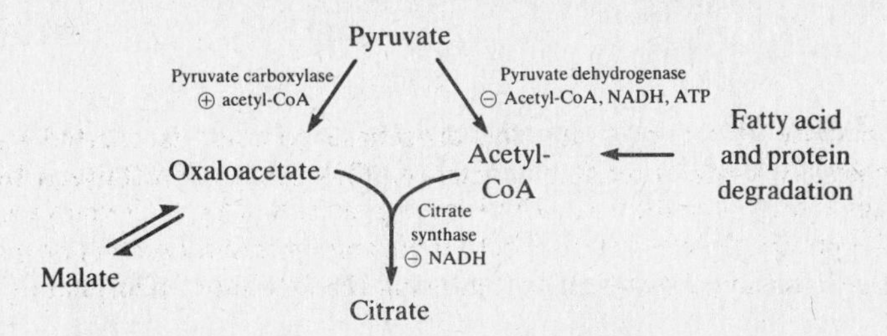

Fig. 12-9 The fates of pyruvate in the mitochondria. ⊕ = activation; ⊖ = inhibition.

EXAMPLE 12.4

The fate of pyruvate within the mitochondria depends on the energy status that exists in the mitochondria. If the levels of high-energy compounds are elevated, i.e., ATP and NADH are plentiful, then pyruvate dehydrogenase will be inhibited. If the concentration of acetyl-CoA is high (perhaps from the degradation of fatty acids or protein), then this will also tend to inhibit pyruvate dehydrogenase and pyruvate will be conserved through inhibition of its conversion to acetyl-CoA. But under these conditions, the acetyl-CoA causes the activation of pyruvate carboxylase, and so oxaloacetate is produced. A lack of oxaloacetate precludes the conversion of acetyl-CoA to citrate, resulting in a high level of acetyl-CoA. By activating pyruvate carboxylase, this mechanism permits the necessary production of oxaloacetate from pyruvate. (Other reasons for a lack of oxaloacetate are discussed in Example 12.6.)

12.7 THE AMPHIBOLIC NATURE OF THE CITRIC ACID CYCLE

Apart from the production of NADH and $FADH_2$, which are the high-energy fuels of electron transport, the citric acid cycle has two other major functions. Several of its intermediate compounds are used to synthesize other cell constituents. This, the provision of molecules for other metabolic or biosynthetic pathways, is the *anabolic* function of the cycle (Table 12.1). Alternatively, certain other processes occurring within the cell may produce intermediates of the citric acid cycle. These compounds enter the reactions of the cycle, and their degradation involves the *catabolic* role of the cycle. These two major capabilities classify the citric acid cycle as an *amphibolic* pathway (Greek: *amphi* meaning "both").

Table 12.1. Anabolic and Catabolic Reactions of the Citric Acid Cycle

Anabolic Reactions	
Intermediate Removed	**Metabolic Fate**
Citrate	Fatty acid biosynthesis
2-Oxoglutarate	Synthesis of glutamate
Succinyl-CoA	Heme biosynthesis
Malate	Gluconeogenesis
Oxaloacetate	Synthesis of aspartate
Catabolic Reactions	
Intermediate Produced	**Source**
2-Oxoglutarate	Glutamate
Succinyl-CoA, Oxaloacetate, Fumarate, 2-Oxoglutarate	Degradation of some amino acids (Chap. 15)
Oxaloacetate	Aspartate

EXAMPLE 12.5

If intermediates of the citric acid cycle enter the cycle via other reactions in the cell, are they oxidized?

If one molecule of 2-oxoglutarate, for example, were to enter the cycle, its metabolism through one turn of the cycle would be:

$$\text{2-Oxoglutarate}^{2-} + 3\text{NAD}^+ + \text{FAD} + \text{GDP}^{3-} + \text{P}_i^{2-} + 2\text{H}_2\text{O} + \text{Acetyl-CoA} \longrightarrow \text{2-Oxoglutarate}^{2-} + 2\text{CO}_2 + 3\text{NADH} + \text{FADH}_2 + 2\text{H}^+ + \text{GTP}^{4-} + \text{CoA}$$

Only the acetyl group of acetyl-CoA is oxidized to CO_2; the 2-oxoglutarate is regenerated.

EXAMPLE 12.6

If intermediates pass from the cycle and enter other anabolic pathways, why does the cycle continue to operate?

The cycle oxidizes acetyl-CoA, and to perform this task, it must convert acetyl-CoA to citrate. For this to be achieved, oxaloacetate must be available. If the removal of intermediates results in a decrease in the amount of oxaloacetate for this purpose, acetyl-CoA cannot be removed and will accumulate. This will *inhibit* the pyruvate dehydrogenase complex and *activate* pyruvate carboxylase, leading to the conversion of pyruvate to oxaloacetate. This product is now available to condense with the acetyl-CoA to produce citrate, which will restore the status quo. Reactions like that of pyruvate carboxylase that provide molecules for the replacement of intermediates of the citric acid cycle are known as *anaplerotic* reactions (Greek, meaning "to fill up": *ana* = "up" + *plerotikos* from *pleroun* = "to make full").

12.8 THE GLYOXYLATE CYCLE

Some plants and bacteria that can use acetate as their sole source of carbon are able to oxidize acetyl-CoA via the citric acid cycle, or the acetate can be converted to carbohydrates via a pathway that is a modification of the citric acid cycle. This pathway is known as the *glyoxylate cycle* (Fig. 12-10) and in plants exists in organelles called *glyoxysomes*. The enzymes of the cycle are very active

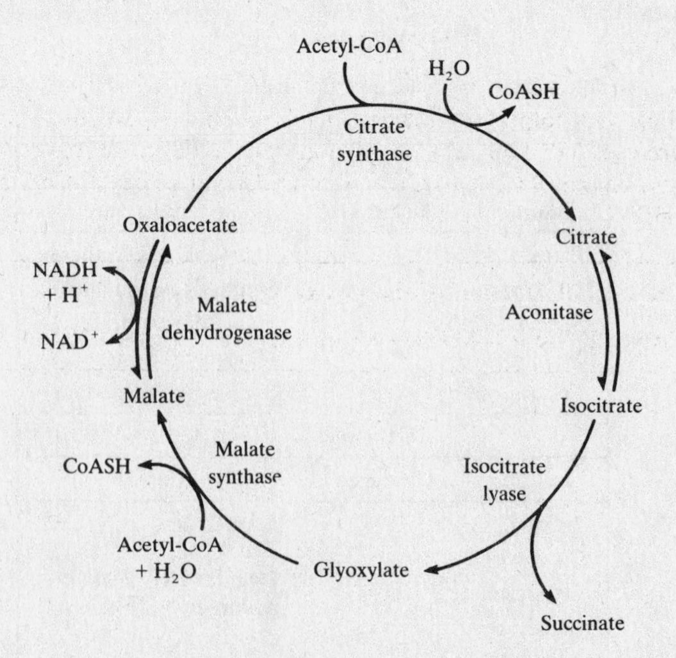

Fig. 12-10 The glyoxylate cycle.

in germinating plant seeds, which are able to convert their stores of fatty acids into carbohydrates during their growth. Animals do not have the enzymes of the glyoxylate cycle and therefore cannot convert fatty acids into carbohydrates. The two reactions unique to this pathway are catalyzed by the enzymes *isocitrate lyase* and *malate synthase*.

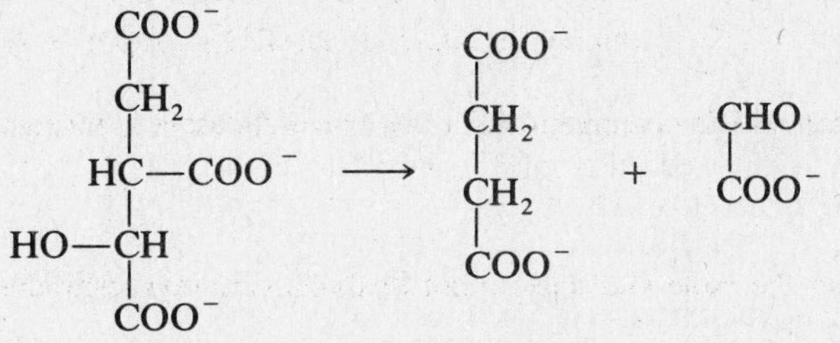

$$\begin{array}{c} COO^- \\ | \\ CH_2 \\ | \\ HC{-}COO^- \\ | \\ HO{-}CH \\ | \\ COO^- \end{array} \longrightarrow \begin{array}{c} COO^- \\ | \\ CH_2 \\ | \\ CH_2 \\ | \\ COO^- \end{array} + \begin{array}{c} CHO \\ | \\ COO^- \end{array}$$

Isocitrate Succinate Glyoxylate

Reaction type: reverse-aldol or aldol cleavage
Enzyme: isocitrate lyase
Coreactants: none

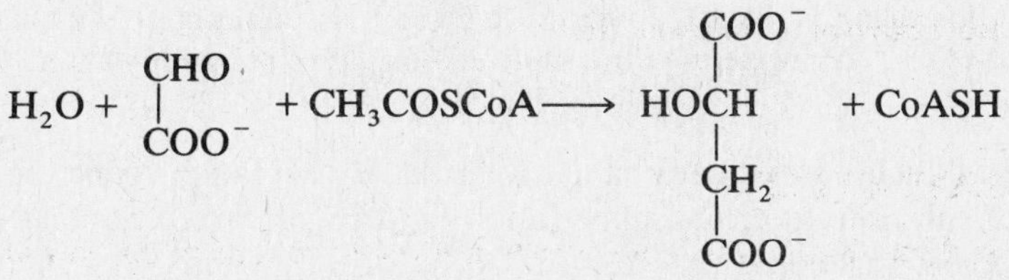

$$H_2O + \begin{array}{c} CHO \\ | \\ COO^- \end{array} + CH_3COSCoA \longrightarrow \begin{array}{c} COO^- \\ | \\ HOCH \\ | \\ CH_2 \\ | \\ COO^- \end{array} + CoASH$$

Glyoxylate Acetyl-CoA Malate Coenzyme A

Reaction type: aldol condensation and hydrolysis
Enzyme: malate synthase
Coreactants: none

These reactions produce two important intermediate compounds, *succinate* and *malate* (which is converted into *oxaloacetate*). The two decarboxylation steps of the citric acid cycle are bypassed, and so there is no oxidation of acetyl-CoA to CO_2. Two molecules of acetyl-CoA are used, but *all* the carbon atoms are retained.

Question: What is the overall reaction of the glyoxylate cycle?

$$2\ \text{Acetyl-CoA} + 2H_2O + NAD^+ \longrightarrow \text{Succinate}^{2-} + NADH + 3H^+ + 2\ CoASH$$

Solved Problems

REACTIONS OF THE CITRIC ACID CYCLE

12.1. What are the overall chemical changes that occur during one complete turn of the citric acid cycle?

SOLUTION

The overall reactions are the complete oxidation of one molecule of acetyl-CoA, the production of two molecules of CO_2, the reduction of three molecules of NAD^+ and one of FAD, and the phosphorylation of one molecule of GDP.

12.2. The overall reaction of the citric acid cycle is

$$CH_3COSCoA + 2H_2O + 3NAD^+ + FAD + GDP + P_i \longrightarrow 2CO_2 + 3NADH + 3H^+ + FADH_2 + GTP$$

Is only one ATP equivalent (GTP) produced for each molecule of acetyl-CoA consumed?

SOLUTION

Only 1 ATP equivalent is *directly* produced, but the reoxidation of NADH and $FADH_2$ yields a further 11 molecules of ATP by electron transport and oxidative phosphorylation (Chap. 14).

12.3. In the citric acid cycle, how many steps involve (*a*) oxidation-reduction, (*b*) hydration-dehydration, (*c*) substrate-level phosphorylation, and (*d*) decarboxylation? List the enzymes responsible for these reactions.

SOLUTION

(*a*) Four steps involve oxidation-reduction. The enzymes involved are isocitrate dehydrogenase, 2-oxoglutarate dehydrogenase, succinate dehydrogenase, and malate dehydrogenase.

(*b*) Two steps involve hydration-dehydration reactions. The enzymes responsible for these reactions are aconitase and fumarase.

(*c*) One step involves substrate-level phosphorylation, and the enzyme is succinyl-CoA synthetase.

(*d*) Two steps involve decarboxylation. The enzymes are isocitrate dehydrogenase and 2-oxoglutarate dehydrogenase.

12.4. Two molecules of CO_2 are produced each time a molecule of acetyl-CoA is oxidized. Do the carbon atoms of *this* acetyl-CoA molecule become the CO_2 in the first turn of the cycle?

SOLUTION

No. Figure 12-11 shows the fate of the carbon atoms of one molecule of acetyl-CoA in *two* turns of the cycle. The two carbon atoms can be followed until step 5, the formation of the *symmetrical* molecule succinate, when they become *randomized*. This means that the two methylene carbons of succinate have equal probability of arising from the methyl group of acetyl-CoA. Steps 3 and 4 of the *second* turn each yield CO_2, which represents, at each step, 50 percent of the original carboxyl carbon of acetyl-CoA. During the *third* turn of the cycle, these two steps collectively liberate 50 percent of the methyl carbon as CO_2.

12.5. Citrate is a symmetrical molecule, yet in step 2 of the citric acid cycle, aconitase catalyzes the removal of the elements of water from only *half* of the molecule and not from the identical other half. How is this rationalized?

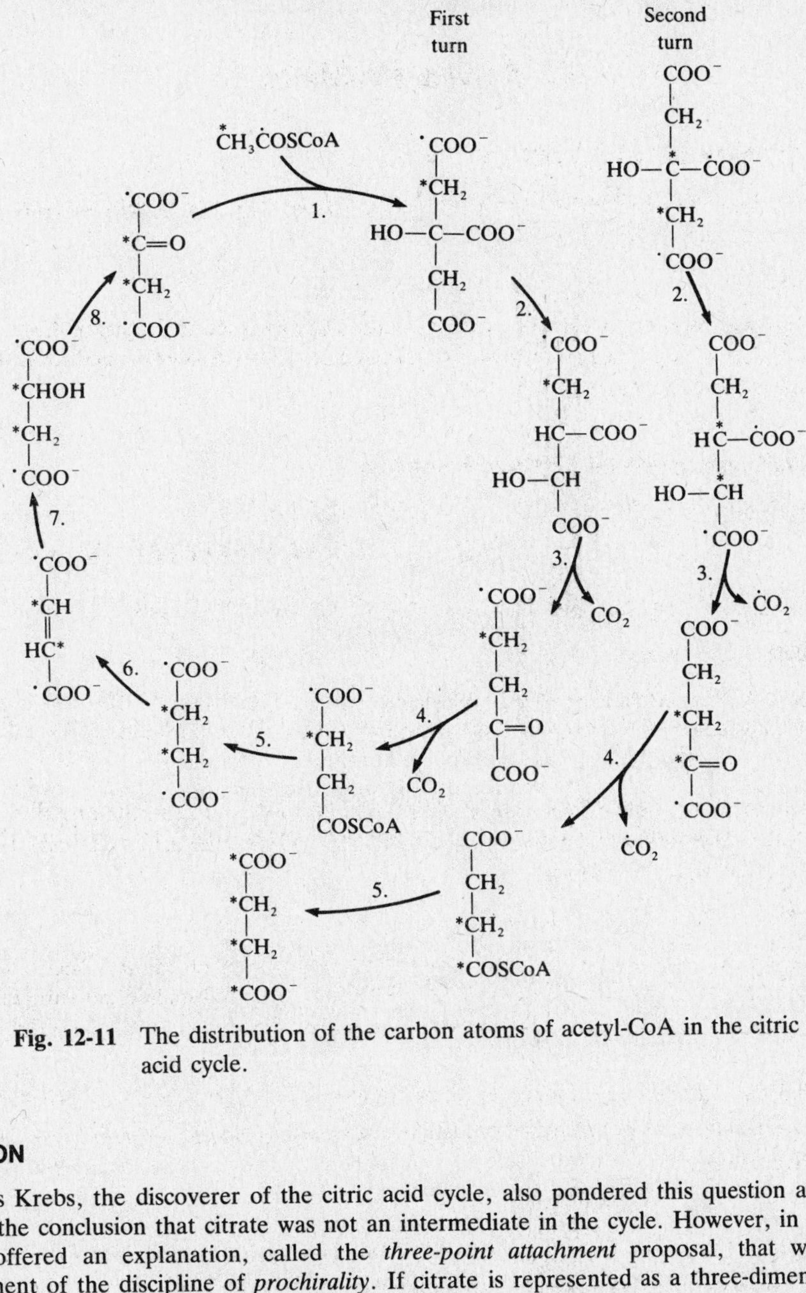

Fig. 12-11 The distribution of the carbon atoms of acetyl-CoA in the citric acid cycle.

SOLUTION

Hans Krebs, the discoverer of the citric acid cycle, also pondered this question and at one stage came to the conclusion that citrate was not an intermediate in the cycle. However, in 1948 Alexander Ogston offered an explanation, called the *three-point attachment* proposal, that was to start the development of the discipline of *prochirality*. If citrate is represented as a three-dimensional structure (Fig. 12-12), then on the assumption that a three-point attachment to aconitase is necessary, it is apparent that citrate can only be accommodated in one way. The removal of the elements of water can then only occur from one particular half of the symmetrical molecule.

12.6. If a cell in which the citric acid cycle operates has the enzyme *phosphoenolpyruvate carboxykinase*, by what processes might it oxidize one molecule of 2-oxoglutarate to five molecules of CO_2?

SOLUTION

The theoretical reaction sequence would be

$$\text{2-Oxoglutarate} \longrightarrow (CO_2) + \text{Succinyl-CoA} \longrightarrow \text{Fumarate} \longrightarrow \text{Malate} \longrightarrow \text{Oxaloacetate} \longrightarrow$$
$$(CO_2) + \text{Phosphoenolpyruvate} \longrightarrow \text{Pyruvate} \longrightarrow (CO_2) + \text{Acetyl-CoA} \longrightarrow 2CO_2$$

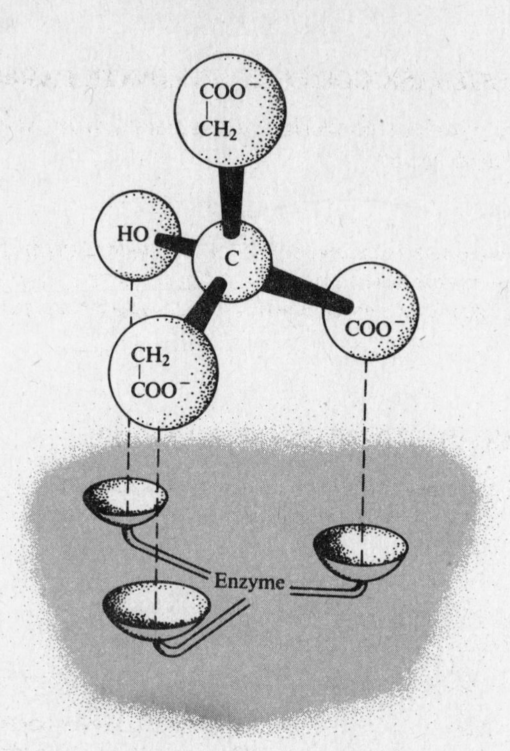

Fig. 12-12 The-three point attachment of citrate to aconitase. This shows that only one *particular* $—CH_2COO^-$ group binds to the enzyme.

12.7. 2-Fluoroacetate is an animal poison found in some South African plants; it is also used in baits to kill wild rabbits (the poison called "1080"). When ingested, it is converted into 2-fluoroacetyl-CoA. What is the fate of 2-fluoroacetyl-CoA in the citric acid cycle and what is the basis of the toxicity of 2-fluoroacetate?

SOLUTION

2-Fluoroacetyl-CoA competes with acetyl-CoA as a substrate for citrate synthase and is converted into 4-fluorocitrate. It is 4-fluorocitrate that blocks the citric acid cycle by being a potent inhibitor of aconitase (Fig. 12-13) and in this lies its toxicity.

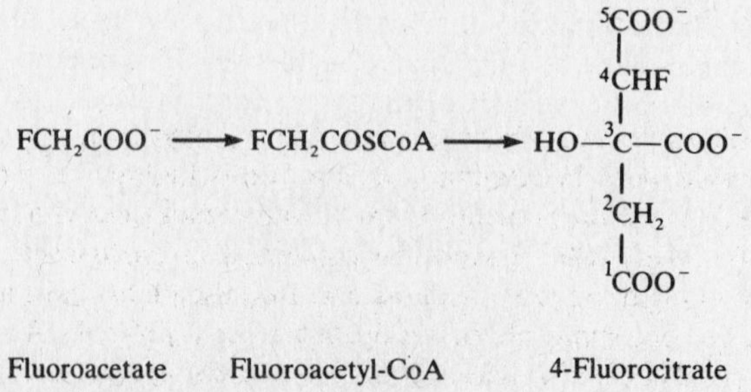

$$
\underset{\text{Fluoroacetate}}{FCH_2COO^-} \longrightarrow \underset{\text{Fluoroacetyl-CoA}}{FCH_2COSCoA} \longrightarrow \underset{\text{4-Fluorocitrate}}{\begin{array}{c} {}^5COO^- \\ | \\ {}^4CHF \\ | \\ HO{-}{}^3C{-}COO^- \\ | \\ {}^2CH_2 \\ | \\ {}^1COO^- \end{array}}
$$

Fig. 12-13 The synthesis of 4-fluorocitrate from 2-fluoroacetate.

THE PYRUVATE DEHYDROGENASE COMPLEX; PYRUVATE CARBOXYLASE

12.8. By which reactions is pyruvate converted to succinate without depleting any of the intermediates of the citric acid cycle?

SOLUTION

Pyruvate can be converted to acetyl-CoA via pyruvate dehydrogenase. Pyruvate can also be carboxylated via pyruvate carboxylase to produce oxaloacetate. So, two molecules of pyruvate can form the precursors of citrate, which can be converted to succinate within the citric acid cycle.

THE AMPHIBOLIC NATURE OF THE CITRIC ACID CYCLE

12.9. Assuming aspartate is the only major carbon source supplied to a preparation of cells, outline all reactions whereby the citric acid cycle would operate in the mitochondria.

SOLUTION

Aspartate would be *transaminated* (Chap. 15) to yield oxaloacetate. Oxaloacetate would then be converted to acetyl-CoA by the following sequence of reactions:

$$\text{Oxaloacetate} \xrightarrow{1.} \text{Malate} \xrightarrow{2.} \text{Oxaloacetate} \xrightarrow{3.} \text{Phosphoenolpyruvate} \xrightarrow{4.} \text{Acetyl-CoA}$$

The enzymes are, respectively: (1) mitochondrial malate dehydrogenase; (2) cytoplasmic malate dehydrogenase; (3) phosphoenolpyruvate carboxykinase; (4) pyruvate kinase and pyruvate dehydrogenase. The acetyl-CoA could then condense with oxaloacetate (produced from a second molecule of aspartate) to yield citrate. Aspartate could, therefore, continue to supply acetyl-CoA, which would continue to fuel the citric acid cycle.

THE GLYOXYLATE CYCLE

12.10. The glyoxylate and citric acid cycles have several reactions in common, but what two enzymes are unique to the glyoxylate cycle?

SOLUTION

Isocitrate lyase and malate synthase. Isocitrate lyase catalyzes the cleavage of isocitrate to succinate and glyoxylate; malate synthase catalyzes the condensation of glyoxylate and acetyl-CoA to yield malate.

12.11. Animals are unable to synthesize carbohydrate from acetyl-CoA, whereas plants can. What is the explanation for this?

SOLUTION

The formation of acetyl-CoA from pyruvate in animals is via pyruvate dehydrogenase, which catalyzes the irreversible decarboxylation reaction. Carbohydrate is synthesized from oxaloacetate, which in turn is synthesized from pyruvate via pyruvate carboxylase. Since the pyruvate dehydrogenase reaction is irreversible, acetyl-CoA cannot be converted to oxaloacetate, and hence animals cannot realize a net gain of carbohydrate from acetyl-CoA. Because plants have a glyoxylate cycle and animals do not, plants synthesize one molecule of succinate and one molecule of malate from two molecules of acetyl-CoA and one of oxaloacetate. The malate is converted to oxaloacetate, which reacts with another molecule of acetyl-CoA and thereby continues the reactions of the glyoxylate cycle. The succinate is also converted to oxaloacetate via the enzymes of the citric acid cycle. Thus, one molecule of oxaloacetate is *diverted* to carbohydrate synthesis and, therefore, plants are able to synthesize carbohydrate from acetyl-CoA.

Supplementary Problems

12.12. Calculate the number of molecules of ATP that can be produced by phosphorylation of ADP from two molecules of acetyl-CoA in the (*a*) citric acid cycle and (*b*) glyoxylate cycle.

12.13. Fumarate contains an *E* (or *trans*) double bond. What would be the result if a cell were supplied with the *Z* (or *cis*) isomer of fumarate (maleate) as the sole carbon source?

12.14. What effect, if any, would a lack of oxygen have on the rate of the citric acid cycle in red blood cells?

12.15. If a potent inhibitor of succinyl-CoA synthetase is applied to liver cells and germinating plant cells, what effect would this have on energy production and the synthesis of carbohydrate in both types of cell?

Chapter 13

Lipid Metabolism

13.1 INTRODUCTION

The major dietary lipids for humans are animal and plant *triacylglycerols*, *sterols*, and *membrane phospholipids*. The process of lipid metabolism fashions and degrades the lipid stores and produces the structural and functional lipids characteristic of individual tissues. For example, the development of a highly organized nervous system has depended on the evolution of specific enzymes to synthesize and degrade (turn over) the lipids of the brain and central nervous system.

13.2 LIPID DIGESTION

People of western culture ingest up to 150 g of triacylglycerol per day. The digestion and absorption of this lipid, together with the ingested phospholipids, depend on secretions from the pancreas (exocrine) and a flow of bile from the gallbladder. The important constituents of the pancreatic secretions are enzymes, and those of the bile are the bile salts (Chap. 6).

The Enzymes

Lipid digestion is accomplished in the small intestine by the action of hydrolytic enzymes, called *lipases* and *phospholipases*, which act on dietary triacylglycerol and phospholipids, respectively.

EXAMPLE 13.1

Consider the action of a lipase:

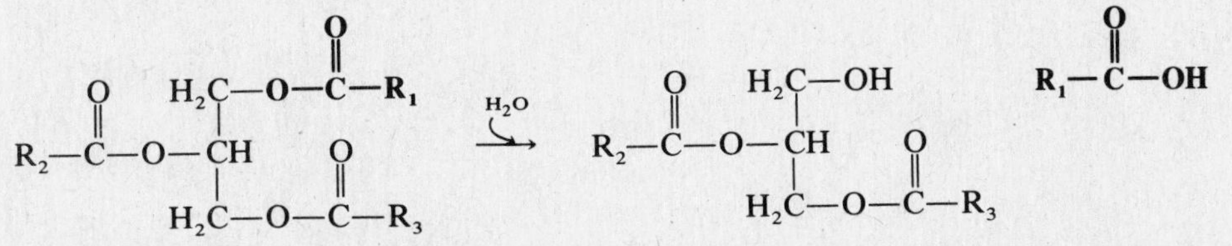

$$\begin{array}{l} H_2C\text{—O—}\overset{O}{\overset{\|}{C}}\text{—}R_1 \\ R_2\text{—}\overset{O}{\overset{\|}{C}}\text{—O—}CH \\ H_2C\text{—O—}\overset{O}{\overset{\|}{C}}\text{—}R_3 \end{array} \xrightarrow{H_2O} \begin{array}{l} H_2C\text{—OH} \\ R_2\text{—}\overset{O}{\overset{\|}{C}}\text{—O—}CH \\ H_2C\text{—O—}\overset{O}{\overset{\|}{C}}\text{—}R_3 \end{array} \quad R_1\text{—}\overset{O}{\overset{\|}{C}}\text{—OH}$$

R_1, R_2, and R_3 are the hydrocarbon chains of fatty acids. Ester bonds between a fatty acid and glycerol are hydrolyzed. In the lumen of the intestine, the complete action of pancreatic lipase on dietary triacylglycerol is to produce the following:

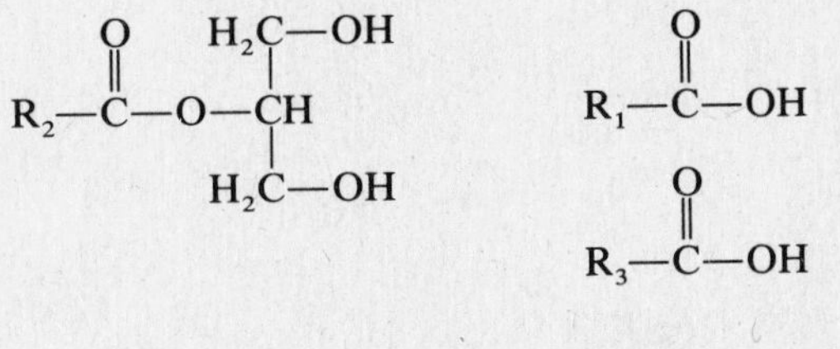

$$\begin{array}{l} H_2C\text{—OH} \\ R_2\text{—}\overset{O}{\overset{\|}{C}}\text{—O—}CH \\ H_2C\text{—OH} \end{array} \qquad \begin{array}{l} R_1\text{—}\overset{O}{\overset{\|}{C}}\text{—OH} \\ R_3\text{—}\overset{O}{\overset{\|}{C}}\text{—OH} \end{array}$$

2-Monoacylglycerol and 2 fatty acids

EXAMPLE 13.2

Phospholipase A_2 catalyzes the following reaction:

$$
\begin{array}{c}
\text{O} \qquad \text{H}_2\text{C—O—}\overset{\text{O}}{\overset{\|}{\text{C}}}\text{—R}_1 \\
\text{R}_2\text{—}\overset{\|}{\text{C}}\text{—O—}\overset{|}{\text{C}}\text{H} \qquad \overset{\text{O}}{\|} \\
\text{H}_2\overset{|}{\text{C}}\text{—O—P—O—R}_4 \\
\underset{\text{O}^-}{|}
\end{array}
\xrightarrow{H_2O}
\begin{array}{c}
\text{H}_2\text{C—O—}\overset{\text{O}}{\overset{\|}{\text{C}}}\text{—R}_1 \\
\text{HO—}\overset{|}{\text{C}}\text{H} \qquad \overset{\text{O}}{\|} \\
\text{H}_2\overset{|}{\text{C}}\text{—O—P—O—R}_4 \\
\underset{\text{O}}{|}
\end{array}
\qquad
\text{R}_2\text{—}\overset{\text{O}}{\overset{\|}{\text{C}}}\text{—OH}
$$

R_1 and R_2 are the carbon chains of fatty acids, R_4 is an alcohol. Only one ester bond between fatty acid and glycerol is hydrolyzed, specifically at position 2 of the glycerol carbon chain.

Phospholipase A_1 hydrolyzes the ester bond between a fatty acid and glycerol in position 1 of the carbon chain of a phosphoglyceride.

Question: How can hydrolytic enzymes act on lipids that repel water?

It is necessary for these enzymes to act at a water-lipid interface. Digestive lipases secreted into the lumen of the small intestine associate with the surface of large fat droplets. The initial products of digestion, by lipases and phospholipases, are fatty acids and lysophosphoglycerides, which are strong *detergents*. These hasten the digestive process because they emulsify the large fat droplets into myriads of tiny ones. As the concentration of fatty acids increases and 2-monoacylglycerol is produced, these are incorporated into micelles (Chap. 6) of bile salts. Monoacylglycerol also increases the detergent action of the bile salts, thus also facilitating emulsification of triacylglycerol. The mixed micelles migrate in large numbers to the surface of the intestinal epithelial cells, where the fatty acids and 2-monoacylglycerol are released from the micelle (Fig. 13-1).

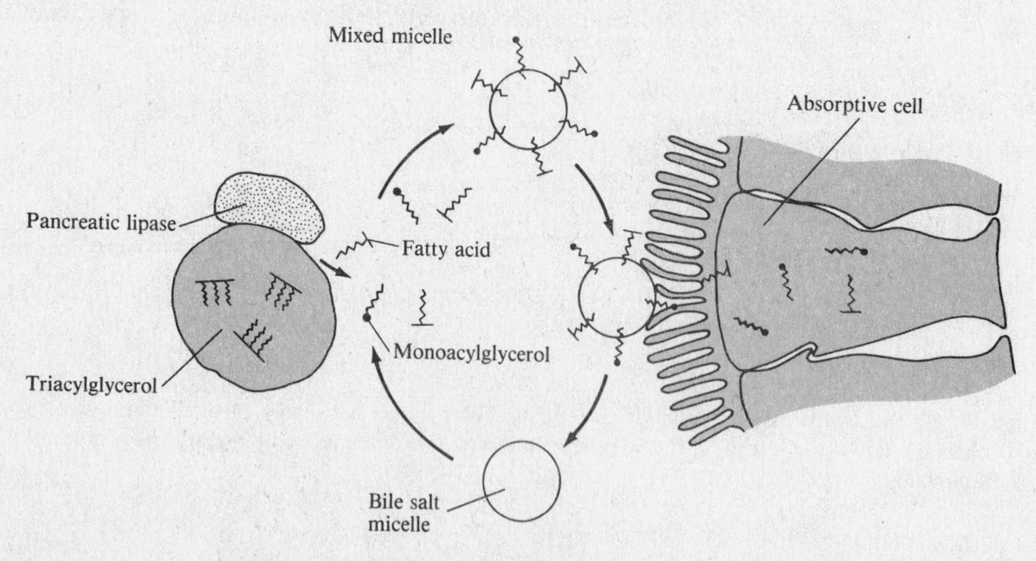

Fig. 13-1 Digestion of triacylglycerol in the intestine.

Absorption of Fatty Acids

Fatty acids of carbon chain length equal to, or greater than, 14 (long-chain acids) diffuse passively into the intestinal epithelial cells. They enter the cells down a concentration gradient because the concentration of free fatty acids in the mixed micelle is high and that of the cell is low. The cell membrane is no barrier to the lipophilic fatty acid. Entry of fatty acid into the cell is immediately followed by binding to a *binding protein*, which has a high affinity for long-chain fatty acids. Simultaneously the 2-monoacylglycerol passively diffuses into the epithelial cell and, with the fatty acids, is converted rapidly to triacylglycerol.

Question: By what mechanism do the fatty acids and 2-monoacylglycerol enter the circulation?

The newly synthesized triacylglycerol becomes organized into *chylomicrons* (a type of lipoprotein; see next section), which are secreted by the intestinal epithelial cell into the *lacteals*, small lymph vessels in the villi of the small intestine. Then from the *lymphatic circulation*, the chylomicrons pass into the *thoracic duct*, from which they enter the blood and thus contribute to the transport of lipid fuel to various tissues.

13.3 LIPOPROTEIN METABOLISM

Role of Lipoproteins

Lipoproteins serve to transport hydrophobic fats in plasma (Fig. 13-2). The major lipoproteins (Chap. 6) circulating in the blood are chylomicrons, VLDLs (very low density lipoproteins), LDLs (low-density lipoproteins), and HDLs (high-density lipoproteins). IDLs (intermediate density lipoproteins) are derived from VLDLs in the formation of LDLs. Fatty acids are important cellular fuels and are stored as triacylglycerols in *adipose tissue*. Fatty acids destined for storage as depot fat are transported to adipose tissue principally as triacylglycerol in chylomicrons and VLDLs. In adipose tissue, chylomicrons are rapidly degraded, and the *remnant* particles reenter the circulation and are removed by the liver. VLDLs are degraded in adipose tissue to LDLs which then circulate as the major transport lipoproteins for cholesterol. HDLs are lipoproteins that continuously circulate; they contain an enzyme, *phosphatidylcholine:cholesterol acyltransferase* (LCAT), that converts free cholesterol to cholesteryl esters.

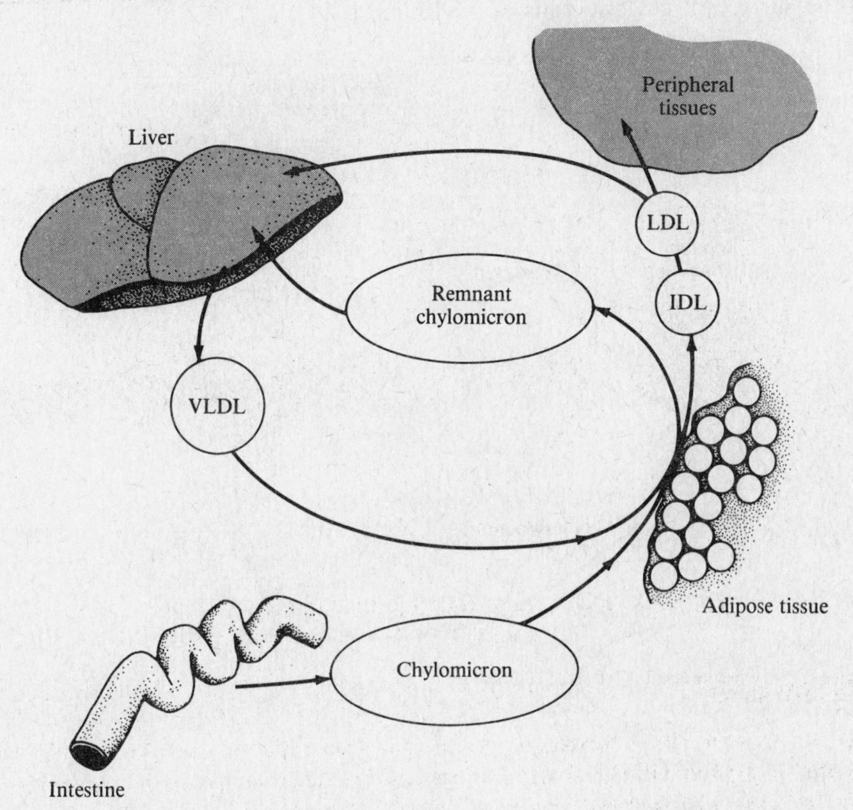

Fig. 13-2 Circulation of lipoproteins in the blood and tissues.

EXAMPLE 13.3

The reaction catalyzed by phosphatidylcholine:cholesterol acyltransferase is:

$$\text{Cholesterol} + \text{Phosphatidylcholine} \rightarrow \text{Cholesteryl ester} + \text{Lysophosphatidylcholine}$$

Linoleic acid is the fatty acid most commonly transferred from phosphatidylcholine to cholesterol, forming linoleoylcholesterol:

$$CH_3{-}(CH_2)_4{-}CH{=}CH{-}CH_2{-}CH{=}CH{-}(CH_2)_7{-}\overset{\displaystyle O}{\overset{\|}{C}}{-}O{-}$$

Linoleoylcholesterol

Release of Fatty Acids from Lipoproteins

The action of an enzyme *lipoprotein lipase* in adipose tissue depletes chylomicrons and VLDLs of their triglyceride (Fig. 13-3). The enzyme is activated by *apolipoprotein C*, with which it specifically interacts, on the surface of chylomicrons and VLDLs.

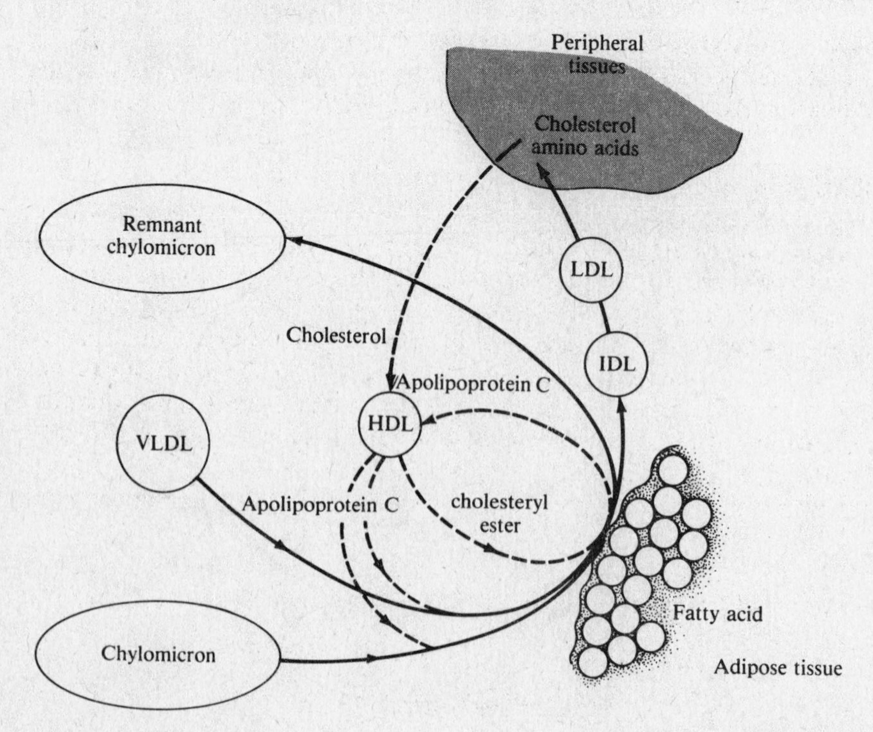

Fig. 13-3 The formation of HDL and LDL from chylomicrons and VLDL, respectively.

Question: What is the reaction catalyzed by lipoprotein lipase?

Lipoprotein lipase is an *extracellular* enzyme that hydrolyzes triacylglycerol into 2-monoacylglycerol and fatty acids. The fatty acids then enter the cell passively down a concentration gradient.

$$\underset{\text{Triacylglycerol}}{\begin{array}{c} H_2C{-}O{-}\overset{O}{\overset{\|}{C}}{-}R_1 \\ | \\ R_2{-}\overset{O}{\overset{\|}{C}}{-}O{-}CH \\ | \\ H_2C{-}O{-}\overset{O}{\overset{\|}{C}}{-}R_3 \end{array}} \longrightarrow \underset{\text{2-Monoacylglycerol}}{\begin{array}{c} H_2C{-}OH \\ | \\ R_2{-}\overset{O}{\overset{\|}{C}}{-}O{-}CH \\ | \\ H_2C{-}OH \end{array}} + \underset{\text{Free fatty acids}}{\begin{array}{c} R_1{-}\overset{O}{\overset{\|}{C}}{-}OH \\ \\ R_3{-}\overset{O}{\overset{\|}{C}}{-}OH \end{array}}$$

In adipose tissue, triacylglycerol is resynthesized from the fatty acids and stored as a large fat droplet, occupying 96 percent of the cellular space in a fat cell (Chap. 1). The fat stores in adipose tissue of an average 70-kg person are sufficient to satisfy the body's energy needs over a long period of starvation (~40 days). Lipoprotein lipase also occurs as an extracellular enzyme in other tissues, and the fatty acids, obtained from the hydrolysis of lipoprotein triglyceride, serve as immediate cellular fuel, or may be stored as triglyceride.

Release of Cholesterol from LDL

When LDL is abundant in the circulation, it provides tissues with an exogenous source of cholesterol.

Question: How is the cholesterol in LDL transported into cells?

LDL binds specifically to protein receptors on the cell surface. The resulting complexes become clustered in regions of the plasma membrane called *coated pits*. *Endocytosis* follows. (See Fig. 13-4.)

Question: What is the next event after endocytosis of the LDL?

The *clathrin coat* (Fig. 13-4) dissociates from the endocytic vesicles, which may recycle the receptors to the plasma membrane or fuse with *lysosomes*. The lysosomal proteinases and lipases then catalyze the hydrolysis of the LDL-receptor complexes; the protein is degraded completely to amino acids, and cholesteryl esters are hydrolyzed to free cholesterol and fatty acid. New LDL receptors are synthesized on the endoplasmic reticulum (ER) membrane and are reintroduced into the plasma membrane. The cholesterol is incorporated into the endoplasmic reticulum membrane or may be stored after esterification as cholesteryl ester in the cytosol; this occurs if the supply of cholesterol exceeds its utilization in membranes. Normally, only very small amounts of cholesteryl esters occur inside cells.

EXAMPLE 13.4

Acyl-CoA:cholesterol acyltransferase catalyzes the esterification of cholesterol in the cell cytoplasm:

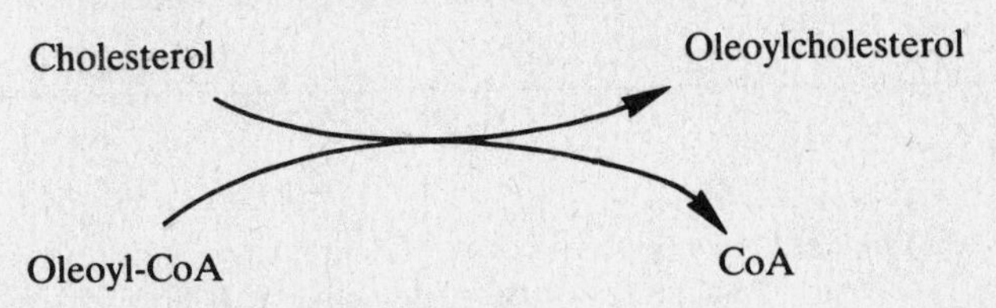

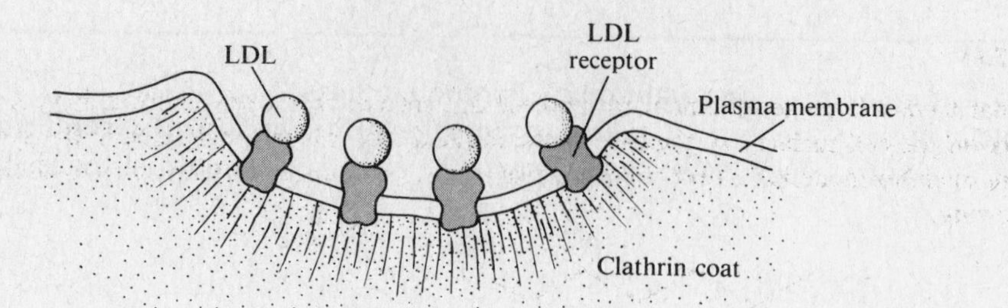

(*a*) Formation of LDL-receptor complexes

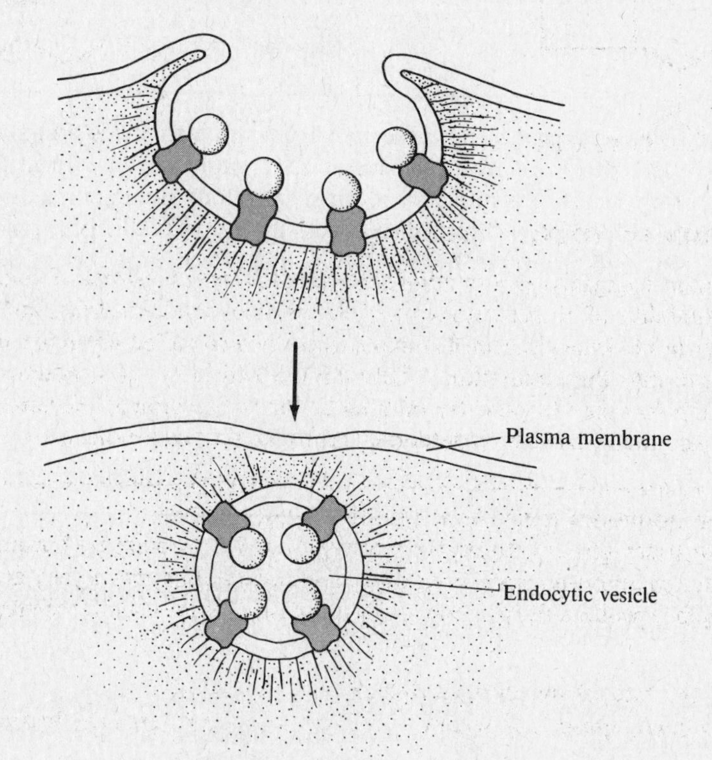

(*b*) Endocytosis

Fig. 13-4 The fate of LDL-receptor complexes in cholesterol uptake into cells.

Question: What chemical mechanism prevents the excessive accumulation of intracellular cholesterol?

The synthesis of *LDL receptors* is inhibited by an excess of intracellular cholesterol, thus preventing the appearance of new receptors on the cell surface. This may lead to high circulating levels of cholesterol.

EXAMPLE 13.5

In the congenital disease *familial hypercholesterolemia*, the high circulating level of cholesterol is due to the complete absence of LDL receptors or to the presence of defective receptors on cell surfaces.

EXAMPLE 13.6

Tissues that have a large requirement for cholesterol, such as the adrenal cortex, have a large number of LDL receptors on the cell surface. In the case of the adrenal, the cholesterol is used in the synthesis of steroid hormones. One such hormone is *cortisol* (hydrocortisone); the similarity of its structure to cholesterol is evident from the following:

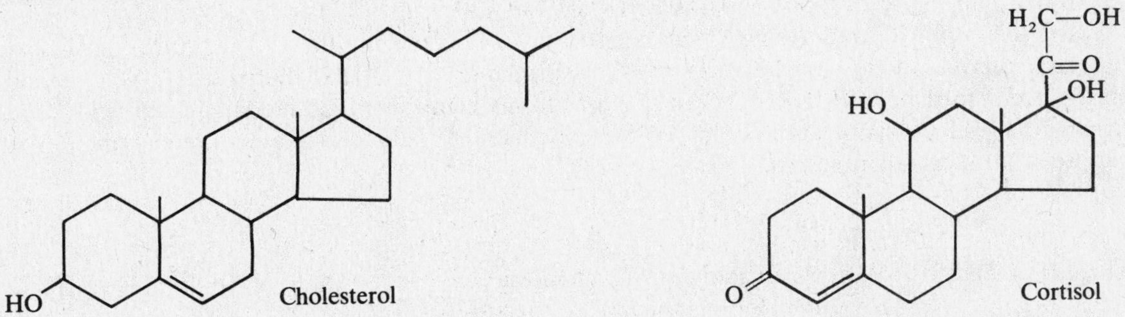

HO Cholesterol

$H_2C—OH$, $C=O$, OH, HO, O Cortisol

13.4 MOBILIZATION OF DEPOT LIPID

Fatty acids stored as triacylglycerol in adipose tissue are a major source of energy for a variety of tissues when the availability of glucose is reduced. Stress, prolonged exercise, and starvation lead to mobilization of depot lipid. Triacylglycerol is hydrolyzed by a so-called *hormone-sensitive lipase*, and fatty acids are released into the circulation. These unesterified fatty acids, bound to *serum albumin*, pass via the circulation to other tissues. By binding to them, the albumin facilitates their entry into the circulation and also minimizes their detergent action during transport.

Question: To which hormones is hormone-sensitive lipase sensitive?

The hydrolysis of triacylglycerol to monoacylglycerol and fatty acids by hormone-sensitive lipase can be stimulated by *epinephrine*, *norepinephrine*, *adrenal steroids*, *glucagon*, and the *hypophysial hormones*—luteotropin (prolactin), β- and α-lipotropins, somatotropin, thyrotropin, and vasopressin.

Question: What is the fate of the monoacylglycerol?

A second intracellular lipase, not sensitive to hormones, completes the hydrolysis of monoacylglycerol.

13.5 OXIDATION OF FATTY ACIDS

Oxidation of fatty acids occurs in three well-defined steps; namely, activation, transport into mitochondria, and oxidation to acetyl-CoA.

Fatty Acid Activation

In general, the entry of a fatty acid into a metabolic pathway is preceded by its conversion to its coenzyme A (CoASH) derivative; this acyl derivative is called an *alkanoyl-* or *alkenoyl-CoA*, and in this form the fatty acid is said to have been *activated*.

$$\underset{\substack{\text{Hexadecanoate}\\ \text{(palmitate)}}}{CH_3\text{-}(CH_2)_{14}\text{-}COO^-} \longrightarrow \underset{\substack{\text{Hexadecanoyl-CoA}\\ \text{(palmitoyl-CoA)}}}{CH_3\text{-}(CH_2)_{14}\text{-}COSCoA}$$

$$\underset{\substack{\text{Octadecenoate}\\ \text{(oleate)}}}{CH_3\text{-}(CH_2)_7CH=CH\text{-}(CH_2)_7\text{-}COO^-} \longrightarrow \underset{\substack{\text{Octadecenoyl-CoA}\\ \text{(oleoyl-CoA)}}}{CH_3\text{-}(CH_2)_7CH=CH\text{-}(CH_2)_7\text{-}COSCoA}$$

The activation of a fatty acid induces the formation of a thioester of fatty acid and CoA. The process is coupled to the hydrolysis of ATP to AMP. For palmitic acid, the reaction is:

$$CH_3\text{-}(CH_2)_{14}\text{-}COO^- + CoASH + ATP \longrightarrow CH_3\text{-}(CH_2)_{14}\text{-}COSCoA + AMP + PP_i$$

The enzyme that catalyzes the reaction is *acyl-CoA synthetase*.

Fatty acids of widely differing chain length may be activated, there being three acyl-CoA synthetase enzymes. One activates acetate (C_2), propionate (C_3), and butyrate (C_4); a second activates medium-chain-length acids (C_4–C_{12}); a third activates long- and medium-chain acids. The long-chain acyl-CoA synthetase occurs in the mitochondria and endoplasmic reticulum and is widespread in mammalian tissues.

Transport of Activated Fatty Acids into the Mitochondria

The enzymes that oxidize fatty acids are located in the mitochondrial matrix. Acyl-CoA derivatives do not permeate the inner mitochondrial membrane, but a specific transport protein allows entry of the acyl chains to the matrix.

EXAMPLE 13.7

Carnitine is an *acyl-group carrier* that transports fatty acids into and out of the mitochondrial matrix (Fig. 13-5). Acyl groups are linked by esterification to the hydroxyl group of carnitine by the action of *carnitine acyltransferase* in the inner membrane of the mitochondrion.

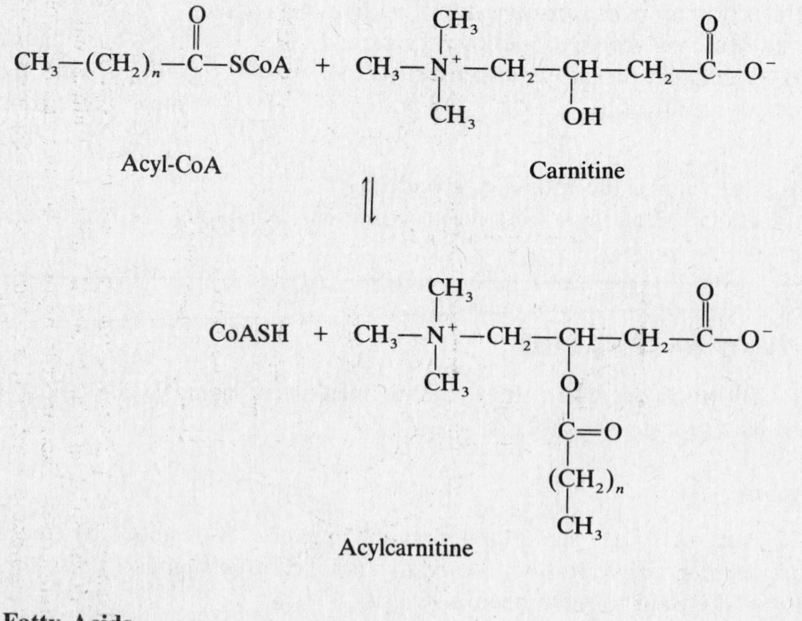

$$CH_3\text{—}(CH_2)_n\text{—}\overset{\overset{\large O}{\|}}{C}\text{—}SCoA \quad + \quad CH_3\text{—}\underset{\underset{\large CH_3}{|}}{\overset{\overset{\large CH_3}{|}}{N^+}}\text{—}CH_2\text{—}\underset{\underset{\large OH}{|}}{CH}\text{—}CH_2\text{—}\overset{\overset{\large O}{\|}}{C}\text{—}O^-$$

Acyl-CoA Carnitine

$$\rightleftharpoons$$

$$CoASH \quad + \quad CH_3\text{—}\underset{\underset{\large CH_3}{|}}{\overset{\overset{\large CH_3}{|}}{N^+}}\text{—}CH_2\text{—}\underset{\underset{\underset{\underset{\underset{\large CH_3}{|}}{(CH_2)_n}}{|}}{C=O}}{\underset{|}{O}}}{\underset{|}{CH}}\text{—}CH_2\text{—}\overset{\overset{\large O}{\|}}{C}\text{—}O^-$$

Acylcarnitine

β Oxidation of Fatty Acids

β oxidation of acyl-CoA derivatives of fatty acids occurs so that fatty acids are sequentially shortened by two carbon units at a time by a process that yields acetyl-CoA as the only product (Fig. 13-6). The acyl chains are cleaved at the bond between C-2 and C-3 of the chain, the so-called β *bond*, by a process that induces oxidation of this part of the molecule. Table 13.1 lists the reactions and enzymes for the β oxidation of fatty acids shown in Fig. 13-6.

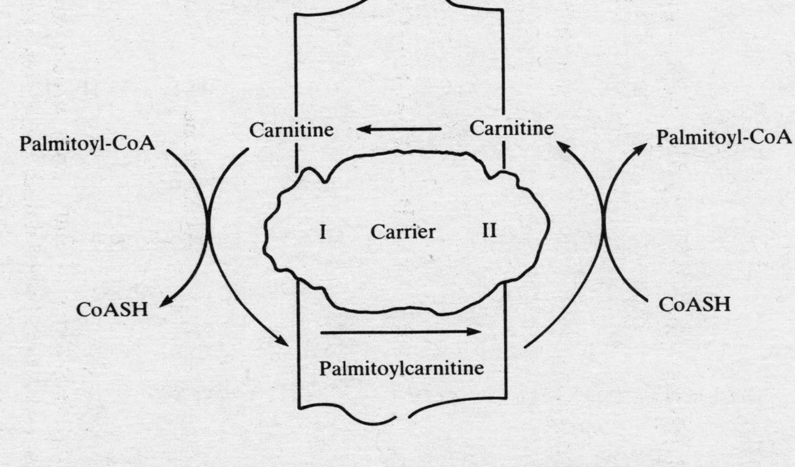

Fig. 13-5 Transport of palmitoyl groups between the cytoplasm and the mitochondrial matrix. I and II are two carnitine palmitoyltransferases. The carrier transports palmitoylcarnitine across the membrane inward, and free carnitine is transported across the membrane outward. Free CoA is not transported.

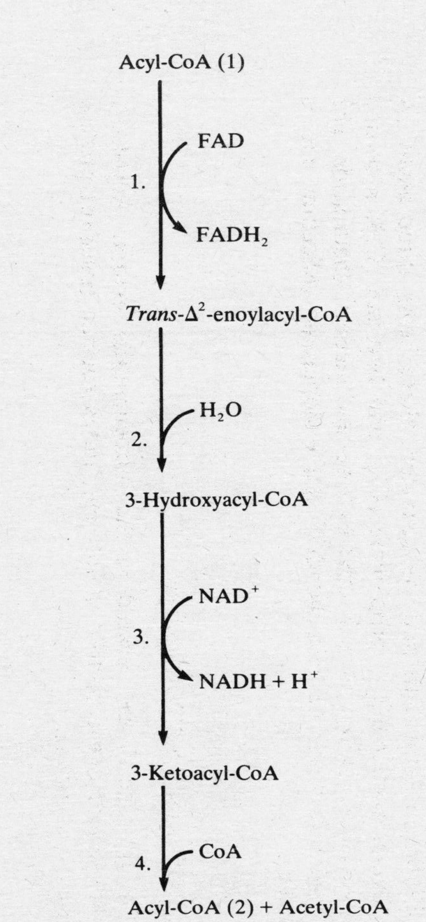

Fig. 13-6 Metabolic pathway of β oxidation, removal of the first two carbon units of a fatty acid. If the acyl-CoA (1) is palmitoyl-CoA ($C_{16:0}$), then acyl-CoA (2) is myristoyl-CoA ($C_{14:0}$). Hence, the complete β oxidation of palmitoyl-CoA requires seven such cleavages and produces eight molecules of acetyl-CoA. The numbers to the left of the arrows correspond to the numbered reactions in Table 13.1.

Table 13.1. The Reactions of β Oxidation of Fatty Acids

Reaction	Enzyme
1. $CH_3\text{-}(CH_2)_n\text{-}\underset{③}{CH_2}\text{—}\underset{②}{CH_2}\text{-}C(=O)\text{-}SCoA + FAD \longrightarrow CH_3\text{-}(CH_2)_n\text{-}\underset{③}{CH}\text{=}\underset{②}{CH}\text{-}C(=O)\text{-}SCoA + FADH_2$	Acyl-CoA dehydrogenase
2. $CH_3\text{-}(CH_2)_n\text{-}CH\text{=}CH\text{-}C(=O)\text{-}SCoA + H_2O \longrightarrow CH_3\text{-}(CH_2)_n\text{-}CHOH\text{—}CH_2\text{-}C(=O)\text{-}SCoA$	Enoyl-CoA hydratase
3. $CH_3\text{-}(CH_2)_n\text{-}CHOH\text{—}CH_2\text{-}C(=O)\text{-}SCoA + NAD^+ \longrightarrow CH_3\text{-}(CH_2)_n\text{-}CO\text{—}CH_2\text{-}C(=O)\text{-}SCoA + NADH + H^+$	3-Hydroxyacyl-CoA dehydrogenase
4. $CH_3\text{-}(CH_2)_n\text{-}CO\text{—}CH_2\text{-}C(=O)\text{-}SCoA + CoASH \longrightarrow CH_3\text{-}(CH_2)_n\text{-}C(=O)\text{-}SCoA + CH_3\text{-}C(=O)\text{-}SCoA$	Thiolase

Question: What happens to the $FADH_2$ and NADH produced during β oxidation?

Acyl-CoA dehydrogenase is a *flavin-linked*, membrane-bound enzyme, associated with the mitochondrial respiratory complexes. When $FADH_2$ is produced, it is oxidized by the respiratory chain (Chap. 14). Hydroxyacyl-CoA dehydrogenase is in the mitochondrial matrix, and the NADH produced by the action of this enzyme on hydroxyacyl-CoA compounds contributes to the pool of NADH in the matrix. It is also oxidized by the respiratory chain (Chap. 14).

13.6 THE FATE OF ACETYL-CoA FROM FATTY ACIDS: KETOGENESIS

Acetyl-CoA is oxidized to carbon dioxide via the citric acid cycle (Chap. 12), thus transforming additional energy to that which has been transformed via β oxidation. In liver mitochondria only, acetyl-CoA may also be converted to *ketone bodies*:

$CH_3\text{-CO-}CH_2\text{-}COO^-$	$CH_3\text{-CO-}CH_3$	$CH_3\text{-CHOH-}CH_2\text{-}COO^-$
Acetoacetate	Acetone	3-Hydroxybutyrate

The reactions and the biosynthetic pathway for these compounds are shown in Table 13.2 and Fig. 13-7.

Table 13.2. Ketogenesis

Reaction	Enzyme
1. $2CH_3\text{-CO-SCoA} \longrightarrow CH_3\text{-CO-}CH_2\text{-CO-SCoA}$	Acetyl-CoA acetyltransferase
2. $CH_3\text{-CO-}CH_2\text{-CO-SCoA} + CH_3\text{-CO-SCoA} \longrightarrow$ $^-OOC\text{-}CH_2\text{-}C(CH_3)(OH)\text{-}CH_2\text{-CO-SCoA} + \text{CoASH}$	Hydroxymethylglutaryl-CoA (HMG-CoA) synthase
3. $^-OOC\text{-}CH_2\text{-}C(CH_3)(OH)\text{-}CH_2\text{-CO-SCoA} \longrightarrow$ $^-OOC\text{-}CH_2\text{-CO-}CH_3 + CH_3\text{-CO-SCoA}$	HMG-CoA lyase
4. $CH_3\text{-CO-}CH_2\text{-}COO^- + \text{NADH} + H^+ \longrightarrow$ $CH_3\text{-CHOH-}CH_2\text{-}COO^- + \text{NAD}^+$	3-Hydroxybutyrate dehydrogenase

Ketone bodies are water-soluble lipid fuels continually being released from the liver. When carbohydrate is plentiful and glucose is readily available to the tissues, the amount of circulating ketone bodies is low ($\sim$0.1 mmol L^{-1}). When large amounts of triacylglycerol are being hydrolyzed in adipose tissues, in response to an increase in whole-body energy demand, the oxidation of fatty acid increases in the liver and other tissues. In the liver, this increases ketogenesis and thus increases the ketone body concentration in the circulation. During exercise, for example, the blood ketone-body concentration rises steadily and can reach 2 to 3 mmol L^{-1}, which is comparable to the blood glucose concentration.

Normally some acetoacetate is converted to 3-hydroxybutyrate. Acetoacetate and 3-hydroxybutyrate are valuable fuels for skeletal and cardiac muscle, and it is estimated that they supply 10 percent of the daily energy consumption of these tissues.

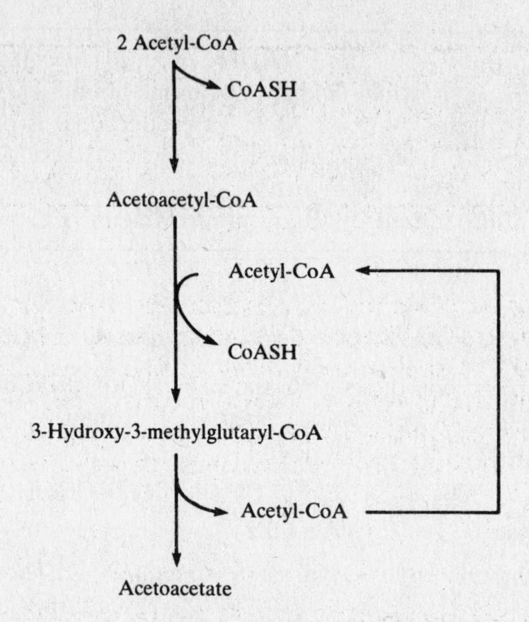

3-Hydroxybutyrate and acetone are derived from acetoacetate:

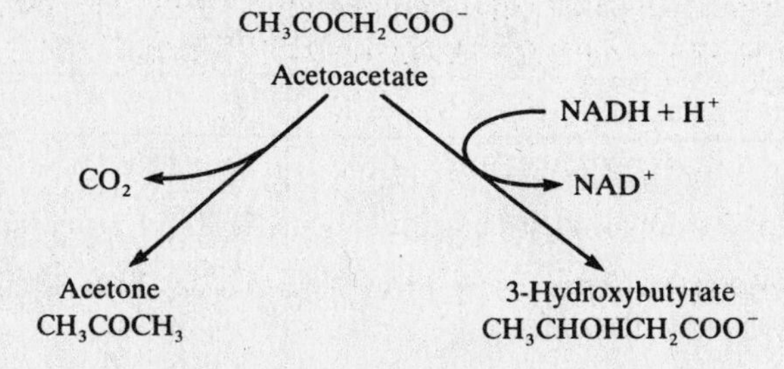

Fig. 13-7 Pathway of ketone body synthesis: ketogenesis. Table 13.2 shows the reactions and enzymes involved in the production of acetoacetate and 3-hydroxybutyrate shown here.

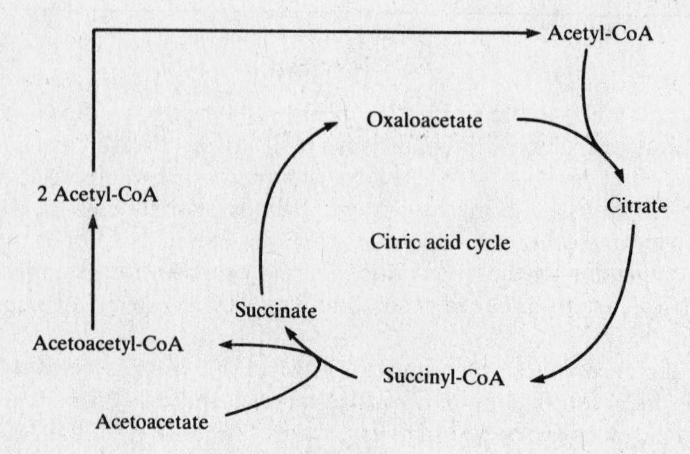

Fig. 13-8

Question: How is the energy of acetoacetate made available for mechanical work in muscles?

The enzyme *3-oxoacid transferase* is in mitochondria of muscle and converts acetoacetate to acetoacetyl-CoA; however, it is absent from liver mitochondria. The acetoacetyl-CoA is cleaved to acetyl-CoA by *thiolase*, which is in the mitochondria of all tissues. See Fig. 13-8.

13.7 LIPOGENESIS

When there is an oversupply of dietary carbohydrate, the excess carbohydrate is converted to triacylglycerol; individuals on low-fat diets also convert glucose to triacylglycerol for storage. This involves the synthesis of fatty acids from acetyl-CoA and the esterification of fatty acids in the production of triacylglycerol. The process is called *lipogenesis*. The major lipogenic tissues are intestine, liver, and adipose tissue. During lactation, the mammary gland, too, becomes a major site of lipogenesis and places a heavy demand on a continuing supply of glucose for the synthesis of milk lipids.

Synthesis of Palmitic Acid

The synthesis of palmitic acid occurs in the cytosol, from acetyl-CoA. When glucose is abundant and the amount of citrate in the mitochondrial matrix exceeds the demand by the citric acid cycle, the excess citrate is transported out of the mitochondria into the cytosol (Fig. 13-9). Citrate in the cytosol is the source of acetyl groups for fatty acid synthesis, and its metabolism there involves the following enzyme reactions:

(*a*) *Citrate lyase*

$$\begin{matrix} H_2C\text{-}COO^- \\ | \\ HOC\text{-}COO^- \\ | \\ H_2C\text{-}COO^- \\ \text{Citrate} \end{matrix} + \text{CoASH} + \text{ATP}^{4-} \longrightarrow \underset{\text{Oxaloacetate}}{^-OOC\text{-}CH_2\text{-}CO\text{-}COO^-} + \underset{\text{Acetyl-CoA}}{CH_3\text{-}CO\text{-}SCoA} + \text{ADP}^{3-} + P_i^{2-}$$

(*b*) *Malate dehydrogenase* (*1*)

$$\underset{\text{Oxaloacetate}}{^-OOC\text{-}CH_2\text{-}CO\text{-}COO^-} + \text{NADH} + H^+ \longrightarrow \underset{\text{Malate}}{^-OOC\text{-}CH_2\text{-}HCOH\text{-}COO^-} + \text{NAD}^+$$

(*c*) *Malate dehydrogenase* (*2*) (oxaloacetate-decarboxylating)($NADP^+$)

$$\underset{\text{Malate}}{^-OOC\text{-}CH_2\text{-}CHOH\text{-}COO^-} + \text{NADP}^+ \longrightarrow \underset{\text{Pyruvate}}{CH_3\text{-}CO\text{-}COO^-} + CO_2 + \text{NADPH}$$

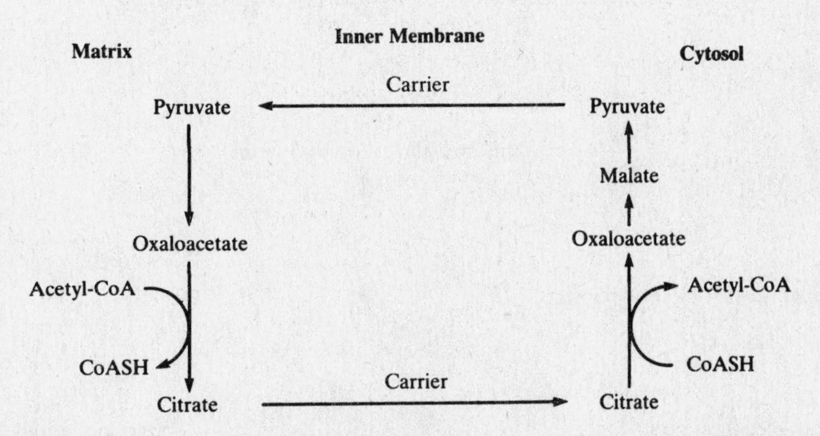

Fig. 13-9 Transfer of acetyl groups from the mitochondrion to the cytosol.

For the conversion of pyruvate to oxaloacetate and the formation of citrate in the mitochondrion, see Chap. 12. Acetyl-CoA for fatty acid synthesis is converted to malonyl-CoA; this reaction is catalyzed by *acetyl-CoA carboxylase*. Seven molecules of acetyl-CoA are converted to malonyl-CoA for the synthesis of one molecule of palmitic acid.

EXAMPLE 13.8

In the formation of malonyl-CoA via acetyl-CoA carboxylase, *biotin* is tightly bound to the enzyme, and it acts as a carrier of a carboxyl group that is transferred to acetyl-CoA.

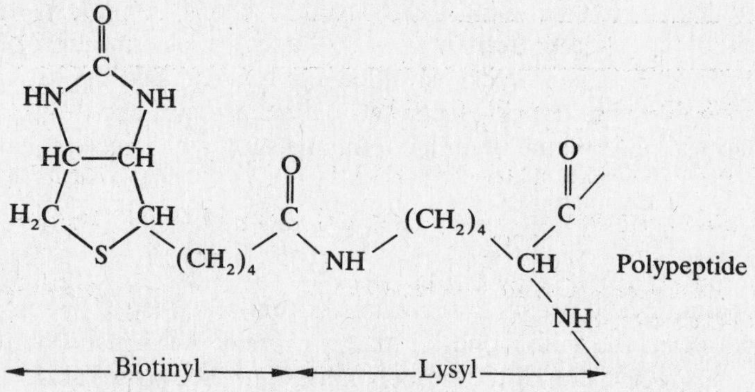

Polypeptide

← Biotinyl → ← Lysyl →

Biotinylenzyme

Normally, acetyl-CoA carboxylase exists in the cytosol as an inactive protomer, $M_r = 2 \times 10^5$. Citrate induces polymerization of the protomer into long filaments of an active, multifunctional enzyme, $M_r = 4.8 \times 10^6$. The active acetyl-CoA carboxylase activates carbon dioxide and transfers a carboxyl group to acetyl-CoA to form malonyl-CoA:

$$HCO_3^- + ATP + \text{Biotinylenzyme} \longrightarrow {}^-OOC\text{-biotinylenzyme} + ADP + P_i$$

$$CH_3\text{-CO-SCoA} + {}^-OOC\text{-biotinylenzyme} \longrightarrow {}^-OOC\text{-}CH_2\text{-CO-SCoA} + \text{Biotinylenzyme}$$

$$\text{Sum: Acetyl-CoA} + HCO_3^- + ATP \longrightarrow \text{Malonyl-CoA} + ADP + P_i$$

The formation of malonyl-CoA signals the beginning of the synthesis of palmitic acid ($C_{16:0}$). This occurs on a multifunctional enzyme complex, the *fatty acid synthase*. In mammalian liver, the enzyme complex consists of two identical polypeptides, each with specific binding sites for malonyl and alkanoyl groups, and eight different enzyme activities.

The malonyl and alkanoyl binding sites on the fatty acid synthase are as follows:

(*a*) Phosphopantotheine (PP) binds the malonyl or acetyl group:

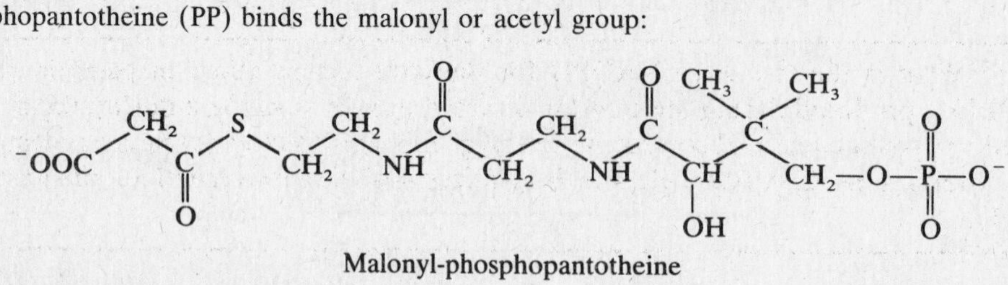

Malonyl-phosphopantotheine

(*b*) Cysteine (Cys) binds alkanoyl groups only:

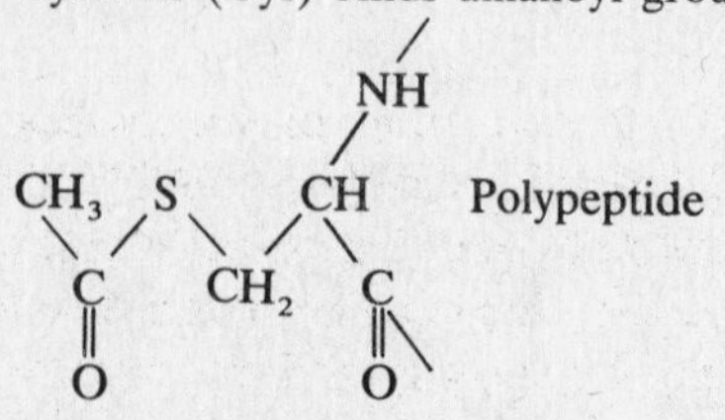

Polypeptide

Acetylcysteinyl-

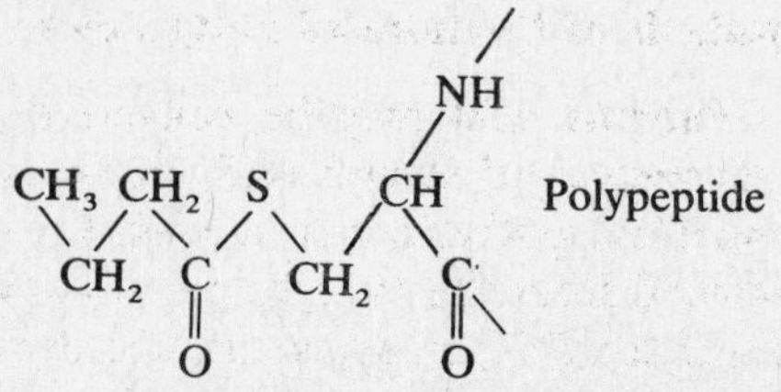

Polypeptide

Butyrylcysteinyl-

Once an acetyl group and a malonyl group are bound to the fatty acid synthase, seven rounds of enzymic reactions proceed for the synthesis of palmitic acid, which is then released from the complex. The overall reaction is:

$$1 \text{ Acetyl-CoA} + 7 \text{ Malonyl-CoA} + 14\text{NADPH} + 14\text{H}^+ \longrightarrow \text{Palmitic acid} + 8 \text{ CoASH} + 14\text{NADP}^+ + 7\text{H}_2\text{O} + 7\text{CO}_2$$

The individual enzymatic reactions are given in Table 13.3.

Table 13.3. Enzymatic Reactions of the Fatty Acid Synthetase

Reaction	Enzyme
1. $CH_3\text{-CO-SCoA} + \text{HS-PP} \xrightarrow[\text{CoASH}]{} CH_3\text{-CO-S-PP}$	Acetyl transacylase
2. $CH_3\text{-CO-S-PP} + \text{HS-Cys} \longrightarrow CH_3\text{-CO-S-Cys} + \text{HS-PP}$	Acetyl transacylase
3. $^-\text{OOC-}CH_2\text{-CO-SCoA} + \text{HS-PP} \xrightarrow[\text{CoASH}]{} {}^-\text{OOC-}CH_2\text{-CO-S-PP}$	Malonyl transacylase
4. $CH_3\text{-CO-S-Cys} + {}^-\text{OOC-}CH_2\text{-CO-S-PP} \longrightarrow \text{HS-Cys} + CH_3\text{-CO-}CH_2\text{-CO-S-PP} + CO_2$	β-Ketoacyl synthase
5. $CH_3\text{-CO-}CH_2\text{-CO-S-PP} \xrightarrow[\text{NADPH+H}^+ \quad \text{NADP}^+]{} CH_3\text{-CHOH-}CH_2\text{-CO-S-PP}$	β-Ketoacyl reductase
6. $CH_3\text{-CHOH-}CH_2\text{-CO-S-PP} \xrightarrow[H_2O]{} CH_3\text{-CH=CH-CO-S-PP}$	β-Hydroxyacyl dehydratase
7. $CH_3\text{-CH=CH-CO-S-PP} \xrightarrow[\text{NADPH+H}^+ \quad \text{NADP}^+]{} CH_3\text{-}CH_2\text{-}CH_2\text{-CO-S-PP}$	Enoyl reductase
8. $CH_3\text{-}CH_2\text{-}CH_2\text{-CO-S-PP} \longrightarrow CH_3\text{-}CH_2\text{-}CH_2\text{-CO-S-Cys}$	Acyl transacylase
Return to step 4 for condensation with malonyl-SPP six times.	
9. $CH_3\text{-}(CH_2)_{14}\text{-CO-S-PP} + H_2O \longrightarrow CH_3\text{-}(CH_2)_{14}\text{-COO}^- + \text{H}^+\text{-HS-Cys}$	Palmitoyl thioesterase

---

Question: What is the source of NADPH for the reduction reactions in palmitic acid synthesis?

NADPH is produced during the transfer of acetyl groups from the mitochondrion, when malate is oxidized to pyruvate and carbon dioxide [see Fig. 13-9 and reaction (*c*) in the accompanying text]. NADPH is also produced when glucose is oxidized and decarboxylated to ribulose 5-phosphate (Chap. 11).

Synthesis of Unsaturated Fatty Acids

Palmitic acid may be converted to stearic acid ($C_{18:0}$) by elongation of the carbon chain. *Desaturation* of stearic acid produces oleic acid ($C_{18:1}\Delta 9$). Linoleic acid ($C_{18:2}\Delta 9,12$), however, cannot be synthesized in mammalian tissues. It is an *essential* fatty acid for animals and must be obtained from the diet; it has two important metabolic roles. One is to maintain the fluid state of membrane lipid, lipoproteins, and storage lipid. The other role is as precursor for *arachidonic* acid, which has a specialized role in the formation of prostaglandins (Sec. 13.9).

Question: What are the steps involved in the conversion of linoleic acid to arachidonic acid?

See Fig. 13-10. The carbon chain of linoleic acid is desaturated at position 6. γ-Linolenic acid is elongated by two carbon units, and then another double bond is introduced in the C_{20} chain at position 5.

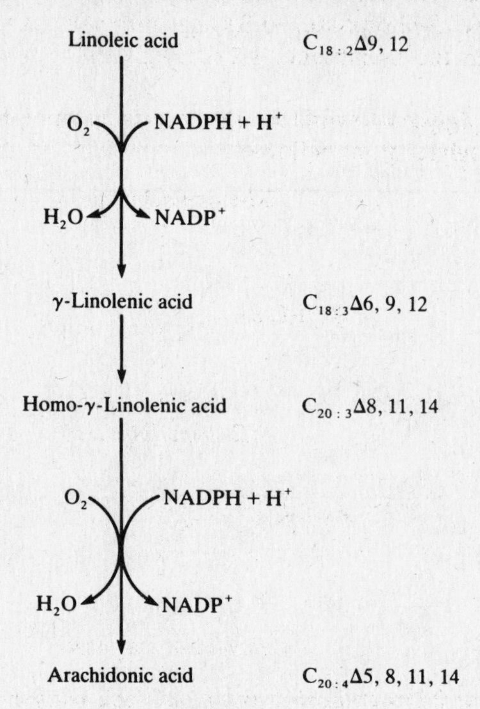

Fig. 13-10

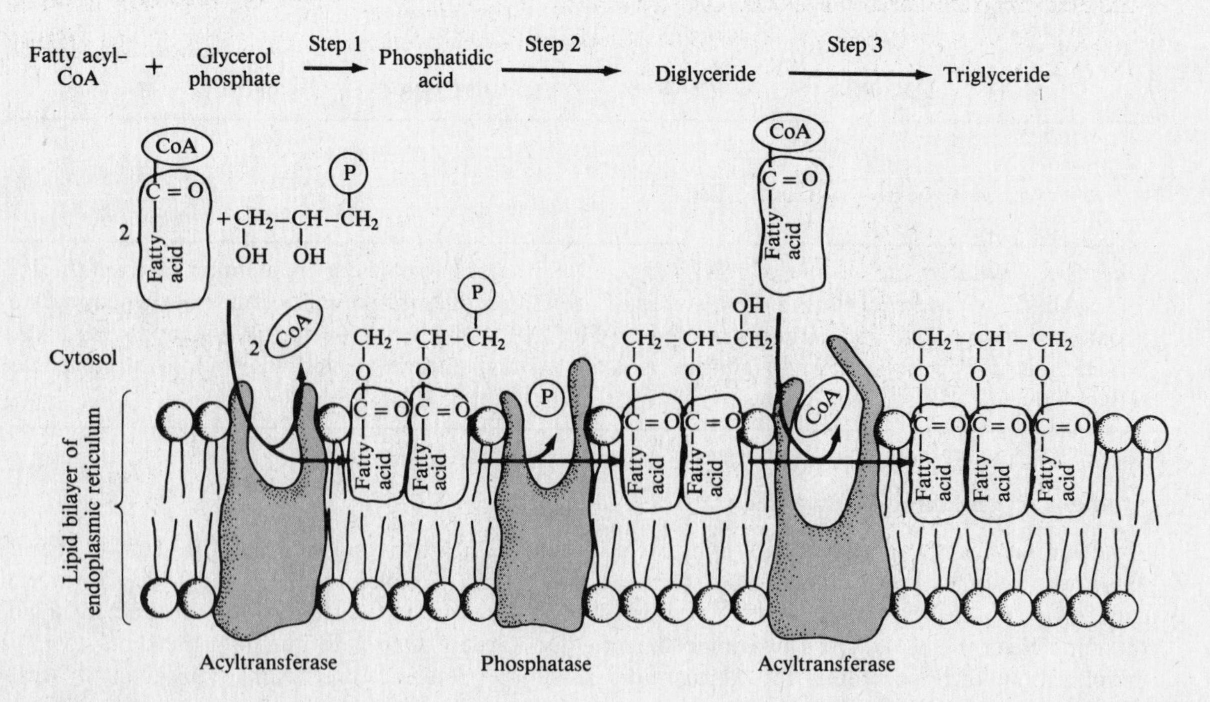

Fig. 13-11 The synthesis of triacylglycerol in the endoplasmic reticulum of liver and fat cells.

Triacylglycerol Synthesis

The synthesis of triacylglycerol takes place in the endoplasmic reticulum (ER). In liver and adipose tissue, fatty acids in the cytosol obtained from the diet or from de novo synthesis of palmitic acid become inserted into the ER membrane. The reactions are shown in Fig. 13-11. Membrane-bound *acyl-CoA synthetase* activates two fatty acids, and membrane-bound *acyl-CoA transferase* esterifies them with glycerol 3-phosphate, to form *phosphatidic acid*. *Phosphatidic acid phosphatase* releases phosphate, and in the membrane, 1,2-diacylglycerol is esterified with a third molecule of fatty acid.

In the intestine, triacylglycerol synthesis also occurs in the ER membrane, but fatty acids are esterified with 2-monoacylglycerol, as follows:

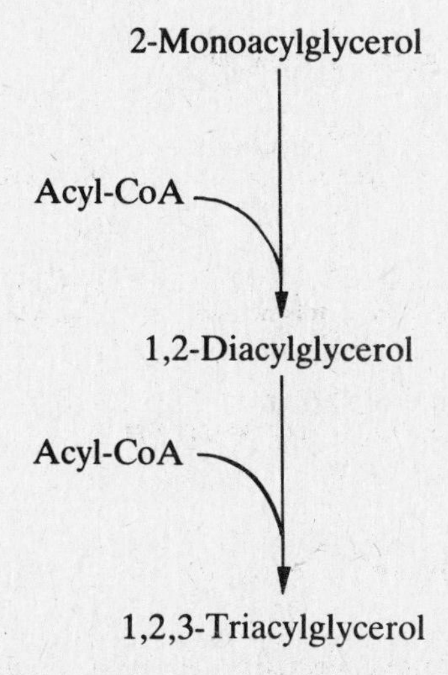

Triacylglycerol has no polar interaction with the membrane phospholipids and is either released into the cytosol as tiny lipid droplets or into the lumen of the ER. In fat cells, oil droplets in the cytosol coalesce, migrate toward and fuse with the large central oil droplets. In the liver and intestine, triacylglycerol is packaged into lipoproteins (VLDL and chylomicrons, respectively), which then are secreted into the circulation. See Fig. 13-12.

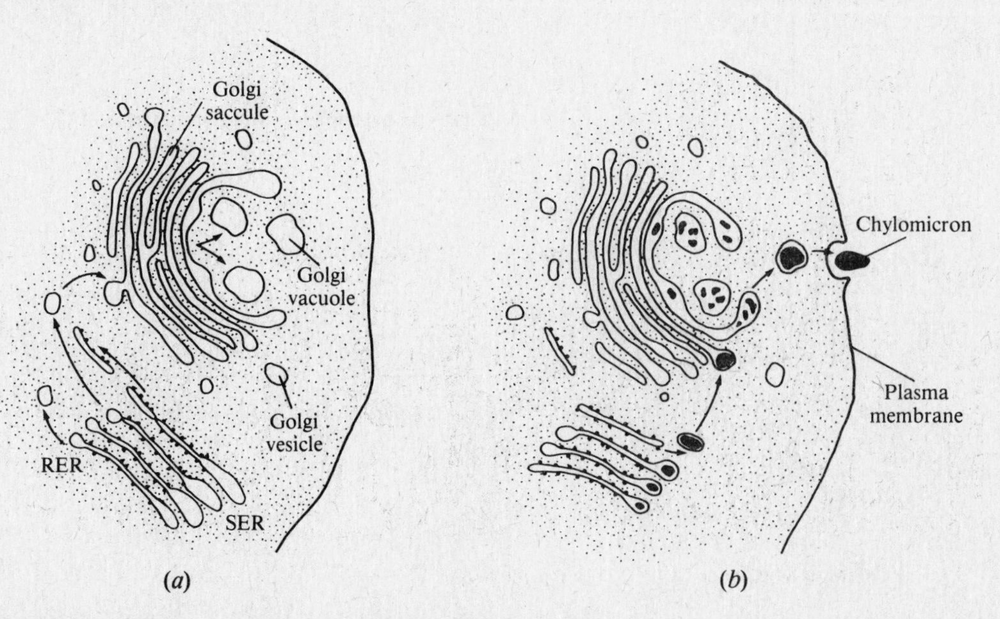

Fig. 13-12 Secretion of chylomicrons and VLDL: (*a*) from rats fasted for 24 h; (*b*) from rats 15–60 min after fat feeding.

13.8 SYNTHESIS OF PHOSPHOLIPIDS AND SPHINGOLIPIDS

Phospholipids

Phosphatidylcholine, a major phospholipid constituent of membranes and lipoproteins, is synthesized de novo in liver cells. The synthesis occurs on the ER and is linked, through 1,2-diacylglycerol, with the synthesis of triacylglycerol. Three compounds specifically involved in the synthesis of phosphatidylcholine are:

1. Choline

$$HO\text{-}CH_2\text{-}CH_2\text{-}N^+(CH_3)_3$$

2. Choline phosphate

$$^{2-}O_3P\text{-}O\text{-}CH_2\text{-}CH_2\text{-}N^+(CH_3)_3$$

3. Cytidine diphosphocholine (CDP-choline)

$$\text{Cytidine-O-}\overset{\overset{\displaystyle O}{\|}}{\underset{\underset{\displaystyle O^-}{|}}{P}}\text{-O-}\overset{\overset{\displaystyle O}{\|}}{\underset{\underset{\displaystyle O^-}{|}}{P}}\text{-O-}CH_2\text{-}CH_2\text{-}N^+(CH_3)_3$$

The synthesis of phosphatidylcholine is shown in Fig. 13-13.

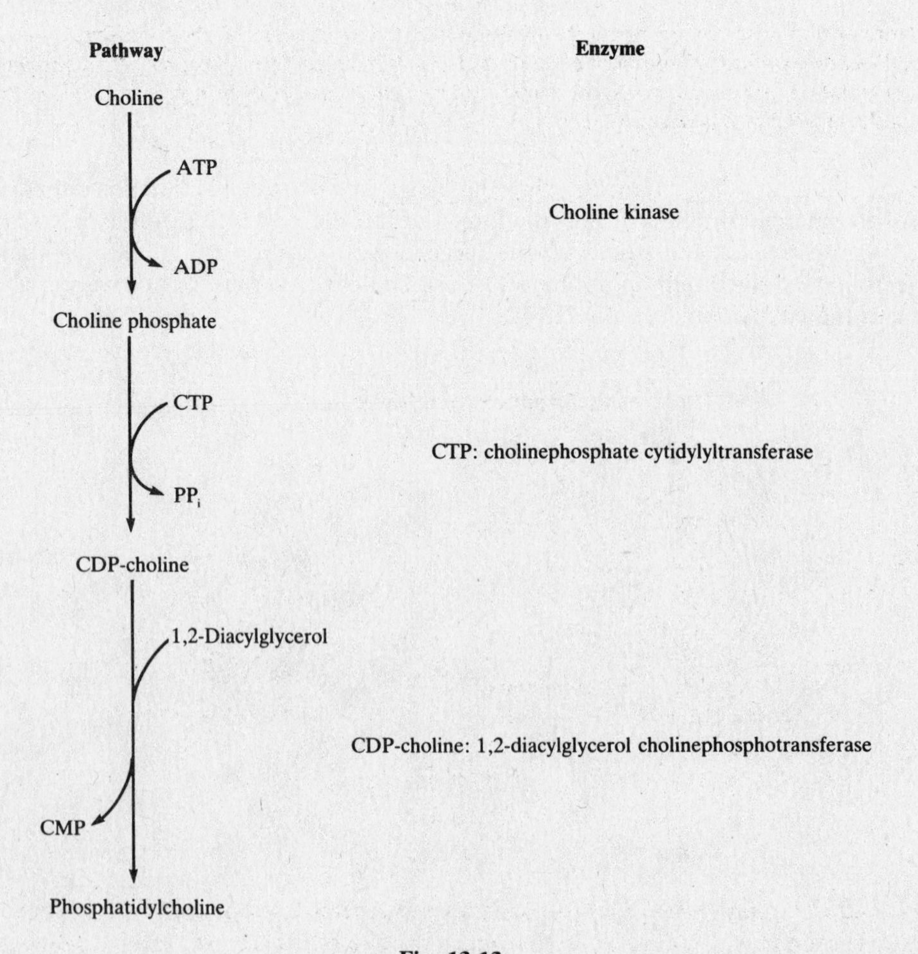

Fig. 13-13

EXAMPLE 13.9

Note that the enzymes CDP-choline:1,2-diacylglycerol transferase in phospholipid synthesis and acyl-CoA transferase in triglyceride synthesis have the common substrate 1,2-diacylglycerol; thus we have the synthesis shown in Fig. 13-14.

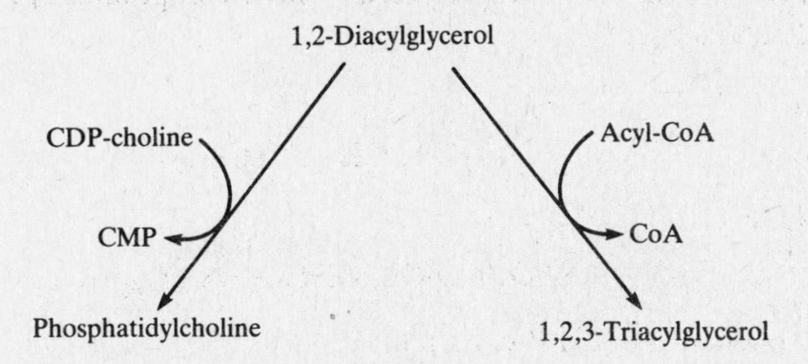

Fig. 13-14

Phosphatidylethanolamine is synthesized de novo in a similar way to phosphatidylcholine. Phosphatidylserine and phosphatidylcholine may then be formed from phosphatidylethanolamine.

EXAMPLE 13.10

Phosphatidylserine arises by an exchange of the ethanolamine residue of phosphatidylethanolamine for a seryl group. Decarboxylation of the serine of phosphatidylserine reforms phosphatidylethanolamine. Three successive methylation reactions convert phosphatidylethanolamine to phosphatidylcholine. *S*-Adenosylmethionine is the methyl-group donor (Chap. 15). See Fig. 13-15.

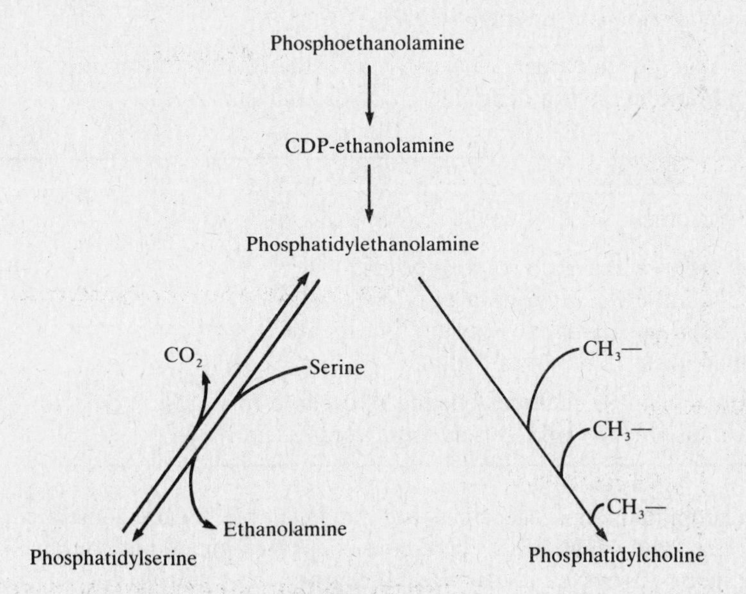

Fig. 13-15

Sphingolipids

Sphingolipids comprise *glycolipids* (gangliosides and cerebrosides) and the phospholipid *sphingomyelin*. These compounds, too, are important membrane constituents. The biosynthesis of sphingolipids involves the common intermediate *ceramide* (Fig. 13-16).

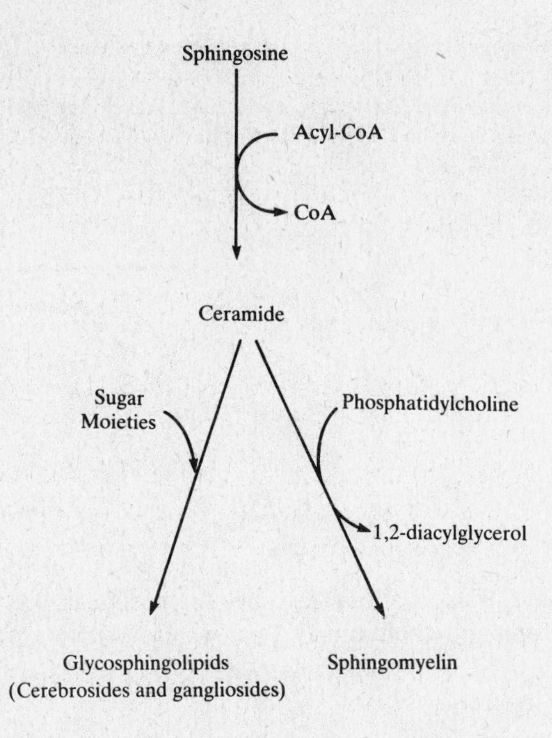

Fig. 13-16 Pathway of synthesis of sphingolipids.

Degradation of Phospholipids and Sphingolipids

During growth and development of tissues, cell material is constantly being degraded and resynthesized. Membrane lipids are degraded by lysosomal enzymes.

Question: What is a sequence of events in the hydrolysis of (*a*) phosphatidylcholine and (*b*) sphingomyelin in lysosomes?

(*a*) 1. Phospholipase A_1 removes a fatty acid.
2. Phospholipase A_2 removes a fatty acid.
3. Phospholipase C removes choline phosphate.
4. Phospholipase D removes choline.

(*b*) 1. Sphingomyelinase removes choline phosphate from sphingomyelin.
2. Ceramidase removes fatty acid from ceramide.

Cerebrosides (ceramide monosaccharides) are important membrane constituents of the brain and central nervous system, and gangliosides (ceramide oligosaccharides that contain sialic acid) have an important role in many tissues as cell-surface receptors for a variety of environmental cues. In humans, there are 10 known classes of lipid storage disease, in which the degradation of sphingolipids does not take place. This results in the accumulation of sphingolipid, followed by swelling and malfunction of tissues; it affects the young and usually leads to an early death. The most common symptom in all these diseases is mental retardation (see Table 13.4); this emphasizes the importance of breakdown and resynthesis of sphingolipids in nervous tissue. The retina, the liver, and the spleen may also be affected. The diseases are inherited and are due to genetic defects that result in reduced hydrolytic enzyme activity.

Table 13.4. Symptoms of Some Lipid-Storage Diseases

Disease	Signs and Symptoms
Gaucher's disease	Spleen and liver enlargement Erosion of long bones and pelvis Mental retardation (only in infantile form)
Niemann-Pick disease	Liver and spleen enlargement Mental retardation About 30% with red spot in retina
Metachromatic leukodystrophy	Mental retardation Psychological disturbances in adult form Yellow-brown staining of nerves, with cresyl violet dye

EXAMPLE 13.11

The degradative pathways for the major gangliosides of the 30 different gangliosides found in humans are shown in Fig. 13-17, together with the common names of the enzymes involved at the various degradation levels and the names of the diseases that result from a defect in the particular enzyme at each of these levels.

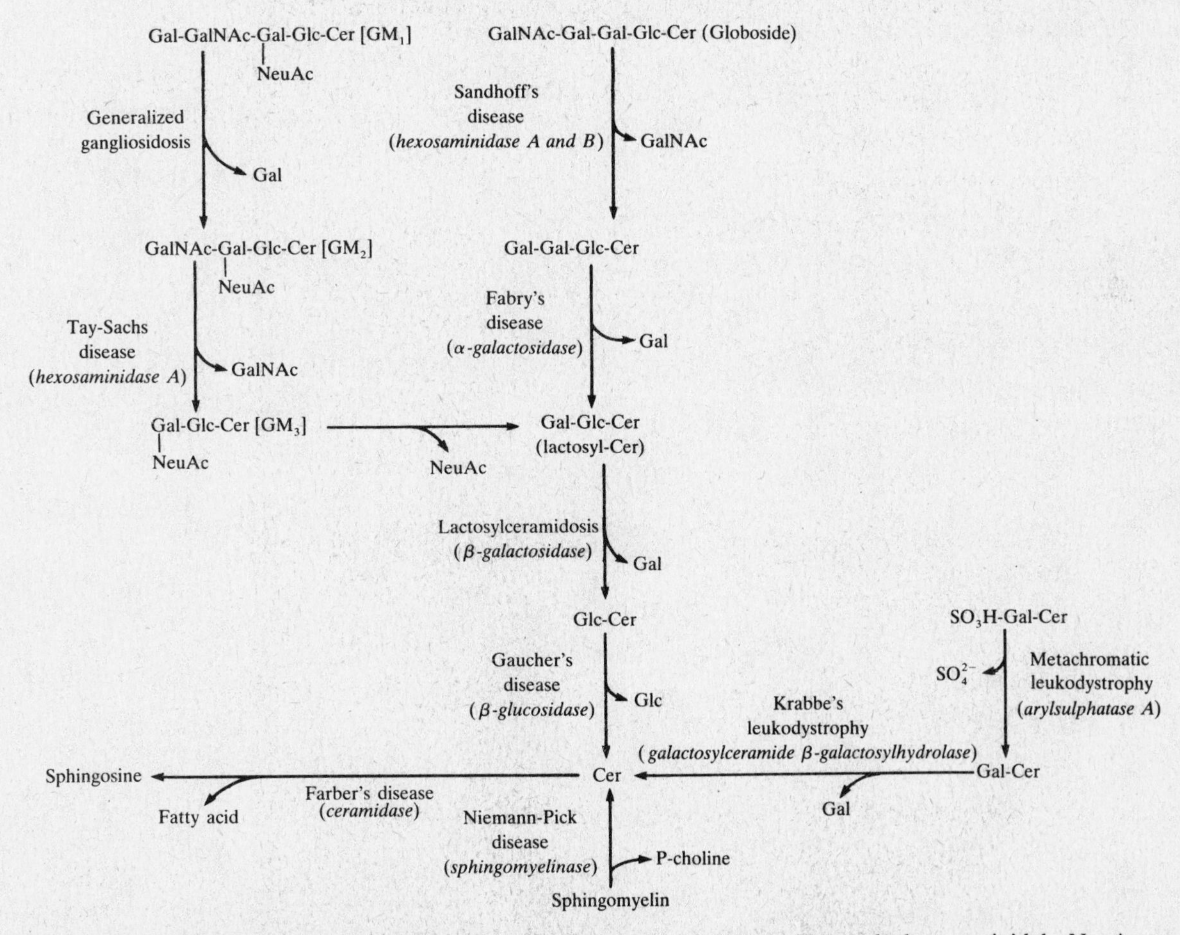

Fig. 13-17 Cer = ceramide; Gal = galactosyl; Glc = glucosyl; GalNAc = N-acetylgalactosaminidyl; NeuAc = N-acetylneuraminidyl; GM_1, GM_2, and GM_3 = type 1, type 2, and type 3 monosialoganglioside, respectively. The specific enzyme deficiency is listed in parentheses for each disease.

13.9 PROSTAGLANDINS

Structure and Nomenclature

The prostaglandins are C20 unsaturated hydroxy acids with a substituted cyclopentane ring and two aliphatic side chains.

Consider the carbon skeleton of the prostaglandins:

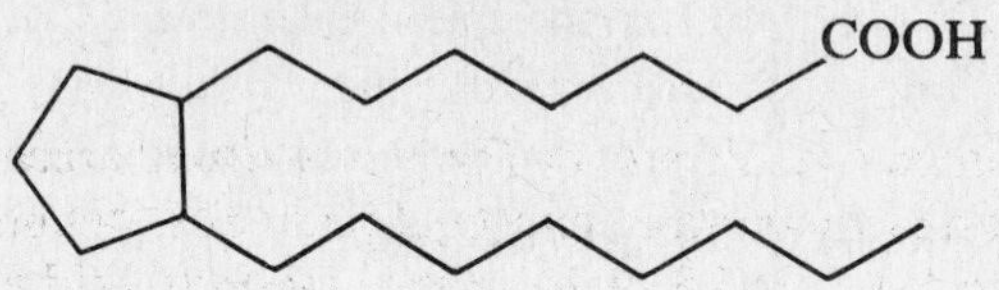

COOH

Prostanoic acid

Individual prostaglandins are described by an abbreviated system of nomenclature, in which the name *prostaglandin* is designated PG, followed by a third letter (A–I) that indicates the nature of the substituents on the cyclopentane ring. A numerical subscript indicates the number of double bonds in the aliphatic chains.

EXAMPLE 13.12

The structures and abbreviated names of some prostaglandins are:

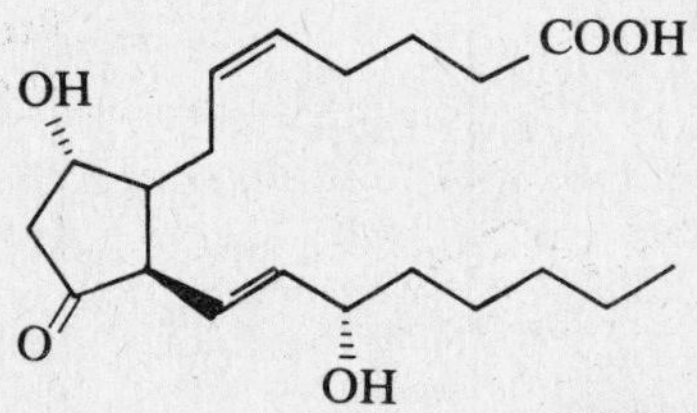

COOH OH O OH

PGD_2

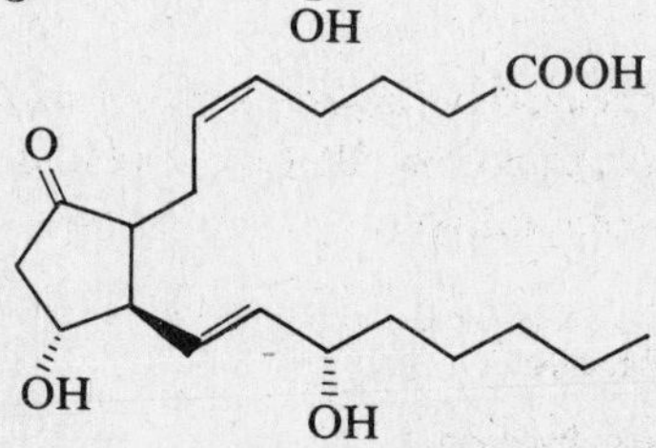

COOH O OH OH

PGE_2

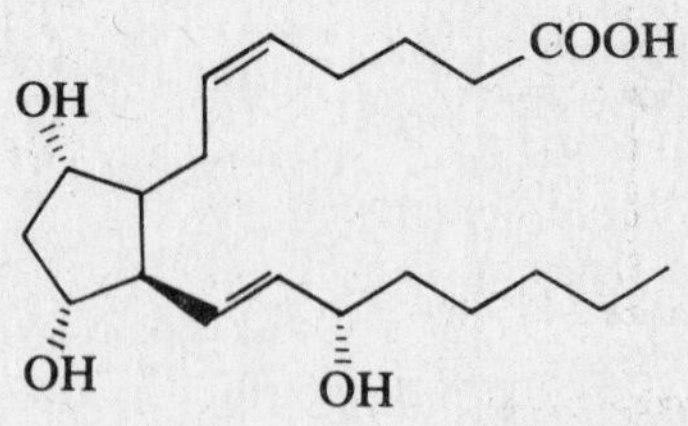

COOH OH OH OH

PGF_2

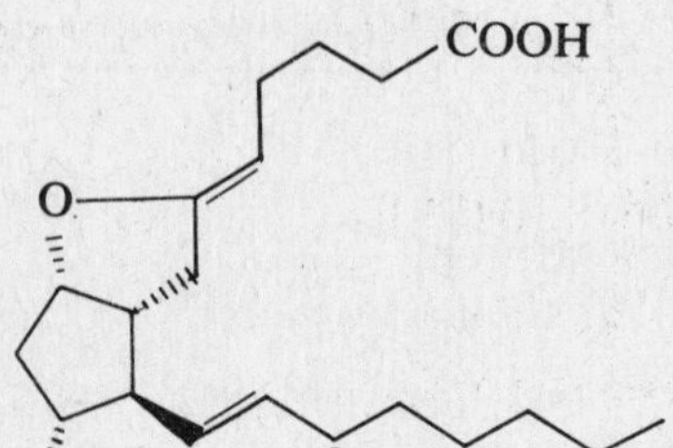

COOH O OH OH

PGI_2
(also called prostacyclin)

Biological Role of the Prostaglandins

The prostaglandins occur in all tissues but in very small amounts. They act on loci in the same cells as those in which they are synthesized, and their biological roles are diverse; e.g., they function in the female reproductive system during ovulation, menstruation, pregnancy, and parturition, and they stimulate uterine muscle contraction.

Synthesis of Prostaglandins

Prostaglandins are synthesized as shown in Fig. 13-18 from arachidonic acid in a metabolic pathway that begins with plasma membrane phospholipids. The double-bond arrangement in the carbon chain of arachidonic acid, $C_{20:4}\Delta5,8,11,14$, makes the fatty acid very susceptible to oxidation by molecular oxygen. The enzyme *cyclooxygenase* catalyzes the introduction of oxygen and the cyclization of the carbon chain of arachidonic acid in the region of the double-bond positions at C-8 and C-11, as follows:

COOH

Arachidonic acid

COOH

O

O

OOH

Prostaglandin G_2
(PGG_2)

Cyclooxygenase is a component of *prostaglandin synthase*, located in the ER membrane.

The biosynthesis of the primary prostaglandin, PGG_2, leads to the biosynthesis of a large number of chemically related secondary compounds.

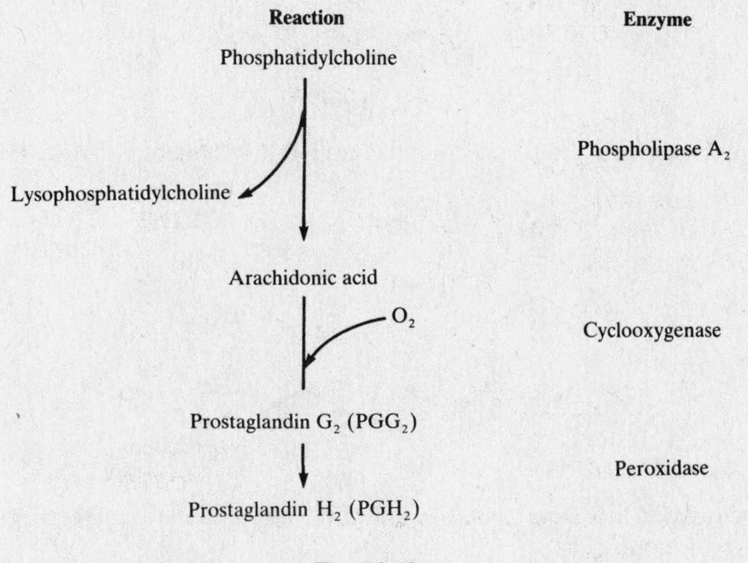

Fig. 13-18

EXAMPLE 13.13

The interrelationships of the prostaglandins are shown in Fig. 13-19.

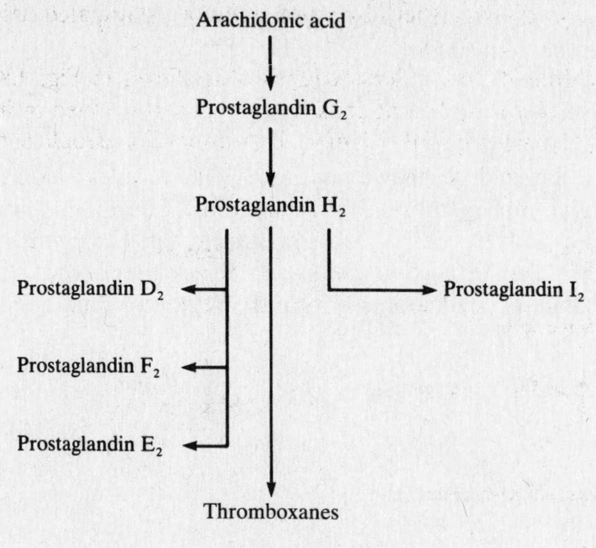

Fig. 13-19

Question: What are the structural characteristics that distinguish prostaglandin H_2 from thromboxanes?

Prostaglandin H_2 (PGH_2) is an *endoperoxide*, the oxygen atoms being attached to the cyclopentane ring:

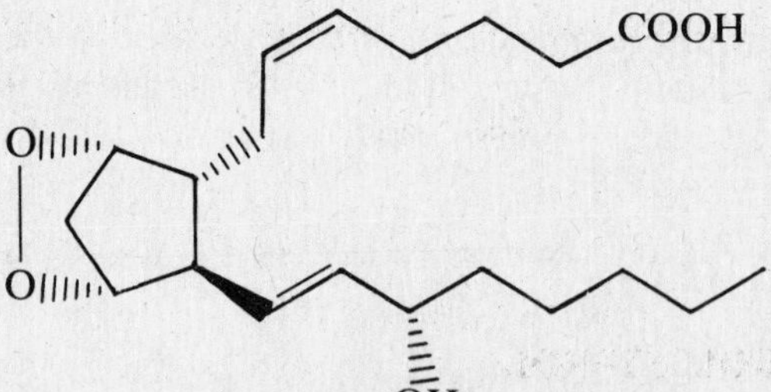

Thromboxanes have an oxygen atom incorporated into the cyclopentane ring, which produces a six-membered ring:

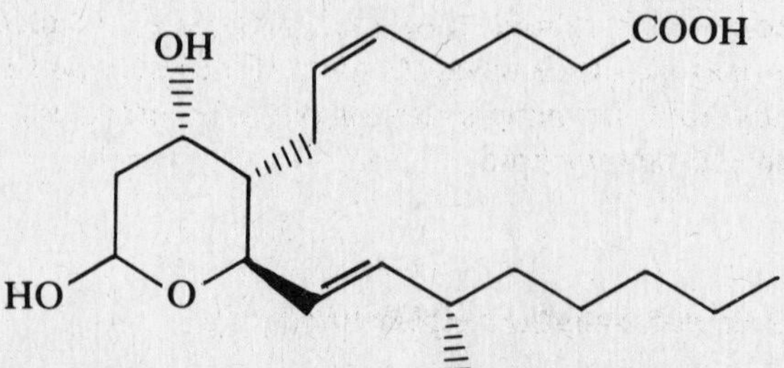

Thromboxanes were isolated first from platelets and were shown to cause the aggregation of platelets during the formation of blood clots.

Leukotrienes

Another group of compounds chemically related to the prostaglandins are the *leukotrienes*. These compounds are also derived from arachidonic acid and are linear oxidation products found in leukocytes. Leukotrienes are distinguished by containing a conjugated triene double-bond arrangement.

The conversion of arachidonic acid to leukotrienes is as shown in Fig. 13-20.

Leukotriene B_4 is one of the most potent chemotactic agents involved in inflammatory response in mammalian tissues; a concentration of $1.0\,\text{nmol}\,L^{-1}$ directs neutrophils to a site of injury or infection.

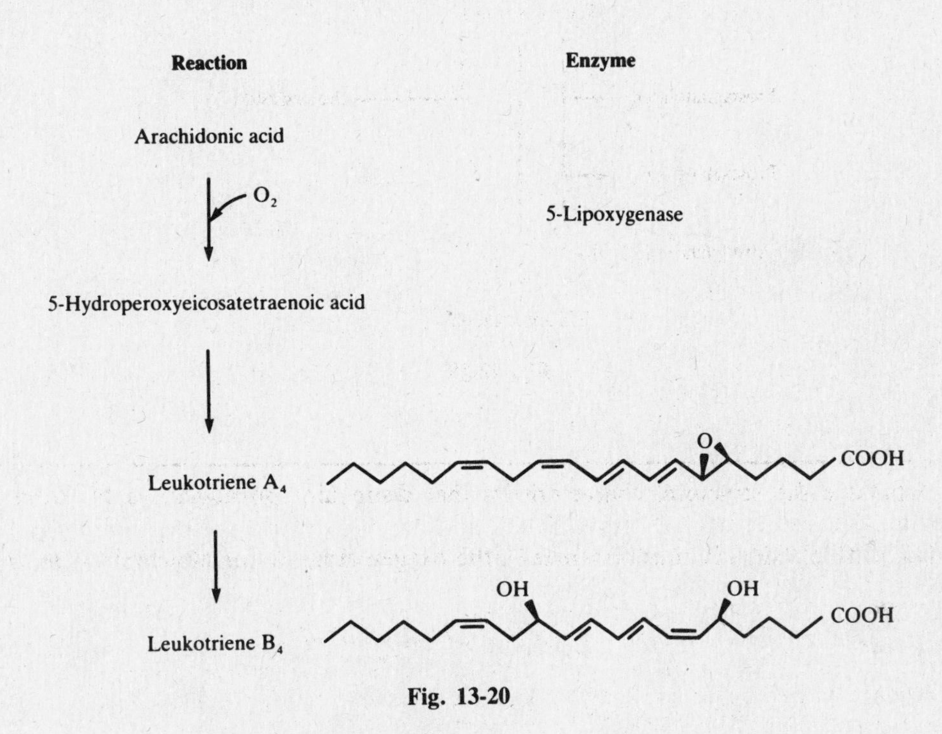

Fig. 13-20

13.10 METABOLISM OF CHOLESTEROL

Biological Role of Cholesterol

Cholesterol fulfills two biological roles. It is a structural component of cell membranes (Chap. 6) and the parent compound from which *steroid hormones*, *vitamin* D_3 (cholecalciferol), and the *bile salts* are derived. Cholesterol may be synthesized de novo in the liver and intestinal epithelial cells and is also derived from dietary lipid. De novo synthesis of cholesterol is regulated by the amount of cholesterol and triglyceride in the dietary lipid.

Cholesterol Biosynthesis in Liver and Intestinal Epithelium

The biosynthesis of cholesterol begins with acetyl-CoA in what is a very complex process involving 32 different enzymes, some of which are soluble in the cytosol and others of which are bound to the ER membrane. The basic carbon building block of cholesterol is *isoprene* (Chap. 6).

The key intermediates in the biosynthesis of cholesterol are shown in Fig. 13-21.

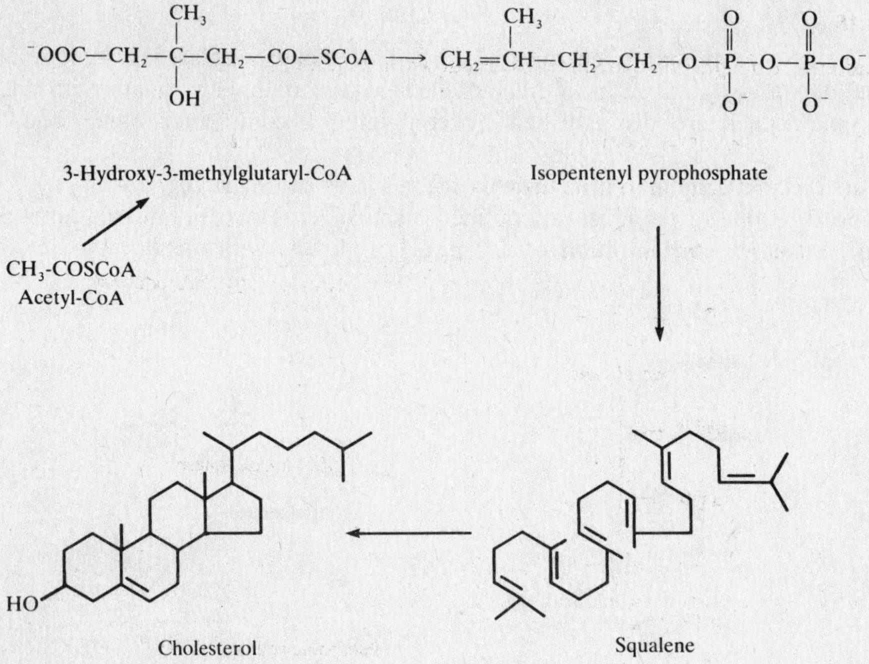

$$^{-}OOC\text{—}CH_2\text{—}\underset{OH}{\overset{CH_3}{C}}\text{—}CH_2\text{—}CO\text{—}SCoA \longrightarrow CH_2{=}\overset{CH_3}{C}H\text{—}CH_2\text{—}CH_2\text{—}O\text{—}\underset{O^-}{\overset{O}{P}}\text{—}O\text{—}\underset{O^-}{\overset{O}{P}}\text{—}O^-$$

3-Hydroxy-3-methylglutaryl-CoA   Isopentenyl pyrophosphate

CH_3-COSCoA
Acetyl-CoA

HO

Cholesterol   Squalene

Fig. 13-21

In the initial reactions of the biosynthetic pathway, 3-hydroxy-3-methylglutaryl-CoA (HMG-CoA) is formed from acetyl-CoA by the action of *thiolase* and *HMG-CoA synthase* in the cytosol of the liver cells.

EXAMPLE 13.14

The reactions catalyzed by thiolase and HMG-CoA synthase are:

(*a*) Thiolase

$$\underset{\text{Acetyl-CoA}}{2CH_3CO\text{—}SCoA} \longrightarrow \underset{\text{Acetoacetyl-CoA}}{CH_3COCH_2CO\text{—}SCoA} + CoASH$$

(*b*) HMG-CoA synthase

$$H_2O + \underset{\text{Acetoacetyl-CoA}}{CH_3COCH_2CO\text{—}SCoA} + \underset{\text{Acetyl-CoA}}{CH_3CO\text{—}SCoA} \longrightarrow \underset{\text{3-Hydroxy-3-methylglutaryl-CoA}}{^{-}OOC\text{-}CH_2\text{-}\underset{OH}{\overset{CH_3}{C}}\text{-}CH_2\text{-}CO\text{—}SCoA} + H^+ + CoASH$$

Question: By what biochemical means is 3-hydroxy-3-methylglutaryl-CoA reduced to the isoprene compound, isopentenyl pyrophosphate?

NADPH is the electron donor in a reaction catalyzed by *hydroxymethylglutaryl-CoA reductase*, followed by two phosphorylation reactions involving ATP, two specific kinases, and an ATP-dependent decarboxylation. The process is summarized in Example 13.15 and Fig. 13-22.

EXAMPLE 13.15

The formation of isopentenyl pyrophosphate is shown in Fig. 13-22.

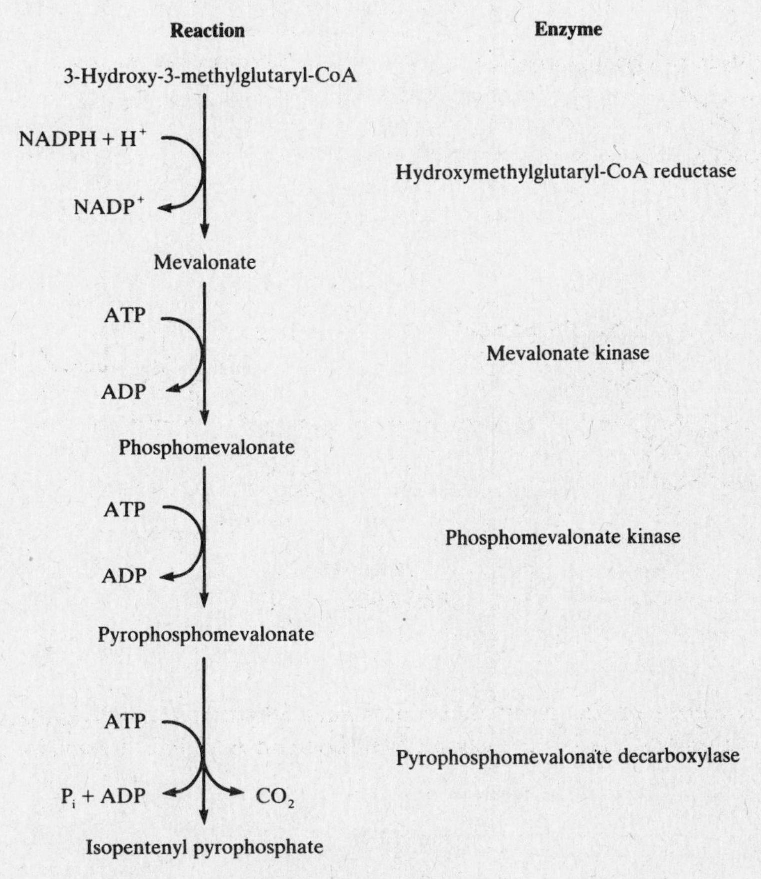

Fig. 13-22

Some isopentenyl pyrophosphate is converted to the isomer *dimethylallyl pyrophosphate*, by an isomerase that produces a mixture, *isopentenyl pyrophosphate* $\rightleftharpoons$ *dimethylallyl pyrophosphate*. From this point on, the carbon chain length of the intermediates progressively increases; this is followed by reduction to *squalene*, which has 30 carbon atoms in a folded chain and no oxygen atoms (Chap. 6). The conversion of isopentenyl pyrophosphate to squalene is shown in Fig. 13-23.

The final stage of the biosynthesis of cholesterol requires molecular oxygen, and the chain of squalene is cyclized to produce the primary steroid lanosterol, which is subsequently modified to cholesterol. The cyclization of squalene is shown in Fig. 13-24.

Question: How is the biosynthesis of cholesterol regulated by the amount of cholesterol in the diet?

A feedback mechanism operates in which intracellular free cholesterol inhibits *HMG-CoA reductase*. When the diet is rich in cholesterol, intracellular cholesterol increases in the liver and the biosynthesis of cholesterol is suppressed. Conversely, a low-cholesterol diet, but one with adequate triglyceride, stimulates cholesterol biosynthesis.

The metabolism of cholesterol involves modifications to the alkyl side chain and the introduction of substituents onto the phenanthrene ring. A group of enzymes, *monooxygenases*, plays an important role in catalyzing these reactions in the formation of bile salts and steroid hormones.

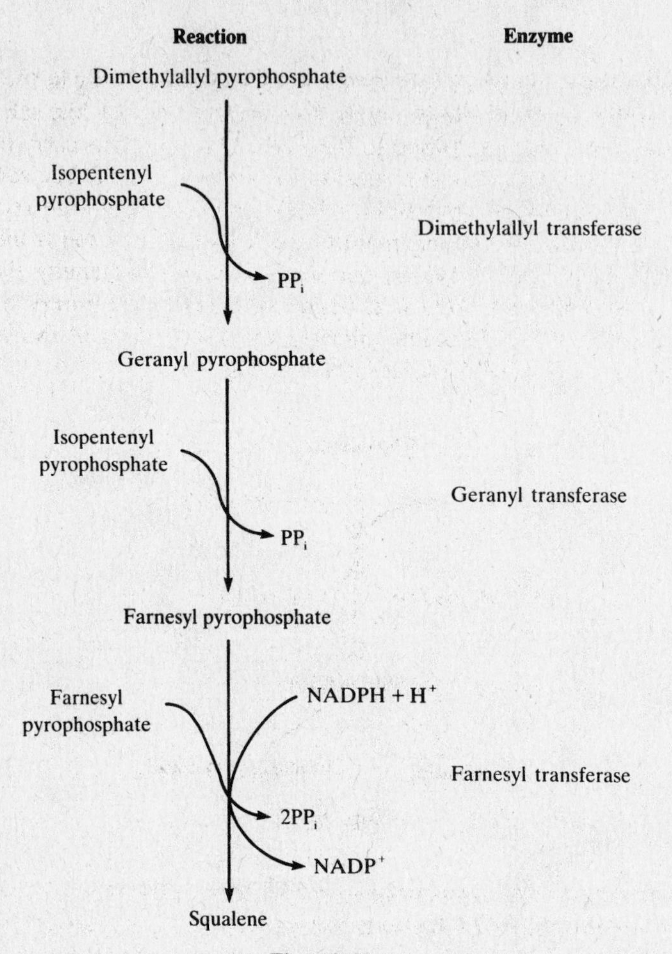

Fig. 13-23

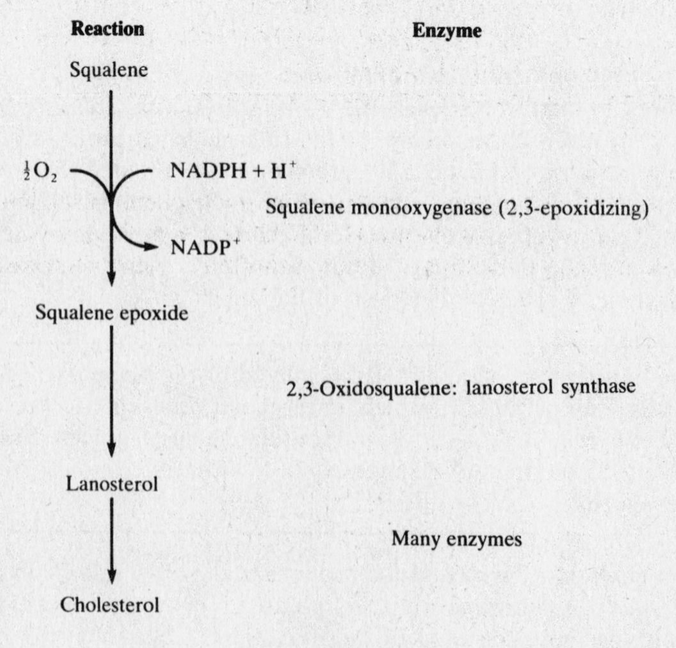

Fig. 13-24

The Bile Salts

The digestion and absorption of dietary lipid can be completed only in the presence of adequate amounts of *bile salts* in the lumen of the jejunum. Reabsorption of the bile salt micelles occurs in the ileum, from which they return via the blood to the liver. The bile ducts carry bile salts from the liver to the gallbladder, where they are stored; cholesterol is dissolved in the bile salt micelles. Overall, 90 percent of the bile salts involved in absorption of lipid in the jejunum are recycled, and 10 percent are lost in excretion. Replacement of this amount necessitates conversion from cholesterol. Thus, de novo synthesis of cholesterol itself plays an important part in maintaining the supply of bile salts.

The conversion of cholesterol to bile salts begins when hydroxyl groups are introduced into the phenanthrene ring structure of cholesterol, followed by modification of the side chain. *Cholic acid* and *chenodeoxycholic acid* are produced, as shown in Fig. 13-25.

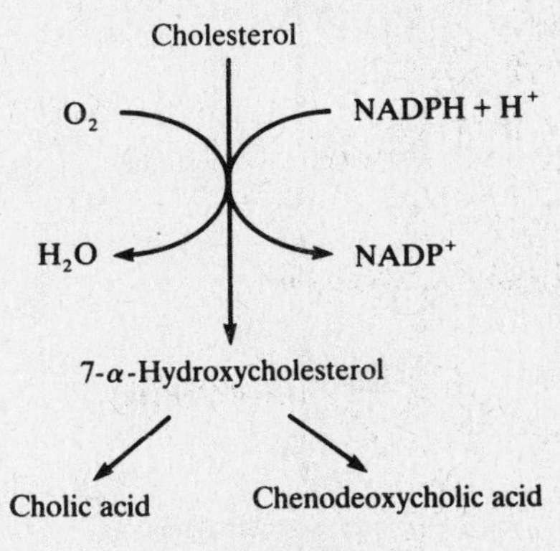

Fig. 13-25

Question: Why do monooxygenase reactions require NADPH as a cosubstrate?

Monooxygenase reactions catalyze the introduction of a hydroxyl or keto group into the substrate. Molecular oxygen is the source of the oxygen atom introduced into the substrate, and the second atom of molecular oxygen is reduced to H_2O. Both the substrate and the NADPH act as proton and electron donors. Monooxygenase reactions occur in the ER membrane and involve iron-sulfur proteins, ferredoxin, and cytochrome P_{450}.

Steroid Hormones

The synthesis of all steroid hormones begins with the conversion of cholesterol to *pregnenolone*. The pathway of this conversion is shown in Fig. 13-26. The side chain of cholesterol is cleaved by three successive monooxygenase reactions, which introduce a keto group at the site of cleavage of the side chain.

Pregnenolone

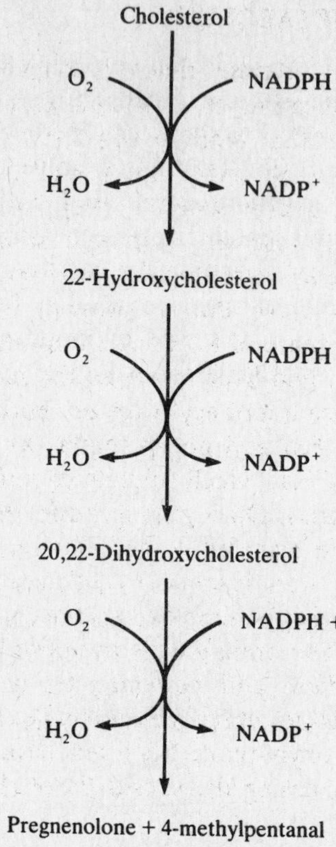

Cholesterol

O_2 → NADPH + H^+

H_2O ← NADP$^+$

↓

22-Hydroxycholesterol

O_2 → NADPH + H^+

H_2O ← NADP$^+$

↓

20,22-Dihydroxycholesterol

O_2 → NADPH + H^+

H_2O ← NADP$^+$

↓

Pregnenolone + 4-methylpentanal

Fig. 13-26

Subsequent molecular changes to pregnenolone give rise to the other steroid hormones. All these changes, catalyzed by monooxygenases, involve the introduction of oxygen atoms, as either hydroxyl or keto groups, at specific sites on the phenanthrene ring of the sterol and further removal of the side chain (Fig. 13-27).

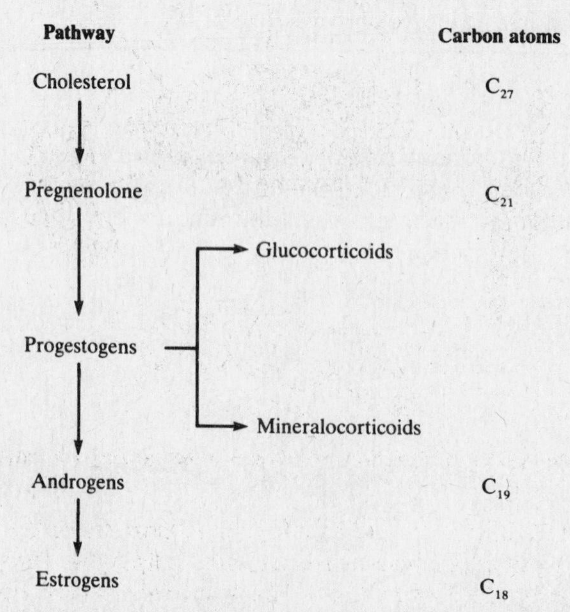

Fig. 13-27 The interrelationships between cholesterol and steroid hormones.

13.11 REGULATION OF LIPID METABOLISM

The release of fatty acids from adipose tissue is regulated by the rate of hydrolysis of triacylglycerol and the rate of esterification of acyl-CoA with glycerol 3-phosphate. The rate of hydrolysis is stimulated by hormones that bind to cell-surface receptors and stimulate *adenylate cyclase* (which catalyzes the production of cAMP from ATP). The *hormone-sensitive lipase* (Sec. 13.4) can exist in two forms, one which exhibits very low activity and a second which is phosphorylated and has high activity. Before hormonal stimulation of adenylate cyclase, the low-activity lipase predominates in the fat cell. Stimulation of protein kinase by an increase in cAMP concentration leads to phosphorylation of the low-activity lipase. An increase in the rate of hydrolysis of triacylglycerol and the release of fatty acids from the fat cell follows. This is a signal to tissues such as heart, skeletal muscle, and liver to utilize the fatty acids.

In liver, β oxidation and reesterification of acyl-CoA are both possible. The rate of β oxidation is determined initially by the rate at which acyl groups enter the mitochondrial matrix. This rate of entry may be decreased by *malonyl-CoA*, which inhibits *carnitine palmitoyltransferase* (an enzyme enabling acyl groups to enter the matrix). During lipogenesis, the concentration of malonyl-CoA in the cytosol is sufficient to inhibit the transferase and thus maintain esterification of fatty acids. During starvation, the release of fatty acids from adipose tissues increases; the rate of ketogenesis in liver also increases. During a 1- to 24-day period of starvation, the concentration of ketone bodies in the blood increases to $\sim 8\ \mathrm{mmol\ L^{-1}}$. The brain adapts to utilize ketone bodies as fuel; 70 percent of its energy requirement is satisfied by them, and the remainder by glucose. The whole-body utilization of glucose decreases as a number of tissues utilize ketone bodies (Fig. 13-28). In very long starvation periods (greater than 24 days), 3-hydroxybutyrate has a regulatory effect on the release of fatty acids from adipose tissues, probably by increasing the sensitivity of these tissues to insulin.

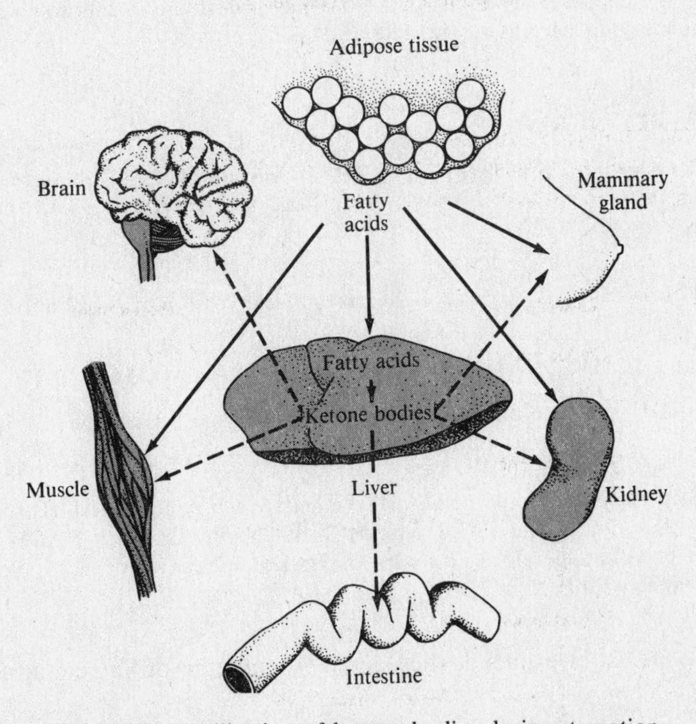

Fig. 13-28 Utilization of ketone bodies during starvation.

Insulin is an *antilipolytic* hormone, and its effect on adipose tissue is to increase the transport of glucose into the fat cell, to stimulate lipogenesis and inhibit lipolysis. Thus, pyruvate dehydrogenase and acetyl-CoA carboxylase are activated, and the hormone-sensitive lipase is inactivated. In the normal, well-fed state, insulin stimulates the deposition of fat.

Insulin also exerts a stimulatory effect on the synthesis of cholesterol in liver. In this tissue, HMG-CoA reductase is activated. HMG-CoA reductase, like hormone-sensitive lipase, can exist in two forms; one is phosphorylated (inactive) and the other is dephosphorylated (active). Phosphorylation of the enzyme depends on an increase in the cellular concentration of cAMP and activation of protein kinase. The dephosphorylation (activation) is catalyzed by a phosphatase. In fat cells, a similar phosphatase dephosphorylates (inactivates) hormone-sensitive lipase. Insulin stimulates the activity of the phosphatase in both liver and fat cells. In this way, active HMG-CoA reductase predominates in the liver cell and directs HMG-CoA into cholesterol synthesis, and in the fat cell hormone-sensitive lipase is inactivated.

Solved Problems

LIPID DIGESTION

13.1. What are the consequences of abnormal deconjugation of bile salts by bacteria in the small intestine?

SOLUTION

The pH of the contents of the lumen of the intestine is between 6.0 and 8.0. Bile salts in the lumen are, thus, ionized: e.g., taurocholate has a $pK_a \simeq 1.5$, owing to the conjugation of taurine with the cholate, and glycocholate has a $pK_a \simeq 3.7$, owing to the glycine conjugate. Deconjugation of either bile salt leaves cholate, $pK_a = 5.0$. The higher pK_a value of cholate reduces its solubility in the aqueous environment of the lumen, compared with the bile salts, which are readily soluble. Reabsorption of bile salts in the ileum is thus decreased, and bile acids are excreted.

OXIDATION OF FATTY ACIDS

13.2. How much energy, in the form of ATP, is obtained from β oxidation of 1 mole of palmitoyl-CoA?

SOLUTION

One mole of palmitoyl-CoA yields eight moles of acetyl-CoA by β oxidation. The overall equation is:

$$\text{Palmitoyl-CoA} + 7\text{FAD} + 7\text{NAD}^+ + 7\text{CoA} + 7\text{H}_2\text{O} \longrightarrow 8\text{Acetyl-CoA} + 7\text{FADH}_2 + 7\text{NADH} + 7\text{H}^+$$

$FADH_2$ and $NADH + H^+$ are oxidized in the electron-transport assemblies of the mitochondria:

$$7\text{FADH}_2 + 7\text{NADH} + 7\text{H}^+ + 7\text{O}_2 \longrightarrow 7\text{FAD} + 7\text{NAD}^+ + 14\text{H}_2\text{O}$$

During electron transport, coupled with ATP synthesis, each mole of $FADH_2$ that is oxidized yields 2 moles of ATP; thus, 7 moles of $FADH_2$ yields 14 moles of ATP. Each NADH mole that is oxidized yields 3 moles of ATP; therefore, 7 moles of NADH + 7 moles of H^+ yields 21 moles of ATP. Therefore, the total yield is 35 moles of ATP.

13.3. What is the sequence of events in the complete oxidation of linoleic acid?

SOLUTION

β oxidation proceeds normally to shorten the 18-carbon chain to dodecadienoic acid, $C_{12:2}\Delta 3,6$; three molecules of acetyl-CoA are produced as follows:

$$\text{CH}_3\text{-(CH}_2)_4\text{-CH=CH-CH}_2\text{-CH=CH-(CH}_2)_7\text{-COO}^- \longrightarrow$$
$$\text{CH}_3\text{-(CH}_2)_4\text{-CH=CH-CH}_2\text{-CH=CH-CH}_2\text{-COO}^- + 3\text{CH}_3\text{-COSCoA}$$

Before the next round of β oxidation can proceed, the double bond *cis* Δ^3 is converted to *trans* Δ^2 by the enzyme Δ^3-cis-Δ^2-trans-*enoyl-CoA isomerase* as follows:

$$CH_3\text{-}(CH_2)_4\text{-}CH{=}CH\text{-}CH_2\text{-}CH\underset{cis}{=}CH\text{-}CH_2\text{-}COO^- \longrightarrow$$
$$CH_3\text{-}(CH_2)_4\text{-}CH{=}CH\text{-}(CH_2)_2\text{-}CH\underset{trans}{=}CH\text{-}COO^-$$

The round of β oxidation is completed, and a further round of β oxidation proceeds, yielding two molecules of acetyl-CoA. In the next round, the configuration of the double-bond Δ^2 in *octenoic acid* (CH_3-$(CH_2)_4$-CH=CH-COO^-) is altered from the cis configuration to the trans configuration, and from a D to an L stereoisomer, by two enzymes Δ^2-*cis*-enoyl-CoA hydratase and 3-hydroxyacyl-CoA epimerase, respectively. Another acetyl-CoA molecule is released, and β oxidation then completely oxidizes the remaining hexanoic acid. Thus, nine molecules of acetyl-CoA are produced from the oxidation of linoleic acid.

THE FATE OF ACETYL-CoA FROM FATTY ACIDS: KETOGENESIS

13.4. What determines the relative concentrations of acetoacetate and 3-hydroxybutyrate in the blood?

SOLUTION

Acetoacetate is converted to 3-hydroxybutyrate by an *NAD-dependent dehydrogenase*. The reaction is reversible, and the equilibrium constant is approximately 1.

$$K_e = \frac{[\text{3-hydroxybutyrate}][\text{NAD}^+]}{[\text{acetoacetate}][\text{NADH}][\text{H}^+]} \simeq 1$$

13.5. 3-Hydroxybutyrate is a higher-energy-content fuel than acetoacetate. How is this energy released in cardiac and skeletal muscle?

SOLUTION

3-Hydroxybutyrate is activated to 3-hydroxybutyryl-CoA and oxidized via a specific 3-hydroxybutyryl-CoA dehydrogenase as in the β-oxidation pathway to acetoacetyl-CoA. This is then converted to acetyl-CoA, which is oxidized in the citric acid cycle.

LIPOGENESIS

13.6. How much glucose is converted to ribulose 5-phosphate when one molecule of palmitic acid is synthesized from acetyl-CoA?

SOLUTION

When glucose is converted to ribulose 5-phosphate, NADPH is produced in the redox reactions. For every molecule of glucose entering the pathway, two $NADP^+$ molecules are reduced. Some of this NADPH is utilized in the synthesis of palmitic acid. To calculate the amount consider:

1. Seven rounds of enzymatic reactions occur on the *fatty acid synthase* to synthesize one molecule of palmitic acid.
2. In each round of enzymatic reactions, there are two reduction reactions that use NADPH as the electron donor. Therefore, the synthesis of 1 molecule of palmitic acid requires 14 molecules of NADPH.
3. Eight acetyl groups are transferred from the mitochondrion to the cytosol to provide the carbon for the synthesis of one molecule of palmitic acid. During the transfer of the acetyl groups, 8 molecules of malate are oxidatively decarboxylated to 8 pyruvate and 8 carbon dioxide molecules, and 8 $NADP^+$ molecules are reduced. Thus, 8 of the 14 molecules of NADPH required are provided during the transfer of acetyl groups; the remaining 6 NADPH molecules come from the pentose phosphate pathway from 3 molecules of glucose.

13.7. What are the characteristics of the fatty acid synthase in the bacterium *Escherichia coli* that distinguish it from the mammalian multienzyme complex?

SOLUTION

Each of the enzymatic activities located in a single polypeptide chain of the mammalian fatty acid synthetase exists as a distinct protein in *E. coli*. The acyl-carrier protein (ACP) of *E. coli* has an $M_r = 8{,}847$ and contains 4-phosphopantotheine. The dehydratase has a molecular weight of $M_r = 28{,}000$ and catalyzes either trans 2—3 or cis 3—4 dehydration of the hydroxy acid intermediates in the biosynthesis of palmitic acid. When the chain length of the hydroxy fatty acid is C_{10}, the synthesis of palmitoleic acid is achieved as follows:

$$\underset{\text{3-Hydroxydecanoyl-ACP}}{CH_3\text{-}(CH_2)_6\text{-}CHOH\text{-}CH_2\text{-}CO\text{-}ACP} \longrightarrow \underset{\Delta^3\text{-Decenoyl-ACP}}{CH_3\text{-}(CH_2)_5\text{-}CH{=}CH\text{-}CH_2\text{-}CO\text{-}ACP}$$

Three more rounds of normal synthase activity elongate Δ^3-Decenoyl-ACP to yield

$$\underset{\text{Palmitoleic acid}}{CH_3\text{-}(CH_2)_5\text{-}CH{=}CH(CH_2)_7COOH}$$

13.8. (*a*) What is the enzymatic reaction that converts saturated fatty acids to their monounsaturated derivatives in mammalian tissues, and (*b*) in which cellular location does the reaction take place?

SOLUTION

(*a*) An enzyme complex consisting of *acyl-CoA desaturase* and *cytochrome* b_5 catalyzes the oxidation of the alkanoyl chain as follows:

$$CH_3{-}(CH_2)_n{-}CH_2{-}CH_2{-}(CH_2)_n{-}COOH$$

$$\tfrac{1}{2}O_2 \rightarrow H_2O \qquad NADPH + H^+ \rightarrow NADP^+$$

$$CH_3{-}(CH_2)_n{-}CH{=}CH{-}(CH_2)_n{-}COOH$$

(*b*) This electron-transport chain occurs in the ER.

SYNTHESIS OF PHOSPHOLIPIDS AND SPHINGOLIPIDS

13.9. Why does a dietary deficiency of choline in humans and laboratory animals induce a *fatty* liver, i.e., a liver in which the hepatocytes contain large lipid spheres?

SOLUTION

The de novo synthesis of phosphatidylcholine requires choline. In liver, phosphatidylcholine is synthesized and enters the membranes and lipoproteins. The de novo synthesis of phosphatidylcholine also requires 1,2-diacylglycerol. If there is insufficient choline available, phosphatidylcholine production by the de novo pathway cannot occur. The 1,2-diacylglycerol is then converted to triacylglycerol, which accumulates, as it is not secreted in lipoproteins. Therefore, the liver cells fill with triacylglycerol.

13.10. How does chronic ingestion of ethanol lead to the development of a fatty liver, if you are told that ethanol stimulates the activity of phosphatidic acid phosphatase?

SOLUTION

The stimulation of phosphatidic acid phosphatase by ethanol in turn stimulates the synthesis of triacylglycerol. Therefore, the triacylglycerol concentration increases in the liver cells.

13.11. Tay-Sachs disease, the lipid storage disease that afflicts 1 in 900 children of Ashkenazic Jewish parents, results in the accumulation of the ganglioside ceramide-β-glucose-β-galactose-*N*-acetylgalactosamine *N*-acetylneuramic acid in the brain. Why does this occur?

SOLUTION

Brain lysosomes are deficient in the enzyme hexosaminidase A; the enzyme has two subunits, A and B. Because of a lack of the enzyme, hydrolysis of the terminal N-acetylgalactosamine from the ganglioside cannot occur and the membrane lipid that is *turned over* in the lysosomes accumulates there.

PROSTAGLANDINS

13.12. What is the biochemical basis of the anti-inflammatory action of the drug aspirin?

SOLUTION

Prostaglandins such as PGE_2 are potent *vasodilators*, so they increase blood flow. Local vasodilation occurs at inflammatory sites. Aspirin irreversibly *inhibits* cyclooxygenase. A single serine residue in the enzyme is acetylated as follows:

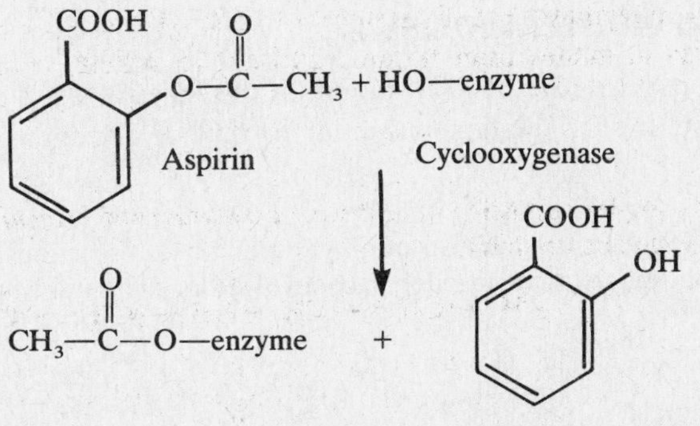

Aspirin + HO—enzyme —(Cyclooxygenase)→ CH_3—C(=O)—O—enzyme + 2-Hydroxybenzoic acid

This inhibition of cyclooxygenase blocks the synthesis of prostaglandin, which in turn reduces the inflammatory response of the tissue.

13.13. Homo-γ-linolenic acid, like arachidonic acid, is a precursor of prostaglandin. How are the compounds that are derived from homo-γ-linolenic acid named?

$$CH_3\text{—}(CH_2)_4\text{—}CH{=}CH\text{—}CH_2\text{—}CH{=}CH\text{—}CH_2\text{—}CH{=}CH(CH_2)_6COOH$$

Homo-γ-linolenic acid

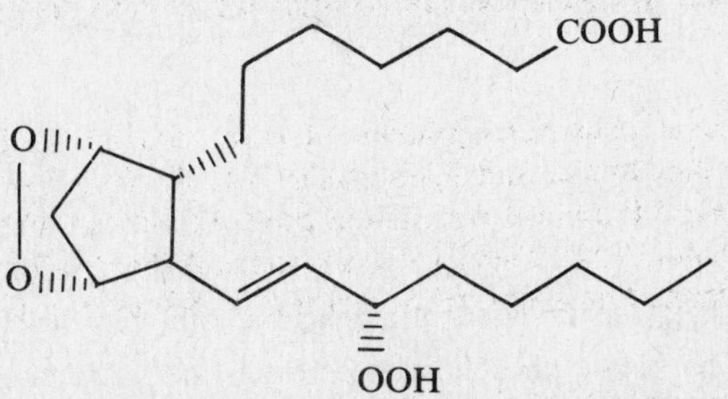

COOH, O, O, OOH

PGG_1

SOLUTION

Because homo-γ-linolenic acid lacks a double bond in position 5 but has a double bond in position 13, all derivatives of PGG_1, the primary prostaglandin of the homo-γ-linolenic acid series, are designated PG, followed by a letter A–I and the number 1 which denotes only one double bond in the molecule.

CHOLESTEROL AND RELATED COMPOUNDS

13.14. Cholecalciferol (vitamin D_3) is derived from cholesterol. How and in which tissue does the conversion occur?

SOLUTION

With normal exposure to sunlight, 7-dehydrocholesterol is converted to cholecalciferol in the skin.

13.15. Cholecalciferol is metabolized to produce steroid hormones. How and in which tissue does this occur?

SOLUTION

Cholecalciferol is hydroxylated at three positions in the carbon skeleton—1, 24, and 25. In the liver, cholecalciferol is hydroxylated to 25-hydroxycholecalciferol. Further hydroxylation reactions occur in the kidney, resulting in the formation of three new metabolites. These are 1,25-dihydroxycholecalciferol; 24,25-dihydroxycholecalciferol; and 1,24,25-trihydroxycholecalciferol. 1,25-Dihydroxy- and 1,24,25-trihydroxycholecalciferol are active hormones involved in calcium uptake from the intestine.

REGULATION OF LIPID METABOLISM

13.16. How does the expenditure of ATP maintain the inactive form of liver HMG-CoA reductase? (Hint: The system of regulation is similar to that of glycogen phosphorylase, p. 332.)

SOLUTION

See Fig. 13-29. ATP is a cosubstrate in phosphorylation reactions that covalently modify enzymatic proteins. Two phosphorylation reactions are involved in the modifications of HMG-CoA reductase:

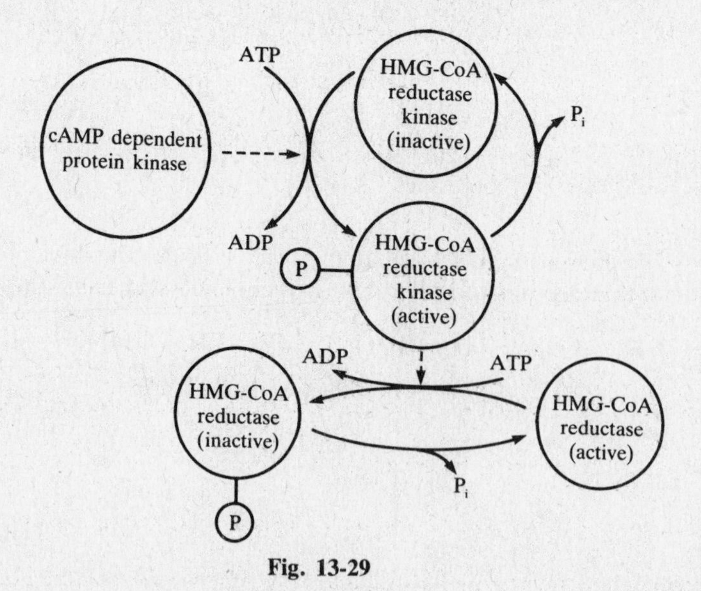

Fig. 13-29

1. HMG-CoA reductase is phosphorylated in a reaction catalyzed by HMG-CoA reductase kinase.
2. cAMP-dependent protein kinase phosphorylates HMG-CoA reductase kinase, converting it to its active form. Subsequent dephosphorylation of HMG-CoA reductase and HMG-CoA reductase kinase releases inorganic phosphate. Each HMG-CoA reductase molecule is phosphorylated at the expense of one molecule of ATP. Similarly, one molecule of ATP is utilized in the initial phosphorylation of HMG-CoA reductase kinase. However, one molecule of active HMG-CoA reductase kinase can catalyze the phosphorylation of many molecules of HMG-CoA reductase.

Supplementary Problems

13.17. What is the mechanism of the stimulation, by triglyceride in the diet, of the synthesis of cholesterol?

13.18. Why does feeding compounds that act as methyl-group donors to patients with fatty liver alleviate the condition?

13.19. How much energy, as ATP, is obtained from the complete oxidation of palmitic acid to carbon dioxide and water?

13.20. What is the yield of ATP from the oxidation of (*a*) 1 mole of acetoacetate and (*b*) 1 mole of 3-hydroxybutyrate?

13.21. What will be the contents of bile salt micelles after feeding: (*a*) 1,3-dipalmitoyl-2-linoleoylglycerol; (*b*) 1-stearoyl-2-arachidonylphosphatidylcholine?

13.22. How much glucose would be utilized to produce NADPH for the synthesis of one molecule of cholesterol from acetyl-CoA?

13.23. How is oleic acid converted to acetyl-CoA?

13.24. How does adipose tissue obtain the glycerol 3-phosphate necessary for triacylglycerol synthesis?

13.25. What properties of acetyl-CoA carboxylase are consistent with a regulatory role for the enzyme in fatty acid synthesis?

13.26. How is it possible for amino acids to act as precursors for the synthesis of palmitic acid?

13.27. What enzymes are involved in the oxidation of oleic acid to acetyl-CoA?

13.28. Explain why fat depots in mammalian adipose tissues are sources of intracellular water.

13.29. What effect would a deficiency of carnitine palmitoyltransferase in skeletal muscle have on the ability of a person to perform prolonged exercise?

13.30. Would aspirin be effective in preventing platelet aggregation?

13.31. If it were possible to provide a person with a diet free of cholesterol and triacylglycerol, how would this affect the deposition of triacylglycerol in adipose tissue?

13.32. 3-Hydroxy-3-methylglutaryl-CoA is an intermediate in the synthesis of acetoacetate and cholesterol in liver. How does this intermediate become available to each biosynthetic pathway?

13.33. How are cholesteryl esters synthesized?

13.34. Why does the concentration of ketone bodies in the blood increase during prolonged starvation?

13.35. What are the physical-chemical reasons that fatty acids are transported in lipoproteins?

13.36. In which cells may arachidonic acid be released into the cytosol?

Chapter 14

Oxidative Phosphorylation

14.1 INTRODUCTION

Bioenergetics is the study of (1) the process by which *reduced nicotinamide* and *flavin nucleotides*—generated primarily from the oxidation of carbohydrates (Chap. 11) and lipids (Chap. 13)—are oxidized ultimately by molecular oxygen via the mitochondrial *electron-transport chain*, and (2) the mechanism by which this oxidation is coupled to *ATP synthesis*. The synthesis of ATP in this way is referred to as *oxidative phosphorylation*, in contrast to phosphorylation of ADP via soluble enzymes. The latter involves intermediate phosphate derivatives of the substrate and is known as *substrate-level phosphorylation* (Chap. 11).

EXAMPLE 14.1

In glycolysis, ADP is phosphorylated to ATP during the oxidation of glyceraldehyde 3-phosphate to 3-phosphoglycerate. The phosphorylated intermediate that receives the energy of the oxidation is 1,3-diphosphoglycerate.

$$\underset{\text{Glyceraldehyde 3-phosphate}}{\begin{array}{c} CHO \\ | \\ HCOH \\ | \\ CH_2OPO_3^{2-} \end{array}} + NAD^+ + P_i \longrightarrow \underset{\text{1,3-Diphosphoglycerate}}{\begin{array}{l} \;\; O \\ \; /\!/ \\ C \\ | \;\diagdown O{-}PO_3H_2 \\ HCOH \\ | \\ CH_2OPO_3^{2-} \end{array}} + NADH + H^+$$

$$\begin{array}{l} \;\; O \\ \; /\!/ \\ C \\ | \;\diagdown O{-}PO_3H_2 \\ HCOH \\ | \\ CH_2OPO_3^{2-} \end{array} + ADP \longrightarrow \underset{\text{3-Phosphoglycerate}}{\begin{array}{c} COO^- \\ | \\ HCOH \\ | \\ CH_2OPO_3^{2-} \end{array}} + ATP$$

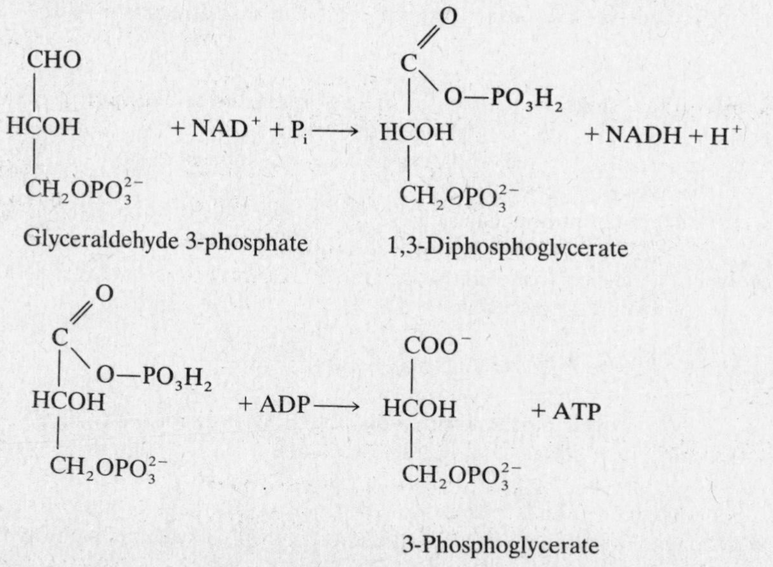

This is an example of *substrate-level phosphorylation*.

Oxidative phosphorylation is central to the metabolism of all higher organisms, because the *free energy* of hydrolysis of the ATP so generated is used in the synthesis of, inter alia, nucleic acids (Chaps. 7 and 16), proteins (Chaps. 4, 9, and 17), and complex lipids (Chap. 6), as well as in processes as diverse as muscle contraction (Chap. 5) and the transmission of nerve impulses.

14.2 COMPONENTS OF THE ELECTRON-TRANSPORT CHAIN

The *electron-transport chain*, or *respiratory chain* in mitochondria forms the means by which electrons, from the reduced electron carriers of intermediary metabolism, are channeled to oxygen and protons to yield H_2O. The main components of the chain are as follows.

$NAD^+/NADH$

The electron-transport reaction for the $NAD^+/NADH$ *conjugate redox pair* is:

$$NAD^+ + H^+ + 2e^- \longrightarrow NADH \qquad E^{\circ\prime} = 0.32\ V$$

where $E^{\circ\prime}$ is the standard redox potential (Chap. 10). In effect, electrons are transported as *hydride ions* (H^-), which are *formally* equivalent to ($H^+ + 2e^-$).

Question: How can electron transport by $NAD^+/NADH$ be measured?

NADH has a characteristic light absorbance maximum at 340 nm, which is absent in NAD^+. Hence electron transport involving $NAD^+/NADH$ can be monitored by measuring the change in absorbance of a sample at 340 nm.

EXAMPLE 14.2

NADH is oxidized by the electron-transport chain inside mitochondria. The inner mitochondrial membrane is not permeable to nucleotides. How can NADH generated in the cytoplasm (for example, during glycolysis) participate in the electron-transport chain?

The reducing equivalents of cytosolic NADH are transferred into mitochondria via *shuttle mechanisms*, such as the one involving oxaloacetate and malate, shown in Fig. 14-1. The net effect of this shuttle is the transport of NADH *into* the mitochondrion.

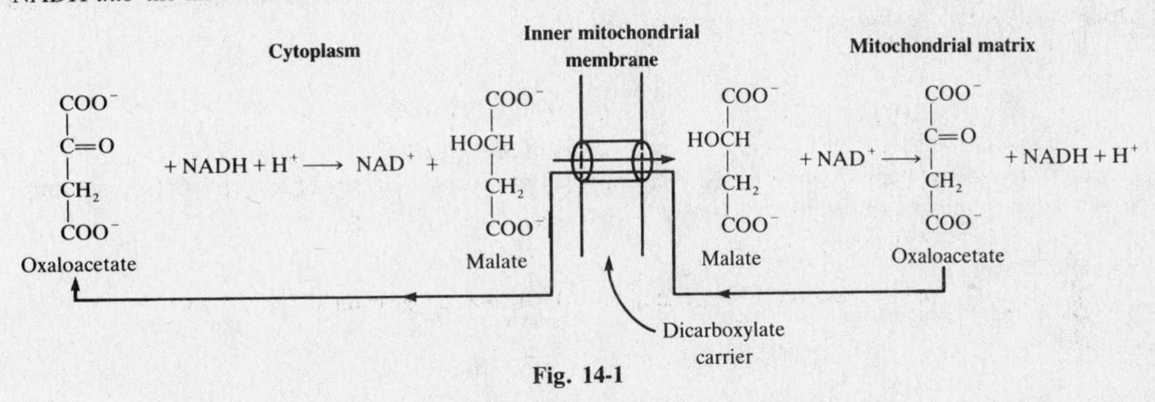

Fig. 14-1

Flavin Nucleotides

The electron-transport reactions for FAD and FMN are:

$$FAD + 2H^+ + 2e^- \longrightarrow FADH_2$$
$$FMN + 2H^+ + 2e^- \longrightarrow FMNH_2$$

Electrons are effectively transported as H atoms by these nucleotides [$H \equiv (H^+ + e^-)$].

These carriers transfer electrons into the electron-transport chain independently of and bypassing the $NAD^+/NADH$ couple.

EXAMPLE 14.3

Electrons from succinate, glycerol 3-phosphate, and the flavin-dependent steps of fatty acid oxidation enter the chain in the following way:

(*a*) Succinate + FAD $\longrightarrow$ Fumarate + $FADH_2$
(*b*) Glycerol 3-phosphate + FAD $\longrightarrow$ Dihydroxyacetone phosphate + $FADH_2$
(*c*) Acyl-CoA + FAD $\longrightarrow$ *trans*-Δ^2-Enoyl-CoA + $FADH_2$

Unlike the $NAD^+/NADH$ couple, the flavin coenzymes are not in free solution and are covalently linked to the dehydrogenases of their respective substrates. These enzymes are *membrane-bound* in close association with the respiratory chain.

Coenzyme Q

Coenzyme Q (alternatively known as *ubiquinone* or *CoQ*) is a *benzoquinone* derivative with a long hydrocarbon side chain made up of repeating isoprene units. The number of units generally categorizes the CoQ molecule. The form of CoQ in mammalian mitochondria contains 10 such units; hence, it is designated CoQ_{10}.

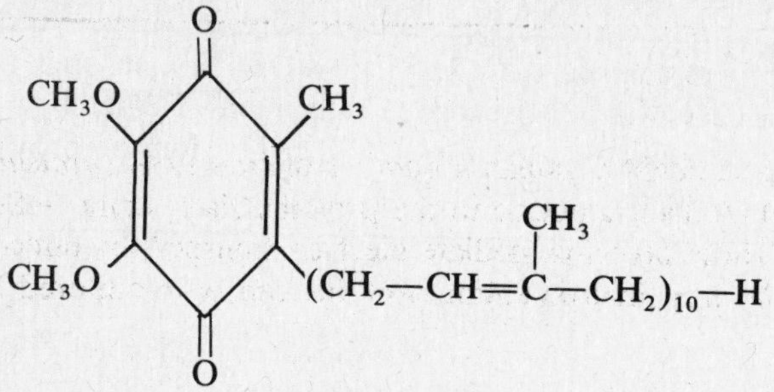
O
CH₃O
CH₃
CH₃
CH₃O
(CH₂—CH=C—CH₂)₁₀—H
O

CoQ_{10}

The molecule undergoes a $(2H^+ + 2e^-)$ reduction to form $CoQH_2$ (alternatively known as *ubiquinol*). The reduction may take place in two stages, yielding an intermediate, half-reduced, *free-radical* form (*semiquinone*), designated CoQH.

Cytochromes

The *cytochromes* (from the Greek: "cell colors") are a family of proteins containing prosthetic *heme groups* (see Chaps. 5 and 15). Mitochondria contain *three* classes of cytochromes: *a*, *b*, and *c*, which have hemes of different structures.

EXAMPLE 14.4

The general *cyclic tetrapyrrole* structure of a heme ring is shown below:

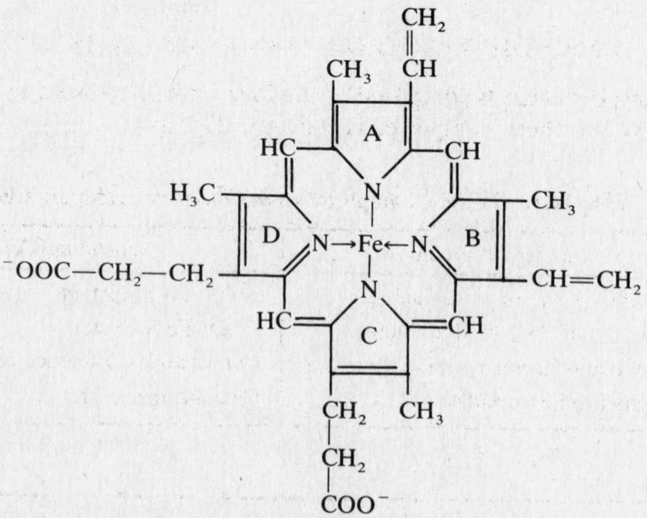
CH₂
CH₃
CH
A
HC
CH
H₃C
N
CH₃
D
N→Fe←N
B
⁻OOC—CH₂—CH₂
N
CH=CH₂
HC=
=CH
C
CH₂
CH₃
CH₂
COO⁻

In cytochromes *c* and c_1, the heme ring is covalently attached to the protein via *thioether bonds*, formed by reaction of the *vinyl* groups ($—CH{=}CH_2$) on pyrrole rings A and B and cysteine residues of the protein. These thioether bonds are absent in cytochrome *b*. In cytochromes *a* and a_3, the vinyl group on ring A is replaced by a hydrocarbon chain, and the methyl group on ring D is replaced by a *formyl* (—CHO) group. In addition, cytochromes *a* and a_3 contain bound Cu ions.

Question: How can cytochrome-mediated electron transport be measured?

Electron transport in cytochromes occurs by *direct electron transfer* between Fe^{2+} and Cu^{+} in cytochromes a and a_3. These changes in metal-ion oxidation state lead to changes in the visible absorption spectra of the cytochromes; spectrophotometric measurement of these changes allows quantification of the electron flow.

Iron-Sulfur Proteins

The electron-transport chain contains a number of *iron-sulfur proteins* (also known as *nonheme iron proteins*). The iron atoms are bound to the proteins via cysteine —S— groups and sulfide ions, as shown in Fig. 14-2. These proteins mediate electron transport by direct electron transfer; changes in oxidation state of the iron in iron-sulfur proteins can be monitored by *electron spin resonance spectroscopy* (*ESR*).

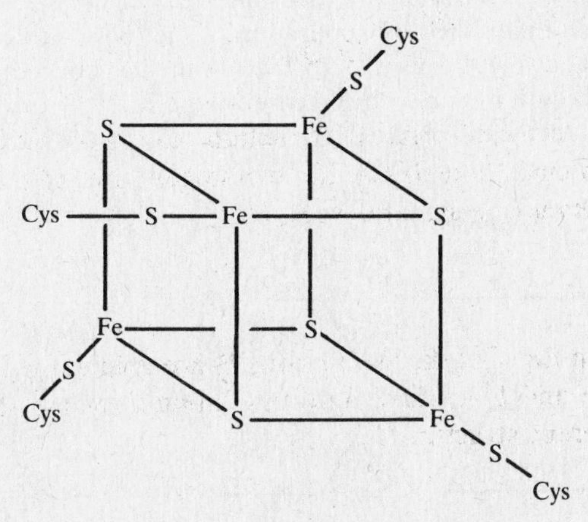

Fig. 14-2

14.3 ORGANIZATION OF THE ELECTRON-TRANSPORT CHAIN

The electron-transport chain is composed of the four complexes listed in Table 14.1. The pattern of electron transfer within these complexes is shown in Fig. 14-3.

Table 14.1. Protein Complexes of the Electron-Transport Chain

Complex	Enzymatic Function	Functional Components
I	NADH/CoQ oxidoreductase	FMN; Fe-S clusters
II	Succinate/CoQ oxidoreductase	FAD; Fe-S clusters
III	CoQ-cytochrome c oxidoreductase	Cytochromes b, cytochrome c_1; Fe-S clusters
IV	Cytochrome c oxidase	Cytochromes a and a_3

Question: By what experimental means has this pattern of electron transfer been determined?

Two broad experimental approaches have been used: examination of the effects of *specific inhibitors*, which block electron flow through a particular complex; and use of synthetic *redox couples*, which are able to yield electrons to specific complexes, depending on the relative $E^{\circ\prime}$ values of the complex and the redox couple.

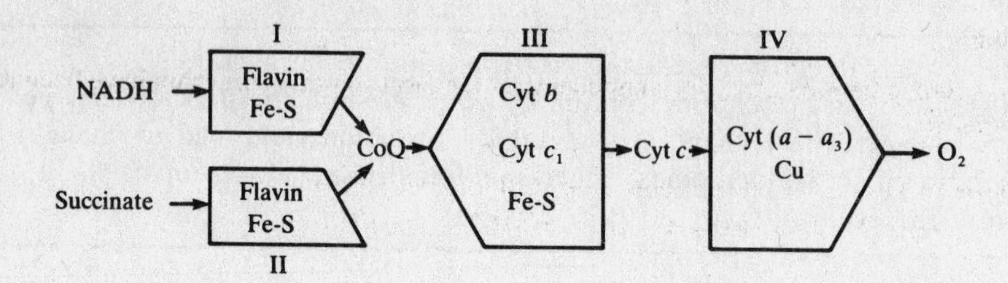

Fig. 14-3 Schematic organization of the electron–transport chain in mitochondria.

EXAMPLE 14.5

The sites of action of some of the commonly used inhibitors of the electron-transport chain are shown in Fig. 14-4. These sites have been established by application of the *crossover theorem* (Chap. 10). For example, the fungus-derived antibiotic *antimycin A* causes an *increase* in the level of reduced cytochrome b and a *decrease* in the level of reduced cytochrome c_1 (i.e., an increase in the level of oxidized cytochrome c_1); thus, it is inferred that antimycin A interacts with complex III.

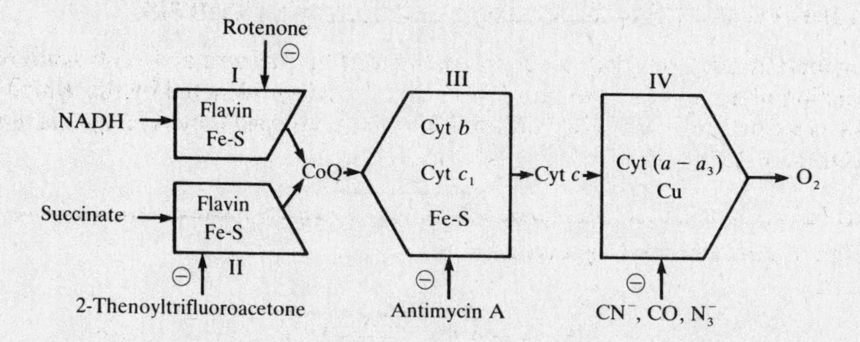

Fig. 14-4 Sites of action of inhibitors of the electron-transport chain.

Question: How can the $E^{\circ\prime}$ values of electron-transport-chain components be measured?

Measurement can be done using the technique of *redox potentiometry*. In experiments of this type, mitochondria are incubated anaerobically in the presence of a reference electrode [for example, a hydrogen electrode (Chap. 10)] and a platinum electrode and with *secondary redox mediators*. These mediators form redox pairs with $E^{\circ\prime}$ values intermediate between the reference electrode and the electron-transport-chain component of interest; they permit rapid equilibration of electrons between the electrode and the electron-transport-chain component. The experimental system is allowed to reach equilibrium at a particular E value. This value can then be changed by addition of a reducing agent (such as reduced ascorbate or NADH), and the relationship between E and the levels of oxidized and reduced electron-transport-chain components is measured. The $E^{\circ\prime}$ values can then be calculated using the Nernst equation (Chap. 10):

$$E = E^{\circ\prime} - \frac{RT}{nF} \ln \frac{[\mathrm{A_{red}}]}{[\mathrm{A_{ox}}]} \qquad (14.1)$$

where $[\mathrm{A_{red}}]$ and $[\mathrm{A_{ox}}]$ are the concentrations of reduced and oxidized electron-chain component A; these are obtained from spectroscopic measurements of the mitochondrial preparation.

EXAMPLE 14.6

Studied by redox potentiometry, the components of the electron-transport chain have been assigned the $E^{\circ\prime}$ values shown below:

$$\begin{array}{ccccccccc} \text{NADH} & \xrightarrow{e} & \text{FMN} & \xrightarrow{e} & \text{Ubiquinone} & \xrightarrow{e} & \text{Cyt } b & \xrightarrow{e} & \text{Cyt } c_1 \\ -320\text{ mV} & & -30\text{ mV} & & +100\text{ mV} & & +70\text{ mV} & & +215\text{ mV} \\ & & & & \uparrow e & & & & \downarrow e \\ & & & & \text{FAD} & & & & \text{Cyt } c \xrightarrow{e} \text{Cyt } a \text{ and } a_3 \xrightarrow{e} O_2 \\ & & & & \uparrow & & & & +210\text{ mV} \quad +210\text{ mV} \quad +820\text{ mV} \\ & & & & \text{Succinate} & & & & \quad\quad +385\text{ mV} \\ & & & & +30\text{ mV} & & & & \end{array}$$

This technique also indicates (1) whether a component is a $1e^-$ carrier or a $2e^-$ carrier, which is evident from the slope of a plot of E versus $\ln \dfrac{[A_{red}]}{[A_{ox}]}$; and (2) whether a component is an $(H^+ + e^-)$ carrier, in which case the value of $E^{\circ\prime}$ is pH-dependent.

14.4 COUPLING OF ELECTRON TRANSPORT AND ATP SYNTHESIS

It is now generally accepted that the coupling of electron transport and ATP synthesis is brought about by the action of a *proton electrochemical-potential gradient*, denoted by the symbol $\Delta\mu_{H^+}$. This gradient arises as a consequence of electron transport and is dissipated by *ATP synthetase* to generate ATP from ADP and P_i.

Question: How is $\Delta\mu_{H^+}$ defined experimentally?

It is defined as

$$\Delta\mu_{H^+} = \Delta\psi - \frac{2.3RT}{F}\Delta\text{pH} \tag{14.2}$$

where $\Delta\psi$ and ΔpH are, respectively, the electrical potential (in volts) and pH difference across the inner mitochondrial membrane; R, T, and F are the gas constant, absolute temperature, and Faraday constant, respectively. Both $\Delta\psi$ and ΔpH can be measured experimentally.

EXAMPLE 14.7

The most common way in which $\Delta\psi$ is determined is from measurement of the concentrations inside and outside mitochondria, at equilibrium, of an ionizable compound that is permeable to the inner mitochondrial membrane. $\Delta\psi$ can then be calculated using the Nernst equation (*14.1*), written in the form:

$$\Delta\psi = \frac{RT}{nF}\ln\frac{[X^{n+}]_{out}}{[X^{n+}]_{in}} \tag{14.3}$$

for an ion X of charge $+n$. It can also be calculated from changes in the spectral properties of membrane constituents arising from $\Delta\psi$ (*electrochromism*); a value of 200 mV for $\Delta\psi$ corresponds to an electric field across the membrane of $\sim 3 \times 10^5$ V cm^{-1}.

EXAMPLE 14.8

The value of ΔpH across the inner mitochondrial membrane can be estimated from the equilibrium distribution of *electroneutrally permeant weak acids* (or *weak bases*). The logic underlying such experiments is illustrated below.

For the reaction

$$H^+_{out} + A^-_{out} \overset{K_a}{\rightleftharpoons} HA_{out} \rightleftharpoons HA_{in} \overset{K_a}{\rightleftharpoons} H^+_{in} + A^-_{in}$$

the neutral species (HA) will equilibrate across the membrane independently of $\Delta\psi$. At equilibrium

$$[HA]_{in} = [HA]_{out} \tag{14.4}$$

Hence, assuming that the equilibrium constant K_a is the same for ionization of HA inside and outside the mitochondrion and the value of K_a is sufficiently low that $[HA] \simeq 0$,

$$\frac{[H^+]_{out}[A^-]_{out}}{[HA]} = \frac{[H^+]_{in}[A^-]_{in}}{[HA]} \tag{14.5}$$

and

$$\frac{[A^-]_{out}}{[A^-]_{in}} = \frac{[H^+]_{in}}{[H^+]_{out}} \tag{14.6}$$

Therefore, ΔpH can be calculated from measurements of $[A^-]_{out}$ and $[A^-]_{in}$.

14.5 THE RATIO OF PROTONS EXTRUDED FROM THE MITOCHONDRION TO ELECTRONS TRANSFERRED TO OXYGEN

The *chemiosmotic model* requires that flow of electrons through the electron-transport chain lead to *extrusion* of protons from mitochondria, thus generating the proton electrochemical-potential gradient. Measurements of the number of H^+ ions extruded per O atom reduced by complex IV of the electron-transport chain (the H^+/O ratio) are experimentally important because the ratio can be used to test the validity of *mechanistic models* of proton translocation (Sec. 14.6).

EXAMPLE 14.9

The H^+/O ratio can be measured by incubating mitochondria with an appropriate substrate (for example, NADH or succinate) under *anaerobic* conditions. The reaction is initiated by addition of a known amount of O_2, and H^+ extrusion is measured (as ΔpH) by using a pH electrode. The results of such an experiment are shown in Fig. 14-5.

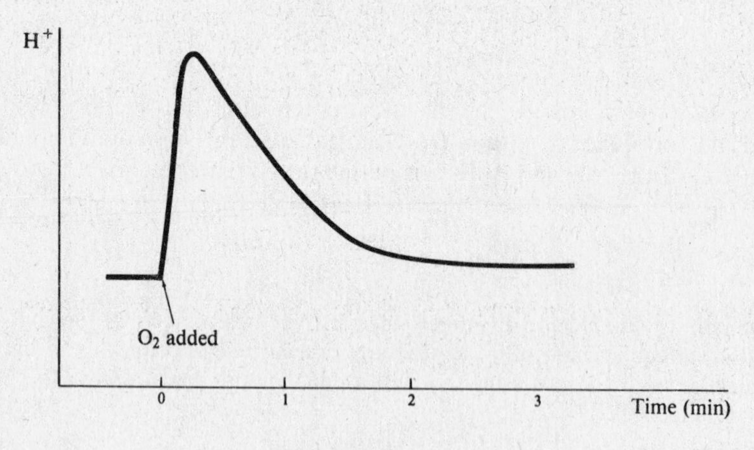

Fig. 14-5

There are a number of complications in the interpretation of such data. The mitochondria must be permeable to cations other than H^+. If this were not the case, then H^+ extrusion would lead to an increase in $\Delta\psi$, which would diminish further H^+ extrusion. Incorporation of K^+ and the *ionophore* valinomycin in the mitochondrial preparation prevents increases in $\Delta\psi$.

Other complications in analyzing the data of this experiment include the necessity to correct for movement of H^+ back into the mitchondria; this may occur *directly* or via the *Na^+/H^+ antiport* translocator of the mitchondrial inner membrane, or via the *H^+/phosphate symport* translocator (see Prob. 14.7).

Question: How does valinomycin render mitochondrial membranes permeable to K^+?

Valinomycin, an antibiotic, is a *mobile* K^+ carrier with an interior rich in polar amino acid residues (providing a binding site for K^+) and hydrophobic valine residues on the outside. The latter allow valinomycin, bearing K^+ in its interior, to diffuse across membranes, thus rendering them permeable to K^+. K^+ is unable to diffuse at appreciable rates across membranes in the absence of valinomycin.

14.6 MECHANISTIC MODELS OF PROTON TRANSLOCATION

Loop Mechanisms

In the chemiosmotic model, as first developed by Mitchell, proton translocation arises from transfer of electrons from an ($H^+ + e^-$) carrier (such as $FMNH_2$) to an electron carrier (such as an iron-sulfur protein), with expulsion of protons to the outer compartment of the inner mitochondrial membrane. This process is followed by electron transfer to an ($H^+ + e^-$) carrier, with uptake of protons from the matrix. In this model, the electron-transport chain is organized into *three* such loops, as shown in Fig. 14-6.

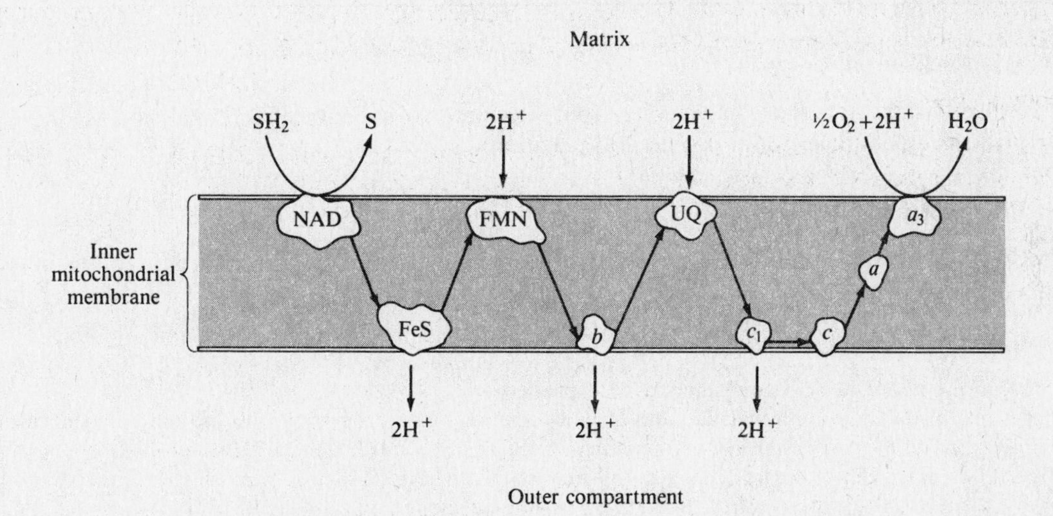

Fig. 14-16 Mitchell loop mechanism for proton translocation. SH_2 = reduced substrate, S = oxidized substrate; FeS = iron-sulfur protein; a, a_3, b, c, c_1 = cytochromes; UQ = coenzyme Q (ubiquinone).

EXAMPLE 14.10

What experimentally testable predictions does the Mitchell loop mechanism make?

This scheme makes two experimentally testable predictions:

1. There should be a ratio of 2 H^+ ions translocated per 2 electrons transported in each loop.
2. The electron-transport chain should be organized to give alternating ($H^+ + e^-$) and pure e^- carriers.

The experimental evidence pertaining to prediction (1) is discussed later in this section, while problems relating to (2) are outlined in Example 14.11.

Proton-Pump Mechanisms

In these mechanisms, electron transport through the various components of the electron-transport chain leads to structural changes in the proteins of the chain, such that there are changes in the pK_a values (Chap. 3) of ionizable amino acid residues in the proteins. For example, an *increase* in the pK_a of a residue adjacent to the *matrix* side of the membrane would lead to proton *uptake* from the matrix, while a *decrease* in the pK_a of a residue adjacent to the *cytosolic* side of the membrane could lead to *release* of a proton. The net effect of these processes is transfer of protons from the matrix to the cytosolic side of the membrane. Proton-pump mechanisms do *not* make strong predictions of the H^+/e^- stoichiometries.

Proton Translocation by Complex I

Complex I mediates electron transfer from $NADH + H^+$ to coenzyme Q. The consequent reduction of coenzyme Q is coupled to *vectorial* proton transport; the H^+/e^- stoichiometry (ratio) has been variously estimated to be 1.0, 1.5, and 2.0.

Question: Is an H^+/e^- stoichiometry of 2 consistent with the simple chemiosmotic hypothesis?

No, it is inconsistent with this hypothesis, which predicts an H^+/e^- value of 1.0. Such stoichiometries may, however, be explained by *proton-pump mechanisms*, in which electron transfer is coupled to changes in the pK_a values of proteins within complex I.

Proton Translocation by Complex II

Complex II mediates electron transfer from succinate to coenzyme Q. This process does *not* appear to be coupled to vectorial proton translocation.

Proton Translocation by Complex III

Complex III catalyzes electron transfer from reduced coenzyme Q to cytochrome *c*; this process is coupled to vectorial proton translocation with an H^+/e^- stoichiometry of 2.

EXAMPLE 14.11

How can the above H^+/e^- stoichiometry be explained?

This is one of the most controversial areas of bioenergetics and is concerned with the role of coenzyme Q. The simplest view of the role of this coenzyme is that it acts as a *mobile* $(2H^+ + 2e^-)$ *carrier*, linking complexes I and II with complex III. However, coenzyme Q may be involved in $(H^+ + e^-)$ transfer within complex III. One model for this is the *proton-motive Q cycle* (Fig. 14-7), developed by Mitchell in 1975. This model satisfies prediction (2) of Example 14.10, in that coenzyme Q acts as an $(H^+ + e^-)$ carrier in *two* loops. In this model, *reduced* coenzyme Q (QH_2) is linked to *oxidized* coenzyme Q (Q) via the free-radical *semiquinone* ($QH\cdot$). This model provides an explanation for the H^+/e^- stoichiometry.

Proton Translocation by Complex IV

Complex IV catalyzes electron transfer from cytochrome *c* to O_2; this process appears to be coupled to proton translocation, with H^+/e^- values of either 1 or 2. Two models have been developed to account for these values (Fig. 14-8).

Question: How can the Mitchell model give an H^+/e^- value of 1 when no protons traverse the membrane?

The Mitchell model can be described by the equation

$$2e^- + 2H^+ + \tfrac{1}{2}O_2 \longrightarrow H_2O$$

Thus, there is a net decrease in $[H^+]$ on the matrix side of the mitochondrial inner membrane.

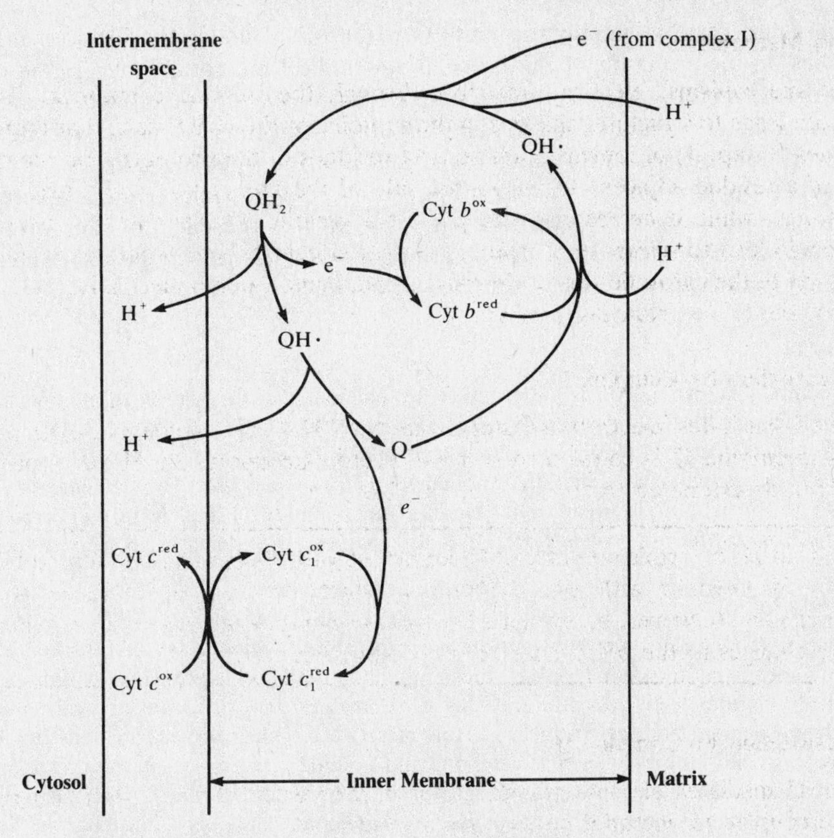

Fig. 14-7 Representation of the proton-motive Q cycle.

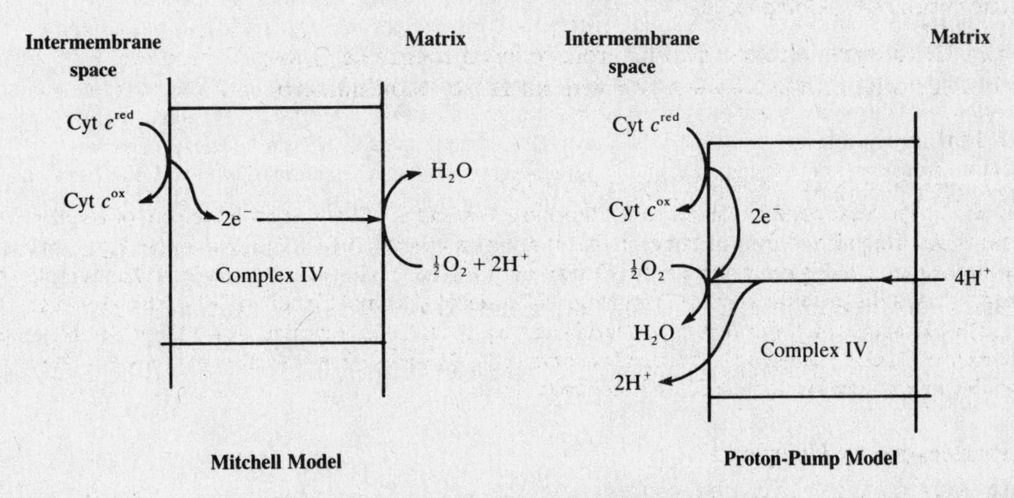

Fig. 14-8 Two models for proton translocation.

14.7 ATP SYNTHETASE

ATP synthetase is found in *all* energy-transducing membranes, including those of *mitochondria*, *chloroplasts*, and *bacteria*. The enzyme complex most probably evolved in primitive anaerobic life forms, before the appearance of oxygen-based respiration. In those early organisms, it possibly functioned in reverse, utilizing ATP generated by *fermentations*, to create proton gradients that were coupled to ion-transport systems.

Tripartite (three-region) membrane units were observed under the electron microscope in the earliest studies on the structure of the inner mitochondrial membrane. While the original ideas of repeating tripartite membrane units—containing a *basal segment* of the four respiratory complexes, a *stalk* of coupling factors, and a *knob* of ATP synthetase—have been superseded, the ATP synthetase itself, however, does consist of transmembrane and matrix-projecting units.

The matrix-projecting unit in mitochondria, termed the F_1 component, is involved in hydrolysis of ATP in vitro and in ATP synthesis in vivo in the presence of the proton-translocating transmembrane subunits, termed F_o (the subscript is a reference to oligomycin). The structures of the F_1 and F_o complexes have now been described in detail.

EXAMPLE 14.12

The F_1 component can be detached from F_o by chemicals such as urea or at low ionic strength and comprises five subunits: these are generally designated α ($M_r = 56{,}000$), β ($M_r = 53{,}000$), γ ($M_r = 33{,}000$), δ ($M_r = 16{,}000$), and ε ($M_r = 11{,}000$). The number of copies of each subunit may vary from one system to another, but the arrangement is, generally, either $\alpha_2\beta_2\gamma_2\delta_2\varepsilon_2$ or $\alpha_3\beta_3\gamma\delta\varepsilon$. The catalytic site resides on the β subunit, while the δ subunit is involved in binding the complex to F_o. A further protein known as the *oligomycin-sensitivity-conferring protein* (OSCP) is also involved in binding F_1 to F_o in mitochondria.

EXAMPLE 14.13

Oligomycin is an antibiotic long known as an inhibitor of respiration in intact mitochondria. Respiration is not inhibited in *uncoupled* mitochondria, i.e., those mitochondria in which O_2 consumption occurs but no ATP is synthesized; the hormone *thyroxin* produces this state, as does *dinitrophenol*. Physical damage to mitochondria during their isolation from cells can cause this effect too. Thus, oligomycin does not block respiratory carriers, in contrast with inhibitors such as rotenone and cyanide. Oligomycin in the presence of OSCP blocks proton translocation to the F_1 component through specific interaction with an F_o subunit in the membrane.

The peptides of the F_o component are highly *hydrophobic*. They form the channel linking the proton gradient to ATP synthesis by F_1 and their importance cannot be overemphasized. Proton movement occurs through the smallest of the three F_o peptides, referred to as a *proteolipid* because of the high proportion of phospholipid normally bound to it. The peptide ($M_r = 5{,}400$) is present in up to six copies per complex. Also present are two other peptides, one of $M_r = 13{,}000$ and the other the OSCP protein.

The proteolipid, now sequenced from a number of sources, forms a *hairpinlike loop* that traverses the inner membrane and has unique hydrophobic amino acid sequences flanking a central, very short, highly charged, hydrophilic segment that interacts with the OSCP or with F_1 component δ. *Dicyclohexylcarbodiimide* (DCCD) inhibits proton translocation through F_o and interacts with a single essential *glutamate carboxyl* in the hydrophobic loop sequence. The loop sequences also include a number of serine and threonine hydroxyls.

The six-times repeating proteolipid unit is analogous in structure to the sevenfold, membrane-traversing polypeptide complex IV of the cytochrome oxidase complex. The latter is also DCCD-sensitive and is involved in proton pumping during the oxidoreduction of cytochrome *a*.

Proton movement through the inner mitochondrial membrane may thus (according to the examples above) occur via a transmembrane hydrogen-bonded network in clustered helices buried inside the membrane.

14.8 THE MECHANISM OF ATP SYNTHESIS

The main tenet of the chemiosmotic hypothesis is that a transmembrane electrochemical gradient acts as the *intermediate* in the transfer of energy to ATP; this energy is available from the difference in redox potential between the $NAD^+/NADH$ couple and the $O_2/2H_2O$ couple in the respiratory chain. Energy coupling occurs in *each* of the three respiratory complexes I, III, and IV, where in each case there is sufficient difference in *midpoint potential* of the donor and acceptor carriers to drive proton transport *against* the electrochemical gradient. Thus, oxidation of NADH results in the production of three ATP molecules per atom of O reduced to water (a P/O ratio of 3). Oxidation of succinate, on the other hand, yields only two ATP molecules, a P/O ratio of 2.

While these regions of the chain were defined by the early crossover experiments as the sites of oxidative phosphorylation, the actual physical site of phosphorylation is now known to be the β subunit of the F_1 component of the ATP synthetase. The electrochemical gradient is the driving force in the reaction:

$$ADP + P_i + \Delta\mu_{H^+} \rightleftharpoons ATP + H_2O$$

The molecular mechanism of ATP synthesis is not yet understood. As with other aspects of the chemiosmotic theory, there continue to be many models proposed for the reaction.

EXAMPLE 14.14

Protonic removal of oxygen from P_i was originally proposed by Mitchell as a mechanism for the above reaction; the mechanism is shown in Fig. 14-9, but is no longer generally accepted.

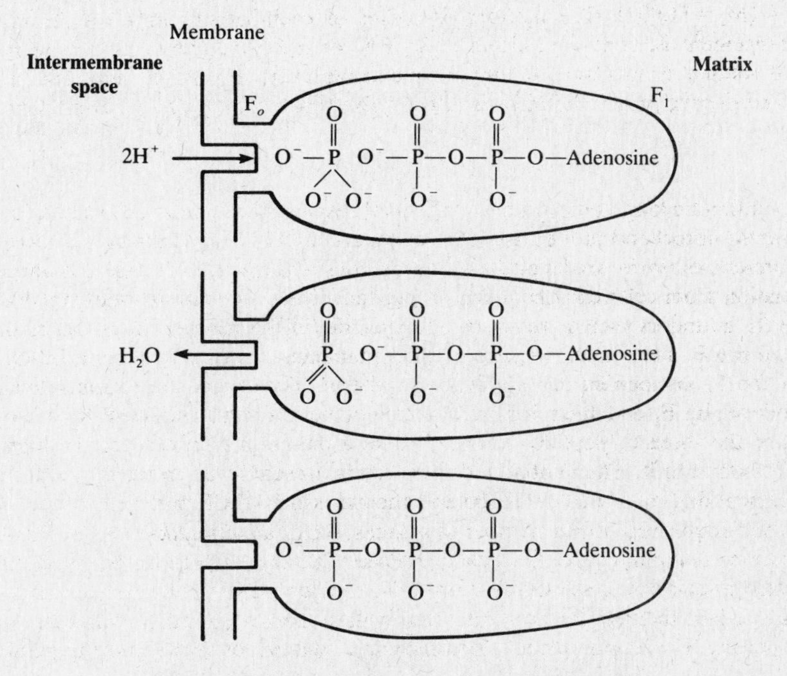

Fig. 14-9 Protonic removal of oxygen from P_i.

The concept of a high-energy phosphorylated intermediate at "coupling sites" was a feature of all earlier concepts of the mechanism of ADP phosphorylation. The *possibility* of such a precursor, engendered by the electrochemical gradient on F_1, has however been eliminated by studies involving the use of ^{18}O-labeled components of the reaction:

$$\text{Adenine—ribose—O—}\overset{\overset{\displaystyle O}{\|}}{\underset{\underset{\displaystyle O^-}{|}}{P}}\text{—O—}\overset{\overset{\displaystyle ^{18}O}{\|}}{\underset{\underset{\displaystyle ^{18}O^-}{|}}{P}}\text{—}^{18}OH \quad + \quad \text{Phosphate} \longrightarrow$$

$$\text{Adenine—ribose—O—}\overset{\overset{\displaystyle O}{\|}}{\underset{\underset{\displaystyle O^-}{|}}{P}}\text{—O—}\overset{\overset{\displaystyle ^{18}O}{\|}}{\underset{\underset{\displaystyle ^{18}O^-}{|}}{P}}\text{—}^{18}O\text{—}\overset{\overset{\displaystyle O}{\|}}{\underset{\underset{\displaystyle O^-}{|}}{P}}\text{—O}^-$$

A possibility of a high-energy derivative of an F_1 component involving an *anhydride of ADP* has been excluded by the demonstration that the oxygen bridge joining the β and γ phosphates of ATP is derived from the ADP-OH and not the P_i-OH. A high-energy derivative of an F_1 component involving an *anhydride of phosphate* has also been excluded as a possibility by the demonstration that exchange of ^{18}O between P_i and water, as indicated below, is enhanced and not reduced by addition of ADP.

$$H_2O + H\overset{18}{O}-\underset{\underset{^{18}OH}{|}}{\overset{\overset{^{18}O}{\|}}{P}}-{}^{18}O^- \rightleftharpoons H_2{}^{18}O + H\overset{18}{O}-\underset{\underset{^{18}O^-}{|}}{\overset{\overset{^{18}OH}{|}}{P}}=O$$

Covalent bond formation during ATP synthesis is thus restricted to the one linking the β and γ phosphates of ATP itself.

The energy of the electrochemical $\Delta\mu_{H^+}$ appears not to be utilized *directly* in the making of this covalent linkage but is used, rather, in the binding of ADP and P_i to F_1 and in the subsequent release of ATP.

EXAMPLE 14.15

Models based on the above concept have been proposed, such as the one shown in Fig. 14-10. Evidence for this model is based on studies of the hydrolysis of ATP labeled in the γ phosphate with ^{18}O. The label is lost from the released P_i to water ($H^{18}OH$) because of the rapid reversibility of reactions 2 and −2. Uncoupling agents do *not* influence this loss, indicating μ_{H^+} is not involved in reactions 2 or −2, nor in reactions −3 or −1.

Energy is thus required in the *binding* of ADP and P_i, which converts the loose (L) binding complex to a tight (T) binding complex. Reversible loss of water (ATP synthesis) or gain of water (ATP hydrolysis) in the tight binding complex may occur without involvement of the μ_{H^+}. Energy is also required in the conversion of the T-bound ATP complex to an L-bound ATP complex, from which ATP is released without additional energy involvement.

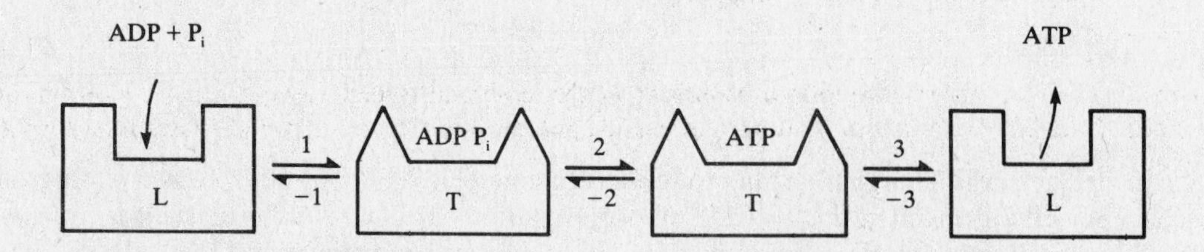

Fig. 14-10 A model for binding of ADP and P_i, with subsequent release of ATP. L = loose binding complex; T = tight binding complex.

As stated in Example 14.12, the catalytic site of F_1 is the β subunit of the complex. A number of amino acid residues essential to the site's reactivity have been identified on the basis of chemical-modification studies. These residues include arginine, lysine, tyrosine, aspartate, and glutamate. Models of the catalytic site have been proposed; conformational changes leading to the formation of the T-bound nucleotide and P_i are envisaged, following μ_{H^+}-dependent protonation of the essential aspartate and glutamate carboxyls.

14.9 TRANSPORT OF ADENINE NUCLEOTIDES TO AND FROM MITOCHONDRIA

Although the mechanism of ATP synthesis is not fully understood, there is no doubt that the ATP is synthesized in the matrix of the mitochondrion. Mitochondrial ATP is then *exported* to the cytoplasm. A specific carrier ($M_r = 40{,}000$) is involved in the simultaneous transport of ATP *out* of and ADP *into* the mitochondria. *Translocation* is inhibited by two well-known toxins, *atractyloside* and *bongkrekic acid*, the former a glucoside found in the Mediterranean thistle and the latter produced by a *Pseudomonas* bacterium.

EXAMPLE 14.16

The translocator moves the two nucleotides in either direction. However, ATP is transported as ATP^{4-} and ADP as ADP^{3-}. Thus the equilibrium position of the exchange is dependent on the electrical potential difference across the membrane. When the potential difference is 160 mV, the ratio of ATP/ADP in the medium outside the mitochondrion is 125/1.

Solved Problems

COMPONENTS OF THE ELECTRON-TRANSPORT CHAIN

14.1. A suspension of mitochondria was incubated under anaerobic conditions in the presence of NADH. Oxidative phosphorylation was initiated by a pulse of oxygen, and it was found that the ATP concentration in the suspension increased by 1.5×10^{-4} mol L^{-1}, with a corresponding decrease in the absorbance of the suspension of 0.30 at 340 nm when using a 1-cm path length. Given that the absorbance coefficient of NADH at 340 nm is 6.2×10^3 L mol^{-1} cm^{-1}, calculate the number of ATP molecules produced per molecule of NADH oxidized.

SOLUTION

The Beer-Lambert equation (Chap. 3) is

$$A = \varepsilon C l$$

where A is the absorbance of the solution; ε, the absorbance coefficient; C, the concentration in mol L^{-1}; and l, the length of light path in cm. By rearranging the equation and substituting the values given above, we get

$$\Delta C = \frac{\Delta A}{\varepsilon l}$$

$$\Delta[\text{NADH}] = \frac{\Delta A_{340\text{ nm}}}{\varepsilon_{340\text{ nm}} \times 1\text{ cm}}$$

$$= \frac{0.30}{6.2 \times 10^3\text{ L mol}^{-1}\text{ cm}^{-1} \times 1\text{ cm}} = 4.84 \times 10^{-5}\text{ mol L}^{-1}$$

Comparison of this value with the change in [ATP] (1.5×10^{-4} mol L^{-1} $\div$ 4.84×10^{-5} mol L^{-1}) gives a value of 3.1 ATP produced per NADH oxidized; this is close to the theoretically predicted value of 3.0.

14.2. Muscle tissue contains two glycerol 3-phosphate dehydrogenases: a cytosolic enzyme, which uses NADH, and a flavin nucleotide–dependent mitochondrial enzyme. What is the metabolic significance of these two enzymes?

SOLUTION

These two enzymes provide a *shuttle mechanism* for transport of reducing equivalents of NADH (generated during glycolysis in the cytosol) into the mitochondria. The cytosolic enzyme catalyzes the following reaction:

$$\text{Dihydroxyacetone phosphate} + \text{NADH} + \text{H}^+ \rightleftharpoons \text{NAD}^+ + \text{Glycerol 3-phosphate}$$

The glycerol 3-phosphate traverses the mitochondrial membrane and is then oxidized back to dihydroxyacetone phosphate by the mitochondrial enzyme:

$$\text{Glycerol 3-phosphate} + \text{FAD} \rightleftharpoons \text{Dihydroxyacetone phosphate} + \text{FADH}_2$$

The $FADH_2$ enters the electron-transport chain at coenzyme Q, while the dihydroxyacetone phosphate can return to the cytoplasm. Although this shuttle is generally *inefficient*, in the sense that only two ATP molecules are produced per $FADH_2$ molecule oxidized, compared with three for NADH oxidation, it provides a mechanism for regeneration of NAD^+ in the cytosol. The presence of cytosolic NAD^+ is essential for continued glycolytic breakdown of glucose (see Chap. 11).

14.3. How can the two types of sulfur atoms in an iron-sulfur protein be distinguished experimentally?

SOLUTION

The sulfur atoms linking the Fe atoms can be removed from the iron-sulfur protein at low pH; they are *acid-labile*, while the sulfur atoms derived from cysteine residues are *not*.

14.4. In the succinate dehydrogenase–catalyzed reaction, why is the appropriate electron acceptor FAD rather than NAD^+, which is used in the other redox reactions of the citric acid cycle (Chap. 12)?

SOLUTION

For FAD-mediated oxidation of succinate, the half-reactions are:

$$\begin{aligned} \text{FAD} + 2\text{H}^+ + 2\text{e}^- &\longrightarrow \text{FADH}_2 \qquad E^{\circ\prime} = 0.05\ \text{V} \\ \text{Succinate} &\longrightarrow \text{Fumarate} + 2\text{H}^+ + 2\text{e}^- \qquad E^{\circ\prime} = -0.03\ \text{V} \end{aligned}$$

Hence, for the overall reaction

$$\text{Succinate} + \text{FAD} \longrightarrow \text{Fumarate} + \text{FADH}_2$$

$E^{\circ\prime}$ is 0.02 V. The value of $\Delta G^{\circ\prime}$ can be calculated as follows:

$$\begin{aligned} \Delta G &= -nFE^{\circ\prime} \\ &= -2 \times 96.5 \times 0.02 = -3.86\ \text{kJ mol}^{-1} \end{aligned}$$

For NAD^+-mediated oxidation of succinate, the nucleotide $E^{\circ\prime}$ value is -0.32 V, giving an $E^{\circ\prime}$ value for the whole reaction of -0.35 V. Hence the $\Delta G^{\circ\prime}$ for NAD^+-mediated oxidation of succinate is unfavorable, being $+67.6\ \text{kJ mol}^{-1}$. Recalling that

$$\Delta G^{\circ\prime} = -RT \ln K$$

the equilibrium constant for FAD-mediated oxidation of succinate is $\sim 10^{12}$-fold more favorable than that for the NAD^+-mediated oxidation.

COUPLING OF ELECTRON TRANSPORT AND ATP SYNTHESIS

14.5. *Uncoupling agents* are compounds that prevent ATP synthesis in mitochondria but allow electron transport to proceed. They generally act by increasing the permeability of the inner mitochondrial membrane to H^+, thus preventing generation of the H^+ gradient. A widely used uncoupling agent is 2,4-dinitrophenol. How could this compound increase the permeability of the inner mitochondrial membrane to H^+?

SOLUTION

At physiological pH, 2,4-dinitrophenol exists predominately as the anion, $C_6H_4(NO_2)_2O^-$. The membrane is permeable both to this anion and to the protonated form, $C_6H_4(NO)_2OH$. The latter form can carry protons across the membrane and return in the anionic form to be reloaded with a proton. 2,4-Dinitrophenol can thus dissipate the H^+ gradient.

14.6. Addition of DCCD (dicyclohexylcarbodiimide) to mitochondrial preparations *decreases* the rates of both ATP synthesis and electron transport. Only the latter process can be restored to normal levels upon addition of 2,4-dinitrophenol. How can these observations be explained?

SOLUTION

DCCD inhibits proton translocation through the F_o subunit of the ATP synthetase. Thus, the value of $\Delta\mu_{H^+}$ increases to a point where proton translocation, and hence electron transport, becomes thermodynamically unfavorable. In addition, DCCD inactivates the ATP synthesis function of the ATP synthetase. The uncoupler, 2,4-dinitrophenol, renders the inner mitochondrial membrane permeable to protons, leading to a decrease in the value of $\Delta\mu_{H^+}$ and restoration of electron transport. However, 2,4-dinitrophenol cannot restore the activity of the DCCD-treated ATP synthetase.

14.7. Most mitochondria contain an active H^+/phosphate symport protein containing a sulfhydryl group essential for its translocator activity. (*a*) How could this translocator complicate measurements of the H^+/O ratio (as in Example 14.9), and (*b*) how could the presence of the sulfhydryl group be used to obviate these complications?

SOLUTION

(*a*) The protons extruded from the mitochondria may be transported, together with phosphate, back into the mitochondria via the H^+/phosphate translocator. This would lead to an underestimate of the number of protons extruded and, hence, an underestimation of the H^+/O ratio.

(*b*) Because the H^+/phosphate translocator has an essential sulfhydryl group, treatment of mitochondrial preparations with *group-selective reagents*, such as N-*ethylmaleimide*, that react with sulfhydryl groups will lead to selective inactivation of the H^+/phosphate translocator. Hence, reentry of protons mediated by this translocator will not occur. Experimentally, higher H^+/O ratios are obtained from mitochondria treated with *N*-ethylmaleimide than from untreated mitochondria.

14.8. *Brown fat* is a form of adipose tissue found in the backs of many young animals. Mitochondria from this tissue have a P/O ratio of less than 1 for ATP synthesis arising from oxidation of NADH. What may be the physiological function of brown fat tissue?

SOLUTION

From the above P/O ratio, it is clear that ATP synthesis in brown fat mitochondria is *naturally uncoupled* from electron transport. Hence, protons extruded from the mitochondria during electron transport must reenter without concomitant ATP synthesis. The energy released as heat during this reentry may help to keep the young animals warm. Such small organisms have a high *surface-to-volume ratio* and therefore readily lose heat through convective and radiative processes.

14.9. In an experiment, purified F_o protein was incorporated into synthetic phospholipid vesicles. When these vesicles were preloaded with K^+ and then valinomycin was added to the suspension, they were able to *take up* protons. No H^+ uptake was seen with control phospholipid vesicles containing no F_o protein. What might be the basis of these observations?

SOLUTION

The addition of *valinomycin*, a K^+ ionophore, leads to generation of a *diffusion potential* through efflux of K^+ down a concentration gradient. The F_o protein permits uptake of H^+ by the vesicles (which are impermeable to H^+ in the absence of F_o) in response to this diffusion-generated gradient.

14.10. In an experiment, a suspension of mitochondria provided with adequate oxygen and pyruvate, but no ADP, consumed oxygen at a very low rate. The relative states of reduction of components of the electron-transport chain were determined: NAD, 100 percent; coenzyme Q, 40 percent; cytochrome *b*, 38 percent; cytochrome *c*, 14 percent; and cytochrome *a*, 0 percent. How could these data allow definition of the sites of oxidative phosphorylation?

SOLUTION

The *absence* of ADP is acting, in effect, as an *inhibitor* of electron transport, for reasons discussed in Prob. 14.6. Hence, by application of the *crossover theorem* (Chap. 10), there are large differences in the reduction sites of the electron-transport-chain components between NAD and coenzyme Q, between cytochrome *b* and cytochrome *c*, and between cytochrome *c* and cytochrome *a*. Therefore, the absence of ADP must be inhibiting electron transport at these points; in fact, these are the sites of proton extrusion leading to ATP synthesis during electron transport.

Supplementary Problems

14.11. How many molecules of ATP are produced per molecule of (*a*) pyruvate, (*b*) NADH, (*c*) glucose, and (*d*) phosphoenolpyruvate, by a cell homogenate in which glycolysis, the citric acid cycle, and oxidative phosphorylation are all completely active?

14.12. What effect would 2,4-dinitrophenol have on the P/O ratio for ATP synthesis using NADH as an electron donor?

14.13. Given that the internal pH of a mitochondrion is 7.8, and assuming that a mitochondrion is a sphere of diameter 1.4×10^{-6} m, calculate the number of protons inside the mitochondrion.

14.14. Why should the ingestion of an uncoupler, such as 2,4-dinitrophenol, lead to sweating, an increase in body temperature, and, in the long term (weeks), weight loss? (Note: this is an extremely dangerous method of weight control.)

14.15. How is it possible to distinguish between the different cytochromes in the electron-transport chain?

14.16. Fe^{3+} in iron-sulfur proteins has an electron spin resonance signal, while Fe^{2+} does not. Assume that you have a preparation of mitochondria able to synthesize ATP via oxidation of NADH; supplies of rotenone, antimycin A, and KCN; and access to an esr spectrometer. How could you establish which of the complexes of the electron-transport chain contain iron-sulfur proteins?

14.17. Calculate $\Delta G^{\circ\prime}$ for electron flow from reduced cytochrome *c* to oxygen.

14.18. Nitrite is an effective antidote to cyanide poisoning. Why? (*Hint*: Nitrite is an oxidizing agent that is able to convert Fe^{2+} to Fe^{3+} in heme groups.)

14.19. Addition of oxygen to cells metabolizing glucose under anaerobic conditions leads to (*a*) a decrease in the rate of glucose consumption and (*b*) cessation of lactate accumulation. The latter phenomenon is known as the *Pasteur effect*. Explain why these changes occur in glucose and lactate metabolism.

Chapter 15

Nitrogen Metabolism

15.1 SYNTHESIS AND DIETARY SOURCES OF AMINO ACIDS

Animals are dependent for growth on a source of fixed nitrogen from other animals or plants; plants in turn are dependent on bacteria for fixing nitrogen. Humans need fixed nitrogen, which must come from the diet (normally as protein), particularly for protein and nucleic acid synthesis but also for synthesizing many specialized metabolites such as porphyrins and phospholipids.

The *amount* of protein (or fixed nitrogen) we ingest determines the state of *nitrogen balance*. Humans, like other animals, will excrete nitrogenous compounds even when they are fed on a protein-free diet, because not all nitrogenous compounds can be recycled. They are then in *negative nitrogen balance*. The amount of protein an adult needs in order to stay in nitrogen balance is not easy to define, because not all amino acids found in proteins, particularly plant proteins, are equally important for animal metabolism.

EXAMPLE 15.1

Why are plant proteins not as useful as animal proteins for dietary purposes?

Cereal proteins are only about 70 percent efficient for replacement purposes. The reason is that cereal proteins are deficient in lysine, an *essential amino acid* for humans (see "Amino Acid Synthesis" in this section). Thus a diet based on one source of protein (e.g., corn) can lead to malnutrition. A partial solution to the problem has been the breeding of high-lysine corn. Other plant proteins, particularly those from pod seeds (e.g., peas and beans) are deficient in the sulfur-containing amino acids. A successful vegetarian diet will be balanced in cereals and pod seeds.

The accepted amount of protein required to maintain nitrogen balance is 28 g per day for a 70-kg man, i.e., about 3.8 g of nitrogen. This is estimated by measuring the N excretion over 6 to 7 days on a protein-free diet. If the protein source is from cereal, then the daily intake would have to be increased to about 40 g per day for a 70-kg man. The difference is due to the variable amounts of essential amino acids found in proteins. The amount required by growing children is larger; the accepted figure is about 0.6 g per kilogram per day.

Question: Is too much protein in the diet harmful?

Eskimos, who have a high protein intake, do have a shorter life span than Europeans, but there is no clear correlation of this with dietary protein.

Nitrogen Fixation

The fixation of nitrogen is the most fundamental biochemical process after photosynthesis. It is the process whereby *atmospheric nitrogen* is reduced to *ammonia*. Nitrogen fixation can be carried out by blue-green algae, some yeasts, and especially bacteria. The reduction of nitrogen

$$N_2 + 3H_2 \longrightarrow 2NH_3 \qquad \Delta G^{\circ\prime} = -33.5 \text{ kJ mol}^{-1}$$

is an *exergonic* reaction. Due to the chemical unreactivity of N_2, this process is accomplished industrially by using efficient catalysts, high temperatures (600°C) and pressures (1,000 atm). The biological process occurs at 1 atm and ~25°C. In bacterial systems, the reaction is catalyzed by the enzyme *nitrogenase*.

EXAMPLE 15.2

What is the source of hydrogen and how is the chemical reactivity of N_2 handled in biological systems?

Protons are used as a source of hydrogen and the hydrolysis of ATP is presumed to overcome the chemical unreactivity of N_2 by assisting in the formation of thermodynamically unfavored intermediates. Eight electrons are required for the reduction of nitrogen, which is always accompanied by the evolution of H_2.

$$N_2 + 8e^- + 8H^+ \xrightarrow{16(ATP + H_2O) \quad 16(ADP + P_i)} 2NH_3 + H_2$$

The electrons can be supplied by several donors, including NADH, flavoproteins, or NADPH.

Ultimately all higher organisms are dependent on bacterially produced ammonia for their nitrogen metabolism.

Question: How do higher organisms obtain ammonia?

Many plants, particularly legumes (peas and beans), have a symbiotic relationship with nitrogen-fixing bacteria, which live in special nodules on the roots. There are ~13,000 species of leguminous plants, all of which have symbiotic bacteria of the genus *Rhizobium*. Some insects (termites and cockroaches) also have symbiotic nitrogen-fixing bacteria in their intestines.

Assimilation of Ammonia

Ammonia is normally condensed with 2-oxoglutarate and thus converted to glutamate via the enzyme *glutamate dehydrogenase*; this enzyme is of highest activity in the liver and kidney. Glutamate is produced from 2-oxoglutarate and ammonia as follows:

$$NH_4^+ + \text{2-Oxoglutarate}^{2-} + NADPH + H^+ \rightleftharpoons \text{Glutamate}^- + NADP^+ + H_2O$$

Glutamate dehydrogenase can also use NADH. This reaction is freely reversible: the direction of *net flux* through the reaction is determined solely by the relative concentrations of the reactants. Thus, this reaction has two equally important functions: the assimilation of ammonia or the removal of ammonia from metabolites.

Glutamate is also produced in some bacteria via reactions catalyzed by the enzymes *glutamine synthetase* and *glutamate synthetase* acting together. Glutamine synthetase, as the name implies, catalyzes the synthesis of glutamine in virtually all organisms. In humans, it is particularly active in the liver; glutamine is transported from the liver to other tissues in the blood plasma.

$$NH_3 + \begin{array}{c} COO^- \\ | \\ CH_2 \\ | \\ CH_2 \\ | \\ HC{-}NH_3^+ \\ | \\ COO^- \end{array} \xrightarrow{ATP \quad ADP + P_i} \begin{array}{c} CONH_2 \\ | \\ CH_2 \\ | \\ CH_2 \\ | \\ HC{-}NH_3^+ \\ | \\ COO^- \end{array}$$

Glutamate Glutamine

Glutamate synthetase, which is not present in humans but is found in bacteria, catalyzes the formation of glutamate:

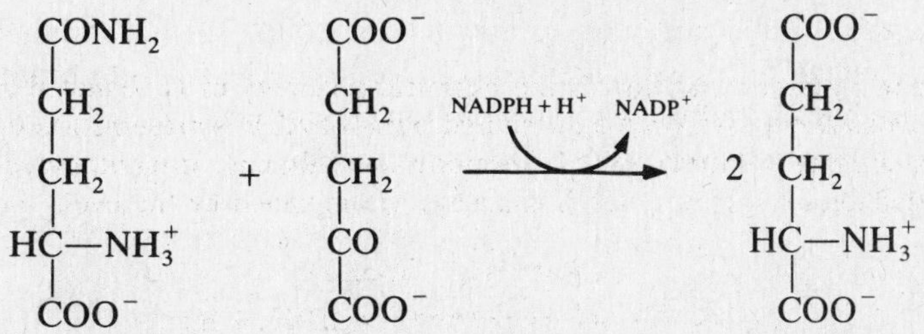

$$\begin{array}{c} CONH_2 \\ | \\ CH_2 \\ | \\ CH_2 \\ | \\ HC{-}NH_3^+ \\ | \\ COO^- \end{array} + \begin{array}{c} COO^- \\ | \\ CH_2 \\ | \\ CH_2 \\ | \\ CO \\ | \\ COO^- \end{array} \xrightarrow{NADPH + H^+ \quad NADP^+} 2\ \begin{array}{c} COO^- \\ | \\ CH_2 \\ | \\ CH_2 \\ | \\ HC{-}NH_3^+ \\ | \\ COO^- \end{array}$$

Glutamine 2-Oxoglutarate Glutamate

This coupled enzyme system is used by the blue-green algae and by *Rhizobium*.

The amide group of glutamine provides the ammonia for the synthesis of many N-containing compounds, e.g., purines and pyrimidines (Sec. 15.6).

Glutamate provides the amino group for the synthesis of many other amino acids through *transamination* reactions. This is true in all cells. These amino acids are then used for protein synthesis and other aspects of nitrogen metabolism. The majority of animals are dependent on plant or animal proteins for fixed nitrogen for their nitrogen metabolism.

Transamination

Transamination is the process whereby ammonia is reversibly transferred between amino acids and 2-oxoacids. The reaction is catalyzed by *aminotransferases*. These enzymes bind *pyridoxal phosphate* as a coenzyme. Pyridoxal phosphate and *pyridoxamine phosphate* are the coenzyme forms of vitamin B_6 (Fig. 15-1).

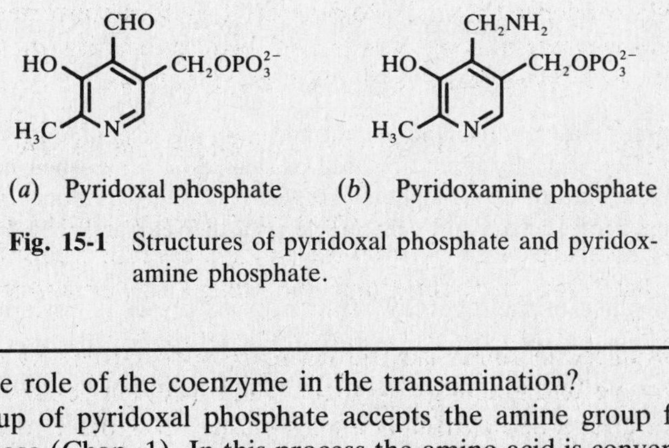

(*a*) Pyridoxal phosphate (*b*) Pyridoxamine phosphate

Fig. 15-1 Structures of pyridoxal phosphate and pyridoxamine phosphate.

Question: What is the role of the coenzyme in the transamination?

The aldehyde group of pyridoxal phosphate accepts the amine group from an amino acid by formation of a Schiff base (Chap. 1). In this process the amino acid is converted to a 2-oxoacid, and pyridoxal phosphate is converted to pyridoxamine phosphate. The amine group on pyridoxamine phosphate can now be transferred to another 2-oxoacid, converting it to an amino acid. In this second reaction, the pyridoxamine phosphate is reconverted to pyridoxal phosphate.

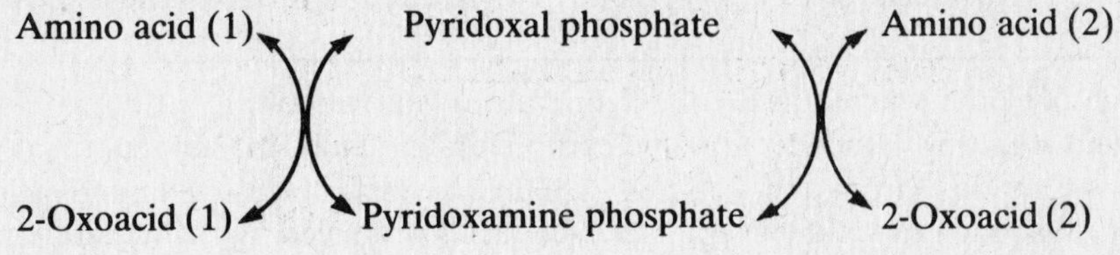

The overall reaction is

$$\text{Amino acid (1)} + \text{2-Oxoacid (2)} \rightleftharpoons \text{2-Oxoacid (1)} + \text{Amino acid (2)}$$

2-Oxoglutarate is the normal acceptor of the amine group. In the aminotransferase reaction, 2-oxoglutarate is transaminated to give glutamate. There are at least 13 different *aminotransferases*, but their specificities are not all known. The most important are (*a*) *aspartate aminotransferase*, which has a high affinity for the compounds in the following reaction:

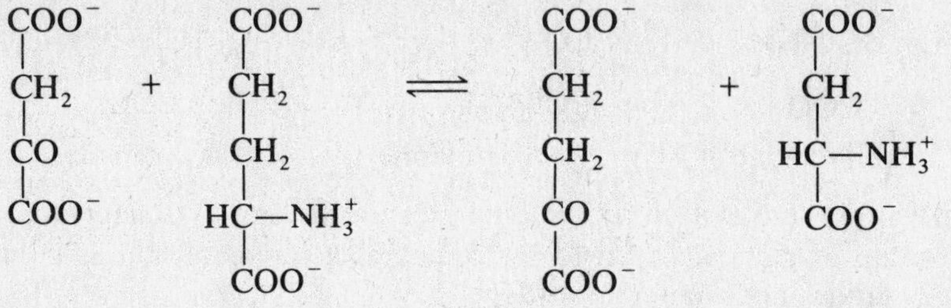

$$
\begin{array}{ccccccc}
\begin{array}{c} COO^- \\ CH_2 \\ CO \\ COO^- \\ \\ \end{array} & + & \begin{array}{c} COO^- \\ CH_2 \\ CH_2 \\ HC{-}NH_3^+ \\ COO^- \end{array} & \rightleftharpoons & \begin{array}{c} COO^- \\ CH_2 \\ CH_2 \\ CO \\ COO^- \end{array} & + & \begin{array}{c} COO^- \\ CH_2 \\ HC{-}NH_3^+ \\ COO^- \\ \\ \end{array} \\
\text{Oxaloacetate} & & \text{Glutamate} & & \text{2-Oxoglutarate} & & \text{Aspartate}
\end{array}
$$

and (*b*) *alanine aminotransferase*, which selectively catalyzes the following reaction:

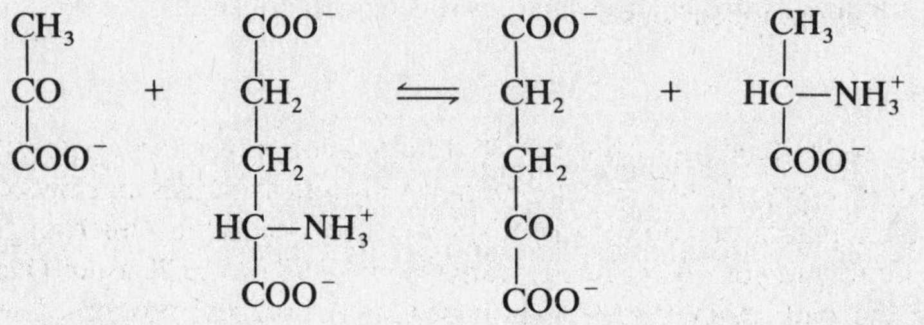

$$
\begin{array}{ccccccc}
\begin{array}{c} CH_3 \\ CO \\ COO^- \\ \\ \\ \end{array} & + & \begin{array}{c} COO^- \\ CH_2 \\ CH_2 \\ HC{-}NH_3^+ \\ COO^- \end{array} & \rightleftharpoons & \begin{array}{c} COO^- \\ CH_2 \\ CH_2 \\ CO \\ COO^- \end{array} & + & \begin{array}{c} CH_3 \\ HC{-}NH_3^+ \\ COO^- \\ \\ \\ \end{array} \\
\text{Pyruvate} & & \text{Glutamate} & & \text{2-Oxoglutarate} & & \text{Alanine}
\end{array}
$$

EXAMPLE 15.3

Both aspartate aminotransferase and alanine aminotransferase are released into the blood after damage to tissues or after cell death. Consequently, they are used as diagnostic tools when heart or liver damage has occurred, such as after a heart attack or in hepatitis, respectively. Other enzymes are also released into the blood at such times. For example, damage to heart muscle is further characterized by the presence of *creatine kinase* in the plasma.

The metabolic significance of the diversity of aminotransferases is not always understood. Thus, although specific aminotransferases exist for histidine, serine, phenylalanine, and methionine, none of the carbon backbones of these amino acids is metabolized to any significant extent in vivo.

The highest concentration of the aminotransferases is in the cytoplasm, but they are also located in mitochondria, where *glutamate dehydrogenase* is located exclusively. The aminotransferases and glutamate dehydrogenase catalyze central reactions in amino acid metabolism. The major aminotransferases and glutamate dehydrogenase are present in all tissues in relatively high concentrations compared with other enzymes, such as those involved in glycolysis. The reversible nature of both reactions allows a rapid exchange of amino groups and formation of 2-oxoacids, shown in Fig. 15-2.

Question: How does the scheme in Fig. 15-2 operate in starvation?

In starvation (or fasting), both protein and carbohydrate may be in short supply. The *net* effect is that endogenous proteins (from the muscles) are hydrolyzed, releasing amino acids for protein synthesis and for oxidation to yield energy. The 2-oxoacids produced by the aminotransferases either enter gluconeogenesis (Chap. 11) or undergo respiration (Chap. 12) and are metabolized to CO_2 and H_2O. The amount of stored lipid and glycogen determines the extent of the net degradation of endogenous protein. Glutamate dehydrogenase, again, catalyzes the formation of ammonia from amino groups as the amino acids are broken down.

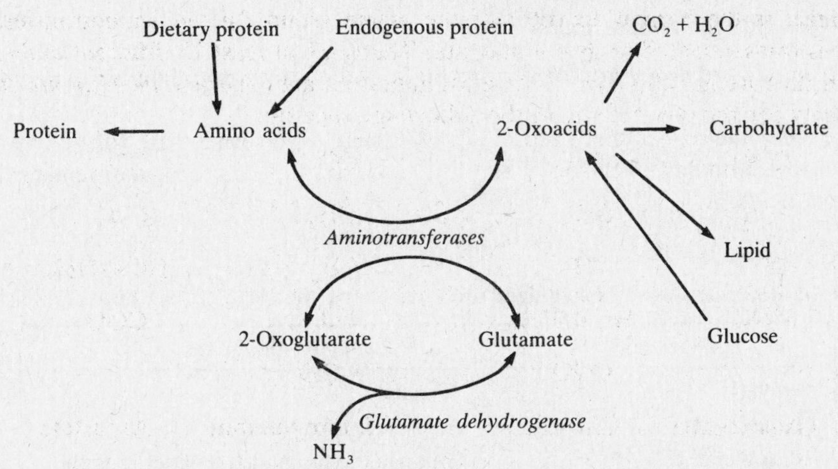

Fig. 15-2 Central role of the *aminotransferases* and *glutamate dehydrogenase* in nitrogen metabolism.

Amino Acid Synthesis

RNA base triplet codes (codons) exist for 20 amino acids in protein synthesis (Chap. 17). The ability of an organism to live and grow is dependent on protein synthesis and hence on a supply of all 20 amino acids. Higher plants are able to synthesize all 20, but many microorganisms and higher animals make considerably fewer. Specifically, humans make 10 of the 20 amino acids; the remainder must be supplied in the diet, usually in the form of plant or animal protein. The amino acids that humans cannot synthesize de novo but are essential for life are termed the *essential* amino acids. Those that we can synthesize are called the *nonessential* amino acids. The essential and nonessential amino acids are listed in Table 15.1.

Question: On what is the synthesis of the nonessential amino acids dependent?

The synthesis depends on the availability of the appropriate *carbon skeletons* and a source of *ammonia*. Glucose is ultimately the source of the carbon skeleton for most of the nonessential amino acids. Two of the essential amino acids are used to form nonessential amino acids. These are *phenylalanine* and *methionine*, which are used to synthesize *tyrosine* and *cysteine*. Since ammonia is available in the fed state, amino acids become essential to our diet when we are not able to synthesize their carbon skeletons.

Table 15.1. Nonessential and Essential Amino Acids for Humans

Nonessential	Essential
Glutamate	Isoleucine
Glutamine	Leucine
Proline	Lysine
Aspartate	Methionine
Asparagine	Phenylalanine
Alanine	Threonine
Glycine	Tryptophan
Serine	Valine
Tyrosine	Arginine*
Cysteine	Histidine

*Essential only in infants and children.

Synthesis of 2-Oxoacids for Amino Acids

Certain 2-oxoacids are necessary for the synthesis of the nonessential amino acids; they are listed in Table 15.2.

Four of the amino acids—alanine, aspartate, glutamate, and serine—are formed by the transamination of their corresponding oxoacids. The other nonessential amino acids are then derived from these four amino acids.

Table 15.2. 2-Oxoacids Required for Synthesis of the Nonessential Amino Acids

2-Oxoacid	Amino Acids
Pyruvate	Alanine
Oxaloacetate	Aspartate, asparagine
2-Oxoglutarate	Glutamate, glutamine, proline, arginine*
Pyruvate, 3-hydroxypyruvate	Serine

*Essential only in infants and children.

Arginine

Arginine is synthesized from aspartate and ornithine during urea formation. *Argininosuccinate synthetase* and *lyase* catalyze the condensation and cleavage reactions, respectively, that result in the formation of arginine (Sec. 15.5).

Glutamine

Glutamine is synthesized via glutamine synthetase, as has already been mentioned earlier in Sec. 15.1.

Asparagine

Asparagine derives its amide group from glutamine; it is formed from aspartate and glutamine in a reaction catalyzed by *asparagine synthetase*:

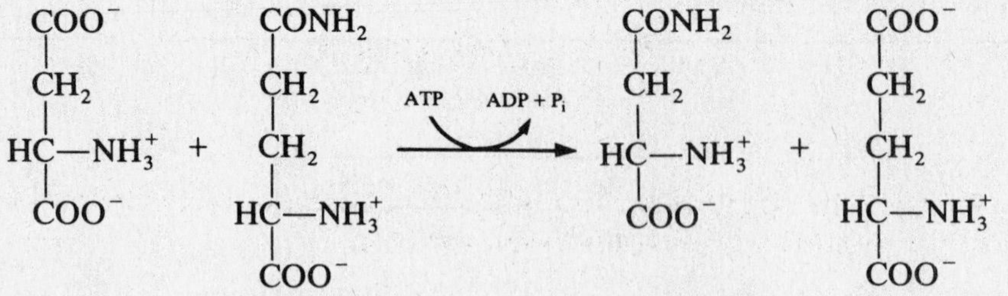

$$\underset{\text{Aspartate}}{\begin{array}{c}COO^-\\|\\CH_2\\|\\HC{-}NH_3^+\\|\\COO^-\end{array}} + \underset{\text{Glutamine}}{\begin{array}{c}CONH_2\\|\\CH_2\\|\\CH_2\\|\\HC{-}NH_3^+\\|\\COO^-\end{array}} \xrightarrow{\text{ATP} \quad \text{ADP} + P_i} \underset{\text{Asparagine}}{\begin{array}{c}CONH_2\\|\\CH_2\\|\\HC{-}NH_3^+\\|\\COO^-\end{array}} + \underset{\text{Glutamate}}{\begin{array}{c}COO^-\\|\\CH_2\\|\\CH_2\\|\\HC{-}NH_3^+\\|\\COO^-\end{array}}$$

Proline

Proline is synthesized from glutamate (Fig. 15-3).

Alanine

Alanine and *serine* are two amino acids that are formed directly from glycolytic intermediates. Alanine is formed from pyruvate via *alanine aminotransferase*.

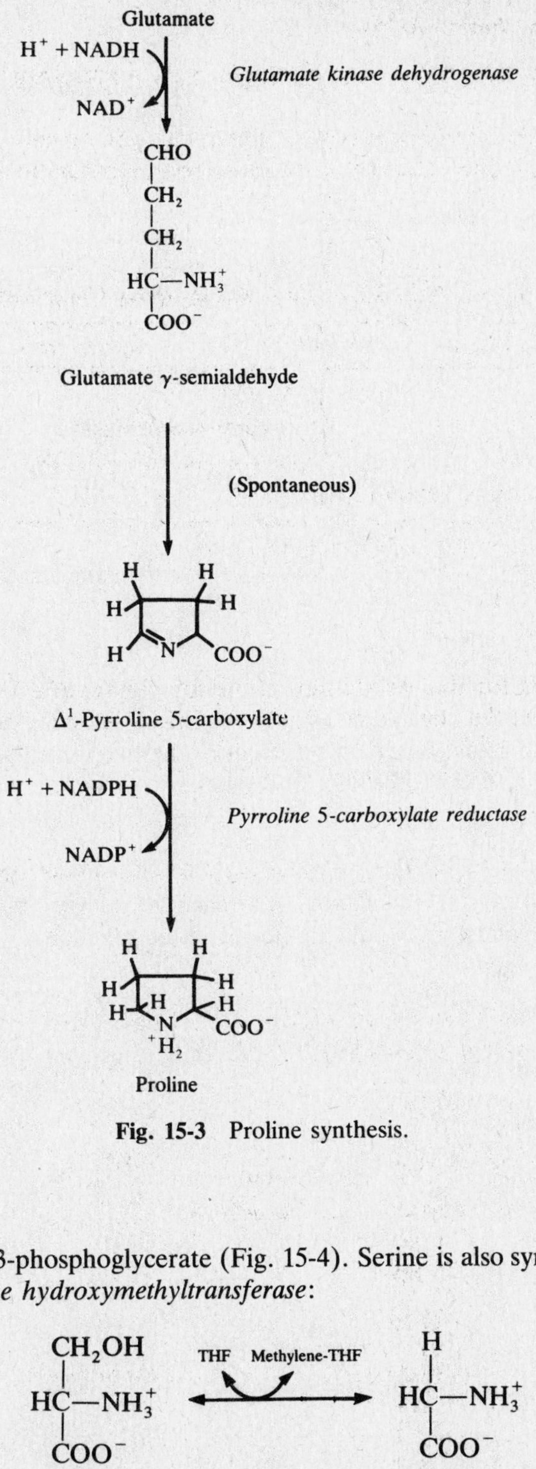

Glutamate

$H^+ + NADH$ → NAD^+ *Glutamate kinase dehydrogenase*

$$\begin{array}{c} CHO \\ | \\ CH_2 \\ | \\ CH_2 \\ | \\ HC—NH_3^+ \\ | \\ COO^- \end{array}$$

Glutamate γ-semialdehyde

(Spontaneous)

Δ^1-Pyrroline 5-carboxylate

$H^+ + NADPH$ → $NADP^+$ *Pyrroline 5-carboxylate reductase*

Proline

Fig. 15-3 Proline synthesis.

Serine

Serine is formed from 3-phosphoglycerate (Fig. 15-4). Serine is also synthesized from glycine in a reaction catalyzed by *serine hydroxymethyltransferase*:

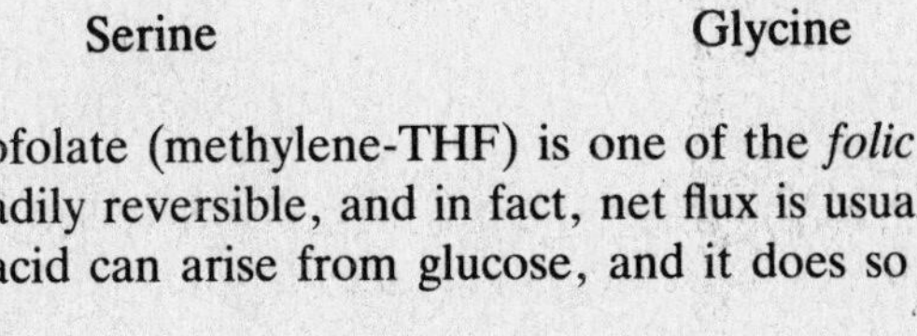

$$\begin{array}{c} CH_2OH \\ | \\ HC—NH_3^+ \\ | \\ COO^- \end{array} \xrightleftharpoons[]{\text{THF} \quad \text{Methylene-THF}} \begin{array}{c} H \\ | \\ HC—NH_3^+ \\ | \\ COO^- \end{array}$$

Serine　　　　Glycine

N^5,N^{10}-Methylenetetrahydrofolate (methylene-THF) is one of the *folic acid* coenzymes (Sec. 15.7). Note that this reaction is readily reversible, and in fact, net flux is usually in the direction of *glycine* synthesis. Thus this amino acid can arise from glucose, and it does so via serine.

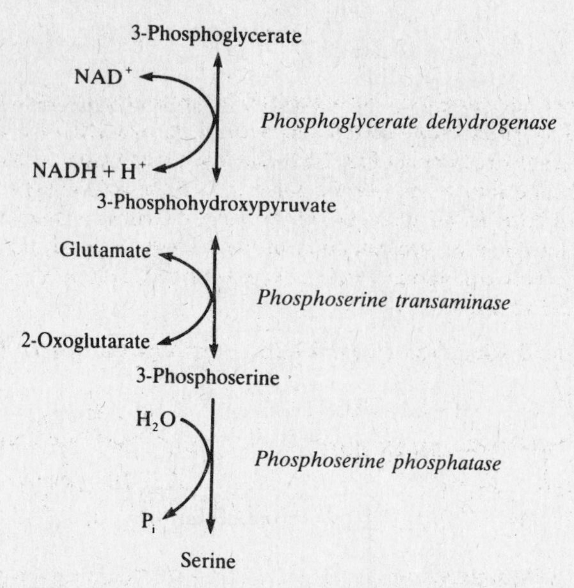

Fig. 15-4 Major pathway for serine synthesis.

Synthesis of Tyrosine and Cysteine

Two of the nonessential amino acids, tyrosine and cysteine, are derived from essential amino acids and may be considered to be breakdown products, as they are intermediates in the normal degradation of these amino acids. Provided sufficient of the two essential amino acids phenylalanine and methionine is available through the diet, *net synthesis* of tyrosine and cysteine can occur.

Tyrosine

Tyrosine is synthesized from phenylalanine in a reaction catalyzed by *phenylalanine hydroxylase*, which has two enzyme activities. The reducing power in the reaction comes from NADPH, and the oxygen from molecular oxygen.

$NADP^+$ ← $H^+ + NADPH$; Dihydrobiopterin → Tetrahydrobiopterin

Tetrahydrobiopterin + O_2 + Phenylalanine → Dihydrobiopterin + H_2O + Tyrosine

The overall reaction is

$$\text{Phenylalanine} + O_2 \xrightarrow{NADPH + H^+ \;\; NADP^+} \text{Tyrosine} + H_2O$$

Phenylalanine: $C_6H_5-CH_2-HC(NH_3^+)-COO^-$

Tyrosine: $HO-C_6H_4-CH_2-HC(NH_3^+)-COO^-$

EXAMPLE 15.4

The first enzyme activity (*dihydrobiopterin reductase*) catalyzes the transfer of hydrogen to *dihydrobiopterin*, which is thus reduced to *tetrahydrobiopterin*. The second enzyme activity is a *hydroxylase* containing two Fe^{3+} atoms, and this catalyzes the reduction of O_2 such that one oxygen atom is incorporated into phenylalanine to form tyrosine and the second into water. At the same time tetrahydrobiopterin is oxidized to dihydrobiopterin. Phenylalanine hydroxylase is an example of a *mixed-function oxidase*. An inherited deficiency of phenylalanine hydroxylase results in an accumulation of phenylalanine that is not converted to tyrosine but is excreted as *phenylpyruvate*. This condition, which affects young infants, is known as *phenylketonuria* and is associated with severe mental retardation.

Biopterin, like folic acid (Sec. 15.7, Fig. 15-26), contains the pterin ring (Fig. 15-5).

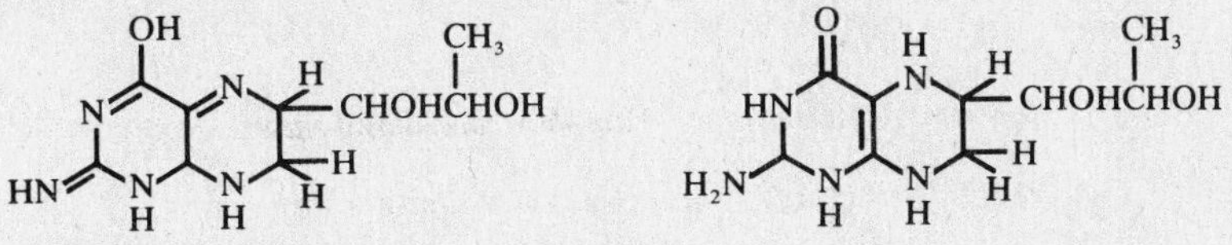

(*a*) Dihydrobiopterin (*b*) Tetrahydrobiopterin

Fig. 15-5 Structures of biopterin derivatives.

Cysteine

Cysteine is synthesized in several steps from methionine. In summary, the S is from methionine, and the carbon skeleton is derived from serine. The pathway is shown in Fig. 15-6.

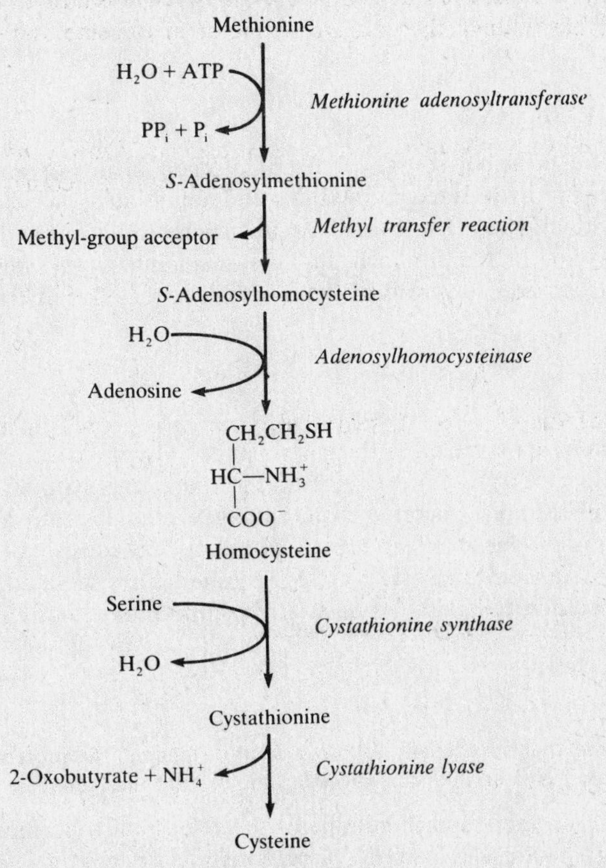

Fig. 15-6 Degradation of methionine and synthesis of cysteine.

Question: What is *S*-adenosylmethionine (Fig. 15-6)?

It is an activated form of methionine (Fig. 15-7) involved in cysteine synthesis and in biological methylations (Sec. 15.7).

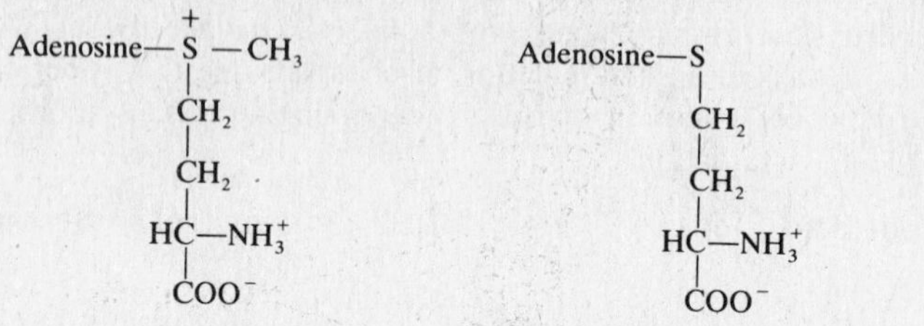

$$\text{Adenosine}-\overset{+}{S}-CH_3 \quad | \quad CH_2 - CH_2 - HC(NH_3^+) - COO^-$$

(*a*) *S*-Adenosylmethionine

$$\text{Adenosine}-S \quad | \quad CH_2 - CH_2 - HC(NH_3^+) - COO^-$$

(*b*) *S*-Adenosylhomocysteine

Fig. 15-7 Structures of *S*-adenosylmethionine and *S*-adenosylhomocysteine.

The synthesis of *S*-adenosylmethionine (SAM) is the only biochemical reaction in which all three phosphates from ATP are hydrolyzed in a single reaction.

EXAMPLE 15.5

SAM owes its activity to the highly reactive positively charged sulphonium ion, which makes the adjacent methyl group electron deficient and susceptible to nucleophilic attack. The sulphonium ion is more stable than the oxonium ion and is highly reactive. The formation of the more stable *S*-adenosylhomocysteine (Fig. 15-7) is the driving force in biological methylations.

15.2 DIGESTION OF PROTEINS

Dietary protein is the principal source of fixed nitrogen in higher animals. Digestion is the process whereby proteins are hydrolyzed to peptides and amino acids, which are absorbed from the lumen of the gastrointestinal tract. The hydrolysis is enzyme-catalyzed and is brought about by a series of hydrolytic enzymes in the stomach and the small intestine. These enzymes are known collectively as *proteolytic enzymes*, or *proteases*, and belong to the class of enzymes called *hydrolases* (Chap. 8).

Zymogens

The proteolytic enzymes are secreted in the gastric juice or by the pancreas as inactive precursors called *zymogens*. In the case of *trypsin*, the zymogen, *trypsinogen*, is synthesized on the endoplasmic reticulum of pancreatic cells and is secreted from zymogen granules into a duct that leads to the duodenum. The granule is produced in the Golgi apparatus and consists of trypsinogen molecules surrounded by a lipid-protein membrane. The zymogen granules are secreted into a duct leading into the duodenum. The pancreatic cells also produce a *trypsin inhibitor* that ensures that they are not autodigested.

EXAMPLE 15.6

In the disease *pancreatitis* that occasionally follows a bout of mumps, the proteolytic enzymes secreted by the pancreas are prematurely activated and start digesting the cells of the pancreas.

The entry of protein into the stomach stimulates the release of a hormone, *gastrin*, which then causes the release of hydrochloric acid from the *parietal cells* and *pepsinogen* from the *chief cells* (Fig. 15-8). Pepsinogen is another zymogen (they all start with *pro-* or end in *-ogen*) that is converted in the gastric juice to the active form *pepsin*.

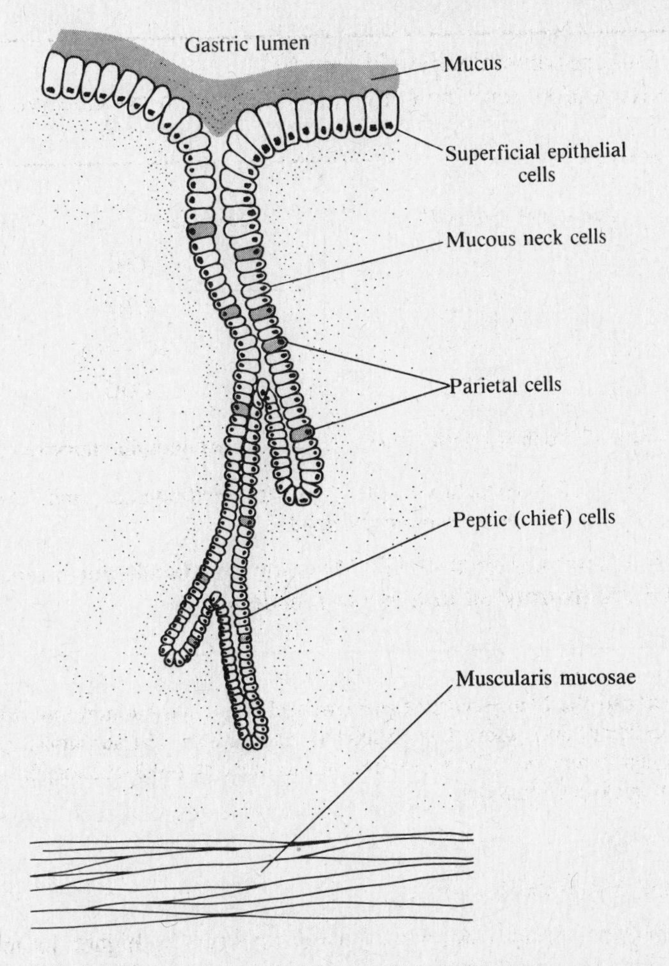

Fig. 15-8 Schematic portrayal of a gastric gland.

Question: What is the function of hydrochloric acid in digestion?

The hydrochloric acid lowers the pH of the stomach contents to ~pH 2; this low pH kills most microbiota entering the stomach and denatures globular proteins, making their peptide bonds more accessible to enzymatic hydrolysis.

The low pH causes the release from the small intestine cells of *secretin*, a hormone, when the stomach contents pass into the small intestine. Secretin causes the release of bicarbonate from the pancreas which neutralizes the hydrochloric acid. This allows the main hydrolytic enzymes—*trypsin*, *chymotrypsin*, *elastase*, and *carboxypeptidase*—to function optimally between pH 7 and 8.

EXAMPLE 15.7

There are a variety of *peptide hormones* acting in the gut: the *gastrins* stimulate gastric acid secretion; *secretin* and *somatostatin* inhibit the production of gastrins. Several hormones can inhibit gastric acid secretion directly. These include *cholecystokinin* and *somatostatin*.

In the duodenum, the pancreatic zymogens—*trypsinogen*, *chymotrypsinogen*, *proelastase*, and *procarboxypeptidase*—are converted into active enzymes by trypsin, as shown in Fig. 15-9. The activation of all the zymogens involves removing peptides or some of the amino acid residues. This allows the remaining polypeptide chain(s) to undergo conformational changes so that an active site can form, thus rendering the protein functional.

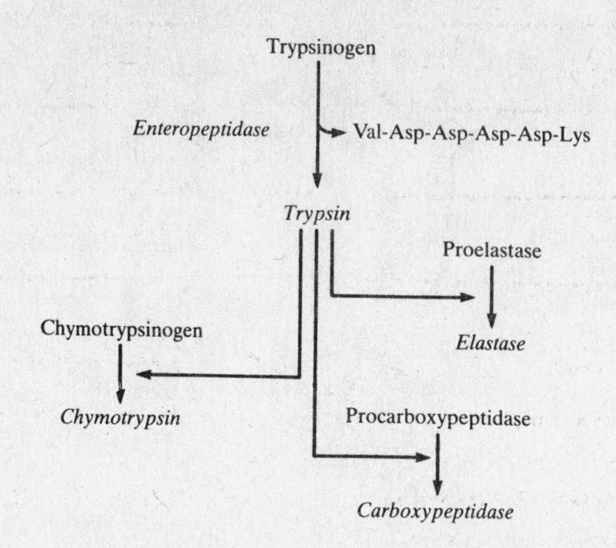

Fig. 15-9 Activation of pancreatic zymogens.

Question: What is the actual sequence of chemical events involved in zymogen activation?

Trypsinogen, chymotrypsinogen, proelastase, and procarboxypeptidase are all synthesized as single polypeptide chains with an M_r around 25,000–30,000. The initial step in the activation is the hydrolysis of a hexapeptide from the N terminus of trypsinogen. This hydrolysis produces trypsin and is catalyzed by *enteropeptidase*, an enzyme on the membranes of brush border cells of the small intestine.

Question: How are zymogens other than trypsinogen activated?

The activation has been studied in detail; that of chymotrypsinogen is shown in Fig. 15-10.

Chymotrypsinogen, a single polypeptide chain of 245 amino acid residues, is eventually converted to α-chymotrypsin, which has three polypeptide chains linked by two of the five disulfide bonds present in the *primary* structure of chymotrypsinogen. π- and δ-Chymotrypsin also have proteolytic activity. In contrast, the conversion of *procarboxypeptidase* to *carboxypeptidase* involves the hydrolytic removal of a single amino acid.

Specificity of Proteases

Theoretically, there are 20×20 possible different combinations of amino acid residues adjacent to one another in a polypeptide. If each possible combination needed a specific protease, then 400 different proteolytic enzymes would be required. However, the proteolytic enzymes have broad specificities, largely confined to groups of amino acids with similar side-chain characteristics, and therefore only a few different types of enzymes are encountered.

Question: What determines the substrate specificities of the proteolytic enzymes?

All the proteolytic enzymes catalyze the hydrolysis of peptide bonds:

$$R\text{—CO—NH—}R' + H_2O \rightleftharpoons R\text{—COO}^- + NH_3^+\text{—}R'$$

The specificities are determined by the side chains of the amino acids on either side of the peptide bond that is hydrolyzed in the polypeptide chain. For the endopeptidases, it is the side chain of the amino acid contributing the carbonyl group of the peptide bond that determines whether the substrate will bind or not. Thus, chymotrypsin hydrolyzes peptide bonds where the carbonyl group is

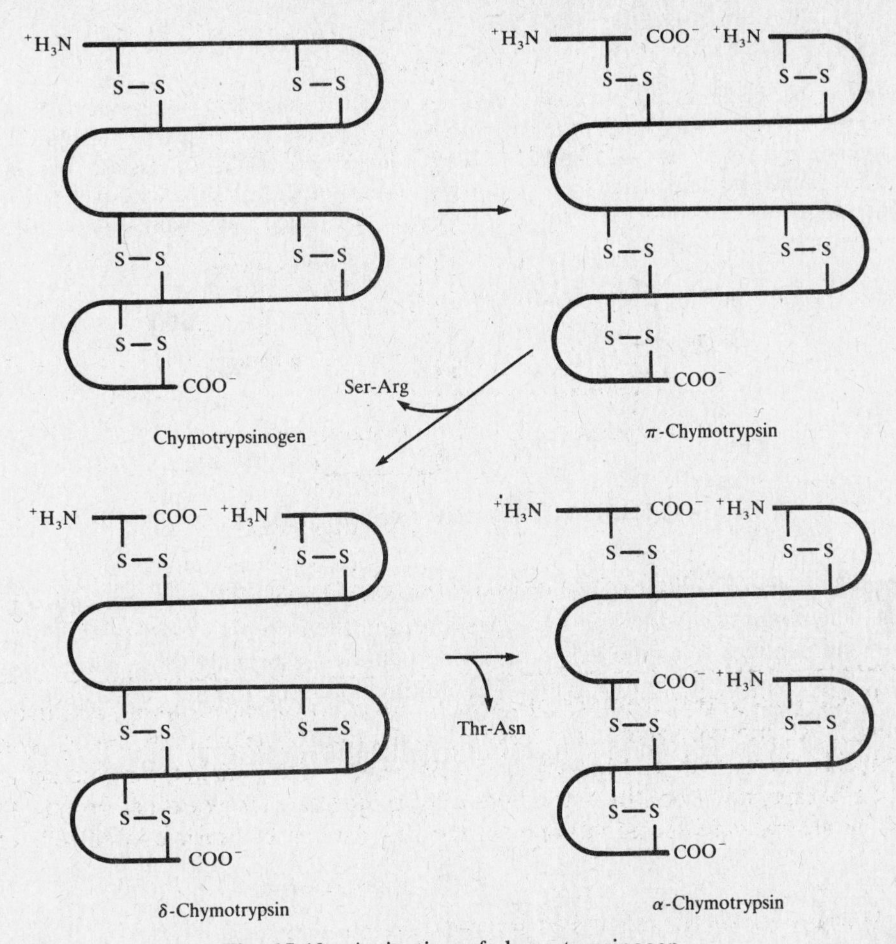

Fig. 15-10 Activation of chymotrypsinogen.

from one of the *aromatic* amino acids, namely, phenylalanine, tyrosine, or tryptophan. The specificities are listed in Table 15.3.

Question: Are there any similarities in the mechanism of catalysis of the pancreatic proteases?

Three of the four pancreatic proteases (trypsin, chymotrypsin, and elastase) are called *serine proteases* because they are all dependent for activity on the side chain of a serine residue in the active site. This serine residue attacks the carbonyl group of the peptide bond to cleave the peptide, giving an ester attached to the serine (Chap. 8). This ester bond is then hydrolyzed.

$$\text{R—CO—NH—R}' + \text{Enz—CH}_2\text{OH} \rightleftharpoons \text{R—CO—OCH}_2\text{—Enz} + \text{R}'\text{NH}_2$$

$$\text{R—CO—OCH}_2\text{—Enz} + \text{H}_2\text{O} \rightleftharpoons \text{R—COOH} + \text{Enz—CH}_2\text{OH}$$

Table 15.3. Specificities of Proteolytic Enzymes

Enzyme	Specificity
Pepsin	Phe, Tyr, Trp; also, Leu, Glu, Gln
Trypsin	Lys, Arg
Chymotrypsin	Phe, Tyr, Trp
Carboxypeptidase	Any carboxy-terminal residue
Elastase	Ala, Gly, Ser
Aminopeptidase	Any amino-terminal residue

EXAMPLE 15.8

The different specificities of the proteolytic enzymes are due to *specificity pockets* at the binding site (Fig. 15-11). In trypsin, a serine residue present in chymotrypsin is replaced by an aspartate residue. This allows the binding of arginine and lysine residues instead of bulky aromatic side chains. In elastase, two glycine residues are replaced by valine and threonine. Their bulky side chains block the specificity pocket so that elastase hydrolyzes peptide bonds in peptides containing the smaller, uncharged side chains.

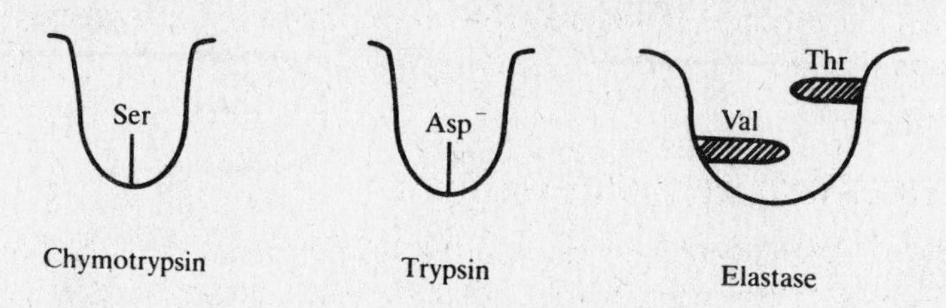

Fig. 15-11 Substrate specificity pockets.

Pepsin and the pancreatic proteases catalyze the conversion of dietary protein to peptides and amino acids. The *aminopeptidases* and the *dipeptidases* in the intestinal mucosa almost complete the hydrolysis of the peptides to amino acids, but some peptides, especially those containing glutamate, pass into the blood with the amino acids. The aminopeptidases remove amino acids from the N terminus of a peptide.

The hydrolysis of proteins in the process of digestion is summarized in Fig. 15-12.

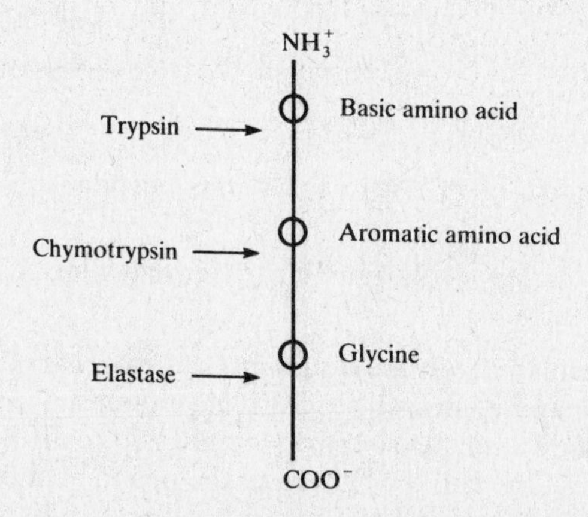

Fig. 15-12 Breakdown of proteins to amino acids in digestion.

Amino Acid Transport

Amino acids, dipeptides, and some tripeptides are transported from the lumen of the intestine through the membrane of the absorptive cells of the brush border, where the peptides are hydrolyzed to amino acids. Transport of peptides and amino acids is *active* and analogous to glucose transport; i.e., they are transported, together with Na^+, across the cell membrane by specific protein carriers called *Na^+ symports*. Between the gut lumen and the cell there is a concentration gradient of Na^+ that is maintained by Na^+/K^+ ATPase at the base of the cell adjacent to the blood capillaries; this Na^+/K ATPase pumps Na^+ into the blood. The Na^+ concentration is lower in the cell than in the gut lumen, ensuring that the amino acids and peptides will be transported across the membrane as Na^+ moves in. There are seven different protein carriers involved in the transport of the amino acids.

EXAMPLE 15.9

How do 7 carriers transport 20 different amino acids?

The carriers have overlapping specificity. Thus, there is one carrier (called *system L*) for leucine and neutral amino acids with branched or aromatic side chains, another for basic amino acids (the *Ly system*), and a low-activity carrier (the *dicarboxylate system*) for dicarboxylic amino acids.

Some of the amino acids diffuse into the bloodstream, where they are transported to the liver and elsewhere. Others, particularly glutamate, glutamine, aspartate, and asparagine are metabolized by the cells for energy.

15.3 DYNAMICS OF AMINO ACID METABOLISM

In addition to being synthesized or produced by the hydrolysis of dietary protein, amino acids can come from hydrolysis of tissue proteins, e.g., intestinal mucosa or, during starvation, muscle. Amino acids enter protein synthesis (Chap. 17); they also enter gluconeogenesis and lipogenesis; are degraded to provide energy; and are used for synthesizing compounds such as purines, pyrimidines, and porphyrins. They are also precursors of specialized metabolites such as epinephrine and creatine.

This metabolic activity is achieved by a turnover of amino acids and proteins that is as rapid as that of lipids and carbohydrates. In an adult human male, 400 g of body proteins is turned over each day. Of this, 50 g is used to replace digestive enzymes (Sec. 15.2), and 16 g to replace hemoglobin (Sec. 15.8). The amino acid concentration in plasma is small (total ~3.2 mmol L^{-1}, of which 25 percent is glutamine), but the turnover of 400 g per day of protein is equivalent to the uptake, and release back into the plasma, of 4.6 moles of α-amino-N, so that the average lifetime of an amino acid in the plasma is about 5 min. Plasma amino acids are turned over with the same kind of rapidity as plasma glucose or free fatty acids. Like that of plasma glucose, the plasma amino acid concentration is remarkably constant, but it is not understood how this is maintained.

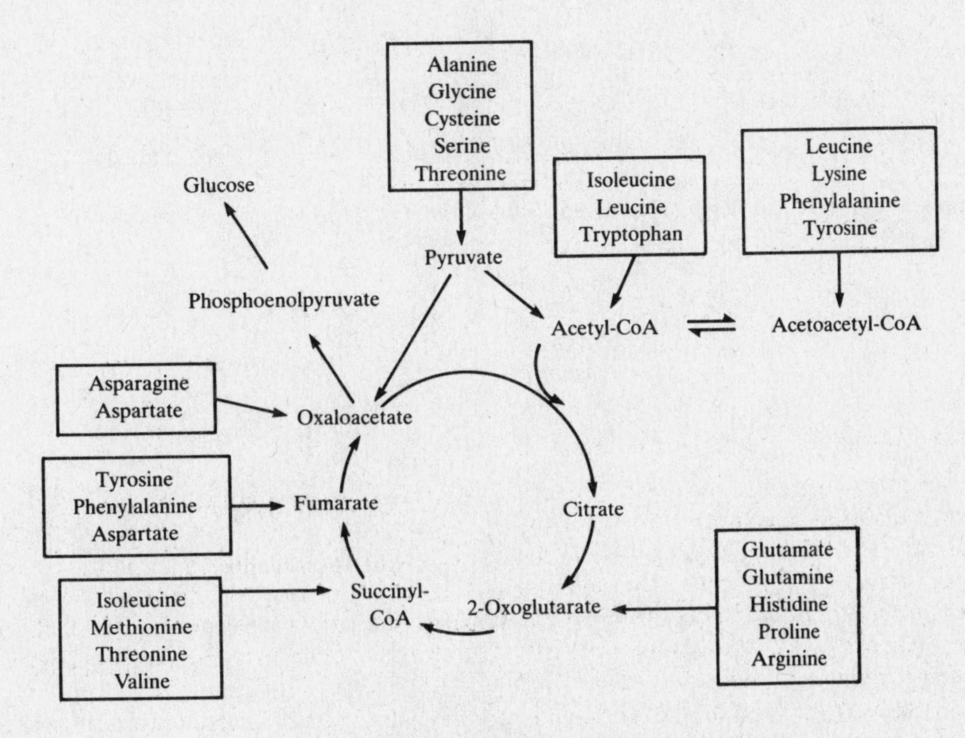

Fig. 15-13 Fates of the carbon skeletons of amino acids.

15.4 AMINO ACID CATABOLISM

The catabolism of the amino acids is complex; there are too many differences between the amino acids for any useful generalization to be made.

The carbon skeletons of the amino acids, with the exception of leucine, can be used for gluconeogenesis. The fates of the carbon atoms are summarized in Fig. 15-13.

All those amino acids which can eventually give rise to pyruvate, which in turn can be used for gluconeogenesis, are called *glucogenic amino acids*. The one amino acid, leucine, that does not give any intermediates of gluconeogenesis (i.e., it is only ketogenic) is degraded to acetoacetate and acetyl-CoA. Several amino acids—phenylalanine, tyrosine, tryptophan, and isoleucine—are both glucogenic and ketogenic. Thus the *majority* of the amino acids are glucogenic. The pathways involved in the catabolism of the individual amino acids range from one-step reactions—such as with aspartate, glutamate, and alanine, which use the appropriate amino transferases—to multistep pathways, as with the aromatic amino acids and lysine; e.g., tyrosine is degraded in four steps to acetoacetate and fumarate.

EXAMPLE 15.10

Tyrosine, itself a degradation product of phenylalanine (Sec. 15.1), is initially converted to *3,4-dihydroxyphenylalanine* (*dopa*) and the corresponding *dopa quinone* by the copper-containing enzyme *tyrosinase*. Tyrosinase is found in *melanocytes* and is a mixed-function oxidase. It catalyzes the following reaction:

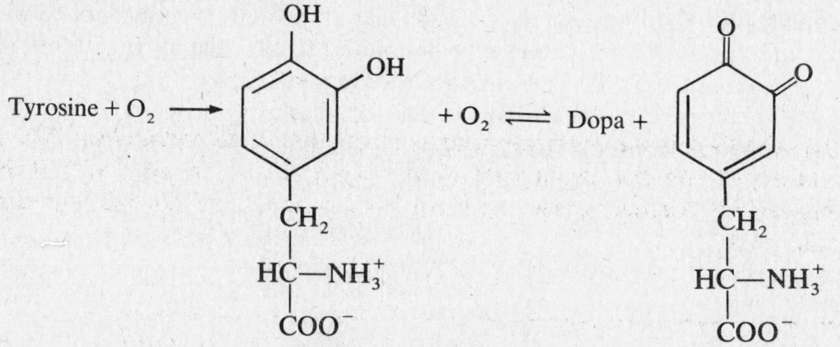

Tyrosine + O_2 ⟶ [Dopa structure: benzene ring with OH, OH; CH_2—HC—NH_3^+—COO^-] + O_2 ⇌ Dopa + [Dopa quinone structure: ring with O, O; CH_2—HC—NH_3^+—COO^-]

Dopa   Dopa quinone

Dopa quinone is converted to *norepinephrine* and *epinephrine* (Fig. 15-14) in the adrenal medulla.

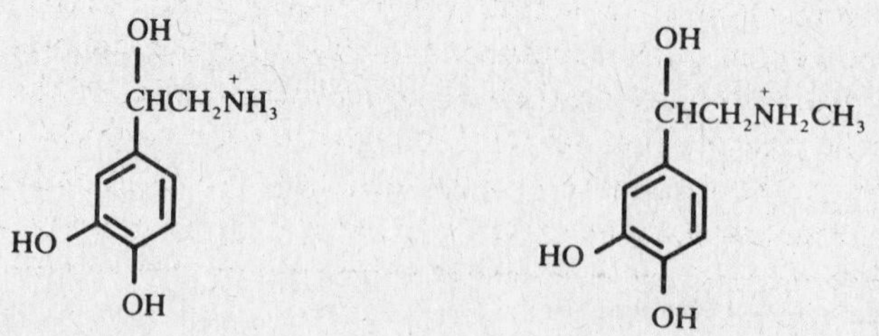

(a) Norepinephrine: OH, $CH\overset{+}{C}H_2NH_3$ (CHCH$_2\overset{+}{N}$H$_3$), ring with HO, OH

(b) Epinephrine: OH, CHCH$_2\overset{+}{N}$H$_2$CH$_3$, ring with HO, OH

Fig. 15-14 Structure of norepinephrine and epinephrine.

Four tissues, liver, muscle, intestine, and kidney, are particularly involved in amino acid metabolism, as summarized in Fig. 15-15.

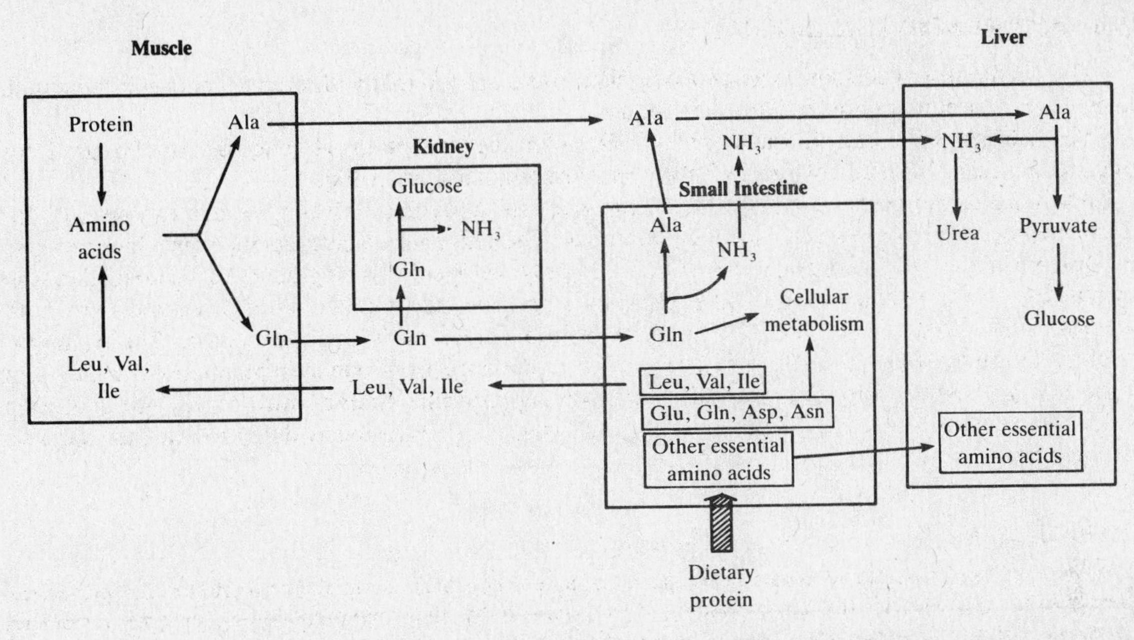

Fig. 15-15 Major reactions of amino acid metabolism in muscle, kidney, intestine, and liver.

Liver

Quantitatively, the liver is the most important site of amino acid metabolism. The liver plays an important role in regulating the concentration of the essential amino acids; it is the site of degradation of most of the essential amino acids (exceptions are valine, leucine, isoleucine) and most of the nonessential ones.

Question: How does the liver control the rate of amino acid degradation?

Most proteins, including enzymes, are constantly being synthesized and degraded, i.e., turned over. With enzymes this represents an additional way of controlling enzyme activity (besides inhibition and activation). Structural proteins can be mobilized in times of need and their amino acids used for gluconeogenesis. The turnover, normally expressed as the half-life of the protein, varies enormously from a few hours (the digestive proteins) to three months (collagen). The liver can vary the turnover of the key enzymes in nitrogen metabolism. For example, on a minimal protein diet on which *nitrogen balance* is just maintained, the concentrations of the amino transferases are low. This ensures that the essential amino acids are spared from degradation and are available for protein synthesis. If the intake of dietary protein increases, the degradative enzymes are induced, and in about 24 h their concentration increases markedly; e.g., by a factor of ~2.

Muscle

Amino acid metabolism in muscle serves two functions. It provides the muscle with energy through partial oxidation of the branched-chain amino acids leucine, isoleucine, and valine, and it exports alanine and glutamine to other tissues (Fig. 15-15). The branched-chain amino acids are transaminated to their 2-oxoacids. The 2-oxoacids are then oxidized by specific pathways eventually to give oxaloacetate, which can be converted to pyruvate (Fig. 15-16).

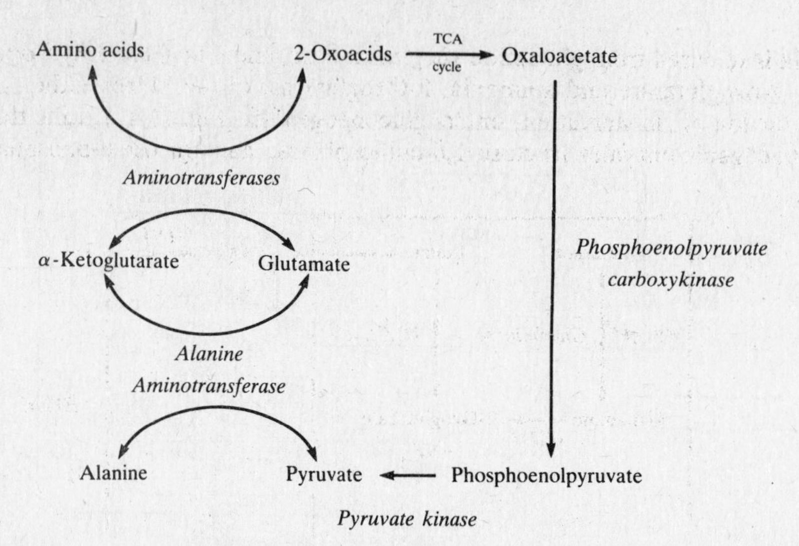

Fig. 15-16 Alanine synthesis in muscle.

Question: How is oxaloacetate converted to pyruvate?

The enzyme *phosphoenolpyruvate carboxykinase* catalyzes the conversion of oxaloacetate to phosphoenolpyruvate, and this is hydrolyzed by pyruvate kinase to give pyruvate. Pyruvate is either oxidized in the mitochondria to form acetyl-CoA or transaminated via the enzyme *alanine aminotransferase*, with glutamate as a source of ammonia, to form alanine, which can then be exported from the cell. In muscle, the glucogenic amino acids are degraded during starvation to yield either pyruvate or 2-oxoglutarate.

Question: What is the route of synthesis of 2-oxoglutarate?

Since all the amino acids metabolized by muscle can give rise to acetyl-CoA and all the amino acids except leucine can ultimately form oxaloacetate, the carbon skeletons of any two amino acids could form citrate (from acetyl-CoA and oxaloacetate), which could be converted to 2-oxoglutarate via the citric acid cycle.

Glutamine, like alanine, is released from skeletal muscle during starvation in amounts greater than its occurrence in muscle protein. Alanine and glutamine account for 60 percent of the amino acids exported from the muscle during starvation. Glutamine is synthesized from glutamate and ammonia. The reaction (see Sec. 15.1) is catalyzed by the enzyme *glutamine synthetase*.

Question: What is the significance of producing alanine and glutamine?

Alanine and glutamine are nontoxic carriers of ammonia in the blood. Alanine is an important precursor of gluconeogenesis in the liver, and so its rate of production by muscle during starvation may permit muscle to regulate hepatic gluconeogenesis. Glutamine is an obligatory fuel of intestinal and rapidly dividing cells; it is also used in the kidney for ammonia production, which is important in the maintenance of whole-body acid-base balance.

Kidney

Glutamine in the kidney is oxidized to NH_3 and CO_2. The physiological role of this pathway is to regulate the body's pH. The overall reaction is

$$\text{Glutamine} + 4\tfrac{1}{2}O_2 \longrightarrow 2NH_3 + 5CO_2 + 2H_2O$$

Ammonia is released from glutamine via *glutaminase*, and *glutamate dehydrogenase* catalyzes the formation of 2-oxoglutarate and ammonia. 2-Oxoglutarate in the kidney either is fully oxidized in energy transduction or, in starvation, enters gluconeogenesis (Fig. 15-17). Like the liver, the kidney has the gluconeogenic enzymes fructose 1,6-diphosphatase and glucose 6-phosphatase.

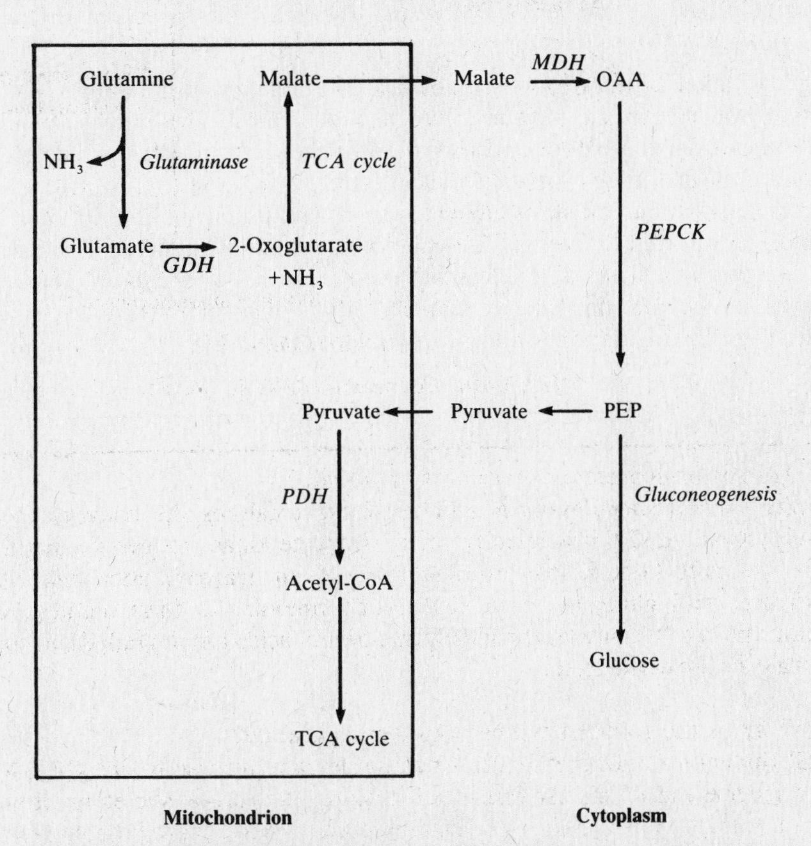

Fig. 15-17 Metabolism of glutamine in the kidney. GDH = glutamate dehydrogenase; MDH = malate dehydrogenase; OAA = oxaloacetic acid; PEPCK = phosphoenolpyruvate carboxykinase; PEP = phosphoenolpyruvate; PDH = pyruvate dehydrogenase.

Question: What determines whether 2-oxoglutarate is oxidized or is converted into glucose?

The activity of pyruvate dehydrogenase is the determining factor. In the normal fed state, this enzyme is active and hence will catalyze pyruvate oxidation. During starvation, when fatty acids are mobilized, the oxidation of pyruvate is severely inhibited since pyruvate dehydrogenase is present in the inactive form.

Intestine

During digestion the epithelial cells of the intestine are constantly sloughed off and are normally replaced with a turnover time of four days. Thus the intestine is the site of considerable metabolic activity. Glutamine, glutamate, asparagine, and aspartate are oxidized to CO_2, lactate, alanine, and citrulline (see "Urea Synthesis," Sec. 15.5). Any nitrogen derived from metabolism of these four amino acids and not retained in alanine or citrulline is released into the blood as ammonia and transported to the liver.

EXAMPLE 15.11

Why is citrulline synthesized in the gut?

Citrulline is an intermediate in arginine synthesis (Sec. 15.5). The gut lacks the enzymes for citrulline utilization and thus exports citrulline to tissues such as kidney for arginine synthesis; the kidney cannot synthesize citrulline and has low levels of *arginase*. The gut is the only tissue that is a net synthesizer of citrulline.

During starvation or after an overnight fast, the amount of glutamine, glutamate, asparagine, and aspartate available in the intestinal lumen is decreased. Under these conditions, most of the glutamine comes from the blood, and oxidation in the mitochondria of this glutamine and ketone bodies provides all the energy for the small intestine.

The branched-chain amino acids leucine, valine, and isoleucine are not metabolized in the small intestine and are exported to the muscle, while other essential amino acids are exported to the liver.

There are two significant differences between intestinal and kidney metabolism. Glutamate is transaminated in the intestine rather than deaminated as in the kidney, and *malic enzyme*, a mitochondrial enzyme, is the only route to pyruvate (Fig. 15-18).

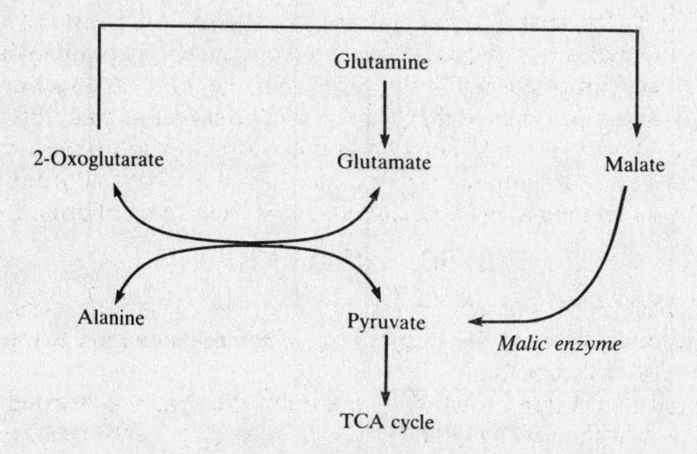

Fig. 15-18 Metabolism of glutamine in the small intestine.

15.5 DISPOSAL OF EXCESS NITROGEN

Question: Why is nitrogen excreted?

There are no stores of nitrogen in the body comparable in form to lipid and glycogen; i.e., any nitrogen in excess of our growth requirements is excreted. If we ingest less nitrogen than we require for normal growth and tissue repair, we utilize the nitrogen stored in our muscle proteins.

Amino acids in excess of metabolic requirements are degraded to their carbon skeletons, which, as discussed in the preceding section, enter energy metabolism or are converted to other compounds and ammonia; the ammonia is excreted as such or converted to urea and then excreted. This is the situation with humans, but there are major species differences as to the means by which excess ammonia is removed.

In aquatic animals, ammonia diffuses out of the body through the skin, but land animals excrete excess ammonia either as urea or uric acid. Ammonia is excreted by humans on high meat diets as a strategy to conserve Na^+ and K^+. Excess PO_4^{3-} and $SO_4^{=}$ produced from phosphorproteins and S-containing amino acids are excreted as ammonium salts. The overall process in the kidney is that Na^+ and K^+ are exchanged for NH_4^+. The excretion of urea requires a plentiful supply of water, as it is normally excreted in solution, whereas uric acid is very insoluble and is excreted as a solid. Thus, in animals in which weight or the conservation of water is important (e.g., birds), excess ammonia is excreted as uric acid.

Urea, NH_2CONH_2, is highly soluble ($10\ mol\ L^{-1}$), nontoxic, high in nitrogen content (47 percent), and requires little energy for synthesis (0.5 ATP per atom of nitrogen). Normal human subjects excrete about 30 g per day on a western diet, but on a high-protein diet, this can increase to 100 g per day. Humans, and other primates, excrete a small amount of uric acid as an end product of purine metabolism. Also, unlike many animals, we do not produce *uricase*, the enzyme that catalyzes the degradation of uric acid. We excrete excess nitrogen as ammonia, urea, and uric acid. Several other nitrogen-containing metabolites, notably the bile pigments, are also excreted. These are degradation products of hemoglobin and other porphyrin-containing molecules.

Formation of Ammonia

The major enzyme involved in the formation of ammonia in the liver, brain, muscle, and kidney is *glutamate dehydrogenase*, which catalyzes the reaction in which ammonia is condensed with 2-oxoglutarate to form glutamate (Sec. 15.1). Small amounts of ammonia are produced from important amine metabolites such as epinephrine, norepinephrine, and histamine via *amine oxidase* reactions. It is also produced in the degradation of purines and pyrimidines (Sec. 15.6) and in the small intestine from the hydrolysis of glutamine. The concentration of ammonia is regulated within narrow limits; the upper limit of normal in the blood in humans is $\sim 70\ \mu mol\ L^{-1}$. It is toxic to most cells at quite low concentrations; hence there are specific chemical mechanisms for its removal. The reasons for ammonia toxicity are still not understood. The activity of the urea cycle in the liver maintains the concentration of ammonia in peripheral blood at about $20\ \mu mol\ L^{-1}$.

In biological tissues, ammonia and the ammonium ion are in equilibrium:

$$NH_3 + H^+ \rightleftharpoons NH_4^+$$

At physiological pH, about 99 percent of the ammonia is in the ionic form. But it is only the un-ionized form that diffuses across cell membranes. The ammonium ion is transported much more slowly via a carrier-mediated process.

It should be noted that a large fraction of the ammonia that is converted to urea in the liver comes from metabolism in the extrahepatic tissues, though only a small fraction leaves these tissues as ammonia. The absorptive cells of the small intestine are exceptional in that they release ammonia into the portal vein; there the ammonia concentration may reach $0.26\ mmol\ L^{-1}$, and this ammonia accounts for 30 percent of the urea synthesized in the liver. The flow of nitrogen compounds to the liver for urea synthesis is shown in Fig. 15-19.

Urea Synthesis

Urea is formed in the liver by a series of reactions known as the *urea cycle*. One nitrogen is derived from ammonia, the second from aspartate; the carbon is derived from CO_2. The synthesis of urea requires the formation of *carbamoyl phosphate* and the four enzymatic reactions of the urea cycle. These take place partly in the mitochondria and partly in the cytoplasm. The enzymes involved in the synthesis of urea are discussed below.

Carbamoyl Phosphate Synthetase I

The formation of carbamoyl phosphate ($NH_2COOPO_3^{2-}$) takes place in the matrix of mitochondria:

$$NH_3 + HCO_3^- + 2ATP \longrightarrow NH_2COOPO_3^{2-} + 2ADP + P_i$$

The ammonia can come from glutamate via glutamate dehydrogenase, and the HCO_3^- comes from respiration.

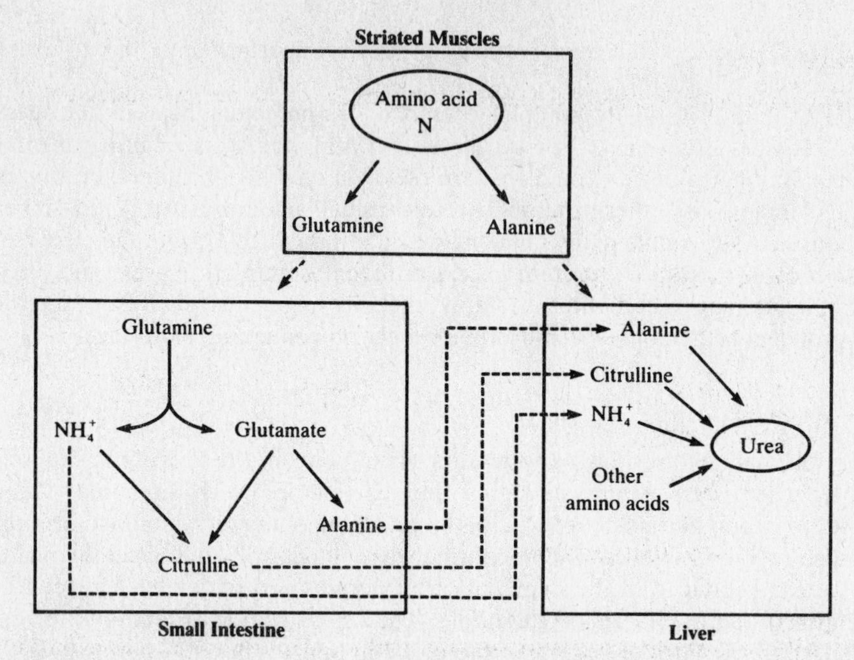

Fig. 15-19 Transfer of nitrogen compounds from the tissues to the liver for urea synthesis.

Ornithine Carbamoyltransferase

The first reaction of the cycle takes place in the matrix of the mitochondria and is catalyzed by ornithine carbamoyltransferase.

$$\underset{\text{L-Ornithine}}{\begin{array}{c}(CH_2)_3NH_3\\|\\HC{-}NH_3^+\\|\\COO^-\end{array}} + \underset{\text{Carbamoyl phosphate}}{NH_2COOPO_3^{2-}} \longrightarrow \underset{\text{Citrulline}}{\begin{array}{c}(CH_2)_3NHCONH_2\\|\\HC{-}NH_3^+\\|\\COO^-\end{array}} + P_i$$

Both ornithine, which is a homologue of lysine, and citrulline are L-amino acids, but neither has a genetic codon, and both are found only as *posttranslational modifications* of arginine residues in some proteins such as keratin. Citrulline leaves the mitochondria by the same transport system that facilitates ornithine's entry from the cytoplasm.

EXAMPLE 15.12

Citrulline takes its name from the watermelon genus (*Citrillus*) in which it was first found in 1930. It was also discovered the same year as a bacterial degradation product of arginine. Krebs, who elucidated the urea cycle, realized that citrulline was probably the intermediate between ornithine and arginine, and he showed that this was so. The urea cycle was the first metabolic cycle to be discovered. In Krebs's words, "it revealed a new pattern of the organization of metabolic processes."

Argininosuccinate Synthetase

Argininosuccinate synthetase and the remaining two enzymes of the urea cycle are found in the cytoplasm. Argininosuccinate synthetase catalyzes the condensation of citrulline with aspartate to form *argininosuccinate*. The reaction requires one molecule of ATP, which is hydrolyzed to AMP and PP_i. Pyrophosphate is a strong inhibitor of the reaction ($K_i = 6.2 \times 10^{-5}$ *M*); inhibition is normally not evident because of pyrophosphatase.

$$\text{Citrulline} + \text{Aspartate} + \text{ATP} \xrightleftharpoons{Mg^{2+}} \text{Argininosuccinate} + \text{AMP} + PP_i$$

Argininosuccinate Lyase

Argininosuccinate lyase reversibly catalyzes the cleavage of argininosuccinate to arginine and fumarate.

$$
\begin{array}{lllll}
NH_2^+ \quad\quad COO^- & & NH_2^+ & & COO^- \\
\| \quad\quad\quad\quad\; | & & \| & & | \\
C-NH-CH & & C-NH_2^+ & & CH \\
| \quad\quad\quad\quad\;\; | & & | & & \| \\
NH \quad\quad CH_2COO^- & \rightleftharpoons & NH & + & HC \\
| & & | & & | \\
(CH_2)_3 & & (CH_2)_3 & & COO^- \\
| & & | & & \\
HC-NH_3^+ & & HC-NH_3^+ & & \\
| & & | & & \\
COO^- & & COO^- & & \\
\text{Argininosuccinate} & & \text{Arginine} & & \text{Fumarate}
\end{array}
$$

This reaction is also involved in the synthesis of arginine for protein synthesis. The fumarate can be converted to oxaloacetate (via coupling with the citric acid cycle) and then transaminated by *aspartate aminotransferase* to aspartate, which can reenter reactions of the urea cycle.

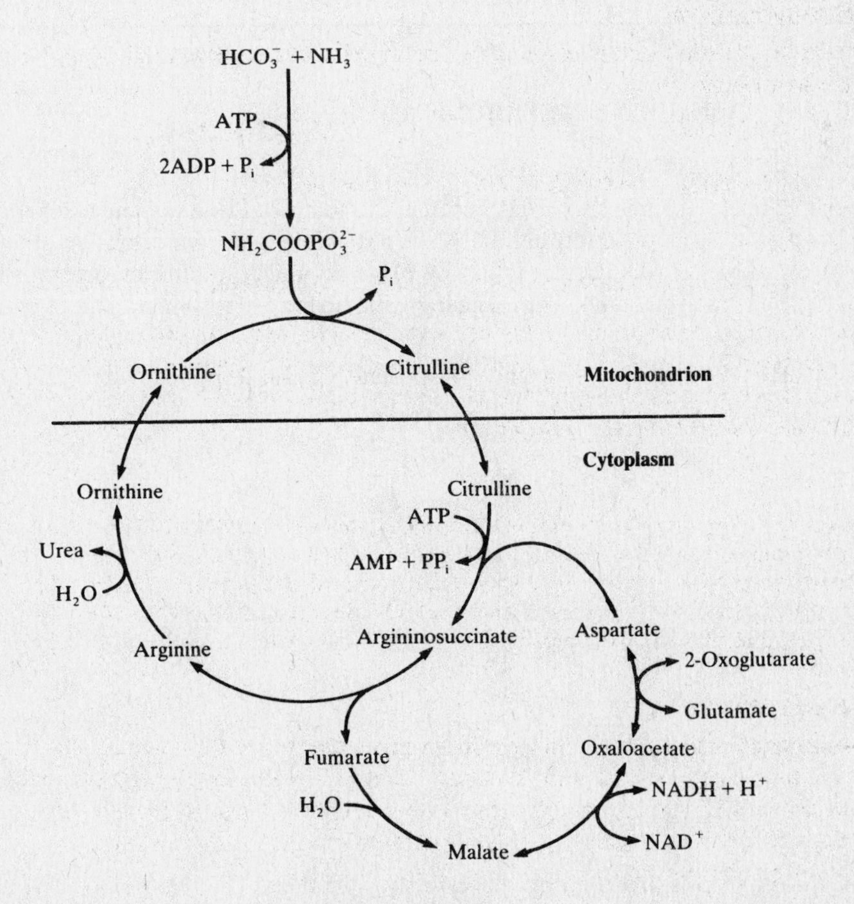

Fig. 15-20 The urea cycle.

Arginase

Arginase, the last enzyme in the urea cycle, catalyzes the hydrolytic cleavage of arginine to urea and ornithine.

$$\text{Arginine} + H_2O \longrightarrow \text{Urea} + \text{Ornithine}$$

The urea diffuses into the blood and is carried to the kidneys. There it passes into the glomerular filtrate, from which it is excreted in the urine.

The reactions of the urea cycle are summarized in Fig. 15-20.

The overall reaction of the urea cycle is;

$$3ATP + NH_3 + CO_2 + 2H_2O + \text{Aspartate} \longrightarrow$$
$$2ADP + 4P_i + AMP + \text{Fumarate} + \text{Urea}$$

Since the regeneration of ATP from AMP requires a molecule of ATP to convert AMP to ADP (a reaction catalyzed by the enzyme *adenylate kinase*), a total of four molecules of ATP are hydrolyzed in the synthesis of one molecule of urea. However, when fumarate is converted to aspartate, one molecule of NADH is produced, which will give three molecules of ATP by oxidative phosphorylation. Thus, overall, only one molecule of ATP is hydrolyzed for each molecule of urea produced.

Question: Why is the urea cycle compartmentalized?

The main reason is that the system evolved to keep the fumarate concentration low, because fumarate (and arginine) readily inhibits argininosuccinate lyase. Thus, this enzyme is cytoplasmic; it is not inhibited by the high concentrations of fumarate from the citric acid cycle since this fumarate is in the mitochondria.

15.6 PURINE AND PYRIMIDINE METABOLISM

Purine Synthesis

The synthesis of nucleotides is important because of the crucial role nucleic acids play in protein synthesis and storage of genetic information. Nucleotides are also important in the synthesis of derivatives such as cAMP, FAD, CoASH, $NAD(P)^+$, and UDPG. Purines are synthesized by a pathway involving eight intermediates, commencing with ribose 5-phosphate. The overall reaction is

$$2NH_3 + 2HCOOH + CO_2 + \text{Glycine} + \text{Aspartate} + \text{Ribose 5-phosphate} + 9ATP$$
$$\downarrow$$

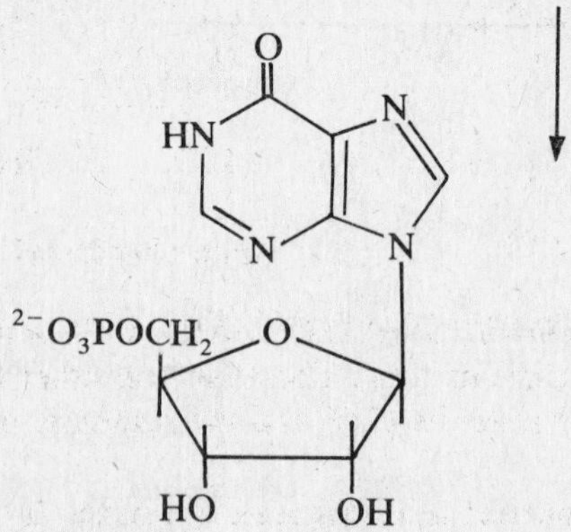

O, HN, N, N, N, $^{2-}O_3POCH_2$, O, HO, OH

$$+ \text{Fumarate} + 8ADP + AMP + PP_i + P_i$$

Inosine monophosphate

The energy for this process is provided by the hydrolysis of the ATP. The purine ring is built up as shown in Fig. 15-21.

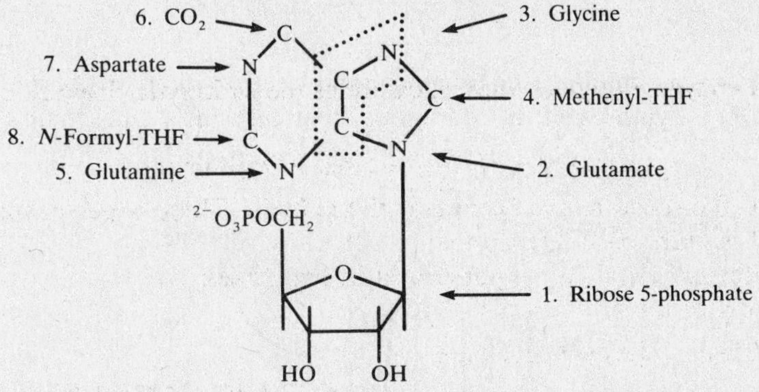
6. CO2
7. Aspartate
8. N-Formyl-THF
5. Glutamine
3. Glycine
4. Methenyl-THF
2. Glutamate
1. Ribose 5-phosphate
2− O3POCH2
HO
OH

Fig. 15-21 Summary of synthesis of the purine ring.

Both adenosine and guanosine phosphates (purine nucleotides) are derived from inosine monophosphate (IMP), as shown in Fig. 15-22.

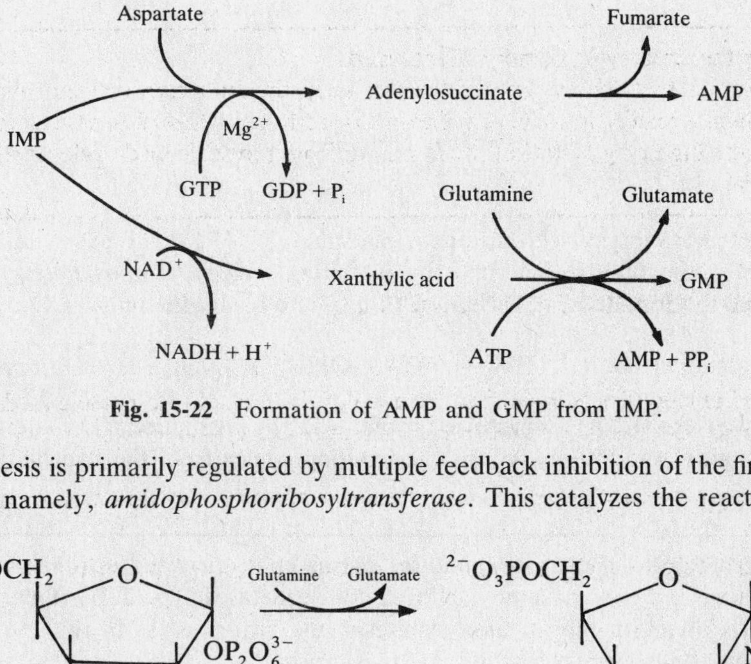
Aspartate
Fumarate
Adenylosuccinate
AMP
IMP
Mg2+
GTP
GDP + Pi
Glutamine
Glutamate
NAD+
Xanthylic acid
GMP
NADH + H+
ATP
AMP + PPi

Fig. 15-22 Formation of AMP and GMP from IMP.

Purine synthesis is primarily regulated by multiple feedback inhibition of the first enzyme unique to the pathway, namely, *amidophosphoribosyltransferase*. This catalyzes the reaction

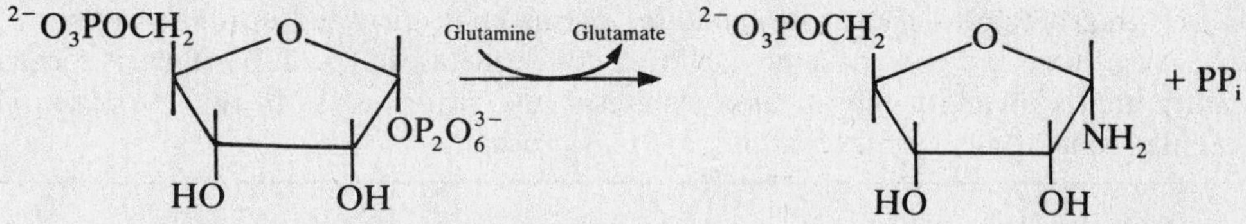
2− O3POCH2
OP2O6 3−
HO
OH
Glutamine
Glutamate
2− O3POCH2
NH2
HO
OH
+ PPi

Phosphoribosyl pyrophosphate (PRPP) 5-Phosphoribosyl-1-amine

The reaction requires Mg^{2+} and is inhibited by GMP, GDP, GTP, AMP, ADP, and ATP. Each series of nucleotides acts at a *separate* control site on the enzyme. There are two other kinds of control in the pathway: the synthesis of AMP from IMP requires GTP and the synthesis of GMP from IMP requires ATP.

The triphosphates that are required for nucleic acid synthesis are made from the monophosphates in two successive phosphorylations involving ATP. For example, the synthesis of GTP from GMP is as follows:

$$\text{GMP} \xrightarrow{\text{ATP} \quad \text{ADP}} \text{GDP} \xrightarrow{\text{ATP} \quad \text{ADP}} \text{GTP}$$

Pyrimidine Synthesis

The synthesis of the pyrimidine ring is much simpler than that of the purine ring. *Uridine monophosphate* (*UMP*) is synthesized in three steps from carbamoyl phosphate (Fig. 15-23).

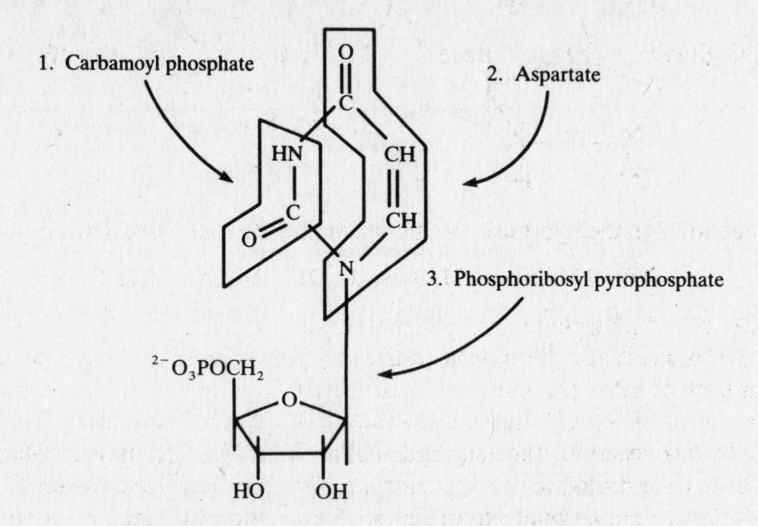

Fig. 15-23 Synthesis of uridine monophosphate.

UMP is further phosphorylated to uridine triphosphate (UTP). UTP exerts the primary control on the synthesis of pyrimidines in mammals by inhibiting *carbamoyl phosphate synthetase II*. This synthetase catalyzes the formation of carbamoyl phosphate from glutamine, CO_2, and ATP:

$$\text{Glutamine} + CO_2 + 2\text{ATP} \longrightarrow NH_2COOPO_3^{2-} + 2\text{ADP} + P_i + \text{Glutamate}$$

Cytidine triphosphate (*CTP*) is produced from UTP by amination. The amino group is also derived from glutamine, and the energy for the reaction comes from the hydrolysis of ATP.

Question: Is the carbamoyl phosphate made for pyrimidines also used in urea synthesis?

No, because there are two carbamoyl phosphate synthetases (I and II). One is located in the mitochondria and is involved only in urea synthesis; the other occurs in the cytoplasm. In other words, cellular compartmentation separates the two processes.

Deoxyribonucleotide Synthesis

The synthesis of DNA is dependent on a ready supply of deoxyribonucleotides. The substrates for these are the ribonucleoside diphosphates ADP, GDP, CDP, and UDP; the enzyme responsible for the reduction of these substrates to their corresponding deoxy derivatives is *ribonucleoside diphosphate reductase*, which uses thioredoxin as a cosubstrate.

EXAMPLE 15.13

Thioredoxin is a protein of molecular weight 12,000 that can donate two electrons by the oxidation of two cysteine sulfhydryl groups to cystine; oxidized thioredoxin is reduced by NADPH.

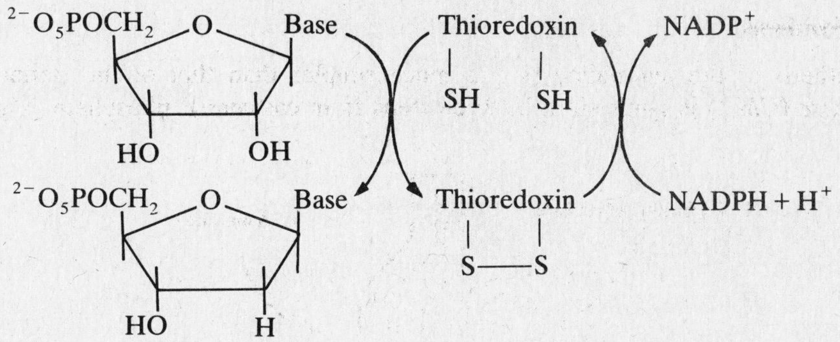

$^{2-}O_5POCH_2$ O Base — Thioredoxin (SH SH) — $NADP^+$
HO OH
$^{2-}O_5POCH_2$ O Base — Thioredoxin (S—S) — $NADPH + H^+$
HO H

The overall reaction for the synthesis of, for example, deoxyadenosine diphosphate (dADP) is

$$ADP + NADPH + H^+ \longrightarrow dADP + NADP^+ + H_2O$$

The deoxyribonucleotide diphosphates are phosphorylated by ATP.

Cells making DNA must also be able to make deoxythymidine triphosphate (dTTP). The key step in the synthesis of dTTP is the conversion of dUMP to dTMP via *thymidylate synthetase*. The reaction requires a source of N^5,N^{10}-methylenetetrahydrofolate (see Sec. 15.7, Fig. 15-26) to provide the methyl group. In this reaction, the tetrahydrofolate is oxidized to dihydrofolate. Dihydrofolate must be reduced to tetrahydrofolate via the enzyme *dihydrofolate reductase* so that more N^5,N^{10}-methylenetetrahydrofolate can be made from serine in a reaction catalyzed by *serine hydroxymethyltransferase*. These three reactions, which are essential for the formation of dTMP, are shown below.

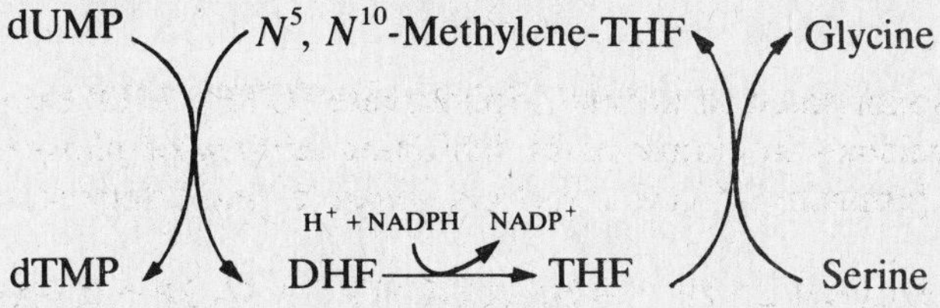

dUMP → dTMP; N^5, N^{10}-Methylene-THF → DHF; DHF → THF (H^+ + NADPH → $NADP^+$); THF → N^5, N^{10}-Methylene-THF (Serine → Glycine)

EXAMPLE 15.14

Inhibition of thymidylate synthetase and dihydrofolate reductase offers a means of inhibiting the growth of some rapidly dividing cancer cells. 5-Fluorouracil (Fig. 15-24), an analog of uracil, is converted in many cells into fluorodeoxyuridine monophosphate, which is an inhibitor of thymidylate synthetase; dihydrofolate reductase is strongly inhibited by the drug *methotrexate* (Fig. 15-24).

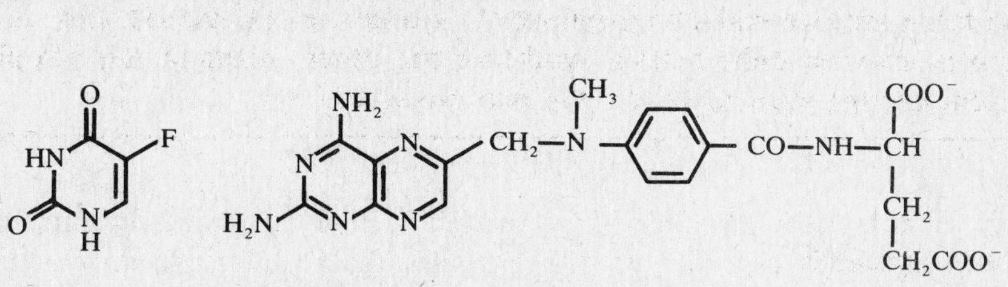

5-Fluorouracil

Methotrexate

Fig. 15-24 Structure of 5-fluorouracil and methotrexate.

Nucleic Acid Degradation

The general scheme of the degradation of nucleic acids has much in common with that of proteins. That is, nucleotides are produced by hydrolysis of both dietary and endogenous nucleic acids. The endogenous polynucleotides are broken down in lysosomes. DNA is not normally broken down, except after cell death and during DNA repair. RNA is turned over in much the same way as protein. The enzymes involved are the *nucleases* (analogs of the proteases); *deoxyribonucleases* and *ribonucleases* hydrolyze DNA and RNA, respectively, to polynucleotides. These can be further hydrolyzed (Fig. 15-25), and eventually purines and pyrimidines are formed.

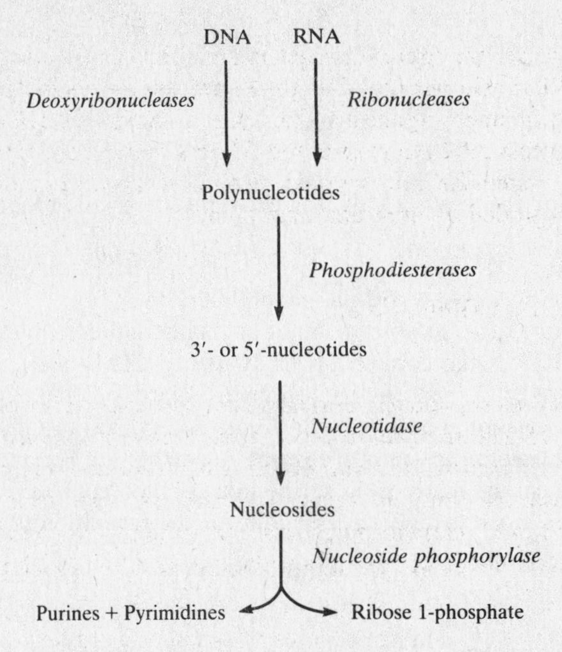

Fig. 15-25 Hydrolysis of DNA and RNA to purines and pyrimidines.

Most of the enzymes involved in the hydrolysis of DNA and RNA are secreted in the intestine. Thus, ribonucleases are found in the lumen of the small intestine, and the phosphodiesterases and the nucleotidases are secreted in the intestinal mucosa.

Question: The synthesis of the purine and pyrimidine bases, particularly the purines, is energetically expensive. Can the bases produced by the breakdown of nucleic acids be reused?

The majority (80 percent) of the purines and pyrimidines are reused by what are called *salvage pathways*. The salvage pathway for purines consists of a single reaction. For example, adenine is salvaged in a reaction with phosphoribosyl pyrophosphate (PRPP), as follows:

$$\text{Adenine} + \text{PRPP} \longrightarrow \text{AMP} + \text{PP}_i$$

Guanine and hypoxanthine (the deamination product of adenine) are similarly converted by a second enzyme to GMP and IMP, respectively. The enzymes that catalyze these reactions are *purine phosphoribosyltransferases*. The pyrimidines are salvaged by a similar mechanism.

Degradation of Purines and Pyrimidines

Purines and pyrimidines in excess of cellular requirements can be further degraded. The degree of degradation depends on the organism. Humans cannot degrade purines beyond uric acid because we lack the enzyme *uricase*, which splits the purine ring to form *allantoin*. In humans excess AMP is deaminated to IMP by the action of a specific deaminase. IMP is then hydrolyzed by a 5′-nucleotidase to form inosine. Inosine and guanine can be oxidized to urate as follows.

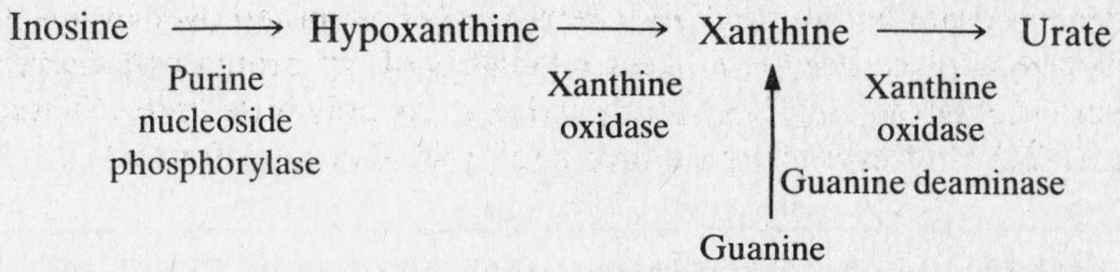

Inosine ⟶ Hypoxanthine ⟶ Xanthine ⟶ Urate
(Purine nucleoside phosphorylase; Xanthine oxidase; Xanthine oxidase)
Guanine ⟶ Xanthine (Guanine deaminase)

Thus, because we lack uricase, we excrete, albeit in small amounts, uric acid every day. The liver synthesizes about 0.8 g of uric acid per day, but 20–50 percent enters the gut in the gastric juice and in the bile and is degraded by microorganisms. For some animals (the *uricoteles*, such as birds), uric acid is the form in which excess nitrogen is excreted. Unless there are enzymes missing, as is the case of humans, nonuricotelic organisms can degrade purines to urea, ammonia, and carbon dioxide. Pyrimidines can also be degraded to urea and ammonia.

15.7 METABOLISM OF C_1 COMPOUNDS

Folic Acid Derivatives

Several processes already described use derivatives of *tetrahydrofolate* (Fig. 15-26). The synthesis of the purine ring (Fig. 15-21) requires two different derivatives, *methenyltetrahydrofolate* and N^{10}*-formyltetrahydrofolate*. Thymidylate synthetase, a key enzyme in pyrimidine synthesis, uses N^5,N^{10}*-methylenetetrahydrofolate* both as a substrate and as a reducing agent. This compound, perhaps the most important in C_1 metabolism, is also involved in the synthesis of serine and glycine. All these compounds are derivatives of 5,6,7,8-tetrahydrofolic acid, which is the reduced form of the vitamin *folate*.

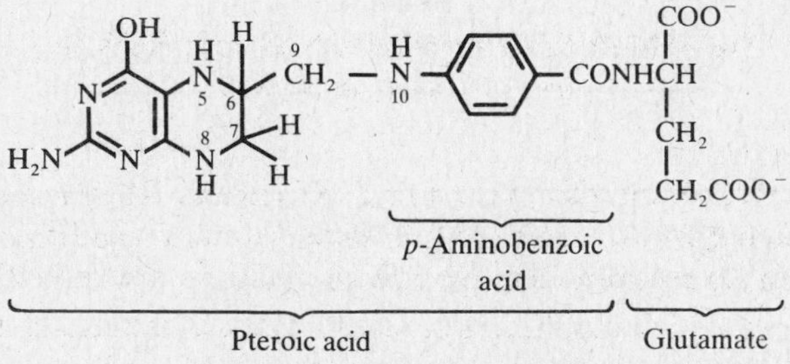

Fig. 15-26 Tetrahydrofolic acid.

The pteroic acid moiety of tetrahydrofolate consists of a reduced *pteridine ring* and p-aminobenzoic acid. Folate from the diet is absorbed in the intestinal mucosa and in two enzymatic steps is reduced to tetrahydrofolate, the active form of the coenzyme. Mammals cannot synthesize folate; this normally does not present a problem as microorganisms of the intestinal tract readily do so.

The two steps in the reduction of folic acid to tetrahydrofolate are catalyzed by *folate reductase* and *dihydrofolate reductase*. Both of these reactions require NADPH as a source of electrons.

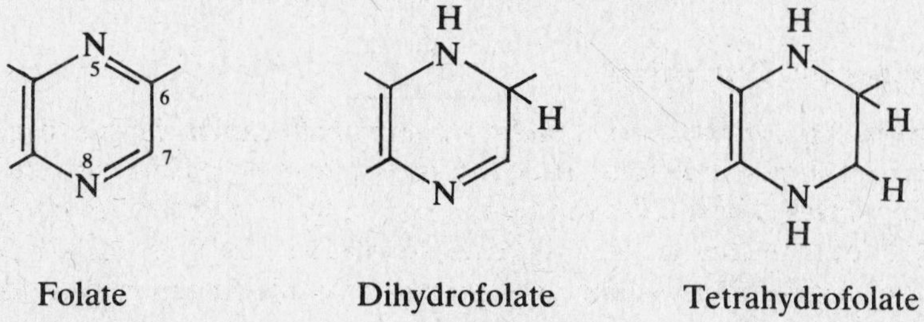

Folate Dihydrofolate Tetrahydrofolate

In order to understand the bewildering variety of reactions involving tetrahydrofolate, it is essential to realize that in biological systems, carbon compounds exist in *five* different oxidation states. The most *reduced* form is methane, CH_4, and the most *oxidized* form is CO_2. In between these two extremes are *methanol* (CH_3OH), *formaldehyde* (CH_2O), and *formate* ($HCOO^-$).

Question: Are all the above one-carbon compounds involved in "C_1 metabolism"?

Methane and carbon dioxide are the exceptions. Methane is an end product of anaerobic metabolism of many microorganisms, and carbon dioxide (for carboxylation) is handled by biotin-containing enzymes.

Table 15.4 lists the various groups carried by tetrahydrofolate derivatives.

The various C_1 groups may be attached to N atoms in positions 5 or 10 (Fig. 15-26) or may form a bridge between the two positions. N^5-methyl-THF is formed in mammals by a virtually irreversible reaction that is catalyzed by the enzyme *methylene-THF reductase*; the other THF derivatives are interconverted through a series of oxidation-reduction and hydration-dehydration reactions.

$$N^5\text{-Methyl-THF} \longleftarrow N^5,N^{10}\text{-Methylene-THF} \rightleftharpoons N^5,N^{10}\text{-Methenyl-THF} \longrightarrow N^{10}\text{-Formyl-THF}$$

Table 15.4. Tetrahydrofolate (THF) Derivatives

Group Carried	THF Derivative
$-CH_3$	N^5-Methyl-THF
$-CH_2OH$	N^5,N^{10}-Methylene-THF
$-CHO$	N-Formyl-THF
$-CH=$	N^5,N^{10}-Methenyl-THF

Question: What is the major reaction that replenishes C_1 units in THF?

With the exception of N^5-methyl-THF, the THF derivatives are directly synthesized from a C_1 unit in the appropriate oxidation state and THF. The major *anaplerotic* reaction is that catalyzed by *serine hydroxymethyltransferase* (Sec. 15.1).

Biological Methylations

The transfer of methyl groups is a common biochemical reaction, and the introduction of a methyl group into a molecule is an important way of modifying biological activity, as in the case of epinephrine versus norepinephrine. The methyl groups originate from N^5-methyltetrahydrofolate, although this compound is involved directly in only one methylation reaction. The simplest form of this reaction occurs in plants, where it is catalyzed by the enzyme *homocysteine transmethylase*:

$$\underset{\text{Homocysteine}}{\begin{matrix} SH \\ | \\ (CH_2)_2 \\ | \\ HC-NH_3^+ \\ | \\ COO^- \end{matrix}} \xrightarrow{N^5\text{-Methyl-THF} \quad \text{THF}} \underset{\text{Methionine}}{\begin{matrix} SCH_3 \\ | \\ (CH_2)_2 \\ | \\ HC-NH_3^+ \\ | \\ COO^- \end{matrix}}$$

The mammalian synthesis of methionine is more complex and requires cobalamin, a coenzyme form of vitamin B_{12}. Note that because methionine is an essential amino acid, it must be supplied in the diet; methionine that is used for methylation (Fig. 15-27) is degraded to homocysteine, and this is remethylated to re-form methionine. These reactions merely *recycle* methionine and do *not* represent a *net* synthesis.

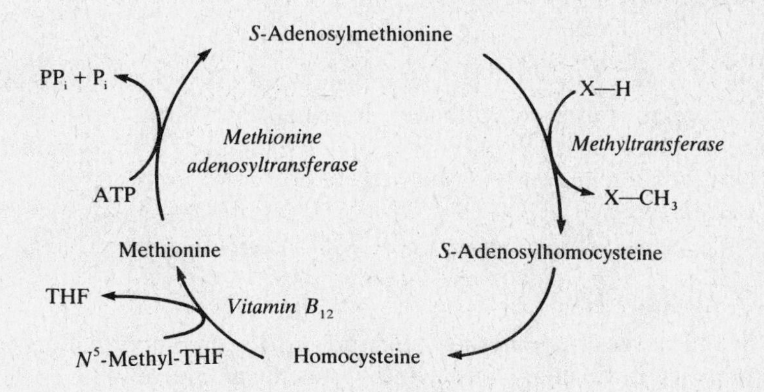

Fig. 15-27 Methionine metabolism.

EXAMPLE 15.15

Vitamin B_{12} does not exist in plants, and strict vegetarians risk suffering from vitamin B_{12} deficiency. Thus, we are dependent on animal or bacterial sources for our vitamin B_{12}.

Cobalamin is a complex molecule containing a Co atom. In the mammalian synthesis of methionine, cobalamin acts as a coenzyme by accepting the methyl group from N^5-methyltetrahydrofolate and transferring it to homocysteine. The reaction is catalyzed by *cobalamin–N^5-methyl-THF:homocysteine methyltransferase*. The overall reaction is

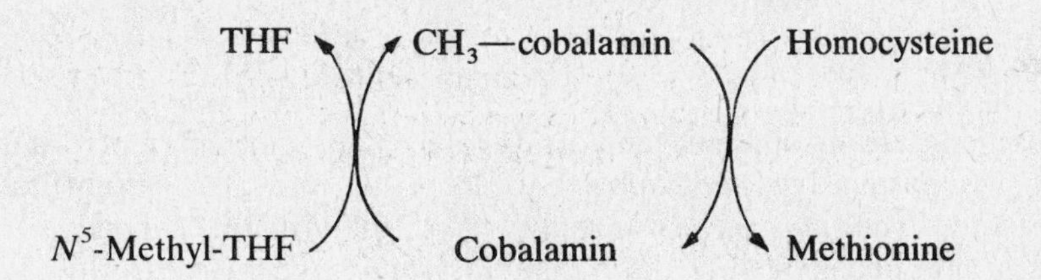

EXAMPLE 15.16

One aspect of vitamin B_{12} deficiency is that it results in an accumulation of N^5-methyl-THF. N^5-Methyl-THF is synthesized in mammals by an irreversible reaction (as seen earlier in this section); if it cannot be utilized (see above reaction) because of a deficiency of vitamin B_{12}, then it accumulates. This causes a depletion of the other forms of THF, resulting in a deficiency of THF. *Megaloblastic anemia* is associated with deficiencies of cobalamin (pernicious anemia) and THF.

The methyl group on methionine is activated when methionine is converted to S-*adenosylmethionine* (Fig. 15-6). It is the methyl group of *S*-adenosylmethionine that is the immediate donor in biological methylations. Important reactions in which *S*-adenosylmethionine acts as the methyl donor are the synthesis of creatine, epinephrine, and phosphatidylcholine.

EXAMPLE 15.17

Creatine is synthesized from guanidinoacetate (which is made from glycine and arginine).

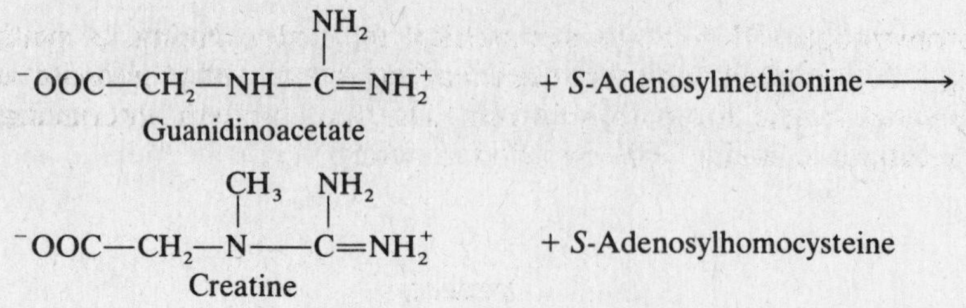

$$^{-}OOC{-}CH_2{-}NH{-}\overset{\displaystyle NH_2}{\overset{|}{C}}{=}NH_2^+ \quad + S\text{-Adenosylmethionine} \longrightarrow$$

Guanidinoacetate

$$^{-}OOC{-}CH_2{-}\overset{\displaystyle CH_3}{\overset{|}{N}}{-}\overset{\displaystyle NH_2}{\overset{|}{C}}{=}NH_2^+ \quad + S\text{-Adenosylhomocysteine}$$

Creatine

The carbon skeleton of homocysteine is salvaged and used to synthesize methionine (Fig. 15-27). Alternatively, homocysteine is used to synthesize cysteine (Fig. 15-6).

The importance of the cycle in Fig. 15-27 is that it conserves homocysteine. Methionine and cysteine that are used for protein synthesis will deplete homocysteine from the cycle (see Fig. 15-6) and must be replaced in the diet.

15.8 PORPHYRIN METABOLISM

The synthesis and turnover of porphyrins, heme precursors, are important because of the central roles of the heme proteins, hemoglobin, and cytochromes. Quantitatively, hemoglobin synthesis is a significant part of the nitrogen economy.

Porphyrin Synthesis

The first step of porphyrin synthesis is the condensation of succinyl-CoA and glycine to form *δ-aminolevulinate*. The reaction takes place in mitochondria, where succinyl-CoA is available. The reaction is irreversible and requires pyridoxal phosphate and Mg^{2+}. It is catalyzed by the enzyme *δ-aminolevulinate synthase*.

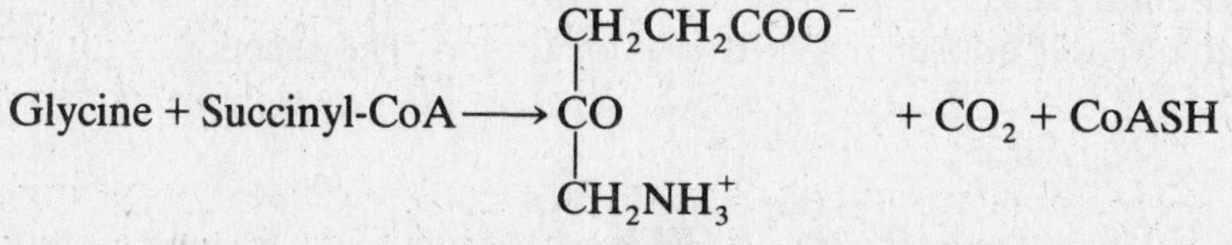

$$\text{Glycine} + \text{Succinyl-CoA} \longrightarrow \begin{array}{c} CH_2CH_2COO^- \\ | \\ CO \\ | \\ CH_2NH_3^+ \end{array} + CO_2 + \text{CoASH}$$

δ-Aminolevulinate

Subsequent reactions are cytoplasmic and irreversible. Two molecules of δ-aminolevulinate are condensed by the enzyme *porphobilinogen synthase* to form the trisubstituted pyrrole *porphobilinogen*. Two enzymes, *uroporphyrinogen synthase* and *uroporphyrinogen cosynthase* condense four molecules of porphobilinogen to the porphyrin *uroporphyrinogen III*.

$^{-}OOCCH_2$ $CH_2CH_2COO^-$

4 $^{+}H_3NCH_2$ N H $\longrightarrow$ $4NH_4^+$

$^{-}OOCH_2C$ $CH_2CH_2COO^-$

(A)

$^{-}OOCH_2C$ N H CH_2COO^-

(D) NH HN (B)

$^{-}OOCCH_2CH_2$ H N $CH_2CH_2COO^-$

(C)

$^{-}OOCCH_2CH_2$ CH_2COO^-

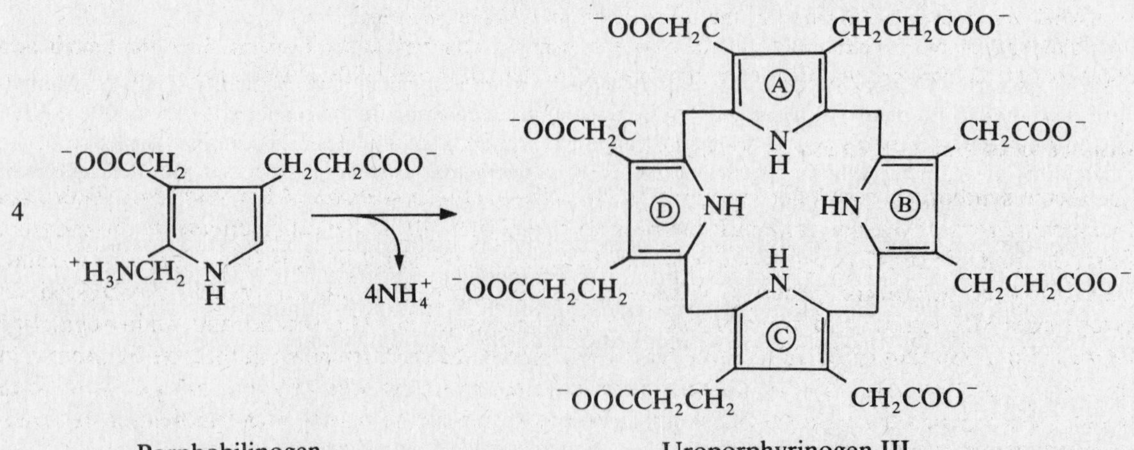

Porphobilinogen Uroporphyrinogen III

Note that uroporphyrinogen III is not a symmetrical molecule. During its synthesis, one of the pyrrole rings (ring D) is reversed, with the result that the acetate and propionate side chains are not symmetrically arranged around the porphyrin ring. The key porphyrin intermediate in cytochrome and hemoglobin synthesis is *protoporphyrin IX* (Fig. 15-28).

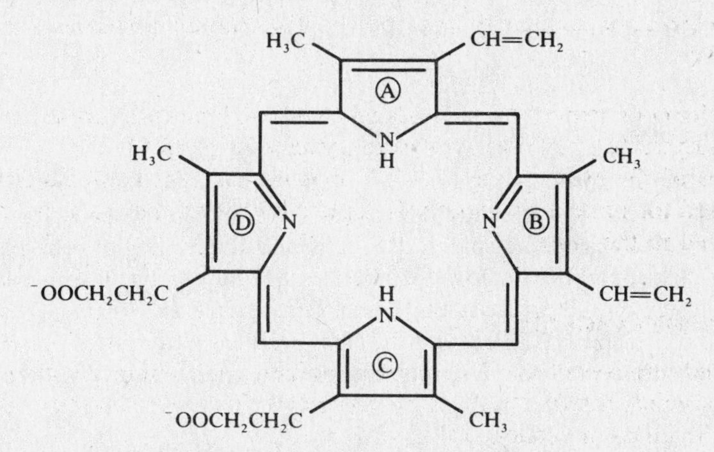

Fig. 15-28 Structure of protoporphyrin IX.

The synthesis of protoporphyrin IX involves two alterations to the side chains of uroporphyrinogen III: decarboxylation of the acetate groups to methyl groups and decarboxylation of the propionate residues in rings A and B to vinyl groups ($-CH{=}CH_2$). The first decarboxylations take place in the cytoplasm, while the formation of the vinyl groups and the conversion of the methylene bridges ($-CH_2-$) to the unsaturated methene ($=CH-$) is mitochondrial. The final product from these reactions is the fully aromatic, planar protoporphyrin IX. The final reaction in the mitochondria is the chelation of Fe^{2+} to form heme, a reaction which occurs spontaneously, although the enzyme *ferrochelatase* accelerates the reaction.

Heme is the functional group in hemoglobin and myoglobin, the cytochromes, and the enzymes catalase and peroxidase. These molecules all have quite different functions: hemoglobin carries oxygen; myoglobin stores oxygen; the cytochromes transfer electrons; and catalase and peroxidase are enzymes that catalyze the decomposition of hydrogen peroxide and oxidation by peroxides, respectively.

EXAMPLE 15.18

What determines the function of the heme group in different proteins?

Although there are differences in the way the heme is attached to the proteins, it is the amino acid sequences of the proteins that determine the function of the porphyrins in these molecules.

Regulation of Heme Synthesis in Reticulocytes

Heme synthesis is controlled primarily by *δ-aminolevulinate synthase* (ALA synthase). There are two mechanisms of control, and each involves a process that affects the *concentration* of the enzyme. First, the half-life of ALA synthase, as shown by experiments in rat liver, is very short (60–70 min). Like many mitochondrial proteins, ALA synthase is encoded by nuclear genes, synthesized on cytoplasmic ribosomes, and translocated into the mitochondria. The second and main regulating factor is the inhibition of ALA synthase by *hemin*. Hemin differs from heme in that the Fe atom is in the Fe^{3+} oxidation state. Heme spontaneously oxidizes to hemin when there is no globin to form hemoglobin. Hemin serves a second function in the regulation of hemoglobin synthesis in reticulocytes. It controls the synthesis of globin.

High concentrations of hemin inhibit the transport of ALA synthase into the mitochondria, where one of the substrates, succinyl-CoA, is formed. Thus, heme synthesis is inhibited until enough globin is made to react with any heme already formed. Low concentrations or the absence of hemin, the signal that globin is not needed, inhibits protein (and, therefore, globin) synthesis. In the absence of hemin, a *protein kinase* is activated. This in turn phosphorylates an *initiation factor* of eukaryotic protein synthesis, *eIF-2*, which then inhibits polypeptide chain initiation (Chap. 17) and hence inhibits globin synthesis.

Degradation of Hemoglobin

The life span of the erythrocyte is about 120 days with about 0.85 percent of the total being broken down each day in the reticuloendothelial (RE) cells of the spleen, liver, and bone marrow. The erythrocytes are lysed inside the splenic RE cells and the hemoglobin released. The globin is hydrolyzed to amino acids, and the heme is metabolized as follows. The porphyrin ring is oxidatively cleaved between rings A and B to form the linear tetrapyrrole *biliverdin*. The complete reaction requires molecular oxygen and NADPH, and the final product is bilirubin. The Fe^{2+} is salvaged via transferrin and stored within the protein *apoferritin*, and the methene bridge between rings A and B is removed as CO.

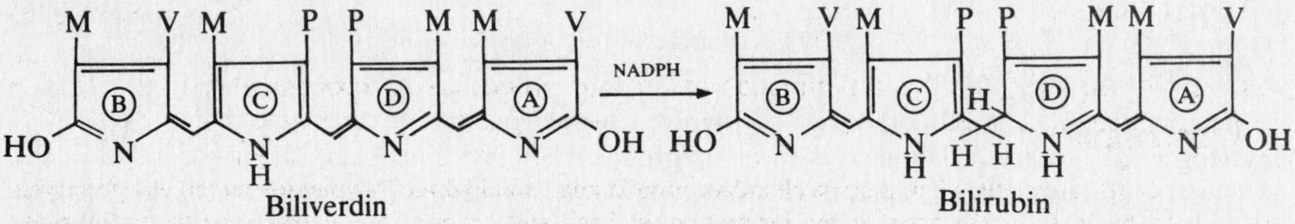

M V M P P M M V — Biliverdin (rings B, C, D, A; HO ... OH) —NADPH→ M V M P P M M V — Bilirubin (rings B, C, D, A; HO ... OH)

Biliverdin **Bilirubin**

In the structures shown, M = methyl group, V = vinyl group, and P = propionate. *Biliverdin* and *bilirubin* are bile pigments and are familiar as the green and orange coloring of bruises. Bilirubin, a water-insoluble molecule, is released into the plasma complexed with albumin and transported to the liver. In the liver it is solubilized by being converted to *bilirubin diglucuronide* (90 percent) and *bilirubin sulfate* (10 percent).

EXAMPLE 15.19

Glucuronic acid is a derivative of glucose. The —CH_2OH group on C-6 has been oxidized to —COOH. The active form of glucuronic acid is UDP-glucuronate and this is used in creating glucuronides, such as bilirubin diglucuronide. Glucuronide formation (and sulfate formation) is a common method of increasing solubility because of the polarity of the —OH and —COO^- groups of glucuronic acid. It is particularly important in the excretion of many insoluble drugs.

Bilirubin diglucuronide is excreted from the liver via the bile into the intestine. Within the bowel, it is hydrolyzed, and the bilirubin is reduced to *urobilinogen* and *stercobilinogen*. These are excreted in urine as *urobilin* and in feces as *stercobilin* and give urine and feces their characteristic colors.

Solved Problems

SYNTHESIS AND DIETARY SOURCES OF AMINO ACIDS

15.1. What effect will high concentrations of NADPH and 2-oxoglutarate have on the assimilation of ammonia?

SOLUTION

Either or both of these compounds in high concentration will favor the products in the reaction catalyzed by glutamate dehydrogenase. This shift of the chemical reaction to the right will result in assimilation of ammonia.

$$NH_4^+ + \text{2-Oxoglutarate} + NADPH \rightleftharpoons \text{Glutamate} + NADP^+ + H_2O$$

15.2. How does the TCA cycle operate if oxaloacetate and 2-oxoglutarate are removed for amino acid synthesis?

SOLUTION

Any oxaloacetate or 2-oxoglutarate removed from the TCA cycle must be replaced. Pyruvate is converted by pyruvate carboxylase to oxaloacetate, which can then enter the TCA cycle to yield 2-oxoglutarate.

15.3. In the fed state, what is the direction of the net carbon flux in transamination?

SOLUTION

In the fed state, when there is abundant protein and carbohydrate, the dietary protein is hydrolyzed to amino acids. Those not required for protein synthesis are converted to 2-oxoacids by the aminotransferases. The 2-oxoacids are then converted into lipids and carbohydrate for storage. Glutamate dehydrogenase catalyzes the formation of ammonia from the excess amino groups; this ammonia is excreted as urea.

15.4. What determines the fate of pyruvate produced in muscle from amino acid metabolism?

SOLUTION

In the normal fed state, pyruvate is oxidized via pyruvate dehydrogenase (PDH), but in starvation, PDH is inactivated; thus, pyruvate is converted into alanine (Fig. 15-16) which enters the blood and is conveyed to the liver, where gluconeogenesis takes place.

DIGESTION OF PROTEINS

15.5. What is the fate of the proteases after protein digestion?

SOLUTION

The proteases and other pancreatic enzymes such as pancreatic lipase are eventually degraded (by proteases). Digestion is responsible for the turnover of about 50 g of endogenous protein per day. This comes from the breakdown of the pancreatic enzymes as well as the epithelial cells of the gut, which are replaced every 24 hours.

AMINO ACID CATABOLISM

15.6. Amino acid oxidases exist for both D- and L-amino acids. They catalyze the following reaction:

$$R{-}CH(COO^-){-}NH_3^+ + \tfrac{1}{2}O_2 \longrightarrow R{-}CO{-}COO^- + NH_4^+$$

What function do they serve?

SOLUTION

The L-amino acid oxidase occurs in low activity in liver and kidney; but D-amino acid oxidase is of high activity in the peroxisomes of liver, kidney, and brain. D-amino acid oxidase is *glycine oxidase*; it contains FAD and catalyzes the formation of glyoxylate:

$$\underset{\text{Glycine}}{^{+}NH_3{-}CH_2{-}COO^-} \xrightarrow[\text{FAD}\ \ \text{FADH}_2]{} [NH{=}CH{-}COO^-] \xrightarrow{H_2O} NH_3 + \underset{\text{Glyoxylate}}{CHO{-}COO^-}$$

Glyoxylate is further oxidized to CO_2 and formate ($HCOO^-$), which is used in C_1 metabolism.

15.7. Pyridoxal phosphate is a coenzyme in amino acid decarboxylations. What is a likely mechanism of the decarboxylation, and what are the products?

SOLUTION

As with transaminations, the first step is the formation of a Schiff base between the amino acid and pyridoxal phosphate:

$$R{-}C(H)(COO^-){-}N{=}CH{-}[\text{pyridinium } N^+H] \xrightarrow{-CO_2} R{-}CH{=}N{-}CH{=}[\text{dihydropyridine } NH]\ \ H^+ \rightleftharpoons R{-}CH_2{-}N{=}CH{-}[\text{pyridinium } N^+H]$$

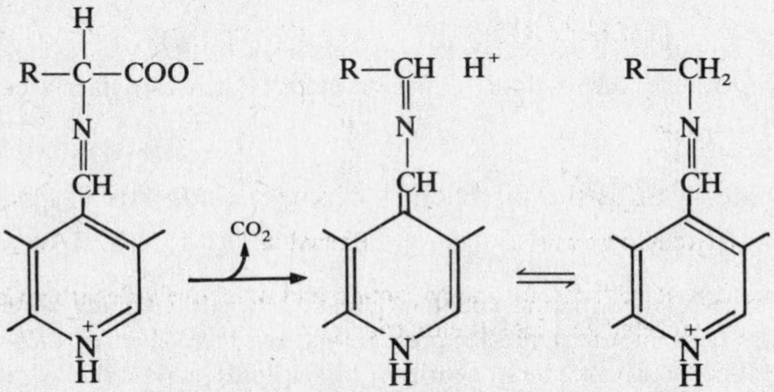

The quaternary nitrogen acts as an electron sink, which facilitates the decarboxylation. Further electron and proton shifts produce a Schiff base between an amine and pyridoxal phosphate, which can then be hydrolyzed.

Amino acid decarboxylations are involved in the synthesis of several physiologically important amines, e.g., *5-hydroxytryptamine* (serotonin) from tryptophan, *histamine* from histidine, and *γ-aminobutyric acid* (GABA) from glutamate.

15.8. Phenylalanine is ultimately degraded to acetoacetate and fumarate; hence it is both ketogenic and glucogenic. Initially, phenylalanine is hydroxylated to tyrosine, then transaminated to *p*-hydroxyphenylpyruvate. This is further oxidized to homogentisic acid and then to fumarate and acetoacetate:

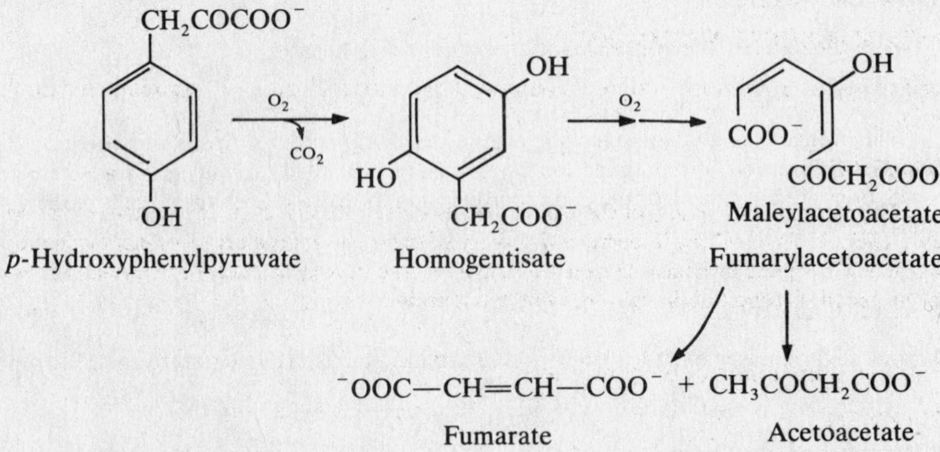

$$\underset{p\text{-Hydroxyphenylpyruvate}}{HO{-}C_6H_4{-}CH_2COCOO^-} \xrightarrow[CO_2]{O_2} \underset{\text{Homogentisate}}{(HO)_2C_6H_3{-}CH_2COO^-} \xrightarrow{O_2} \underset{\text{Maleylacetoacetate} \to \text{Fumarylacetoacetate}}{HOC(=CH{-}CH{=}CH{-}COO^-)COCH_2COO^-}$$

$$\to \underset{\text{Fumarate}}{^{-}OOC{-}CH{=}CH{-}COO^-} + \underset{\text{Acetoacetate}}{CH_3COCH_2COO^-}$$

What happens if the enzyme needed to cleave the aromatic ring is absent?

SOLUTION

Alkaptonuria is a rare inborn error of metabolism caused by a lack of the enzyme *homogentisate oxidase*, resulting in a failure to oxidize homogentisate, which is excreted in the urine. Cartilage and other connective tissues become pigmented by polymerized oxidized homogentisate, and in later years, arthritis develops.

15.9. The disease phenylketonuria, which causes severe mental retardation, is characterized by the urinary excretion of phenylpyruvate. Why is this formed?

SOLUTION

Phenylketonuria is due to an inborn error of phenylalanine metabolism. Typically, it is due to a deficiency of phenylalanine hydroxylase. Atypically, it can be caused by a deficiency of *dihydrobiopterin reductase* and a resultant inability to synthesize *biopterin*. All these conditions cause an accumulation of phenylalanine, which can be transaminated to phenylpyruvic acid.

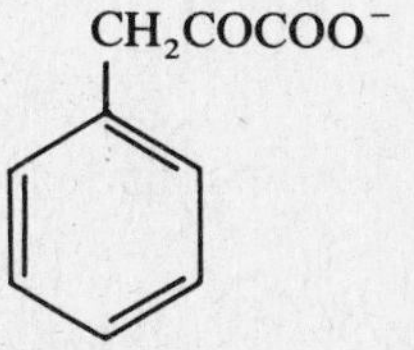

CH_2COCOO^- (Phenylpyruvate) — $CH_2CH(OH)COO^-$ (Phenyl lactate) — CH_2COO^- (Phenyl acetate)

Phenylpyruvate 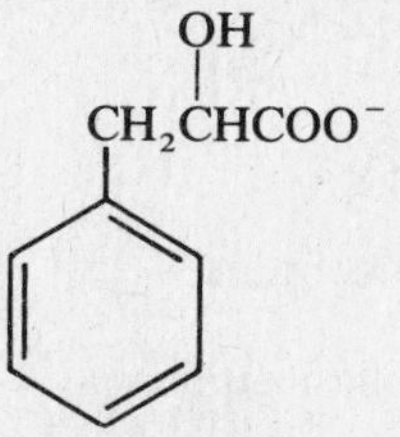 | Phenyl lactate 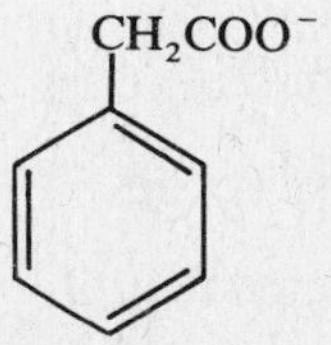 | Phenyl acetate

Phenylpyruvate can be reduced to phenyl lactate and oxidatively decarboxylated to phenylacetate, both of which are also excreted in phenylketonuria.

15.10. Normally, homocysteine is either remethylated to methionine with N^5-methyltetrahydrofolate (N^5-Me-THF) or converted into cysteine via cystathionine, but abnormally it can be excreted as homocystine. What does urinary excretion of homocystine indicate?

SOLUTION

Homocystinuria is a biochemical abnormality caused either by a deficiency of cystathionine β-synthase or by impaired activity of cobalamin-N^5-methyltetrahydrofolate:homocysteine methyltransferase. The classical homocystinuria occurs when the conversion of homocysteine to cystathionine is limited by a deficiency of cystathionine β-synthase, with accumulation of methionine and homocysteine and a decrease in cysteine.

DISPOSAL OF EXCESS NITROGEN

15.11. Under what conditions would citrulline, argininosuccinate, or arginine accumulate in the liver?

SOLUTION

The accumulation of any of these amino acids is due to a lack of their respective enzymes in the urea cycle (Sec. 15.5), resulting in decreased activity of the cycle. Inborn errors of metabolism are known for deficiencies in these enzymes. Decreased activity of the urea cycle results in elevated levels of ammonia in the blood, a condition known as *hyperammonemia*.

15.12. Can hyperammonemia be controlled?

SOLUTION

Feeding low-nitrogen diets that contain the 2-oxoacid counterparts of the essential amino acids will reduce ammonia concentrations.

15.13. Predict the form in which excess ammonia would be excreted in the following organisms: tadpoles, frogs, birds, and mammals.

SOLUTION

As aquatic animals, tadpoles can use diffusion to remove excess ammonia, but the amphibian frog and the reptiles, which have skins designed to retain water and prevent dehydration, excrete urea. Birds do not have any spare water (it represents extra weight) and excrete uric acid. In most mammals, excess ammonia is excreted as urea.

15.14. Creatinine is an important nitrogenous constituent of urine. It is formed nonenzymatically from phosphocreatine by spontaneous cyclization, resulting from phosphorylosis.

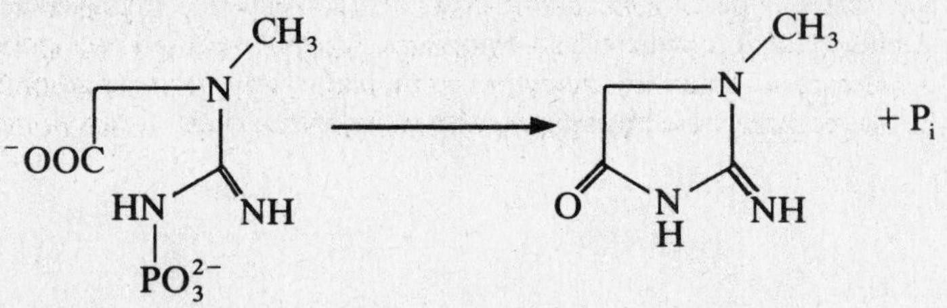

CH_3 N ^{-}OOC HN NH PO_3^{2-} → CH_3 N O N H NH $+ P_i$

Phosphocreatine Creatinine

Is the formation of creatinine significant?

SOLUTION

The rate of formation of creatinine is constant; the amount is dependent on muscle mass. Hence, an individual excretes a constant amount (1–1.5 g per day for an adult man). Its excretion represents a net loss of methyl groups.

PURINE AND PYRIMIDINE METABOLISM

15.15. Uric acid crystals can be deposited in the joints, causing the painful condition known as *gout*. What, in turn, causes the deposition of the uric acid crystals?

SOLUTION

Gout is caused by an overproduction of purines, which leads to the overproduction of uric acid. Because of its insolubility, uric acid precipitates in the joints and causes inflammation.

15.16. The *Lesch-Nyhan syndrome* is a distressing disorder that includes neurological abnormalities, self-mutilation, and overproduction of uric acid. Why is uric acid overproduced?

SOLUTION

The Lesch-Nyhan syndrome is caused by a deficiency of the phosphoribosyltransferase that is involved in the salvage pathway for hypoxanthine and guanine. The absence of the enzyme results in insufficient GMP to stimulate purine biosynthesis.

METABOLISM OF C_1 COMPOUNDS

15.17. The sulfonamides are antibacterial agents which are S analogs of *p*-aminobenzoate. The simplest of these is sulfanilamide.

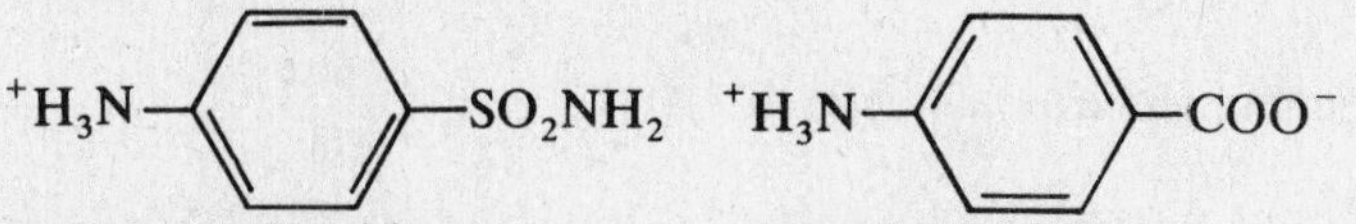

$^{+}H_3N-C_6H_4-SO_2NH_2$ $^{+}H_3N-C_6H_4-COO^{-}$

Sulfanilamide *p*-Aminobenozate

How do the sulfonamides work?

SOLUTION

They act as competitive inhibitors of the incorporation of *p*-aminobenzoate into folic acid by bacteria. Without folic acid, the bacteria cannot grow.

PORPHYRIN METABOLISM

15.18. Inborn errors of metabolism are normally associated with a deficiency or absence of an enzyme. Can excessive production of an enzyme also result in an inborn error of metabolism?

SOLUTION

Excessive production of liver δ-aminolevulinate synthetase causes two forms of congenital *porphyria*. These diseases are characterized by overproduction of porphyrins and excretion of large amounts of δ-aminolevulinate and porphobilinogen. Some ethnic groups are susceptible to this disease, and in these people, acute attacks are brought on by barbiturates and other compounds that induce synthesis of the enzyme.

Supplementary Problems

15.19. Why would a diet rich in energy but low in nitrogen result in malnutrition?

15.20. Would you expect a diet in which meat had been replaced by cheese to be nutritionally adequate in nitrogen?

15.21. Which proteases would be needed to hydrolyze the following peptides completely?

(*a*) Tyr-Phe-Gly-Ala

(*b*) Ala-Arg-Tyr-Glu

(*c*) Leu-Trp-Lys-Ser

15.22. Glutamine from the diet and from the muscle during fasting can be converted into alanine in the intestine. Which enzymes are involved in this conversion?

15.23. Why is arginine classified as an essential amino acid when it is synthesized in the urea cycle?

15.24. Pyridoxal phosphate is the coenzyme involved in amino acid epimerases. Write a mechanism for this reaction, using the mechanism for amino acid decarboxylation in Prob. 15.7 as a guide.

Chapter 16

Replication and Maintenance of the Genetic Material

16.1 INTRODUCTION

DNA is the genetic material of cells; i.e., it contains the genes. An individual *gene* is represented by a segment of DNA (up to several kb), that is eventually *expressed* as a polypeptide, frequently a structural protein or enzyme. The DNA is present in the chromosomes of a cell, with one or more chromosomes constituting the *genome*. The genome represents a single complement of the genetic information of a particular cell. Each chromosome contains a large number of genes. During cell division, the chromosomal DNA must produce exact replicas of itself for *partitioning*, through the process of *segregation*, into daughter cells. This production of copies of the DNA is known as *replication* and involves the synthesis of new DNA chains.

For normal cell growth and proliferation, the DNA must be protected from various types of damage. Such damage, induced, for example, by uv irradiation, can involve the chemical alteration of the DNA and, consequently, deleterious *mutation*. Cells are able to correct or repair such damage. One of the best-understood mechanisms of repair involves the synthesis of new DNA, which replaces the damaged portion. This is called *repair synthesis* of DNA. The extent of repair synthesis is very small in comparison with the DNA synthesis accompanying replication of the chromosomes.

Most cells (bacteria, plants, and animals) contain one, two, or a small number of most genes. However, in response to certain situations, some cells will produce many copies of a particular gene. This is known as *gene* or *DNA amplification* and involves severalfold replication of specific segments of the DNA.

EXAMPLE 16.1

The drug *methotrexate* is an inhibitor of the enzyme dihydrofolate reductase (Chap. 15). When mammalian cells are grown in the presence of this drug, they can develop resistance to it. Examination of the DNA from such resistant cells shows that frequently the number of copies of the gene for dihydrofolate reductase has been increased. The consequent increase in the level of dihydrofolate reductase activity endows resistance to the methotrexate.

16.2 SEMICONSERVATIVE REPLICATION OF DNA

DNA in the chromosomes of most organisms is double-helical; i.e., it consists of two polydeoxynucleotide chains (or strands) twisted around one another in the form of a helix. The genetic information is contained in the *sequence* of nucleotides along one of the chains, with the sequence in one being *complementary* to that in the other (Chap. 7). A *replica* of DNA is one that is an exact copy of itself.

Question: What feature of the DNA structure provides the basis for reproduction of the original nucleotide sequence in a replica?

Each chain of double-helical DNA is bound to the other through complementary base pairs, with adenine (A) in one being hydrogen-bonded to thymine (T) in the other and guanine (G) to cytosine (C). Watson and Crick proposed that, to achieve precise copying of a nucleotide (base) sequence, the two chains of the DNA must *unwind* from one another to allow each single chain to act as a *template* for the synthesis of a new one. Thus, the assembly of the sequence in the newly synthesized chain is determined by the base-pairing specificity of the sequence in the template. This is illustrated in Fig. 16-1 (the DNA *duplex* is shown as a ladder-type structure).

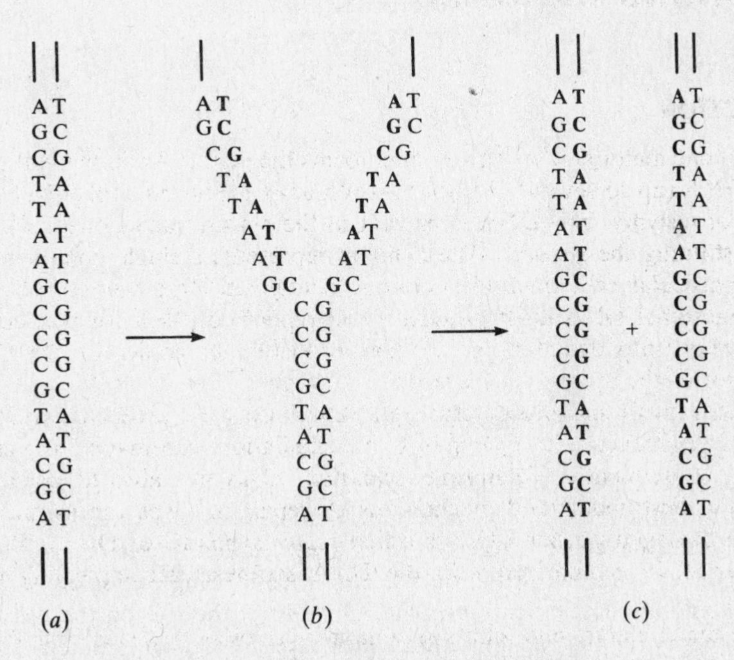

Fig. 16-1 Unwound DNA as a template for replication. The bold letters indicate newly synthesized DNA.

In Fig. 16-1, the DNA to be replicated (*a*) is shown to unwind from the top, and the partially replicated structure (*b*) appears Y-shaped and contains a *replication fork*. Within each arm of this structure, the newly synthesized chains (shown in bold type) are assembled according to the base-pairing instructions of the unwound template chains. The length of sequence shown in (*a*) represents only a very minor segment of a giant DNA molecule in a chromosome. The completely replicated segments are shown in (*c*); they are now separated from each other. Each contains one of the original chains and one new chain. The *parental* segment of DNA gives rise to two *daughter* segments, which will remain intact through subsequent generations.

Question: What is meant by the term *semiconservative replication* of DNA?

Semiconservative replication refers to the conservation of just one half of the parental DNA structure when it undergoes replication to give two daughter molecules. Thus, in Fig. 16-1, each chain of the parental DNA acts as a template and remains intact through the doubling process.

16.3 TOPOLOGY OF DNA REPLICATION

A chromosome contains a single DNA molecule, which is generally very large; e.g., the bacterial chromosome is composed of $\sim 4 \times 10^6$ base pairs. Furthermore, in many cases, the DNA is a closed or *circular* structure. We will concentrate first on the topology of replication of the circular bacterial (*Escherichia coli*) chromosome.

Question: What is the form taken by a replicating bacterial chromosome (circular DNA molecule)?

The replicating bacterial chromosome remains in a closed form, with a portion of its length duplicated and joined to the rest of the DNA at replication *forks* (Fig. 16-2).

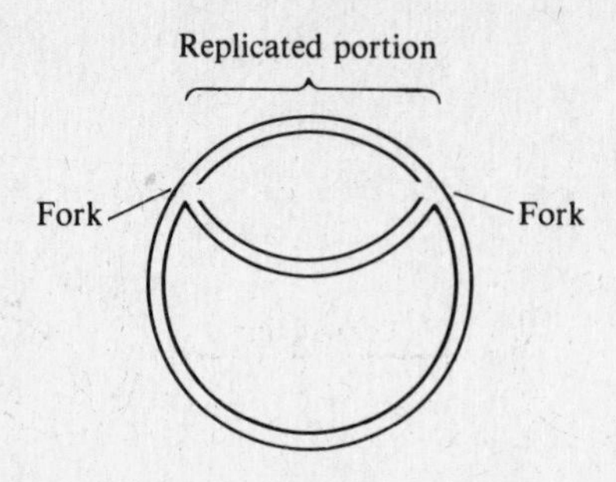

Fig. 16-2 Replicating circular bacterial chromosome.

Question: The replicating bacterial chromosome shown in Fig. 16-2 contains *two* forks. Are both of these replication forks?

Yes, both are actively involved in DNA replication and move at approximately equal rates in opposite directions around the circular molecule. This is known as *bidirectional replication*. The replicated portion of the molecule is referred to as a *replication bubble* or *eye form* (because of its appearance in diagrams). The size of the bubble varies from being extremely small to being nearly twice the size of the nonreplicating chromosome. Obviously, the site on the circular molecule of a very small bubble represents the region within which replication was initiated.

Question: Does initiation of replication of the bacterial chromosome occur at a fixed or a variable position?

Genetic studies have established that initiation of replication occurs at a fixed site, called the *initiation site* or *origin of the chromosome* (*oriC*). The nucleotide sequence in this region binds to various proteins to initiate the two forks.

Question: What happens when the two replication forks meet one another at the position opposite *oriC*?

The approaching forks meet, thus releasing two completed circular molecules (daughters). This is called *termination of replication*; at present, it is not known if the two forks meet at a fixed site on the bacterial chromosome.

EXAMPLE 16.2

Initiation of replication at a fixed site on the chromosome, the origin, and movement of replication forks away from this site toward the terminus imply a fixed order of replication of genes during this process. This has been confirmed through measurements of the relative *copy number* of genes in DNA isolated from growing cells. Genes near the origin of replication occur at a higher average frequency than those near the terminus.

Replication of the bacterial chromosome can be divided into three stages: initiation, elongation, and termination (Fig. 16-3). *Initiation* refers to the generation of replication forks at the origin. *Elongation* describes the progression of these forks around the chromosome, with concomitant DNA synthesis or *chain growth*. *Termination* refers to the fusion of the approaching forks, which results in two completed chromosomes that can separate from one another.

The term *replicon* is used to describe a *unit of replication* that is under the control of a single origin. Thus, the bacterial chromosome is a replicon; its replication is under the control of *oriC*.

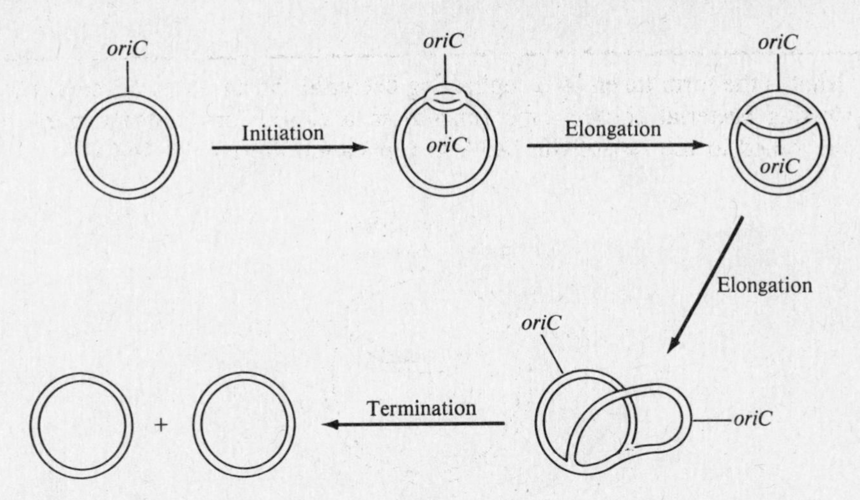

Fig. 16-3 Stages in the replication of a bacterial chromosome.

Question: Do all circular, double-stranded DNA molecules replicate in the same manner as the bacterial chromosome?

Yes, in that there is a defined origin of replication, but in some cases, e.g., some plasmids, replication proceeds in a single unique direction around the DNA molecule. This is called *unidirectional replication*. In other situations, e.g., mitochondrial DNA (Chap. 1), there is a more complex sequence of events that involves the unidirectional copying of only one strand over most of the DNA length, in a different type of eye form, referred to as a *D loop*.

Replication of the 4,000-kb bacterial chromosome takes about 40 min and occurs *throughout* the bacterial division cycle. Thus, *each* fork replicates about 50 kb of DNA per minute. In eukaryotic cells, DNA replication is restricted to the portion of the mitotic cell division cycle called the *S* (synthesis) *phase*, which can extend for several hours.

EXAMPLE 16.3

The cell cycle in eukaryotes is divided into four phases, G_1, S, G_2, and M (Fig. 16-4). The cycle starts at the beginning of G_1 (G = gap), which often represents the major phase in duration. DNA replication commences at the beginning of the S phase and is completed before entry into G_2. Mitosis, or the M phase, is relatively short and includes those steps that lead to chromosome segregation and partitioning into two daughter cells.

Question: A eukaryotic cell contains ~1,000 times the amount of DNA of a bacterial chromosome. How can replication of eukaryotic DNA be accomplished within the period of the S phase (a few hours)?

Examination of replicating eukaryotic DNA by electron microscopy shows the presence of many *tandemly* arranged bubbles separated by only 30–300 kb and clustered in groups of 20–80 in various regions of the DNA. Both forks of each bubble represent sites of replication, which move in opposite directions, until they fuse with an approaching fork from an adjacent bubble. Thus, eukaryotic DNA contains many tandemly arranged *replicons*, each with an origin of replication. Actually the rate of fork movement in eukaryotes is less than 10 percent of the rate in bacteria, but the genome can be replicated in the time available because of the multitude or origins.

EXAMPLE 16.4

A cultured mammalian cell contains about 20,000 origins. As shown in Fig. 16-5, the replication forks initiated at each origin, O, have only a relatively short distance to traverse before fusion (or termination) with an approaching fork from an adjacent replicon.

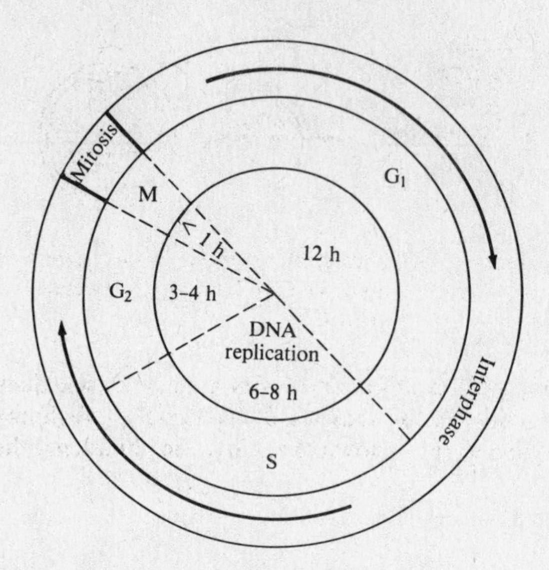

Fig. 16-4 The eukaryotic cell cycle. The durations of the four phases as shown are typical for higher eukaryotic cells growing in tissue culture.

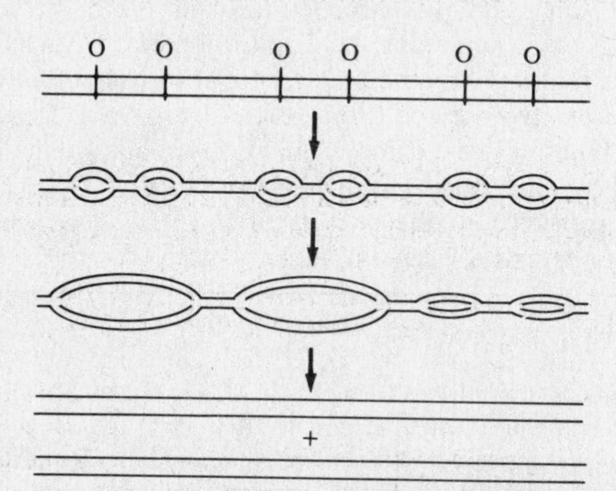

Fig. 16-5 Tandemly arranged replicons in eukaryotic DNA generate a chain of replication bubbles. O denotes an origin of replication.

16.4 CONTROL OF DNA REPLICATION

The rate of DNA replication is coordinated with the rate of cell division. Thus, a bacterial culture growing in a rich medium has a short generation time and must accomplish chromosome replication more quickly than one in a poor medium, where the generation time might be three to four times longer.

Question: By what mechanism is the rate of DNA replication in bacteria altered?

Under all nutrient conditions at a fixed temperature, the rate of replication-fork movement remains fairly constant. The overall rate of DNA replication is determined by the *frequency* of initiation at *oriC*. To accelerate replication, initiation at daughter origins occurs before the ongoing cycle of replication is completed. This results in the formation of a *multiforked chromosome*, as shown in Fig. 16-6. Note that a multiforked chromosome contains at least four copies of *oriC*.

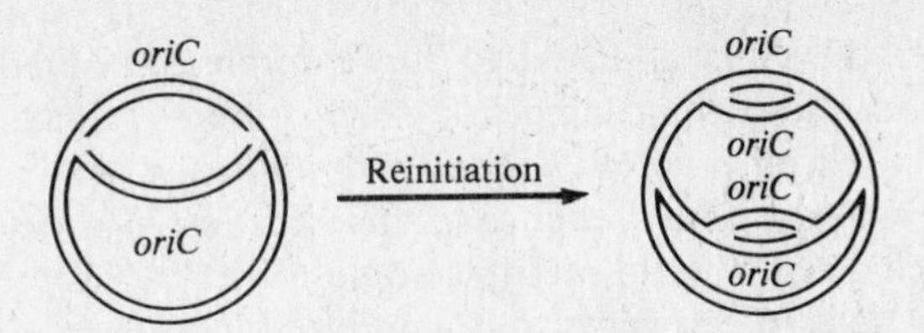

Figure 16-6 A multiforked chromosome generated by reinitiation of replication at *oriC*.

The control of DNA replication in eukaryotes is more complex than in bacteria. Once a cell enters the S phase, the multitude of individual replicons throughout the genome commence replication in a defined sequence. It is known that, in some situations, the duration of the S phase can vary greatly.

EXAMPLE 16.5

The first two nuclei formed after fertilization of an egg of *Drosophila melanogaster* undergo division in just a few minutes; thus, the length of the S phase is *very* short. In this situation, the origins of adjacent replicons are spaced at much shorter than normal intervals, 7–8 kb, as compared with a norm of 30–300 kb. The whole genome is replicated quickly because each replication fork traverses only a few kb before fusion with an approaching one.

EXAMPLE 16.6

Can individual replicons in eukaryotes undergo reinitiation, as in bacteria, to yield multiforked structures?

No, but a situation very similar to it can arise in some specialized cells in which specific genes are *amplified*. For example, the developing egg of many animals contains at least a thousand copies of the rRNA genes (Chaps. 7 and 17). In *Drosophila*, the DNA sequences coding for the *chorion* (egg) *proteins* are amplified about 30-fold just before they are needed. To achieve amplification, a small region of the DNA traversing the particular gene (or genes) is replicated many times during a single cell generation, as shown in Fig. 16-7.

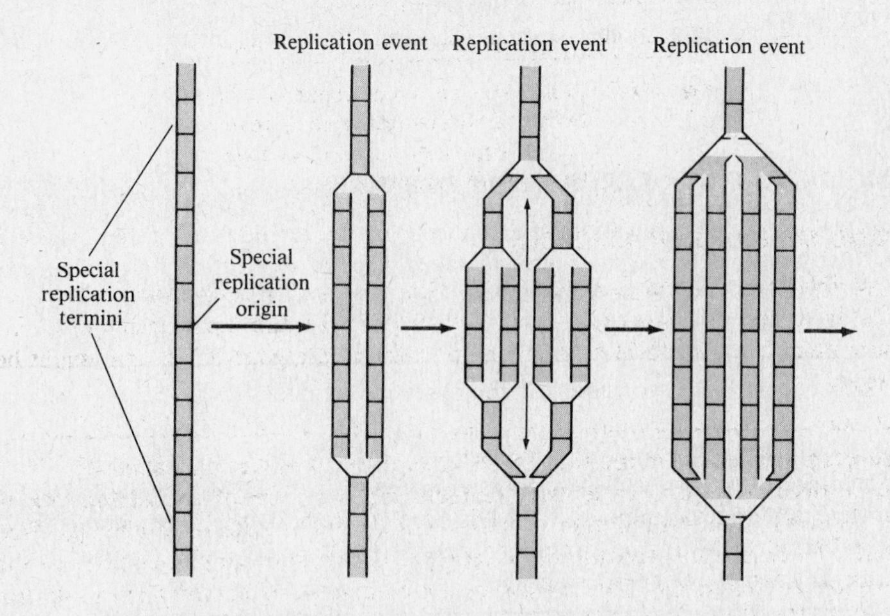

Fig. 16-7 Amplification of a segment of DNA through replication.

Note that the amplified DNA is attached to the chromosomal DNA (Fig. 16-7). When the amplified DNA is *freed* from the chromosome, the freed form is referred to as *extrachromosomal DNA*. Such DNA is generated by *nonhomologous recombination* within an amplified segment to produce a *circular excised sequence*. The sequence in this circular DNA can then be amplified further through a *rolling-circle* mechanism to yield multiple copies of it that are tandemly arranged within a long linear DNA molecule. It is in this form that amplified rRNA genes exist.

EXAMPLE 16.7

How does the rolling-circle mechanism operate?

A cut or *nick* is made in one of the two chains, and the circle rolls, peeling away one end of the cut chain, to yield the equivalent of two single chains, which function as templates for DNA synthesis at a fork, as shown in Fig. 16-8. As the circle rolls through cycle after cycle, each newly formed daughter chain built on the circular template begins to peel off at the beginning of each new cycle and become the new linear template. Thus, a linear duplex molecule containing tandem copies of the sequence (concatemer) is generated. The rolling-circle mechanism is also used as a stage in the replication of some viral DNA molecules.

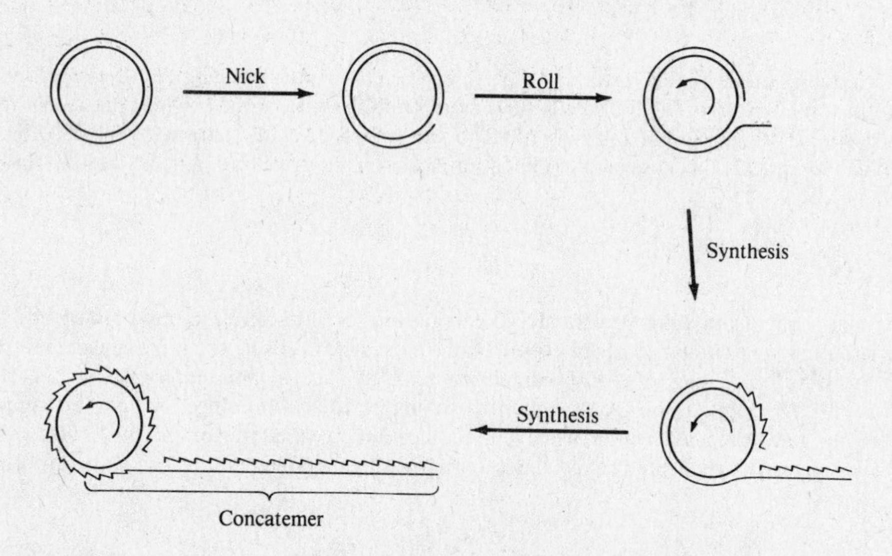

Fig. 16-8 Rolling-circle mechanism of DNA replication.

16.5 ENZYMOLOGY OF DNA REPLICATION IN BACTERIA

As already discussed, the replication of a replicon can be divided into three stages: initiation, elongation, and termination. DNA chain growth takes place at replication forks. The process as it occurs in the bacterium *E. coli* is best understood and serves as the prototype for other systems.

Question: Which enzyme is responsible for the synthesis of new DNA chains at the replication fork?

The enzyme, which uses unwound, single-stranded DNA as a *template*, is called a *DNA polymerase*. There are three distinct DNA polymerases in *E. coli*: DNA polymerase I, II, and III. DNA polymerase I is the most abundant, and DNA polymerase III the least abundant. These two enzymes have important roles in the overall process of DNA replication. A specific function for DNA polymerase II has not yet been established.

DNA Polymerase I

DNA polymerase I of *E. coli* has been studied in considerable detail. It uses the four deoxynucleoside triphosphates (dNTPs) and Mg^{2+} to catalyze the template-directed polymerization of nucleotides (Fig. 16-9). In the reaction shown in Fig. 16-9, an incoming dTTP is positioned opposite adenine (A) in the template (through base pairing), and a phosphodiester bond is formed through a *nucleophilic attack* by the 3′ hydroxyl of the growing chain on the α phosphorus of the incoming triphosphate. Pyrophosphate (PP_i) is released. The next nucleotide is then incorporated to extend the chain farther, and so on. Growth is exclusively in the 5′⟶3′ direction. Another important feature is that the enzyme can add only to a *preexisting* chain (*primer*); it cannot start a new one. Rapid hydrolysis of the released PP_i by *pyrophosphatase* in the cell serves to drive the reaction in the direction of DNA chain growth. The overall reaction catalyzed by DNA polymerase I is:

$$(\text{dNMP})_n + \text{dNTP} \xrightleftharpoons{\text{Mg}^{2+}} (\text{dNMP})_{n+1} + \text{PP}_i$$

However, this is only one of *three* types of reactions catalyzed by this enzyme. Its two other activities involve the *hydrolysis* of phosphodiester bonds. In one it acts as a *5′⟶3′ exonuclease* on *double-stranded* DNA, and in the other, it acts as a *3′⟶5′* exonuclease on a *frayed* or *mismatched* terminus of double-stranded DNA. DNA polymerase I is thus a multifunctional enzyme. It consists of a single polypeptide chain ($M_r = 110{,}000$). Its cleavage by proteases yields a *large* fragment (*Klenow fragment*; $M_r = 75{,}000$) that demonstrates the polymerase and 3′⟶5′ exonuclease activities and a *small* fragment ($M_r = 35{,}000$) that demonstrates only the 5′⟶3′ exonuclease activity.

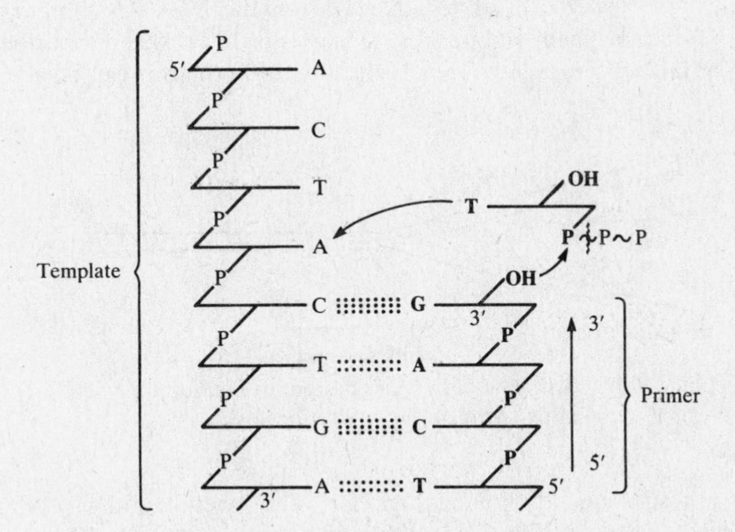

Fig. 16-9 Template-directed addition of a nucleotide unit to a growing DNA chain by DNA polymerase I. The bold letters indicate newly synthesized DNA.

DNA Polymerase III

DNA polymerase III is also a multifunctional enzyme. It resembles DNA polymerase I in catalytic properties; however, there are slight differences with respect to the type of template primer preferred for DNA synthesis, as well as the preferred substrates for the two exonuclease activities.

The DNA polymerase III discussed so far is more accurately known as *DNA polymerase III core enzyme*. It has an $M_r = 180{,}000$ and contains three polypeptide subunits (α, ε, and θ). The *holoenzyme* is a more complex form of DNA polymerase III that contains additional subunits (τ, γ, δ, and β). The holoenzyme functions to carry out most of the DNA synthesis at the replication fork in vivo.

Recall that the two chains in double-stranded DNA have opposite polarities (Chap. 7); one is $5' \longrightarrow 3'$ in direction and the other is $3' \longrightarrow 5'$. But, the DNA polymerases can extend a chain only in the $5' \longrightarrow 3'$ direction, so at a replication fork, only one of the new chains can be made $5' \longrightarrow 3'$ and move in the direction of fork movement. In Fig. 16-10, this occurs in the upper arm.

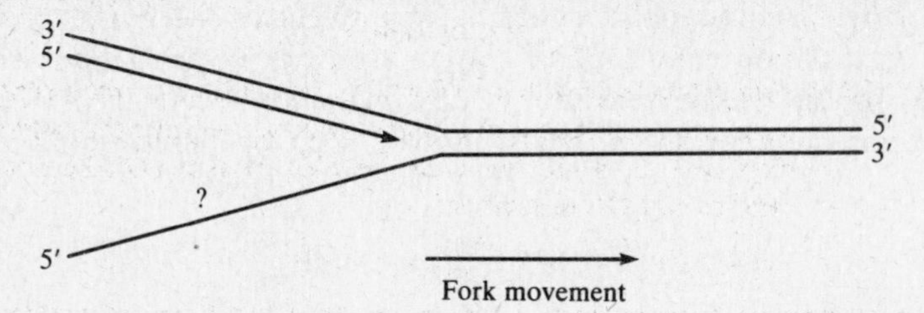

Fig. 16-10 Synthesis of DNA in one arm of the fork is in the same direction as fork movement.

Okazaki Fragments

Question: Figure 16-10 shows that the new chain synthesized in the lower arm must be $3' \longrightarrow 5'$ if synthesized in the direction of fork movement. How is synthesis achieved?

DNA synthesis in the lower arm of the diagram (on the $5' \longrightarrow 3'$ template) is made in the direction *opposite* to fork movement and in short segments of 100–1,000 nucleotides in length. These are called *Okazaki* or *nascent fragments*. Thus replication of both arms can be represented as shown in Fig. 16-11.

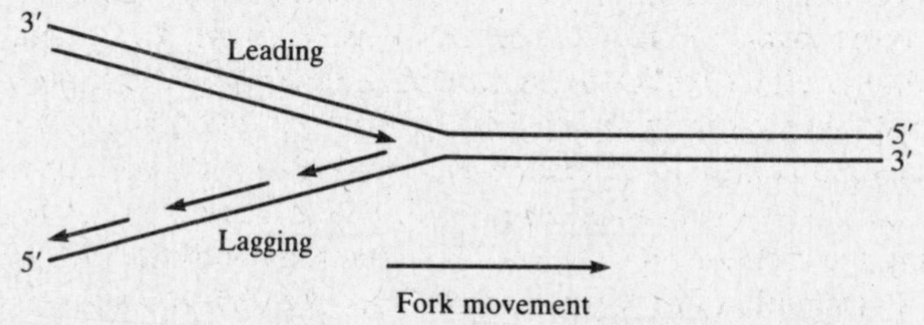

Fig. 16-11 Synthesis of DNA as nascent or Okazaki fragments in one arm of the replication fork.

The fragments are subsequently joined together. This overall pathway of assembly of new strands at the replication fork is called *discontinuous DNA replication*. The strand that is made continuously in the direction of fork movement is called the *leading* strand, and the other the *lagging* strand.

DNA polymerases are unable to initiate synthesis of a new chain; they can only *extend* a chain. Therefore, the Okazaki fragments are initiated by a special type of RNA polymerase, called *primase*, which makes short *RNA primers* opposite in direction to that of fork movement; the primers are made at many regions along the $5' \longrightarrow 3'$ template as it is progressively exposed by unwinding of the helix. Following release of the primase after completing each primer, the DNA chain is then extended (from $5' \longrightarrow 3'$) by DNA polymerase III holoenzyme. (The ability of the RNA polymerases to start chains de novo is a significant difference from the DNA polymerases.) Chain extension by DNA polymerase III continues until the newly synthesized DNA fragment comes up to the $5'$ end of an RNA primer in the adjacent fragment, as shown in Fig. 16-12.

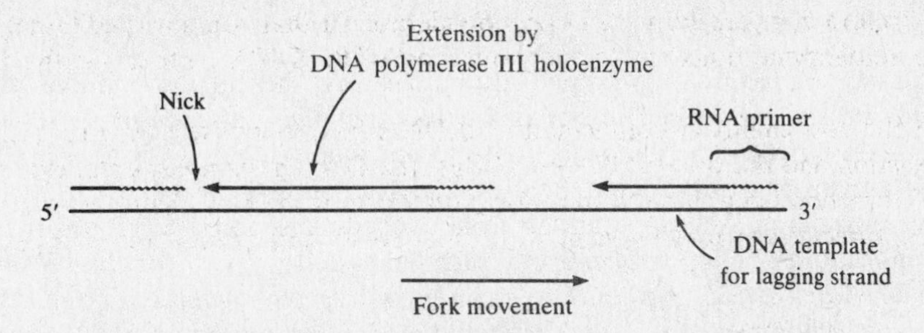

Fig. 16-12 Priming and growth of Okazaki fragments. The *nick* represents the unformed phosphodiester bond between the polynucleotide chains—in this case the RNA primer and the newly synthesized DNA segment.

Question: What determines the sites of initiation of RNA primer formation along the 5′⟶3′ template?

RNA formation, from a DNA template, is initiated at sites on DNA known as *promoters* (Chap. 17). For primer formation at the moving replication fork, the cell provides a mobile promoter (or *primosome*) that moves along the 5′⟶3′ template and functions at regular intervals to promote the initiation of RNA primer synthesis by the primase. The primosome is a complex of six proteins (n, n′, n″, i, *dna*C, and *dna*B proteins), and its movement along the DNA strand uses the energy of ATP hydrolysis. The *dna*B protein recognizes appropriate, regularly spaced sequences and induces the binding of primase to allow initiation of RNA primer synthesis.

The RNA primer must be removed and replaced by DNA. This is accomplished as follows. The DNA polymerase III holoenzyme is released when it reaches the 5′ end of the RNA primer, and DNA polymerase I takes over. A feature of DNA polymerase I, in contrast to other DNA polymerases, is its ability to effect replication at a nick. It doesn't matter if the polynucleotide *ahead* of the nick is DNA or RNA. Referring to Fig. 16-12, we would find that DNA polymerase I binds at the solid left arrowhead and continues to extend the DNA chain. At the same time, it removes the RNA primer through its 5′⟶3′ exonuclease activity. The overall effect is for it to replace the RNA by DNA, pushing the nick ahead of it, and this continues until the nick has shifted beyond the RNA section. This is an example of *nick translation*. Once the nick is bounded by DNA, it is sealed.

Question: How is a nick sealed in a DNA strand?

DNA ligase is responsible for sealing nicks. This enzyme catalyzes the formation of a phosphodiester link between adjacent 5′-phosphoryl and 3′-hydroxyl groups in double-stranded DNA. DNA ligase from *E. coli* requires NAD^+ as a coreactant, and the reaction mechanism involves the formation of an intermediate in which the adenyl group of NAD^+ is covalently attached to the enzyme. The 5′-phosphoryl terminus at the nick is activated by transfer of the adenyl group to form a *DNA-adenylate*. The phosphodiester bond is then formed by attack of the 3′ hydroxyl on the activated 5′ phosphate. The resultant reaction is shown in Fig. 16-13.

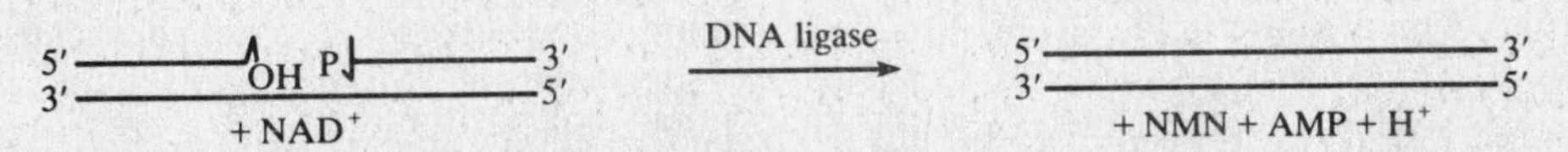

Fig. 16-13 The reaction catalyzed by DNA ligase in *E. coli*. (NMN = nicotinamide mononucleotide.) Eukaryotic ligases use ATP instead of NAD^+ as the coreactant.

Unwinding the Double Helix

The two strands of DNA are wrapped around one another in the form of a helix, and the discussion so far has not addressed in detail the fact that these strands must unwind in order to expose the template for copying. The main route to unwinding DNA is through the action of *DNA helicases*. These are enzymes that couple the unwinding of double-helical DNA to hydrolysis of ATP. *E. coli* has several different helicases, and at least two are considered to be involved in unwinding DNA during replication. The *rep protein* is one such helicase. It moves along the $3' \longrightarrow 5'$ template, unwinding the DNA just ahead of the site of chain growth of the *continuous* strand. It is assisted by *DNA helicase II*, which moves in the same overall direction but on the $5' \longrightarrow 3'$ template strand. It is possible that unwinding is facilitated in other ways: by the binding of *single-stranded DNA binding protein* (SSB protein) to DNA ahead of the fork, as well as by the introduction of *negative supercoiling* (Chap. 7) in this region by the enzyme *DNA gyrase*.

Question: What is the nature of the interaction of SSB protein with DNA?

SSB protein, also called *helix-destabilizing protein*, binds tightly and cooperatively to single-stranded DNA; i.e., when one molecule of SSB protein binds to the DNA, it *facilitates* the binding of more molecules of SSB protein. DNA is not a static structure; transient and localized strand separation occurs continuously, and may therefore occur ahead of the replication fork. Thus, SSB protein, by binding cooperatively to the transient single-stranded regions, actually lowers the melting temperature of DNA; in other words, it aids DNA separation into component strands. As well as facilitating unwinding, SSB protein functions in another way. A single strand to which SSB protein is bound is a more rapidly processed template for DNA polymerase; this is an important factor in promoting efficient DNA chain growth.

Question: What is the mechanism of DNA gyrase action?

DNA gyrase belongs to a class of enzymes known as *topoisomerases*. Such enzymes, by catalyzing the concerted cutting and closing of strands in DNA, convert one topological form (*topoisomer*, Chap. 7) to another. There are two types of topoisomerases: *topoisomerase I* cuts just one strand of duplex DNA; *topoisomerase II* cuts across two strands of a DNA duplex.

Topoisomerase I catalyzes the *relaxation* of *negatively supercoiled* DNA; no coreactants are needed. Topoisomerase II catalyzes the negative supercoiling of DNA, using ATP as a coreactant, and it is also able to relax *positive supercoiling* (Chap. 7). DNA gyrase is an example of a topoisomerase II.

Topoisomeric forms of DNA are important in DNA replication in the sense that the introduction of negative supercoiling facilitates unwinding at the fork. Also, in a closed molecule such as the bacterial chromosome, unwinding at the fork would cause positive supercoiling in the DNA ahead of it, and this would hinder further unwinding. DNA gyrase has the ability to induce negative supercoiling as well as to relax positive supercoiling.

EXAMPLE 16.8

Summarize the events at the replication fork.

A model that summarizes the multitude of events involving the proteins and enzymes described in the foregoing discussion is shown in Fig. 16-14. The leading strand is synthesized in a continuous manner by DNA polymerase III holoenzyme. Its action is facilitated by the binding of SSB protein to the single-stranded template, made available through unwinding of the unreplicated duplex. This unwinding is accomplished by two helicases, the *rep* protein, which moves along the $3' \longrightarrow 5'$ template, and helicase II, which is associated with the $5' \longrightarrow 3'$ template. The lagging strand is synthesized (discontinuously) as short fragments. The *primosome* moves along the $5' \longrightarrow 3'$ template and at regular intervals promotes the binding of primase, which makes a short piece of RNA as primer. This is extended by DNA polymerase III holoenzyme in the direction opposite to fork movement. When DNA polymerase III reaches the primer, which has initiated an adjoining fragment,

DNA polymerase I takes its place and, through the process of nick translation, removes the RNA and replaces it by DNA. The nick that remains is sealed by DNA ligase. Finally, it is possible that the overall process is facilitated through the action of *DNA gyrase* in inducing negative supercoiling or relieving positive supercoiling in the DNA ahead of the fork.

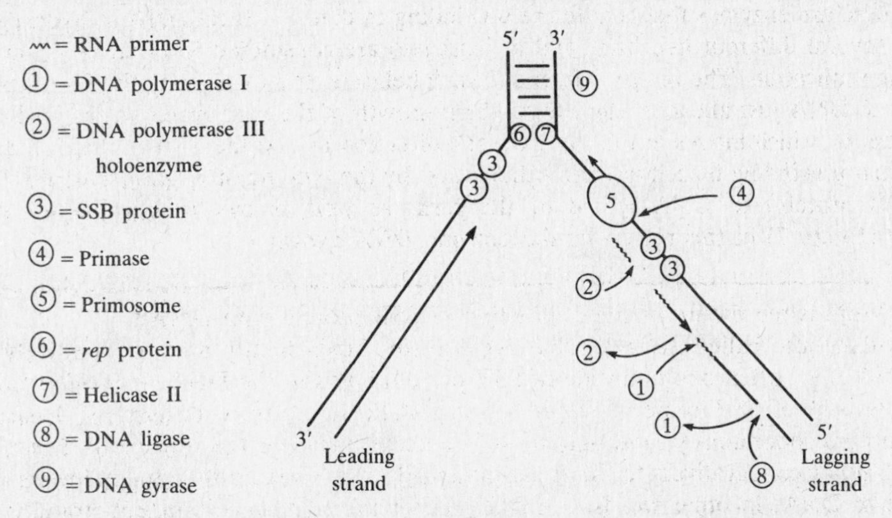

Fig. 16-14 Events involved in DNA chain growth at the replication fork.

EXAMPLE 16.9

DNA replication must be an accurate process. What happens if an incorrect base is incorporated by either of the DNA polymerases?

Mistakes in base incorporation can be made; this is largely a result of the transient existence of *tautomeric* forms of the bases (Chap. 7). If at the instant of insertion of a new nucleotide by DNA polymerase, the base in the template shifts to its rare tautomeric form, which has an altered base-pairing specificity, an incorrect nucleotide may be added to the chain; e.g., guanine instead of adenine opposite the enol form of thymine.

The $3' \longrightarrow 5'$ exonuclease activity of DNA polymerase I, at least, functions to *proofread* for such mistakes. After the incorrect base is incorporated, it will not remain hydrogen-bonded to the tautomeric base in the template once the latter returns, almost immediately, to its more stable form. The $3' \longrightarrow 5'$ exonuclease activity shows a strong preference for a *frayed* or non-hydrogen-bonded terminus and removes the misincorporated nucleotide before chain growth proceeds further. DNA polymerase III holoenzyme also has the potential to proofread by the same mechanism.

16.6 ENZYMOLOGY OF DNA REPLICATION IN EUKARYOTES

The model presented for events at the replication fork in bacteria is considered to apply, in general terms, to other systems. In eukaryotes it is known, for example, that DNA chain growth occurs exclusively in the $5' \longrightarrow 3'$ direction and through a discontinuous mechanism.

EXAMPLE 16.10

What enzymes are involved in DNA replication in eukaryotes?

Three DNA polymerases have been identified: two (*DNA polymerase* α and β) in the nucleus and one (*DNA polymerase* γ) in the mitochondria. The mechanism of DNA chain growth is the same as for the bacterial enzymes, but unlike the latter, the DNA polymerases have *no* associated exonuclease activities. DNA polymerase α, the most abundant of the three, is associated with chromosome replication in the nucleus, while DNA polymerase β is associated mainly with DNA repair. DNA polymerase γ is involved in replication of mitochondrial DNA.

Some other enzymes and proteins that could perform functions analogous to some of the well-characterized ones in bacteria have been identified in eukaryotes, but our overall understanding of their roles is much less clear.

Question: In what phase of the eukaryotic cell cycle does nucleosome (Chap. 7) assembly occur?

The histones needed in this assembly are synthesized mainly during the S phase. It appears that within a few minutes of a fork traversing a section of DNA, both new daughter arms associate with histones to give nucleosomes. It is not clear how the nucleosome structures ahead of the fork are removed. However, there is evidence that the *old* histone octamers associate preferentially with the leading-strand arm of the fork and *new* histones associate with the lagging-strand arm.

16.7 *dna* MUTATIONS AND INITIATION OF REPLICATION AT THE CHROMOSOME ORIGIN

The elongation phase of DNA replication in bacteria has been seen to involve many enzymes and proteins; some are associated with discrete functional complexes, such as the DNA polymerase III holoenzyme. Thus, a mutation in a gene for one of several proteins could lead to impairment of elongation.

EXAMPLE 16.11

Temperature-sensitive mutations, in particular, have been very valuable in helping to define many of the proteins involved in replication. Such mutations take effect at a certain temperature, e.g., 42–47°C, and not at another, e.g., 30°C or less. Mutations that affect replication are called *dna* mutations. Most that have been identified in *E. coli* so far (13 altogether) code for various proteins associated with DNA chain growth at the replication fork. For example, the gene *dna*G codes for primase. Some genes, however, appear to code for proteins involved also or exclusively with the *initiation of a cycle of replication*. Examples of the latter are *dna*A, C, J, K, and P.

EXAMPLE 16.12

What is known of the mechanism of initiation of a cycle of replication in *E. coli*?

Enzymological details of this process are limited. Unlike the initiation of Okazaki fragments during elongation, initiation at *oriC* requires RNA polymerase (unlike primase, it is sensitive to rifampicin; see Chap. 17) and the *dna*A product, which binds to the *oriC* sequence. DNA gyrase is also needed; negative supercoiling at the origin facilitates the binding of enzymes and other factors involved in initiation. Initiation of a cycle of replication is the point of control of DNA replication in bacteria. At the molecular level, it is likely to be a complex process and could involve other cell components, such as the membrane.

EXAMPLE 16.13

What is known of the mechanism of termination of a cycle of replication?

Very little definitive information is available. However, it is obviously an important aspect of the overall process of replication and is a prerequisite for chromosome segregation. It is a process that is coordinated with *septation* in bacteria, i.e., the formation of new cell wall and membrane between the segregated chromosomes to yield the daughter cells.

16.8 INHIBITORS OF DNA REPLICATION

Inhibitors of DNA replication are sometimes valuable in the treatment of various types of disease. They fall into several classes, listed below.

Inhibitors of Nucleotide Biosynthesis

Examples of inhibitors of nucleotide biosynthesis are *methotrexate* and *fluorodeoxyuridylate*. They interfere with the production of dTTP, which has been discussed in Chap. 15.

Inhibitors That Interact with the DNA Template

Actinomycin D is a commonly used inhibitor of both DNA and RNA synthesis. Its planar structure binds noncovalently between the stacked base pairs of duplex DNA; this is called *intercalation*. In this situation the DNA functions as a poor template. Compounds that bind in a similar way include *acridine* and *ethidium*. These affect the fidelity of DNA replication.

Nucleotide Analog Inhibitors

A number of nucleotide analog compounds function by blocking further chain growth at the replication fork. The *2′3′-dideoxynucleosides* can be converted to the triphosphates (ddNTPs). In the case of bacteria, these are incorporated onto the 3′-hydroxyl end of a growing DNA chain, and because the new end now lacks a 3′ hydroxyl, no further additions can occur. They are used in conjunction with DNA polymerase I in the dideoxy method of Sanger for determining DNA sequences (Example 16.14).

EXAMPLE 16.14

The *dideoxy method* of Sanger for sequencing DNA involves the copying of a single strand of the DNA by DNA polymerase I to yield new strands (radioactively labeled for identification purposes), which are terminated at certain positions through the incorporation of a dideoxy nucleotide at the 3′ end. A short primer, complementary to a sequence at one end of the single-strand template, is used to start synthesis from a fixed position. Termination at variable sites corresponding to the incorporation of a particular nucleotide is achieved by the inclusion of the ddNTP for that nucleotide. Four complete reactions (substrates plus enzyme) are set up, each containing, in addition, just one of the four ddNTPs. Thus, for a reaction using ddATP, the synthesis will proceed as in Fig. 16-15. The new strands of varying length are analyzed by gel electrophoresis (Chap. 4), which can resolve chains differing by as little as one base. From the collection of strand lengths produced by use of each of the four ddNTPs, the nucleotide sequence can be determined.

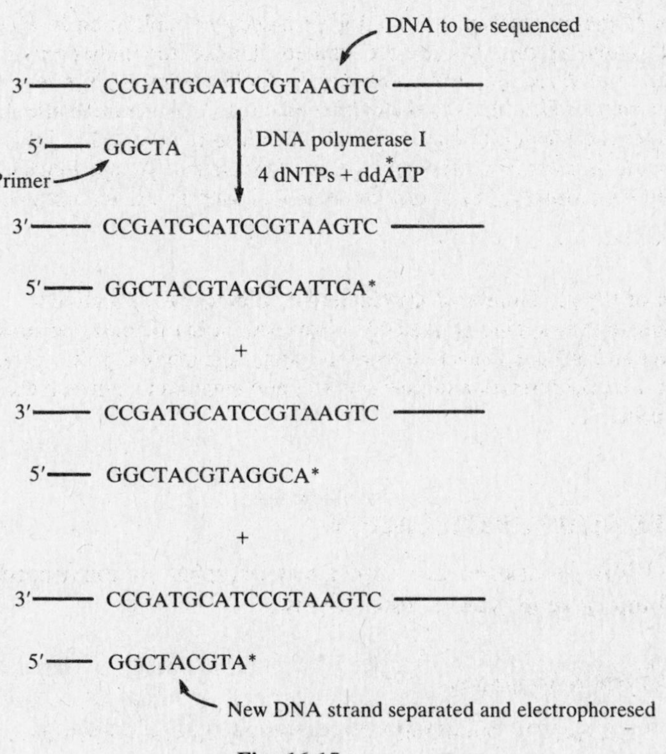

Fig. 16-15

Inhibitors That Bind to Replication Proteins

The *arylhydrazinopyrimidines* are potent inhibitors of DNA polymerase III from gram-positive bacteria (e.g., *Staphyllococcus*, *Streptococcus*, and *Lactobaccilus*; *E. coli* is a gram-*negative* organism). These compounds form *ternary complexes* with the polymerase and the DNA template. *Aphidicolin*, a tetracyclic diterpenoid, is a potent inhibitor of mammalian DNA polymerase α; it does not affect the β and γ polymerases. *Nalidixic acid* and *novobiocin* bind to the A and B subunits, respectively, of *E. coli* DNA gyrase to inhibit its action and hence DNA replication.

EXAMPLE 16.15

Acridine drugs have been used to treat malignant and parasitic diseases. The structure of acridine is

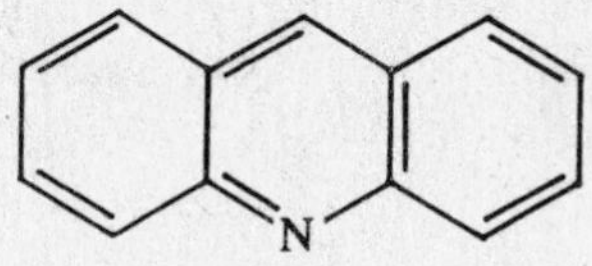

Acridine

Nalidixic acid is an example of a clinically useful antibacterial agent, but *oxolinic acid*, which acts in the same way, is 10 times more potent. The structures of these two antibiotics are

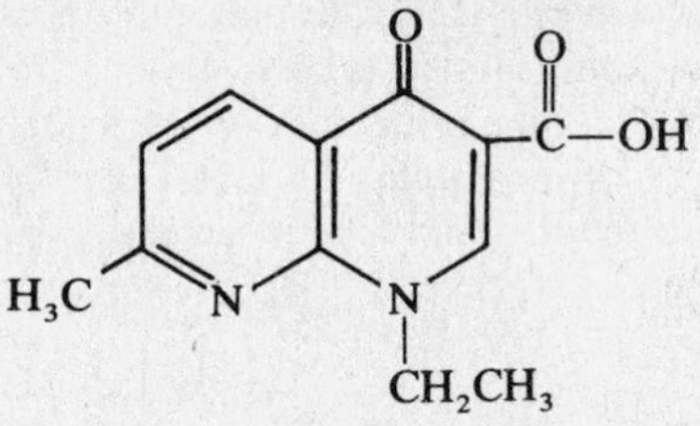

Nalidixic acid

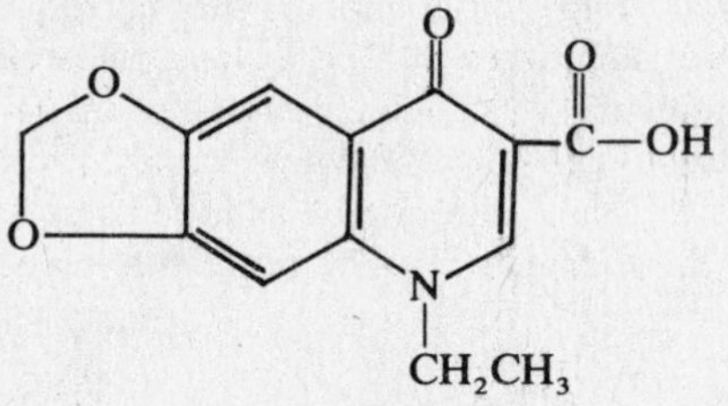

Oxolinic acid

Aphidicolin has been used to treat herpes virus infections of the eye. Its structure is

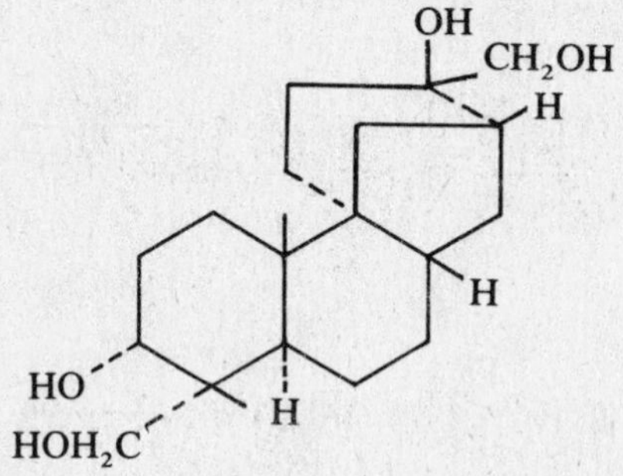

Aphidicolin

16.9 REPAIR OF DNA DAMAGE

Damage to DNA is caused by a number of agents, including uv irradiation, ionizing radiation, and various chemicals. Such damage can be deleterious or lethal to an organism, and a number of mechanisms exist for removing it. The best understood is that known as *excision repair* of damage caused by uv irradiation.

EXAMPLE 16.16

What is the nature of the damage to DNA caused by uv irradiation?

Pyrimidine bases in adjacent nucleotides of a DNA strand become covalently cross-linked after irradiation by uv light. The cross-linking occurs mostly between two thymines to form a *thymine dimer* (Fig. 16-16). The thymine dimer causes a structural distortion within the DNA chain and represents a physical impediment to replication and transcription.

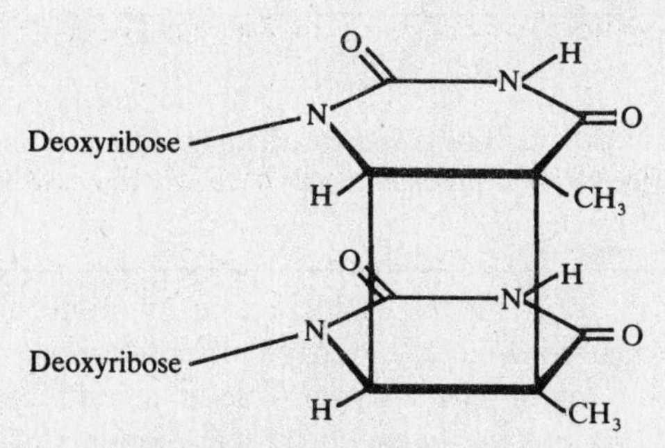

Fig. 16-16 Structure of a thymine dimer. The dimer is attached at two positions to the sugar deoxyribose.

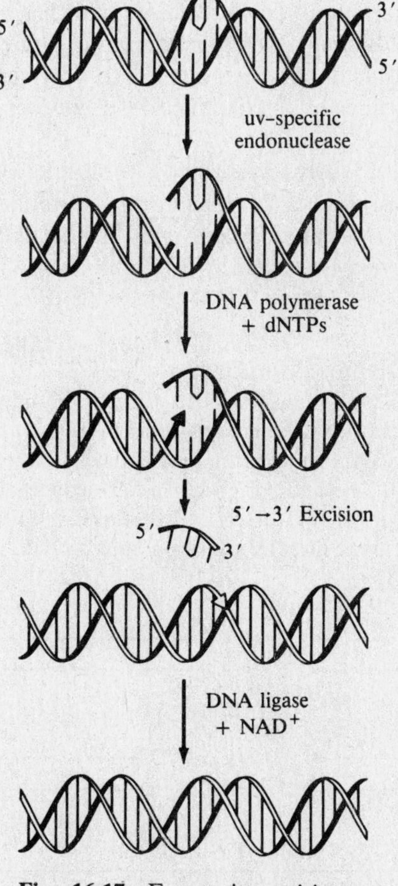

Fig. 16-17 Events in excision repair of uv-induced damage to DNA.

Question: How, if at all, is damaged DNA repaired?

In excision repair (Fig. 16-17), the damaged portion is cut out and replaced by new DNA. The process in *E. coli* is well understood. It occurs in three steps. First, an endonuclease called *uv specific endonuclease* cleaves the strand on the 5′ side of the dimer and within a few nucleotides of it. Second, DNA polymerase I binds at this site and, by means of *nick translation*, removes the damaged portion and replaces it with new DNA. Third, DNA ligase seals the nick.

EXAMPLE 16.17

Excision repair in mammalian cells is considered to occur by a similar mechanism. Defects in this repair pathway can lead to the disease *xeroderma pigmentosum*, in which the skin is very sensitive to sunlight; this results in a high incidence of skin cancer.

Question: Is excision repair the only means of correcting uv-induced damage to DNA?

No. Dimers can also be repaired directly by enzymatic photoreactivation. The *photoreactivating enzyme* binds to the DNA in the region of the dimer to form a complex that absorbs visible light and catalyzes cleavage of the covalent linkage between the components of the dimer. Photoreactivation occurs in both bacterial and mammalian cells.

16.10 RECOMBINANT DNA AND ISOLATION OF GENES

With the development of *recombinant DNA technology*, it has been possible to isolate particular segments of DNA, sometimes containing all or part of a gene, and examine their structure and other properties in considerable molecular detail. This has led to very significant advances in understanding, at a very refined level of molecular detail, many aspects of replication and expression of the genetic material.

Recombinant DNA is an artificially constructed molecule containing DNA segments from different organisms. If one of the segments carries an origin of replication, the recombinant, or *joint*, molecule has the potential to be replicated in an organism that is able to recognize that origin. In such a situation, the segment carrying the origin, which enables replication, is referred to as a *vector*.

EXAMPLE 16.18

How are recombinant DNA molecules constructed?

There are several methods available for this purpose. One of the most commonly used approaches takes advantage of the single-stranded ends of DNA segments generated by *type II restriction endonucleases* such as *Eco*RI. Fragments produced by this enzyme contain overlapping 5′ termini (Chap. 7). These short single-stranded termini spontaneously join to one another under suitable conditions of temperature and ionic strength through complementary base pairing. This process is called *annealing* (Fig. 16-18). The rejoined molecule is held together through four base pairs, but there is a nick on each strand. These can be sealed by DNA ligase.

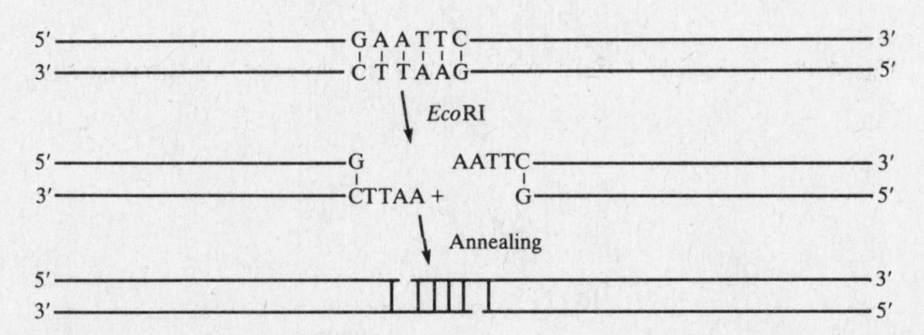

Fig. 16-18 Annealing of DNA through base pairing of overlapping termini generated by *Eco*RI.

EXAMPLE 16.19

How is a segment of DNA linked to a suitable vector?

Commonly used vectors are *bacterial plasmids*. These are small circular DNA molecules of 3–100 kb. Circularity is essential for their replication. If a plasmid is enzymatically *cut* at only one site by, say, *Eco*RI, the linear fragment obtained can be annealed with a fragment that has been formed by cutting other DNA with *Eco*RI. This generates a circular joint molecule, which is sealed by DNA ligase (Fig. 16-19).

The circular joint molecule is introduced into an appropriate bacterial host, such as *E. coli*, through a process known as *transformation*; this involves treatment of the bacterial cells with $CaCl_2$ at low temperature (to allow uptake of DNA), followed by incubation of the bacteria with the DNA under appropriate conditions. The foreign DNA replicates under the control of the vector. If the foreign DNA carries an intact gene, it is possible that this gene will be expressed in the bacterium. This type of approach has led to the mass production of human insulin in bacterial cells. The overall technique is sometimes referred to as *gene cloning*.

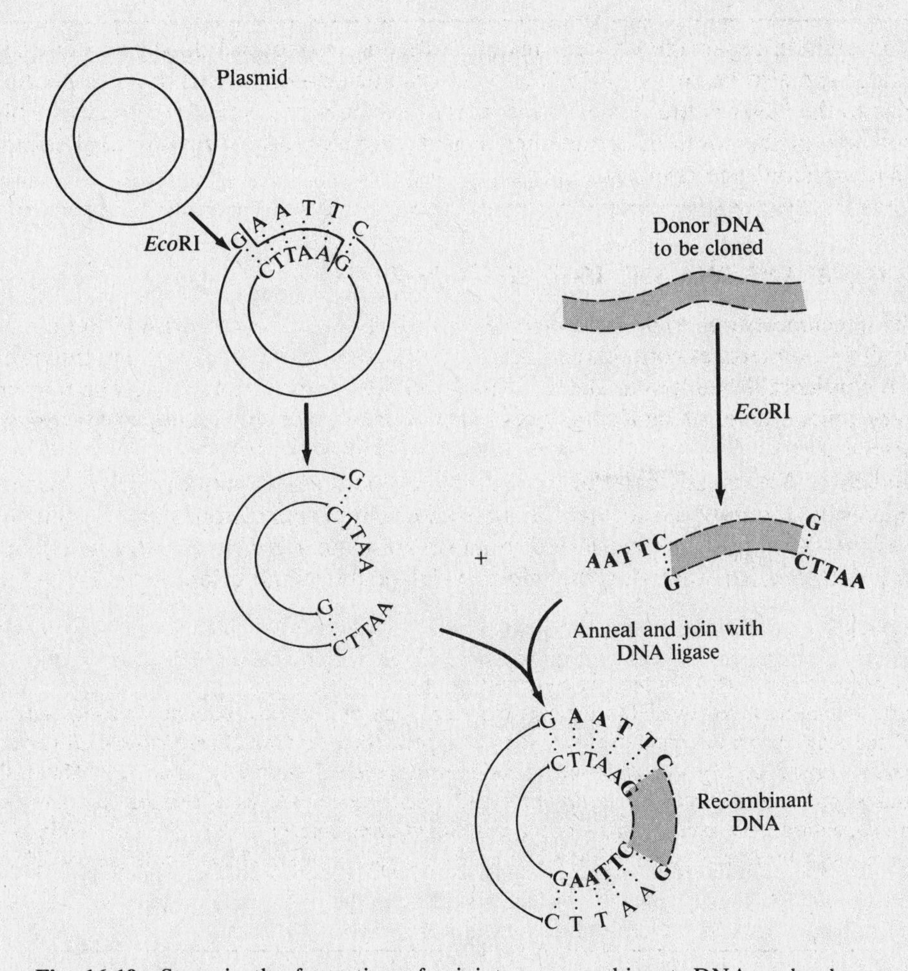

Fig. 16-19 Steps in the formation of a joint, or recombinant, DNA molecule.

Solved Problems

SEMICONSERVATIVE REPLICATION OF DNA

16.1. Define the term *template* as it applies to DNA replication.

SOLUTION

The template is the DNA strand that binds directly to the various replication enzymes and that defines the sequence of the newly synthesized DNA strand. The DNA duplex cannot be copied per se. Rather, it must unwind into its component single strands, in which the nucleotide sequence is accessible to copying through base pairing.

16.2. Consider a linear duplex DNA molecule that undergoes replication through five successive generations. What would be the proportion of the *original* DNA within the *total* DNA population?

SOLUTION

The amount of DNA doubles in each generation. Thus, five successive generations result in a 32-fold (2^5) increase in the DNA. Therefore, the original DNA represents 3.125 percent of the total.

16.3. With respect to the situation outlined in Prob. 16.2, how is the original DNA molecule distributed among the progeny molecules?

SOLUTION

The progeny consists of 32 molecules. Because DNA replication is semiconservative, the single strands of the original (or parental) DNA, which remain intact, each compose half of two of the progeny molecules.

TOPOLOGY OF DNA REPLICATION

16.4. A replicating bacterial chromosome exists as a so-called θ *structure* (because of its appearance in electron micrographs and diagrammatically), in which the upper closed section represents the duplicated portion, or replication bubble. Why is there now an origin of replication situated at the middle of *each* arm of the bubble?

SOLUTION

Replication of the bacterial chromosome commences at a unique site, the origin, and replication forks diverge from this site at approximately equal rates. Consequently, the origin is duplicated completely and each daughter origin remains at the central portion of the expanding bubble. This is in contrast to unidirectional replication, which would leave the origin only partially replicated and located at the stationary fork as the replication fork moves away.

16.5. What is the major difference, with respect to the topology of replication, between bacterial and eukaryotic chromosomes?

SOLUTION

The major difference arises from the existence of only *one* origin of replication in the bacterial chromosome compared with a *multitude* of origins in a eukaryotic chromosome. This gives rise to a single replication bubble in the former, and clusters of tandemly arranged bubbles in a eukaryotic chromosome. In the case of bacteria, the two replication forks that meet at termination arise from the same origin; in eukaryotes they arise from neighboring origins.

CONTROL OF DNA REPLICATION

16.6. What conditions give rise to *multiforked chromosomes* in bacteria?

SOLUTION

A multiforked bacterial chromosome is one that contains more than two replication forks and results from *reinitiation* at the daughter origins within a replication bubble. In this situation, cycles of replication are completed at more frequent intervals, and this gives *shorter* generation times. It occurs under conditions of fast growth, induced by nutritionally rich media.

16.7. Under what conditions do sections of eukaryotic chromosomes exist as *multiforked* structures?

SOLUTION

Unlike the situation in bacteria, multiforked structures in eukaryotic chromosomes do *not* form in response to a demand for increased rate of DNA replication. But, multiforked DNA does appear over restricted regions of eukaryotic chromosomes; this gives DNA (or gene) amplification and occurs in response to certain conditions or treatments. For example, it occurs at the dihydrofolate reductase gene during prolonged treatment of certain cancers with the inhibitor of this enzyme, methotrexate.

16.8. Chromosome replication in bacteria can be divided into three stages. Which of these stages is central to the overall control of the process?

SOLUTION

The stages of replication in bacteria are *initiation*, *elongation*, and *termination*. Control of replication is effected at the level of initiation. Thus, the frequency of initiation determines the frequency of completion of cycles of replication. The rate of replication-fork movement (elongation) remains fairly constant under conditions that change the overall rate of replication. There is no evidence to suggest that termination has any control over the rate of replication.

ENZYMOLOGY OF DNA REPLICATION

16.9. Distinguish between the *substrates* and *template* of DNA polymerase.

SOLUTION

The substrates undergo chemical modification and include the four deoxynucleoside triphosphates. The template undergoes no alteration but provides the instructions for the order of assembly of nucleotides into the growing chain. The template for DNA synthesis is the parental, single-stranded DNA.

16.10. Do DNA polymerase I and III of *E. coli* extend a DNA chain by different mechanisms?

SOLUTION

No. All DNA polymerases use the deoxynucleoside 5′-triphosphates to add new nucleotide units, one at a time, onto the 3′-hydroxyl terminus of the growing chain. The main difference between DNA polymerase I and III relates to the type and relative contribution each makes to replication and repair synthesis of DNA.

16.11. DNA polymerase I of *E. coli* is a multifunctional enzyme. What is meant by this?

SOLUTION

DNA polymerase I, which consists of a single polypeptide chain, contains three distinct activities. These are polymerase, $5' \longrightarrow 3'$ exonuclease, and $3' \longrightarrow 5'$ exonuclease. Each activity has an important biological role. The $5' \longrightarrow 3'$ exonuclease activity can be separated from the other two by proteolytic cleavage of the enzyme.

16.12. What constitutes the DNA polymerase III core enzyme of *E. coli*?

SOLUTION

DNA polymerase III occurs within the cell as a functional complex of at least seven polypeptide chains. This is called the *holoenzyme*. A subcomplex containing three of these chains (α, ε, and θ) is readily isolated and exhibits polymerase activity. It is called the DNA polymerase III core enzyme.

16.13. What is understood by *discontinuous* DNA replication?

SOLUTION

DNA chain growth occurs on both daughter arms at a replication fork. On one arm, chain growth occurs continuously ($5' \longrightarrow 3'$), in the same direction as fork movement. On the other arm, chain growth occurs in separate short pieces ($5' \longrightarrow 3'$) and in the direction opposite to fork movement. The short pieces (nascent or Okazaki fragments) subsequently join. Replication in the latter fashion is known as *discontinuous* DNA replication.

16.14. What is meant by the terms *leading* and *lagging strands* in DNA replication?

SOLUTION

The leading strand is that which is synthesized continuously and in the same direction as fork movement. The lagging strand is synthesized discontinuously, i.e., in short pieces that are subsequently joined, and in the direction opposite to fork movement.

16.15. What is the role of the $3' \longrightarrow 5'$ exonuclease of DNA polymerase I of *E. coli*?

SOLUTION

This enzyme has a proofreading role. At a low random frequency, incorrect bases (in the form of nucleotides) are inserted into the growing DNA chain. This results from the existence of rare tautomeric forms of the four bases, which, if occurring transiently in the template at the moment of insertion of an incoming nucleotide, will cause a mistake in base pairing. When such a template nucleotide shifts back to its preponderant form, a base pair mismatch results. The $3' \longrightarrow 5'$ exonuclease recognizes the mismatch and catalyzes the hydrolytic removal of the nucleotide from the end of the chain before elongation resumes.

16.16. A characteristic feature of the DNA polymerases is their inability to initiate synthesis of a polydeoxynucleotide chain; they can only *extend* an existing chain. How is initiation of the nascent fragments within the discontinuously synthesized DNA strand achieved?

SOLUTION

Unlike the DNA polymerases, RNA polymerase is able to initiate a new RNA chain, using DNA as a template (Chap. 17). The DNA polymerases are able to extend the DNA from an RNA primer. In discontinuous DNA chain growth, a particular type of RNA polymerase, called *primase* in *E. coli*, lays down short RNA primers at fairly regular base intervals, as unwinding of the helix at the replication fork proceeds. These primers are involved in the initiation of synthesis of nascent DNA chains by DNA polymerase.

16.17. In *E. coli*, what is the function of the $5' \longrightarrow 3'$ exonuclease activity of DNA polymerase I in the overall replication process?

SOLUTION

DNA polymerase I has an important role in the assembly of new DNA chains via discontinuous DNA synthesis in the lagging strand at a replication fork. DNA polymerase III is responsible for the bulk of the synthesis of each nascent DNA fragment. When chain growth by this enzyme reaches the 5′ RNA-primer end of an adjoining fragment of DNA, polymerase I takes over and continues DNA extension through the RNA-primer region. In this case the segment of RNA is removed through the $5' \longrightarrow 3'$ exonuclease activity of the multifunctional DNA polymerase I.

16.18. What reaction is catalyzed by DNA ligase, and what is its role in DNA replication?

SOLUTION

DNA ligase catalyzes the covalent linkage of two segments of DNA. A phosphodiester link is formed between adjacent 5′-phosphoryl and 3′-hydroxyl groups within *duplex* DNA. In other words, DNA ligase is able to *seal* a nick. DNA ligase has an important role in several reactions involving DNA. In replication it functions to join the nascent DNA fragments of the lagging strand. This follows replacement of the RNA primer by DNA through the action of DNA polymerase I.

16.19. What are topoisomerases, and how might they be involved in the process of DNA replication?

SOLUTION

Topoisomerases catalyze the conversion of one topological form of DNA to another. There are two general enzyme types, I and II. The former acts by cutting just one strand of duplex DNA, while type II cuts across both strands of DNA. DNA gyrase, from bacteria, is an example of a type II topoisomerase and is known to be involved in DNA replication. It can induce negative supercoiling in DNA as well as being able to relax positive supercoiling. It is required for initiation of a cycle of replication in bacteria. In this case, it probably introduces negative supercoiling at the origin, thus allowing the binding of enzymes and other factors involved in initiation. Also, it facilitates the unwinding of the duplex at the fork by inducing negative supercoiling ahead of the fork or by relaxing the positive supercoiling that may build up in this region.

dna MUTATIONS AND INITIATION OF REPLICATION AT THE CHROMOSOME ORIGIN

16.20. What is understood by the term *initiation of a cycle of replication* in bacteria?

SOLUTION

DNA replication in bacteria is a cyclic process in the sense that at regular time intervals, depending on the richness of the growth medium, initiation of replication is effected at the chromosome origin (*oriC*). A number of specific proteins and enzymes are directly involved in the initiation of replication, and they are not involved in the subsequent growth of the DNA. Once initiated, a cycle of replication runs to completion in the absence of initiation proteins or enzymes; but in the absence of these enzymes and factors, no new cycles start.

16.21. Distinguish between the roles of *RNA polymerase* and *primase* in the DNA replication process in bacteria.

SOLUTION

Both of these enzymes synthesize RNA. RNA polymerase is inhibited by the drug rifampicin; primase is not. Initiation of a cycle of replication, from *oriC*, is blocked by rifampicin, thus showing that RNA polymerase is required for initiation. Primase, on the other hand, is involved in initiation of nascent fragments in the lagging strand during the subsequent elongation or chain-growth stage. This stage is unaffected by rifampicin. Thus, these two enzymes, both of which catalyze the synthesis of RNA, have distinct roles in the process of DNA replication.

INHIBITORS OF DNA REPLICATION

16.22. Why does 2′,3′-ddCTP inhibit DNA polymerase I?

SOLUTION

DNA polymerase I catalyzes the addition of the monophosphate from 2′,3′-ddCTP onto the 3′-hydroxyl end of the growing chain and opposite guanosine on a template DNA strand. The growing chain is then devoid of a 3′ hydroxyl at the terminus, and the addition of further nucleotide units is blocked.

16.23. Would 2′,3′-ddCTP block the replication of $d(AT)_n$ by DNA polymerase I?

SOLUTION

$d(AT)_n$ is a polydeoxynucleotide containing alternating A and T residues in each strand of a duplex structure. 2′,3′-ddCTP has no effect on its replication by DNA polymerase I because G residues, to which this chain-terminating inhibitor would base-pair, are absent from the template.

16.24. You wish to measure the level of *DNA polymerase* α activity in a mammalian cell extract, which would also contain DNA polymerases β and γ. What replication inhibitor could be used for this purpose?

SOLUTION

Aphidicolin inhibits only DNA polymerase α. DNA polymerase activity would first be measured in the *absence* of this inhibitor. Assay of the extract in the presence of aphidicolin would allow estimation of the level of DNA polymerase $\beta + \gamma$. The difference between the activities in the two experiments would give the level of DNA polymerase α.

REPAIR OF DNA DAMAGE

16.25. Distinguish between *replication* and *repair synthesis* of DNA.

SOLUTION

Replication refers to the reproduction, through the copying of template DNA strands, of complete chromosomes. In this case DNA synthesis is extensive. *Repair synthesis* of DNA is involved in correcting the damage (caused by physical or chemical treatments) within isolated portions of DNA. In this case, DNA synthesis is restricted to the immediate vicinity of the damage and is usually only minor in extent.

16.26. What enzymes are involved in removing the damaged portion of DNA during excision repair of uv-induced damage in bacteria?

SOLUTION

Two enzymes are directly involved. They are uv-specific endonuclease and $5' \longrightarrow 3'$ exonuclease of DNA polymerase I. The former cleaves the DNA strand on the 5′ side of the pyrimidine dimer. The second enzyme catalyzes the removal of the dimer, as well as a region of several nucleotides on both sides of it. The associated DNA polymerase activity of DNA polymerase I leads to filling of the gap with new DNA.

RECOMBINANT DNA AND ISOLATION OF GENES

16.27. What features of a *recombinant* DNA molecule are essential for its replication in a host cell?

SOLUTION

First, the recombinant DNA must carry an origin of replication that is *functional* in the particular host. Second, the molecule must be *circular*, or be capable of circularization.

16.28. A piece of human DNA carrying a particular gene has been *cloned* in a bacterial cell. Will the human gene be *expressed*?

SOLUTION

The human gene will be replicated, along with the *vector* in the recombinant DNA, but it won't necessarily be expressed. This only occurs if its disposition within the recombinant molecule is such that it is transcribed as mRNA and then translated into protein. For *transcription*, the gene must be positioned appropriately with respect to a bacterial *promoter*. For successful translation, the gene must be free of introns (Chap. 17), and its transcript must contain a bacterial ribosomal binding site (Chap. 17) at the correct location.

Supplementary Problems

16.29. (*a*) What is meant by *unidirectional* replication of DNA? (*b*) Is the rolling-circle mechanism of replication unidirectional or bidirectional?

16.30. The DNA synthesis (S) phase of the cell cycle is very short (<20 min) in embryonic cells of some organisms. What process takes place that allows such rapid replication of all the chromosomal DNA? (*Hint*: Consider the size of the replicons.)

16.31. All three enzymatic activities of DNA polymerase I of *E. coli* are utilized in the overall replication process. Describe these.

16.32. Priming of DNA chain growth occurs at more than one fixed position on the bacterial chromosome. What is that position?

16.33. By what means can DNA gyrase assist in unwinding DNA at the replication fork?

16.34. Is DNA helicase a topoisomerase?

16.35. What structural features of the ends of DNA chains at a nick are necessary for sealing by DNA ligase?

16.36. What is meant by the proofreading function of DNA polymerase?

16.37. Devise a possible sequence of protein- (enzyme)-mediated events for the generation of a replication bubble at an origin of replication in double-helical DNA.

16.38. What is meant by *dna mutation*?

16.39. Would fluorodeoxyuridylate be expected to block initiation of DNA chain growth? Its structure is

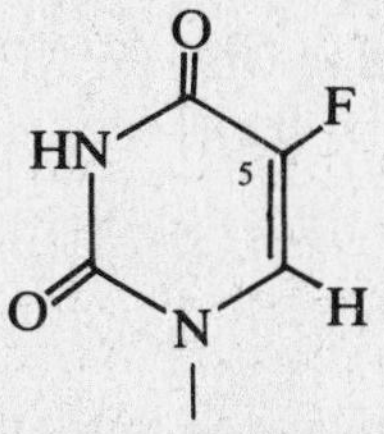

O
HN F
5
O N H

Deoxyribose 5′-phosphate

16.40. What effect would an inhibitor of RNA polymerase be expected to have on DNA replication in bacteria?

16.41. Irradiation by uv light can cause the formation of thymine dimers. Does this bring about cross-linking of complementary strands in double-helical DNA?

16.42. What enzymes are common to the processes of replication of DNA and excision repair of DNA damage in *E. coli.*?

16.43. Is excision repair the only means of removing thymine dimers from DNA?

16.44. What feature of circular, double-helical DNA is essential for it to function as a vector for DNA cloning?

Chapter 17

Gene Expression and Protein Synthesis

17.1 INTRODUCTION

Most genes are expressed as protein (Chap. 16). The process by which this is accomplished is called gene expression. In this process a sequence of deoxynucleotides in DNA (which defines the gene) is first *transcribed* into a sequence of ribonucleotides in RNA (*messenger RNA*, or *mRNA*). This is then *translated* into a sequence of amino acids to give a polypeptide of defined length. The amino acid sequence within the latter determines the manner in which the molecule folds upon itself to yield the biologically active protein.

In bacterial cells, there is no membrane surrounding the DNA *nucleoid*, and both *DNA transcription* and *RNA translation* proceed within the single cell compartment. In eukaryotes, the nucleus is bounded by a membrane. Transcription occurs within the nucleus, and the mRNA must pass into the cytoplasm, where it is translated. Frequently, the immediate polypeptide product of translation is subsequently modified, sometimes in a process that enables it to be transported out of the cell in which it was made.

The sequence of nucleotides within the *single-stranded* mRNA is assembled according to the complementary-base-pairing (Chap. 7) instructions of one of the strands of duplex DNA, which contains the gene. Since this DNA strand provides the *template* for transcription, it is called the *sense* strand.

The *genetic code* is the basis for converting a nucleotide sequence of mRNA into an amino acid sequence of a polypeptide. It is the genetic code that describes how various *combinations* of nucleotides (of which there are only 4 types in DNA or RNA) can be read as individual amino acids (of which there are 20 types). The nature of the genetic code was elucidated in the 1960s.

17.2 THE GENETIC CODE

Because there are 20 amino acids and only 4 nucleotides, there must be a combination of at least 3 nucleotides to define each amino acid. A code based on two nucleotides would provide only 4^2 or 16 combinations, which is insufficient. Proof that the *codon* for each amino acid consists of 3 nucleotides was provided by genetic studies of the effects on the polypeptide product of nucleotide addition to or deletion from a gene.

Question: A trinucleotide-based code would provide 4^3 or 64 codons. Are the extra codons used?

Yes, it turns out that they are all used. In the vast majority of cases, a single amino acid has more than one codon. For this reason the code is said to be *degenerate*.

EXAMPLE 17.1

Degeneracy of the code is very obvious from an examination of the codon assignments shown in Table 17.1. For example, there are six codons for leucine. It should be noted that the nucleotide components of each triplet are written in terms of the ribonucleotides (containing A, U, C, G; see Chap. 7) in mRNA. The first position refers to the initial nucleotide of the triplet, positioned at its 5′ end. The third nucleotide is at the 3′ end.

Table 17.1. Codon–Amino Acid Assignments of the Genetic Code

First Position	Second Position				Third Position
	U	C	A	G	
U	Phe	Ser	Tyr	Cys	U
	Phe	Ser	Tyr	Cys	C
	Leu	Ser	(CT)*	(CT)*	A
	Leu	Ser	(CT)*	Trp	G
C	Leu	Pro	His	Arg	U
	Leu	Pro	His	Arg	C
	Leu	Pro	Gln	Arg	A
	Leu	Pro	Gln	Arg	G
A	Ile	Thr	Asn	Ser	U
	Ile	Thr	Asn	Ser	C
	Ile	Thr	Lys	Arg	A
	Met(CI)†	Thr	Lys	Arg	G
G	Val	Ala	Asp	Gly	U
	Val	Ala	Asp	Gly	C
	Val	Ala	Glu	Gly	A
	Val(CI)†	Ala	Glu	Gly	G

*CT = chain termination.
†CI = chain initiation.

Question: What is the codon for methionine?

From Table 17.1, the codon for methionine is seen to be AUG. Methionine is one of two amino acids for which there is only one codon.

Question: What is the significance of the term *CI* in Table 17.1?

This stands for *chain initiation* and indicates that the codon for methionine (AUG) defines the beginning of the translated portion of an mRNA; i.e., methionine is the first amino acid to be incorporated into a polypeptide chain. Less frequently GUG, normally for valine, can function in place of AUG as the chain-initiation codon incorporating methionine (see Table 17.1).

Question: At what end of a growing polypeptide chain, N or C terminus, is the initiating amino acid found?

It is always incorporated at the N terminus. Thus, the direction of assembly of a chain is $N \longrightarrow C$.

Question: Is there an analogous codon that defines the termination site on the mRNA with respect to polypeptide chain formation?

There are actually three codons that function in this way. They are UAA, UGA, and UAG, and are referred to as the *chain termination* (*CT*) *triplets* (see Table 17.1).

EXAMPLE 17.2

Write a sequence of ribonucleotides that would define a short polypeptide with the amino acid sequence Met-Leu-Arg-Asn-Ala-Val-Glu-Ser-Ile-Cys-Phe-Thr.

A possible sequence is as follows:

5′ AUG UUA CGU AAU GCU GUC GAA UCU AUU UGC UUU ACA UAA 3′

Note the presence of chain initiation and termination codons, respectively, at the beginning and end of the sequence. In translating this sequence, which would occur within a longer mRNA molecule, into a sequence of amino acids, it has been assumed that the codons do not overlap. This has been established experimentally, and for this reason, the triplet genetic code is said to be *nonoverlapping*.

Question: In the ribonucleotide sequence shown in Example 17.2, there is an AUG triplet that can be formed by joining portions of two adjacent triplets, starting from A at the eleventh position. Is it possible for this to function as an alternative initiation codon?

For this to be so, the overall *frame of reading* would be different, i.e., an alternative sequence of triplets would be read to the right of this position. As will be seen below, there are sequences in mRNA, upstream (to the left) of a potential start site, e.g., AUG, that define the frame of reading.

Question: Is the dictionary of amino acid codons shown in Table 17.1 the same for all organisms?

The genetic code dictionary was originally established from studies on the bacterium *E. coli*. It is now known to be the same for all organisms; i.e., it is universal. The only exceptions occur for a few codons in mitochondria from a number of species.

17.3 DNA TRANSCRIPTION IN BACTERIA

As with the establishment of the genetic code and information on the molecular mechanism of DNA replication (Chap. 16), the present detailed knowledge of the mechanism of DNA transcription to produce RNA rests largely upon studies with bacteria, particularly *E. coli*. It is convenient to treat transcription in bacteria first.

Most of the DNA sequences which are transcribed give rise to mRNA, which is subsequently translated into protein. However, the most abundant species of RNA are *ribosomal RNA* (*rRNA*) and *transfer RNA* (*tRNA*), which do not code for protein but function in the process of translation. They are formed by a high level of transcription of a relatively small number of genes (called *rRNA* and *tRNA genes*). In bacteria, transcription of *all* genes is brought about by the enzyme *RNA polymerase*.

Question: What is the nature of the chemical reaction catalyzed by RNA polymerase?

The overall reaction is

$$n\text{NTP} \xrightarrow{\text{Mg}^{2+}} (\text{NMP})_n + n\text{PP}_\text{i}$$

Thus, it uses the four ribonucleoside triphosphates (ATP, GTP, UTP, and CTP) to assemble an RNA chain, the sequence of which is determined by the sense strand of DNA. Nucleotide addition occurs *sequentially*, the phosphodiester bond being formed through the same mechanism as described for DNA polymerase (see Chap. 16, Fig. 16-9). RNA chain growth is in the $5' \longrightarrow 3'$ direction. An important distinction between RNA polymerase and DNA polymerase, however, is the ability of the former to start a new chain de novo; i.e., it does not have an obligatory requirement for a primer. The first nucleotide to be incorporated into the chain of RNA contains either adenine or guanine and retains its 5′ triphosphate.

Question: Within a continuous nucleotide sequence, a single DNA molecule contains a multitude of genes, and when a particular one is to be transcribed, the RNA polymerase must know where to start. How is this achieved?

To transcribe a particular stretch of sequence the RNA polymerase binds to the DNA at a site called a *promoter*, just upstream (i.e., on the 5′ side) of the *transcriptional start site* defined by the sense strand.

Many bacterial promoters have been sequenced, and it has been found that, while the sequence of the region is not the same in all cases, there are two segments situated around nucleotide positions 10 and 35 from the start site that vary only slightly from one another, so that a *consensus* sequence can be defined for each. Figure 17-1 shows sequences (complementary to the sense strand) from a number of promoter regions, which extend through the first or -10 *region* (which is underlined). The -10 region is also called a *Pribnow box* (after the person who discovered it).

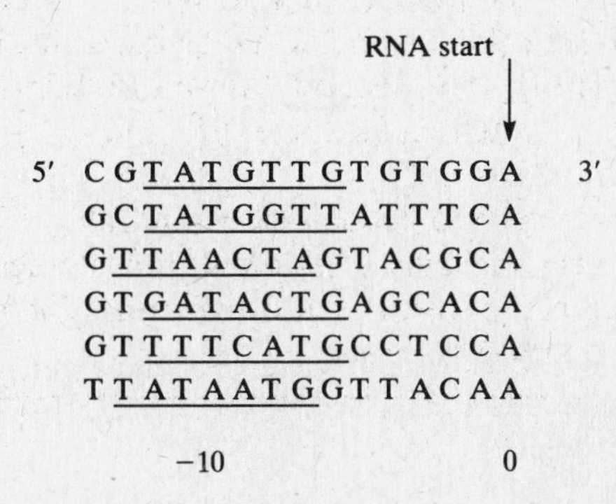

Fig. 17-1 Pribnow box sequences from a number of promoters in bacteria. The regions of homology are underlined.

Question: How does RNA polymerase recognize the promoter?

RNA polymerase is a multisubunit enzyme of molecular weight 480,000. The four major subunits, β, β', α, and σ ($M_r = 150{,}000$; 160,000; 36,500; and 86,000, respectively) are present in the ratio 1:1:2:1, and the total complex is more correctly called the *RNA polymerase holoenzyme*. The σ subunit is involved directly in *promoter recognition*. The complex lacking the σ subunit is called the *RNA polymerase core enzyme*. To start transcription, the σ subunit directs the holoenzyme to a promoter site to form a *binary complex* in which there is a limited unwinding of the DNA duplex to generate an *open promoter complex*. This is the first step in the overall transcription cycle and is called *template binding*.

EXAMPLE 17.3

The transcription cycle is shown in Fig. 17-2.

The template-binding step involves interaction of the holoenzyme, through its σ subunit, with the promoter to give the open promoter complex, as already described. Initiation of the RNA chain can then proceed through the formation of the first phosphodiester bond between ATP (or GTP) and the next nucleotide defined by the template to yield a dinucleoside tetraphosphate:

$$\text{ppp}_{\text{G}}^{\text{A}} + \text{pppN} \longrightarrow \text{ppp}_{\text{G}}^{\text{A}}\text{pN} + \text{PP}_i$$

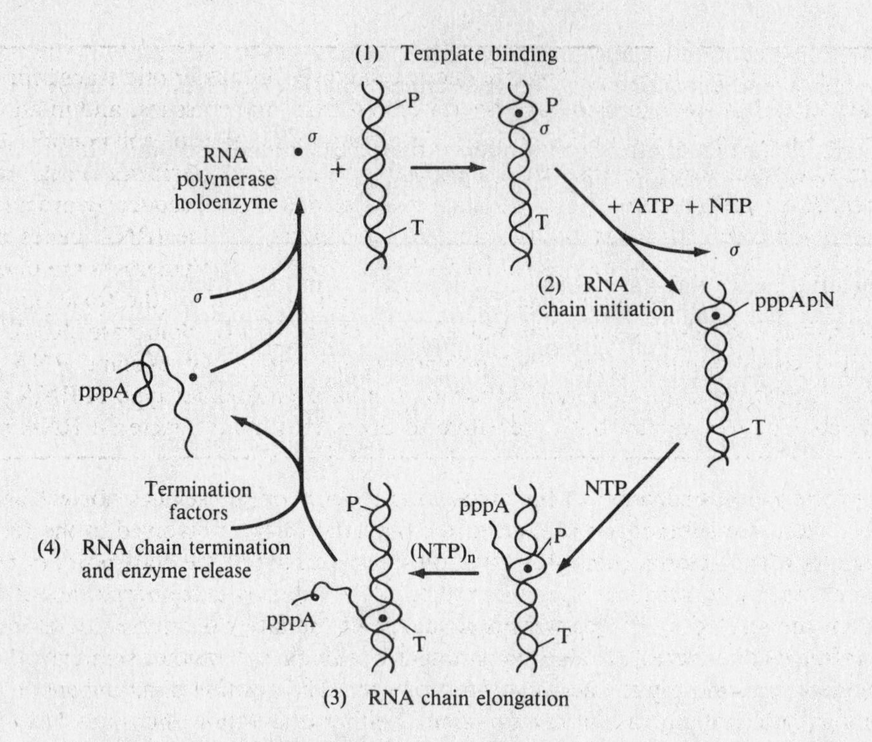

Fig. 17-2 The transcription cycle in bacteria. P and T refer to promoter and termination sites, respectively, for a single RNA transcript, and pppA denotes the triphosphate ATP.

Following initiation, the σ subunit is released, and RNA chain elongation proceeds by the sequential addition of nucleotide units to the 3′OH of the previously incorporated nucleotide, as described for the DNA polymerases (Chap. 16). Termination of transcription is effected when the core enzyme reaches a *termination* sequence. In *E. coli* two types of termination sequence have been identified: one requires an additional protein, called *rho*, to effect termination; the other does not. Termination sequences are relatively long (up to ~50 nucleotides) and function through the formation of *hairpins* in the single-strand RNA transcript. Hairpins reflect the presence of *inverted repeat* sequences that allow the RNA chain to bend back on itself and be stabilized through complementary base pairs. Following release of the RNA transcript, the core enzyme is also released. It is then available, after interaction with the σ subunit, for a further cycle of transcription. The RNA released upon termination is called a *primary transcript* because, in some cases, it is modified before being used in some subsequent process.

Question: Can a single RNA transcript in bacteria carry information from more than one gene?

Yes. Very frequently groups of adjacent genes are transcribed from a single promoter to give an RNA molecule carrying information for all of them. If this information is to be expressed as protein, as is normally the case, the single RNA molecule is called *polycistronic mRNA*. (A *cistron* is a genetic unit equivalent to a gene.) *Collections* of rRNA and tRNA genes are usually transcribed as single units. In these cases, the primary transcript is modified and subsequently cut by nucleases to yield individual rRNA and tRNA molecules. This is one example of *processing* of primary transcripts, which is much more common in eukaryotes.

17.4 DNA TRANSCRIPTION IN EUKARYOTES

The process of transcription in eukaryotes is similar to that in bacteria, but there are important differences.

Question: Is there a single type of RNA polymerase involved in eukaryotic transcription?

No. Eukaryotic RNA polymerases have been isolated from many tissues, and in all cases, three distinct enzymes have been found in the nucleus. All contain a number of polypeptide subunits and are complex in structure. *RNA polymerase I* is known to be involved specifically in the transcription of rRNA genes. *RNA polymerase II* gives rise to transcripts that are subsequently processed to yield mRNA. *RNA polymerase III* is responsible for the transcription of the tRNA genes and a small ribosomal RNA gene that yields a species called *5S RNA*. The three polymerases are distinguishable from one another by their differential sensitivity to the drug *α-amanitin* (the toxic principle of the mushroom *Amanita phalloides*), which does not affect bacterial RNA polymerase. RNA polymerase II is very sensitive to α-amanitin, while RNA polymerase I is completely resistant. RNA polymerase III is moderately sensitive to this inhibitor. Mitochondria have yet another type of RNA polymerase, which is unaffected by α-amanitin but is sensitive to drugs that inhibit bacterial RNA polymerase.

The role of the various subunits of the eukaryotic RNA polymerases has not yet been defined, but presumably there are subunits equivalent to the bacterial σ factor involved in the recognition of promoter sites. As with bacterial promoters, homologies upstream of the start point of transcription have been identified and are restricted to rather short sequences centered about 75 and 25 nucleotides from the start site. The consensus sequence of the latter is referred to as the *TATA* or *Hogness box* (after its discoverer). It also appears that in eukaryotes another sequence, known as an *enhancer*, can affect enormously the activity of a promoter. The position of an enhancer relative to a promoter can vary substantially, and it can function in either orientation and up to 1,000–3,000 base pairs on either side of the promoter. Much less is known about termination of transcription in eukaryotes when compared with bacteria.

Question: How does histone (Chap. 17) interaction with DNA (to give nucleosomes) affect transcription?

It is generally agreed that the nucleosome structure must be "dissolved" in order for transcription to occur. This dissolution, or removal, of histones must be specific in the case of transcription from restricted regions. The histones interact with the DNA through the positive charge on their basic amino acids, and it has been established that chemical modification to reduce the extent of this charge is one mechanism used to remove histones. *Acetylation* of arginine residues, for example, removes positive charges, while *phosphorylation* and *polyadenosine diphosphate ribosylation* add negative charges. In the last process, the enzyme poly(ADP-ribose) synthetase catalyzes the transfer of the ADP-ribose portion of NAD^+ to histones (as well as to nonhistone chromosomal proteins). Exactly how such modifications are effected in response to situations such as hormone stimulation has not been elucidated.

Question: Does transcription in eukaryotes yield polycistronic mRNA?

No, and this is in marked contrast to the situation in bacteria. In eukaryotes, the mRNA, which must be transported out of the nucleus for the purpose of translation, is always *mono*cistronic.

EXAMPLE 17.4

In bacteria, and prokaryotes in general, the primary transcript provides functional mRNA, ready for translation. In eukaryotes, the vast majority of primary transcripts are chemically modified and have sequences removed from within them before maturing as *functional* mRNAs. This is because eukaryotic genes, which will be expressed as protein, contain nontranslated *intervening* sequences, or *introns*. These are excised, or spliced out, at the primary transcript level to leave what corresponds to the translated segments, or *exons*, in the mRNA. A diagrammatic representation of the β-globin gene is shown in Fig. 17-3.

A primary transcript corresponding to the full length of the gene is first made. This is then chemically modified, and introns (two in the case of β-globin) are removed by splicing. The mixture of primary transcripts present in the nucleus is known as *heterogeneous nuclear RNA* (hnRNA).

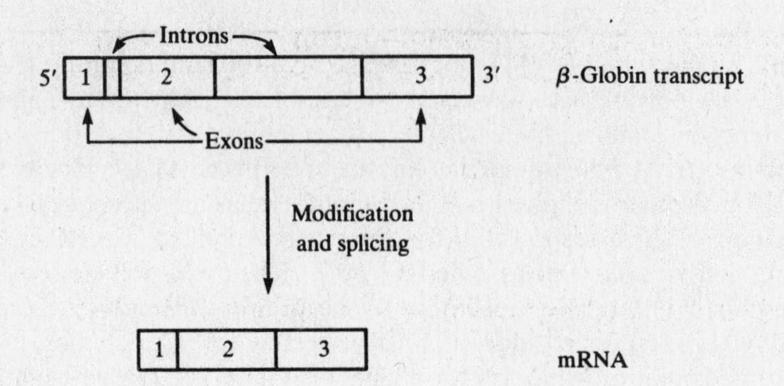

Fig. 17-3 The arrangement of introns and exons in the β-globin gene.

Question: What modifications are effected in the primary mRNA transcript?

1. So-called *capping* occurs at the 5′ end of the transcript shortly after its initiation. In the first step of capping, GTP is used to add a guanine-containing nucleotide, which is linked to the chain through a triphosphate bridge; this is catalyzed by *guanyltransferase*. In subsequent reactions, both the added guanine and the first two nucleotides in the primary transcript are methylated. The 5′ cap is thus relatively complex in structure. It has an important role in the subsequent initiation of translation.
2. *Polyadenylation* results in the addition of a *poly*(*A*) *tail* of 40–200 residues at the 3′ end of the transcript. The enzyme responsible for this addition is *poly*(*A*) *polymerase*. The function of the poly(A) tail is unknown.

Although the excision of introns must be a very precise reaction if functional messages are to result, there is no *unique* mechanism for RNA splicing. Exon-intron junctions within nuclear genes have a *consensus sequence*, and this suggests that there may be a common mechanism for splicing in these cases. There are some indications that small RNAs restricted to the nucleus (snRNAs), at least one of which has a sequence complementary to the consensus sequences at the splice junctions, could form a secondary structure across two adjacent splice junctions to juxtapose the ends of neighboring exons and provide a framework for cutting and sealing by *processing enzymes*. In the case of an rRNA from *Tetrahymena thermophila*, a single intron is spliced out in the presence of only GTP and certain cations.

EXAMPLE 17.5

The steps involved in the production of a functional mRNA from a typical eukaryotic gene are shown in Fig. 17-4. For this illustration, we assume that the gene has six introns, which are identified as A–F. Exons are represented as 1–7. Not shown in the diagram is the fact that there is probably a defined order for excision of the individual introns.

Question: What is the reason for the presence of introns in most eukaryotic genes?

The precise function of introns is unknown. Nevertheless, there is some evidence that the exons, or coding sequences, give rise to structural domains in the final protein product. It is therefore considered that *interrupted genes* (the exons), as they are called, may reflect an evolutionary process in which combinations of various exons gave rise to different proteins through the joining of different protein *domains*.

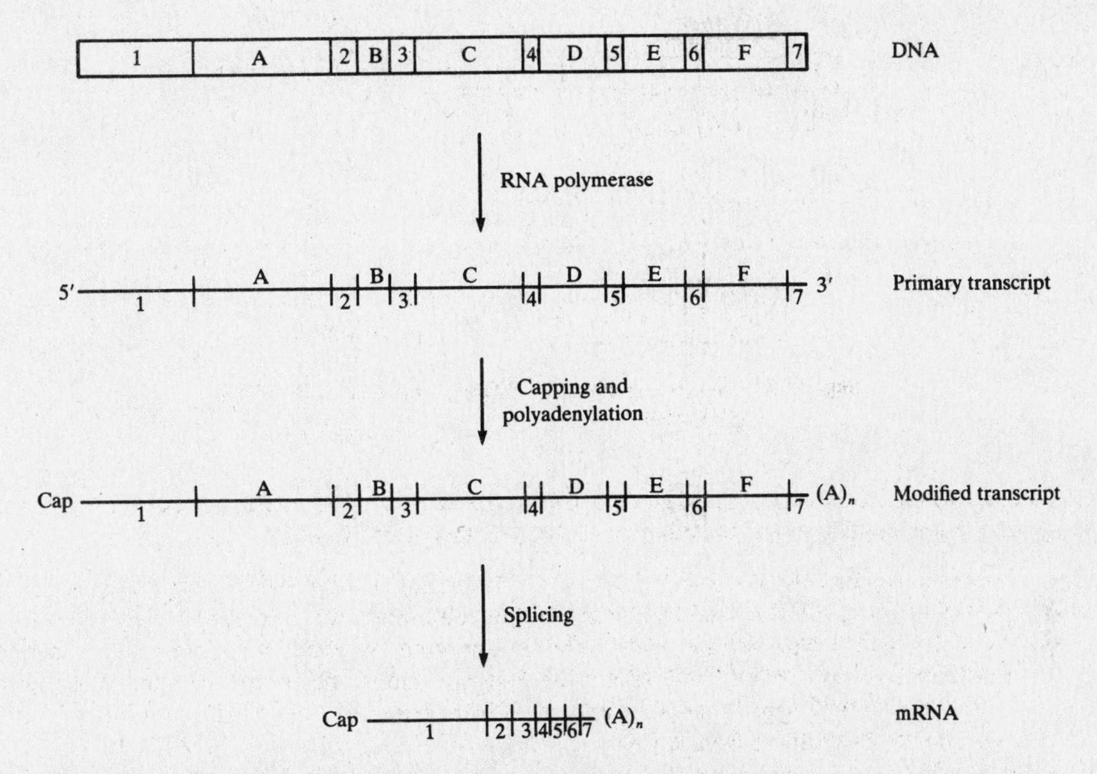

Fig. 17-4 Steps involved in the transcription and processing of a eukaryotic RNA transcript into mRNA.

17.5 INHIBITORS OF TRANSCRIPTION

A number of antibiotics function by inhibiting transcription. *Actinomycin D* (see Chap. 16) is an example of one that exerts its effect by binding to the DNA template; it can also block DNA replication.

Question: Would actinomycin D be expected to inhibit transcription in both bacteria and eukaryotes?

Yes, because in binding to the template, it recognizes a structural feature of the DNA double helix and therefore cannot discriminate between the two types of organisms.

Question: Are there inhibitors that discriminate between transcription in bacteria and eukaryotes?

Yes. Examples are *rifampicin* and *streptolydigin*, which bind only to bacterial RNA polymerase and block its action; *α-amanitin* binds only to eukaryotic RNA polymerase II and, to a lesser extent, to RNA polymerase III to block their action. The structures of these three inhibitors are shown in Fig. 17-5.

Question: Bacterial RNA polymerase is a multisubunit enzyme. Are there particular subunits to which rifampicin and streptolydigin bind?

Yes, both bind only to the β subunit. But rifampicin blocks only *initiation* of RNA synthesis, while streptolydigin preferentially blocks *elongation*. This shows that the β subunit is involved in both initiation and elongation of RNA chains.

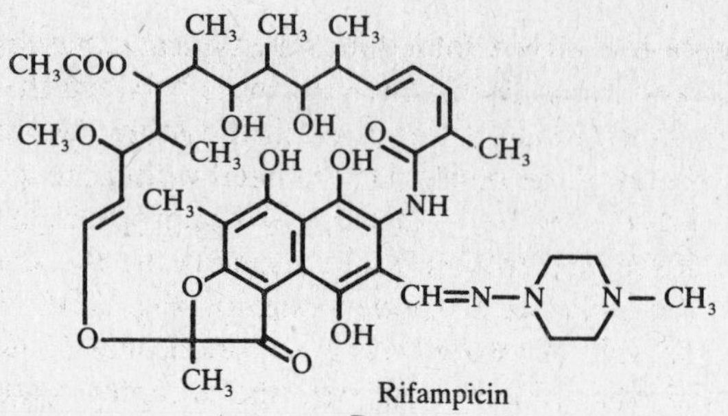

Rifampicin

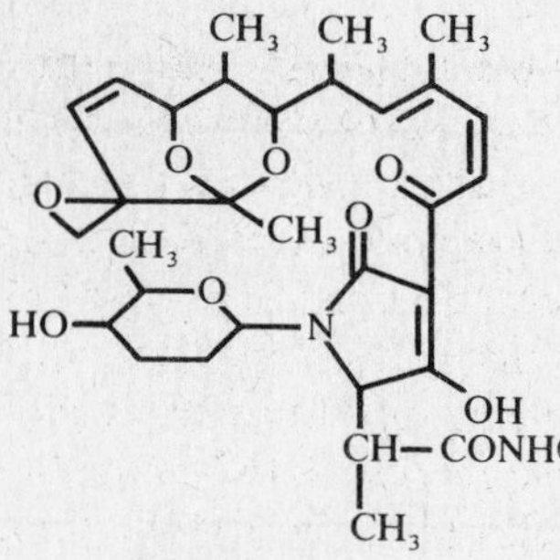

Streptolydigin

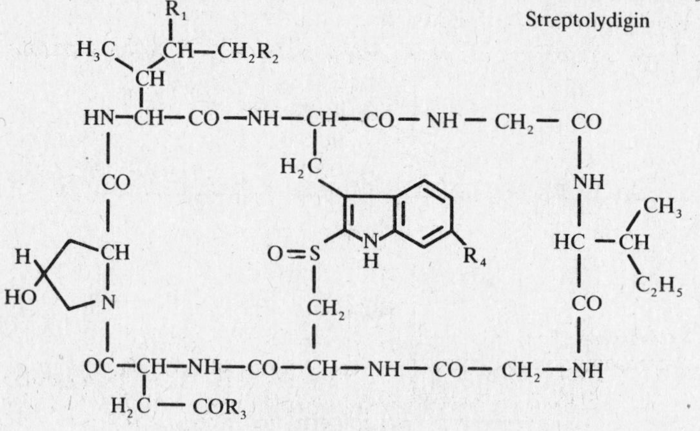

$R_1 = OH; R_2 = OH; R_3 = NH_2; R_4 = OH$

α-Amanitin

Fig. 17-5 Some inhibitors of DNA transcription.

17.6 THE mRNA TRANSLATION MACHINERY

The sequence of nucleotides in mRNA is converted through the translation "machinery" into a sequence of amino acids that constitutes a polypeptide. This machinery includes tRNA and ribosomes (which contain rRNA and a collection of unique proteins). The function of tRNA is to act as an adapter between the nucleotide sequence (defining the order of codons) and the amino acid sequence to be assembled into a polypeptide.

EXAMPLE 17.6

In *E. coli*, at any one time mRNA constitutes only 3–4 percent of the total cell RNA. Only 0.2 percent of the DNA genome is used to code for the >20 tRNAs, and 0.5 percent for rRNA. Thus, more than 99 percent of the genome serves as a template for mRNA synthesis.

Question: How does a tRNA molecule function as an adapter between a codon and an amino acid?

The nucleotide sequences of many tRNAs from a wide variety of organisms have been determined. All contain approximately 80 nucleotides, many of them of unusual structure (see Chap. 7). There is at least one tRNA corresponding to each amino acid, and while the sequences within individual tRNAs vary, they all form a common type of secondary structure (cloverleaf) in which the RNA chain folds back on itself to give a maximum amount of base pairing. One part of this structure is involved in the binding of an amino acid, and another part contains a sequence of 3 nucleotides complementary to one (or more) of the codons for this amino acid. This sequence of 3 nucleotides interacts with codons in the mRNA during the translational process. There are other features of the structure that are essential for the action of tRNA.

Figure 17-6 is a diagrammatic representation of tRNA folded into the typical cloverleaf structure, containing a number of stems (base-paired) and loops. While the sequences between the tRNAs are different, there are regions that remain *invariant*. Most of these are in the *loops*, within which the unusual bases are concentrated, and at the 3′ end of the molecule contained within the *acceptor stem*. The sequence at this end is always CCA, and it is to the 3′ OH that the appropriate amino acid is attached through its carboxyl group. The three nucleotides complementary to the codon for the amino acid make up what is known as the *anticodon* (shaded part of Fig. 17-6). The three-dimensional structure of tRNA is known. In this structure, there are additional H bonds, which stabilize the cloverleaf in a more elongated L-shaped structure, with the acceptor sequence at one end and the anticodon loop at the other.

While there is at least one tRNA for each amino acid, there is *not* a separate one for each codon.

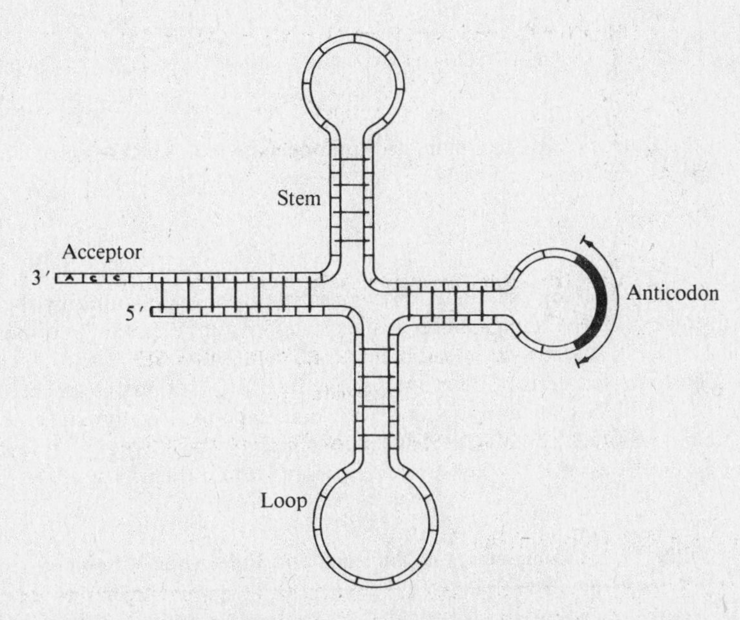

Fig. 17-6 A diagrammatic representation of the folded cloverleaf structure of tRNA.

Question: How can a single tRNA molecule accommodate more than one type of codon?

This can be accounted for by the *wobble hypothesis*: It appears that when a codon in mRNA interacts with the anticodon, unconventional pairing can form between the base in the third position of the codon (3′ end of triplet) and the first position of the anticodon. The unusual nucleoside *inosine* (Chap. 7) frequently occurs in the latter position, and it can pair with A, U, or C. The possibility of more than one type of pairing in this position accounts for the fact that when there is more than one codon for a single amino acid (called *synonyms*, see Table 17.1), the differences are usually in the third position only.

The attachment of an amino acid to an appropriate tRNA is accomplished via *aminoacyl-tRNA synthetase* and the hydrolysis of ATP. There is a separate enzyme specific for each amino acid, and it will recognize all tRNAs for that amino acid. The reaction proceeds in two steps and requires Mg^{2+} (Fig. 17-7). The first step, *amino acid activation*, results in the formation of an aminoacyl-AMP-enzyme intermediate. In the second step, the aminoacyl group is transferred to its appropriate (*cognate*) tRNA, the amino acid being linked to tRNA through an ester bond. (It appears that recognition between the synthetase and tRNA is achieved through very precise contact between the two molecules, with single contact points distinguishing one tRNA from another.) The first reaction is driven to the right by the hydrolysis of PP_i. Thus, in the overall activation and attachment of an amino acid, two high-energy phosphate bonds are "consumed."

mRNA and aminoacylated tRNAs (charged tRNAs) interact on *ribosomes*. The initial interaction occurs in such a way as to allow the codon for the initiating amino acid (methionine) to interact with its appropriately charged tRNA to commence polypeptide synthesis.

Amino acid activation

$$\text{ATP} + \text{R}'\text{CH}(\text{NH}_2)\text{—COOH} + \text{Enz}^{\text{R}'} \xrightleftharpoons{\text{Mg}^{2+}} \left[\text{R}'\text{CH}(\text{NH}_2)\text{—C}(=\text{O})\text{—O—P}(=\text{O})(\text{O}^-)\text{—O—Ribose—A}\right]\text{—Enz}^{\text{R}'} + \text{PP}_i$$

Transfer of aminoacyl group to its cognate tRNA

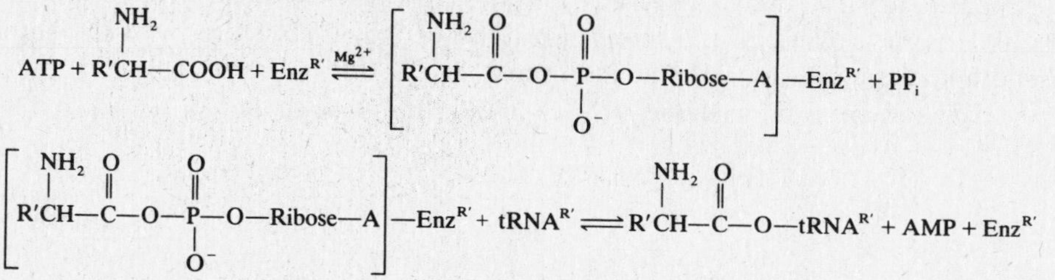

$$\left[\text{R}'\text{CH}(\text{NH}_2)\text{—C}(=\text{O})\text{—O—P}(=\text{O})(\text{O}^-)\text{—O—Ribose—A}\right]\text{—Enz}^{\text{R}'} + \text{tRNA}^{\text{R}'} \rightleftharpoons \text{R}'\text{CH}(\text{NH}_2)\text{—C}(=\text{O})\text{—O—tRNA}^{\text{R}'} + \text{AMP} + \text{Enz}^{\text{R}'}$$

Fig. 17-7 Reactions in the attachment of an amino acid to its cognate tRNA. R′ refers to the amino acid, $\text{Enz}^{\text{R}'}$ to the appropriate synthetase, and $\text{tRNA}^{\text{R}'}$ to the cognate tRNA.

EXAMPLE 17.7

Ribosomes comprise small and large subunits, distinguishable from one another by their different rates of sedimentation in a centrifuge cell. The small subunit has a special role in the initiation of polypeptide synthesis. In bacteria, the small and large subunits have sedimentation coefficients of 30S and 50S, respectively. They interact to give a 70S ribosome. In polypeptide synthesis, the interaction occurs as an early step in the overall process. In eukaryotes, the subunit makeup is similar, with both subunits slightly larger. The small (40S) and large (60S) subunits yield an 80S ribosome. The individual subunits in both types of ribosome have the same functions, determined by the types of RNA and proteins present within them. These are listed in Table 17-2.

Table 17.2. Components of Bacterial and Eukaryotic Ribosomes

Bacteria	Eukaryotes
70S ribosome:	*80S ribosome*:
30S subunit	40S subunit
= 16S RNA + 21 proteins	= 18S RNA + ~30 proteins
50S subunit	60S subunit
= 23S RNA + 5S RNA + 34 proteins	= 28S RNA + 5.8S RNA + 5S RNA + ~50 proteins

A considerable amount of information on the precise architecture of the small and large subunits of bacterial ribosomes, defining the surface location of the many proteins, and on the manner of interaction of the subunits is available. Some of this is illustrated in Fig. 17-8; refer to subsequent text for further explanation of this figure.

17.7 RNA TRANSLATION IN BACTERIA

Because we have a clearer picture of the way in which ribosomes interact with mRNA and assemble polypeptides in bacteria, this will be considered in some detail first. The overall process in eukaryotes is very similar, and the special features of eukaryotes will be treated in the section that follows.

Translation of an RNA message into a polypeptide occurs in three stages: *initiation*, *elongation*, and *termination*. As already mentioned, initiation in bacteria involves the interaction of the 30S ribosomal subunit at the appropriate location on mRNA.

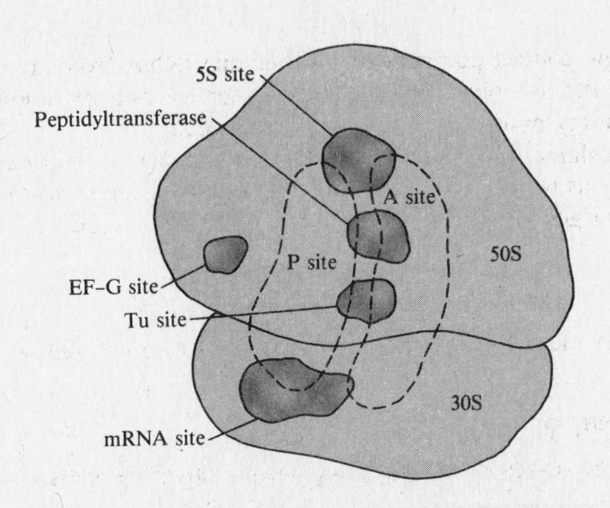

Fig. 17-8 A diagrammatic representation of some of the sites on the *E. coli* ribosome. (The peptidyl (P) site accommodates fMet-tRNA$_f^{Met}$, which is involved in initiation of the polypeptide chain. The A site accommodates aminoacyl-tRNA. The protein Tu (*T* for transfer, and *u* for unstable when heated) releases the aminoacyl-tRNA to the A site. EF-G refers to elongation factor G.

Question: What features of mRNA structure enable interaction with the 30S subunit?

Toward the 5′ end of mRNA, there is a region of 20 or so nucleotides before the initiation codon AUG is reached. This *leader* region contains a sequence responsible for interaction of the mRNA with the 30S subunit. It is known as the *Shine-Dalgarno* (S-D) sequence, and it can bind to a complementary sequence at the 3′ end of the 16S rRNA to position the 30S subunit appropriately for initiation. Other sequences in the leader region are possibly involved in the overall process of initiation of translation, which involves also the binding of the appropriately charged methionyl-tRNA opposite the AUG codon.

Question: Is there any special feature of methionyl-tRNA, in addition to the presence of an anticodon for AUG, that is required for its participation in initiation of translation?

Yes. There are two tRNAs for methionine, distinguishable by their capacity when charged with methionine to be *formylated* by a *transformylase*. The two species are called $tRNA_f^{Met}$ and $tRNA_m^{Met}$. The former is capable of being formylated to yield *N*-formylMet-tRNA$_f^{Met}$ (or fMet-tRNA$_f^{Met}$, for short) and is the species involved exclusively in initiation of the polypeptide chain. Presumably, unique features of the structure of the RNA in this case are required for the initiation process.

In addition to mRNA, fMet-tRNA$_f^{Met}$, and the ribosome subunits, three *initiation factors* (proteins) and GTP are involved in initiation of polypeptide synthesis. The process is depicted in Example 17.8.

EXAMPLE 17.8

The first step of RNA translation begins with the initiation of polypeptide synthesis (Fig. 17-9). GTP is bound into the 30S initiation complex and is subsequently hydrolyzed and released upon binding to the 50S subunit. The fMet-tRNA$_f^{Met}$ occupies what is known as the peptidyl (P) site of the ribosome (Fig. 17-8); another site (A), capable of accommodating an aminoacyl-tRNA, is empty at this stage. It is aligned with the next codon (shown as XXX in Fig. 17-9) in the mRNA.

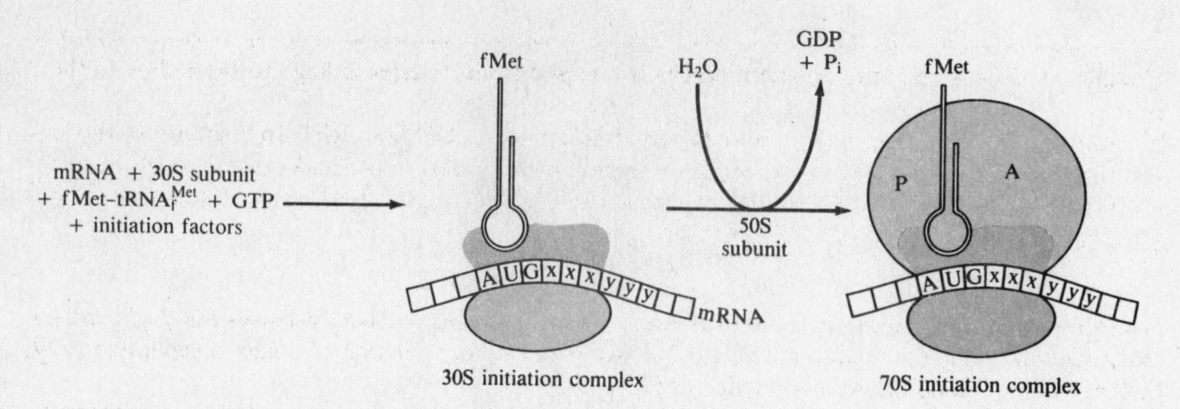

Fig. 17-9 Initiation of polypeptide synthesis.

Question: Is there any special mechanism for filling the A site with the appropriate aminoacyl-tRNA?

Yes. Transport of the appropriately charged tRNA to the A site requires association with a protein-GTP complex. The protein, called *Tu* (so-named because it is a *transfer* factor and is *unstable* when heated) is known as an *elongation factor*. Upon releasing the aminoacyl-tRNA to the A site, the Tu-GTP is hydrolyzed to Tu-GDP + P_i. Tu-GTP is then regenerated from Tu-GDP through reactions involving another protein, *Ts* (a heat-*stable transfer* factor), and GTP. Thus, one high-energy phosphate bond is consumed in binding the aminoacyl-tRNA into the A site. Everything is now ready for peptide bond formation (elongation).

The elongation phase of polypeptide synthesis and its termination are depicted in Example 17.9. The A site is shown to be filled by AA_2-tRNA where the codon XXX is located.

EXAMPLE 17.9

The second step of RNA translation involves elongation of the polypeptide chain (Fig. 17-10). One of the protein components of the 50S subunit is a *peptidyltransferase*. As the name implies, it transfers the fMet (and in later reactions, a peptide) from the P site to the A site. To do this, the ester bond linking fMet to its tRNA is broken and the aminoacyl is transferred to the amino group of the adjacent aminoacyl-tRNA (AA_2-tRNA in Fig. 17-10) to form the first peptide bond. In the next step, a *translocase* called *elongation factor G* (EF-G), in association with GTP hydrolysis, shifts (or translocates) the ribosome by one codon to position the dipeptidyl-tRNA in the P site, leaving the A site available for the binding of another aminoacyl-tRNA. This process of aminoacyl-tRNA binding, peptide bond formation, and translocation continues until a *stop codon*, which defines the completion of the polypeptide chain, is aligned with the empty A site.

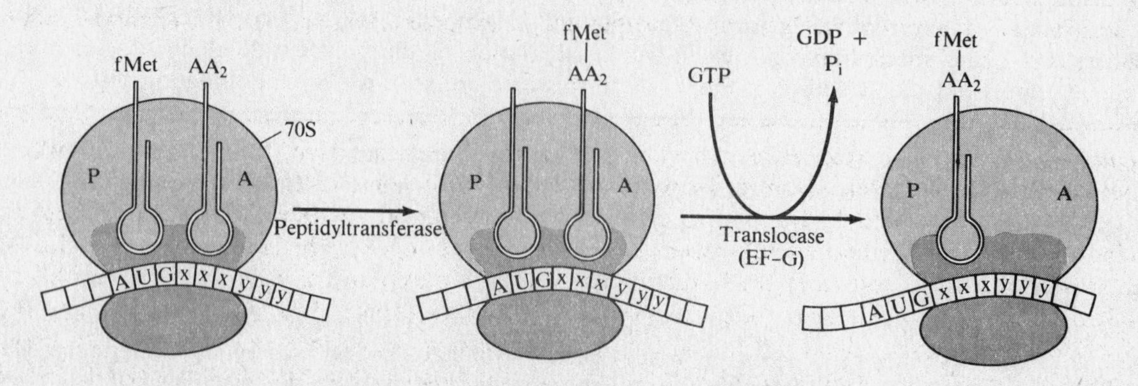

Fig. 17-10 Elongation step of polypeptide synthesis.

Question: How is the long polypeptide chain released from its ester linkage to the tRNA in the P site?

In bacteria, there are three *release-factor proteins*: RF1, RF2, and RF3. In response to the stop codons, they presumably bind (in various combinations) to the A site and cause hydrolysis of the ester bond to release the chain; in so doing, they generate the free carboxyl terminus of the polypeptide.

Question: During translation, as the ribosome moves along the mRNA, it leaves the leader region (containing the *ribosome binding site*) empty. Is it possible for initiation of a new polypeptide chain to occur before completion of a previous one?

Yes. It is common for any single mRNA to be translated simultaneously by many ribosomes. They give rise to structures called *polyribosomes* or *polysomes*.

17.8 RNA TRANSLATION IN EUKARYOTES

The molecular mechanism of translation in eukaryotes is very similar to that in bacteria. The activation of amino acids and attachment to tRNAs and the steps of initiation, elongation, and termination of polypeptide chains are essentially the same in overall terms. The small and large ribosomal subunits of bacteria and eukaryotes are equivalent with respect to their roles in initiation and elongation of chains.

There are two *significant* differences between the mechanism of translation in bacteria and in eukaryotes, and they relate to initiation of translation. First, while there are two forms of tRNA for methionine in eukaryotes, one of which is used in initiation, neither charged form is formylated; the transformylase is not present in eukaryotes. (Eukaryotic Met-$tRNA_t^{Met}$, however, can be formylated by the bacterial enzyme.) A second difference, of more significance, is the involvement of the methylated 5′ cap of the mRNA in initiation of translation. If the cap is missing, translation is inefficient. It has been established that binding of the 40S ribosomal subunit to the leader region of mRNA requires additional factors called *cap-binding proteins*. There is evidence to suggest that the cap is the major structural feature needed for 40S subunit binding.

17.9 POSTTRANSLATIONAL MODIFICATION OF PROTEINS

Most polypeptides synthesized on ribosomes are later chemically modified. Thus the formyl group on the N-terminal methionine in polypeptides of bacteria is removed by a *deformylase*. In both bacteria and eukaryotes, the N-terminal methionine, sometimes along with a few additional amino acids, is removed by *aminopeptidases*.

Question: The amino acids *hydroxyproline* and *hydroxylysine* are absent from Table 17.1, which describes the genetic code. How do these amino acids arise in some proteins?

Hydroxyproline and hydroxylysine occur most noticeably in collagen. These are formed by modification of proline and lysine residues by specific enzymes *after* synthesis of the collagen chains. It is interesting to note that *prolylhydroxylase*, which hydroxylates proline, requires *ascorbate* (vitamin C) as a coreactant. Other chemical modifications known to occur commonly are the attachment of sugars (glycosylation) to asparagine, serine, and threonine residues and the phosphorylation of serine. Chemical modifications are also associated with the transport of proteins out of the cells in which they are synthesized.

Question: How can proteins be transported through a hydrophobic membrane and out of a cell?

Proteins destined for *secretion* from cells (in bacteria and eukaryotes) are usually synthesized in a *precursor form*. This form contains what is called a *signal sequence*, which is relatively hydrophobic and consists of 15–30 amino acid residues, at the N terminus. This sequence, as it is formed on the ribosome, somehow attaches to a membrane and penetrates it. As the polypeptide chain continues to elongate, it passes through to the other side of the membrane. In the meantime, the signal sequence is removed by a *signal peptidase*. In the case of bacterial cells, such a process can lead to transport of the protein into the *periplasmic space* (between the inner and outer membranes at the surface) or out of the cell altogether. In the case of eukaryotes, the signal peptide enables the transfer of the protein into the lumen of the endoplasmic reticulum. From here it is transported by other mechanisms out of the cell. Insulin provides a good example of the transport or secretion of a protein out of a eukaryotic cell.

EXAMPLE 17.10

Insulin is formed in special cells of the pancreas. The immediate product of mRNA translation is a single polypeptide called *preproinsulin*. The modifications associated with the conversion of preproinsulin to insulin are shown in Fig. 17-11.

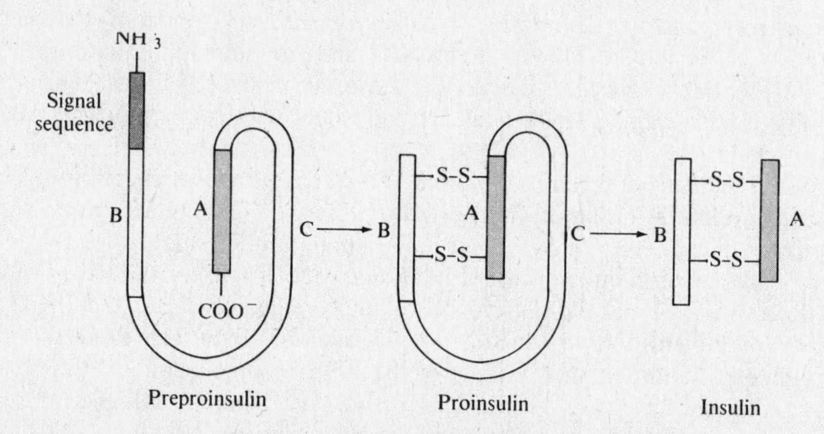

Fig. 17-11 Conversion of preproinsulin into insulin.

Porcine preproinsulin is a single polypeptide containing 107 amino acid residues. After the signal peptide is synthesized on the ribosome, it recognizes a receptor on the endoplasmic reticulum and attaches to it. (The attachment of ribosomes to the endoplasmic reticulum through peptide chains is responsible for what is known as the *rough endoplasmic reticulum*, Chap. 1.) The polypeptide chain passes through the membrane and into the lumen, where the signal peptide of 23 residues is cleaved off to yield a shortened chain of 84 residues; the latter folds on itself to form intramolecular disulfide bridges joining cysteine residues. This folded molecule is called *proinsulin*.

Proinsulin is packaged into membrane vesicles and transported to the *Golgi apparatus*, where conversion to insulin begins. This includes cleavage of the uncross-linked chain (C), thus removing the connecting peptide. Conversion is completed as the molecule is transported via *secretory granules* from the Golgi to the *cytoplasmic membrane*, with which the granules fuse to release the mature insulin, now consisting of two disulfide-cross-linked polypeptide chains, into the circulation.

17.10 INHIBITORS OF TRANSLATION

Because of the large number of steps associated with the translation of mRNA into protein, there are numerous opportunities available for blocking it with inhibitors. The action of many antibiotics is based on blocking translation in bacteria.

Question: Which inhibitors will block translation in both bacteria and eukaryotes?

Good examples are *fusidic acid* and *puromycin*. The former inhibits the binding of charged tRNA to the A site of the ribosome. Puromycin acts by virtue of its similarity in structure to an aminoacyl-tRNA (see below).

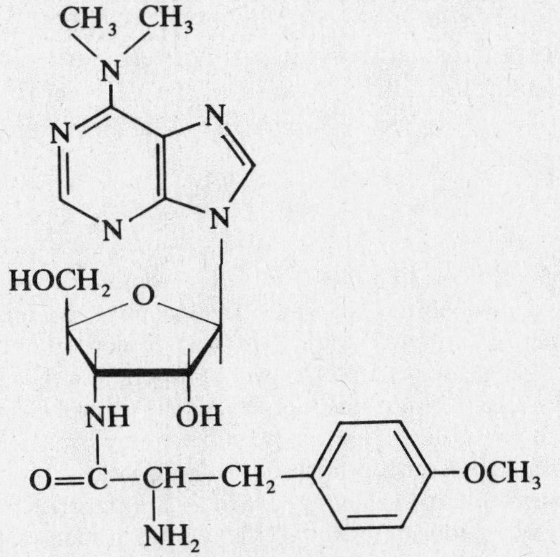

Puromycin

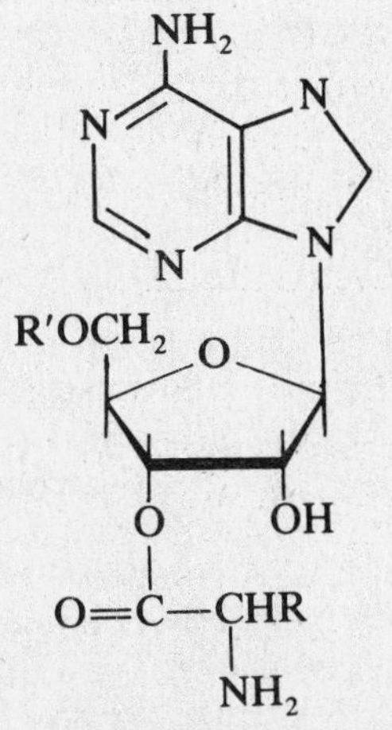

Aminoacyl-tRNA

It competes with the latter as an acceptor in the peptidyl transfer reaction. The growing chain is transferred to the NH_2 group of puromycin and is prematurely terminated.

Question: Are there different inhibitors that act on analogous targets in bacteria and eukaryotes?

Yes. *Chloramphenicol* inhibits the peptidyltransferase of the 50S ribosomal subunit of bacteria, while *cycloheximide* inhibits the analogous enzyme in the 60S subunit of eukaryotic ribosomes. Their structures are shown below.

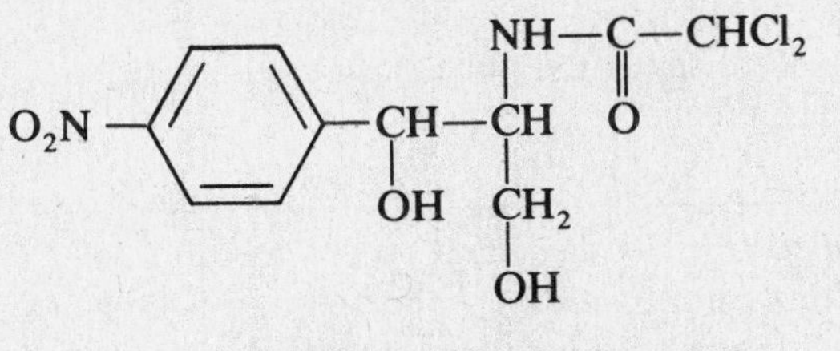

Chloramphenicol

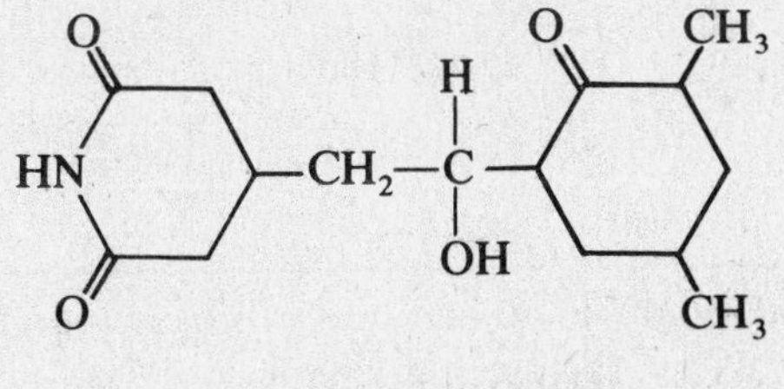

Cycloheximide

It is interesting to note that chloramphenicol blocks translation in mitochondria, further indicating the similarity of mechanisms of gene expression in mitochondria and bacteria.

There are many other inhibitors specific for just bacterial or eukaryotic cells. An interesting one is *diphtheria toxin*, which is effective only in eukaryotes.

EXAMPLE 17.11

Diphtheria toxin, produced by *Corynebacterium diphtheriae*, is a single polypeptide ($M_r = 63{,}000$) with two intrachain disulfide bridges. A portion of the molecule, the A fragment ($M_r = 21{,}000$), must enter the cytoplasm of a cell to exert the toxic effect. Through a fairly complex mechanism, the larger molecule is cleaved into two on the membrane surface to yield A and B fragments. The latter facilitates entry of the A fragment into the cytoplasm of the cell, where it specifically catalyzes the chemical modification (ADP-ribosylation) of the translocase (called EF2) to inactivate it and block polypeptide chain growth.

17.11 CONTROL OF GENE EXPRESSION

The end products of gene expression are proteins, mainly enzymes, and it is essential that their levels be strictly controlled. There are many potential sites of control in both bacteria and eukaryotes. DNA or gene amplification (Chap. 16) in eukaryotes is one way of responding to the need for more of the protein product; if there are more copies of the gene, then transcription can occur at a faster rate. More often, control is effected at the level of either transcription or translation, with the former probably being more important for both bacteria and eukaryotes. Transcriptional control in bacteria is particularly effective because of the very short half-life (a few minutes) of mRNA in such cells; the half-life is longer in eukaryotes. The prototype for transcriptional control is the *lactose operon* in *E. coli*.

EXAMPLE 17.12

The lactose operon (or *lac* operon), a region of ~5.3 kb of the *E. coli* chromosome, contains the genes coding for the enzymes responsible for lactose metabolism. The genes for the three enzymes involved—*β-galactosidase*, *galactoside permease*, and *thiogalactoside transacetylase*—are situated next to one another in this segment of DNA. They are transcribed as a single unit of RNA, with transcription being controlled by sequences (control elements) toward the 5′ end of the 5.3-kb segment of DNA. The control elements comprise a *promoter* (to which RNA polymerase binds) and an adjacent *operator* (to which a *repressor protein* can bind to block transcription by the RNA polymerase). The repressor is produced by a gene located on the 5′ side of the promoter. This is illustrated in Fig. 17-12. Because all three genes—*z*, *y*, and *a*—are transcribed as a single unit (polycistronic mRNA), they are said to be expressed *coordinately*. When transcription is blocked by the repressor, none of the genes is expressed.

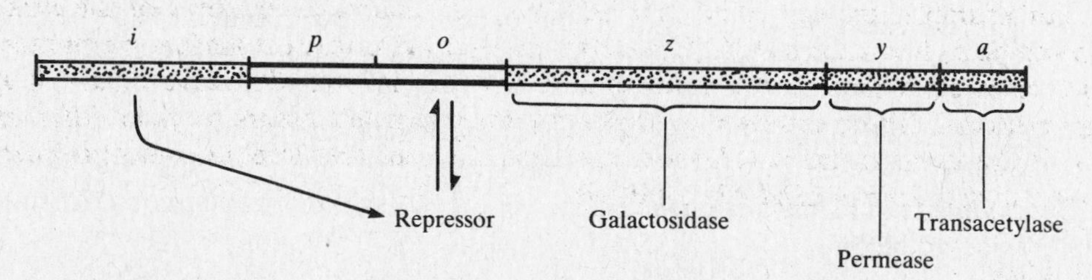

Fig. 17-12 The lactose operon of *E. coli*. Abbreviations: i = repressor gene; p = promoter; o = operator; z = β-galactosidase gene; y = permease gene; and a = transacetylase gene. The i gene product is the repressor.

Question: How does the cell overcome repression to express the genes of the *lac* operon?

When lactose is added to a cell culture, a small amount of it enters the cell and is converted to *allolactose*. This metabolite binds to the repressor, altering its conformation and causing it to be released from the operator. RNA polymerase is then free to transcribe the three genes, and the allolactose is said to be acting as an *inducer*. This relief of repression by an inducer is an example of *negative control* of expression.

There is also an aspect of *positive control* in the lac operon. The *catabolite activator protein* (CAP), carrying bound cAMP, is needed for the binding of RNA polymerase to the promoter; i.e., it has a direct, positive effect on transcription. However, relief of repression (i.e., induction) will not occur in the presence of glucose, because glucose lowers the level of cAMP, so that CAP is unable to exert its effect. This reflects the preference of the cell to use glucose rather than lactose as a carbon source. Thus it can be seen that the cell stringently controls expression of the *lac* genes; it expresses them only if it needs to metabolize lactose.

There are several other examples, in addition to the *lac* operon, of transcriptional control in bacteria, and some show additional features.

EXAMPLE 17.13

The *trp* operon defines a cluster of several genes involved in tryptophan synthesis. The *trp* operon is a single transcriptional unit under negative control. Unlike the *lac* operon, however, the *trp* repressor (or *aporepressor*) must first form a complex with tryptophan or it cannot bind to the operator. Thus, tryptophan functions as a *corepressor* and switches off transcription when the cell has made a sufficient level of it. In addition to the promoter-operator region, the *trp* operon contains a second regulatory site called the *attenuator*. This is a sequence within the leader region of the mRNA, between the promoter and the initiation codon of the first gene, *trp E*. Through a complex mechanism utilizing the level of charged $tRNA^{Trp}$ in the cell, transcription can be terminated at the attenuator region in the presence of excess tryptophan.

There are also examples of *positive* transcriptional control in bacteria. In these situations, relief of repression is not involved; instead, a regulatory compound or gene product binds to a promoter region to induce transcription.

Question: Are there good examples of transcriptional control in eukaryotes?

It has been established that many of the *steroid hormones* (Chap. 13) act by stimulating transcription. Through a series of complex steps, involving interaction with a cytoplasmic receptor, the hormone enters the nucleus, where it binds, either by itself or in association with the receptor or other proteins, to a specific site on the DNA. Exactly how it then induces transcription is still not clear, but both positive and negative control mechanisms seem to be involved. A well-studied example is the estrogen-mediated induction of ovalbumin mRNA synthesis in the chicken oviduct.

Eukaryotes potentially have many more opportunities for control of gene expression than do bacteria. For example, the cell could take advantage of control at the level of the *processing of primary transcripts*. It is known that RNA is not transported across the nuclear membrane until all introns are excised. A more subtle form of control could involve *alternative modes of splicing* a particular transcript. There are now examples known where this occurs to yield different mRNA molecules. Perhaps one of the best-known examples of yet another level of control in eukaryotes is that of translational control of *globin* synthesis.

EXAMPLE 17.14

Globin is synthesized in reticulocytes (see Chap. 1, Prob. 1.1), which have no nucleus and therefore cannot utilize transcriptional and other potential modes of control. Control of globin synthesis from the pool of globin-enriched mRNA is geared to the level of *hemin* [Fe(III)-protoporphyrin], which has the ability to inactivate a translational inhibitor of protein synthesis. The inhibitor is a *protein kinase* that phosphorylates and inactivates one of the initiation factors involved in initiation of translation. When the level of hemin is high, it binds to a regulatory subunit of the kinase and, as a result, initiation of globin synthesis can proceed.

Solved Problems

THE GENETIC CODE

17.1. Assuming that the first three nucleotides define the first codon, what is the peptide sequence coded for by (*a*) UAAUAGUGAUAA, (*b*) UUAUUGCUUCUCCUACUG?

SOLUTION

(*a*) None. These four codons are not translatable; they are signals for chain termination.

(*b*) $(\text{Leu})_6$. This illustrates the degeneracy of the genetic code for leucine; there are six codons for leucine, the most for any of the amino acids.

17.2. From within the sequence AAUUAUGUUUCCAUGUCCACCU, identify two possible sites from which initiation of translation could commence and write the sequence of the first three amino acids.

SOLUTION

AUG is the most common initiation codon. The first one encountered in the sequence would give the following sequence for nine nucleotides from this position: AUGUUUCCA. This codes for Met-Phe-Pro. Farther along there is the sequence AUGUCCACC, which codes for Met-Ser-Thr.

17.3. Within the antisense strand of the DNA in a gene, the codon sequence ATA is changed by mutation to ATG. Following replication of the DNA, what change in the polypeptide product would this cause?

SOLUTION

The antisense strand of DNA gives the same sequence as the mRNA, except for T being present in the DNA in place of U. Thus, the codon will change from AUA to AUG. This causes isoleucine to be replaced by methionine.

17.4. An mRNA contains the following translated sequence, with the reading frame defined by the grouping of triplets:

AUG.CUC.ACU.UCA.GGG.AGA.AGC

(*a*) What amino acids would result from this sequence? (*b*) If the first C (nucleotide containing cytosine) encountered is deleted from the sequence, what new amino acid sequence would result?

SOLUTION

(*a*) The original sequence would give Met-Leu-Thr-Ser-Gly-Arg-Ser.

(*b*) The new sequence would be

AUG.UCA.CUU.CAG.GGA.GAA.GC

which would give Met-Ser-Leu-Gln-Gly-Glu.

17.5. The DNA of the virus ϕX174 appears to contain insufficient nucleotide residues to code for the nine different proteins that are its gene products. How might this arise?

SOLUTION

In this case, the same sequence of nucleotides codes for more than one protein, by the use of more than one reading frame. That is, the sequence of codons for one protein is out of phase with the sequence of the overlapping gene. See the figure on the following page.

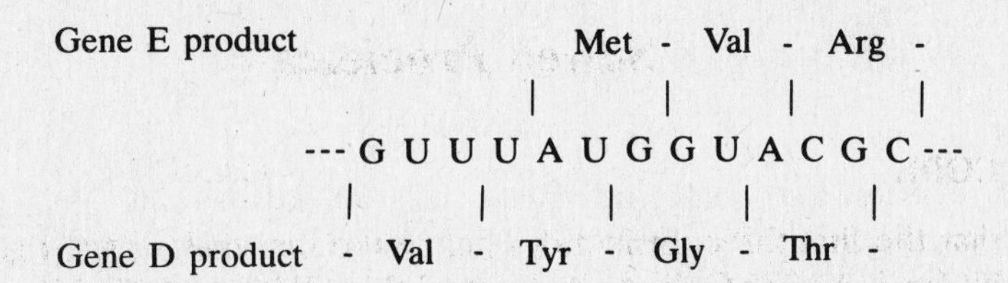

17.6. The fidelity of DNA replication is enhanced by a "proofreading" function, whereby errors in the complementary sequence are excised and repaired (Chap. 16). Why is a similar mechanism not found in protein synthesis?

SOLUTION

The consequences of errors in protein synthesis are not as serious. A single defective protein molecule will, in general, not cause deleterious effects; such a protein may not function properly or may be unstable, and may represent an energy wastage to the cell; however, such errors do not become perpetuated in future generations.

17.7. The sense strand of a double-helical segment of DNA contains the sequence:

$$5' \text{ GCTACGGTAGCGCAA } 3'$$

(*a*) What sequence of mRNA can be transcribed from this strand?

(*b*) What amino acid sequence would be coded for, assuming that the entire transcript could be translated?

SOLUTION

(*a*) The transcribed RNA would be complementary to the above strand, and would have U replacing T:

$$3' \text{ CGAUGCCAUCGCGUU } 5'$$

Written in the $5' \longrightarrow 3'$ direction this becomes:

$$5' \text{ UUGCGCUACCGUAGC } 3'$$

(*b*) Translation occurs in the $5' \longrightarrow 3'$ direction, resulting in the following sequence:

-Leu-Arg-Tyr-Arg-Ser-

DNA TRANSCRIPTION IN BACTERIA

17.8. Would it be possible for a single strand of DNA to function both as a sense and antisense strand with regard to transcription?

SOLUTION

Yes. For this to occur, transcription by RNA polymerase would proceed through the segment of duplex DNA in *opposite directions* and from different promoters. Each strand of the duplex DNA would be functioning as both sense and antisense strands, but for two different RNA transcripts.

17.9. Which of the following could not represent the 5′ end of an RNA transcript?

(*a*) pppApGpCpU . . . . .

(*b*) pppUpCpGpA . . . . .

(*c*) pppCpApGpA . . . . .

SOLUTION

Only (*a*) is possible, because in initiating transcription, RNA polymerase always incorporates a pppA or a pppG in the first position.

17.10. Is the Pribnow box the same as a bacterial promoter?

SOLUTION

No. The Pribnow box is only *part* of the sequence that defines the promoter. It corresponds to the -10 region of ~ 7 nucleotides. There is another region centered around the -35 position that is an essential part of the promoter.

17.11. Which subunits of bacterial RNA polymerase are needed for initiation of transcription from a promoter?

SOLUTION

First, the σ subunit is needed for promoter binding and the formation of an open promoter complex. Subsequent to this, the β subunit (which binds the inhibitor rifampicin) is essential for the formation of the first phosphodiester bond.

17.12. Is the transcription termination sequence incorporated into the RNA transcript?

SOLUTION

Yes. After the RNA polymerase transcribes the inverted repeat sequence, it is the ability of this region to form a hairpin in the single-strand transcript that is responsible, at least in part, for termination of the chain.

17.13. How many transcription termination sequences would be present in polycistronic mRNA?

SOLUTION

Polycistronic mRNA, which forms as a primary transcript in bacteria, is a continuous length of RNA transcribed from a single promoter. It will therefore contain only one *normal* termination sequence (i.e., ignoring a possible *attenuator* sequence before the first initiation codon).

17.14. Does processing of RNA transcripts occur in bacteria?

SOLUTION

Processing refers to modification of the primary transcripts formed by RNA polymerase. In bacteria, it is restricted to the transcripts that contain rRNA and tRNA. In these cases, larger transcripts are chemically modified and then cut down to the smaller, mature forms of rRNA and tRNA by nucleases.

DNA TRANSCRIPTION IN EUKARYOTES

17.15. How many types of RNA polymerase might you expect to find in a eukaryotic cell?

SOLUTION

Four. In the nucleus, there are the RNA polymerases I, II, and III. In mitochondria, there is another RNA polymerase, which is similar to bacterial RNA polymerase.

17.16. With respect to the general mechanism of RNA chain growth, are there any differences among the various types of RNA polymerase in eukaryotic cells?

SOLUTION

No. All RNA polymerases use duplex DNA as a template and copy one of the strands; they synthesize RNA in the $5' \longrightarrow 3'$ direction and use ribonucleoside triphosphates as substrates.

17.17. Would the major transcriptional activity of a eukaryotic cell be affected by α-amanitin?

SOLUTION

No. The most abundant RNA component of a cell is always rRNA (Chap. 7). The rRNA genes are transcribed by RNA polymerase I, which is resistant to α-amanitin.

17.18. Is it possible that α-amanitin exerts its inhibitory effect on certain eukaryotic RNA polymerases by interfering with the availability of the substrates?

SOLUTION

No. All RNA polymerases use the same substrates, the α-amanitin inhibits some RNA polymerases but not others.

17.19. Does monocistronic mRNA of eukaryotes generally represent a primary RNA transcript?

SOLUTION

No. In the vast majority of cases, the mRNA, which is always monocistronic in eukaryotes, is formed by the processing (modification and splicing) of primary transcripts.

INHIBITORS OF TRANSCRIPTION

17.20. Would actinomycin D be expected to block transcription of rRNA and tRNA genes and those that code for protein products?

SOLUTION

Yes. Actinomycin D blocks transcription by binding to the DNA template. In doing so, it recognizes a common structural feature of all duplex DNAs, binding by intercalation between stacked base pairs (Chap. 16).

17.21. The bacterial RNA polymerase inhibitors rifampicin and streptolydigin each bind to the same subunit of the enzyme, but their overall effect on the enzyme's activity is different. Why is this so?

SOLUTION

Each of these inhibitors binds exclusively to the β subunit of RNA polymerase. This subunit is involved in both initiation and elongation of RNA chain growth. Rifampicin must bind to the subunit in a way that affects only the initiation step; it has no effect on elongation. Streptolydigin, on the other hand, binds in a manner that blocks both activities.

THE mRNA TRANSLATION MACHINERY

17.22. How many varieties of rRNA and tRNA species are present in a cell actively involved in translation?

SOLUTION

At the most, there are 3–4 types of rRNA, depending on whether they are bacterial or eukaryotic. On the other hand, there will be *at least* 20 types of tRNA (at least 1 for each amino acid) in any type of cell.

17.23. Why would it be necessary for all tRNAs to have similar overall dimensions?

SOLUTION

All tRNAs must have similar overall dimensions because, during translation, all charged tRNAs interact singly and very precisely with the same sites on the ribosome. The anticodon is positioned at one "end" to allow it to interact with the bound mRNA, and the amino acid is precisely located on the surface of the ribosome with respect to the location of the bound peptidyltransferase.

17.24. Would it be possible for a single tRNA to accommodate all the codons for leucine?

SOLUTION

Leucine has six codons, the largest number for any amino acid. According to the *wobble hypothesis*, a single tRNA can accommodate more than one codon. The wobble hypothesis allows for up to 3 different nucleotides—but only at the third position in the codon—to interact with a single nucleotide in the anticodon. The fact that there are six codons in leucine means that they must differ at positions other than the third, and they therefore could not be accommodated by a single anticodon in a particular tRNA molecule.

17.25. Aminoacyl-tRNA synthetase functions in two steps to bring about the attachment of an amino acid to its cognate tRNA. Does it show specificity to either the amino acid or the tRNA at each of these steps?

SOLUTION

In the first step (amino acid activation), the enzyme recognizes its appropriate amino acid. In the second step, the amino acid must be attached to the correct tRNA, and it is therefore essential that the latter be recognized by the enzyme. (It does happen that sometimes, at a low frequency, a similar but incorrect amino acid is incorporated in the first step; the incorrect amino acid is released when the enzyme recognizes its appropriate tRNA.)

RNA TRANSLATION IN BACTERIA

17.26. What is the "leader region" of bacterial mRNA?

SOLUTION

The leader is the region between the 5′ end of the RNA and the initiation codon. It contains untranslated sequences that are involved in ribosome binding, a step essential for translation of the message into a polypeptide.

17.27. Why can't $tRNA_m^{Met}$ function in initiation of polypeptide synthesis from an appropriate AUG codon?

SOLUTION

In initiation of translation, the methionyl-tRNA species $tRNA_f^{Met}$ is used exclusively. Presumably, the unique nucleotide sequence of this species is required for the initial interaction with the small ribosomal subunit, the leader region of mRNA, initiation factors, and GTP to give the first initiation complex. It is unlikely that formylation (which can occur only with methionyl-$tRNA_f^{Met}$) is essential for initiation, as this is not the case in eukaryotes. It thus appears that $tRNA_m^{Met}$ is excluded from initiation because essential structural requirements in the RNA are not met.

17.28. How many ATP *equivalents* are consumed with the incorporation of an amino acid into a polypeptide?

SOLUTION

One ATP is used for charging of the tRNA, and then one GTP at each of the steps of binding aminoacyl-tRNA to the A site of the ribosome and translocation. Thus, ignoring initiation, the equivalent of three ATPs are used for each amino acid incorporated. But remember that in amino acid activation, the products are AMP and PP_i, the latter being hydrolyzed to P_i to drive the reaction to completion. Thus, the equivalent of four high-energy phosphate bonds are used for each amino acid incorporated.

17.29. In translocation, the peptidyl-tRNA is shifted from the A site to the P site on the ribosome. What happens to the peptidyl-tRNA anticodon-codon interaction?

SOLUTION

This must remain undisturbed, and it is probably clearer to say that the whole ribosome moves by one codon position with respect to the peptidyl-tRNA–mRNA complex.

17.30. What major structural requirement for initiation of translation of eukaryotic mRNA does not exist for bacterial mRNA?

SOLUTION

During the processing of primary transcripts destined for mRNA formation in eukaryotes, a methylated guanine "cap" is attached to the 5′ end. This is essential for efficient initiation of translation of the mature eukaryotic mRNA and does not occur in prokaryotes.

17.31. Is there any obvious reason why polysomes could not form in eukaryotes?

SOLUTION

Polysomes form as the result of the loading of sequential ribosomes at the 5′ end of mRNA such that many ribosomes are progressing through a single mRNA at any one time. Clearly, polysomes do form in eukaryotes. Once the first ribosome has moved a significant distance from the ribosome binding site, there is no reason why a subsequent one should not be loaded through the normal sequence of events.

17.32. *Coupled transcription-translation* in prokaryotes refers to the commencement of translation of an RNA molecule before its transcription from the DNA template is complete. Could such a situation arise in eukaryotes?

SOLUTION

No. Transcription in eukaryotes must be completed within the nucleus. The mRNA finally produced is transported to the cytoplasm for translation into a polypeptide.

POSTTRANSLATIONAL MODIFICATION OF PROTEINS

17.33. When proteins are hydrolyzed with acid at high temperature, they are broken down to their constituent amino acids. Commonly, the amino acid cystine is found among them. But cystine is not included among the amino acids listed in the dictionary of codons. Why?

SOLUTION

Cystine is composed of two molecules of cysteine linked through oxidation of —SH groups to give a disulfide bond. Such oxidation, which is important in stabilizing the folded structure of some proteins, represents a posttranslational modification of a protein. Thus, cystine is never incorporated as such into a polypeptide during translation, and there is no codon that corresponds to it.

17.34. The signal sequence on a protein destined for export from a cell is never located at the C-terminal end of the polypeptide chain. Why?

SOLUTION

This is because the signal sequence is required for passage of the polypeptide chain, *as it is being assembled*, through a membrane. The C-terminal portion is always synthesized last, and a signal sequence at this location could not allow the "threading" of the growing chain through the membrane.

INHIBITORS OF TRANSLATION

17.35. Why is puromycin able to function as an inhibitor of translation in both bacteria and eukaryotes?

SOLUTION

Puromycin is similar in structure to the 3′ end of aminoacyl-tRNAs, with which it competes during translocation. The 3′ end of aminoacyl-tRNAs is the same in all organisms.

17.36. Chloramphenicol, when used as an antibiotic to treat bacterial infection in animals, can have side effects. What might contribute to these?

SOLUTION

Chloramphenicol blocks translation in bacteria by inhibiting peptidyltransferase of the large ribosomal subunit. It does not interfere with peptidyltransferase in the large subunit of eukaryotic ribosomes. However, the mitochondrion of animal cells contains ribosomes that are similar to bacterial ribosomes, and chloramphenicol can block protein synthesis in this organelle. This could contribute to the side effects of this drug when used in the treatment of animals.

CONTROL OF GENE EXPRESSION

17.37. Why is control of expression of the *lac* operon in *E. coli* said to be an example of *negative* control?

SOLUTION

This is because control of expression is brought about through modulating the effectiveness of a negatively acting agent, the repressor. In other words, expression of the *lac* operon is achieved through negating the effect of the repressor. In *positive* control systems, on the other hand, expression is achieved through the immediate effect of an agent to positively induce or increase expression.

17.38. What would be needed for the coordinate control of expression (at the transcriptional level) of a number of genes not located next to one another on the chromosome?

SOLUTION

Each gene would need to have common regulatory elements—promoters, operators, or both—associated with it. In this way, a single control factor, such as a repressor or positively acting substance, could influence all genes simultaneously.

17.39. Transcriptional control of globin synthesis in reticulocytes is not possible because transcription does not occur in these cells. Does this mean that the overall control of globin synthesis is completely lacking an aspect of transcriptional control?

SOLUTION

No. Prior to the formation of the reticulocyte from its precursor cells during the process of erythropoiesis (Chap. 1), there must have been a stage of preferential transcription of the globin genes to yield the globin-enriched mRNA of the reticulocyte.

Supplementary Problems

17.40. Decode the following RNA sequence into the corresponding amino acid sequence:

CAU AUU ACU CAU GAA CGU GAA

17.41. The following occurs at the start of the coding sequence of a eukaryotic mRNA:

GUG UUU UUU GUG UUU

For what amino acid sequence does it code?

17.42. The following segment of duplex DNA contains the region defining the start codon for a protein:

5′-GATGTCTCCT- 3′
3′-CTACAGAGGA- 5′

Identify the sense strand.

17.43. Which is the most abundant RNA species in a cell?

17.44. DNA polymerase needs a primer on which to attach a new nucleotide unit for chain growth. Is a primer obligatory for RNA polymerase action?

17.45. For which is a promoter needed: initiation of RNA synthesis or of polypeptide synthesis?

17.46. The Pribnow box occurs about 35 nucleotides upstream of the initiation codon. (*a*) True; (*b*) false.

17.47. The σ subunit of RNA polymerase is needed only for initiation of transcription. (*a*) True; (*b*) false.

17.48. Is the termination sequence for transcription near the 5′ or the 3′ end of the transcript?

17.49. Can monocistronic mRNA represent a primary transcription product of bacterial cells?

17.50. In which are enhancer sequences found, bacteria or eukaryotes?

17.51. Would phosphorylation of histones decrease or increase the net positive charge on such molecules?

17.52. Genes of eukaryotes are generally made up of introns and exons. Are these sequences both transcribed and translated?

17.53. Does splicing occur in DNA, RNA, or polypeptides?

17.54. When does capping of an RNA molecule occur, before or after splicing?

17.55. Can an exon contain an initiation codon?

17.56. Does rifampicin inhibit the transcription of histone genes?

17.57. In what species of RNA do codons and anticodons occur?

17.58. Why is the tRNA molecule so large, when the codon is only three nucleotide residues long?

17.59. tRNA is a single-stranded molecule, but much of its structure is double-helical. How is this possible?

17.60. Does the acceptor stem of a tRNA molecule contain the 3′ or the 5′ end?

17.61. Are there more species of aminoacyl-tRNA synthetase or tRNA in a cell?

17.62. Which of the subunits of a ribosome recognizes the nontranslated leader sequence of mRNA?

17.63. Subsequent to the formation of the 70S initiation complex in bacteria, at what steps of translation is energy consumed in the form of high-energy phosphate bonds?

17.64. Are termination codons associated with termination of transcription or termination of translation?

17.65. Is the methylated cap on eukaryotic mRNA attached to the RNA while it is in the nucleus or in the cytoplasm?

17.66. What is the first translational product of the insulin gene?

17.67. Does a signal peptidase cleave the nontranslated leader sequence of mRNA?

17.68. Does a bacterial repressor molecule bind to the promoter or operator site adjacent to a gene in DNA?

17.69. What is the difference between a repressor and a corepressor in relation to transcription?

Answers to Supplementary Problems

Chapter 1

1.15. The membranes of all living cells are *selectively* permeable to ions and other chemical species. This selectivity is in many cases linked to the supply of ATP (Chap. 10), and one aspect of cell death is low levels of ATP. In this state, the cells no longer exclude foreign compounds, such as toluidine dye.

1.16. Glutaraldehyde forms a Schiff base between side-chain amino groups of neighboring protein molecules, thus cross-linking them (Chaps. 3 and 4).

1.17. The arylsulfatase substrate *p*-nitrophenyl sulfate is used together with lead nitrate in a manner analogous to the Gomori reaction (Example 1.5).

1.18. Yes, problems would arise in interpreting the autoradiograph because the [^{3}H]glucose would not only be incorporated into glycogen but would also be metabolized via glycolysis (Chap. 11) to yield amino acids and fatty acids; these could appear in a whole array of cellular organelles.

1.19. Fragments of endoplasmic reticulum are transformed from lipid bilayer sheets, with attached ribosomes, into spherical vesicles. This is a result of the homogenization used in preparing the samples and also the tendency of lipid bilayers (Fig. 1-4) to *spontaneously reseal.*

1.20. It enables *separate control* over urea and pyrimidine synthesis (Chap. 15).

1.21. Incubate the reticulocytes with [^{3}H]leucine, which will be incorporated into proteins. Prepare electron microscope autoradiographs and count silver grains per cell and the number of polysomes. The latter appear as rosettes of five ribosomes in these cells. A statistical comparison between the number of polysomes and the amount of protein synthesized during the incubation time (proportional to the number of silver grains) indicates whether there are nonactive polysomes. In fact, many of the polysomes are inactive; i.e., they are "switched off" (see Chap. 17 for the control of protein synthesis).

1.22. (*a*) Mother. (*b*) If a defect exists in a mitochondrial gene, all progeny from that *female* will carry the defect. No well-defined cases of such a defect have, as yet, been described.

1.23. In fact, bacteria do not have mitochondria, but some types do have membranous intrusions into the cytoplasm called *mesosomes*. These are similar in function to the inner membrane of mitochondria (Chap. 14). The reason mitochondria are distinct from other membranous structures in higher cells is possibly due to their evolutionary origin as intracellular symbionts and to the fact that the spatial separation of functions leads to more advantageous (in terms of natural selection) control of the various metabolic processes.

1.24. (*a*) The release of peptidases, in particular, leads to tissue protein hydrolysis and hence breakdown. (*b*) Treatment is aimed at reducing inflammation with anti-inflammatory steroid drugs, which also serve to stabilize the lysosomal membranes.

1.25. A fragment, usually the short arm, of chromosome 21 is translocated onto another chromosome; thus, there are three copies of a fragment of the short arm in any one cell. This is a rare occurrence.

Chapter 2

2.10. (*a*) Pentose, D-xylose, β; (*b*) pentose, D-ribose, β; (*c*) aminohexose, *N*-acetyl-D-glucosamine, β; (*d*) hexose, D-glucose, β.

2.11. (*a*), (*c*), (*d*), (*e*), (*f*), (*g*), (*h*), (*i*)

2.12. (*a*) (*b*)

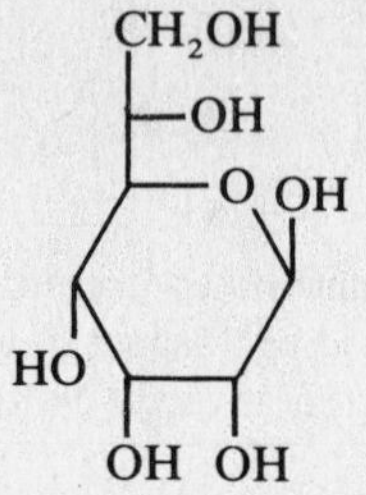

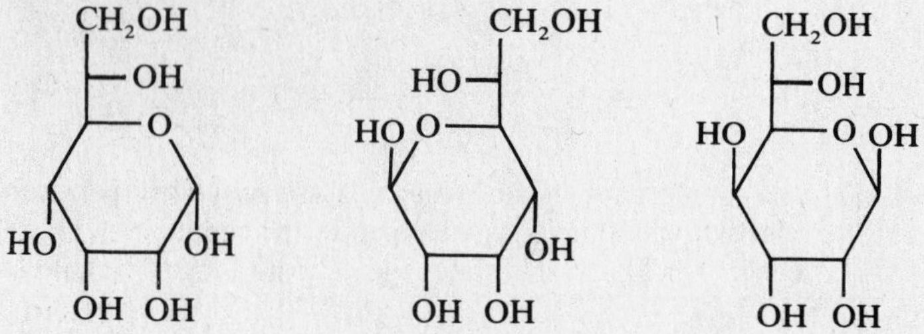

Anomer

Enantiomer

An epimer

2.13.

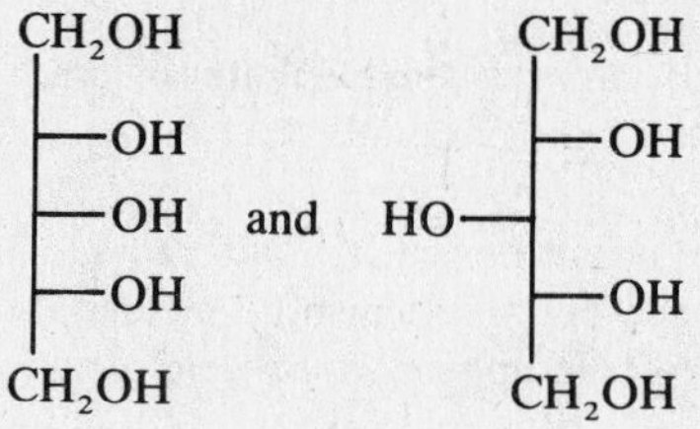

and

2.14. α (28 percent); β (72 percent)

2.15. 0.167 g cm^{-3}

2.16. Fructose is (−)

2.17.

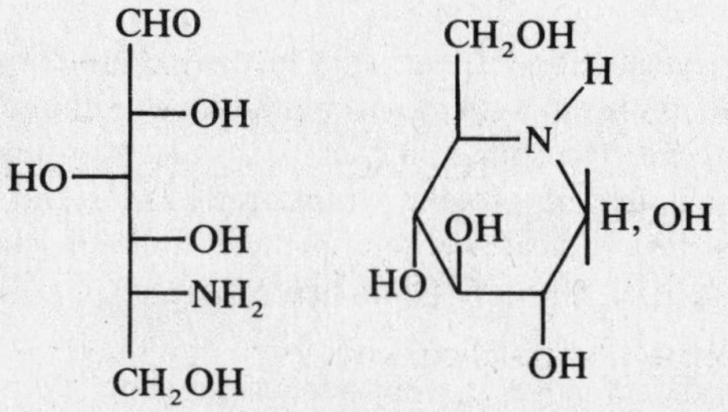

2.18. Six

2.19. The substituents of β-L-glucopyranose are all on the opposite side of the Haworth structure from those of β-D-glucopyranose (i.e., they are in the positions occupied by the H's in β-D-glucopyranose). Therefore, from Example 2.13, it may be seen that in the 1C chair conformer, all the substituents of β-L-glucopyranose would be equatorial.

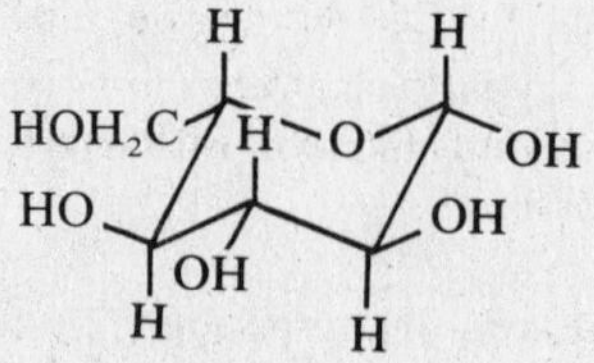

2.20.

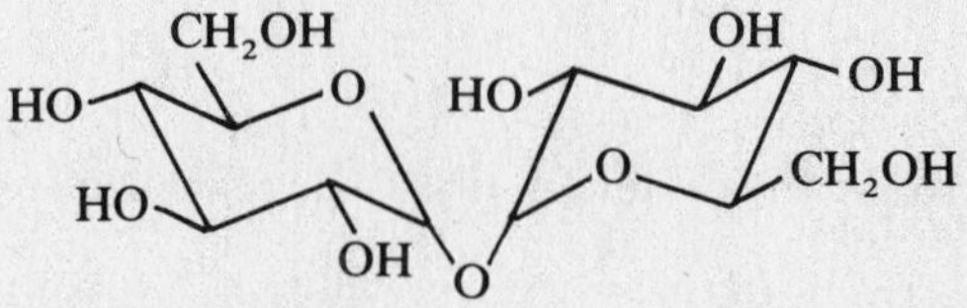

2.21. Erythritol is symmetrical, and both chiral carbon atoms are equivalent.

2.22.

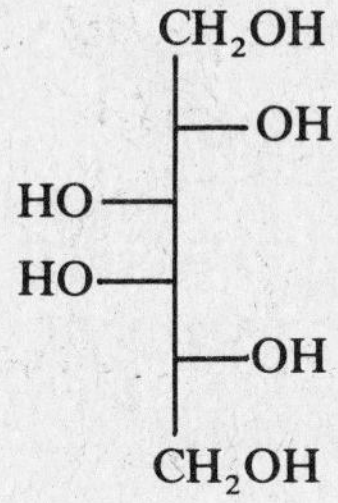

```
       CH2OH
         |—OH
    HO—|
    HO—|
         |—OH
       CH2OH
```

2.23.

```
         COOH                    CHO
           |—OH                    |—OH
     HO—|                    HO—|
(a)        |—OH        (b)         |—OH
           |—OH                    |—OH
         CH2OH                   COOH
```

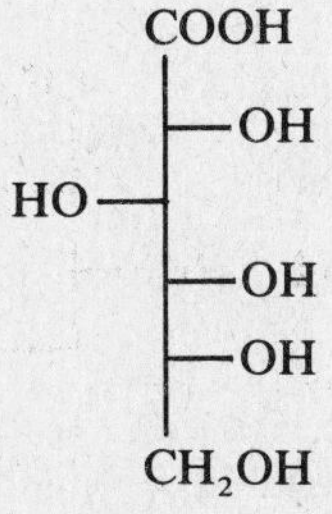

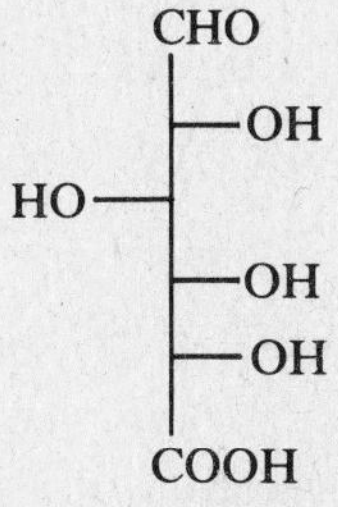

2.24. By inversion of C-5

2.25.

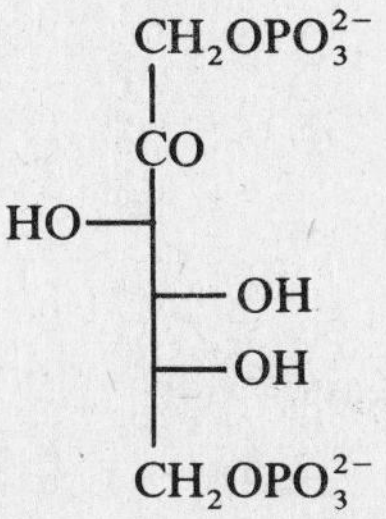

```
     CH2OPO3 2-        COO-              CHO               CHO
       CO                |—OH              |—OH              |—NHCOCH3
 HO—|              HO—|              HO—|              HO—|
       |—OH              |—OH      -O3SO—|           -O3SO—|
       |—OH              |—OH              |—OH              |—OH
     CH2OPO3 2-        CH2OPO3 2-        CH2OH             CH2OH
```

Fructose 1,6-diphosphate; 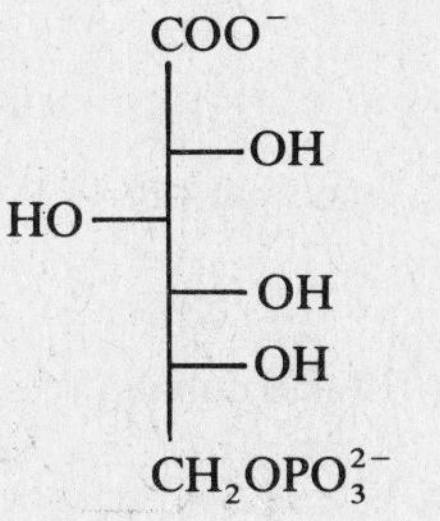 6-Phosphogluconate; 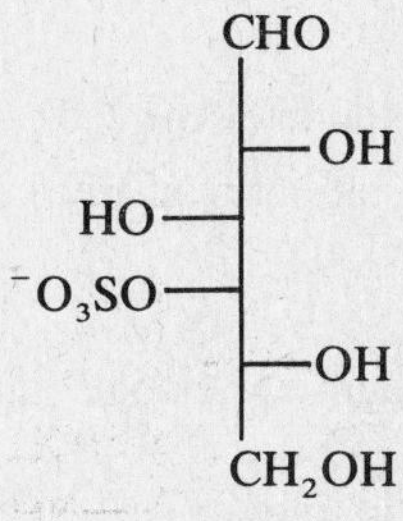 D-Galactose 4-sulfate; 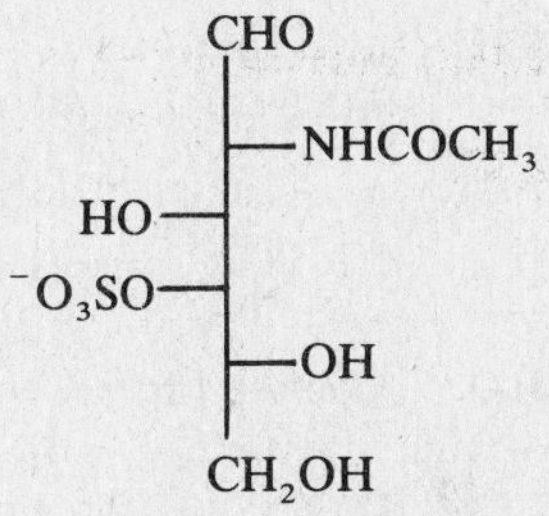 *N*-acetylgalactosamine 4-sulfate

2.26. No; L-fucose and L-rhamnose are deoxyhexoses.

2.27. Isomers

Chapter 3

3.21. (*a*) Serine, threonine (—OH); asparagine, glutamine (—NH_2); cysteine (—SH); tyrosine (—OH); aspartic acid, glutamic acid (—COOH); lysine, histidine, arginine (—NH).

(*b*) Serine, threonine (—OH); asparagine, glutamine (—C(=O)—); cysteine (—S—); tyrosine (—O—); aspartate, glutamate (—C(=O)—O^-); histidine, lysine, arginine (usually the unprotonated form)

3.22. Phenylalanine, tyrosine, tryptophan

3.23. Glycine

3.24. (*a*) $HO—CH_2—CH_2—SH$; (*b*) [pyridinium ring: H—N^+ in a six-membered aromatic ring] (*c*) [imidazolium: $H—\overset{+}{N}$=CH—N—H, CH=CH]

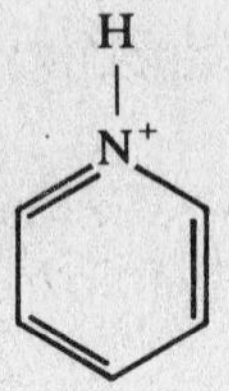

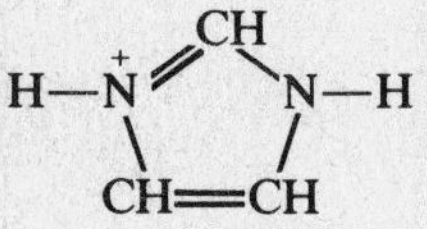

3.25. I.P. = 10.75

3.26. I.P. = 7.6

3.27. pH = 2.35

3.28. (*a*) 1.61; (*b*) 1.0; (*c*) −0.15

3.29. (*a*) 0.98; (*b*) 0.06; (*c*) −0.94

Chapter 4

4.19. (*a*) Actin, myosin, tropomyosin, troponin, myoglobin; (*b*) collagen; (*c*) keratin.

4.20. 17,000

4.21. (*a*) Anode
(*b*) Cathode at pH 3.0; anode at pH 9.0
(*c*) Cathode at pH 4.5; stationary at pH 9.5; anode at pH 11.0
(*d*) Anode at pH 3.5, 7.0, 9.5

4.22. Aldolase (first), serum albumin, hemoglobin, β-lactoglobulin, ribonuclease.

4.23. See Sec. 4.6 for these definitions.

4.24. (*a*) Surface: histidine, arginine, glutamine, glutamic acid; these are polar or charged.
(*b*) Interior: methionine, phenylalanine, valine; these are nonpolar. In addition, glutamine and uncharged histidine may be found in the interior if they can form hydrogen bonds.

4.25. (*a*) Nonpolar groups at the surface may participate in the binding sites for other molecules. (*b*) Charged groups in the interior may be important in the catalytic mechanisms of some enzymes.

4.26. A domain is an independently folded region of a protein; e.g., the NAD^+-binding domain of glyceraldehyde phosphate dehydrogenase.

4.27. Such a conformation leads to clashes between atoms.

4.28. (*a*) Urea in high concentrations (6–8 mol L^{-1}) weakens hydrogen bonds and hydrophobic interactions. (*b*) Urinary urea is very dilute (~0.2 mol L^{-1}).

4.29. (*a*) Hydrogen bonds, hydrophobic interactions, van der Waals interactions, charge-charge interactions. (*b*) These act cooperatively.

4.30. (*a*) $p = 0.54$ nm, $d = 0.15$ nm
(*b*) 153×0.15 nm = 22.95 nm
(*c*) 153×0.35 nm = 53.55 nm
(*d*) 153×0.36 nm = 55.08 nm

4.31. (*a*) α helix
(*b*) α helix (weak)
(*c*) disordered (charge repulsion)
(*d*) α helix (weak)
(*e*) α helix
(*f*) probably disordered

4.32. Hydrogen bonds, hydrophobic interactions, charge-charge interactions, van der Waals interactions.

4.33. Proline is too bulky to accommodate the close approach of the chains at every third residue.

4.34. Glycine is small enough to accommodate the close approach at every third residue.

4.35. (*a*) Insulin has two chains, formed by cleavage from a single-chain precursor; it is stabilized by disulfide bonds.
(*b*) Hemoglobin has four chains—two α and two β; it is held together by noncovalent interactions.
(*c*) Collagen has three chains and forms a triple helix; it is stabilized by hydrogen bonds and by additional chemical cross-links.

4.36. Maximal hydrogen-bond formation; all bond lengths and angles are normal; no clashes occur between atoms of the backbone (i.e., the α helix maps to a favorable region of the Ramachandran plot).

4.37. (*a*) Differences: H bonds in the α helix are formed between peptide groups of the same chain; in β structures, H bonds may be formed between different chains. H bonds are approximately parallel to the α-helix axis, and perpendicular to the direction of the chain in the β structures.
(*b*) Similarities: both the α helix and β structures are regular, repeating structures, are stabilized by hydrogen bonding, and map to favorable regions of the Ramachandran plot.

Chapter 5

5.19. (*a*) 10^{-3} mol L^{-1}; (*b*) 1 mol L^{-1}; (*c*) 10^{-2} mol L^{-1}

5.20. (*a*) 24 percent; (*b*) 98.4 percent
N.B.: The percentage of dimer increases as the equilibrium constant increases.

5.21. (*a*) 0.65
(*b*) 0.025
(*c*) 0.04
(*d*) 0.74
(*e*) 0.48

5.22. The oxygen affinity of hemoglobin is lowered by the chloride ion strengthening the salt link. Thus, in the presence of chloride, more oxygen would be released.

$$Hb(O_2)_4 + Cl^- \rightleftharpoons Hb(Cl^-) + 4O_2$$

5.23. The oxygen affinity is reduced (p_{50} is 70 torr, and the arterial blood is only 70 percent saturated).

5.24. Polypeptides (*b*) and (*c*) would form triple helices, and (*b*) would be the more stable.

5.25. Since the polypeptide chains of collagen are almost completely extended, they cannot be stretched much farther without breaking covalent bonds. The α helix can almost double its length by breaking weak, noncovalent hydrogen bonds.

5.26. (*a*) 6
(*b*) 4

5.27. Hydroxyapatite binds in the gap between the collagen molecules in a way similar to the heavy-metal stains used in electron microscopy.

5.28. The proteases cleave the link and core proteins in the region nearest the hyaluronate, where there is least protection from the glycosaminoglycans.

5.29. 9,400

5.30. The intermediate filaments are cell-specific and do not change when the cell is transformed. Thus, by identifying the type of intermediate filament present in a tumor, the origin and characteristics of even a micrometastasis can be determined.

Chapter 6

6.10.

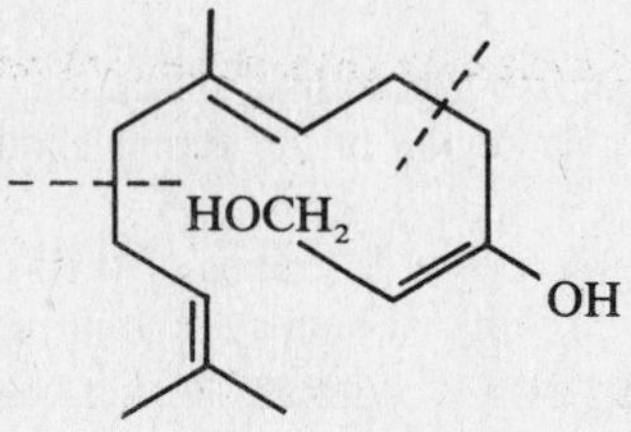

HOCH$_2$ OH

(*a*)

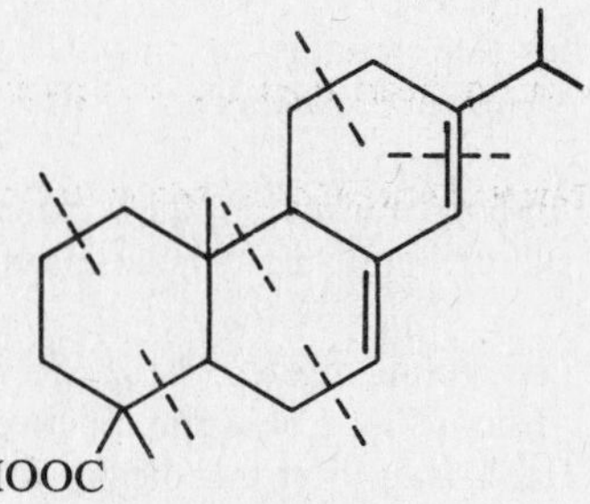

HOOC

(*b*)

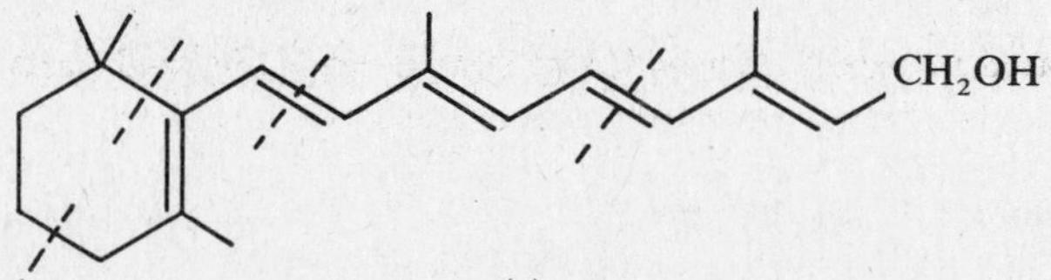

CH$_2$OH

(*c*)

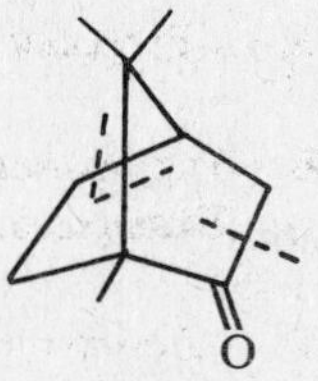

O

(*d*)

6.11. (*a*) A lipid that contains carbohydrate; (*b*) 6-α-D-galactopyranosyl-β-D-galactopyranosyldiglyceride (a glycoglycerolipid) and galactosylceramide (a glycosphingolipid)

6.12. Because they are polymers of the two-carbon compound acetate

6.13. (*a*) $CH_3(CH_2)_{12}COOH$
(*b*) $CH_3(CH_2)_3CH{=}CH(CH_2)_7COOH$
(*c*) $CH_3(CH_2)_5CH(OH)CH_2CH{=}CH(CH_2)_7COOH$

6.14. 64 (24 if the three fatty acids are always different)

6.15.

(*a*)
$CH_2O{\cdot}OC(CH_2)_{14}CH_3$
$CH_3(CH_2)_{14}CO{\cdot}OCH$
$CH_2O{\cdot}OC(CH_2)_{14}CH_3$

(Solid)

(*b*)
$CH_2O{\cdot}OC(CH_2)_7CH{=}CH(CH_2)_7CH_3$
$CH_3(CH_2)_{14}CO{\cdot}OCH$
$CH_2O{\cdot}OC(CH_2)_7CH{=}CH(CH_2)_7CH_3$

(Liquid)

6.16. To the anode: PG, DPG, PS; stationary: PE

6.17.

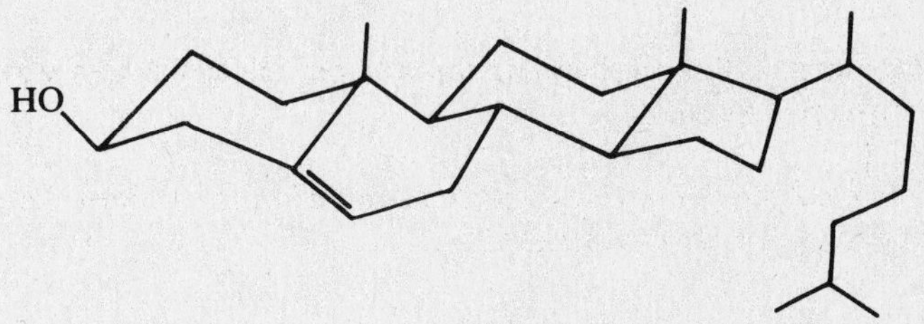

6.18. Phosphatidic acid

6.19. 2.86×10^6

6.20. Calculating the surface area of one red blood cell and multiplying this by the 4.74×10^9 cells yields a total area that is roughly only half the area found empirically. Therefore, the actual lipid content must be roughly twice that calculated for a monolayer—as would be the case for a surface bilayer.

6.21. PE has a smaller polar head than PC and packs the concave inner surface better.

6.22. (*a*) Lowers; (*b*) raises; (*c*) raises transition temperature. (*a*) Increases; (*b*) reduces; (*c*) reduces mobility of phospholipids

6.23. 1.04 g cm^{-3}

6.24. 120

6.25. 1-Propanol (fastest), 1,3 propanediol, propionamide, propionic acid, alanine

6.26. The answer should be based on material provided by Example 6.9, Fig. 6-10, and the introduction to Example 6.12.

6.27. $3{,}300 \text{ cpm min}^{-1}$

6.28. 35.5 kJ mol^{-1}

6.29. (*a*) No effect; (*b*) transport inhibited

Chapter 7

7.35. Structure (*a*) is a pyrimidine, and (*c*) is a purine.

7.36. (*a*) RNA: adenine, guanine, uracil, and cytosine
(*b*) DNA: adenine, guanine, thymine, and cytosine

7.37.

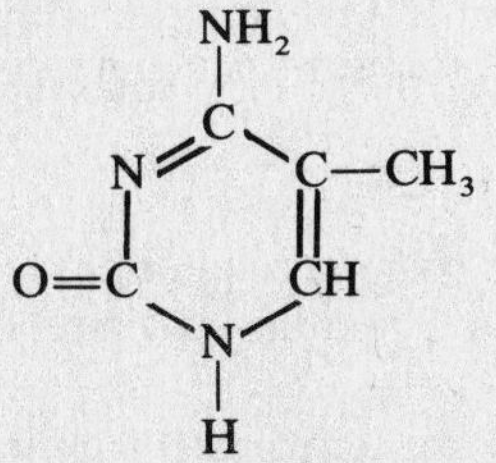

(*b*) Most DNAs contain very small amounts of 5-methylcytosine.

7.38.

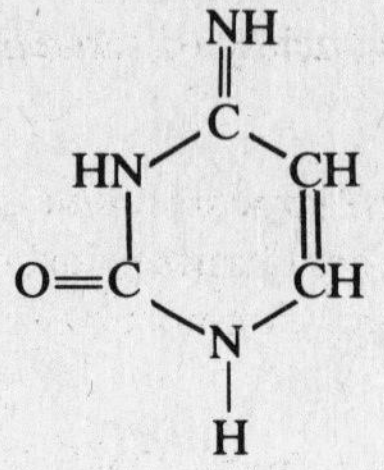

7.39.

(*a*) Deoxythymidine

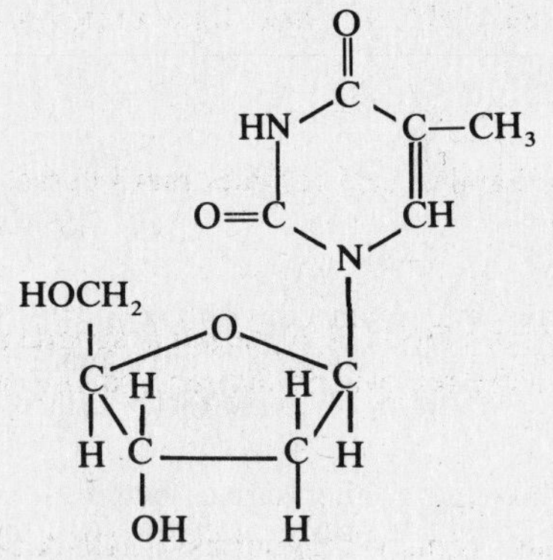

(*b*) Ribothymidine

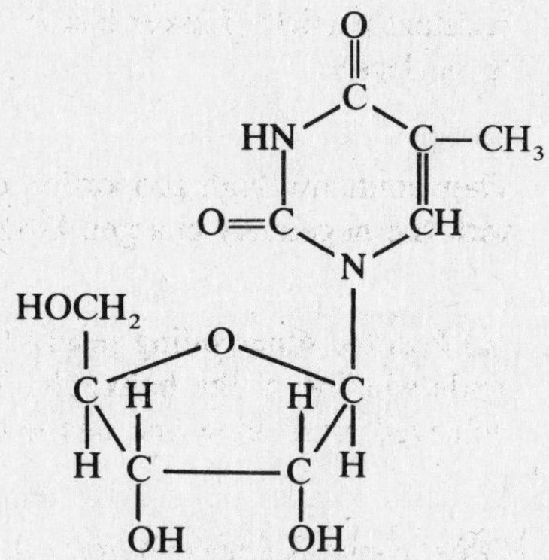

7.40. Adenosine

7.41. A nucleotide contains a phosphate group; a nucleoside does not.

7.42.

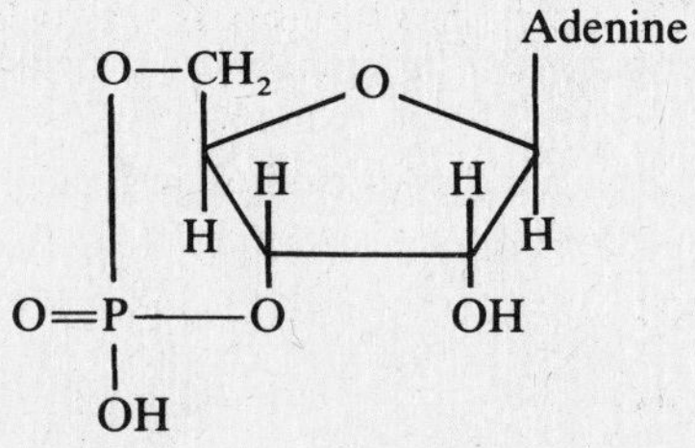

7.43. (*b*) Deoxyguanosine monophosphate

7.44. Because it contains uracil, which does not normally occur in DNA

7.45. Less resistant

7.46. *Sequence complementarity* refers to the matching of a sequence in a strand of DNA with that in another in terms of the ability of the two to form complementary base pairs.

7.47. The B form is a right-handed helix; the Z form is left-handed.

The repeating unit in the B form is a mononucleotide and a dinucleotide in the Z form.

The sugar phosphate backbone of the Z form follows a zigzag course, whereas that of the B form is relatively smooth.

7.48. At high ionic strength, positively charged counterions reduce the negative charge on the DNA, and this lowers the electrostatic repulsion within the structure, making it more stable.

7.49. Because the bacterial genome is smaller in size than the eukaryotic genome, nonrepeated sequences in DNA from the former will be present at a higher concentration. Renaturation is a bimolecular reaction and will proceed more quickly when the reactants are at a higher concentration.

7.50. Sequence complexity refers to the variety of sequences present in a genome.

7.51. The term *chromosome* refers to the structural unit within which the genetic material of a cell (usually DNA) is organized. The term *genome* describes a single complement of genetic material of the cell and could be made up of more than one chromosome.

7.52. Core histones are those around which the DNA in eukaryotic cells is wrapped to yield a nuclease-resistant particle. Linker histone is associated with the DNA that links these particles and is accessible to nucleases.

7.53. They contain a high proportion of positively charged amino acid residues that interact electrostatically with the negatively charged DNA.

7.54. (*a*) Positive supercoiling results from overwinding of the DNA helix; negative supercoiling results from underwinding of the helix. (*b*) Relaxation refers to the removal of the supercoiling, and (*c*) it can be achieved by nicking one of the two strands of DNA.

7.55. DNA contains deoxyribose as the sugar and thymine as one of the four bases. RNA contains ribose as the sugar and uracil as one of the four bases.

7.56. Because it is transcribed (formed) from a much larger portion of the genome than the other types of RNA.

7.57. (*c*) The sequence GTAATC would be the least likely to be cut because it does not have two-fold rotational symmetry.

Chapter 8

8.15. (*a*) Glucose

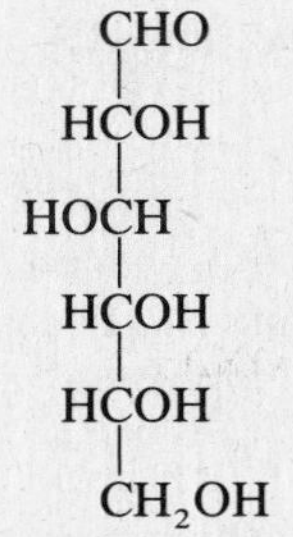

CHO
|
HCOH
|
HOCH
|
HCOH
|
HCOH
|
CH_2OH

Glucose 6-phosphatase is a hydrolase, EC 3.1.3.9.

(*b*) Pyruvate

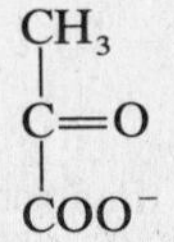

CH_3
|
C=O
|
COO^-

Lactate dehydrogenase is an oxidoreductase, EC 1.1.1.27.

(*c*) Arginine

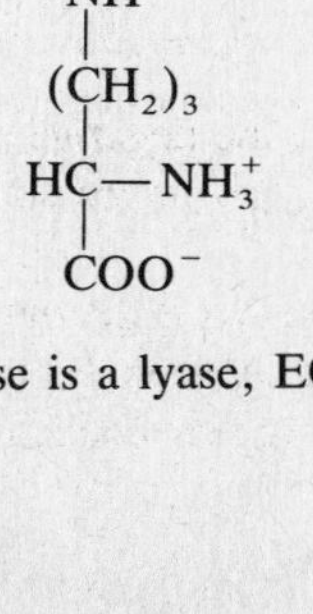

NH_2
|
$C \overset{+}{=} NH_2$
|
NH
|
$(CH_2)_3$
|
$HC—NH_3^+$
|
COO^-

Arginosuccinase is a lyase, EC 4.3.2.1.

(*d*) Malate

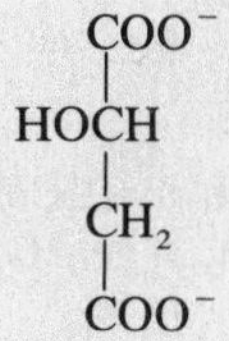

```
  COO⁻
   |
 HOCH
   |
  CH₂
   |
  COO⁻
```

Fumarase is a hydratase, EC 4.2.1.2.

8.16. (*a*) Ethanol > methanol > propanol > butanol > cyclohexanol > phenol.

(*b*) Ethanol is the "best" substrate; methanol is relatively too small, while the rest are too large or too hydrophobic to be accommodated well in the active set.

8.17. (*a*) Thiohemiacetal

(*b*)

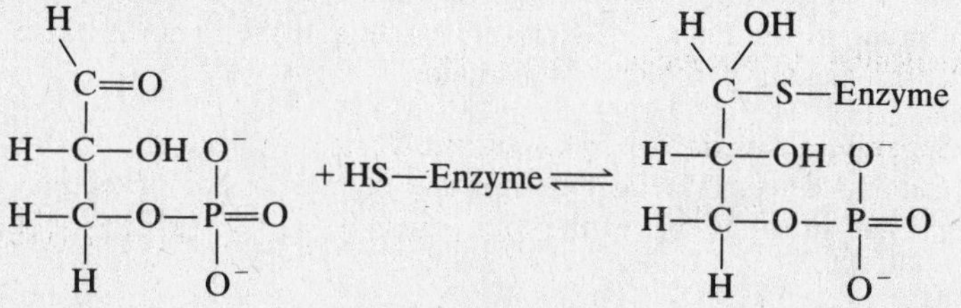

```
     H                                            H  OH
      \                                            \ /
       C=O                                          C—S—Enzyme
       |                                            |
   H—C—OH O⁻                                    H—C—OH  O⁻
       |    |      + HS—Enzyme ⇌                    |     |
   H—C—O—P=O                                    H—C—O—P=O
       |    |                                       |     |
       H    O⁻                                      H     O⁻
```

8.18. (*a*) Mutases are members of the EC class 5, i.e., isomerases. The enzymes of class 5 catalyze geometrical structural changes *within* one molecule. When the isomerization consists of an intramolecular transfer of a *group*, the enzyme is called a mutase. Examples include phosphoglucomutase and chorismate mutase.

(*b*) These three enzyme types are all members of EC class 1: oxidoreductases. The enzymes of this class catalyze oxidation/reduction reactions, and the systematic name is *hydrogen donor* : *acceptor oxidoreductase*. The recommended name is *dehydrogenase* or, alternatively, *reductase*. *Oxidase* is used only in cases where O_2 is the acceptor. Examples are lactate dehydrogenase and choline oxidase (EC 1.1.3.4), which catalyze the formation of betaine and H_2O_2 from choline, respectively.

8.19. In humans, the time spent by red blood cells flowing through capillaries of the lung alveoli is ~0.3 s. In that time, HCO_3^- in the plasma must reenter the red cells and be dehydrated to yield CO_2, which then diffuses across the membranes of the red cells and capillaries into the alveoli (and is expelled into the atmosphere). The spontaneous dehydration reaction is simply too slow.

8.20. A substrate analog on which an enzyme operates but which then covalently modifies the active site, permanently inhibiting it.

8.21. No.

8.22. There is no single, simple answer to the question. Possible reasons are to: (1) provide the "correct" chemical environment for binding and catalysis, e.g., lower the pK_a of the group, (2) absorb energy of bombardment of Brownian motion of water and "funnel" it into the active site to enhance the catalytic rate, (3) allow for *control* of catalysis via conformational changes induced by effectors binding to other sites on the enzyme, (4) allow the fixing of enzymes in membranes or in large organized complexes, (5) prevent their loss by filtration through membranes, e.g., in the kidney.

8.23. DNA polymerase (Chap. 16), glycogen phosphorylase (Chap. 11).

Chapter 9

9.35.

$$v_0 = \frac{V_{max}[S]_0}{K_m + [S]_0}$$

$$v_0(K_m + [S]_0) = V_{max}[S]_0$$

$$\frac{v_0(K_m + [S]_0)}{[S]_0} = \frac{V_{max}[S]_0}{[S]_0}$$

$$\frac{v_0 K_m}{[S]_0} + v_0 = V_{max}$$

$$v_0 = V_{max} - \frac{v_0 K_m}{[S]_0}$$

9.36. Multiply the Lineweaver-Burk equation by $[S]_0$ on both sides of the equals sign.

9.37. (*a*) Remember to use reciprocals. (*b*) No.

9.38. (*a*) Separate the variables and integrate. Then

$$V_{max}t = [S]_0 - [S] + \left(K_m + \frac{[I]}{K_I}\right)\ln([S]_0/[S])$$

(*b*) Use the graphical analysis given in Example 9.13 for a fixed $[S]_0$ value and a range of [I] values, to obtain a series of slopes. A secondary plot of the modulus of these slopes versus [I] has a slope equal to $1/K_I$.

9.39. (*a*) Separate the variables, noting that $[I] = [P] = [S]_0 - [S]$; then integrate directly:

$$V_{max}t = \left(1 + \frac{[I]}{K_I}\right)\left([S]_0 - [S]\right) + K_m \ln([S]_0/[S])$$

(*b*) Use the graphical analysis given in Example 9.12 for a fixed $[S]_0$ value and a range of [I] values, to obtain a series of lines with the same slope but different abscissal and ordinate intercepts. A secondary plot of the reciprocal of the abscissal intercept versus [I] has the slope $1/K_I$.

9.40. Consider

$$\frac{d[S]}{dt} = -\frac{V_{max}[S]}{K_m \dfrac{[1 + ([S]_0 - [S])}{K_I} + [S]}$$

If $K_I = K_m$, then

$$\frac{d[S]}{dt} = -\frac{V_{max}[S]}{K_m \dfrac{[(K_m + [S]_0 - [S])]}{K_m} + [S]}$$

from which the result follows, after cancellation of the K_m's in the numerator and direct integration.

9.41. See Fig. A-1. (1) In the presence of a *pure noncompetitive* inhibitor, the lines intersect at a point lower than in the absence of inhibitor, but with no change in the abscissal value (K_m). (2) For *pure competitive* inhibition, the intersection points move out on a horizontal line with no change in the coordinate of the dependent variable (v_0). (3) For an *anticompetitive* inhibitor, the intersection point of the lines moves in along a line of constant slope as the inhibition concentration is increased; the increment along the line is a constant proportion of the inhibitor concentration. (4) For *mixed* inhibition, the situation is similar to (3), but there is *no constant proportion*.

9.42. Use any of the graphical procedures. $K_m = 0.8\ \mu\text{mol L}^{-1}$; $V_{max} = 3\ \mu\text{mol L}^{-1}\text{ min}^{-1}$; and from the expression $V_{max} = k_{cat}[E]_0$, $k_{cat} = 300\text{ s}^{-1}$.

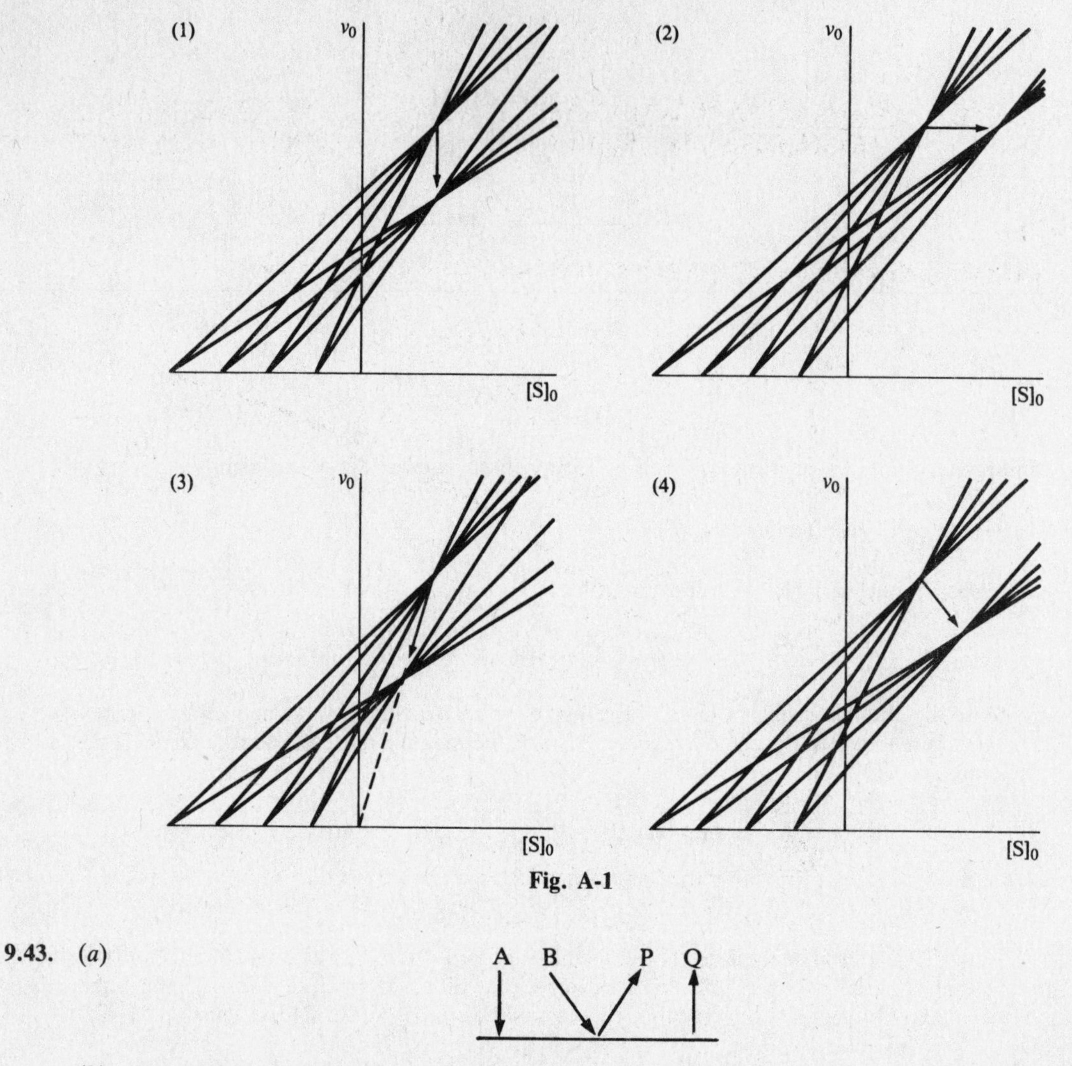

Fig. A-1

9.43. (*a*)

(*b*) Derive the rate expression using the King-Altman or Indge and Childs procedure. The relationships follow from arguments based on those used in Example 9.8.

9.44. This is an important question that has been addressed by many enzyme kineticists over the years. For the correct application of the Briggs-Haldane steady-state analysis, in a *closed* system, $[S]_0$ must be $\gg$ $[E]_0$, where the $\gg$ sign implies a factor of at least 1,000. M. F. Chaplin in 1981 noted that the expression $v_0 = V_{max}[S]_0/(K_m + [S]_0 + [E]_0)$ yields, for example, only a 1 percent error in the estimate of v_0 for $[S]_0 = 10 \times [E]_0$ and $[S]_0 = 0.1\ K_m$; the expression thus applies under much less stringent conditions than does the simple Michaelis-Menten equation. In open systems $[S]_0$ can approximate $[E_0]$ and a steady state of enzyme-substrate complexes can pertain; computer simulation of both types of system is the best way to gain insight into the conditions necessary for a steady state of the complex.

9.45. (*a*) $\frac{1}{9}\ K_m$; (*b*) $\frac{1}{3}\ K_m$; (*c*) K_m; (*d*) $9\ K_m$

9.46. Differentiate Eq. (*9.41*).

9.47. (*a*) ~7; (*b*) ~10; (*c*) 81

9.48. Yes. If $0 < n < 1$, the data conform to a negatively cooperative enzyme or binding protein; this is shown by using the analysis of Prob. 9.26, and in the present case $1/n > 1$, so $(81)^{1/n} > 81$.

9.49.
$$Y = \frac{K_1[X] + K_1K_2[X]^2}{1 + 2K_1[X] + K_1K_2[X]^2}$$

9.50.
$$Y = \frac{K_{AB}^3(K_sK_t[S]) + 3K_{AB}^4K_{BB}(K_sK_t[S])^2 + 3K_{AB}^3K_{BB}(K_sK_t[S])^3 + K_{BB}^6(K_sK_t[S])^4}{1 + 4K_{AB}^3(K_sK_t[S]) + 6K_{AB}^4K_{BB}(K_sK_t[S])^2 + 4K_{AB}^3K_{BB}^3(K_sK_t[S])^3 + K_{BB}^6(K_sK_t[S])^4}$$

Chapter 10

10.12. Use the equations for activity and activity coefficient in Prob. 10.3. See Fig. A-2.

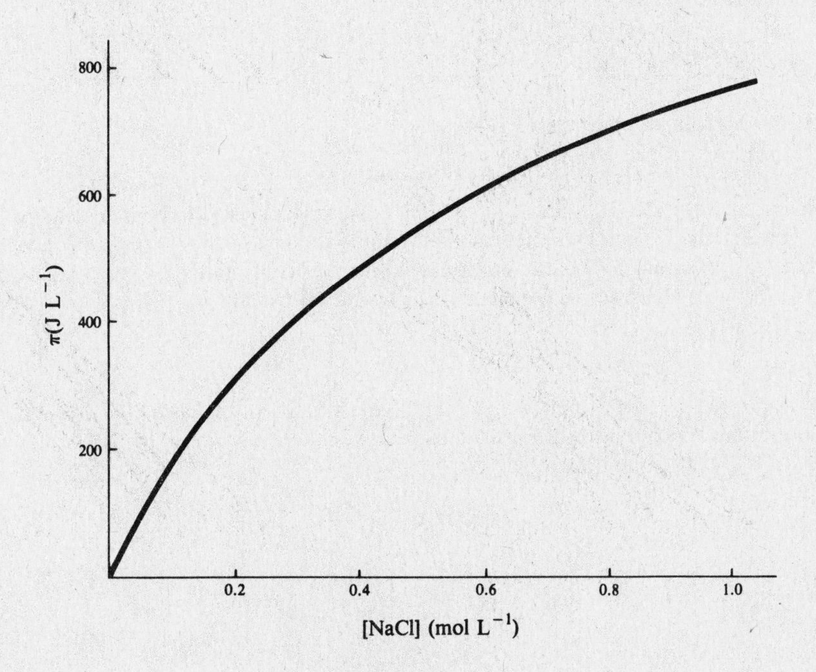

Fig. A-2

10.13. $\Delta G^{\circ\prime} = 219.9 \text{ kJ mol}^{-1}$

10.14. Under biochemical standard-state conditions, the free energy of ATP synthesis is given by the expression

$$\Delta G^{\circ\prime} = -nF\,\Delta p$$

where n is the number of protons translocated per ATP molecule synthesized and Δp is the proton-motive force. Substituting values for $\Delta G^{\circ\prime}$ and n gives

$$\Delta p = -0.158 \text{ V}$$

Δp is given by the expression

$$\Delta p = \Delta\psi - \frac{2.3RT}{F}\,\Delta\text{pH}$$

which is Eq. (*10.35*) in which both sides have been divided by F.

Hence, $$\Delta\mu = \Delta P \cdot \text{F}$$

Therefore, $$\Delta\mu = -15.25 \text{ kJ mol}^{-1}$$

10.15. (*a*) 0.03; (*b*) 0.15; (*c*) 0.3; (*d*) 3.0

10.16. Use Eq. (*10.10*).

10.17. (*a*) $\Delta G^{\circ\prime}_{\mathrm{pH}=x} = \Delta G^\circ + RT \ln(10^{-x})$

(*b*) Using this equation to calculate ΔG° for the value for $\Delta G^{\circ\prime}_{\mathrm{pH}=7}$, we get $\Delta G^{\circ\prime}_{\mathrm{pH}=6} = 23.2 \text{ kJ mol}^{-1}$.

10.18. (*a*) $\Delta S^\circ = 92.2 \text{ J K}^{-1} \text{ mol}^{-1}$

(*b*) The equilibrium constant at 323 K is 1.756×10^{-5}, if we assume that ΔH° and ΔS° are independent of temperature over the range 298–323 K.

10.19. (*a*) and (*b*)

10.20. At high $[S_i]$, $v_i = k_{\mathrm{cat}_i}[E_i]$.

10.21. Consider a chain of enzymes E_1 to E_n,

$$S_1 \xrightarrow{E_1} S_2 \xrightarrow{E_2} S_3 \cdots\cdots S_n \xrightarrow{E_n} S_{n+1}$$

with various values of $V_{\max}$ and K_m, etc. The system is assumed to be in a steady state with constant flux of v_g through it. If, in a hypothetical experiment, we increased all enzyme concentrations (an enzyme's activity is assumed to be proportional to its concentration) by the same small factor α (i.e., $\Delta[E_i]/[E_i] = \alpha$) the fractional change in v_g would be exactly α:

$$\frac{\Delta v_g}{v_g} = \alpha$$

Now, provided α is very small, $\Delta v_g/v_g$ can be the sum of all the *individual changes* that would be caused by alterations to each of the *separate* enzymes. For the ith enzyme, the control coefficient is given by C_i (see Prob. 10.20); therefore,

$$C_i = \left(\frac{\Delta v_g}{v_g}\right)_i \Big/ \frac{\Delta[E_i]}{[E_i]}$$

$$= \left(\frac{\Delta v_g}{v_g}\right)_i \Big/ \alpha$$

$$\alpha C_i = \left(\frac{\Delta v_g}{v_g}\right)_i$$

The summation of this equation gives

$$\frac{\Delta v_g}{v_g} = \sum_{i=1}^{n} \left(\frac{\Delta v_g}{v_g}\right)_i = \alpha \sum_{i=1}^{n} C_i$$

But, from the first equation, above, $\Delta v_g/v_g = \alpha$; therefore $\sum_{i=1}^{n} C_i = 1$.

Chapter 11

11.12. During glycolysis, glyceraldehyde 3-phosphate is converted to 1,3-diphosphoglycerate, so that the equilibrium is *not* attained.

11.13. 2,3-Diphosphoglycerate is synthesized from 1,3-diphosphoglycerate, a reaction catalyzed by diphosphoglycerate mutase. The enzyme 2,3-diphosphoglycerate phosphatase catalyzes the hydrolysis of 2,3-diphosphoglycerate to 3-phosphoglycerate.

11.14. Two of the carbon atoms are converted to carbon dioxide, and the other four carbon atoms yield two molecules of ethyl alcohol.

11.15. Four

11.16. It can be incorporated into glycogen, used in the pentose phosphate pathway, or hydrolyzed to glucose.

11.17. The gut microbiota of termites synthesize and secrete the enzyme cellulase, which hydrolyzes the cellulose.

11.18. (*a*) Alcoholic fermentation operates: the glucose is converted to pyruvate, which forms CO_2 and ethanol and regenerates NAD^+. (*b*) The inhibitor would block glycolysis, and the yeast would die.

11.19. (*a*) Insulin promotes the uptake of glucose from the blood to all cells. Excess glucose is converted to glucose 6-phosphate (via glucokinase) in the liver and is then stored as glycogen.

(*b*) Glucagon mobilizes the breakdown of glycogen in the liver, which produces glucose to maintain an adequate concentration in the blood.

11.20. Blood glucose concentrations would remain high after meals, and the majority of the glucose would be excreted in the urine. Glucose metabolism would operate, therefore, at a lower level than normal.

11.21. The pentose phosphate pathway would produce the necessary NADPH, and the ribose 5-phosphate would be converted into the glycolytic intermediates fructose 6-phosphate and glyceraldehyde 3-phosphate.

11.22. Glycolysis, the pentose phosphate pathway, the production of glycogen, and the formation of glucose (glucose 6-phosphatase).

Chapter 12

12.12. (*a*) Citric acid cycle: two; (*b*) glyoxylate cycle: zero

12.13. Fumarate is the substrate for fumarase, whereas maleate is not. The cell could not use maleate as a carbon source.

12.14. Red blood cells do not have mitochondria and so do not have the enzymes to operate the citric acid cycle.

12.15. The inhibitor would prevent the citric acid cycle from operating, but in germinating plant cells, the glyoxylate cycle would be unaffected. Thus, energy production would be decreased in both cells, but their ability to synthesize glucose would be unimpaired.

Chapter 13

13.17. Cholesterol is the precursor of bile salts, and their secretion into the intestine is stimulated during the digestion and absorption of triglyceride.

13.18. Methyl groups are utilized in the de novo synthesis of phosphatidylcholine, a process that decreases the availability of 1,2-diacylglycerol for triacylglycerol synthesis.

13.19. The complete oxidation of 1 mole of palmitic acid to CO_2 and H_2O produces NADH and $FADH_2$ by β oxidation and by citric acid cycle activity. ATP synthesis is coupled to the oxidation of NADH and $FADH_2$ produced in these processes (Chap. 14). Per mole of palmitoyl-CoA oxidized, 131 moles of ATP are synthesized.

13.20. (*a*) Acetoacetate is completely oxidized by entry of two molecules of acetyl-CoA into the citric acid cycle; 22 moles of ATP per mole of acetoacetate are synthesized from NADH and $FADH_2$ produced by the cycle. Two additional moles of ATP could arise from two moles of GTP, also produced in the citric acid cycle.

(*b*) 3-Hydroxybutyrate is oxidized by β oxidation to acetyl-CoA, which then enters the citric acid cycle; 25 moles of ATP per mole of 3-hydroxybutyrate are synthesized from the NADH and $FADH_2$ produced, and 2 moles of ATP from GTP may also be synthesized.

13.21. (*a*) Palmitic acid and 2-linoleoylglycerol; (*b*) arachidonic acid

13.22. Glucose is oxidized in the pentose phosphate pathway (Chap. 11), to produce NADPH for cholesterol synthesis. One molecule of glucose is required per molecule of lanosterol synthesized. (The reactions that convert lanosterol to cholesterol are outside the scope of Chap. 13.)

13.23. β oxidation

13.24. Glucose is converted to dihydroxyacetone phosphate (Chap. 10), which is then reduced to glycerol 3-phosphate by NADH-dependent glycerol 3-phosphate dehydrogenase.

13.25. Acetyl-CoA carboxylase exists in the cytosol in two forms. Citrate converts the inactive form to the active form, and long-chain acyl-CoA is a feedback inhibitor that converts the active form back to the inactive form.

13.26. Transamination reactions produce pyruvate, and deamination followed by degradation of the carbon skeleton of ketogenic amino acids produces acetyl-CoA (see Chap. 15).

13.27. Acyl-CoA dehydrogenase, enoyl-CoA hydratase, 3-hydroxyacyl-CoA dehydrogenase, thiolase, and Δ^3-*cis*-Δ^2-*trans*-enoyl-CoA isomerase

13.28. Fat depots provide fatty acids for cellular fuel. The oxidation of fatty acids produces NADH and $FADH_2$, which are oxidized (by the respiratory assemblies in the inner membrane of mitochondria) with the concomitant production of water.

13.29. Fatty acid oxidation would be inhibited because acyl groups could not be transported into the mitochondrial matrix. This would restrict long-term muscle activity.

13.30. Yes, because platelet aggregation depends on the release of thromboxanes from the cells. Aspirin inhibits cyclooxygenase, and thus synthesis of prostaglandins and thromboxanes is inhibited.

13.31. Triacylglycerol would be synthesized from dietary carbohydrate and protein.

13.32. Cytosolic 3-hydroxy-3-methylglutaryl-CoA synthase produces the 3-hydroxy-3-methylglutaryl-CoA for cholesterol synthesis, and mitochondrial HMG-CoA synthase produces the HMG-CoA, which is converted to acetoacetate.

13.33. Cholesterylesters arise from the activity of acyl-CoA:cholesterol acyltransferase, which catalyzes the formation of the esters from acyl-CoA's, and also from the activity of phosphatidylcholine:cholesterol transferase, which catalyzes the formation of esters from phosphatidylcholine.

13.34. Depot fat is depleted, and fatty acid oxidation in the liver increases. This provides the acetyl-CoA for the synthesis of ketone bodies.

13.35. Long-chain fatty acids are relatively insoluble in ionic media, but still form salts, some of which act as detergents (soaps) or form insoluble precipitates, such as Ca^{2+} and Mg^{2+} salts.

13.36. Cells that synthesize prostaglandins, thromboxanes, and leukotrienes

Chapter 14

14.11. (*a*) 15; (*b*) 3; (*c*) 36; (*d*) 16

14.12. The P/O ratio will decrease.

14.13. Approximately 14

14.14. Review Prob. 14.8, and think about the metabolic sources of NADH and $FADH_2$. Uncoupling agents, such as 2,4-dinitrophenol, render the inner mitochondrial membrane permeable to the protons extruded during electron transport. The energy released as these protons return to the mitochondrial matrix leads to an increase in body temperature; this is relieved by sweating. Because, under these conditions, ATP synthesis is no longer completely coupled to electron transport, the oxidation of NADH and $FADH_2$ must increase to maintain a supply of ATP. As these compounds are produced during catabolism of carbohydrates, lipids, and proteins, there is an increased turnover of these body constituents, leading to loss of body weight.

14.15. On the basis of their visible absorbance spectra. Many textbooks give examples of such spectra.

14.16. Apply the crossover theorem, with reference to the figure in Example 14.5. The proportions of the iron-sulfur proteins in complexes I, II, and III with iron atoms in the Fe^{3+} and Fe^{2+} states can, in principle, be estimated from the electron spin resonance signals of the preparation of mitochondria. In the case of mitochondria oxidizing NADH in the presence of rotenone, application of the crossover theorem reveals that the iron atoms in the iron-sulfur proteins in complex I will become more reduced, while those in complex III will become more oxidized, in comparison to their states in the absence of rotenone. In the case of KCN-treated mitochondria, both sets of iron-sulfur proteins will become more reduced. With antimycin A–treated mitochondria, the complex I iron-sulfur proteins will become more reduced, while those in complex III may, or may not, depending on the site of action of antimycin A within this complex.

14.17. 114 kJ mol^{-1}

14.18 Nitrite converts some ferrohemoglobin to ferrihemoglobin. The latter is able to bind cyanide, preventing it from inactivating cytochromes a and a_3 by binding to their ferriheme rings.

14.19. (*a*) Oxidative phosphorylation is more efficient than substrate-level phosphorylation.

(*b*) The NADH needed for lactate production will be oxidized to NAD^+ in oxidative phosphorylation.

Chapter 15

15.19. A person on this diet would be in negative nitrogen balance and could not sustain normal metabolism.

15.20. Yes, because cheese contains animal protein.

15.21. (*a*) Chymotrypsin and elastase; (*b*) Elastase, trypsin, and chymotrypsin; (*c*) Pepsin, chymotrypsin, and trypsin

15.22. Glutaminase, aspartate aminotransferase, TCA cycle enzymes, citrate lyase, malate dehydrogenase, and $NADP^+$-linked malate enzyme

15.23. The *rate* of synthesis of arginine is insufficient to sustain growth in children, so a dietary source of arginine is essential.

15.24.

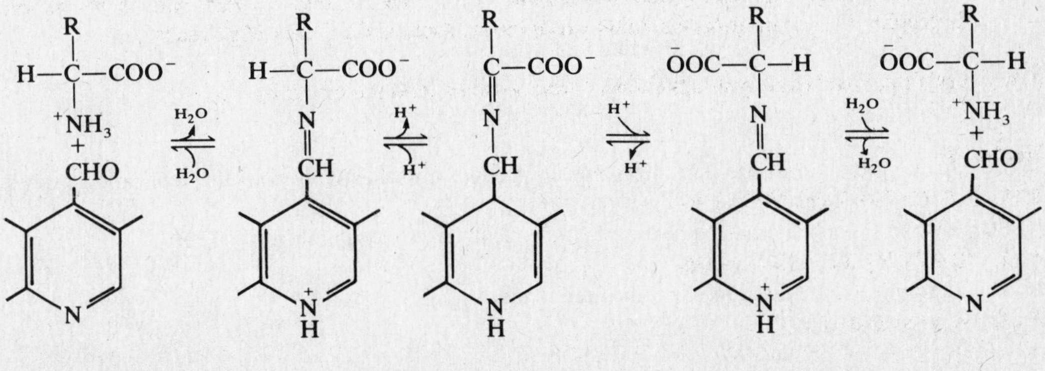

Chapter 16

16.29. (*a*) In unidirectional replication, a single replication fork generated at the origin moves in a unique direction along the DNA. (*b*) The rolling-circle mechanism is an example of undirectional replication.

16.30. In embryonic cells, replication is initiated at sites (origins) along the DNA that are more closely spaced than in terminally differentiated cells. Thus, the distance over which each replication fork travels to replicate the complement of DNA is very short, and the overall process is rapid.

16.31. The $5' \rightarrow 3'$ exonuclease and polymerase activities are utilized on the lagging strand arm to remove the RNA primers and replace them with DNA, respectively. The $3' \rightarrow 5'$ exonuclease activity is used in proofreading during this process.

16.32. At the origin of replication

16.33. It could aid unwinding in one of two ways: by introduction of negative supercoiling or removal of positive supercoiling ahead of the replication fork.

16.34. No.

16.35. There must be present a 3′ hydroxyl and a 5′ phosphate, one at each end of the chains to be joined.

16.36. This refers to the $3' \rightarrow 5'$ exonuclease activity associated with the enzyme. It acts to remove mismatched nucleotides incorporated infrequently by the polymerase.

16.37.

Origin of replication

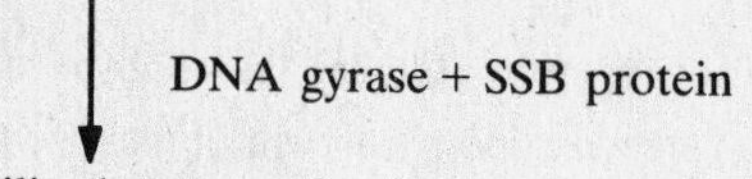

Stabilized, unwound
DNA at origin

↓ Primosome + primose + rNTPs + DNA polymerase III + dNTPs

Leading strands extending
from primer and away from
origin on opposite templates

↓ Movement of primosome away from origin on opposite strands

Priming and synthesis of
lagging strands. Generation
of replication bubble.

16.38. A *dna* mutation is one that alters a gene product directly involved in the replication process.

16.39. Yes, to the extent that replication is dependent on the availability of dTTP.

16.40. It would block initiation of replication at the origin.

16.41. No.

16.42. DNA polymerase I (with its associated 5′→3′ exonuclease activity) and DNA ligase.

16.43. No.

16.44. It must have an origin of replication that is functional in the host cell.

Chapter 17

17.40. His-Ile-Thr-His-Glu-Arg-Glu.
(By using the single letter notation for amino acids (Table 3.1), this sequence reads: HI THERE!)

17.41. Met-Phe-Phe-Val-Phe

17.42. The lower strand

17.43. Ribosomal RNA (rRNA)

17.44. No. The RNA polymerases can start chains de novo.

17.45. RNA synthesis

17.46. (*b*)

17.47. (*a*)

17.48. Near the 3′ end

17.49. Yes. (The *lac* repressor gene of *E. coli* would give rise to monocistronic mRNA.)

17.50. Eukaryotes

17.51. Phosphorylation would *decrease* the net positive charge.

17.52. No. Introns and exons are both transcribed, but only exons are translated.

17.53. RNA

17.54. *Before* splicing

17.55. Yes. The first exon (closest to the 5′ end) of the mRNA must contain the initiation codon.

17.56. No. Histones occur in eukaryotes, and rifampicin blocks transcription only in bacteria.

17.57. Codons occur in mRNA; anticodons, in tRNA.

17.58. The tRNA molecule must not only recognize the codon, but also recognize specific regions of the ribosome, elongation factors, and the appropriate aminoacyl-tRNA synthetase, each of which involves cooperative interactions through noncovalent bonds.

17.59. The single tRNA chain folds back on itself to allow the formation of several segments of double helix through intrastrand base pairing.

17.60. It contains both.

17.61. There are more species of tRNA.

17.62. The small ribosomal subunit

17.63. At two steps: the binding of each charged tRNA and translocation after the formation of each peptide.

17.64. Termination of *translation*

17.65. Capping of the RNA occurs *before* splicing and must occur in the nucleus.

17.66. Preproinsulin

17.67. No.

17.68. It binds to the operator site.

17.69. A *repressor* molecule can bind to an operator and block transcription; a *corepressor* is a compound which binds to a protein to give a functional repressor, which can then bind to an operator.

Index

The letters *d* and *t* following page numbers stand for diagram and table, respectively.

Isotope-exchange methods, 273